THE PHYSICS TEXTBOOK 85% OF STUDENTS PREFER COMES WITH THE MOST EFFECTIVE AND VALUABLE MULTIMEDIA AVAILABLE TO HELP YOU SUCCEED IN YOUR COURSE.

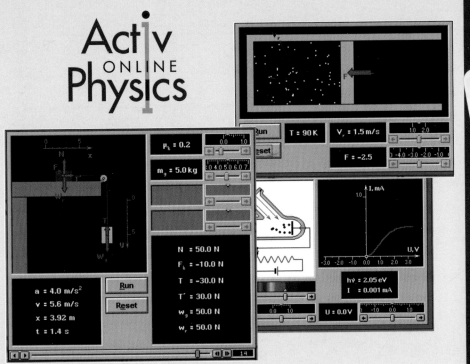

ActivPhysics™ OnLine utilizes visualization, simulation and multiple representations to help you better understand key physical processes, experiment quantitatively and develop your critical-thinking skills. The carefully designed interactive simulations provide thought-provok-ing questions and problems to guide your understanding of physics.

Website: www.aw-bc.com/knight

Minimum System Requirements:
Windows: 250 MHz; OS 98, NT, ME, 2000, XP
Macintosh: 233 MHz; OS 9.2, 10
Both:
- 64 RAM installed
- 1024 x 768 screen resolution
- Browsers: Internet Explorer 5.5, 6.0; Netscape 6.2.3, 7.0
- Plug Ins: Macromedia's Flash 6.2.3

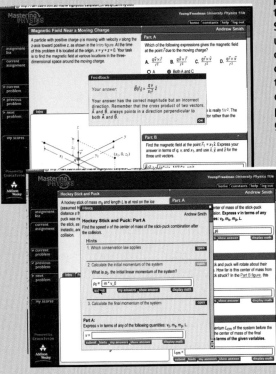

MasteringPhysics™ is the first Socratic tutoring system developed specifically for physics students like you. It is the result of years of detailed studies of how students work physics problems, and where they get stuck and need help. Studies show students who used MasteringPhysics significantly improved their scores on traditional final exams and the Force Concept Inventory (a conceptual test) when compared with traditional hand-graded homework.

With your purchase of a new copy of Knight's *Physics for Scientists and Engineers*, you should have received a Student Access Kit for **MasteringPhysics™** if your professor required it as a component of your course. The kit contains instructions and a code for you to access MasteringPhysics.

If you did not purchase a new textbook and your professor requires you to enroll in the **MasteringPhysics** online homework and tutorial program, you may purchase an online subscription with a major credit card. Go to www.masteringphysics.com and follow the links to purchasing online.

Minimum System Requirements:
Windows: 250 MHz; OS 98, NT, ME, 2000, XP
Macintosh: 233 MHz; OS 9.2, 10
RedHat Linux 8.0
All:
- 64 RAM installed
- 1024 x 768 screen resolution
- Browsers: Internet Explorer 5.0, 5.5, 6.0; Netscape 6.2.3, 7.0; Mozilla 1.2, 1.3

MasteringPhysics™ is powered by MyCyberTutor by Effective Educational Technologies

AW003

Table of Problem-Solving Strategies

Note for users of the five-volume edition:
Volume 1 (pp. 1–481) includes chapters 1–15.
Volume 2 (pp. 482–607) includes chapters 16–19.
Volume 3 (pp. 608–779) includes chapters 20–24.
Volume 4 (pp. 780–1194) includes chapters 25–36.
Volume 5 (pp. 1148–1383) includes chapters 36–42.

Chapters 37–42 are not in the Standard Edition.

Chapter	Problem-Solving Strategy		Page
Chapter 1	**1.1**	**Motion diagrams**	17
Chapter 1	**1.2**	**Problem-solving strategy**	24
Chapter 2	**2.1**	**Kinematics with constant acceleration**	57
Chapter 5	**5.1**	**Equilibrium problems**	123
Chapter 5	**5.2**	**Dynamics problems**	126
Chapter 6	**6.1**	**Projectile motion problems**	161
Chapter 7	**7.1**	**Circular motion problems**	200
Chapter 8	**8.1**	**Interacting-system problems**	217
Chapter 9	**9.1**	**Conservation of momentum**	250
Chapter 10	**10.1**	**Conservation of mechanical energy**	279
Chapter 11	**11.1**	**Solving energy problems**	326
Chapter 13	**13.1**	**Rotational dynamics problems**	387
Chapter 13	**13.2**	**Rigid-body equilibrium problems**	390
Chapter 17	**17.1**	**Work in ideal-gas processes**	518
Chapter 17	**17.2**	**Calorimetry problems**	531
Chapter 19	**19.1**	**Heat-engine problems**	584
Chapter 21	**21.1**	**Interference of two waves**	669
Chapter 25	**25.1**	**Electrostatic forces and Coulomb's law**	799
Chapter 26	**26.1**	**The electric field of multiple point charges**	819
Chapter 26	**26.2**	**The electric field of a continuous distribution of charge**	826
Chapter 27	**27.1**	**Gauss's law**	866
Chapter 29	**29.1**	**Conservation of energy in charge interactions**	912
Chapter 29	**29.2**	**The electric potential of a continuous distribution of charge**	922
Chapter 31	**31.1**	**Resistor circuits**	981
Chapter 32	**32.1**	**The magnetic field of a current**	1005
Chapter 33	**33.1**	**Electromagnetic induction**	1058
Chapter 36	**36.1**	**Relativity**	1176
Chapter 40	**40.1**	**Quantum mechanics problems**	1283

Physics for Scientists and Engineers

A Strategic Approach

ActivPhysics™ OnLine Activities www.aw-bc.com/knight

1.1 Analyzing Motion Using Diagrams
1.2 Analyzing Motion Using Graphs
1.3 Predicting Motion from Graphs
1.4 Predicting Motion from Equations
1.5 Problem-Solving Strategies for Kinematics
1.6 Skier Races Downhill
1.7 Balloonist Drops Lemonade
1.8 Seat Belts Save Lives
1.9 Screeching to a Halt
1.10 Pole-Vaulter Lands
1.11 Car Starts, Then Stops
1.12 Solving Two-Vehicle Problems
1.13 Car Catches Truck
1.14 Avoiding a Rear-End Collision
2.1.1 Force Magnitudes
2.1.2 Skydiver
2.1.3 Tension Change
2.1.4 Sliding on an Incline
2.1.5 Car Race
2.2 Lifting a Crate
2.3 Lowering a Crate
2.4 Rocket Blasts Off
2.5 Truck Pulls Crate
2.6 Pushing a Crate Up a Wall
2.7 Skier Goes Down a Slope
2.8 Skier and Rope Tow
2.9 Pole-Vaulter Vaults
2.10 Truck Pulls Two Crates
2.11 Modified Atwood Machine
3.1 Solving Projectile Motion Problems
3.2 Two Balls Falling
3.3 Changing the x-Velocity
3.4 Projectile x- and y-Accelerations
3.5 Initial Velocity Components
3.6 Target Practice I
3.7 Target Practice II
4.1 Magnitude of Centripetal Acceleration
4.2 Circular Motion Problem Solving
4.3 Cart Goes Over Circular Path
4.4 Ball Swings on a String
4.5 Car Circles a Track
4.6 Satellites Orbit
5.1 Work Calculations
5.2 Upward-Moving Elevator Stops
5.3 Stopping a Downward-Moving Elevator
5.4 Inverse Bungee Jumper
5.5 Spring-Launched Bowler
5.6 Skier Speed
5.7 Modified Atwood Machine
6.1 Momentum and Energy Change
6.2 Collisions and Elasticity
6.3 Momentum Conservation and Collisions
6.4 Collision Problems
6.5 Car Collision: Two Dimensions
6.6 Saving an Astronaut
6.7 Explosion Problems
6.8 Skier and Cart
6.9 Pendulum Bashes Box
6.10 Pendulum Person-Projectile Bowling
7.1 Calculating Torques
7.2 A Tilted Beam: Torques and Equilibrium
7.3 Arm Levers
7.4 Two Painters on a Beam
7.5 Lecturing from a Beam

7.6 Rotational Inertia
7.7 Rotational Kinematics
7.8 Rotoride: Dynamics Approach
7.9 Falling Ladder
7.10 Woman and Flywheel Elevator: Dynamics Approach
7.11 Race Between a Block and a Disk
7.12 Woman and Flywheel Elevator: Energy Approach
7.13 Rotoride: Energy Approach
7.14 Ball Hits Bat
8.1 Characteristics of a Gas
8.2 Maxwell-Boltzmann Distribution: Conceptual Analysis
8.3 Maxwell-Boltzmann Distribution: Quantitative Analysis
8.4 State Variables and Ideal Gas Law
8.5 Work Done by a Gas
8.6 Heat, Internal Energy, and First Law of Thermodynamics
8.7 Heat Capacity
8.8 Isochoric Process
8.9 Isobaric Process
8.10 Isothermal Process
8.11 Adiabatic Process
8.12 Cyclic Process: Strategies
8.13 Cyclic Process: Problems
8.14 Carnot Cycle
9.1 Position Graphs and Equations
9.2 Describing Vibrational Motion
9.3 Vibrational Energy
9.4 Two Ways to Weigh Young Tarzan
9.5 Ape Drops Tarzan
9.6 Releasing a Vibrating Skier I
9.7 Releasing a Vibrating Skier II
9.8 One- and Two-Spring Vibrating Systems
9.9 Vibro-Ride
9.10 Pendulum Frequency
9.12 Risky Pendulum Walk
9.13 Physical Pendulum
10.1 Properties of Mechanical Waves
10.2 Speed of Waves on a String
10.3 Speed of Sound in a Gas
10.4 Standing Waves on Strings
10.5 Tuning a Stringed Instrument: Standing Waves
10.6 String Mass and Standing Waves
10.7 Beats and Beat Frequency
10.8 Doppler Effect: Conceptual Introduction
10.9 Doppler Effect: Problems
10.10 Complex Waves: Fourier Analysis
11.1 Electric Force: Coulomb's Law
11.2 Electric Force: Superposition Principle
11.3 Electric Force Superposition Principle (Quantitative)
11.4 Electric Field: Point Charge
11.5 Electric Field Due to a Dipole
11.6 Electric Field: Problems
11.7 Electric Flux
11.8 Gauss's Law
11.9 Motion of a Charge in an Electric Field: Introduction
11.10 Motion in an Electric Field: Problems
11.11 Electric Potential: Qualitative Introduction

11.12 Electric Potential, Field, and Force
11.13 Electrical Potential Energy and Potential
12.1 DC Series Circuits (Qualitative)
12.2 DC Parallel Circuits
12.3 DC Circuit Puzzles
12.4 Using Ammeters and Voltmeters
12.5 Using Kirchhoff's Laws
12.6 Capacitance
12.7 Series and Parallel Capacitors
12.8 RC Circuit Time Constants
13.1 Magnetic Field of a Wire
13.2 Magnetic Field of a Loop
13.3 Magnetic Field of a Solenoid
13.4 Magnetic Force on a Particle
13.5 Magnetic Force on a Wire
13.6 Magnetic Torque on a Loop
13.7 Mass Spectrometer
13.8 Velocity Selector
13.9 Electromagnetic Induction
13.10 Motional emf
14.1 The RL Circuit
14.2 The RLC Oscillator
14.3 The Driven Oscillator
15.1 Reflection and Refraction
15.2 Total Internal Reflection
15.3 Refraction Applications
15.4 Plane Mirrors
15.5 Spherical Mirrors: Ray Diagrams
15.6 Spherical Mirror: The Mirror Equation
15.7 Spherical Mirror: Linear Magnification
15.8 Spherical Mirror: Problems
15.9 Thin-Lens Ray Diagrams
15.10 Converging Lens Problems
15.11 Diverging Lens Problems
15.12 Two-Lens Optical Systems
16.1 Two-Source Interference: Introduction
16.2 Two-Source Interference: Qualitative Questions
16.3 Two-Source Interference: Problems
16.4 The Grating: Introduction and Qualitative Questions
16.5 The Grating: Problems
16.6 Single-Slit Diffraction
16.7 Circular Hole Diffraction
16.8 Resolving Power
16.9 Polarization
17.1 Relativity of Time
17.2 Relativity of Length
17.3 Photoelectric Effect
17.4 Compton Scattering
17.5 Electron Interference
17.6 Uncertainty Principle
17.7 Wave Packets
18.1 The Bohr Model
18.2 Spectroscopy
18.3 The Laser
19.1 Particle Scattering
19.2 Nuclear Binding Energy
19.3 Fusion
19.4 Radioactivity
19.5 Particle Physics
20.1 Potential Energy Diagrams
20.2 Particle in a Box
20.3 Potential Wells
20.4 Potential Barriers

Physics for Scientists and Engineers Volume 4

A Strategic Approach

Randall D. Knight

California Polytechnic State University, San Luis Obispo

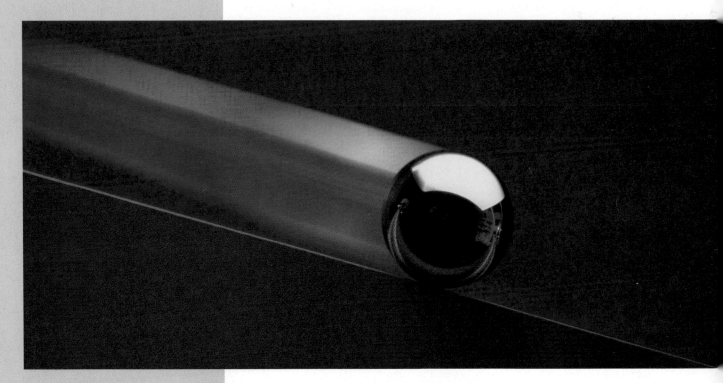

PEARSON

Addison
Wesley

San Francisco   Boston   New York
Cape Town   Hong Kong   London   Madrid   Mexico City
Montreal   Munich   Paris   Singapore   Sydney   Tokyo   Toronto

Executive Editor:	Adam Black, Ph.D.
Development Editor:	Alice Houston, Ph.D.
Project Manager:	Laura Kenney Editorial & Production Services
Associate Editor:	Liana Allday
Media Producer:	Claire Masson
Marketing Manager:	Christy Lawrence
Market Development:	Susan Winslow
Manufacturing Supervisor:	Vivian McDougal
Art Director:	Blakely Kim
Production Service:	Thompson Steele, Inc.
Text Design:	Mark Ong, Side by Side Studios
Cover Design:	Yvo Riezebos Design
Illustrations:	Precision Graphics
Photo Research:	Cypress Integrated Systems
Cover Printer:	Phoenix Color Corporation
Printer and Binder:	R. R. Donnelley & Sons

Cover Image:	Rainbow/PictureQuest
Credits:	see page x

Library of Congress Cataloging-in-Publication Data
Knight, Randall Dewey.
    Physics for scientists and engineers : a strategic approach / Randall D. Knight.
       p. cm.
    Includes index.
    ISBN 0-8053-8960-1 (extended ed. with MasteringPhysics)
      1. Physics   I. Title.

  QC23.2.K65 2004
  530--dc22

                                                                              2003062809
  ISBN 0-8053-8973-3 Volume 4 with MasteringPhysics
  ISBN 0-8053-9015-4 Volume 4 without MasteringPhysics

Many of the designations used by manufacturers and sellers to distinguish their products are claimed
as trademarks. Where those designations appear in this book, and the publisher was aware of a
trademark claim, the designations have been printed in initial caps or all caps.

6 7 8 9 10—DOW—06
www.aw-bc.com

Brief Contents

Part I Newton's Laws

Chapter 1 Concepts of Motion 3
Chapter 2 Kinematics: The Mathematics
of Motion 35
Chapter 3 Vectors and Coordinate Systems 78
Chapter 4 Force and Motion 97
Chapter 5 Dynamics I: Motion Along
a Line 122
Chapter 6 Dynamics II: Motion in a Plane 151
Chapter 7 Dynamics III: Motion in a Circle 177
Chapter 8 Newton's Third Law 207

Part II Conservation Laws

Chapter 9 Impulse and Momentum 239
Chapter 10 Energy 268
Chapter 11 Work 304

Part III Applications of Newtonian Mechanics

Chapter 12 Newton's Theory of Gravity 343
Chapter 13 Rotation of a Rigid Body 369
Chapter 14 Oscillations 413
Chapter 15 Fluids and Elasticity 444

Part IV Thermodynamics

Chapter 16 A Macroscopic Description
of Matter 485
Chapter 17 Work, Heat, and the First Law
of Thermodynamics 512
Chapter 18 The Micro/Macro Connection 547
Chapter 19 Heat Engines and Refrigerators 573

Part V Waves and Optics

Chapter 20 Traveling Waves 611
Chapter 21 Superposition 646
Chapter 22 Wave Optics 684
Chapter 23 Ray Optics 714
Chapter 24 Modern Optics and Matter
Waves 757

Part VI Electricity and Magnetism

Chapter 25 Electric Charges and Forces 783
Chapter 26 The Electric Field 817
Chapter 27 Gauss's Law 849
Chapter 28 Current and Conductivity 879
Chapter 29 The Electric Potential 900
Chapter 30 Potential and Field 932
Chapter 31 Fundamentals of Circuits 961
Chapter 32 The Magnetic Field 996
Chapter 33 Electromagnetic Induction 1041
Chapter 34 Electromagnetic Fields and
Waves 1084
Chapter 35 AC Circuits 1121

Part VII Relativity and Quantum Physics

Chapter 36 Relativity 1151
Chapter 37 The End of Classical Physics 1195
Chapter 38 Quantization 1220
Chapter 39 Wave Functions and
Uncertainty 1253
Chapter 40 One-Dimensional Quantum
Mechanics 1277
Chapter 41 Atomic Physics 1317
Chapter 42 Nuclear Physics 1352

Appendix A Mathematics Review A-1
Appendix B Periodic Table of Elements A-3
Answers to Odd-Numbered Problems A-4

About the Author

Randy Knight has taught introductory physics for over 20 years at Ohio State University and California Polytechnic University, where he is currently Professor of Physics. Professor Knight received a bachelor's degree in physics from Washington University in St. Louis and a Ph.D. in physics from the University of California, Berkeley. He was a post-doctoral fellow at the Harvard-Smithsonian Center for Astrophysics before joining the faculty at Ohio State University. It was at Ohio State that he began to learn about the research in physics education that, many years later, led to this book.

Professor Knight's research interests are in the field of lasers and spectroscopy, and he has published over 25 research papers. He recently led the effort to establish an environmental studies program at Cal Poly, where, in addition to teaching introductory physics, he also teaches classes on energy, oceanography, and environmental issues. When he's not in the classroom or in front of a computer, you can find Randy hiking, sea kayaking, playing the piano, or spending time with his wife Sally and their seven cats.

Credits

Electricity and Magnetism, Volume 4

All Part Overview images are courtesy of Lucasfilm Ltd. Addison Wesley would like to give special thanks to Lucy Wilson, Christopher Holm, and the staff of Lucasfilm Ltd. for granting us permission to use these images and for their help in selecting them.

PART VI
Part VI Overview image: *Star Wars: Episode II – Attack of the Clones* © 2002 Lucasfilm Ltd. & ™. All rights reserved. Used under authorization. Unauthorized duplication is a violation of applicable law. Page **782:** UHB Trust/Getty Images.

CHAPTER 25
Page **783:** Gandee Vasan/Getty Images. Page **784:** Tony Freeman/ PhotoEdit. Page **794:** Courtesy Xerox Corporation. Page **802** T, B: Richard Megna/Fundamental Photographs.

CHAPTER 26
Page **817:** Rachel Epstein/PhotoEdit. Page **832** U: Jody Dole/Getty Images. Page **832** L: Hannu-Pekka Hedman, University of Turku.

CHAPTER 27
Page **849:** Paul A. Souders/Corbis.

CHAPTER 28
Page **879:** Visuals Unlimited. Page **895:** IBM Research/Peter Arnold, Inc.

CHAPTER 29
Page **900:** Corbis Digital Stock. Page **920:** Christopher Johnson, University of Utah.

CHAPTER 30
Page **932:** Martyn F. Chillmaid/Science Library Photo. Page **939:** Paul Silverman/Fundamental Photos. Page **942:** Tom Pantages. Page **946:** Tom Pantages. Page **953:** Adam Hart-Davis/Photo Researchers, Inc.

CHAPTER 31
Page **961:** Courtesy Intel Corporation. Page **963:** Tom Ridley/Dorling Kindersley Media Library. Page **972:** Maya Barnes/The Image Works. Page **983:** Francisco Cruz/SuperStock.

CHAPTER 32
Page **996:** Photodisc Green/Getty Images. Page **1000:** Richard Megna/Fundamental Photographs. Page **1016:** Charles Thatcher/Getty Images. Page **1019:** Richard Megna/Fundamental Photographs. Page **1020:** Courtesy Dr. L.A. Frank, University of Iowa. Page **1021:** CERN Geneva. Page **1022:** Ernest Orlando Lawrence Berkeley National Laboratory.

CHAPTER 33
Page **1041:** Photodisc Green/Getty Images. Page **1042:** Hulton Archive/ Getty Images. Page **1063:** Lester Lefkowitz/Corbis. Page **1064:** Jonathan Nourok/PhotoEdit.

CHAPTER 34
Page **1084:** Lawrence Manning/Corbis. Page **1113** L, M, R: Richard Megna/Fundamental Photographs.

CHAPTER 35
Page **1121:** Inga Spence/Visuals Unlimited. Page **1138:** Courtesy Edwards, Inc.

PART VII
Part VII Overview image: *Star Wars: Episode VI – Return of the Jedi* © 1983 and 1997 Lucasfilm Ltd. & ™. All rights reserved. Used under authorization. Unauthorized duplication is a violation of applicable law. Page **1150:** IBM Research, Almaden Research Center.

CHAPTER 36
Page **1151:** John Y. Fowler. Page **1152:** Topham/The Image Works. Page **1172:** Stanford Linear Accelerator Center. Page **1186:** Science Photo Library/Photo Researchers. Page **1187:** Wellcome Dept. of Cognitive Neurology/Science Photo Library/Photo Researchers

Preface to the Instructor

In 1997 we published *Physics: A Contemporary Perspective*. This was the first comprehensive, calculus-based textbook to make extensive use of results from physics education research. The development and testing that led to this book had been partially funded by the National Science Foundation. In the preface we noted that it was a "work in progress" and that we very much wanted to hear from users—both instructors and students—to help us shape the book into a final form.

And hear from you we did! We received feedback and reviews from roughly 150 professors and, especially important, 4500 of their students. This textbook, the newly titled *Physics for Scientists and Engineers: A Strategic Approach*, is the result of synthesizing that feedback and using it to produce a book that we hope is uniquely tuned to helping today's students succeed. It is the first introductory textbook built from the ground up on research into how students can more effectively learn physics.

Objectives

My primary goals in writing *Physics for Scientists and Engineers: A Strategic Approach* have been:

- To produce a textbook that is more focused and coherent, less encyclopedic.
- To move key results from physics education research into the classroom in a way that allows instructors to use a range of teaching styles.
- To provide a balance of quantitative reasoning and conceptual understanding, with special attention to concepts known to cause student difficulties.
- To develop students' problem-solving skills in a systematic manner.
- To support an active-learning environment.

These goals and the rationale behind them are discussed at length in my small paperback book, *Five Easy Lessons: Strategies for Successful Physics Teaching* (Addison Wesley, 2002). Please request a copy from your local Addison Wesley sales representative if it would be of interest to you (ISBN 0-8053-8702-1).

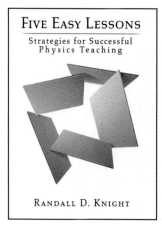

FIVE EASY LESSONS
Strategies for Successful
Physics Teaching

RANDALL D. KNIGHT

Textbook Organization

The 42-chapter extended edition (ISBN 0-8053-8685-8) of *Physics for Scientists and Engineers* is intended for use in a three-semester course. Most of the 36-chapter standard edition (ISBN 0-8053-8982-2), ending with relativity, can be covered in two semesters, but the judicious omission of a few chapters will avoid rushing through the material and give students more time to develop their knowledge and skills.

There's a growing sentiment that quantum physics is quickly becoming the province of engineers, not just scientists, and that even a two–semester course should include a reasonable introduction to quantum ideas. The *Instructor's Guide* outlines a couple of routes through the book that allow most of the quantum physics chapters to be reached in two semesters. I've written the book with the hope that an increasing number of instructors will choose one of these routes.

- **Extended edition**, with modern physics (ISBN 0-8053-8685-8): chapters 1–42.
- **Standard edition** (ISBN 0-8053-8982-2): chapters 1–36.
- **Volume 1** (ISBN 0-8053-8963-6) covers mechanics: chapters 1–15.
- **Volume 2** (ISBN 0-8053-8966-0) covers thermodynamics: chapters 16–19.
- **Volume 3** (ISBN 0-8053-8969-5) covers waves and optics: chapters 20–24.
- **Volume 4** (ISBN 0-8053-8972-5) covers electricity and magnetism, plus relativity: chapters 25–36.
- **Volume 5** (ISBN 0-8053-8975-X) covers relativity and quantum physics: chapters 36–42.
- **Volumes 1–5** boxed set (ISBN 0-8053-8978-4).

The full textbook is divided into seven parts: Part I: *Newton's Laws*, Part II: *Conservation Laws*, Part III: *Applications of Newtonian Mechanics*, Part IV: *Thermodynamics*, Part V: *Waves and Optics*, Part VI: *Electricity and Magnetism*, and Part VII: *Relativity and Quantum Mechanics*. Although I recommend covering the parts in this order (see below), doing so is by no means essential. Each topic is self-contained, and Parts III–VI can be rearranged to suit an instructor's needs. To facilitate a reordering of topics, the full text is available in the five individual volumes listed in the margin.

Organization Rationale: Thermodynamics is placed before waves because it is a continuation of ideas from mechanics. The key idea in thermodynamics is energy, and moving from mechanics into thermodynamics allows the uninterrupted development of this important idea. Further, waves introduce students to functions of two variables, and the mathematics of waves is more akin to electricity and magnetism than to mechanics. Thus moving from waves to fields to quantum physics provides a gradual transition of ideas and skills.

The purpose of placing optics with waves is to provide a coherent presentation of wave physics, one of the two pillars of classical physics. Optics as it is presented in introductory physics makes no use of the properties of electromagnetic fields. There's little reason other than historical tradition to delay optics until after E&M. The documented difficulties that students have with optics are difficulties with waves, not difficulties with electricity and magnetism. However, the optics chapters are easily deferred until the end of Part VI for instructors who prefer that ordering of topics.

More Effective Problem-Solving Instruction

Careful and systematic instruction is provided on all aspects of problem solving. Some of the features that support this approach are described here, and more details are provided in the *Instructor's Guide*.

- An emphasis on using *multiple representations*—descriptions in words, pictures, graphs, and mathematics—to look at a problem from many perspectives.
- The explicit use of *models*, such as the particle model, the wave model, and the field model, to help students recognize and isolate the essential features of a physical process.
- **TACTICS BOXES** for the development of particular skills, such as drawing a free-body diagram or using Lenz's law. Tactics Box steps are explicitly illustrated in subsequent worked examples, and these are often the starting point of a full problem-solving strategy.

TACTICS BOX 4.3 **Drawing a free-body diagram**

❶ **Identify all forces acting on the object.** This step was described in Tactics Box 4.2.

❷ **Draw a coordinate system.** Use the axes defined in your pictorial representation. If those axes are tilted, for motion along an incline, then the axes of the free-body diagram should be similarly tilted.

❸ **Represent the object as a dot at the origin of the coordinate axes.** This is the particle model.

❹ **Draw vectors representing each of the identified forces.** This was described in Tactics Box 4.1. Be sure to label each force vector.

❺ **Draw and label the *net force* vector $\vec{F}_{net}$.** Draw this vector beside the diagram, not on the particle. Or, if appropriate, write $\vec{F}_{net} = \vec{0}$. Then check that $\vec{F}_{net}$ points in the same direction as the acceleration vector $\vec{a}$ on your motion diagram.

TACTICS BOX 32.2 **Evaluating line integrals**

❶ If $\vec{B}$ is everywhere perpendicular to a line, the line integral of $\vec{B}$ is

$$\int_i^f \vec{B} \cdot d\vec{s} = 0$$

❷ If $\vec{B}$ is everywhere tangent to a line of length L *and* has the same magnitude B at every point, the line integral of $\vec{B}$ is

$$\int_i^f \vec{B} \cdot d\vec{s} = BL$$

- **PROBLEM-SOLVING STRATEGIES** that help students develop confidence and more proficient problem-solving skills through the use of a consistent four-step approach: **MODEL, VISUALIZE, SOLVE, ASSESS.** Strategies are provided for each broad class of problems, such as dynamics problems or problems involving electromagnetic induction. The (MP) icon directs students to the specially developed *Skill Builder* tutorial problems in MasteringPhysics™ (see page xi), where they can interactively work through each of these strategies online.

- Worked **EXAMPLES** that illustrate good problem-solving practices through the consistent use of the four-step problem-solving approach and, where appropriate, the Tactics Box steps. The worked examples are often very detailed and carefully lead the student step by step through the *reasoning* behind the solution, not just through the numerical calculations. Steps that are often implicit or omitted in other textbooks, because they seem so obvious to experts, are explicitly discussed since research has shown these are often the points where students become confused.

- **NOTE** ▶ Paragraphs within worked examples caution against common mistakes and point out useful tips for tackling problems.

- The *Student Workbook* (see page xi), a unique component of this text, bridges the gap between worked examples and end-of-chapter problems. It provides qualitative problems and exercises that focus on developing the skills and conceptual understanding necessary to solve problems with confidence.

- Approximately 3000 original and diverse *end-of-chapter problems* have been carefully crafted to exercise and test the full range of qualitative and quantitative problem-solving skills. *Exercises*, which are keyed to specific sections, allow students to practice basic skills and computations. *Problems* require a better understanding of the material and often draw upon multiple representations of knowledge. *Challenge Problems* are more likely to use calculus, utilize ideas from more than one chapter, and sometimes lead students to explore topics that weren't explicitly covered in the chapter.

(MP) **PROBLEM-SOLVING STRATEGY 5.2 Dynamics problems**

MODEL Make simplifying assumptions.

VISUALIZE

Pictorial representation. Show important points in the motion with a sketch, establish a coordinate system, define symbols, and identify what the problem is trying to find. This is the process of translating words to symbols.

Physical representation. Use a motion diagram to determine the object's acceleration vector $\vec{a}$. Then identify all forces acting on the object and show them on a free-body diagram.

It's OK to go back and forth between these two steps as you visualize the situation.

SOLVE The mathematical representation is based on Newton's second law

$$\vec{F}_{net} = \sum_i \vec{F}_i = m\vec{a}$$

The vector sum of the forces is found directly from the free-body diagram. Depending on the problem, either

- Solve for the acceleration, then use kinematics to find velocities and positions, or
- Use kinematics to determine the acceleration, then solve for unknown forces.

ASSESS Check that your result has the correct units, is reasonable, and answers the question.

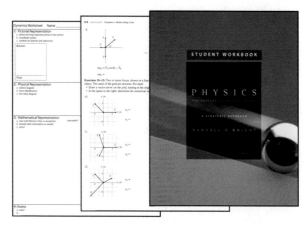

Proven Features to Promote Deeper Understanding

Research has shown that many students taking calculus-based physics arrive with a wealth of misconceptions and subsequently struggle to develop a coherent understanding of the subject. Using a number of unique, reinforcing techniques, this book tackles these issues head-on to enable students to build a solid foundation of understanding.

- A *concrete-to-abstract* approach introduces new concepts through observations about the real world and everyday experience. Step by step, the text then builds up the concepts and principles needed by a theory that will make sense of the observations and make new, testable predictions. This inductive approach better matches how students learn, and it reinforces how physics—and science in general—operates.

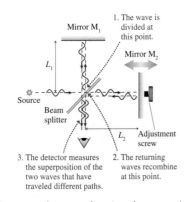

1. The wave is divided at this point.

2. The returning waves recombine at this point.

3. The detector measures the superposition of the two waves that have traveled different paths.

Annotated **FIGURE** showing the operation of the Michelson interferometer.

- **STOP TO THINK** questions embedded in each chapter allow students to assess whether they've understood the main idea of a section. The *Stop to Think* questions, which include concept questions, ratio reasoning, and ranking tasks, are primarily derived from physics education research.

- **NOTE ▶** paragraphs draw attention to common misconceptions, clarify possible confusions in terminology and notation, and provide important links to previous topics.

- Unique *annotated figures*, based on research into visual learning modes, make the artwork a teaching tool on a par with the written text. Commentary in blue—the "instructor's voice"—helps students "read" the figure. Students "learn by viewing" how to interpret a graph, how to translate between multiple representations, how to grasp a difficult concept through a visual analogy, and many other important skills.

- The learning goals and links that begin each chapter outline what the student needs to remember from previous chapters and what to focus on in the chapter ahead.

  - ▶ **Looking Ahead** lists key concepts and skills the student will learn in the coming chapter.

  - ◀ **Looking Back** suggests important topics students should review from previous chapters.

- Unique schematic *Chapter Summaries* help students organize their knowledge in an expert-like hierarchy, from general principles (top) to applications (bottom). Side-by-side pictorial, graphical, textual, and mathematical representations are used to help students with different learning styles and enable them to better translate between these key representations.

- *Part Overviews and Summaries* provide a global framework for the student's learning. Each part begins with an overview of the chapters ahead. It then concludes with a broad summary to help students draw connections between the concepts presented in that set of chapters. **KNOWLEDGE STRUCTURE** tables in the part summaries, similar to the chapter summaries, help students see a forest rather than dozens of individual trees.

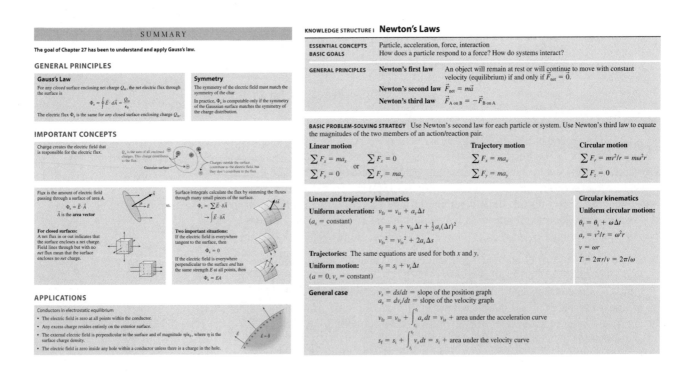

SUMMARY

The goal of Chapter 27 has been to understand and apply Gauss's law.

GENERAL PRINCIPLES

Gauss's Law
For any *closed* surface enclosing net charge Q_{in}, the net electric flux through the surface is

$$\Phi_e = \oint \vec{E} \cdot d\vec{A} = \frac{Q_{in}}{\epsilon_0}$$

The electric flux Φ_e is the same for *any* closed surface enclosing charge Q_{in}.

Symmetry
The symmetry of the electric field must match the symmetry of the charge.

In practice, Φ_e is computable only if the symmetry of the Gaussian surface matches the symmetry of the charge distribution.

IMPORTANT CONCEPTS

Charge creates the electric field that is responsible for the electric flux.

Q_{in} is the sum of all enclosed charges. This charge contributes to the flux.

Charges outside the surface contribute to the electric field, but they don't contribute to the flux.

Flux is the amount of electric field passing through a surface of area A.

$$\Phi_e = \vec{E} \cdot \vec{A}$$

$\vec{A}$ is the **area vector**.

Surface integrals calculate the flux by summing the fluxes through many small pieces of the surface.

$$\Phi_e = \sum \vec{E} \cdot \delta\vec{A}$$
$$\rightarrow \int \vec{E} \cdot \delta\vec{A}$$

For closed surfaces:
A net flux in or out indicates that the surface encloses a net charge. Field lines through but with no *net* flux mean that the surface encloses no *net* charge.

Two important situations:
If the electric field is everywhere tangent to the surface, then
$$\Phi_e = 0$$
If the electric field is everywhere perpendicular to the surface *and* has the same strength E at all points, then
$$\Phi_e = EA$$

APPLICATIONS

Conductors in electrostatic equilibrium
- The electric field is zero at all points within the conductor.
- Any excess charge resides entirely on the exterior surface.
- The external electric field is perpendicular to the surface and of magnitude η/ϵ_0, where η is the surface charge density.
- The electric field is zero inside any hole within a conductor unless there is a charge in the hole.

$\vec{E} = \vec{0}$

KNOWLEDGE STRUCTURE I Newton's Laws

ESSENTIAL CONCEPTS	Particle, acceleration, force, interaction
BASIC GOALS	How does a particle respond to a force? How do systems interact?

GENERAL PRINCIPLES	Newton's first law	An object will remain at rest or will continue to move with constant velocity (equilibrium) if and only if $\vec{F}_{net} = \vec{0}$.
	Newton's second law	$\vec{F}_{net} = m\vec{a}$
	Newton's third law	$\vec{F}_{A \text{ on } B} = -\vec{F}_{B \text{ on } A}$

BASIC PROBLEM-SOLVING STRATEGY Use Newton's second law for each particle or system. Use Newton's third law to equate the magnitudes of the two members of an action/reaction pair.

Linear motion	Trajectory motion	Circular motion
$\sum F_x = ma_x$ or $\sum F_x = 0$	$\sum F_x = ma_x$	$\sum F_r = mv^2/r = m\omega^2 r$
$\sum F_y = 0$ $\sum F_y = ma_y$	$\sum F_y = ma_y$	$\sum F_z = 0$

Linear and trajectory kinematics

Uniform acceleration: $v_{fs} = v_{is} + a_s \Delta t$
(a_s = constant)
$$s_f = s_i + v_{is}\Delta t + \tfrac{1}{2}a_s(\Delta t)^2$$
$$v_{fs}^2 = v_{is}^2 + 2a_s\Delta s$$

Trajectories: The same equations are used for both x and y.

Uniform motion: $s_f = s_i + v_s \Delta t$
($a = 0$, v_s = constant)

Circular kinematics

Uniform circular motion:
$$\theta_f = \theta_i + \omega\Delta t$$
$$a_r = v^2/r = \omega^2 r$$
$$v = \omega r$$
$$T = 2\pi r/v = 2\pi/\omega$$

General case
$v_s = ds/dt$ = slope of the position graph
$a_s = dv_s/dt$ = slope of the velocity graph

$$v_{fs} = v_{is} + \int_{t_i}^{t_f} a_s\,dt = v_{is} + \text{area under the acceleration curve}$$

$$s_f = s_i + \int_{t_i}^{t_f} v_s\,dt = s_i + \text{area under the velocity curve}$$

The Student Workbook

A key component of *Physics for Scientists and Engineers: A Strategic Approach* is the accompanying *Student Workbook*. The workbook bridges the gap between textbook and homework problems by providing students the opportunity to learn and practice skills prior to using those skills in quantitative end-of-chapter problems, much as a musician practices technique separately from performance pieces. The workbook exercises, which are keyed to each section of the textbook, focus on developing specific skills, ranging from identifying forces and drawing free-body diagrams to interpreting wave functions.

The workbook exercises, which are generally qualitative and/or graphical, draw heavily upon the physics education research literature. The exercises deal with issues known to cause student difficulties and employ techniques that have proven to be effective at overcoming those difficulties. The workbook exercises can be used in-class as part of an active-learning teaching strategy, in recitation sections, or as assigned homework. More information about effective use of the *Student Workbook* can be found in the *Instructor's Guide*.

Available versions: Extended (ISBN 0-8053-8961-X), Standard (ISBN 0-8053-8984-9), Volume 1 (ISBN 0-8053-8965-2), Volume 2 (ISBN 0-8053-8968-7), Volume 3 (ISBN 0-8053-8971-7), Volume 4 (ISBN 0-8053-8974-1), and Volume 5 (ISBN 0-8053-8977-6).

Instructor Supplements

- The **Instructor's Guide for Physics for Scientists and Engineers** (ISBN 0-8053-8985-7) offers detailed comments and suggested teaching ideas for every chapter, an extensive review of what has been learned from physics education research, and guidelines for using active-learning techniques in your classroom.

- The **Instructor's Solutions Manuals**, **Chapters 1–19** (ISBN 0-8053-8986-5), and **Chapters 20–42** (ISBN 0-8053-8989-X), written by Professors Pawan Kahol and Donald Foster, at Wichita State University, provide *complete* solutions to all the end-of-chapter problems. The solutions follow the four-step Model/Visualize/Solve/Assess procedure used in the *Problem-Solving Strategies* and all worked examples. Emphasis is placed on the reasoning behind the solution, rather than just the numerical manipulations. The full text of each solution is available as an editable Word document and as a pdf file on the *Instructor's Supplement CD-ROM* for your own use or for posting on your course website.

- The cross-platform **Instructor's Resource CD-ROMs** (ISBN 0-8053-8996-2) consists of the **Simulation and Image Presentation CD-ROM** and the **Instructor's Supplement CD-ROM**. The *Simulation and Image Presentation CD-ROM* provides a comprehensive library of more than 220 applets from *ActivPhysics OnLine*, as well as all the figures from the textbook (excluding photographs) in JPEG format. In addition, all the tables, chapter summaries, and knowledge structures are provided as JPEGs, and the Tactics Boxes, Problem-Solving Strategies, and key (boxed) equations are provided in editable Word format. The *Instructor's Supplement CD-ROM* provides editable Word versions and pdf files of the *Instructor's Guide* and the *Instructor's Solutions Manuals*. Complete *Student Workbook* solutions are also provided as pdf files.

- **MasteringPhysics**™ (www.masteringphysics.com) is a sophisticated, research-proven online tutorial and homework assignment system that provides students with individualized feedback and hints based on their input. It provides a comprehensive library of conceptual tutorials (including one for each

Problem-Solving Strategy in this textbook), multistep self-tutoring problems, and end-of-chapter problems from *Physics for Scientists and Engineers*. *MasteringPhysics*™ provides instructors with a fast and effective way to assign online homework assignments that comprise a range of problem types. The powerful post-assignment diagnostics allow instructors to assess the progress of their class as a whole or to quickly identify individual students' areas of difficulty.

- **ActivPhysics**™ **OnLine** (www.aw-bc.com/knight) provides a comprehensive library of more than 420 tried and tested *ActivPhysics* applets updated for web delivery using the latest online technologies. In addition, it provides a suite of highly regarded applet-based tutorials developed by education pioneers Professors Alan Van Heuvelen and Paul D'Alessandris. The *ActivPhysics* margin icon directs students to specific exercises that complement the textbook discussion.

  The online exercises are designed to encourage students to confront misconceptions, reason qualitatively about physical processes, experiment quantitatively, and learn to think critically. They cover all topics from mechanics to electricity and magnetism and from optics to modern physics. The highly acclaimed *ActivPhysics OnLine* companion workbooks help students work through complex concepts and understand them more clearly. More than 220 applets from the *ActivPhysics OnLine* library are also available on the *Simulation and Image Presentation CD-ROM*.

- The **Printed Test Bank** (ISBN 0-8053-8994-6) and cross-platform **Computerized Test Bank** (ISBN 0-8053-8995-4), prepared by Professor Benjamin Grinstein, at the University of California, San Diego, contain more than 1500 high-quality problems, with a range of multiple-choice, true/false, short-answer, and regular homework-type questions. In the computerized version, more than half of the questions have numerical values that can be randomly assigned for each student.

- The **Transparency Acetates** (ISBN 0-8053-8993-8) provide more than 200 key figures from *Physics for Scientists and Engineers* for classroom presentation.

Student Supplements

- The **Student Solutions Manuals Chapters 1–19** (ISBN 0-8053-8708-0) and **Chapters 20–42** (ISBN 0-8053-8998-9), written by Professors Pawan Kahol and Donald Foster at Wichita State University, provides *detailed* solutions to more than half of the odd-numbered end-of-chapter problems. The solutions follow the four-step Model/Visualize/Solve/Assess procedure used in the *Problem-Solving Strategies* and all worked examples.

- **MasteringPhysics**™ (www.masteringphysics.com) provides students with individualized online tutoring by responding to their wrong answers and providing hints for solving multistep problems. It gives them immediate and up-to-date assessment of their progress, and shows where they need to practice more.

- **ActivPhysics**™ **OnLine** (www.aw-bc.com/knight) provides students with a suite of highly regarded applet-based tutorials (see above). The accompanying workbooks help students work though complex concepts and understand them more clearly. The *ActivPhysics* margin icon directs students to specific exercises that complement the textbook discussion.

- **ActivPhysics OnLine Workbook Volume 1: Mechanics • Thermal Physics • Oscillations & Waves** (ISBN 0-8053-9060-X)

- **ActivPhysics OnLine Workbook Volume 2: Electricity & Magnetism • Optics • Modern Physics** (ISBN 0-8053-9061-8)

■ The **Addison-Wesley Tutor Center** (www.aw.com/tutorcenter) provides one-on-one tutoring via telephone, fax, email, or interactive website during evening hours and on weekends. Qualified college instructors answer questions and provide instruction for *Mastering Physics*™ and for the examples, exercises, and problems in *Physics for Scientists and Engineers*.

Acknowledgments

I have relied upon conversations with and, especially, the written publications of many members of the physics education community. Those who may recognize their influence include Arnold Arons, Uri Ganiel, Ibrahim Halloun, Richard Hake, David Hestenes, Leonard Jossem, Jill Larkin, Priscilla Laws, John Mallinckrodt, Lillian McDermott, Edward "Joe" Redish, Fred Reif, Rachel Scherr, Bruce Sherwood, David Sokoloff, Ronald Thornton, Sheila Tobias, and Alan Van Heuleven. John Rigden, founder and director of the Introductory University Physics Project, provided the impetus that got me started down this path. Early development of the materials was supported by the National Science Foundation as the *Physics for the Year 2000* project; their support is gratefully acknowledged.

I am grateful to Pawan Kahol and Don Foster for the difficult task of writing the *Instructor's Solutions Manuals*; to Jim Andrews and Susan Cable for writing the workbook answers; to Wayne Anderson, Jim Andrews, Dave Ettestad, Stuart Field, Robert Glosser, and Charlie Hibbard for their contributions to the end-of-chapter problems; and to my colleague Matt Moelter for many valuable contributions and suggestions.

I especially want to thank my editor Adam Black, development editor Alice Houston, editorial assistant Liana Allday, and all the other staff at Addison Wesley for their enthusiasm and hard work on this project. Project manager Laura Kenney, Carolyn Field and the team at Thompson Steele, Inc., copy editor Kevin Gleason, photo researcher Brian Donnelly, and page-layout artist Judy Maenle get much of the credit for making this complex project all come together. In addition to the reviewers and classroom testers listed below, who gave invaluable feedback, I am particularly grateful to Wendell Potter and Susan Cable for their close scrutiny of every word and figure.

Finally, I am endlessly grateful to my wife Sally for her love, encouragement, and patience, and to our many cats for their innate abilities to hold down piles of papers and to type qqqqqqqq whenever it was needed.

Randy Knight, September 2003
rknight@calpoly.edu

Reviewers and Classroom Testers

Gary B. Adams, *Arizona State University*

Wayne R. Anderson, *Sacramento City College*

James H. Andrews, *Youngstown State University*

David Balogh, *Fresno City College*

Dewayne Beery, *Buffalo State College*

Joseph Bellina, *Saint Mary's College*

James R. Benbrook, *University of Houston*

David Besson, *University of Kansas*

Randy Bohn, *University of Toledo*

Art Braundmeier, *University of Southern Illinois, Edwardsville*

Carl Bromberg, *Michigan State University*

Douglas Brown, *Cabrillo College*

Ronald Brown, *California Polytechnic State University, San Luis Obispo*

Mike Broyles, *Collin County Community College*

James Carolan, *University of British Columbia*
Michael Crescimanno, *Youngstown State University*
Wei Cui, *Purdue University*
Robert J. Culbertson, *Arizona State University*
Purna C. Das, *Purdue University North Central*
Dwain Desbien, *Estrella Mountain Community College*
John F. Devlin, *University of Michigan, Dearborn*
Alex Dickison, *Seminole Community College*
Chaden Djalali, *University of South Carolina*
Sandra Doty, *Denison University*
Miles J. Dresser, *Washington State University*
Charlotte Elster, *Ohio University*
Robert J. Endorf, *University of Cincinnati*
Tilahun Eneyew, *Embry-Riddle Aeronautical University*
F. Paul Esposito, *University of Cincinnati*
John Evans, *Lee University*
Michael R. Falvo, *University of North Carolina*
Abbas Faridi, *Orange Coast College*
Stuart Field, *Colorado State University*
Daniel Finley, *University of New Mexico*
Jane D. Flood, *Muhlenberg College*
Thomas Furtak, *Colorado School of Mines*
Richard Gass, *University of Cincinnati*
J. David Gavenda, *University of Texas, Austin*
Stuart Gazes, *University of Chicago*
Katherine M. Gietzen, *Southwest Missouri State University*
Robert Glosser, *University of Texas, Dallas*
William Golightly, *University of California, Berkeley*
Paul Gresser, *University of Maryland*
C. Frank Griffin, *University of Akron*
John B. Gruber, *San Jose State University*
Randy Harris, *University of California, Davis*
Stephen Haas, *University of Southern California*
Nicole Herbots, *Arizona State University*
Scott Hildreth, *Chabot College*
David Hobbs, *South Plains College*
Laurent Hodges, *Iowa State University*
John L. Hubisz, *North Carolina State University*
George Igo, *University of California, Los Angeles*
Bob Jacobsen, *University of California, Berkeley*
Rong-Sheng Jin, *Florida Institute of Technology*
Marty Johnston, *University of St. Thomas*
Stanley T. Jones, *University of Alabama*
Darrell Judge, *University of Southern California*
Pawan Kahol, *Wichita State University*
Teruki Kamon, *Texas A&M University*
Richard Karas, *California State University, San Marcos*
Deborah Katz, *U.S. Naval Academy*
Miron Kaufman, *Cleveland State University*
M. Kotlarchyk, *Rochester Institute of Technology*
Cagliyan Kurdak, *University of Michigan*
Fred Krauss, *Delta College*
H. Sarma Lakkaraju, *San Jose State University*

Darrell R. Lamm, *Georgia Institute of Technology*
Robert LaMontagne, *Providence College*
Alessandra Lanzara, *University of California, Berkeley*
Sen-Ben Liao, *Massachusetts Institute of Technology*
Dean Livelybrooks, *University of Oregon*
Chun-Min Lo, *University of South Florida*
Richard McCorkle, *University of Rhode Island*
James McGuire, *Tulane University*
Theresa Moreau, *Amherst College*
Gary Morris, *Rice University*
Michael A. Morrison, *University of Oklahoma*
Richard Mowat, *North Carolina State University*
Taha Mzoughi, *Mississippi State University*
Vaman M. Naik, *University of Michigan, Dearborn*
Craig Ogilvie, *Iowa State University*
Martin Okafor, *Georgia Perimeter College*
Benedict Y. Oh, *University of Wisconsin*
Georgia Papaefthymiou, *Villanova University*
Peggy Perozzo, *Mary Baldwin College*
Brian K. Pickett, *Purdue University, Calumet*
Joe Pifer, *Rutgers University*
Dale Pleticha, *Gordon College*
Robert Pompi, *SUNY-Binghamton*
David Potter, *Austin Community College*
Chandra Prayaga, *University of West Florida*
Didarul Qadir, *Central Michigan University*
Michael Read, *College of the Siskiyous*
Michael Rodman, *Spokane Falls Community College*
Sharon Rosell, *Central Washington University*
Anthony Russo, *Okaloosa-Walton Community College*
Otto F. Sankey, *Arizona State University*
Rachel E. Scherr, *University of Maryland*
Bruce Schumm, *University of California, Santa Cruz*
Douglas Sherman, *San Jose State University*
Elizabeth H. Simmons, *Boston University*
Alan Slavin, *Trent College*
William Smith, *Boise State University*
Paul Sokol, *Pennsylvania State University*
Chris Sorensen, *Kansas State University*
Anna and Ivan Stern, *AW Tutor Center*
Michael Strauss, *University of Oklahoma*
Arthur Viescas, *Pennsylvania State University*
Chris Vuille, *Embry-Riddle Aeronautical University*
Ernst D. Von Meerwall, *University of Akron*
Robert Webb, *Texas A&M University*
Zodiac Webster, *California State University, San Bernardino*
Robert Weidman, *Michigan Technical University*
Jeff Allen Winger, *Mississippi State University*
Ronald Zammit, *California Polytechnic State University, San Luis Obispo*
Darin T. Zimmerman, *Pennsylvania State University, Altoona*

Preface to the Student

From Me to You

The most incomprehensible thing about the universe is that it is comprehensible.
—Albert Einstein

The day I went into physics class it was death.
—Sylvia Plath, *The Bell Jar*

Let's have a little chat before we start. A rather one-sided chat, admittedly, because you can't respond, but that's OK. I've heard from many of your fellow students over the years, so I have a pretty good idea of what's on your mind.

What's your reaction to taking physics? Fear and loathing? Uncertainty? Excitement? All of the above? Let's face it, physics has a bit of an image problem on campus. You've probably heard that it's difficult, maybe downright impossible unless you're an Einstein. Things that you've heard, your experiences in other science courses, and many other factors all color your *expectations* about what this course is going to be like.

It's true that there are many new ideas to be learned in physics and that the course, like college courses in general, is going to be much faster paced than science courses you had in high school. I think it's fair to say that it will be an *intense* course. But we can avoid many potential problems and difficulties if we can establish, here at the beginning, what this course is about and what is expected of you—and of me!

Just what is physics, anyway? Physics is a way of thinking about the physical aspects of nature. Physics is not better than art or biology or poetry or religion, which are also ways to think about nature; it's simply different. One of the things this course will emphasize is that physics is a human endeavor. The information content of this book was not found in a cave or conveyed to us by aliens; it was discovered by real people engaged in a struggle with real issues. I hope to convey to you something of the history and the process by which we have come to accept the principles that form the foundation of today's science and engineering.

You might be surprised to hear that physics is not about "facts." Oh, not that facts are unimportant, but physics is far more focused on discovering *relationships* that exist between facts and *patterns* that exist in nature than on learning facts for their own sake. As a consequence, there's not a lot of memorization when you study physics. Some—there are still definitions and equations to learn—but less than in many other courses. Our emphasis, instead, will be on thinking and reasoning. This is important to factor into your expectations for the course.

Perhaps most important of all, *physics is not math!* Physics is much broader. We're going to look for patterns and relationships in nature, develop the logic that relates different ideas, and search for the reasons *why* things happen as they do. In doing so, we're going to stress qualitative reasoning, pictorial and graphical reasoning, and reasoning by analogy. And yes, we will use math, but it's just one tool among many.

It will save you much frustration if you're aware of this physics–math distinction up front. Many of you, I know, want to find a formula and plug numbers into it—that is, to do a math problem. Maybe that's what you learned in high school science courses, but it is *not* what this course expects of you. We'll certainly do

(a) X-ray diffraction pattern

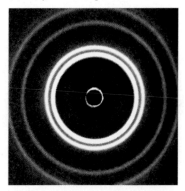

(b) Electron diffraction pattern

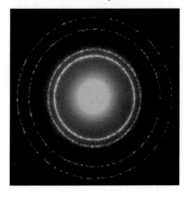

many calculations, but the specific numbers are usually the last and least important step in the analysis.

Physics is about recognizing patterns. The top photograph is an x-ray diffraction pattern that shows how a collimated beam of x rays spreads out after passing through a crystal. The bottom photograph shows what happens when a collimated beam of electrons is shot through the same crystal. What does the obvious similarity in these two photographs tell us about the nature of light and about the nature of matter?

As you study, you'll sometimes be baffled, puzzled, and confused. That's perfectly normal and to be expected. Making mistakes is OK too *if* you're willing to learn from the experience. No one is born knowing how to do physics any more than he or she is born knowing how to play the piano or shoot basketballs. The ability to do physics comes from practice, repetition, and struggling with the ideas until you "own" them and can apply them yourself in new situations. There's no way to make learning effortless, at least for anything worth learning, so expect to have some difficult moments ahead.

But also expect to have some moments of excitement at the joy of discovery. There will be instants at which the pieces suddenly click into place and you *know* that you understand a difficult idea. There will be times when you'll surprise yourself by successfully working a difficult problem that you didn't think you could solve. My hope, as an author, is that the excitement and sense of adventure will far outweigh the difficulties and frustrations.

Many of you, I suspect, would like to know the "best" way to study for this course. There is no best way. People are too different, and what works for one student works less effectively for another. But I do want to stress that *reading the text* is vitally important. Class time will be used to clarify difficulties and to develop tools for using the knowledge, but your instructor will *not* use class time simply to repeat information in the text. The basic knowledge for this course is written down within these pages, and the *number one expectation* is that you will read carefully and thoroughly to find and learn that knowledge.

Despite there being no best way to study, I will suggest *one* way that is successful for many students. It consists of the following four steps:

1. **Read each chapter *before* it is discussed in class.** I cannot stress too highly how important this step is. Class attendance is largely ineffective if you have not prepared. When you first read a chapter, focus on learning new vocabulary, definitions, and notation. There's a list of terms and notations at the end of each chapter. Learn them! You won't understand what's being discussed or how the ideas are being used if you don't know what the terms and symbols mean.
2. **Participate actively in class.** Take notes, ask and answer questions, take part in discussion groups. There is ample scientific evidence that *active participation* is far more effective for learning science than is passive listening.
3. **After class, go back for a *careful* rereading of the chapter.** In your second reading, pay closer attention to the details and the worked examples. Look for the *logic* behind each example (and I've tried to help make this clear), not just at what formula is being used. Do the *Student Workbook* exercises for each section as you finish your reading of it.
4. **Finally, apply what you have learned to the homework problems at the end of each chapter.** I strongly encourage you to form a study group with two or three classmates. There's good evidence that students who study regularly with a group do better than the rugged individualists who try to go it alone.

Did someone mention a workbook? The companion *Student Workbook* is a vital part of this course. It contains questions and exercises that ask you to reason *qualitatively*, to use graphical information, and to give explanations. It is through these exercises that you will learn what the concepts mean and will practice the reasoning skills appropriate to the chapter. You will then have acquired the baseline knowledge that you need *before* turning to the end-of-chapter homework problems. In sports or in music, you would never think of performing before you practice, so why would you want to do so in physics? The workbook is where you practice and work on basic skills.

Many of you, I know, would like to go straight to the homework problems and then thumb through the text looking for a formula that seems like it will work. That approach will not succeed in this course, and it's guaranteed to make you frustrated and discouraged. Very few homework problems are "plug and chug" problems where you simply put numbers into a formula. To work the homework problems successfully, you need a better study strategy—either that outlined above or your own—that helps you learn the concepts and the relationships between the ideas. Many of the chapters in this book have Problem-Solving Strategies to help you develop effective problem-solving skills.

A traditional guideline in college is to study two hours outside of class for every hour spent in class, and this text is designed with that expectation. Of course, two hours is an average. Some chapters are fairly straightforward and will go quickly. Others likely will require much more than two study hours per class hour.

Now that you know more about what is expected of you, what can you expect of me? That's a little trickier, because the book is already written! Nonetheless, it was prepared on the basis of what I think my students throughout the years have expected—and wanted—from their physics textbook.

You should know that these course materials—the text and the workbook—are based upon extensive research about how students learn physics and the challenges they face. The effectiveness of many of the exercises has been demonstrated through extensive class testing. I've written the book in an informal style that I hope you will find appealing and that will encourage you to do the reading. And finally, I have endeavored to make clear not only that physics, as a technical body of knowledge, is relevant to your profession but also that physics is an exciting adventure of the human mind.

I hope you'll enjoy the time we're going to spend together.

Detailed Contents

Volume 1 contains chapters 1–15; Volume 2 contains chapters 16–19; Volume 3 contains chapters 20–24; Volume 4 contains chapters 25–36; Volume 5 contains chapters 36–42.

Part VI Electricity and Magnetism

OVERVIEW Charges, Currents, and Fields   781

Chapter 25 **Electric Charges and Forces**   783
25.1   Developing a Charge Model   784
25.2   Charge   788
25.3   Insulators and Conductors   791
25.4   Coulomb's Law   796
25.5   The Concept of a Field   802
25.6   The Field Model   805
       SUMMARY   810
       EXERCISES AND PROBLEMS   811

Chapter 26 **The Electric Field**   817
26.1   Electric Field Models   817
26.2   The Electric Field of Multiple Point
       Charges   819
26.3   The Electric Field of a Continuous
       Charge Distribution   824
26.4   The Electric Fields of Rings, Planes,
       and Spheres   829
26.5   The Parallel-Plate Capacitor   834
26.6   Motion of a Charged Particle in an
       Electric Field   835
26.7   Motion of a Dipole in an Electric
       Field   839
       SUMMARY   842
       EXERCISES AND PROBLEMS   843

Chapter 27 **Gauss's Law**   849
27.1   Symmetry   850
27.2   The Concept of Flux   853
27.3   Calculating Electric Flux   855
27.4   Gauss's Law   861
27.5   Using Gauss's Law   865
27.6   Conductors in Electrostatic Equilibrium   870
       SUMMARY   873
       EXERCISES AND PROBLEMS   874

Chapter 28 **Current and Conductivity**   879
28.1   The Electron Current   880
28.2   Creating a Current   884
28.3   Batteries   889
28.4   Current and Current Density   890
28.5   Conductivity and Resistivity   893
       SUMMARY   896
       EXERCISES AND PROBLEMS   897

Chapter 29 **The Electric Potential**   900
29.1   Electric Potential Energy   901
29.2   The Potential Energy of Point Charges   904
29.3   The Potential Energy of a Dipole   909
29.4   The Electric Potential   910
29.5   The Electric Potential Inside
       a Parallel-Plate Capacitor   914
29.6   The Electric Potential of a Point Charge   918
29.7   The Electric Potential of Many Charges   920
       SUMMARY   925
       EXERCISES AND PROBLEMS   926

Chapter 30 **Potential and Field**   932
30.1   Connecting Potential and Field   933
30.2   Finding the Electric Field from
       the Potential   935
30.3   A Conductor in Electrostatic
       Equilibrium   939
30.4   Sources of Electric Potential   941
30.5   Connecting Potential and Current   943
30.6   Capacitance and Capacitors   946
30.7   The Energy Stored in a Capacitor   951
       SUMMARY   954
       EXERCISES AND PROBLEMS   955

Chapter 31 **Fundamentals of Circuits 961**
31.1 Resistors and Ohm's Law 962
31.2 Circuit Elements and Diagrams 964
31.3 Kirchhoff's Laws and the Basic Circuit 965
31.4 Energy and Power 970
31.5 Series Resistors 973
31.6 Real Batteries 976
31.7 Parallel Resistors 978
31.8 Resistor Circuits 981
31.9 Getting Grounded 983
31.10 *RC* Circuits 985
SUMMARY 988
EXERCISES AND PROBLEMS 989

Chapter 32 **The Magnetic Field 996**
32.1 Magnetism 997
32.2 The Discovery of the Magnetic Field 998
32.3 The Source of the Magnetic Field: Moving Charges 1001
32.4 The Magnetic Field of a Current 1004
32.5 Magnetic Dipoles 1008
32.6 Ampère's Law and Solenoids 1011
32.7 The Magnetic Force on a Moving Charge 1017
32.8 Magnetic Forces on Current-Carrying Wires 1024
32.9 Forces and Torques on Current Loops 1026
32.10 Magnetic Properties of Matter 1028
SUMMARY 1032
EXERCISES AND PROBLEMS 1033

Chapter 33 **Electromagnetic Induction 1041**
33.1 Induced Currents 1041
33.2 Motional emf 1043
33.3 Magnetic Flux 1050
33.4 Lenz's Law 1053
33.5 Faraday's Law 1056
33.6 Induced Fields and Electromagnetic Waves 1061
33.7 Induced Currents: Three Applications 1063
33.8 Inductors 1065
33.9 *LC* Circuits 1069
33.10 *LR* Circuits 1072
SUMMARY 1075
EXERCISES AND PROBLEMS 1076

Chapter 34 **Electromagnectic Fields and Waves 1084**
34.1 Electromagnetic Fields and Forces 1085
34.2 *E* or *B*? It Depends on Your Perspective 1088
34.3 Faraday's Law Revisited 1094
34.4 The Displacement Current 1099
34.5 Maxwell's Equations 1102
34.6 Electromagnetic Waves 1104
34.7 Properties of Electromagnetic Waves 1109
34.8 Polarization 1112
SUMMARY 1115
EXERCISES AND PROBLEMS 1116

Chapter 35 **AC Circuits 1121**
35.1 AC Sources and Phasors 1122
35.2 Capacitor Circuits 1124
35.3 *RC* Filter Circuits 1126
35.4 Inductor Circuits 1129
35.5 The Series *RLC* Circuit 1131
35.6 Power in AC Circuits 1135
SUMMARY 1040
EXERCISES AND PROBLEMS 1041
PART SUMMARY Electricity and Magnetism 1146

Part VII Relativity

OVERVIEW Contemporary Physics 1149

Chapter 36 **Relativity 1151**
36.1 Relativity: What's It All About? 1152
36.2 Galilean Relativity 1152
36.3 Einstein's Principle of Relativity 1157
36.4 Events and Measurements 1160
36.5 The Relativity of Simultaneity 1164
36.6 Time Dilation 1166
36.7 Length Contraction 1171
36.8 The Lorentz Transformations 1175
36.9 Relativistic Momentum 1180
36.10 Relativistic Energy 1183
SUMMARY 1189
EXERCISES AND PROBLEMS 1190

Appendix A Mathematics Review A-1
Appendix B Periodic Table of Elements A-3
Answers to Odd-Numbered Questions A-4
Index I-1

Anakin's and Obi-Wan's speeder discharges the power coupler on the planet Coruscant by flying through it. If the discharge current through the speeder is 5000 A, what voltage is developed across the speeder? To find out, what properties of the speeder do you need to estimate?

Electricity and Magnetism

Charges, Currents, and Fields

Amber, or fossilized tree resin, has long been prized for the beauty of its lustrous yellow color. Amber is of scientific interest today because biologists have learned how to recover strands of DNA from million-year-old insects that were trapped in the resin. But amber has an ancient scientific connection as well. The Greek word for amber is *elektron*.

It has been known since at least the fifth century B.C. that a piece of amber that has been rubbed briskly can attract feathers or small pieces of straw—seemingly magical powers to a pre-scientific society. It was also known to Greeks of the same time period that certain stones from the region they called *Magnesia* (in present-day Turkey) could pick up pieces of iron. It is from these humble beginnings that we today have high-speed computers, lasers, fiber-optic communications, and magnetic-resonance imaging as well as such mundane modern-day miracles as the light bulb.

The story of electricity and magnetism is vast. The development of a successful electromagnetic theory, which occupied the leading physicists of Europe for most of the nineteenth century, led to sweeping revolutions in both science and technology. The complete formulation of the theory of the electromagnetic field has been called by no less than Einstein "the most important event in physics since Newton's time." Not surprisingly, all that we can do in this text is to develop some of the basic ideas and concepts, leaving many details and advanced applications to later courses. Even so, our study of electricity and magnetism will require learning many new and important ideas. Foremost among these will be the idea of a *field*.

Phenomena and Theories

The basic phenomena of electricity and magnetism are not as familiar to most people as those of mechanics. You have spent your entire life exerting forces on objects and watching them move, but your experience with electricity and magnetism is probably much more limited.

It is hard to motivate the need for a major new theory if you are not aware of what the theory is meant to explain. We will deal with this lack of experience by placing a large emphasis on the basic *phenomena* of electricity and magnetism. We will begin by looking in detail at the fundamental properties of *electric charge* and the process of *charging* an object. It is easy to make systematic observations of how charges behave, and we will be led to consider the forces between charges and how charges behave in different materials. The development of electrical technology and the dawn of the electronic age came about as scientists and

781

engineers learned to *control* the movement of charges. Electric current, whether it be for lighting a light bulb or changing the state of a computer memory element, is simply a controlled motion of charges through conducting materials. One of our goals will be to understand how charges move through electric circuits.

When we turn to magnetic behavior, we will start by observing how magnets stick to some metals but not others and how magnets affect compass needles. But our most important observation, which you may have seen, is that an electric current can affect a compass needle in exactly the same way as a magnet. This observation will suggest to us that there is a close connection between electricity and magnetism, and we will explore this relationship in detail. Our path will eventually lead to the discovery of electromagnetic waves.

Our goal in Part VI is to develop a theory that will explain the phenomena of electricity and magnetism. The linchpin of our theory will be the entirely new concept of a *field*. Electricity and magnetism are about the long-range interactions of charges, both static charges and moving charges, and the field concept will help us understand how these interactions take place. Much of our attention will be focused on the interplay between charges and fields: How fields are created by charges and how charges, in return, respond to the fields. Bit by bit, we will assemble a theory—based on the new concepts of electric and magnetic fields—that will allow us to understand, explain, and predict a wide range of electromagnetic behavior.

The Microscopic Model

There are two different aspects to the theory of electromagnetism. The field theory provides a macroscopic perspective on the phenomena, but we can also take a microscopic view. At the microscopic level, we want to know what charges are, how they are related to atoms and molecules, and how they move about through various kinds of materials. We will develop a microscopic model of electrons and ions moving in response to electric and magnetic forces. This microscopic perspective of electricity and magnetism is analogous to the kinetic theory of gases in our study of thermodynamics.

Likely *the* most important scientific discovery of the modern era is that matter consists of atoms. It was found near the end of the nineteenth century that the atoms themselves are not indestructible objects but, instead, have constituents that are *charged particles*. We know these today as electrons and protons. Much of the time these charged

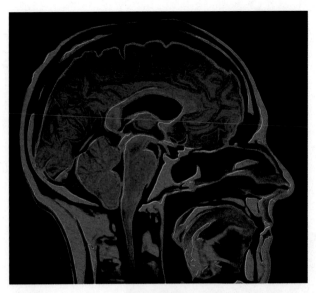

Magnetic-resonance imaging, or MRI, uses the magnetic properties of atoms as a non-invasive probe of the human body.

constituents all balance, and the atoms, as well as macroscopic objects built of these atoms, are electrically neutral. That has been the implicit assumption for all the physics we have done until now. But under some circumstances, the charges can become separated and move about. An important goal of our microscopic model will be to understand how charges become separated and how charged particles move through conductors as what we call a *current*.

But our interest at the microscopic level is more than simply how currents flow. Electricity and magnetism are of particular significance in our quest to understand the atomic structure of matter itself. The electric force is the "glue" that holds the atom together and that binds atoms into molecules and solids. The forces that we see in the macroscopic world as friction, tension, adhesion, and other contact forces are really electric forces acting at the atomic level. Magnetism plays a lesser, but not insignificant, role. While magnetism is not involved in the forces that bind atoms and molecules together, it does figure prominently in the macroscopic behavior of materials. In addition, the magnetic properties of atoms have become important probes into the interior of solids and liquids. Magnetic-resonance imaging in medicine is the most well known example of this technology, but many of the same techniques are used in science and engineering to characterize materials. For all these many reasons, acquiring a knowledge of electricity and magnetism is an essential part of science and engineering education.

25 Electric Charges and Forces

Lightning is a vivid manifestation of electric charges and forces.

▶ Looking Ahead

The goal of Chapter 25 is to develop a basic understanding of electric phenomena in terms of charges, forces, and fields. In this chapter you will learn to:

- Use a charge model to explain basic electric phenomena.
- Understand the electric properties of insulators and conductors.
- Use Coulomb's law to calculate the electric force between charges.
- Use a field model to explain the long-range interaction between charges.
- Calculate and display the electric field of a point charge.

◀ Looking Back

The mathematical analysis of electric forces and fields makes extensive use of vector addition. The electric force is in some ways analogous to gravity. Please review:

- Sections 3.2–3.4 Vector properties and vector addition.
- Sections 12.3–12.4 Newton's theory of gravity.

The electric force is one of the fundamental forces of nature. Sometimes, as in this lightning strike, electric forces can be wild and uncontrolled. On the other hand, controlled electricity is the cornerstone of our modern, technological society. Electric devices range from light bulbs and motors to computers and sophisticated medical equipment. Try imagining what it would be like to live without electricity!

But how do we control and manage this force? What are the properties of electricity and electric forces? How do we generate, transport, and use electricity? These are the questions we will explore throughout Part VI. Electricity is a big topic, and we cannot hope to answer all these questions at once.

A charged object, such as a comb that you've run through your hair, picks up small pieces of paper.

We will begin, in this chapter, by investigating some of the basic phenomena of electricity. It's hard to see what rubbing plastic rods with wool has to do with computers or generators, but only by starting at the very beginning, with simple observations, will we develop the understanding we need to use electricity in a controlled and precise manner.

25.1 Developing a Charge Model

You can receive a mildly unpleasant shock and produce a little spark if you touch a metal doorknob after walking across a carpet. Vigorously brushing your freshly washed hair makes all the hairs fly apart. A plastic comb that you've run through your hair will pick up bits of paper and other small objects, but a metal comb won't.

The common factor in these observations is that two objects are *rubbed* together. Why should rubbing an object cause forces and sparks? What kind of forces are these? Why do metallic objects behave differently from nonmetallic? These are the questions with which we begin our study of electricity.

Our first goal is to develop a model for understanding electric phenomena in terms of *charges* and *forces*. We will later use our contemporary knowledge of atoms to understand electricity on a microscopic level, but the basic concepts of electricity make *no* reference to atoms or electrons. The theory of electricity was well established long before the electron was discovered.

Experimenting with Charges

Let us enter a laboratory where we can make observations of electric phenomena. This is a modest laboratory, much like one you would have found in the year 1800. The major tools in the lab are:

- A variety of plastic, glass, and wood rods, each a few inches long. These can be held in your hand or suspended by threads from a support.
- A few metal rods with wood handles.
- Pieces of wool and silk.
- Small metal spheres, an inch or two in diameter, on wood stands.

We will manipulate and use these tools with the goal of developing a theory to explain the phenomena we see.

Discovering electricity I

| Experiment 1 | Experiment 2 | Experiment 3 | Experiment 4 |

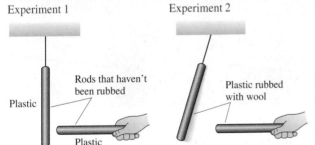
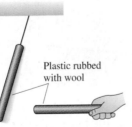
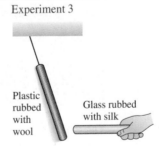
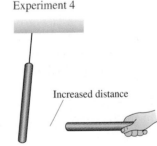

Take a plastic rod that has been undisturbed for a long period of time and hang it by a thread. Pick up another undisturbed plastic rod and bring it close to the hanging rod. Nothing happens to either rod.

Vigorously rub both the hanging plastic rod and the hand-held plastic rod with wool. Now the hanging rod tries to move away from the handheld rod when you bring the two close together. Rubbing two glass rods with silk produces the same result: The two rods repel each other.

Bring a glass rod that has been rubbed with silk close to a hanging plastic rod that has been rubbed with wool. These two rods *attract* each other.

Further observations show that:

- The strength of these forces is greater for rods that have been rubbed more vigorously.
- The strength of the forces decreases as the separation between the rods increases.

No forces were observed in Experiment 1. We will say that the original objects are **neutral.** Rubbing the rods (Experiments 2 and 3) somehow causes forces to be exerted between them. We will call the rubbing process **charging** and say that a rubbed rod is *charged.* For now, these are simply descriptive terms. The terms don't tell us anything about the process itself.

Experiment 2 shows that there is a *long-range repulsive force* between two identical objects that have been charged in the *same* way, such as two plastic rods both rubbed with wool. Furthermore, Experiment 4 finds that the force between two charged objects depends on the distance between them. This is the first long-range force we've encountered since gravity was introduced in Chapter 4. It is also the first time we've observed a repulsive force, so right away we see that new ideas will be needed to understand electricity.

Experiment 3 is a puzzle. Two rods *seem* to have been charged in the same way, by rubbing, but these two rods *attract* each other rather than repel. Why should the outcome of Experiment 3 differ from that of Experiment 2? Back to the lab.

Discovering electricity II

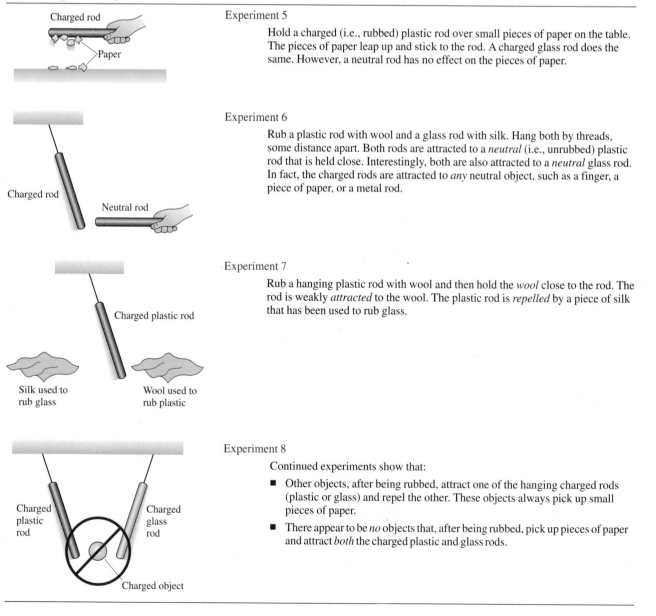

Experiment 5

Hold a charged (i.e., rubbed) plastic rod over small pieces of paper on the table. The pieces of paper leap up and stick to the rod. A charged glass rod does the same. However, a neutral rod has no effect on the pieces of paper.

Experiment 6

Rub a plastic rod with wool and a glass rod with silk. Hang both by threads, some distance apart. Both rods are attracted to a *neutral* (i.e., unrubbed) plastic rod that is held close. Interestingly, both are also attracted to a *neutral* glass rod. In fact, the charged rods are attracted to *any* neutral object, such as a finger, a piece of paper, or a metal rod.

Experiment 7

Rub a hanging plastic rod with wool and then hold the *wool* close to the rod. The rod is weakly *attracted* to the wool. The plastic rod is *repelled* by a piece of silk that has been used to rub glass.

Experiment 8

Continued experiments show that:

- Other objects, after being rubbed, attract one of the hanging charged rods (plastic or glass) and repel the other. These objects always pick up small pieces of paper.
- There appear to be *no* objects that, after being rubbed, pick up pieces of paper and attract *both* the charged plastic and glass rods.

Our first set of experiments found that charged objects exert forces on each other. The forces are sometimes attractive, sometimes repulsive. Experiments 5 and 6 now show that there is an attractive force between a charged object and a *neutral* (uncharged) object. This discovery presents us with a bit of a problem: How can we tell if an object is charged or neutral? Because of the attractive force between a charged and a neutral object, simply observing an electric force does *not* imply that an object is charged.

However, an important characteristic of any *charged* object appears to be that **a charged object picks up small pieces of paper.** This behavior provides a straightforward test to answer the question, "Is this object charged?" An object that passes the test by picking up paper is charged; an object that fails the test is neutral.

These observations let us tentatively advance the first stages of a **charge model.** We can give our model these postulates:

1. Frictional forces, such as rubbing, add something called **charge** to an object or remove it from the object. The process itself is called *charging.* More vigorous rubbing produces a larger quantity of charge.
2. There are two and only two kinds of charge. For now we will call these "plastic charge" and "glass charge." Other objects can sometimes be charged by rubbing, but the charge they receive is either "plastic charge" or "glass charge."
3. Two **like charges** (plastic/plastic or glass/glass) exert repulsive forces on each other. Two **opposite charges** (plastic/glass), exert attractive forces on each other.
4. The force between two charges is a long-range force. The size of the force increases as the quantity of charge increases and decreases as the distance between the charges increases.
5. *Neutral* objects have an *equal mixture* of both "plastic charge" and "glass charge." The rubbing process somehow manages to separate the two.

Postulate 2 is based on Experiment 8. If an object is charged (i.e., picks up paper), it always attracts one charged rod and repels the other. That is, it acts either "like plastic" or "like glass." If there were a third kind of charge, different from the first two, an object with that charge should pick up paper and attract *both* the charged plastic and glass rods. No such objects have ever been found.

The basis for postulate 5 is the observation in Experiment 7 that a charged plastic rod is attracted to the wool used to rub it but repelled by silk that has rubbed glass. It appears that rubbing glass causes the silk to acquire "plastic charge." The easiest way to explain this is to hypothesize that the silk starts out with equal amounts of "glass charge" and "plastic charge" and that the rubbing somehow transfers "glass charge" from the silk to the rod. This leaves an excess of "glass charge" on the rod and an excess of "plastic charge" on the silk.

While the charge model is *consistent* with the observations, it is by no means proved. One could easily imagine other hypotheses that are just as consistent with the limited observations we have made so far. We still have some very large unexplained puzzles, such as why charged objects exert attractive forces on neutral objects.

STOP TO THINK 25.1   To determine if an object has "glass charge," you need to

a. See if the object attracts a charged plastic rod.
b. See if the object repels a charged glass rod.
c. Do both a and b.
d. Do either a or b.

Electric Properties of Materials

We still need to clarify how different types of materials respond to charges.

Discovering electricity III

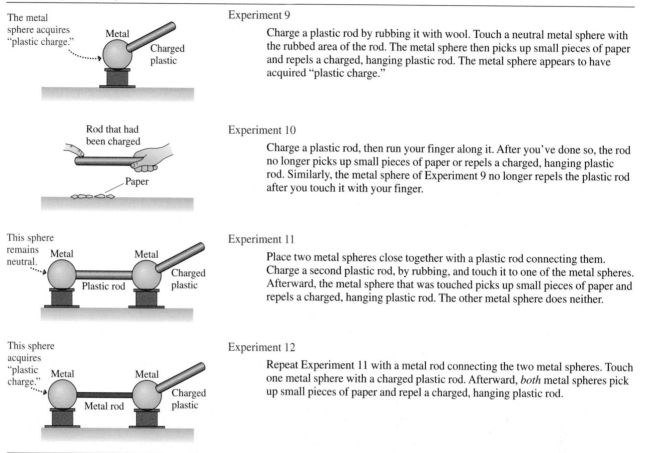

The metal sphere acquires "plastic charge."
Metal
Charged plastic

Experiment 9

Charge a plastic rod by rubbing it with wool. Touch a neutral metal sphere with the rubbed area of the rod. The metal sphere then picks up small pieces of paper and repels a charged, hanging plastic rod. The metal sphere appears to have acquired "plastic charge."

Rod that had been charged
Paper

Experiment 10

Charge a plastic rod, then run your finger along it. After you've done so, the rod no longer picks up small pieces of paper or repels a charged, hanging plastic rod. Similarly, the metal sphere of Experiment 9 no longer repels the plastic rod after you touch it with your finger.

This sphere remains neutral.
Metal   Metal
Plastic rod   Charged plastic

Experiment 11

Place two metal spheres close together with a plastic rod connecting them. Charge a second plastic rod, by rubbing, and touch it to one of the metal spheres. Afterward, the metal sphere that was touched picks up small pieces of paper and repels a charged, hanging plastic rod. The other metal sphere does neither.

This sphere acquires "plastic charge."
Metal   Metal
Metal rod   Charged plastic

Experiment 12

Repeat Experiment 11 with a metal rod connecting the two metal spheres. Touch one metal sphere with a charged plastic rod. Afterward, *both* metal spheres pick up small pieces of paper and repel a charged, hanging plastic rod.

Our final set of experiments has shown that

- Charge can be *transferred* from one object to another, but only when the objects *touch*. Contact is required. Removing charge from an object, which you can do by touching it, is called **discharging.**
- There are two types or classes of materials with very different electric properties. We call these *conductors* and *insulators*.

Experiment 12, in which a metal rod is used, is in sharp contrast to Experiment 11. Charge somehow *moves through* or along a metal rod, from one sphere to the other, but remains *fixed in place* on a plastic or glass rod. Let us define **conductors** as those materials through or along which charge easily moves and **insulators** as those materials on or in which charges remain immobile. Glass and plastic are insulators; metal is a conductor.

This new information allows us to add two more postulates to our charge model:

6. There are two types of materials. Conductors are materials through or along which charge easily moves. Insulators are materials on or in which charges remain fixed in place.

7. Charge can be transferred from one object to another by contact.

NOTE ▶ Both insulators and conductors can be charged. They differ in the *mobility* of the charge. ◀

We have by no means exhausted the number of experiments and observations we might try. Early scientific investigators were faced with all of these results, plus many others. Moreover, many of these experiments are hard to reproduce with much accuracy. How should we make sense of it all? The charge model seems promising, but certainly not proven. We have not yet explained how charged objects exert attractive forces on *neutral* objects, nor have we explained what charge is, how it is transferred, or *why* it should move through some objects but not others. Nonetheless, we will take advantage of our historical hindsight and continue to pursue this model. Homework problems will let you practice using the model to explain other observations.

EXAMPLE 25.1 Transferring charge

In Experiment 12, touching one metal sphere with a charged plastic rod caused a second metal sphere to become charged with the same type of charge as the rod. Use the postulates of the charge model to explain this.

SOLVE We need the following ideas from the charge model:

1. Charge is transferred upon contact.
2. Metal is a conductor.
3. Like charges repel.

The plastic rod was charged by rubbing with wool. The charge doesn't move around on the rod, because it is an insulator, but some of the "plastic charge" is transferred to the metal upon contact. Once in the metal, which is a conductor, the charges are free to move around. Furthermore, because like charges repel, these plastic charges quickly move as far apart as they possibly can. Some move through the connecting metal rod to the second sphere. Consequently, the second sphere acquires "plastic charge."

25.2 Charge

As you probably know, the modern names for the two types of charge are *positive charge* and *negative charge*. You may be surprised to learn that the names were coined by Benjamin Franklin. Franklin was among the first to make quantitative studies of electric phenomena, as opposed to making qualitative observations, and he found that charge behaves like positive and negative numbers. If a plastic rod is charged twice, by rubbing, and twice transfers charge to a metal sphere, the electric forces exerted by the sphere are doubled. That is, $2 + 2 = 4$. But the sphere is found to be neutral after receiving equal amounts of "plastic charge" and "glass charge." This is like $2 + (-2) = 0$. These experiments establish an important property of charge.

So what is positive and what is negative? It's entirely up to us! Franklin established the convention that a glass rod that has been rubbed with silk is *positively* charged. That's it. Any other object that repels a charged glass rod is also positively charged. Any charged object that attracts a charged glass rod is negatively charged. It was only long afterward, with the discovery of electrons and protons, that electrons were found to be attracted to a charged glass rod while protons were repelled. Thus by convention electrons have a negative charge and protons a positive charge.

NOTE ▶ In hindsight, it would have been better had Franklin made the opposite choice. Electrons are the carriers of electric currents in metals, and the convention of assigning a negative charge to electrons will later present us with some sign difficulties that could have been avoided with positive electrons. ◀

Atoms and Electricity

Now let's fast forward to the 21st century. The theory of electricity was developed without knowledge of atoms, but there is no reason for us to continue to overlook this important part of our contemporary perspective. For now, we will assert without proof some of the relevant characteristics of atoms and matter. You will have later opportunities to learn about the experimental evidence supporting these assertions.

Figure 25.1 shows that an atom consists of a very small and dense *nucleus* (diameter $\sim 10^{-14}$ m) surrounded by much less massive orbiting *electrons*. The electron orbital frequencies are so enormous ($\sim 10^{15}$ revolutions per second) that the electrons seem to form an **electron cloud** of diameter $\sim 10^{-10}$ m, a factor 10^4 larger than the nucleus. In fact, the wave-particle duality of quantum physics destroys any notion of a well-defined electron trajectory, and *all* we know of the electrons is the size and shape of the electron cloud.

Experiments at the end of the 19th century—experiments we will study in Part VII—revealed that electrons are *particles* with both mass and a negative charge. The nucleus is a composite structure consisting of *protons*, which are positively charged particles, and neutral *neutrons*. The atom is held together by the attractive electric force between the positive nucleus and the negative electrons.

One of the most important discoveries is that **charge, like mass, is an inherent property of electrons and protons.** It's no more possible to have an electron without charge than it is to have an electron without mass. As far as we know today, electrons and protons have charges of opposite sign but *exactly* equal magnitude. (Very careful experiments have never found any difference.) This atomic-level unit of charge, called the **fundamental unit of charge,** is represented by the symbol e. Table 25.1 shows the masses and charges of protons and electrons. We need to define a unit of charge, which we will do in Section 25.5, before we can specify how much charge e is.

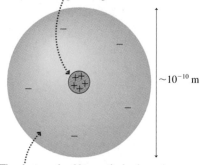

The nucleus, exaggerated for clarity, contains positive protons.

$\sim 10^{-10}$ m

The electron cloud is negatively charged.

FIGURE 25.1 An atom.

TABLE 25.1 Protons and electrons

Particle	Mass (kg)	Charge
Proton	1.67×10^{-27}	$+e$
Electron	9.11×10^{-31}	$-e$

The Micro/Macro Connection

Electrons and protons are the basic charges in nature. **There are no other sources of charge.** Consequently, the various observations we made in Section 25.1 need to be explained in terms of electrons and protons.

> **NOTE** ▶ Electrons and protons are particles of matter. Their motion is governed by Newton's laws. Electrons can move from one object to another when the objects are in contact, but neither electrons nor protons can leap through the air from one object to another. An object does not become charged simply from being close to a charged object. ◀

Charge is represented by the symbol q (or sometimes Q). A macroscopic object, such as a plastic rod, has charge

$$q = N_p e - N_e e = (N_p - N_e)e \qquad (25.1)$$

where N_p and N_e are the number of protons and electrons contained in the object. Most macroscopic objects have an *equal number* of protons and electrons and therefore have $q = 0$. An object with no *net* charge (i.e., $q = 0$) is said to be *electrically neutral.*

> **NOTE** ▶ *Neutral* does *not* mean "no charges" but, instead, means that there is no *net* charge. A typical 1 cm^3 solid contains $\sim 10^{24}$ electrons and an equal number of protons. This is a tremendous number of charges, but most solids are electrically neutral or very close to it. A glass rod loses only $\sim 10^{10}$ electrons as it is charged by rubbing. This corresponds to only 1 electron out of every 10^{14}. ◀

A charged object has an unequal number of protons and electrons. An object is positively charged if $N_p > N_e$. It is negatively charged if $N_p < N_e$. Notice that an object's charge is always an integer multiple of e. That is, the amount of charge on

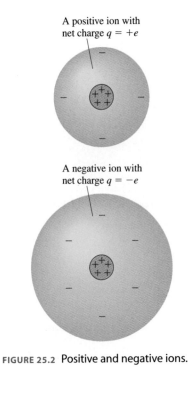

A positive ion with
net charge $q = +e$

A negative ion with
net charge $q = -e$

FIGURE 25.2 Positive and negative ions.

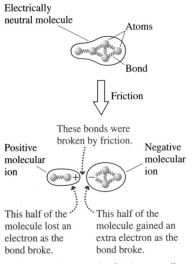

Electrically
neutral molecule

Atoms

Bond

Friction

These bonds were
broken by friction.

Positive
molecular
ion

Negative
molecular
ion

This half of the
molecule lost an
electron as the
bond broke.

This half of the
molecule gained an
extra electron as the
bond broke.

FIGURE 25.3 Charging by friction usually
creates molecular ions as bonds are broken.

an object varies by small but discrete steps, not continuously. This is called **charge quantization.**

In practice, objects acquire a positive charge not by gaining protons, which you might expect, but by losing electrons. Protons are *extremely* tightly bound within the nucleus and cannot be added to or removed from atoms. Electrons, on the other hand, are bound rather loosely and can be removed without great difficulty. The process of removing an electron from the electron cloud of an atom is called **ionization.** An atom that is missing an electron is called a *positive ion.* Its *net* charge is $q = +e$.

It turns out that some atoms can accommodate an *extra* electron and thus become a *negative ion* with net charge $q = -e$. A saltwater solution is a good example. When table salt (the chemical sodium chloride, NaCl) dissolves, it separates into positive sodium ions Na^+ and negative chlorine ions Cl^-. Figure 25.2 shows positive and negative ions.

All the charging processes we observed in Section 25.1 involved rubbing and friction. The forces of friction cause molecular bonds at the surface to break as the two materials slide past each other. Molecules are electrically neutral, but Figure 25.3 shows that *molecular ions* can be created when one of the bonds in a large molecule is broken. The positive molecular ions remain on one material and the negative ions on the other, so one of the objects being rubbed ends up with a net positive charge and the other with a net negative charge. This is the way in which a plastic rod is charged by rubbing with wool or a comb is charged by passing through your hair.

Frictional charging via bond breaking works best with large organic molecules. This explains not only how plastic is charged by rubbing with wool but also such familiar experiences as the production of "static cling" in a clothes dryer. Metals usually can *not* be charged by rubbing them.

Charge Conservation and Charge Diagrams

One of the important discoveries about charge is the **law of conservation of charge:** Charge is neither created nor destroyed. Charge can be transferred from one object to another as electrons and ions move about, but the *total* amount of charge remains constant. For example, charging a plastic rod by rubbing it with wool transfers electrons from the wool to the plastic as the molecular bonds break. The wool is left with a positive charge equal in magnitude but opposite in sign to the negative charge of the rod: $q_{wool} = -q_{plastic}$. The *net* charge remains zero.

Diagrams are going to be an important tool for understanding and explaining charges and the forces on charged objects. As you begin to use diagrams, it will be important to make explicit use of charge conservation. The net number of plusses and minuses drawn on your diagrams should *not* change as you show them moving around.

TACTICS BOX 25.1 **Drawing charge diagrams**

❶ Draw a simplified two-dimensional cross section of the object.
❷ Draw *surface* charges *very close* to the object's boundary.
❸ Draw *interior* charges uniformly within the interior of the object.
❹ Show only the *net* charge. A neutral object should show *no* charges, not a lot of plusses and minuses.
❺ Conserve charge from one diagram to the next if you use a series of diagrams to explain a process.

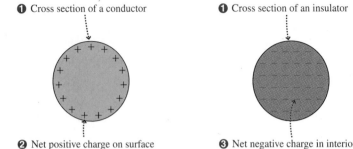

❶ Cross section of a conductor   ❶ Cross section of an insulator

❷ Net positive charge on surface   ❸ Net negative charge in interior

FIGURE 25.4 Charge diagrams.

Figure 25.4 shows two examples of charge diagrams. Step 5 will become clearer as you see it used in examples. Step 4 is especially important. For example, a positively charged object is missing electrons. Regardless of how the object became charged, the charge diagram should show plusses.

> **STOP TO THINK 25.2**   Rank in order, from most positive to most negative, the charges q_a to q_e of these five systems.

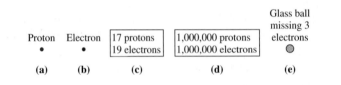

Proton	Electron	17 protons 19 electrons	1,000,000 protons 1,000,000 electrons	Glass ball missing 3 electrons
•	•			●
(a)	**(b)**	**(c)**	**(d)**	**(e)**

25.3 Insulators and Conductors

You have seen that there are two classes of materials as defined by their electrical properties: insulators and conductors. It's time for a closer look at these materials.

Figure 25.5 looks inside an insulator and a metallic conductor. The electrons in the insulator are all tightly bound to the positive nuclei and not free to move around. Charging an insulator by friction leaves patches of molecular ions on the surface, but these patches are immobile.

In metals, the outer atomic electrons (called the *valence electrons* in chemistry) are only weakly bound to the nuclei. As the atoms come together to form a solid, these outer electrons become detached from their parent nuclei and are free to wander about through the entire solid. The solid *as a whole* remains electrically neutral, because we have not added or removed any electrons, but the electrons are now rather like a negatively charged gas or liquid—what physicists like to call a **sea of electrons**—permeating an array of positively charged **ion cores.**

The primary consequence of this structure is that electrons in a metal are highly mobile. They can quickly and easily move through the metal in response to electric forces. The motion of charges through a material is what we will later call a **current,** and the charges that physically move are called the **charge carriers.** The charge carriers in metals are electrons.

Metals aren't the only conductors. Ionic solutions, such as salt water, are also good conductors. But the charge carriers in an ionic solution are the ions, not electrons. We'll focus on metallic conductors because of their importance in applications of electricity.

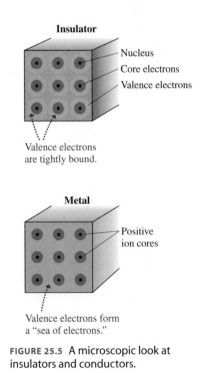

FIGURE 25.5 A microscopic look at insulators and conductors.

Charging

Insulators are often charged by rubbing, as shown in Figure 25.6. The drawing shows that the charges on the rod are right at the surface and that charge is conserved. The charge on the rod is immobile. It can be transferred to another object upon contact, but it doesn't move around on the rod.

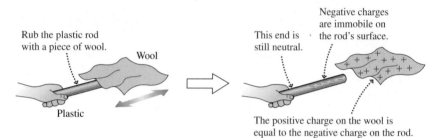

Rub the plastic rod with a piece of wool.

Wool

Plastic

This end is still neutral.

Negative charges are immobile on the rod's surface.

The positive charge on the wool is equal to the negative charge on the rod.

FIGURE 25.6 An insulating rod is charged by rubbing.

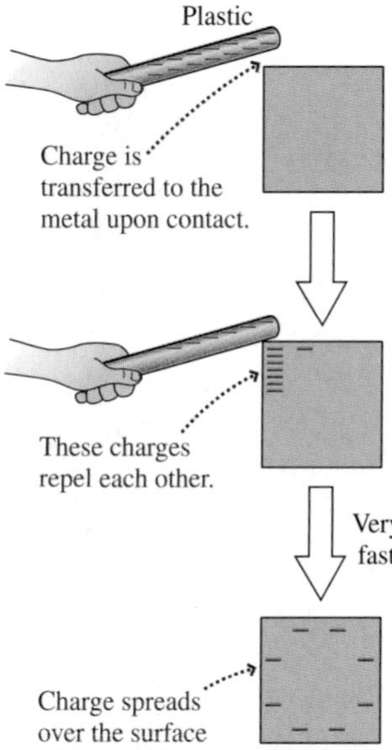

Plastic

Metal

Charge is transferred to the metal upon contact.

These charges repel each other.

Very fast

Charge spreads over the surface of the metal.

FIGURE 25.7 A conductor is charged by contact with a charged plastic rod.

Metals usually cannot be charged by rubbing, but Experiment 9 showed that a metal sphere can be charged by contact with a charged plastic rod. Figure 25.7 gives a pictorial explanation. An essential idea is that **the electrons in a conductor are free to move.** Once charge is transferred to the metal, repulsive forces between the negative charges cause the electrons to move apart from each other.

Note that the newly added electrons do not themselves need to move to the far corners of the metal. Because of the repulsive forces, the newcomers simply "shove" the entire electron sea a little to the side. The electron sea takes an extremely short time to adjust itself to the presence of the added charge, typically less than 10^{-9} s. For all practical purposes, a conductor responds *instantaneously* to the addition or removal of charge.

Other than this very brief interval during which the electron sea is adjusting, the charges in an *isolated* conductor are in static equilibrium. That is, the charges are at rest and there is no *net* force on any charge. This condition is called **electrostatic equilibrium.** If there *were* a net force on one of the charges, it would quickly move to an equilibrium point at which the force was zero.

Electrostatic equilibrium has an important consequence:

In an isolated conductor, any excess charge is located on the surface of the conductor.

To see this, suppose there *were* an excess electron in the interior of an isolated conductor. The extra electron would upset the electrical neutrality of the interior and exert forces on nearby electrons, causing them to move. But their motion would violate the assumption of static equilibrium, so we're forced to conclude that there cannot be any excess electrons in the interior. Because any excess electrons repel each other, they move as far apart as possible and spread out along the surface.

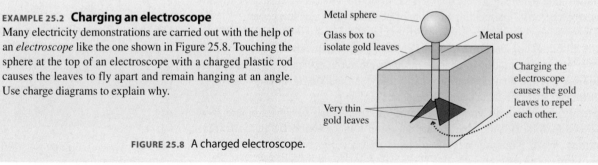

EXAMPLE 25.2 **Charging an electroscope**
Many electricity demonstrations are carried out with the help of an *electroscope* like the one shown in Figure 25.8. Touching the sphere at the top of an electroscope with a charged plastic rod causes the leaves to fly apart and remain hanging at an angle. Use charge diagrams to explain why.

Metal sphere

Glass box to isolate gold leaves

Metal post

Charging the electroscope causes the gold leaves to repel each other.

Very thin gold leaves

FIGURE 25.8 A charged electroscope.

MODEL We'll use the charge model and the model of a conductor as a material through which electrons move.

VISUALIZE Figure 25.9 uses a series of charge diagrams to show the charging of an electroscope.

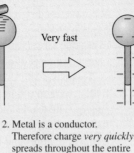
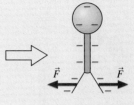

Plastic

Electroscope

Very fast

$\vec{F}$ $\vec{F}$

1. Negative charges (i.e., electrons) are transferred from the rod to the metal sphere upon contact.

2. Metal is a conductor. Therefore charge *very quickly* spreads throughout the entire electroscope. The leaves become negatively charged.

3. Like charges repel. The negatively charged leaves exert repulsive forces on each other, causing them to spread apart.

FIGURE 25.9 The process by which an electroscope is charged.

Discharging

Pure water is not a terribly good conductor, but nearly all water contains a variety of dissolved minerals that float around as ions. Dissolved table salt, as we noted previously, separates into Na^+ and Cl^- ions. These ions are the charge carriers, allowing salt water to be a fairly good conductor.

The human body consists largely of salt water. Consequently, and occasionally tragically, humans are reasonably good conductors. This fact allows us to understand how it is that *touching* a charged object discharges it, as we observed in Experiment 10. Figure 25.10 shows a person touching a positively charged metal, one that is missing electrons. Upon contact, some of the negative Cl^- ions on the skin surface transfer their extra electron to the metal, neutralizing both the metal and the chlorine atoms. This leaves the body with an excess of positive Na^+ ions and, thus, a net positive charge. As in any conductor, these excess positive charges quickly spread as far apart as possible over the surface of the conductor.

The net effect of touching a charged metal is that it and the conducting human together become a much larger conductor than the metal alone. Any excess charge that was initially confined to the metal can now spread over the larger metal + human conductor. This may not entirely discharge the metal, but in typical circumstances, where the human is much larger than the metal, the residual charge remaining on the metal is much reduced from the original charge. The metal, for most practical purposes, is discharged. In essence, two conductors in contact "share" the charge that was originally on just one of them.

Moist air is a conductor, although a rather poor one. Charged objects in air slowly lose their charge as the object shares its charge with the air. The earth itself is a giant conductor because of its water, moist soil, and a variety of ions—not, admittedly, as good a conductor as a piece of copper, but a conductor nonetheless. Any object that is physically connected to the earth through a conductor is said to be **grounded.** The effect of being grounded is that the object shares any excess charge it has with the entire earth! But the earth is so enormous that any conductor attached to the earth will be completely discharged.

The purpose of *grounding* objects, such as circuits and appliances, is to prevent the buildup of any charge on the objects. As you will see later, grounding has the effect of preventing a *voltage difference* between the object and the ground. The third prong on appliances and electronics that have a three-prong plug is the

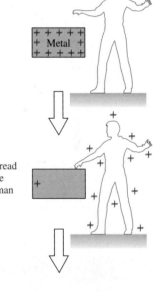

Charges spread through the metal + human system.

Very little charge is left on the metal.

FIGURE 25.10 Touching a charged metal discharges it.

ground connection. The building wiring physically connects that third wire deep into the ground somewhere just outside the building, often by attaching it to a metal water pipe that goes underground.

Charge Polarization

We have made great strides in learning how the atomic structure of matter can explain charging processes and the properties of insulators and conductors. However, one observation from Section 25.1 still needs an explanation. How do charged objects of either sign exert an attractive force on a *neutral* object?

To answer this question, consider an electroscope. Figure 25.11 shows a positively charged rod held close to a *neutral* electroscope *without touching* the metal sphere. The leaves move apart and stay apart as long as you hold the rod near, but they quickly collapse when it is removed. Can we understand this behavior in terms of charges and forces?

We can, and Figure 25.12a shows how. Because the electrons in a conductor are free to move about, the whole electron sea shifts slightly toward an external positive charge. Although the metal as a whole is still electrically neutral, we say that the object has been *polarized*. **Charge polarization** is a slight separation of the positive and negative charge in a neutral object.

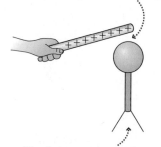

Bring a positively charged glass rod close to an electroscope without touching the sphere.

The electroscope is neutral, yet the leaves repel each other. Why?

FIGURE 25.11 A charged rod held close to an electroscope causes the leaves to repel each other.

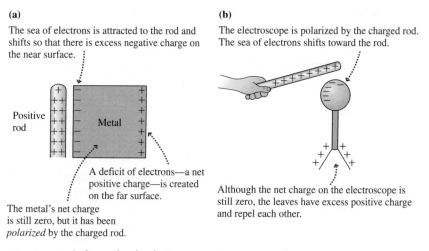

(a)

The sea of electrons is attracted to the rod and shifts so that there is excess negative charge on the near surface.

Positive rod

Metal

A deficit of electrons—a net positive charge—is created on the far surface.

The metal's net charge is still zero, but it has been *polarized* by the charged rod.

(b)

The electroscope is polarized by the charged rod. The sea of electrons shifts toward the rod.

Although the net charge on the electroscope is still zero, the leaves have excess positive charge and repel each other.

FIGURE 25.12 A charged rod polarizes a metal.

Charge polarization produces a net positive charge on the leaves of the electroscope shown in Figure 25.12b, so they repel each other. But because the electroscope has no *net* charge, the electron sea quickly readjusts to balance the positive ions once the rod is removed.

You might think that *all* the electrons in Figure 25.12a would rush to the side near the positive charge, but once the electron sea shifts slightly, the stationary positive ions begin to exert a force, a restoring force, pulling the electrons back to the right. The equilibrium position for the sea of electrons is just far enough to the left that the forces due to the external charge and the positive ions are in balance. In practice, the displacement of the electron sea is usually *less than* 10^{-15} *m*!

Charge polarization explains not only why the electroscope leaves deflect but also how a charged object exerts an attractive force on a neutral object. Figure 25.13 shows a positively charged rod near a neutral piece of metal. Because the electric force decreases with distance, the attractive force on the electrons at the top surface is *slightly greater* than the repulsive force on the ions. The net

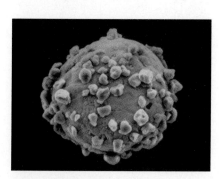

Toner particles in a photocopy machine stick to charged *carrier beads* because of a polarization force. Later, the toner particles will be transferred to charged areas on a sheet of paper to form the photocopied image.

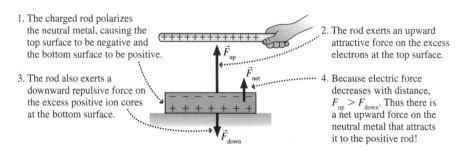

1. The charged rod polarizes the neutral metal, causing the top surface to be negative and the bottom surface to be positive.

2. The rod exerts an upward attractive force on the excess electrons at the top surface.

3. The rod also exerts a downward repulsive force on the excess positive ion cores at the bottom surface.

4. Because electric force decreases with distance, $F_{up} > F_{down}$. Thus there is a net upward force on the neutral metal that attracts it to the positive rod!

$\vec{F}_{up}$

$\vec{F}_{net}$

$\vec{F}_{down}$

FIGURE 25.13 The polarization force on a neutral piece of metal is due to the slight charge separation.

force toward the charged rod is called a **polarization force.** The polarization force arises because the charges in the metal are separated, *not* because the rod and metal are oppositely charged.

A negatively charged rod would push the electron sea slightly away, polarizing the metal to have a positive upper surface charge and negative lower surface charge. Once again, these are the conditions for the charge to exert a *net attractive force* on the metal. Thus our charge model explains how a charged object of *either* sign attracts neutral pieces of metal.

The Electric Dipole

Now let's consider a slightly trickier situation. Why does a charged rod pick up paper, which is an insulator rather than a metal? First consider what happens if we bring a positive charge near an atom. As Figure 25.14a shows, the charge polarizes the atom. The electron cloud doesn't move far, because the force from the positive nucleus pulls it back, but the center of positive charge and the center of negative charge are now slightly separated.

(a)

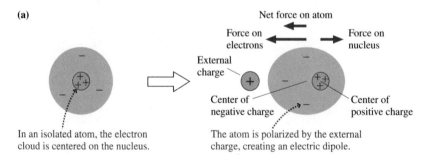

Net force on atom

Force on electrons

Force on nucleus

External charge

Center of negative charge

Center of positive charge

In an isolated atom, the electron cloud is centered on the nucleus.

The atom is polarized by the external charge, creating an electric dipole.

(b)

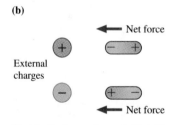

Net force

External charges

Net force

Electric dipoles can be created by either positive or negative charges. In both cases, there is an attractive net force toward the external charge.

FIGURE 25.14 A neutral atom is polarized by an external charge, forming an *electric dipole.*

Two opposite charges with a slight separation between them form what is called an **electric dipole.** Figure 25.14b shows that an external charge of either sign polarizes the atom to produce an electric dipole with the near end opposite in sign to the charge. (The actual distortion from a perfect sphere is minuscule, nothing like the distortion shown in the figure.) The attractive force on the dipole's near end *slightly* exceeds the repulsive force on its far end because the near end is closer to the charge. The net force, which is an *attractive* force between charge and the atom, is another example of a polarization force.

An insulator has no sea of electrons to shift if an external charge is brought close. Instead, as Figure 25.15 shows, all the individual atoms inside the insulator become polarized. The polarization force acting *on each atom* produces a net

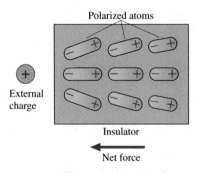

Polarized atoms

External charge

Insulator

Net force

FIGURE 25.15 The atoms in an insulator are polarized by an external charge.

polarization force toward the external charge. This solves the puzzle. A charged rod picks up pieces of paper by

- Polarizing the atoms in the paper,
- Then exerting an attractive polarization force on each atom.

This is a significant conclusion. Make sure you understand all the steps in the reasoning.

STOP TO THINK 25.3   An electroscope is positively charged by *touching* it with a positive glass rod. The electroscope leaves spread apart and the glass rod is removed. Then a negatively charged plastic rod is brought close to the top of the electroscope, but it doesn't touch. What happens to the leaves?

a. The leaves get closer together.   b. The leaves spread farther apart.
c. One leaf moves higher, the other lower.   d. The leaves don't move.

Charging by Induction

Charge polarization is responsible for an interesting and counterintuitive way of charging an electroscope. Figure 25.16 shows a positively charged glass rod held near an electroscope but not touching it, while a person touches the electroscope with a finger. Unlike what happens in Figure 25.11, the electroscope leaves do not move.

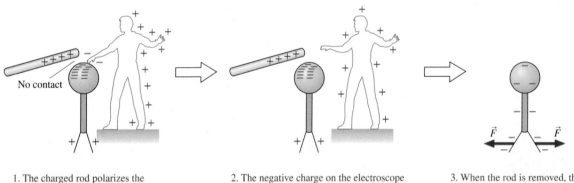

1. The charged rod polarizes the electroscope+person conductor. The leaves repel slightly, due to polarization within the electroscope, but overall the electroscope has an excess of electrons and the person has a deficit of electrons.

2. The negative charge on the electroscope is isolated when contact is broken.

3. When the rod is removed, the leaves first collapse, as the polarization vanishes, then repel as the excess negative charge spreads out. The electroscope has been *negatively* charged.

FIGURE 25.16 Charging by induction.

Charge polarization occurs, as it did in Figure 25.11, but this time in the much larger electroscope + person conductor. If the person removes his or her finger while the system is polarized, the electroscope is left with a *net* negative charge and the person has a net positive charge. When the rod is then removed, the electroscope leaves first collapse, as the polarization vanishes, then spread apart as the excess negative charge spreads out. The electroscope has been charged *opposite to the rod* in a process called **charging by induction.**

25.4 Coulomb's Law

The last few sections have established a *model* of charges and electric forces. This model is very good at explaining electric phenomena and providing a general understanding of electricity. Now we need to become quantitative. Experiment 4 in Section 25.1 found that the electric force increases for objects with more charge and decreases as charged objects are moved farther apart. The force law that describes this behavior is known as *Coulomb's law.*

Charles Coulomb was one of many scientists investigating electricity in the late 18th century. Coulomb had the idea of studying electric forces using the torsion balance scheme by which Cavendish had measured the value of the gravitational constant G (see Section 12.4). This was a difficult experiment. Cavendish's masses could be placed in position and did not change, but Coulomb was constantly having to recharge the ends of his balance. How could he do this reproducibly? How could he know if two objects were "equally charged"? How could he know for sure where the charge was located?

Despite these obstacles, Coulomb announced in 1785 that the electric force obeys an *inverse-square law* analogous to Newton's law of gravity. Historians of science debate whether Coulomb really discovered this law from his data or, perhaps, if he leapt to unwarranted conclusions because he so wanted his discovery to match that of the great Newton. Nonetheless, Coulomb's discovery or lucky guess, whichever it was, was subsequently confirmed, and the basic law of electric force bears his name.

> **Coulomb's law** This law can be stated as follows:
>
> 1. If two charged particles having charges q_1 and q_2 are a distance r apart, the particles exert forces on each other of magnitude
>
> $$F_{1 \text{ on } 2} = F_{2 \text{ on } 1} = \frac{K|q_1||q_2|}{r^2} \qquad (25.2)$$
>
> where K is called the **electrostatic constant.** These forces are an action/reaction pair, equal in magnitude and opposite in direction.
> 2. The forces are directed along the line joining the two particles. The forces are *repulsive* for two like charges and *attractive* for two opposite charges.

We sometimes speak of the "force between charge q_1 and charge q_2," but keep in mind that we are really dealing with charged *objects* that also have a mass, a size, and other properties. Charge is not some disembodied entity that exists apart from matter. Coulomb's law describes the force between charged *particles,* which are also called **point charges.** A charged particle, which is an extension of the particle model we used in Part I, has a mass and a charge but has no size.

Coulomb's law looks much like Newton's law of gravity, but there is one important difference: The charge q can be either positive or negative. Consequently, the absolute value signs in Equation 25.2 are especially important. The first part of Coulomb's law gives only the *magnitude* of the force, which is always positive. The direction must be determined from the second part of the law. Figure 25.17 shows the forces between different combinations of positive and negative charges.

Units of Charge

Coulomb had no *unit* of charge, so he was unable to determine a value for K, whose numerical value depends upon the units of both charge and distance. The SI unit of charge, the **coulomb** (C), is derived from the SI unit of *current,* so we'll have to await the study of current in Chapter 28 before giving a precise definition. For now we'll note that the fundamental unit of charge e has been measured to have the value

$$e = 1.60 \times 10^{-19} \text{ C}$$

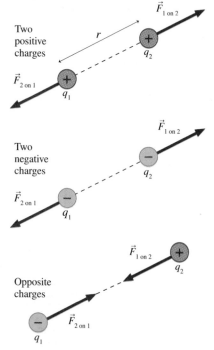

FIGURE 25.17 Attractive and repulsive forces between charges.

This is a very small amount of charge. Stated another way, 1 C is the net charge of roughly 6.25×10^{18} protons.

NOTE ▶ The amount of charge produced by rubbing plastic or glass rods is typically in the range 1 nC (10^{-9} C) to 100 nC (10^{-7} C). This corresponds to an excess or deficit of 10^{10} to 10^{12} electrons. ◀

Once the unit of charge is established, torsion balance experiments such as Coulomb's can be used to measure the electrostatic constant K. In SI units

$$K = 8.99 \times 10^9 \text{ N m}^2/\text{C}^2$$

It is customary to round this to $K = 9.0 \times 10^9$ N m^2/C^2 for all but extremely precise calculations, and we will do so.

Surprisingly, we will find that Coulomb's law is not explicitly used in much of the theory of electricity. While it *is* the basic force law, most of our future discussion and calculations will be of things called *fields* and *potentials*. It turns out that we can make many future equations easier to use if we rewrite Coulomb's law in a somewhat more complicated way. Let's define a new constant, called the **permittivity constant** ϵ_0 (pronounced "epsilon zero" or "epsilon naught"), as

$$\epsilon_0 = \frac{1}{4\pi K} = 8.85 \times 10^{-12} \text{ C}^2/\text{N m}^2$$

Rewriting Coulomb's law in terms of ϵ_0 gives us

$$F = \frac{1}{4\pi\epsilon_0} \frac{|q_1||q_2|}{r^2} \tag{25.3}$$

It will be easiest when using Coulomb's law directly to use the electrostatic constant K. However, in later chapters we will switch to the second version with ϵ_0.

Using Coulomb's Law

11.1–11.3   Activ Physics

Coulomb's law is a force law, and forces are vectors. It has been many chapters since we made much use of vectors and vector addition, but these mathematical techniques will be essential in our study of electricity and magnetism. You may wish to review vector addition in Chapter 3.

There are three important observations regarding Coulomb's law:

1. **Coulomb's law applies only to point charges.** A point charge is an idealized material object with charge and mass but with no size or extension. For practical purposes, two charged objects can be modeled as point charges if they are much smaller than the separation between them.
2. Strictly speaking, **Coulomb's law applies only to electrostatics,** the electric forces between *static charges*. In practice, Coulomb's law is a good approximation to the electric force between two moving charged particles if their relative speed is much less than the speed of light.
3. **Electric forces, like other forces, can be superimposed.** If multiple charges 1, 2, 3, . . . are present, the *net* electric force on charge *j* due to all other charges is

$$\vec{F}_{\text{net}} = \vec{F}_{1 \text{ on } j} + \vec{F}_{2 \text{ on } j} + \vec{F}_{3 \text{ on } j} + \cdots \tag{25.4}$$

where each of the $\vec{F}_{i \text{ on } j}$ is given by Equation 25.2 or 25.3.

These conditions are the basis of a strategy for using Coulomb's law to solve electrostatic force problems.

(MP) PROBLEM-SOLVING STRATEGY 25.1 **Electrostatic forces and Coulomb's law**

MODEL Identify point charges or objects that can be modeled as point charges.

VISUALIZE Use a *pictorial representation* to establish a coordinate system, show the positions of the charges, show the force vectors on the charges, define distances and angles, and identify what the problem is trying to find. This is the process of translating words to symbols.

SOLVE The mathematical representation is based on Coulomb's law

$$F_{1 \text{ on } 2} = F_{2 \text{ on } 1} = \frac{K|q_1||q_2|}{r^2}$$

- Show the directions of the forces—repulsive for like charges, attractive for opposite charges—on the pictorial representation.
- When possible, do graphical vector addition on the pictorial representation. While not exact, it tells you the type of answer you should expect.
- Write each force vector in terms of its *x*- and *y*-components, then add the components to find the net force. Use the pictorial representation to determine which components are positive and which are negative.

ASSESS Check that your result has the correct units, is reasonable, and answers the question.

EXAMPLE 25.3 **The sum of two forces**

Two +10 nC charged particles are 2.0 cm apart on the *x*-axis. What is the net force on a +1.0 nC charge midway between them? What is the net force if the charged particle on the right is replaced by a −10 nC charge?

MODEL Model the charged particles as point charges.

VISUALIZE Figure 25.18 establishes a coordinate system and shows the forces $\vec{F}_{1 \text{ on } 3}$ and $\vec{F}_{2 \text{ on } 3}$.

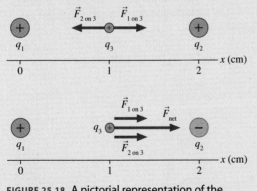

FIGURE 25.18 A pictorial representation of the charges and forces.

SOLVE Electric forces are vectors, and the net force on q_3 is the *vector* sum $\vec{F}_{net} = \vec{F}_{1 \text{ on } 3} + \vec{F}_{2 \text{ on } 3}$. Charges q_1 and q_2 each exert a repulsive force on q_3, but they are equal in magnitude and opposite in direction. Consequently, $\vec{F}_{net} = \vec{0}$. The situation changes if q_2 is negative. Now the two forces are equal in magnitude but in the *same* direction, so $\vec{F}_{net} = 2\vec{F}_{1 \text{ on } 3}$. The magnitude of the force is given by Coulomb's law:

$$F_{1 \text{ on } 3} = \frac{K|q_1||q_3|}{r_{13}^2}$$

$$= \frac{(9.0 \times 10^9 \text{ N m}^2/\text{C}^2)(10 \times 10^{-9} \text{ C})(1 \times 10^{-9} \text{ C})}{(0.010 \text{ m})^2}$$

$$= 9.0 \times 10^{-4} \text{ N}$$

Thus the net force on the 1.0 nC charge is $\vec{F}_{net} = 1.8 \times 10^{-3} \hat{\imath}$ N.

ASSESS This example illustrates the important idea that electric forces are *vectors*.

EXAMPLE 25.4 The point of zero force

Two positively charged particles q_1 and $q_2 = 3q_1$ are 10.0 cm apart. Where (other than at infinity) could a third charge q_3 be placed so as to experience no net force?

MODEL Model the charged particles as point charges.

VISUALIZE Figure 25.19 establishes a coordinate with q_1 at the origin. We first need to identify the region of space in which q_3 must be located. We have no information about the sign of q_3, so apparently the position for which we are looking will work for either sign. You can see from the figure that the forces at point A, above the axis, and at point B, outside the charges, cannot possibly add to zero. However, at some point C on the x-axis *between* the charges, the two forces will be oppositely directed.

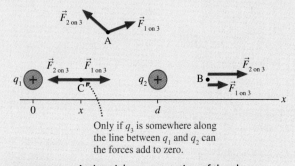

FIGURE 25.19 A pictorial representation of the charges and forces.

SOLVE The mathematical problem is to find the position for which the forces $\vec{F}_{1\,\text{on}\,3}$ and $\vec{F}_{2\,\text{on}\,3}$ are equal in magnitude. If q_3 is distance x from q_1, it is distance $d - x$ from q_2. The *magnitudes* of the forces are

$$F_{1\,\text{on}\,3} = \frac{Kq_1|q_3|}{r_{13}^2} = \frac{Kq_1|q_3|}{x^2}$$

$$F_{2\,\text{on}\,3} = \frac{Kq_2|q_3|}{r_{23}^2} = \frac{K(3q_1)|q_3|}{(d-x)^2}$$

Charges q_1 and q_2 are positive and do not need absolute value signs. Equating the two forces gives

$$\frac{Kq_1|q_3|}{x^2} = \frac{3Kq_1|q_3|}{(d-x)^2}$$

The term $Kq_1|q_3|$ cancels. Multiplying by $x^2(d-x)^2$ gives

$$(d-x)^2 = 3x^2$$

which can be rearranged into the quadratic equation

$$2x^2 + 2dx - d^2 = 2x^2 + 20x - 100 = 0$$

where we used $d = 10$ cm and x is in cm. The solutions to this equation are

$$x = +3.66 \text{ cm} \quad \text{or} \quad -13.66 \text{ cm}$$

Both are points where the *magnitudes* of the two forces are equal, but $x = -13.66$ cm is a point where the magnitudes are equal but the directions are the same. The solution we want, which is between the charges, is $x = 3.66$ cm. Thus the point to place q_3 is 3.66 cm from q_1 along the line joining q_1 and q_2.

ASSESS q_1 is smaller than q_2, so we expect the point at which the forces balance to be closer to q_1 than to q_2. The solution seems reasonable. Note that the problem statement has no coordinates, so "$x = 3.66$ cm" is *not* an acceptable answer. You need to describe the position relative to q_1 and q_2.

EXAMPLE 25.5 Three charges

Three charged particles with $q_1 = -50$ nC, $q_2 = +50$ nC, and $q_3 = +30$ nC are placed on the corners of the 5.0 cm × 10.0 cm rectangle shown in Figure 25.20. What is the net force on charge q_3 due to the other two charges? Give your answer both in component form and as a magnitude and direction.

MODEL Model the charged particles as point charges.

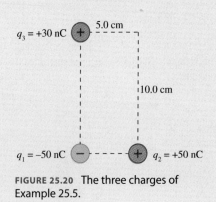

FIGURE 25.20 The three charges of Example 25.5.

VISUALIZE The pictorial representation of Figure 25.21 establishes a coordinate system. q_1 and q_3 are opposite charges, so force vector $\vec{F}_{1\,\text{on}\,3}$ is an attractive force toward q_1. q_2 and q_3 are like charges, so force vector $\vec{F}_{2\,\text{on}\,3}$ is a repulsive force away from q_2. q_1 and q_2 have equal magnitudes, but $\vec{F}_{2\,\text{on}\,3}$ has been drawn shorter than $\vec{F}_{1\,\text{on}\,3}$ because q_2 is farther from q_3. Vector addition has been used to draw the net force vector $\vec{F}_3$ and to define the angle ϕ.

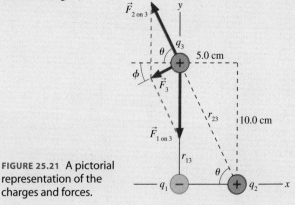

FIGURE 25.21 A pictorial representation of the charges and forces.

SOLVE The question asks for a *force*, so our answer will be the *vector* sum $\vec{F}_3 = \vec{F}_{1\,\text{on}\,3} + \vec{F}_{2\,\text{on}\,3}$. We need to write $\vec{F}_{1\,\text{on}\,3}$ and $\vec{F}_{2\,\text{on}\,3}$ in component form. The magnitude of force $\vec{F}_{1\,\text{on}\,3}$ can be found using Coulomb's law:

$$
\begin{aligned}
F_{1\,\text{on}\,3} &= \frac{K|q_1||q_3|}{r_{13}^2} \\
&= \frac{(9.0 \times 10^9 \text{ N m}^2/\text{C}^2)(50 \times 10^{-9} \text{ C})(30 \times 10^{-9} \text{ C})}{(0.100 \text{ m})^2} \\
&= 1.35 \times 10^{-3} \text{ N}
\end{aligned}
$$

where we used $r_{13} = 10$ cm. The pictorial representation shows that $\vec{F}_{1\,\text{on}\,3}$ points down, in the negative y-direction, so

$$
\vec{F}_{1\,\text{on}\,3} = -1.35 \times 10^{-3}\hat{j} \text{ N}
$$

To calculate $\vec{F}_{2\,\text{on}\,3}$ we first need the distance r_{23} between the charges:

$$
r_{23} = \sqrt{(5.0 \text{ cm})^2 + (10.0 \text{ cm})^2} = 11.18 \text{ cm}
$$

The magnitude of $\vec{F}_{2\,\text{on}\,3}$ is thus

$$
\begin{aligned}
F_{2\,\text{on}\,3} &= \frac{K|q_2||q_3|}{r_{23}^2} \\
&= \frac{(9.0 \times 10^9 \text{ N m}^2/\text{C}^2)(50 \times 10^{-9} \text{ C})(30 \times 10^{-9} \text{ C})}{(0.1118 \text{ m})^2} \\
&= 1.08 \times 10^{-3} \text{ N}
\end{aligned}
$$

This is only a magnitude. The *vector* $\vec{F}_{2\,\text{on}\,3}$ is

$$
\vec{F}_{2\,\text{on}\,3} = -F_{2\,\text{on}\,3}\cos\theta\,\hat{i} + F_{2\,\text{on}\,3}\sin\theta\,\hat{j}
$$

where angle θ is defined in the figure and the signs (negative x-component, positive y-component) were determined from the pictorial representation. From the geometry of the rectangle,

$$
\theta = \tan^{-1}\left(\frac{10.0 \text{ cm}}{5.0 \text{ cm}}\right) = \tan^{-1}(2.0) = 63.4°
$$

Thus $\vec{F}_{2\,\text{on}\,3} = (-4.83\hat{i} + 9.66\hat{j}) \times 10^{-4}$ N. Now we can add $\vec{F}_{1\,\text{on}\,3}$ and $\vec{F}_{2\,\text{on}\,3}$ to find

$$
\vec{F}_3 = \vec{F}_{1\,\text{on}\,3} + \vec{F}_{2\,\text{on}\,3} = (-4.83\hat{i} - 3.84\hat{j}) \times 10^{-4} \text{ N}
$$

This would be an acceptable answer for many problems, but sometimes we need the net force as a magnitude and direction. With angle ϕ as defined in the figure, these are

$$
F_3 = \sqrt{F_{3x}^2 + F_{3y}^2} = 6.17 \times 10^{-4} \text{ N}
$$

$$
\phi = \tan^{-1}\left|\frac{F_{3y}}{F_{3x}}\right| = 38.5°
$$

Thus $\vec{F}_3 = (6.17 \times 10^{-4}$ N, $38.5°$ below the negative x-axis).

ASSESS The forces are not large, but they are typical of electrostatic forces. Even so, you'll soon see that these forces can produce very large accelerations because the masses of the charged objects are usually very small.

EXAMPLE 25.6 Lifting a glass bead

A small plastic sphere is charged to -10 nC. It is held 1.0 cm above a small glass bead at rest on a table. The bead has a mass of 15 mg and a charge of $+10$ nC. Will the glass bead "leap up" to the plastic sphere?

MODEL Model the plastic sphere and glass bead as point charges.

VISUALIZE Figure 25.22 establishes a y-axis, identifies the plastic sphere as q_1 and the glass bead as q_2, and shows a free-body diagram.

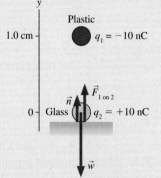

FIGURE 25.22 A pictorial representation of the charges and forces.

SOLVE If $F_{1\,\text{on}\,2}$ is less than the bead's weight $w = m_2 g$, then the bead will remain at rest on the table with $\vec{F}_{1\,\text{on}\,2} + \vec{w} + \vec{n} = \vec{0}$. But if $F_{1\,\text{on}\,2}$ is greater than the bead's weight, the glass bead will accelerate upward from the table. Using the values provided,

$$
F_{1\,\text{on}\,2} = \frac{K|q_1||q_2|}{r^2} = 9.0 \times 10^{-3} \text{ N}
$$

$$
w = m_2 g = 1.5 \times 10^{-4} \text{ N}
$$

$F_{1\,\text{on}\,2}$ exceeds the weight by a factor of 60, so the glass bead will leap upward.

ASSESS The values used in this example are realistic for spheres ≈ 2 mm in diameter. In general, as in this example, electric forces are *significantly* larger than weight forces. Consequently, we can neglect weight forces when working electric-force problems unless the particles are fairly massive.

Charges A and B exert repulsive forces on each other. $q_A = 4q_B$. Which statement is true?

a. $F_{A \text{ on } B} > F_{B \text{ on } A}$
b. $F_{A \text{ on } B} = F_{B \text{ on } A}$
c. $F_{A \text{ on } B} < F_{B \text{ on } A}$

A   B

25.5 The Concept of a Field

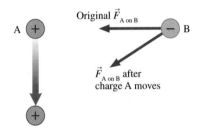

FIGURE 25.23 If charge A moves, how long does it take the force vector on B to respond?

Electric and magnetic forces, like gravity, are *long-range forces.* No contact is required for one charged particle to exert a force on another charged particle. Somehow, in a process called *action at a distance,* the force is transmitted through empty space. The concept of action at a distance greatly troubled many of the leading thinkers of Newton's day, following the publication of his theory of gravity. Force, they believed, should have some *mechanism* by which it is exerted, and the idea of action at a distance, with no apparent mechanism, was more than most scientists could accept. Nonetheless, they could not dispute the success of Newton's theory.

The great prestige and success of Newton was able to keep scientists' doubts and reservations in check until the end of the 18th century, when investigations of electric and magnetic phenomena reopened the issue of action at a distance. For example, consider the charged particles A and B in Figure 25.23. If the particles have been at rest for a long period of time, then we can confidently use Coulomb's law to determine the force that A exerts on B. But suppose that A suddenly starts moving, as shown by the arrow. In response, the force vector on B must pivot to follow A. Does this happen *instantly?* Or is there some *delay* between when A moves and when the force $\vec{F}_{A \text{ on } B}$ responds?

Neither Coulomb's law nor Newton's law of gravity is dependent upon time, so the answer from the perspective of Newtonian physics has to be "instantly." Yet most scientists found this troubling. What if A is 100,000 light years from B? Will B respond *instantly* to an event 100,000 light years away? The idea of instantaneous transmission of forces was becoming unbelievable to most scientists by the beginning of the 19th century. But if there is a delay, how long is it? How does the information to "change force" get sent from A to B? These were the issues when a young Michael Faraday appeared on the scene.

Michael Faraday is one of the most interesting figures in the history of science. Born in 1791, the son of a poor blacksmith near London, Faraday was sent to work at an early age with almost no formal education. As a teenager, he found employment with a printer and bookbinder, and he began to read the books that came through the shop. By happenstance, a customer brought in a copy of the *Encyclopedia Britannica* to be rebound, and there Faraday discovered a lengthy article about electricity. It was all the spark he needed to set him on a course that, by his death in 1867, would make him one of the most esteemed scientists in Europe.

You will learn more about Faraday in later chapters. For now, suffice it to say that Faraday was never able to become fluent in mathematics. Apparently the late age at which he started his studies was too much of a detriment for mathematical learning. In place of mathematics, Faraday's brilliant and insightful mind developed many ingenious *pictorial* methods for thinking about and describing physical phenomena. By far the most important of these was the field.

Faraday was particularly impressed with the pattern that iron filings make when sprinkled around a magnet, as seen in Figure 25.24. The pattern's regularity

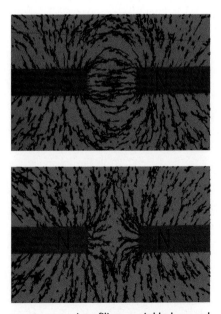

FIGURE 25.24 Iron filings sprinkled around the ends of a magnet suggest that the influence of the magnet extends into the space around it.

and the curved lines suggested to Faraday that the *space itself* around the magnet is filled with some kind of magnetic influence. Perhaps the magnet in some way alters the space around it. In this view, a piece of iron near the magnet responds not directly to the magnet but, instead, to the alteration of space caused by the magnet. This space alteration, whatever it is, is the *mechanism* by which the long-range force is exerted.

Figure 25.25 illustrates Faraday's idea. The Newtonian view was that A and B interact directly. In Faraday's view, A first alters or modifies the space around it, and Object B then comes along and interacts with this altered space. The alter-ation of space becomes the *agent* by which A and B interact. Furthermore, this alteration could easily be imagined to take a finite time to propagate outward from A, perhaps in a wave-like fashion. If A changes, B responds only when the new alteration of space reaches it. The interaction between B and this alteration of space is a *local* interaction, rather like a contact force.

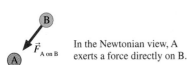

In the Newtonian view, A exerts a force directly on B.

In Faraday's view, A alters the space around it. (The wavy lines are poetic license. We don't know what the alteration looks like.)

Particle B then responds to the altered space. The altered space is the agent that exerts the force on B.

FIGURE 25.25 Newton's and Faraday's ideas about long-range forces.

Faraday's idea came to be called a **field.** The term *field,* which comes from mathematics, describes a function $f(x, y, z)$ that assigns a value to every point in space. When used in physics, a field conveys the idea that the physical entity exists at every point in space. That is, indeed, what Faraday was suggesting about how long-range forces operate. The charge makes an alteration *everywhere* in space. Other charges then respond to the alteration at their position. The alteration of the space around a mass is called the **gravitational field.** Similarly, the space around a charge is altered to create the **electric field.**

> **NOTE** ▶ The concept of a field is in sharp contrast to the concept of a particle. A particle exists at *one* point in space. The purpose of Newton's laws of motion is to determine how the particle moves from point to point along a trajectory. A field exists simultaneously at *all* points in space. A wave is an example of a field, although we didn't use the term during our study of waves. ◀

Faraday proposed a novel way to think about how one object exerts forces on another, but his idea was not taken seriously at first. It seemed too vague and non-mathematical to scientists steeped in the Newtonian tradition of particles and forces. But the significance of the concept of field grew as electromagnetic theory developed during the first half of the 19th century. What seemed at first a pictorial "gimmick" came to be seen as more and more essential for understanding electric and magnetic forces.

Faraday's field ideas were finally placed on a firm mathematical foundation in 1865 by James Clerk Maxwell, a young Scottish physicist possessing both great physical insight and mathematical ability. Maxwell was able to describe com-pletely all the known behaviors of electric and magnetic fields in four equations, known today as Maxwell's equations. We will explore aspects of Maxwell's the-ory as we go along, then look at the full implications of Maxwell's equations in Chapter 34.

I have preferred to seek an explanation [of electric and magnetic phenomena] by supposing them to be produced by actions which go on in the surrounding medium as well as in the excited bodies, and endeavoring to explain the action between distant bodies without assuming the existence of forces capable of acting directly, . . . The theory I propose may therefore be called a theory of the Electromagnetic Field because it has to do with the space in the neighborhood of the electric and magnetic bodies.

James Clerk Maxwell, 1865

An Example of a Field: The Gravitational Field

Before we dive into investigating the electric field, we will take a slight detour to examine the gravitational field. The field concept is not as necessary for under-standing the force of gravity as it will be for electric and magnetic forces.

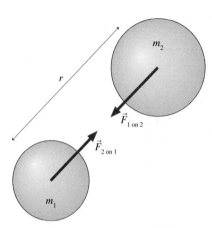

FIGURE 25.26 The gravitational forces between two masses.

Nonetheless, it will be useful to introduce the mathematical statement of the field concept in a familiar context before venturing into new and unexplored territory.

Figure 25.26 reminds you of Newton's law of gravity, which says that masses 1 and 2 exert equal but opposite attractive forces on each other of magnitude

$$F_{1 \text{ on } 2} = F_{2 \text{ on } 1} = \frac{Gm_1 m_2}{r^2} \tag{25.5}$$

where G is the gravitational constant and r is the distance between their centers of mass. Let's focus our attention on the force that mass 1 exerts on mass 2. Because force is a vector, we can write the force on mass 2 as

$$\vec{F}_{1 \text{ on } 2} = \left(\frac{Gm_1 m_2}{r^2}, \text{ toward mass 1} \right) = m_2 \left(\frac{Gm_1}{r^2}, \text{ toward mass 1} \right) \tag{25.6}$$

On the right side we separated out the mass m_2 on which the force is exerted.

Faraday's hypothesis, applied to gravity, is that the *gravitational field* of mass 1 exerts the force on mass 2. Suppose we define the gravitational field, which we will give the symbol $\vec{g}$, so that

$$\vec{F}_{\text{on } m} = m\vec{g} \tag{25.7}$$

In other words, $\vec{F}_{\text{on } m}$ is the gravitational force exerted on a mass m placed at a point in space where, due to the presence of *other* masses, the gravitational field is $\vec{g}$. The field exists at this point in space whether or not mass m is present; mass m simply *responds* to the presence of the field. Equation 25.7 is the mathematical representation of Faraday's idea that **the field is the agent that exerts the force.**

NOTE ▶ Equation 25.7 defines the gravitational field to be a *vector,* a quantity with both a magnitude and a direction. ◀

Comparing Equations 25.6 and 25.7 shows you that the gravitational field of mass 1 is

$$\vec{g}_{\text{mass 1}} = \left(\frac{Gm_1}{r^2}, \text{ toward mass 1} \right) \tag{25.8}$$

Equation 25.8 tells us that there is a gravitational field at *every point* in space around mass 1, regardless of whether or not another mass is present. At any particular point,

1. The field strength g is proportional to the mass m_1. Larger masses create stronger gravitational fields.
2. The field strength is inversely proportional to the square of the distance from mass 1. The field is weaker at larger distances, but there is never a point at which the field is exactly zero. The gravitational field of a mass extends through *all* of space.
3. The field points directly toward mass 1.

NOTE ▶ The gravitational field is determined by mass 1, the object that *creates* the field. The value m_2 of the mass that *experiences* the field does not enter into Equation 25.8. ◀

Figure 25.27 is a pictorial representation of the gravitational field of mass 1. It shows the vector $\vec{g}$ at just a *few* points in space. It is impossible to draw a vector at every point, but you should keep in mind that such a vector *is* at every point, whether shown or not.

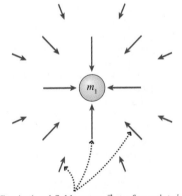

Gravitational field vectors $\vec{g}$ at a few points in space near mass m_1. All point toward A and have a magnitude inversely proportional to the square of the distance from the mass.

FIGURE 25.27 A pictorial representation of the gravitational field around mass 1.

The concept of a gravitational field is not limited to the gravitational field of a single mass. Figure 25.28 shows a mass placed in an arbitrary gravitational field $\vec{g}$ represented by the vector arrows. This is *not* the gravitational field of a single mass. In fact, we have no idea what masses are the source of this field because they are not shown. Nonetheless, knowing the *field* allows us to determine that the gravitational force exerted on this mass is $\vec{F}_{\text{on } m} = m\vec{g}$. The force is in the same direction as $\vec{g}$.

NOTE ▶ The field exists at the point where the mass is placed even though no field vector is drawn at that exact point. The field exists at *all* points in space.◀

Our choice of the symbol $\vec{g}$ to represent the gravitational field was not arbitrary. Consider the situation near the surface of the earth. The gravitational force on mass m is then a simpler

$$\vec{F}_{\text{on } m} = -mg\hat{\jmath} \tag{25.9}$$

where the minus sign indicates a force vector pointing in the downward direction. This g is the familiar acceleration due to gravity, $g = 9.80$ m/s^2. According to Equation 25.7, the gravitational field just above the earth's surface is

$$\vec{g} = \frac{\vec{F}_{\text{on } m}}{m} = -g\hat{\jmath} \tag{25.10}$$

In other words, the gravitational field near the earth's surface is a downward vector with magnitude g. Our old friend g is nothing other than the magnitude of the gravitational field!

Figure 25.29 shows the gravitational field near the surface of the earth. All the vectors are of equal length $g = 9.8$ N/kg $= 9.8$ m/s^2, and they all point straight down. A mass m *responds* to the field by experiencing the force $\vec{F} = m\vec{g}$. It is important to keep in mind that mass m does not *cause* the field. The field is created by the earth and is there whether mass m is present or not.

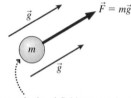

The gravitational field exerts a force on the mass even though no field vector is shown at this point.

FIGURE 25.28 A mass in a gravitational field experiences a force.

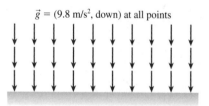

FIGURE 25.29 The gravitational field near the earth's surface.

25.6 The Field Model

Many of the electric phenomena you have seen in this chapter are more complex than gravitational phenomena. One reason is that whereas there is only one kind of mass, there are two kinds of charge, and thus both attractive and repulsive forces. Another is that there are two types of materials, conductors and insulators, with very different mobility for charge. These factors alone will make Faraday's concept of field useful for the study of electricity.

However, there's a more fundamental reason for introducing the electric field. For purely static situations, where electric and magnetic fields do not change with time, we *could* dispense with the concept of a field and base all our calculations on Coulomb's law. But later, when we get to the electro*dynamics* of fields that change with time, we'll find phenomena that can be understood *only* in terms of fields.

We begin our investigation of electric fields by postulating a **field model** that describes how two charges interact:

1. Some charges, which we will call the **source charges,** alter the space around them by creating an *electric field* $\vec{E}$.
2. A separate charge *in* the electric field experiences a force $\vec{F}$ exerted *by the field.*

We must accomplish two tasks to make this a useful model of electric interactions. First, we must learn how to calculate the electric field for a configuration of source charges. Second, we must determine the forces on and the motion of a charge in the electric field.

Suppose charge q experiences an electric force $\vec{F}_{on\,q}$ due to other charges. The strength and direction of this force vary from point to point in space, so $\vec{F}_{on\,q}$ is a continuous function of the charge's coordinates $(x,\ y,\ z)$. This suggests that "something" is present at each point in space to cause the force that charge q experiences. Using Equation 25.7 for the gravitational field as an analogy, let us define the electric field $\vec{E}$ at the point $(x,\ y,\ z)$ as

$$\vec{E}(x, y, z) = \frac{\vec{F}_{on\,q} \text{ at } (x, y, z)}{q} \qquad (25.11)$$

We're *defining* the electric field as a force-to-charge ratio, hence the units of the electric field are newtons per coulomb, or N/C. The magnitude E of the electric field is called the **electric field strength.**

You can think of using charge q as a *probe* to determine if an electric field is present at a point in space. If charge q experiences an electric force at a point in space, as Figure 25.30a shows, we say that there is an electric field at that point causing the force. Further, we *define* the electric field at that point to be the vector given by Equation 25.11. Figure 25.30b shows the electric field at two points, but you can imagine "mapping out" the electric field by moving charge q all through space.

The basic idea of the field model is that **the field is the agent that exerts an electric force on charge q.** Notice three important things about the field:

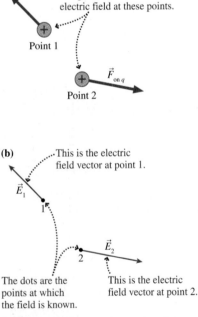

(a) Charge q is placed at two different points in space. The force on the charge tells us that there's an electric field at these points.

Point 1

$\vec{F}_{on\,q}$

Point 2

(b) This is the electric field vector at point 1.

$\vec{E}_1$

1

$\vec{E}_2$

2

The dots are the points at which the field is known.

This is the electric field vector at point 2.

FIGURE 25.30 Charge q is a probe of the electric field.

1. Equation 25.11 assigns a *vector* to *every point* in space. That is, the electric field is a *vector field.* Electric field diagrams will show a sample of the vectors, but there is an electric field vector at every point whether one is shown or not.
2. If q is positive, the electric field vector points in the same direction as the force on the charge.
3. Because q appears in Equation 25.11, it may seem that the electric field depends on the size of the charge used to probe the field. It doesn't. We know from Coulomb's law that the force $\vec{F}_{on\,q}$ is proportional to q. Thus the electric field defined in Equation 25.11 is *independent* of the charge q that probes the field. The electric field depends only on the source charges that create the field.

In practice we will usually want to turn Equation 25.11 around and find the force exerted by a known field. That is, a charge q at a point in space where the electric field is $\vec{E}$ experiences an electric force

$$\vec{F}_{on\,q} = q\vec{E} \qquad (25.12)$$

If q is positive, the force on charge q is in the direction of $\vec{E}$. The force on a negative charge is *opposite* the direction of $\vec{E}$.

STOP TO THINK 25.5   An electron is placed at the position marked by the dot. The force on the electron is

a. Zero.
b. To the right.
c. To the left.
d. There's not enough information to tell.

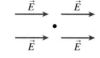

The Electric Field of a Point Charge

We will begin to put the definition of the electric field to full use in the next chapter. For now, to develop the ideas, we will determine the electric field of a single point charge q. Figure 25.31a shows charge q and a point in space at which we would like to know the electric field. We need a second charge, shown as q' in Figure 25.31b, to serve as a probe of the electric field.

For the moment, assume both charges are positive. The force on q', which is repulsive and directed straight away from q, is given by Coulomb's law:

$$\vec{F}_{on\ q'} = \left(\frac{1}{4\pi\epsilon_0} \frac{qq'}{r^2}, \text{away from } q \right) \tag{25.13}$$

It's customary to use $1/4\pi\epsilon_0$ rather than K for field calculations. Equation 25.11 defined the electric field in terms of the force on a probe charge, thus the electric field at this point is

$$\vec{E} = \frac{\vec{F}_{on\ q'}}{q'} = \left(\frac{1}{4\pi\epsilon_0} \frac{q}{r^2}, \text{away from } q \right) \tag{25.14}$$

The electric field is shown in Figure 25.31c.

NOTE ▶ The expression for the electric field is similar to Coulomb's law. To distinguish the two, remember that Coulomb's law has a product of two charges in the numerator. It describes the force between *two* charges. The electric field has a single charge in the numerator. It is the field of *a* charge. ◀

The field *strength* at distance r from a point charge depends inversely on the square of the distance: $E = q/4\pi\epsilon_0 r^2$. In Figure 25.32a, the field strength E_1 is larger than the field strength E_2 because $r_1 < r_2$. If we calculate the field at a sufficient number of points, we can draw a **field diagram** such as the one shown in Figure 25.32b. Notice that the field vectors all point straight away from charge q.

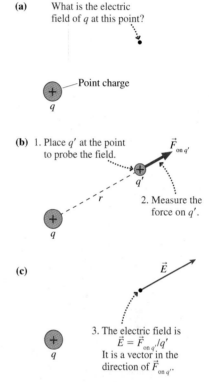
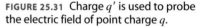

(a) What is the electric field of q at this point?

Point charge

(b) 1. Place q' at the point to probe the field.

$\vec{F}_{on\ q'}$

2. Measure the force on q'.

(c)

$\vec{E}$

3. The electric field is $\vec{E} = \vec{F}_{on\ q'}/q'$ It is a vector in the direction of $\vec{F}_{on\ q'}$.

FIGURE 25.31 Charge q' is used to probe the electric field of point charge q.

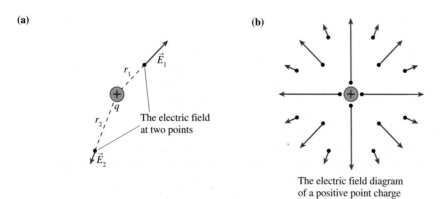

(a)

$\vec{E}_1$

r_1

q

r_2

The electric field at two points

$\vec{E}_2$

(b)

The electric field diagram of a positive point charge

FIGURE 25.32 The electric field near a positive charge.

11.4 Activ Physics

Keep these three important points in mind when using field diagrams:

1. The diagram is just a representative sample of electric field vectors. The field exists at all the other points. A well-drawn diagram can tell you fairly well what the field would be like at a neighboring point.

2. The arrow indicates the direction and the strength of the electric field *at the point to which it is attached.* That is, at the point where the *tail* of the vector is placed. In this chapter, we indicate the point at which the electric field is measured with a dot. The length of any vector is significant only relative to the lengths of other vectors.

3. Although we have to draw a vector across the page, from one point to another, an electric field vector is *not* a spatial quantity. It does not "stretch" from one point to another. Each vector represents the electric field at *one point* in space.

Unit Vector Notation

Equation 25.14 is precise, but it's not terribly convenient. Furthermore, what happens if the source charge q is negative? We need a more concise notation to write the electric field, a notation that will allow q to be either positive or negative.

The basic need is to express "away from q" in mathematical notation. "Away from q" is a *direction* in space. To guide us, recall that we already have a notation for expressing certain directions; namely, the unit vectors $\hat{\imath}$, $\hat{\jmath}$, and $\hat{k}$. For example, unit vector $\hat{\imath}$ means "in the direction of the positive x-axis." With a minus sign, $-\hat{\imath}$ means "in the direction of the negative x-axis." Unit vectors, with a magnitude of 1 and no units, provide purely directional information.

With this in mind, let's define the unit vector $\hat{r}$ to be a vector of length 1 that points from the origin to a point of interest. Unit vector $\hat{r}$ provides no information at all about the distance to the point. It merely specifies the direction.

Figure 25.33a shows unit vectors $\hat{r}_1$, $\hat{r}_2$, and $\hat{r}_3$ pointing toward points 1, 2, and 3. Unlike $\hat{\imath}$ and $\hat{\jmath}$, unit vector $\hat{r}$ does not have a fixed direction. Instead, unit vector $\hat{r}$ specifies the direction "straight outward from this point." But that's just what we need to describe the electric field vector. Figure 25.33b shows the electric fields at points 1, 2, and 3 due to a positive charge at the origin. No matter which point you choose, the electric field at that point is "straight outward" from the charge. In other words, the electric field $\vec{E}$ points in the direction of the unit vector $\hat{r}$.

Using this notation, the electric field at distance r from a point charge q can be written

$$\vec{E} = \frac{1}{4\pi\epsilon_0}\frac{q}{r^2}\hat{r} \qquad \text{(electric field of a point charge)} \qquad (25.15)$$

where $\hat{r}$ is the unit vector from the charge to the point at which we want to know the field. Equation 25.15 is identical to Equation 25.14, but written in a notation in which the unit vector $\hat{r}$ expresses the idea "away from q."

Equation 25.15 works equally well if q is negative. A negative sign in front of a vector simply reverses its direction, so the unit vector $-\hat{r}$ points *toward* charge q. Figure 25.34 shows the electric field of a negative point charge. It looks like the electric field of a positive point charge except that the vectors point inward, toward the charge, instead of outward.

We'll end this chapter with two examples of the electric field of a point charge. Chapter 26 will expand these ideas to the electric fields of multiple charges and of extended objects.

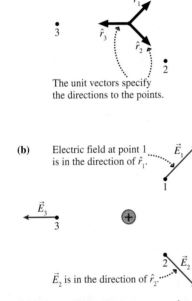

(a)

The unit vectors specify the directions to the points.

(b) Electric field at point 1 is in the direction of $\hat{r}_1$.

$\vec{E}_2$ is in the direction of $\hat{r}_2$.

FIGURE 25.33 Using the unit vector $\hat{r}$.

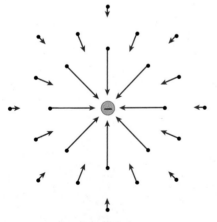

FIGURE 25.34 The electric field of a negative point charge.

EXAMPLE 25.8 Calculating the electric field
A -1.0 nC charged particle is located at the origin. Points 1 to 3 have (x, y) coordinates (1 cm, 0 cm), (0 cm, 1 cm), and (1 cm, 1 cm), respectively. Determine the electric field $\vec{E}$ at these three points, then show the vectors on an electric field diagram.

MODEL The electric field is that of a negative point charge.

VISUALIZE The electric field points straight *toward* the origin. It will be weaker at (1 cm, 1 cm), which is farther from the charge.

SOLVE The electric field is

$$\vec{E} = \frac{1}{4\pi\epsilon_0} \frac{q}{r^2} \hat{r}$$

where $q = -1.0$ nC $= -1.0 \times 10^{-9}$ C. The distance r is 1.0 cm $= 0.010$ m for points 1 and 2 and ($\sqrt{2} \times 1.0$ cm) $= 0.0141$ m for point 3. The *magnitude* of $\vec{E}$ at the three points is

$$E_1 = \frac{1}{4\pi\epsilon_0} \frac{|q|}{r_1^2}$$

$$= \frac{(9.0 \times 10^9 \, N m^2/C^2)(1.0 \times 10^{-9} \, C)}{(0.010 \, m)^2} = 90{,}000 \, N/C$$

$$E_2 = \frac{1}{4\pi\epsilon_0} \frac{|q|}{r_2^2}$$

$$= \frac{(9.0 \times 10^9 \, N m^2/C^2)(1.0 \times 10^{-9} \, C)}{(0.010 \, m)^2} = 90{,}000 \, N/C$$

$$E_3 = \frac{1}{4\pi\epsilon_0} \frac{|q|}{r_3^2}$$

$$= \frac{(9.0 \times 10^9 \, N m^2/C^2)(1.0 \times 10^{-9} \, C)}{(0.0141 \, m)^2} = 45{,}000 \, N/C$$

Because q is negative, the field at each of these positions points directly at charge q. The electric field vectors, in component form, are

$$\vec{E}_1 = -90{,}000\hat{\imath} \, N/C$$

$$\vec{E}_2 = -90{,}000\hat{\jmath} \, N/C$$

$$\vec{E}_3 = -E_3\cos 45°\hat{\imath} - E_3\sin 45°\hat{\jmath}$$

$$= (-31{,}800\hat{\imath} - 31{,}800\hat{\jmath}) \, N/C$$

These vectors are shown on the electric field diagram of Figure 25.35.

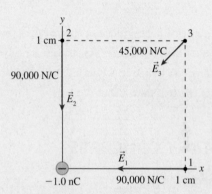

FIGURE 25.35 The electric field diagram of a -1.0 nC charged particle.

EXAMPLE 25.9 The electric field of a proton
The electron in a hydrogen atom orbits the proton at a radius of 0.053 nm.

a. What is the proton's electric field strength at the position of the electron?
b. What is the magnitude of the electric force on the electron?

SOLVE

a. The proton's charge is $q = e$. Its electric field strength at the distance of the electron is

$$E = \frac{1}{4\pi\epsilon_0} \frac{e}{r^2} = \frac{1}{4\pi\epsilon_0} \frac{1.60 \times 10^{-19} \, C}{(5.3 \times 10^{-11} \, m)^2}$$

$$= 5.12 \times 10^{11} \, N/C$$

Notice how large this field is in comparison to the field of Example 25.8.

b. We could use Coulomb's law to find the force on the electron, but the whole point of knowing the electric field is that we can use it directly to find the force on a charge in the field. The magnitude of the force on the electron is

$$F_{\text{on elec}} = |q_e| E_{\text{of proton}}$$

$$= (1.60 \times 10^{-19} \, C)(5.12 \times 10^{11} \, N/C)$$

$$= 8.20 \times 10^{-8} \, N$$

STOP TO THINK 25.6 Rank in order, from largest to smallest, the electric field strengths E_1 to E_4 at points 1 to 4.

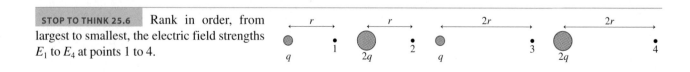

SUMMARY

The goal of Chapter 25 has been to develop a basic understanding of electric phenomena in terms of charges, forces, and fields.

GENERAL PRINCIPLES

Coulomb's Law

The forces between two charged particles q_1 and q_2 separated by distance r are

$$F_{1 \text{ on } 2} = F_{2 \text{ on } 1} = \frac{K|q_1||q_2|}{r^2}$$

These forces are an action/reaction pair directed along the line joining the particles.

- The forces are repulsive for two like charges, attractive for two opposite charges.
- The net force on a charge is the sum of the forces from all other charges.
- The unit of charge is the coulomb (C).

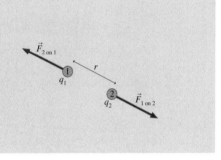

IMPORTANT CONCEPTS

The Charge Model

There are **two kinds of charge,** called *positive* and *negative*.

- Fundamental charges are protons and electrons, with charge $\pm e$ where $e = 1.60 \times 10^{-19}$ C.
- Objects are charged by adding or removing electrons.
- The amount of charge is $q = (N_p - N_e)e$.
- An object with an equal number of protons and electrons is neutral, meaning no *net* charge.

Charged objects exert **electric forces** on each other.

- Like charges repel, opposite charges attract.
- The force increases as the charge increases.
- The force decreases as the distance increases.

There are **two types of material,** insulators and conductors.

- Charge remains fixed in or on an insulator.
- Charge moves easily through or along conductors.
- Charge is transferred by contact between objects.

Charged objects attract neutral objects.

- Charge polarizes metal by shifting the electron sea.
- Charge polarizes atoms, creating electric dipoles.
- The polarization force is always an attractive force.

The Field Model

Charges interact with each other via the electric field $\vec{E}$.

- Charge A alters the space around it by creating an electric field.

- The field is the agent that exerts a force. The force on charge q_B is $\vec{F}_{\text{on B}} = q_B\vec{E}$.

An electric field is identified and measured in terms of the force on a **probe charge** q.

$$\vec{E} = \vec{F}_{\text{on } q}/q$$

- The electric field exists at all points in space.
- An electric field vector shows the field only at one point, the point at the tail of the vector.

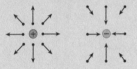

The electric field of a **point charge** is

$$\vec{E} = \frac{1}{4\pi\epsilon_0}\frac{q}{r^2}\hat{r}$$

TERMS AND NOTATION

neutral	ionization	Coulomb's law
charging	law of conservation of charge	electrostatic constant, K
charge model	sea of electrons	point charge
charge, q or Q	ion core	coulomb, C
like charges	current	permittivity constant, ϵ_0
opposite charges	charge carriers	field
discharging	electrostatic equilibrium	gravitational field, $\vec{g}$
conductor	grounded	electric field, $\vec{E}$
insulator	charge polarization	field model
electron cloud	polarization force	source charge
fundamental unit of charge, e	electric dipole	electric field strength, E
charge quantization	charging by induction	field diagram

EXERCISES AND PROBLEMS

Exercises

Section 25.1 Developing a Charge Model

Section 25.2 Charge

1. A glass rod is charged to $+5$ nC by rubbing.
   a. Have electrons been removed from the rod or protons added? Explain.
   b. How many electrons have been removed or protons added?
2. A plastic rod is charged to -20 nC by rubbing.
   a. Have electrons been added to the rod or protons removed? Explain.
   b. How many electrons have been added or protons removed?
3. A 2.0-mm-diameter copper ball is charged to $+50$ nC. What fraction of its electrons have been removed?
4. How many coulombs of positive charge are in 1.0 mol of O_2 gas?

Section 25.3 Insulators and Conductors

5. Figure 25.9 showed how an electroscope becomes negatively charged. The leaves will also repel each other if you touch the electroscope with a positively charged glass rod. Use a series of charge diagrams to explain what happens and why the leaves repel each other.
6. A plastic balloon that has been rubbed with wool will stick to a wall.
   a. Can you conclude that the wall is charged? If not, why not? If so, where does the charge come from?
   b. Draw a series of charge diagrams showing how the balloon is held to the wall.
7. Two neutral metal spheres on wood stands are touching. A negatively charged rod is held directly above the top of the left sphere, not quite touching it. While the rod is there, the right sphere is moved so that the spheres no longer touch. Then the rod is withdrawn. Afterward, what is the charge state of each sphere? Use charge diagrams to explain your answer.
8. You have two neutral metal spheres on wood stands. Devise a procedure for charging the spheres so that they will have opposite charges of *exactly* equal magnitude. Use charge diagrams to explain your procedure.
9. You have two neutral metal spheres on wood stands. Devise a procedure for charging the spheres so that they will have like charges of *exactly* equal magnitude. Use charge diagrams to explain your procedure.
10. An object passes the "Is it charged?" test by picking up small pieces of paper.
    a. Use a series of charge diagrams to explain how a charged object picks up a piece of paper.
    b. Write a paragraph explaining why this test works for both positively and negatively charged objects. Your explanation should be based on the properties of charges, forces, and insulators.

Section 25.4 Coulomb's Law

11. Two 1.0 kg masses are 1.0 m apart on a frictionless table. Each has $+1.0$ C of charge.
    a. What is the magnitude of the electric force on one of the masses?
    b. What is the initial acceleration of the mass if it is released and allowed to move?
12. Two small plastic spheres each have a mass of 2.0 g and a charge of -50.0 nC. They are placed 2.0 cm apart.
    a. What is the magnitude of the electric force between the spheres?
    b. By what factor is the electric force on a sphere larger than its weight?
13. A small glass bead has been charged to $+20$ nC. A metal ball bearing 1.0 cm above the bead feels a 0.018 N downward electric force. What is the charge on the ball bearing?

14. What is the net electric force on charge A in Figure Ex25.14?

FIGURE EX25.14

15. Object A, which has been charged to $+10.0$ nC, is at the origin. Object B, which has been charged to -20.0 nC, is at $(x, y) = (0.0 \text{ cm}, 2.0 \text{ cm})$. Determine the electric force on each object. Write each force vector in component form.

16. A small glass bead has been charged to $+20$ nC. What are the magnitude and direction of the acceleration of (a) a proton and (b) an electron that is 1.0 cm from the center of the bead?

Section 25.5 The Concept of a Field

17. What is the strength of the gravitational field at a point where a 2.0 kg mass experiences a 60.0 N force?

18. Mass m experiences a gravitational force of magnitude 15 N. What will be the magnitude of the gravitational force on mass $3m$ at the same point in the gravitational field?

19. What are the magnitude and direction of the earth's gravitational field $\vec{g}$ at
    a. The surface of the earth?
    b. The radius of the moon's orbit?

20. A satellite orbits a planet. The gravitational field strength at the radius of the orbit is 12 N/kg. What will the gravitational field strength at the position of the satellite be if
    a. The radius of the orbit is doubled?
    b. The planet's density is doubled?
    c. The satellite's mass is doubled?

Section 25.6 The Field Model

21. What magnitude charge creates a 1.0 N/C electric field at a point 1.0 m away?

22. What are the strength and direction of the electric field 2.0 cm from a small glass bead that has been charged to $+6.0$ nC?

23. The electric field 2.0 cm from a small object points toward the object with a strength of 180,000 N/C. What is the object's charge?

24. What are the strength and direction of the electric field 1.0 mm from (a) a proton and (b) an electron?

25. What are the strength and direction of an electric field that will balance the weight of a 1.0 g plastic sphere that has been charged to -3.0 nC?

26. What are the strength and direction of an electric field that will balance the weight of (a) a proton and (b) an electron?

27. A $+10.0$ nC charge is located at the origin.
    a. What are the electric fields at the positions $(x, y) = (5 \text{ cm}, 0 \text{ cm})$, $(-5 \text{ cm}, 5 \text{ cm})$, and $(-5 \text{ cm}, -5 \text{ cm})$? Write each electric field vector in component form.
    b. Draw a field diagram showing the electric field vectors at these points.

28. A -10 nC charge is located at the origin.
    a. What are the electric fields at the positions $(x, y) = (0 \text{ cm}, 5 \text{ cm})$, $(-5 \text{ cm}, -5 \text{ cm})$, and $(-5 \text{ cm}, 5 \text{ cm})$? Write each electric field vector in component form.
    b. Draw a field diagram showing the electric field vectors at these points.

Problems

29. Pennies today are copper-covered zinc, but older pennies are 3.1 g of solid copper. What are the total positive charge and total negative charge in a solid copper penny that is electrically neutral?

30. A plastic rod that has been charged to -15 nC touches a metal sphere. Afterward, the rod's charge is -10 nC.
    a. What kind of charged particle was transferred between the rod and the sphere, and in which direction? That is, did it move from the rod to the sphere or from the sphere to the rod?
    b. How many charged particles were transferred?

31. A glass rod that has been charged to $+12$ nC touches a metal sphere. Afterward, the rod's charge is $+8$ nC.
    a. What kind of charged particle was transferred between the rod and the sphere, and in which direction? That is, did it move from the rod to the sphere or from the sphere to the rod?
    b. How many charged particles were transferred?

32. Two identical metal spheres A and B are connected by a metal rod. Both are initially neutral. 1.0×10^{12} electrons are added to sphere A, then the connecting rod is removed. Afterward, what are the charge of A and the charge of B?

33. Two identical metal spheres A and B are connected by a plastic rod. Both are initially neutral. 1.0×10^{12} electrons are added to sphere A, then the connecting rod is removed. Afterward, what are the charge of A and the charge of B?

34. Two protons are 2.0 fm apart.
    a. What is the magnitude of the electric force on one proton due to the other proton?
    b. What is the magnitude of the gravitational force on one proton due to the other proton?
    c. What is the ratio of the electric force to the gravitational force?

35. The nucleus of a ^{125}Xe atom (an isotope of the element xenon with mass 125 u) is 6.0 fm in diameter. It has 54 protons and charge $q = +54e$.
    a. What is the electric force on a proton 2.0 fm from the surface of the nucleus?
    Hint: Treat the spherical nucleus as a point charge.
    b. What is the proton's acceleration?

36. Two 1.0 g spheres are charged equally and placed 2.0 cm apart. When released, they begin to accelerate at 225 m/s². What is the magnitude of the charge on each sphere?

37. Objects A and B are both positively charged. Both have a mass of 100 g, but A has twice the charge of B. When A and B are placed 10 cm apart, B experiences an electric force of 0.45 N.
    a. How large is the force on A?
    b. What are the charges q_A and q_B?
    c. If the objects are released, what is the initial acceleration of A?

38. What is the force $\vec{F}$ on the 1 nC charge? Give your answer as a magnitude and a direction.

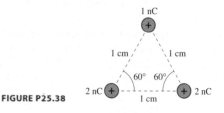

FIGURE P25.38

39. What is the force $\vec{F}$ on the 1 nC charge? Give your answer as a magnitude and a direction.

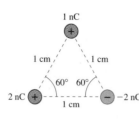

FIGURE P25.39

40. What is the force $\vec{F}$ on the −10 nC charge? Give your answer as a magnitude and a direction.

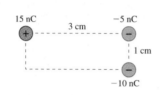

FIGURE P25.40

41. What is the force $\vec{F}$ on the −10 nC charge? Give your answer as a magnitude and a direction.

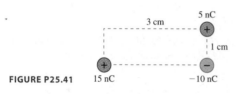

FIGURE P25.41

42. What is the force $\vec{F}$ on the 5 nC charge in Figure P25.42? Give your answer as a magnitude and a direction.

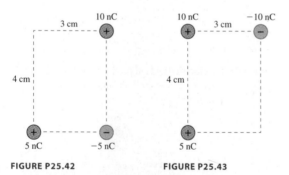

FIGURE P25.42   **FIGURE P25.43**

43. What is the force $\vec{F}$ on the 5 nC charge in Figure P25.43? Give your answer as a magnitude and a direction.

44. What is the force $\vec{F}$ on the 1 nC charge in the middle due to the four other charges? Give your answer in component form.

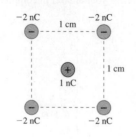

FIGURE P25.44

45. What is the force $\vec{F}$ on the 1 nC charge in the middle due to the four other charges? Give your answer in component form.

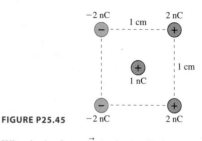

FIGURE P25.45

46. What is the force $\vec{F}$ on the 1 nC charge at the bottom? Give your answer in component form.

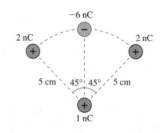

FIGURE P25.46

47. What is the force $\vec{F}$ on the 1 nC charge at the bottom? Give your answer in component form.

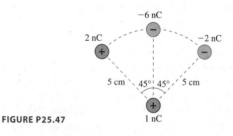

FIGURE P25.47

48. A +2.0 nC charge and a −4.0 nC charge are 1.0 cm apart.
    a. Where could you place a proton so that it would experience no net force?
    b. Would the net force be zero for an electron placed at the same position? Explain.

49. The net force on the 1 nC charge is zero. What is q?

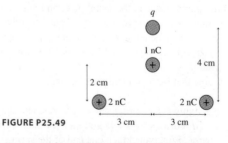

FIGURE P25.49

50. Charge q_2 is in static equilibrium. What is q_1?

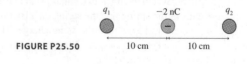

FIGURE P25.50

51. A positive point charge Q is located at $x = a$ and a negative point charge $-Q$ is at $x = -a$. A positive charge q can be placed anywhere on the y-axis. Find an expression for $(F_{net})_x$, the x-component of the net force on q.

52. A positive point charge Q is located at $x = a$ and a negative point charge $-Q$ is at $x = -a$. A positive charge q can be placed anywhere on the x-axis. Find an expression for $(F_{net})_x$, the x-component of the net force on q, when (a) $|x| < a$ and (b) $|x| > a$.

53. Figure P25.53 shows four charges at the corners of a square of side L. Assume q and Q are positive.
    a. Draw a diagram showing the three forces on charge q due to the other charges. Give your vectors the correct relative lengths.
    b. What is the magnitude of the net force on q?

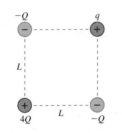

FIGURE P25.53

54. Two point charges q and $4q$ are distance L apart and free to move. A third charge is placed so that the entire three-charge system is in static equilibrium. What are the magnitude, sign, and position of the third charge?

55. Suppose that the magnitudes of the proton charge and the electron charge differ by a mere 1 part in 10^9.
    a. What would be the force between two 2.0-mm-diameter copper spheres 1.0 cm apart? Assume that each copper atom has an equal number of electrons and protons.
    b. Would this amount of force be detectable? What can you conclude from the fact that no such forces are observed?

56. In a simple model of the hydrogen atom, the electron moves in a circular orbit of radius 0.053 nm around a stationary proton. How many revolutions per second does the electron make?

57. As a science project, you've invented an "electron pump" that moves electrons from one object to another. To demonstrate your invention, you bolt a small metal plate to the ceiling, connect the pump between the metal plate and yourself, and start pumping electrons from the metal plate to you. How many electrons must be moved from the metal plate to you in order for you to hang suspended in the air 2.0 m below the ceiling? Your mass is 60 kg.
    Hint: Assume that both you and the plate can be modeled as point charges.

58. You have a lightweight spring whose unstretched length is 4.0 cm. You're curious to see if you can use this spring to measure charge. First, you attach one end of the spring to the ceiling and hang a 1.0 g mass from it. This stretches the spring to a length of 5.0 cm. You then attach two small plastic beads to the opposite ends of the spring, lay the spring on a frictionless table, and give each plastic bead the same charge. This stretches the spring to a length of 4.5 cm. What is the magnitude of the charge (in nC) on each bead?

59. Two 5.0 g spheres on 1.0-m-long threads repel each other after being charged to $+100$ nC. What is the angle θ? You can assume that θ is a small angle.

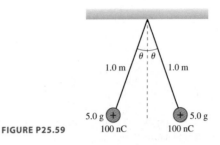

FIGURE P25.59

60. Two 3.0 g spheres on 1.0-m-long threads repel each other after being equally charged. What is the charge q?

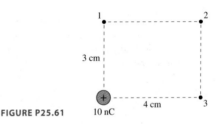

FIGURE P25.60

61. What are the electric fields at points 1, 2, and 3 in Figure P25.61? Give your answer in component form.

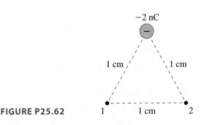

FIGURE P25.61

62. What are the electric fields at points 1 and 2 in Figure P25.62? Give your answer as a magnitude and direction.

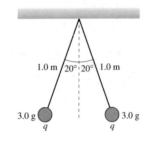

FIGURE P25.62

63. What are the electric fields at points 1, 2, and 3 in Figure P25.63? Give your answer in component form.

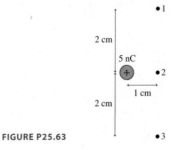

FIGURE P25.63

64. A -10.0 nC charge is located at position $(x, y) = (2.0$ cm, 1.0 cm). At what (x, y) position(s) is the electric field
    a. $-225,000\hat{\imath}$ N/C?
    b. $(161,000\hat{\imath} - 80,500\hat{\jmath})$ N/C?
    c. $(28,800\hat{\imath} + 21,600\hat{\jmath})$ N/C?

65. A 10.0 nC charge is located at position $(x, y) = (1.0$ cm, 2.0 cm). At what (x, y) position(s) is the electric field
    a. $-225,000\hat{\imath}$ N/C?
    b. $(161,000\hat{\imath} + 80,500\hat{\jmath})$ N/C?
    c. $(21,600\hat{\imath} - 28,800\hat{\jmath})$ N/C?

66. Three 1.0 nC charges are placed as shown in Figure P25.66. Each of these charges creates an electric field $\vec{E}$ at a point 3.0 cm in front of the middle charge.
    a. What are the three fields $\vec{E}_1$, $\vec{E}_2$, and $\vec{E}_3$ created by the three charges? Write your answer for each as a vector in component form.
    b. Do you think that electric fields obey a principle of superposition? That is, is there a "net field" at this point given by $\vec{E}_{net} = \vec{E}_1 + \vec{E}_2 + \vec{E}_3$? Use what you learned in this chapter and previously in our study of forces to argue why this is or is not true.
    c. If it is true, what is $\vec{E}_{net}$?

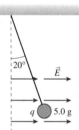

FIGURE P25.66

67. The electric field at a point in space is $\vec{E} = (200\hat{\imath} + 400\hat{\jmath})$ N/C.
    a. What is the electric force on a proton at this point? Give your answer in component form.
    b. What is the electric force on an electron at this point? Give your answer in component form.
    c. What is the magnitude of the proton's acceleration?
    d. What is the magnitude of the electron's acceleration?

68. An electric field $\vec{E} = 100,000\hat{\imath}$ N/C causes the 5.0 g ball in Figure P25.68 to hang at a 20° angle. What is the charge on the ball?

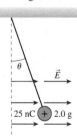

FIGURE P25.68

69. An electric field $\vec{E} = 200,000\hat{\imath}$ N/C causes the ball in Figure P25.69 to hang at an angle. What is θ?

FIGURE P25.69

In Problems 70 through 73 you are given the equation(s) used to solve a problem. For each of these,
    a. Write a realistic problem for which this is the correct equation(s).
    b. Finish the solution of the problem.

70. $\dfrac{(9.0 \times 10^9\,\text{N}\,\text{m}^2/\text{C}^2) \times N \times (1.60 \times 10^{-19}\,\text{C})}{(1.0 \times 10^{-6}\,\text{m})^2}$
    $= 1.5 \times 10^6$ N/C

71. $\dfrac{(9.0 \times 10^9\,\text{N}\,\text{m}^2/\text{C}^2)q^2}{(0.0150\,\text{m})^2} = 0.020$ N

72. $\dfrac{(9.0 \times 10^9\,\text{N}\,\text{m}^2/\text{C}^2)(15 \times 10^{-9}\,\text{C})}{r^2} = 54,000$ N/C

73. $\sum F_x = 2 \times \dfrac{(9.0 \times 10^9\,\text{N}\,\text{m}^2/\text{C}^2)(1.0 \times 10^{-9}\,\text{C})q}{((0.020\,\text{m})/\sin 30°)^2} \times \cos 30°$
    $= 5.0 \times 10^{-5}$ N
    $\sum F_y = 0$ N

Challenge Problems

74. Three 3.0 g balls are tied to 80-cm-long threads and hung from a *single* fixed point. Each of the balls is given the same charge q. At equilibrium, the three balls form an equilateral triangle in a horizontal plane with 20.0 cm sides. What is q?

75. The identical small spheres shown in Figure CP25.75 are charged to $+100$ nC and -100 nC. They hang as shown in a 100,000 N/C electric field. What is the mass of each sphere?

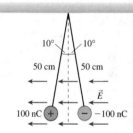

FIGURE CP25.75

76. You sometimes create a spark when you touch a doorknob after shuffling your feet on a carpet. Why? The air always has a few free electrons that have been kicked out of atoms by cosmic rays. If an electric field is present, a free electron is accelerated until it collides with an air molecule. It will transfer its kinetic energy to the molecule, then accelerate, then collide, then accelerate, collide, and so on. If the electron's kinetic energy just before a collision is 2.0×10^{-18} J or more, it has sufficient energy to kick an electron out of the molecule it hits. Where there was one free electron, now there are two! Each of these can then accelerate, hit a molecule, and kick out another electron. Then there will be four free electrons. In other words, as Figure CP25.76 on the next page shows, a sufficiently strong electric field causes a "chain reaction" of electron production. This is called a *breakdown* of the air. The current of moving electrons is what gives you the shock, and a spark is generated when the electrons recombine with the positive ions and give off excess energy as a burst of light.
    a. The average distance an electron travels between collisions is $2.0\ \mu$m. What acceleration must an electron have to gain 2.0×10^{-18} J of kinetic energy in this distance?

b. What force must act on an electron to give it the acceleration found in part a?

c. What strength electric field will exert this much force on an electron? This is the *breakdown field strength*.

d. Suppose a free electron in air is 1.0 cm away from a point charge. What minimum charge q_{min} must this point charge have to cause a breakdown of the air and create a spark?

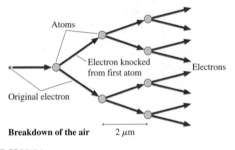

Breakdown of the air   2 μm

FIGURE CP25.76

77. In Section 25.3 we claimed that a charged object exerts a net attractive force on an electric dipole. Let's investigate this. The figure shows a *permanent* electric dipole consisting of charges $+q$ and $-q$ separated by the fixed distance s. Charge $+Q$ is distance r from the center of the dipole. We'll assume, as is usually the case in practice, that $s \ll r$.

a. Write an expression for the net force exerted on the dipole by charge $+Q$.

b. Is this force toward $+Q$ or away from $+Q$? Explain.

c. Use the *binomial approximation* $(1 + x)^{-n} \approx 1 - nx$ if $x \ll 1$ to show that your expression from part a can be written $F_{net} = 2KqQs/r^3$.

d. How can an electric force have an inverse-cube dependence? Doesn't Coulomb's law say that the electric force depends on the inverse square of the distance? Explain.

s Q

$+$ $-$ $+$

q $-q$ r

FIGURE CP25.77

STOP TO THINK ANSWERS

Stop to Think 25.1: b. Charged objects are attracted to neutral objects, so an attractive force is inconclusive. Repulsion is the only sure test.

Stop to Think 25.2: $q_e(+3e) > q_a(+1e) > q_d(0) > q_b(-1e) > q_c(-2e)$.

Stop to Think 25.3: a. The negative plastic rod will polarize the electroscope by pushing electrons down toward the leaves. This will partially neutralize the positive charge the leaves had acquired from the glass rod.

Stop to Think 25.4: b. The two forces are an action/reaction pair, opposite in direction but *equal* in magnitude.

Stop to Think 25.5: c. There's an electric field at *all* points, whether an $\vec{E}$ vector is shown or not. The electric field at the dot is to the right. But an electron is a negative charge, so the force of the electric field on the electron is to the left.

Stop to Think 25.6: $E_2 > E_1 > E_4 > E_3$.

26 The Electric Field

Liquid crystal displays work by using electric fields to align long polymer molecules.

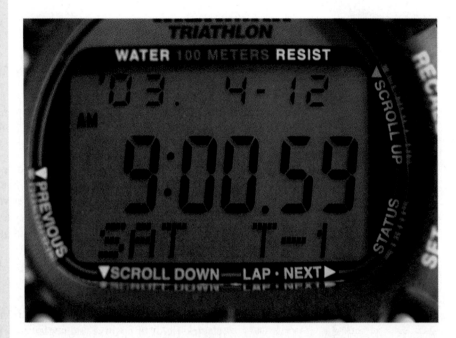

▶ **Looking Ahead**

The goal of Chapter 26 is to learn how to calculate and use the electric field. In this chapter you will learn to:

- Calculate the electric field due to multiple point charges.
- Calculate the electric field due to a continuous distribution of charge.
- Use the electric field of dipoles, lines of charge, and planes of charge.
- Generate a uniform electric field with a parallel-plate capacitor.
- Calculate the motion of charges and dipoles in an electric field.

◀ **Looking Back**

This chapter builds on the ideas about electric forces and fields that were introduced in Chapter 25. Charged-particle motion in an electric field is similar to projectile motion. Please review:

- Section 6.3 Projectile motion.
- Section 25.4 Coulomb's law.
- Sections 25.5–25.7 The electric field of a point charge.

You can't see them, but they are all around you—electric fields, that is. Electric fields line up polymer molecules to form the images in the liquid crystal display (LCD) of a wristwatch or a flat-panel computer screen. Electric fields are responsible for the electric currents that flow in your computer and your stereo, and they are essential to the functioning of your brain, your heart, and your DNA.

We introduced the idea of an electric field, in Chapter 25, in order to understand the long-range electric interaction between charges. The electric field of a point charge is pretty simple, but real-world charged objects have vast numbers of charges arranged in complex patterns. To make practical use of electric fields, we need to know how to calculate the electric field of a complicated distribution of charge. The major goal of this chapter is to develop a procedure for calculating the electric field of a specified configuration or arrangement of charge.

Chapter 25 made a distinction between those charged particles that are the *sources* of an electric field and other charged particles that *experience* and move in the electric field. This is a very important distinction. Most of this chapter will be concerned with the *sources* of the electric field. Only at the end, once we know how to calculate the electric field, will we look at what happens to charges that find themselves *in* an electric field.

26.1 Electric Field Models

The electric fields used in science and engineering are often caused by fairly complicated distributions of charge. Sometimes these fields require exact calculations, but much of the time we can understand the essential physics on the basis of simplified *models* of the electric field.

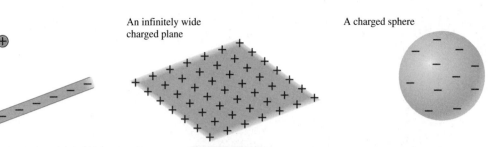

A point charge

An infinitely long
charged wire

An infinitely wide
charged plane

A charged sphere

FIGURE 26.1 Four basic electric field models.

Four widely used electric field models, illustrated in Figure 26.1, are:

- The electric field of a point charge.
- The electric field of an infinitely long charged wire.
- The electric field of an infinitely wide charged plane.
- The electric field of a charged sphere.

Small charged objects can often be modeled as point charges or charged spheres. Real wires aren't infinitely long, but in many practical situations this approximation is perfectly reasonable. As we derive and use these electric fields, we'll consider the conditions under which they are appropriate models.

Our starting point is the electric field of a point charge q. We found in Chapter 25 that

$$\vec{E} = \frac{1}{4\pi\epsilon_0} \frac{q}{r^2} \hat{r} \qquad \text{(electric field of a point charge)} \qquad (26.1)$$

where $\hat{r}$ is a unit vector pointing away from q. Figure 26.2 reminds you how the electric fields of positive and negative point charges look. Although we have to give each vector we draw a length, keep in mind that each arrow represents the electric field *at a point*. The electric field is not a spatial quantity that "stretches" from one end of the arrow to the other.

The electric field was defined as $\vec{E} = \vec{F}_{\text{on } q}/q$, where $\vec{F}_{\text{on } q}$ is the electric force on charge q. Forces add as vectors, so the net force on q due to a group of point charges is the vector sum

$$\vec{F}_{\text{on } q} = \vec{F}_{1 \text{ on } q} + \vec{F}_{2 \text{ on } q} + \cdots$$

Consequently, the net electric field due to a group of point charges is

$$\vec{E}_{\text{net}} = \frac{\vec{F}_{\text{on } q}}{q} = \frac{\vec{F}_{1 \text{ on } q}}{q} + \frac{\vec{F}_{2 \text{ on } q}}{q} + \cdots = \vec{E}_1 + \vec{E}_2 + \cdots = \sum_i \vec{E}_i \qquad (26.2)$$

where $\vec{E}_i$ is the field from point charge i.

Equation 26.2, which is the primary tool for calculating electric fields, tells us that **the net electric field is the *vector sum* of the electric fields due to each charge.** In other words, electric fields obey the *principle of superposition*. Figure 26.3 illustrates this important idea. Much of this chapter will be focused on the mathematical aspects of performing the sum.

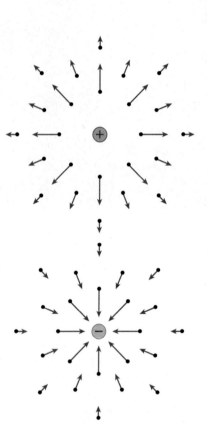

FIGURE 26.2 The electric field of a positive and a negative point charge.

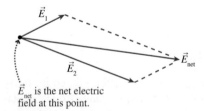

$\vec{E}_{\text{net}}$ is the net electric field at this point.

FIGURE 26.3 Electric fields obey the principle of superposition.

Limiting Cases and Typical Field Strengths

Suppose you have a small, charged object. The electric field very near the object depends on the object's shape and on how the charge is distributed. But from far away, the object appears to be a point charge in the distance. Thus the object's electric field should approach that of a point charge in the *limiting case* that the distance from the object approaches infinity.

We'll have occasion to look at limiting cases both very far from and very close to charged objects. Limiting cases allow us to

- Check a solution by seeing if it has the expected behavior as distances become very large or very small.
- Use simpler expressions for the electric field at points very close to or very far from a charged object.

We'll emphasize limiting cases throughout this chapter as we develop models of the electric field.

Knowing some typical electric field strengths will also be helpful. The values shown in Table 26.1 will help you judge the reasonableness of your solutions to problems.

TABLE 26.1 Typical electric field strengths

Field location	Field strength (N/C)
Inside a current-carrying wire	10^{-2}
Near the earth's surface	10^2–10^4
Near objects charged by rubbing	10^3–10^6
Electric breakdown in air, causing a spark	3×10^6
Inside an atom	10^{11}

26.2 The Electric Field of Multiple Point Charges

As we noted in Chapter 25, it is important to distinguish between those charges that are the sources of an electric field and those that experience and move in the electric field. Suppose the source of an electric field is a group of point charges $q_1, q_2, \ldots$. According to Equation 26.2, the electric field at each point in space is a superposition of the electric fields due to each individual charge.

The *vector* sum of Equation 26.2 can be written as the three simultaneous sums

$$(E_{net})_x = (E_1)_x + (E_2)_x + \cdots = \sum (E_i)_x$$

$$(E_{net})_y = (E_1)_y + (E_2)_y + \cdots = \sum (E_i)_y \qquad (26.3)$$

$$(E_{net})_z = (E_1)_z + (E_2)_z + \cdots = \sum (E_i)_z$$

The problem and the context will sometimes favor expressing $\vec{E}_{net}$ in component form: $\vec{E}_{net} = (E_{net})_x \hat{\imath} + (E_{net})_y \hat{\jmath} + (E_{net})_z \hat{k}$. At other times you will want to write $\vec{E}_{net}$ as a magnitude and a direction.

(MP) **PROBLEM-SOLVING STRATEGY 26.1** The electric field of multiple point charges

MODEL Model charged objects as point charges.

VISUALIZE For the pictorial representation:

- Establish a coordinate system and show the locations of the charges.
- Identify the point P at which you want to calculate the electric field.
- Draw the electric field of each charge at P.
- Use symmetry to determine if any components of $\vec{E}_{net}$ are zero.

SOLVE The mathematical representation is $\vec{E}_{net} = \sum \vec{E}_i$.

- For each charge, determine its distance from P and the angle of $\vec{E}_i$ from the axes.
- Calculate the field strength of each charge's electric field.
- Write each vector $\vec{E}_i$ in component form.
- Sum the vector components to determine $\vec{E}_{net}$.
- If needed, determine the magnitude and direction of $\vec{E}_{net}$.

ASSESS Check that your result has the correct units, is reasonable, and agrees with any known limiting cases.

EXAMPLE 26.1 **The electric field of three equal point charges**

Three equal point charges q are located on the y-axis at $y = 0$ and at $y = \pm d$. What is the electric field at a point on the x-axis?

MODEL This problem is a step along the way to understanding the electric field of a charged wire. We'll assume that q is positive when drawing pictures, but the solution should allow for the possibility that q is negative. The question does not ask about any specific point, so we will be looking for a symbolic expression in terms of the unspecified position x.

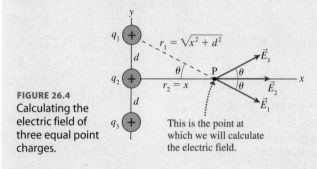

FIGURE 26.4 Calculating the electric field of three equal point charges.

This is the point at which we will calculate the electric field.

VISUALIZE Figure 26.4 shows the charges, the coordinate system, and the three electric field vectors $\vec{E}_1$, $\vec{E}_2$, and $\vec{E}_3$. Each of these fields points *away from* its source charge because of the assumption that q is positive. We need to find the vector sum $\vec{E}_{net} = \vec{E}_1 + \vec{E}_2 + \vec{E}_3$.

Before rushing into a calculation, we can make our task *much* easier by first thinking qualitatively about the situation. For example, the fields $\vec{E}_1$, $\vec{E}_2$, and $\vec{E}_3$ all lie in the xy-plane, hence we can conclude without doing any calculations that $(E_{net})_z = 0$. Next, look at the y-components of the fields. Because the charges are located *symmetrically* on either side of the x-axis and are of equal value, the fields $\vec{E}_1$ and $\vec{E}_3$ have equal magnitudes and are tilted away from the x-axis by the same angle θ. Consequently, the y-components of $\vec{E}_1$ and $\vec{E}_3$ will *cancel* when added. $\vec{E}_2$ has no y-component, so we can conclude that $(E_{net})_y = 0$. The only component we need to calculate is $(E_{net})_x$.

SOLVE We're finally ready to calculate. The x-component of the electric field is

$$(E_{net})_x = (E_1)_x + (E_2)_x + (E_3)_x = 2(E_1)_x + (E_2)_x$$

where we used the fact that fields $\vec{E}_1$ and $\vec{E}_3$ have *equal* x-components. Vector $\vec{E}_2$ has *only* the x-component

$$(E_2)_x = E_2 = \frac{1}{4\pi\epsilon_0}\frac{q_2}{r_2^2} = \frac{1}{4\pi\epsilon_0}\frac{q}{x^2}$$

where $r_2 = x$ is the distance from q_2 to the point at which we are calculating the field. Vector $\vec{E}_1$ is at angle θ from the x-axis, so its x-component is

$$(E_1)_x = E_1\cos\theta = \frac{1}{4\pi\epsilon_0}\frac{q_1}{r_1^2}\cos\theta$$

where r_1 is the distance from q_1. This expression for $(E_1)_x$ is correct, but it is not yet sufficient. Both the distance r_1 and the angle θ vary with the position x and need to be expressed as functions of x. From the Pythagorean theorem, $r_1 = (x^2 + d^2)^{1/2}$. Then from trigonometry,

$$\cos\theta = \frac{x}{r_1} = \frac{x}{(x^2 + d^2)^{1/2}}$$

By combining these pieces, we see that $(E_1)_x$ is

$$(E_1)_x = \frac{1}{4\pi\epsilon_0}\frac{q}{x^2 + d^2}\frac{x}{(x^2 + d^2)^{1/2}} = \frac{1}{4\pi\epsilon_0}\frac{xq}{(x^2 + d^2)^{3/2}}$$

This expression is a bit complex, but notice that the dimensions of $x/(x^2 + d^2)^{3/2}$ are $1/m^2$, as they *must* be for the field of a point charge. Checking dimensions is a good way to verify that you haven't made algebra errors.

We can now combine $(E_1)_x$ and $(E_2)_x$ to write the x-component of $\vec{E}_{net}$ as

$$(E_{net})_x = 2(E_1)_x + (E_2)_x = \frac{q}{4\pi\epsilon_0}\left[\frac{1}{x^2} + \frac{2x}{(x^2 + d^2)^{3/2}}\right]$$

The other two components of $\vec{E}_{net}$ are zero, hence the electric field of the three charges at a point on the x-axis is

$$\vec{E}_{net} = \frac{q}{4\pi\epsilon_0}\left[\frac{1}{x^2} + \frac{2x}{(x^2 + d^2)^{3/2}}\right]\hat{\imath}$$

ASSESS This is the electric field only at points *on the x-axis*. Furthermore, this expression is valid only for $x > 0$. The electric field to the left of the charges points in the opposite direction, but our expression doesn't change sign for negative x. (This is a consequence of how we wrote $(E_2)_x$.) We would need to modify this expression to use it for negative values of x. The good news, though, is that our expression *is* valid for both positive and negative q. A negative value of q makes $(E_{net})_x$ negative, which would be an electric field pointing to the left, toward the negative charges.

Let's explore this example a bit more. If you think about it, there are two limiting cases for which we know what the result should be. First, let x become really, really small. As the point in Figure 26.4 approaches the origin, the fields $\vec{E}_1$ and $\vec{E}_3$ become opposite to each other and cancel. Thus as $x \to 0$, the field should be that of the single point charge q at the origin, a field we already know. Is it? Notice that

$$\lim_{x\to 0}\frac{2x}{(x^2 + d^2)^{3/2}} = 0 \tag{26.4}$$

Thus $E_{net} \to q/4\pi\epsilon_0 x^2$ as $x \to 0$, the expected field of a single point charge.

Now consider the opposite situation, where x becomes extremely large. From very far away, the three source charges will seem to merge into a single charge of size $3q$, just as three very distant light bulbs appear to be a single light. Thus the field for $x \gg d$ *should* be that of a point charge $3q$. Is it?

The field is zero in the limit $x \to \infty$. That doesn't tell us much, so we don't want to go *that* far away. We simply want x to be very large in comparison to the spacing d between the source charges. If $x \gg d$, then the denominator of the second term of $\vec{E}_{\text{net}}$ is well approximated by $(x^2 + d^2)^{3/2} \approx (x^2)^{3/2} = x^3$. Thus

$$\lim_{x \gg d} \left[\frac{1}{x^2} + \frac{2x}{(x^2 + d^2)^{3/2}} \right] = \frac{1}{x^2} + \frac{2x}{x^3} = \frac{3}{x^2} \qquad (26.5)$$

Consequently, the net electric field far from the source charges is

$$\vec{E}_{\text{net}}(x \gg d) = \frac{1}{4\pi\epsilon_0} \frac{(3q)}{x^2} \hat{\imath} \qquad (26.6)$$

As expected, this is the field of a point charge $3q$. These checks of limiting cases provide confidence in the result of the calculation.

It's often useful to represent the electric field strength as a graph. This is especially true when the algebraic expression is complicated. Figure 26.5 is a graph of E_{net} for the three charges of Example 26.1. Although we don't have any numerical values, we can specify x as a multiple of the charge separation d. Notice how the graph matches the field of a single point charge when $x \ll d$ and matches the field of a point charge $3q$ when $x \gg d$.

The Electric Field of a Dipole

Two opposite charges $\pm q$ separated by a small distance s form an *electric dipole*. You learned in Chapter 25 that an electric dipole is created when a neutral atom is polarized by an external charge. That was an example of an *induced electric dipole* because the polarization was caused, or induced, by the electric field of the external charge. There are also *permanent electric dipoles* in which two oppositely charged particles maintain a small permanent separation. Some molecules, such as the water molecule, are permanent dipoles due to the way in which the electrons are distributed between the atoms. Figure 26.6 shows two examples of electric dipoles.

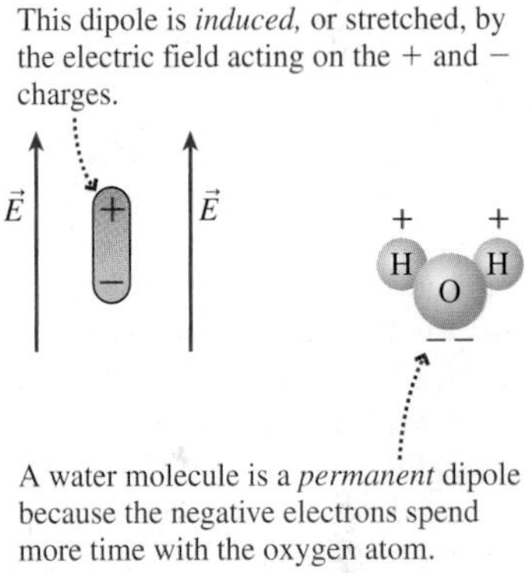

Although a dipole has zero net charge, it *does* have a net electric field. Figure 26.7 shows an electric dipole aligned along the y-axis. It makes no difference whether this dipole is induced or permanent. You can see that $\vec{E}_{\text{dipole}}$ is in the negative y-direction at points along the x-axis because the x-components of $\vec{E}_+$ and $\vec{E}_-$ cancel. $\vec{E}_{\text{dipole}}$ is in the positive y-direction at points along the y-axis because E_+ is slightly larger than E_-.

Let's calculate the electric field of a dipole at a point on the axis of the dipole. This is the y-axis in Figure 26.7. The point is distance $r_+ = y - s/2$ from the positive charge and $r_- = y + s/2$ from the negative charge. The net electric field at this point has only a y-component, and the superposition of the fields of the two point charges gives

$$(E_{\text{dipole}})_y = (E_+)_y + (E_-)_y = \frac{1}{4\pi\epsilon_0} \frac{q}{(y - \frac{1}{2}s)^2} + \frac{1}{4\pi\epsilon_0} \frac{(-q)}{(y + \frac{1}{2}s)^2}$$

$$= \frac{q}{4\pi\epsilon_0} \left[\frac{1}{(y - \frac{1}{2}s)^2} - \frac{1}{(y + \frac{1}{2}s)^2} \right] \qquad (26.7)$$

If we combine the two terms over a common denominator, Equation 26.7 becomes

$$(E_{\text{dipole}})_y = \frac{q}{4\pi\epsilon_0} \left[\frac{2ys}{(y - \frac{1}{2}s)^2(y + \frac{1}{2}s)^2} \right] \qquad (26.8)$$

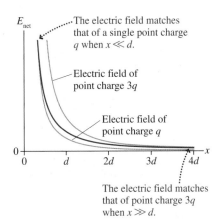

FIGURE 26.5 The electric field strength along a line perpendicular to three equal point charges.

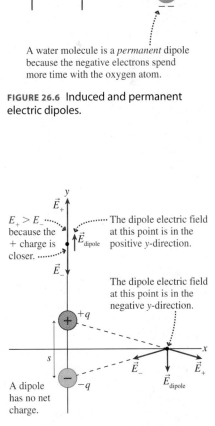

FIGURE 26.6 Induced and permanent electric dipoles.

FIGURE 26.7 The dipole electric field at two points.

We omitted some of the algebraic steps, but be sure you can do this yourself. Some of the homework problems will require similar algebra.

In practice, we almost always observe the electric field of a dipole only for distances $y \gg s$; that is, only for distances much larger than the charge separation. In such cases, the denominator can be approximated $(y - \frac{1}{2}s)^2(y + \frac{1}{2}s)^2 \approx y^4$. With this approximation, Equation 26. 8 becomes

$$(E_{\text{dipole}})_y \approx \frac{1}{4\pi\epsilon_0} \frac{2qs}{y^3} \tag{26.9}$$

It is useful to define the **dipole moment** $\vec{p}$, shown in Figure 26.8, as the vector

$$\vec{p} = (qs, \text{ from the negative to the positive charge}) \tag{26.10}$$

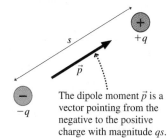

The dipole moment $\vec{p}$ is a vector pointing from the negative to the positive charge with magnitude qs.

FIGURE 26.8 The dipole moment.

The direction of $\vec{p}$ identifies the orientation of the dipole, and the dipole-moment magnitude $p = qs$ determines the electric field strength. The SI units of the dipole moment are C m.

We can use the dipole moment to write a succinct expression for the electric field at a point on the axis of a dipole:

$$\vec{E}_{\text{dipole}} \approx \frac{1}{4\pi\epsilon_0} \frac{2\vec{p}}{r^3} \quad \text{(on the axis of an electric dipole)} \tag{26.11}$$

where r is the distance measured from the *center* of the dipole. We've switched from y to r because we've now specified that Equation 26.11 is valid only along the axis of the dipole. Notice that the electric field along the axis points in the direction of the dipole moment $\vec{p}$.

A homework problem will let you calculate the electric field in the plane that bisects and is perpendicular to the dipole. This is the field shown on the x-axis in Figure 26.7, but it could equally well be the field on the z-axis as it comes out of the page. The field, for $r \gg s$, is

$$\vec{E}_{\text{dipole}} \approx -\frac{1}{4\pi\epsilon_0} \frac{\vec{p}}{r^3} \tag{26.12}$$

(in the plane perpendicular to an electric dipole)

This field is *opposite* to $\vec{p}$, and it is only half the strength of the on-axis field at the same distance.

NOTE ▶ Do these inverse-cube equations violate Coulomb's law? Not at all. Coulomb's law describes the force between two *point charges,* and from Coulomb's law we found that the electric field of a *point charge* varies with the inverse square of the distance. But a dipole is not a point charge. The field of a dipole decreases more rapidly than that of a point charge, which is to be expected because the dipole is, after all, electrically neutral. ◀

EXAMPLE 26.2 The electric field of a water molecule
The water molecule H_2O has a permanent dipole moment of magnitude 6.2×10^{-30} C m. What is the electric field strength 1.0 nm from a water molecule at a point on the axis of the dipole?

MODEL The size of a molecule is ≈ 0.1 nm. Thus $r \gg s$, and we can use Equation 26.11 for the on-axis electric field of the molecule's dipole moment.

SOLVE The electric field strength at $r = 1.0$ nm on the dipole axis is

$$E \approx \frac{1}{4\pi\epsilon_0} \frac{2p}{r^3} = (9.0 \times 10^9 \, \text{N m}^2/\text{C}^2) \frac{2(6.2 \times 10^{-30} \, \text{C m})}{(1.0 \times 10^{-9} \, \text{m})^3}$$

$$= 1.1 \times 10^8 \, \text{N/C}$$

ASSESS By referring to Table 26.1 you can see that the field strength is "strong" compared to our everyday experience with charged objects but "weak" compared to the electric field inside the atoms themselves. This seems reasonable.

Picturing the Electric Field

We can't see the electric field. Consequently, we need pictorial tools to help us visualize it in a region of space. One method, which is introduced in Chapter 25, is to picture the electric field by drawing electric field vectors at various points in space. Another way to picture the field is to draw **electric field lines.** These are imaginary lines drawn through a region of space so that

- The tangent to a field line at any point is in the direction of the electric field $\vec{E}$ at that point, and
- The field lines are closer together where the electric field strength is larger.

Figure 26.9 shows the relationship between electric field lines and electric field vectors in one region of space.

Electric field lines have two other important properties:

- Every field line starts on a positive charge and ends on a negative charge.
- Field lines cannot cross.

The first of these properties follows from the fact that the electric field is created by charges. The second property is required to make sure that $\vec{E}$ has a unique direction at every point in space.

Figure 26.10a represents the electric field of a dipole as a field-vector diagram. Figure 26.10b shows the same field using electric field lines. Notice how the on-axis field points in the direction of $\vec{p}$, both above and below the dipole, while the field in the bisecting plane points opposite to $\vec{p}$. At most points, however, $\vec{E}$ has components both parallel to $\vec{p}$ and perpendicular to $\vec{p}$.

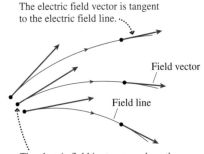

The electric field vector is tangent to the electric field line.

Field vector

Field line

The electric field is stronger where the electric field vectors are longer and where the electric field lines are closer together.

FIGURE 26.9 Electric field lines and electric field vectors.

Activ Physics 11.5, 11.6

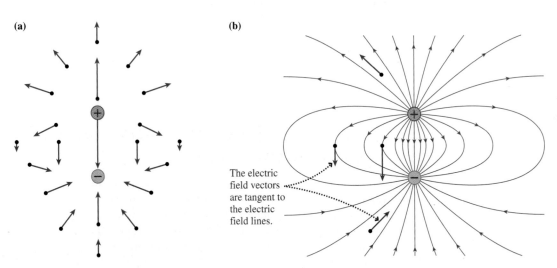

The electric field vectors are tangent to the electric field lines.

FIGURE 26.10 The electric field of a dipole.

A somewhat similar electric field is produced by two same-sign charges, as Figure 26.11 on the next page shows, using two positive charges. Figure 26.11a represents the field as vectors, Figure 26.11b represents it as electric field lines.

Neither field-vector diagrams nor field-line diagrams are perfect pictorial representations of an electric field. The field vectors are somewhat harder to draw, and they show the field at only a few points, but they do clearly indicate the direction and strength of the electric field at those points. Field-line diagrams perhaps look more elegant, and they're sometimes easier to sketch, but there's no formula for knowing where to draw the lines and it's harder to infer the actual direction and strength of the electric field.

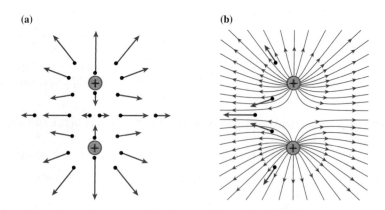

(a)   **(b)**

FIGURE 26.11 The electric field of two equal positive charges.

There simply is no pictorial way to show *exactly* what the field is. Only the mathematical representation is exact. We'll use both field-vector diagrams and field-line diagrams, depending upon the circumstances, but you'll see that the preference of this text is usually to use a field-vector diagram.

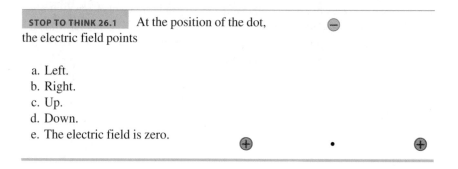

STOP TO THINK 26.1 At the position of the dot, the electric field points

a. Left.
b. Right.
c. Up.
d. Down.
e. The electric field is zero.

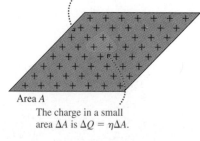

(a) Charge Q on a rod of length L. The linear charge density is $\lambda = Q/L$.

L

ΔL

The charge in a small length ΔL is $\Delta Q = \lambda \Delta L$.

(b) Charge Q on a surface of area A. The surface charge density is $\eta = Q/A$.

Area A

The charge in a small area ΔA is $\Delta Q = \eta \Delta A$.

FIGURE 26.12 One-dimensional and two-dimensional continuous charge distributions.

26.3 The Electric Field of a Continuous Charge Distribution

Ordinary objects—tables, chairs, beakers of water, flowing streams—seem to our senses to be continuous distributions of matter. There is no obvious evidence for an atomic structure, even though we have good reasons to believe that we would find atoms if we subdivided the matter sufficiently far. Thus it is easier, for many practical purposes, to consider matter to be continuous and to talk about the *density* of matter. Density—the number of kilograms of matter per cubic meter—allows us to describe the *distribution* of matter *as if* the matter were continuous rather than atomic.

Much the same situation occurs with charge. If a charged object contains a large number of excess electrons—for example, 10^{12} extra electrons on a metal rod—it is not practical to keep track of every electron. It makes more sense to consider the charge to be *continuous* and to describe how it is *distributed* over the object.

Figure 26.12a shows an object of length L, such as a plastic rod or a metal wire, with charge Q spread uniformly along it. (We will use an uppercase Q for the total charge of an object, reserving lowercase q for individual point charges.) The **linear charge density** λ is defined to be

$$\lambda = \frac{Q}{L} \tag{26.13}$$

Linear charge density, which has units of C/m, is the amount of charge *per meter* of length. The linear charge density of a 20-cm-long wire with 40 nC of charge is 2.0 nC/cm or 2.0×10^{-7} C/m.

NOTE ▶ The linear charge density λ is analogous to the linear mass density μ that you used in Chapter 20 to find the speed of a wave on a string. ◀

We'll also be interested in charged surfaces. Figure 26.12b shows a two-dimensional distribution of charge across a surface of area A. We define the **surface charge density** η (lowercase Greek eta) to be

$$\eta = \frac{Q}{A} \tag{26.14}$$

Surface charge density, with units of C/m^2, measures the amount of charge *per square meter.* A 1.0 mm $\times$ 1.0 mm square on a surface with $\eta = 2.0 \times 10^{-4}$ C/m^2 contains 2.0×10^{-10} C or 0.20 nC of charge.

Figure 26.12 and the definitions of Equations 26.13 and 26.14 assume that the object is **uniformly charged,** meaning that the charges are evenly spread over the object. We will assume objects to be uniformly charged unless noted otherwise.

NOTE ▶ Some textbooks represent the surface charge density with the symbol σ. Because σ is also used to represent *conductivity,* an idea we'll introduce in Chapter 28, we've selected a different symbol for surface charge density. ◀

STOP TO THINK 26.2 A piece of plastic is uniformly charged with surface charge density η_1. The plastic is then broken into a large piece with surface charge density η_2 and a small piece with surface charge density η_3. Rank in order, from largest to smallest, the surface charge densities η_1 to η_3.

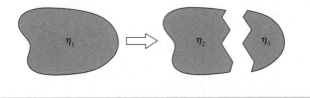

A Problem-Solving Strategy

Our goal is to find the electric field of a continuous distribution of charge, such as a charged rod or a charged disk. We have two basic tools to work with:

- The electric field of a point charge, and
- The principle of superposition.

We can apply these tools to a continuous distribution of charge if we follow a three-part strategy:

1. Divide the total charge Q into many small point-like charges ΔQ.
2. Use our knowledge of the electric field of a point charge to find the electric field of each ΔQ.
3. Calculate the net field $\vec{E}_{net}$ by summing the fields of all the ΔQ.

In practice, as you may have guessed, we'll let the sum become an integral.

The difficulty with electric field calculations is not the summation or integration itself, which is the last step, but in setting up the calculation and knowing *what* to integrate. We will go step-by-step through several examples to illustrate the procedures. However, we first need to flesh out the steps of the problem-solving strategy. The aim of this problem-solving strategy is to break a difficult problem down into small steps that are individually manageable.

(MP) **PROBLEM-SOLVING STRATEGY 26.2** **The electric field of a continuous distribution of charge**

MODEL Model the distribution as a simple shape, such as a line of charge or a disk of charge. Assume the charge is uniformly distributed.

VISUALIZE For the pictorial representation:

❶ Draw a picture and establish a coordinate system.
❷ Identify the point P at which you want to calculate the electric field.
❸ Divide the total charge Q into small pieces of charge ΔQ, using shapes for which you *already know* how to determine $\vec{E}$. This is often, but not always, a division into point charges.
❹ Draw the electric field vector at P for one or two small pieces of charge. This will help you identify distances and angles that need to be calculated.
❺ Look for symmetries of the charge distribution that simplify the field. You may conclude that some components of $\vec{E}$ are zero.

SOLVE The mathematical representation is $\vec{E}_{net} = \sum \vec{E}_i$.

■ Use superposition to form an algebraic expression for *each* of the three components of $\vec{E}$ (unless you are sure one or more is zero) at point P.
■ Let the (x, y, z) coordinates of the point remain as variables.
■ Replace the small charge ΔQ with an equivalent expression involving a charge density and a coordinate, such as dx, that describes the shape of charge ΔQ. **This is the critical step in making the transition from a sum to an integral** because you need a coordinate to serve as the integration variable.
■ All angles and distances must be expressed in terms of the coordinates.
■ Let the sum become an integral. The integration will be over the coordinate variable that is related to ΔQ. The integration limits for this variable will depend on your choice of coordinate system. Carry out the integration and simplify the result as much as possible.

ASSESS Check that your result is consistent with any limits for which you know what the field should be.

EXAMPLE 26.3 The electric field of a line of charge

Figure 26.13 shows a thin, uniformly charged rod of length L with total charge Q that can be either positive or negative. Find the electric field strength at distance r in the plane that bisects the rod.

MODEL The rod is thin, so we'll assume the charge lies along a line and forms what we call a *line of charge*. This is an important charge distribution that models the electric field of a charged rod or a charged metal wire. The rod's linear charge density is $\lambda = Q/L$.

VISUALIZE Figure 26.14 illustrates the five steps of the problem-solving strategy. We've chosen a coordinate system in which the rod lies along the y-axis and point P, in the bisecting plane, is on the x-axis. We've then divided the rod into N small segments of charge ΔQ, each of which can be modeled as a point charge. For

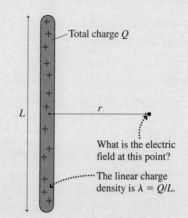

FIGURE 26.13 A thin, uniformly charged rod.

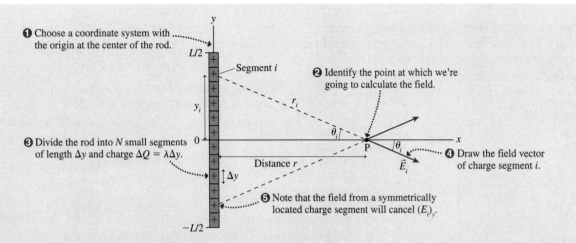

① Choose a coordinate system with the origin at the center of the rod.

② Identify the point at which we're going to calculate the field.

③ Divide the rod into N small segments of length Δy and charge $\Delta Q = \lambda \Delta y$.

④ Draw the field vector of charge segment i.

⑤ Note that the field from a symmetrically located charge segment will cancel $(E_i)_y$.

FIGURE 26.14 Calculating the electric field of a line of charge.

every ΔQ in the bottom half of the wire with a field that points to the right and up, there's a matching ΔQ in the top half whose field points to the right and down. The y-components of these two fields cancel, hence the net electric field on the x-axis points straight away from the rod. The only component we need to calculate is E_x. (This is the same reasoning on the basis of symmetry that we used in Example 26.1.)

SOLVE Each of the little segments of charge can be modeled as a point charge. We know the electric field of a point charge, so we can write the x-component of $\vec{E}_i$, the electric field of segment i, as

$$(E_i)_x = E_i \cos\theta_i = \frac{1}{4\pi\epsilon_0} \frac{\Delta Q}{r_i^2} \cos\theta_i$$

where r_i is the distance from charge i to point P. You can see from the figure that $r_i = (y_i^2 + r^2)^{1/2}$ and $\cos\theta_i = r/r_i = r/(y_i^2 + r^2)^{1/2}$. With these, $(E_i)_x$ is

$$(E_i)_x = \frac{1}{4\pi\epsilon_0} \frac{\Delta Q}{y_i^2 + r^2} \frac{r}{\sqrt{y_i^2 + r^2}}$$
$$= \frac{1}{4\pi\epsilon_0} \frac{r\Delta Q}{(y_i^2 + r^2)^{3/2}}$$

Compare this result to the very similar calculation we did in Example 26.1. If we now sum this expression over all the charge segments, the net x-component of the electric field is

$$E_x = \sum_{i=1}^{N} (E_i)_x = \frac{1}{4\pi\epsilon_0} \sum_{i=1}^{N} \frac{r\Delta Q}{(y_i^2 + r^2)^{3/2}}$$

This is the same superposition we did for the $N = 3$ case in Example 26.1. The only difference is that we have now written the result as an explicit summation so that N can have any value. We want to let $N \rightarrow \infty$ and to replace the sum with an integral, but we can't integrate over Q; it's not a geometric quantity. This is where the linear charge density enters. The quantity of charge in each segment is related to its length Δy by $\Delta Q = \lambda \Delta y = (Q/L)\Delta y$. In terms of the linear charge density, the electric field is

$$E_x = \frac{Q/L}{4\pi\epsilon_0} \sum_{i=1}^{N} \frac{r\Delta y}{(y_i^2 + r^2)^{3/2}}$$

Now we're ready to let the sum become an integral. If we let $N \rightarrow \infty$, then each segment becomes an infinitesimal length $\Delta y \rightarrow dy$ while the discrete position variable y_i becomes the continuous integration variable y. The sum from $i = 1$ at the bottom end of the line of charge to $i = N$ at the top end will be replaced with an integral from $y = -L/2$ to $y = +L/2$. Thus in the limit $N \rightarrow \infty$,

$$E_x = \frac{Q/L}{4\pi\epsilon_0} \int_{-L/2}^{L/2} \frac{r\,dy}{(y^2 + r^2)^{3/2}}$$

This is a standard integral that you have learned to do in calculus and that can be found in any table of integrals. Note that r is a *constant* as far as this integral is concerned. Integrating gives

$$E_x = \frac{Q/L}{4\pi\epsilon_0} \frac{y}{r\sqrt{y^2 + r^2}} \Big|_{-L/2}^{L/2}$$
$$= \frac{Q/L}{4\pi\epsilon_0} \left[\frac{L/2}{r\sqrt{(L/2)^2 + r^2}} - \frac{-L/2}{r\sqrt{(-L/2)^2 + r^2}} \right]$$
$$= \frac{1}{4\pi\epsilon_0} \frac{Q}{r\sqrt{r^2 + (L/2)^2}}$$

Because E_x is the *only* component of the field, the electric field strength E_{rod} at distance r from the center of a charged rod is

$$E_{\text{rod}} = \frac{1}{4\pi\epsilon_0} \frac{|Q|}{r\sqrt{r^2 + (L/2)^2}}$$

The field strength must be positive, so we added absolute value signs to Q to allow for the possibility that the charge could be negative. The only restriction is to remember that this is the electric field at a point in the plane that bisects the rod.

ASSESS Suppose we are at a point *very* far from the rod. If $r \gg L$, the length of the rod is not relevant and the rod appears to be a point charge Q in the distance. Thus in the *limiting case* $r \gg L$, we expect the rod's electric field to be that of a point charge. If $r \gg L$, the square root becomes $(r^2 + (L/2)^2)^{1/2} \approx (r^2)^{1/2} = r$ and the electric field strength at distance r becomes $E_{\text{rod}} \approx Q/4\pi\epsilon_0 r^2$. The fact that our expression of E_{rod} has the correct limiting behavior gives us confidence that we haven't made any mistakes in its derivation.

EXAMPLE 26.4 The electric field of a charged rod

What is the electric field strength 1.0 cm from the middle of an 8.0-cm-long glass rod that has been charged to 10 nC?

SOLVE Example 26.3 found that the electric field strength in the plane that bisects a charged rod is

$$E_{rod} = \frac{1}{4\pi\epsilon_0} \frac{|Q|}{r\sqrt{r^2 + (L/2)^2}}$$

Using $L = 0.080$ m, $r = 0.010$ m, and $Q = 1.0 \times 10^{-8}$ C, we can calculate

$$E_{rod} = 2.18 \times 10^5 \text{ N/C}.$$

For comparison, the field 1.0 cm from a 10 nC *point* charge would be a somewhat larger 9.0×10^5 N/C.

ASSESS This result is consistent with the values in Table 26.1.

An Infinite Line of Charge

What happens if the rod or wire becomes very long while the linear charge density λ remains constant? That is, more charge is added so that the ratio $\lambda = |Q|/L$ stays constant as L increases. In the limit that L approaches infinity, the electric field strength becomes

$$E_{line} = \lim_{L\to\infty} \frac{1}{4\pi\epsilon_0} \frac{|Q|}{r\sqrt{r^2 + (L/2)^2}} = \frac{1}{4\pi\epsilon_0} \frac{|Q|}{rL/2} = \frac{1}{4\pi\epsilon_0} \frac{2|\lambda|}{r} \qquad (26.15)$$

This is the field strength of an infinitely long **line of charge** having linear charge density λ. The linear charge density measures how close together or far apart the charges are on the rod, and this distance did not change as the rod's length L was increased.

NOTE ▶ Unlike a point charge, for which the field decreases as $1/r^2$, the field of an infinitely long charged wire decreases more slowly—as only $1/r$. ◀

Equation 26.15 is of considerable practical significance. Although no real wire is infinitely long, the fact that the field of a point charge decreases inversely with the square of the distance means that the electric field at a point near the wire is determined primarily by the nearest charges on the wire. Over most of the length of a wire, the ends of the wire are too far away to make any significant contribution. Consequently, the field of a realistic finite-length wire is well approximated by Equation 26.15, the field of an infinitely long line of charge, except at points near the end of wire.

Figure 26.15 shows the electric field vectors of an infinite line of positive charge. Notice that the field points straight away from the line at all points and the field strength decreases as the distance increases. The vectors would point inward for a negative line of charge.

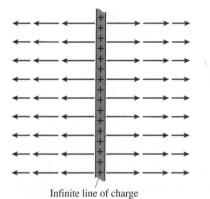

Infinite line of charge

FIGURE 26.15 The electric field of an infinite line of charge.

STOP TO THINK 26.3   Which of the following actions will increase the electric field strength at the position of the dot?

a. Make the rod longer without changing the charge.
b. Make the rod shorter without changing the charge.
c. Make the rod wider without changing the charge.
d. Make the rod narrower without changing the charge.
e. Add charge to the rod.
f. Remove charge from the rod.
g. Move the dot farther from the rod.
h. Move the dot closer to the rod.

Charged rod

26.4 The Electric Fields of Rings, Planes, and Spheres

In this section we'll derive the electric fields for three important and related charge distributions: a ring of charge, a disk of charge, and a plane of charge. The ring of charge is the most fundamental and will be the basis for determining the other two. We'll also look at the electric field of a sphere of charge.

EXAMPLE 26.5 The electric field of a ring of charge
Figure 26.16 shows a thin, uniformly charged ring of radius R with total charge Q. Find the electric field at a point on the axis of the ring (perpendicular to the page).

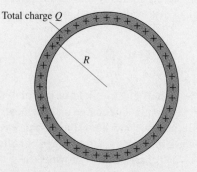

Total charge Q

R

FIGURE 26.16 A thin, uniformly charged ring.

MODEL Because the ring is thin, we'll assume the charge lies along circle of radius R. You can think of this as a line of charge of length $2\pi R$ wrapped into a circle. The linear charge density along the ring is $\lambda = Q/2\pi R$.

VISUALIZE Figure 26.17 illustrates the five steps of the problem-solving strategy. We've chosen a coordinate system in which

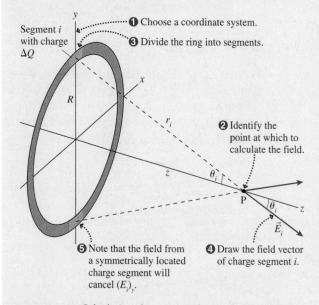

❶ Choose a coordinate system.
❸ Divide the ring into segments.
Segment i with charge ΔQ
y
x
R
r_i
z
θ_i
P
θ_i
z
$\vec{E}_i$
❷ Identify the point at which to calculate the field.
❹ Draw the field vector of charge segment i.
❺ Note that the field from a symmetrically located charge segment will cancel $(E_i)_y$.

FIGURE 26.17 Calculating the on-axis electric field of a ring of charge.

the ring lies in the xy-plane and point P is on the z-axis. We've then divided the ring into N small segments of charge ΔQ, each of which can be modeled as a point charge. For every pair of segments that are diametrically opposed, such as segments i and j in the figure, the components of their fields perpendicular to the axis will cancel. The symmetry of the ring tells us that the field at a point on the axis will point along the z-axis; outward for a positive ring, inward for a negative ring. Thus we need to calculate only the z-component E_z.

SOLVE The z-component of the electric field due to segment i is

$$(E_i)_z = E_i \cos\theta_i = \frac{1}{4\pi\epsilon_0}\frac{\Delta Q}{r_i^2}\cos\theta_i$$

You can see from the figure that *every* segment of the ring, independent of i, has

$$r_i = \sqrt{z^2 + R^2}$$
$$\cos\theta_i = \frac{z}{r_i} = \frac{z}{\sqrt{z^2 + R^2}}$$

Consequently, the field of segment i is

$$(E_i)_z = \frac{1}{4\pi\epsilon_0}\frac{\Delta Q}{z^2+R^2}\frac{z}{\sqrt{z^2+R^2}} = \frac{1}{4\pi\epsilon_0}\frac{z}{(z^2+R^2)^{3/2}}\Delta Q$$

The net electric field is found by summing $(E_i)_z$ due to all N segments:

$$E_z = \sum_{i=1}^{N}(E_i)_z = \frac{1}{4\pi\epsilon_0}\frac{z}{(z^2+R^2)^{3/2}}\sum_{i=1}^{N}\Delta Q$$

We were able to bring all terms involving z to the front because z is a constant as far as the summation is concerned. Surprisingly, we don't need to convert the sum to an integral to complete this calculation. The sum of all the ΔQ around the ring is simply the ring's total charge, $\sum\Delta Q = Q$, hence the electric field on the axis of a charged ring is

$$(E_{ring})_z = \frac{1}{4\pi\epsilon_0}\frac{zQ}{(z^2+R^2)^{3/2}}$$

This expression is valid for both positive and negative z (i.e., on either side of the ring) and for both positive and negative charge.

ASSESS It will be left as a homework problem to show that this result gives the expected limit when $z \gg R$.

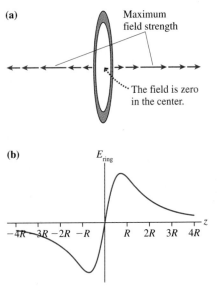

(a)

Maximum field strength

The field is zero in the center.

(b)

E_{ring}

$-4R$ $-3R$ $-2R$ $-R$ R $2R$ $3R$ $4R$ z

FIGURE 26.18 The on-axis electric field of a ring of charge.

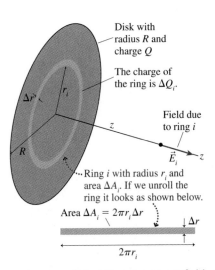

Disk with radius R and charge Q

The charge of the ring is ΔQ_i.

Field due to ring i

$\vec{E}_i$ z

···Ring i with radius r_i and area ΔA_i. If we unroll the ring it looks as shown below.

Area $\Delta A_i = 2\pi r_i \Delta r$

$\downarrow \Delta r$
$\uparrow$

$2\pi r_i$

FIGURE 26.19 Calculating the on-axis field of a charged disk.

Figure 26.18 shows two representations of the on-axis electric field of a positively charged ring. Figure 26.18a shows that the electric field vectors point away from the ring, increasing in length until reaching a maximum when $|z| \approx R$, then decreasing. The graph of $(E_{\text{ring}})_z$ in Figure 26.18b confirms that the field strength has a maximum on either side of the ring. Notice that the electric field at the center of the ring is zero, even though this point is surrounded by charge. You might want to spend a minute thinking about why this has to be the case.

A Disk of Charge

Our goal is to find the electric field of an infinitely wide charged plane because it is one of our basic electric field models. Most of the work is now done. We'll first use the result for a ring of charge to find the on-axis electric field of a *disk* of charge. Then we'll let the disk expand until it becomes a plane of charge.

Figure 26.19 shows a disk having radius R that is uniformly charged with charge Q. This is a mathematical disk, with no thickness, and its surface charge density is

$$\eta = \frac{Q}{A} = \frac{Q}{\pi R^2} \tag{26.16}$$

We would like to calculate the on-axis electric field of this disk. Our problem-solving strategy tells us to divide a continuous charge into segments for which we already know how to find $\vec{E}$. Because we now know the on-axis electric field of a ring of charge, let's divide the disk into N very narrow rings of radius r and width Δr. One such ring, with radius r_i and charge ΔQ_i, is shown.

We need to be careful with notation. The R in Example 26.5 was the radius of the ring. Now we have many rings, and the radius of ring i is r_i. Similarly, Q was the charge on the ring. Now the charge on ring i is ΔQ_i, a small fraction of the total charge on the disk. With these changes, the electric field of ring i, with radius r_i, is

$$(E_i)_z = \frac{1}{4\pi\epsilon_0} \frac{z\Delta Q_i}{(z^2 + r_i^2)^{3/2}} \tag{26.17}$$

The on-axis electric field of the charged disk is the sum of the electric fields of all of the rings:

$$(E_{\text{disk}})_z = \sum_{i=1}^{N}(E_i)_z = \frac{z}{4\pi\epsilon_0}\sum_{i=1}^{N}\frac{\Delta Q_i}{(z^2 + r_i^2)^{3/2}} \tag{26.18}$$

The critical step, as always, is to relate ΔQ to a coordinate. Because we now have a surface, rather than a line, the charge in ring i is $\Delta Q = \eta\Delta A_i$, where ΔA_i is the area of ring i. We can find ΔA_i, as you've learned to do in calculus, by "unrolling" the ring to form a narrow rectangle of length $2\pi r_i$ and height Δr. Thus the area of ring i is $\Delta A_i = 2\pi r_i \Delta r$ and the charge is $\Delta Q_i = 2\pi \eta r_i \Delta r$. With this substitution, Equation 26.18 becomes

$$(E_{\text{disk}})_z = \frac{\eta z}{2\epsilon_0}\sum_{i=1}^{N}\frac{r_i \Delta r}{(z^2 + r_i^2)^{3/2}} \tag{26.19}$$

As $N \to \infty$, $\Delta r \to dr$ and the sum becomes an integral. Adding all the rings means integrating from $r = 0$ to $r = R$, thus

$$(E_{\text{disk}})_z = \frac{\eta z}{2\epsilon_0}\int_0^R \frac{r\,dr}{(z^2 + r^2)^{3/2}} \tag{26.20}$$

All that remains is to carry out the integration. This is straightforward if we make the variable change $u = z^2 + r^2$. Then $du = 2r\,dr$ or, equivalently, $r\,dr = \frac{1}{2}du$. At the lower integration limit $r = 0$, our new variable is $u = z^2$. At the upper limit $r = R$, the new variable is $u = z^2 + R^2$.

NOTE ▶ When changing variables in a definite integral, you *must* also change the limits of integration. ◀

With this variable change the integral becomes

$$(E_{disk})_z = \frac{\eta z}{2\epsilon_0} \frac{1}{2} \int_{z^2}^{z^2+R^2} \frac{du}{u^{3/2}} = \frac{\eta z}{4\epsilon_0} \frac{-2}{u^{1/2}} \Big|_{z^2}^{z^2+R^2} = \frac{\eta z}{2\epsilon_0} \left[\frac{1}{z} - \frac{1}{\sqrt{z^2 + R^2}} \right] \quad (26.21)$$

If we multiply through by z, the on-axis electric field of a charged disk with $\eta = Q/\pi R^2$ is

$$(E_{disk})_z = \frac{\eta}{2\epsilon_0} \left[1 - \frac{z}{\sqrt{z^2 + R^2}} \right] \quad (26.22)$$

It's a bit difficult see what Equation 26.22 is telling us, so let's compare it to what we already know. First, you can see that the quantity in square brackets is dimensionless. The surface charge density $\eta = Q/A$ has the same units as q/r^2, so $\eta/2\epsilon_0$ has the same units as $q/4\pi\epsilon_0 r^2$. This tells us that $\eta/2\epsilon_0$ really is an electric field.

Next, let's move very far away from the disk. At distance $z \gg R$, the disk appears to be a point charge Q in the distance and the field of the disk should approach that of a point charge. If we let $z \to \infty$ in Equation 26.22, so that $z^2 + R^2 \approx z^2$, we find $(E_{disk})_z \to 0$. This is true, but not quite what we wanted. We need to let z be very large in comparison to R, but not so large as to make E_{disk} vanish. That requires a little more care in taking the limit.

We can cast Equation 26.22 into a somewhat more useful form by factoring the z^2 out of the square root to give

$$(E_{disk})_z = \frac{\eta}{2\epsilon_0} \left[1 - \frac{1}{\sqrt{1 + R^2/z^2}} \right] \quad (26.23)$$

Now $R^2/z^2 \ll 1$ if $z \gg R$, so the second term in the square brackets is of the form $(1 + x)^{-1/2}$ where $x \ll 1$. We can then use the *binomial approximation*

$$(1 + x)^n \approx 1 + nx \quad \text{if} \quad x \ll 1 \quad \text{(binomial approximation)}$$

to simplify the expression in square brackets:

$$1 - \frac{1}{\sqrt{1 + R^2/z^2}} = 1 - (1 + R^2/z^2)^{-1/2} \approx 1 - \left(1 + \left(-\frac{1}{2} \right) \frac{R^2}{z^2} \right) = \frac{R^2}{2z^2} \quad (26.24)$$

This is a good approximation when $z \gg R$. Substituting this approximation into Equation 26.23, we find that the electric field of the disk for $z \gg R$ is

$$(E_{disk})_z \approx \frac{\eta}{2\epsilon_0} \frac{R^2}{2z^2} = \frac{Q/\pi R^2}{4\epsilon_0} \frac{R^2}{z^2} = \frac{1}{4\pi\epsilon_0} \frac{Q}{z^2} \quad \text{if} \quad z \gg R \quad (26.25)$$

This is, indeed, the field of a point charge Q, giving us confidence in Equation 26.22 for the on-axis electric field of a disk of charge.

NOTE ▶ The binomial approximation is an important tool for looking at the limiting cases of electric fields. ◀

EXAMPLE 26.6 The electric field of a charged disk
A 10-cm-diameter plastic disk is charged uniformly with an extra 10^{11} electrons. What is the electric field 1.0 mm above the surface at a point near the center?

MODEL Model the plastic disk as a uniformly charged disk. We are seeking the on-axis electric field. Because the charge is negative, the field will point *toward* the disk.

SOLVE The total charge on the plastic square is $Q = N(-e) = -1.60 \times 10^{-8}$ C. The surface charge density is

$$\eta = \frac{Q}{A} = \frac{Q}{\pi R^2} = \frac{-1.60 \times 10^{-8} \text{ C}}{\pi (0.050 \text{ m})^2} = -2.04 \times 10^{-6} \text{ C/m}^2$$

The electric field at $z = 0.0010$ m, given by Equation 26.22, is

$$E_z = \frac{\eta}{2\epsilon_0}\left[1 - \frac{1}{\sqrt{1 + R^2/z^2}}\right] = \frac{-2.04 \times 10^{-6} \text{ C/m}^2}{2 \times 8.85 \times 10^{-12} \text{ C}^2/\text{N m}^2}$$

$$\times \left[1 - \frac{1}{\sqrt{1 + (0.050 \text{ m}/0.001 \text{ m})^2}}\right]$$

$$= -1.13 \times 10^5 \text{ N/C}$$

The minus sign indicates that the field points *toward*, rather than away from, the disk. As a vector,

$$\vec{E} = (1.13 \times 10^5 \text{ N/C, toward the disk})$$

ASSESS The total charge, -16 nC, is typical of the amount of charge produced on a small plastic object by rubbing or friction. Thus 10^5 N/C is a typical electric field strength near an object that has been charged by rubbing.

Electrodes a few millimeters in size guided electrons through old-fashioned vacuum tubes. Modern field-effect transistors use an electrode, called a *gate,* that is only about 1 μm wide.

A Plane of Charge

Many electronic devices used charged, flat surfaces—disks, squares, rectangles, and so on—to steer electrons along the proper paths. These charged surfaces are called **electrodes.** Although any real electrode is finite in extent, we can often model an electrode as an infinite **plane of charge.** As long as the distance z to the plane is small in comparison to the distance to the edges, we can reasonably treat the edges *as if* they are infinitely far away.

The electric field of a plane of charge is found from the on-axis field of a charged disk by letting the radius $R \to \infty$. That is, a disk with infinite radius is an infinite plane. From Equation 26.22, we see that the electric field of a plane of charge with surface charge density η is:

$$E_{\text{plane}} = \frac{\eta}{2\epsilon_0} = \text{constant} \tag{26.26}$$

This is a simple result, but what does it tell us? First, the field strength is directly proportional to the charge density η: More charge, bigger field. Second, and more interesting, the field strength is the same at *all* points in space, independent of the distance z. The field strength 1000 m from the plane is the same as the field strength 1 mm from the plane.

How can this be? It seems that the field should get weaker as you move away from the plane of charge. But remember that we are dealing with an *infinite* plane of charge. What does it mean to be "close to" or "far from" an infinite object? For a disk of finite radius R, whether a point at distance z is "close to" or "far from" the disk is a comparison of z to R. If $z \ll R$, the point is close to the disk. If $z \gg R$, the point is far from the disk. But as $R \to \infty$, we have no *scale* for distinguishing near and far. In essence, *every* point in space is "close to" a disk of infinite radius.

No real plane is infinite in extent, but we can interpret Equation 26.26 as saying that the field of a surface of charge, regardless of its shape, is a constant $\eta/2\epsilon_0$ for those points whose distance z to the surface is much smaller than their distance to the edge. Eventually, when $z \gg R$, the charged surface will begin to look like a point charge Q and the field will have to decrease as $1/z^2$.

We do need to note that the derivation leading to Equation 26.26 considered only $z > 0$. For a positively charged plane, with $\eta > 0$, the electric field points *away from* the plane on both sides of the plane. This requires $E_z < 0$ ($\vec{E}$ pointing in the negative z-direction) on the side with $z < 0$. Thus a complete description of the electric field, valid for both sides of the plane and for either sign of η, is

$$(E_{\text{plane}})_z = \begin{cases} +\dfrac{\eta}{2\epsilon_0} & z > 0 \\[2mm] -\dfrac{\eta}{2\epsilon_0} & z < 0 \end{cases} \qquad (26.27)$$

Figure 26.20 shows two views of the electric field of a positively charged plane. All the arrows would be reversed for a negatively charged plane. Notice that the electric field on either side of the plane looks like the gravitational field near the surface of the earth. It would have been very difficult to anticipate this result from Coulomb's law or from the electric field of a single point charge, but step by step we have been able to use the concept of the electric field to look at increasingly complex distributions of charge.

The Electric Field of a Sphere of Charge

The one last charge distribution for which we need to know the electric field is a **sphere of charge.** This problem is analogous to wanting to know the gravitational field of a spherical planet or star. The procedure for calculating the field of a sphere of charge is the same as we used for lines and planes, but the integrations are significantly more difficult. We will skip the details of the calculations and, for now, simply assert the result without proof. In Chapter 27 we'll use an alternative procedure to find the field of a sphere of charge.

A sphere of charge Q and radius R, be it a uniformly charged sphere or just a spherical shell, has an electric field *outside* the sphere ($r \geq R$) that is exactly the same as that of a point charge Q located at the center of the sphere:

$$\vec{E}_{\text{sphere}} = \frac{Q}{4\pi\epsilon_0 r^2}\hat{r} \qquad \text{for } r \geq R \qquad (26.28)$$

This assertion is analogous to our earlier assertion that the gravitational force between stars and planets can be computed as if all the mass is at the center.

Figure 26.21 shows the electric field of a sphere of positive charge. The field of a negative sphere would point inward. Note that the field inside the sphere ($r < R$) is *not* given by Equation 26.28.

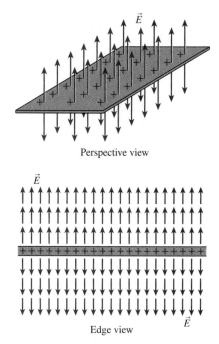

FIGURE 26.20 Perspective and edge views of the electric field of a positive plane of charge.

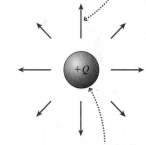

The electric field outside the sphere is the same as the field of a point charge Q at the center.

A positive sphere or spherical shell. We don't know what the electric field inside the sphere is.

FIGURE 26.21 The electric field of a sphere of positive charge.

STOP TO THINK 26.4 Rank in order, from largest to smallest, the electric field strengths E_a to E_e at these five points near a plane of charge.

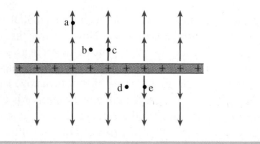

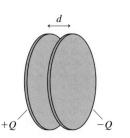

FIGURE 26.22 A parallel-plate capacitor.

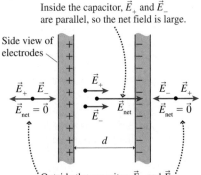

FIGURE 26.23 The electric fields inside and outside a parallel-plate capacitor.

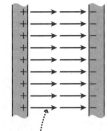

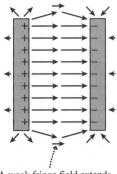

FIGURE 26.24 The electric field of a capacitor.

26.5 The Parallel-Plate Capacitor

Figure 26.22 shows two electrodes, one with charge $+Q$ and the other with $-Q$, placed face-to-face a distance d apart. This arrangement of two electrodes, charged equally but oppositely, is called a **parallel-plate capacitor.** Capacitors play important roles in many electric circuits. Our goal is to find the electric field both inside the capacitor (i.e., between the plates) and outside the capacitor.

> **NOTE** ▶ The *net* charge of a capacitor is zero. Capacitors are charged by transferring electrons from one plate to the other. The plate that gains N electrons has charge $-Q = N(-e)$. The plate that loses electrons has charge $+Q$. ◀

Let's begin with a qualitative investigation. Figure 26.23 is an enlarged view of the capacitor plates, seen from the side. Because opposite charges attract, all of the charge is on the *inner* surfaces of the two plates. Thus the inner surfaces can be modeled as *charged planes* with equal but opposite surface charge densities. The electric field $\vec{E}_+$ of the positive plate points away from the charged surface. The field $\vec{E}_-$ of the negative plate points toward the surface. The figure shows the fields both between the plates and to the left and right of the capacitor.

> **NOTE** ▶ You might think the right capacitor plate would somehow "block" the electric field created by the positive plate and prevent the presence of an $\vec{E}_+$ field to the right of the capacitor. To see that it doesn't have this effect, consider an analogous situation with gravity. The strength of gravity above a table is the same as its strength below it. Just as the table doesn't block the earth's gravitational field, intervening matter or charges do not alter or block an object's electric field. ◀

Inside the capacitor, $\vec{E}_+$ and $\vec{E}_-$ are parallel and of equal strength. Their superposition creates a net electric field inside the capacitor that points from the positive plate to the negative plate. Outside the capacitor, $\vec{E}_+$ and $\vec{E}_-$ point in opposite directions and, because the field of a plane of charge is independent of the distance from the plane, have equal magnitudes. Consequently, the fields $\vec{E}_+$ and $\vec{E}_-$ add to zero outside the capacitor plates.

We can calculate the fields between the capacitor plates from the field of an infinite charged plane. Between the electrodes, $\vec{E}_+$ is of magnitude $\eta/2\epsilon_0$ and points from the positive toward the negative side. The field $\vec{E}_-$ is *also* of magnitude $\eta/2\epsilon_0$ and *also* points from positive to negative. Thus the electric field inside the capacitor is

$$\vec{E}_{\text{capacitor}} = \vec{E}_+ + \vec{E}_- = \left(\frac{\eta}{\epsilon_0}, \text{from positive to negative}\right)$$

$$= \left(\frac{Q}{\epsilon_0 A}, \text{from positive to negative}\right)$$

(26.29)

Outside the capacitor plates, where $\vec{E}_+$ and $\vec{E}_-$ have equal magnitudes but *opposite* directions, $\vec{E} = \vec{0}$.

Figure 26.24a shows the electric field of an ideal parallel-plate capacitor constructed from two infinite charged planes. Now, it's true that no real capacitor is infinite in extent, but the ideal parallel-plate capacitor is a very good approximation for all but the most precise calculations as long as the electrode separation d is much smaller than the electrodes' size—that is, their edge length or radius. Figure 26.24b shows that the interior field of a real capacitor is virtually identical to that of an ideal capacitor but that the exterior field isn't quite zero. This weak field outside the capacitor is called the **fringe field.** We will keep things simple by always assuming the plates are very close together and using Equation 26.29 for the field inside a parallel-plate electrode.

NOTE ▶ The shape of the electrodes—circular or square or any other shape— is not relevant as long as the electrodes are very close together. ◀

Uniform Electric Fields

Figure 26.25 shows an electric field that is the *same*—in strength and direction— at every point in a region of space. This is called a **uniform electric field.** A uniform electric field is analogous to the uniform gravitational field near the surface of the earth. Uniform fields are of great practical significance because, as you will see in the next section, computing the trajectory of a charged particle moving in a uniform electric field is a straightforward process.

The easiest way to produce a uniform electric field is with a parallel-plate capacitor. Indeed, our interest in capacitors is due in large measure to the fact that the electric field is uniform. Many electric field problems refer to a uniform electric field. Such problems carry an implicit assumption that the action is taking place *inside* a parallel-plate capacitor.

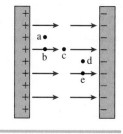

FIGURE 26.25 A uniform electric field.

EXAMPLE 26.7 The electric field inside a capacitor
Two 1.0 cm × 2.0 cm rectangular electrodes are 1.0 mm apart. What charge must be placed on each electrode to create a uniform electric field of strength 2.0×10^6 N/C? How many electrons must be moved from one electrode to the other to accomplish this?

MODEL The electrodes can be modeled as a parallel-plate capacitor because the spacing between them is much smaller than their lateral dimensions.

SOLVE The electric field strength inside the capacitor is $E = Q/\epsilon_0 A$. Thus the charge needed to produce a field of strength E is

$$Q = \epsilon_0 AE$$

$$= (8.85 \times 10^{-12} \text{ C}^2/\text{Nm}^2)(2.0 \times 10^{-4} \text{ m}^2)(2.0 \times 10^6 \text{ N/C})$$

$$= 3.5 \times 10^{-9} \text{ C} = 3.5 \text{ nC}$$

The positive plate must be charged to +3.5 nC and the negative plate to −3.5 nC. In practice, the plates are charged by using a *battery* to move electrons from one plate to the other. The number of electrons in 3.5 nC is

$$N = \frac{Q}{e} = \frac{3.5 \times 10^{-9} \text{ C}}{1.60 \times 10^{-19} \text{ C/electron}} = 2.2 \times 10^{10} \text{ electrons}$$

Thus 2.2×10^{10} electrons are moved from one electrode to the other. Note that the capacitor *as a whole* has no net charge.

ASSESS The plate spacing does not enter the result. As long as the spacing is much smaller than the plate dimensions, as is true in this example, the field is independent of the spacing.

STOP TO THINK 26.5   Rank in order, from largest to smallest, the forces F_a to F_e a proton would experience if placed at points a to e in this parallel-plate capacitor.

26.6 Motion of a Charged Particle in an Electric Field

Our motivation for introducing the concept of the electric field was to understand the long-range electric interaction of charges. We said that some charges, the *source charges,* create an electric field. Other charges then respond to that electric field. The first five sections of this chapter have focused on the electric field of the source charges. Now we turn our attention to the second half of the interaction.

Activ Physics 11.9, 11.10

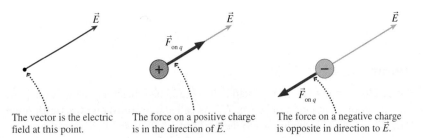

The vector is the electric field at this point.

The force on a positive charge is in the direction of $\vec{E}$.

The force on a negative charge is opposite in direction to $\vec{E}$.

FIGURE 26.26 The electric field exerts a force on a charged particle.

Figure 26.26 shows a particle of charge q and mass m at a point where an electric field $\vec{E}$ has been produced by *other* charges, the source charges. The electric field exerts a force

$$\vec{F}_{\text{on } q} = q\vec{E}$$

on the charged particle. This relationship between field and force was the *definition* of the electric field. Notice that the force on a negatively charged particle is *opposite* in direction to the electric field vector. Signs are important!

The force on the charged particle causes an acceleration

$$\vec{a} = \frac{\vec{F}_{\text{on } q}}{m} = \frac{q}{m}\vec{E} \tag{26.30}$$

This acceleration is the *response* of the charged particle to the source charges that created the electric field. The ratio q/m is especially important for the dynamics of charged-particle motion. It is called the **charge-to-mass ratio.** Two *equal* charges, say a proton and a Na$^+$ ion, will experience *equal* forces $\vec{F} = q\vec{E}$ if placed at the same point in an electric field, but their accelerations will be *different* because they have different masses and thus different charge-to-mass ratios. Two particles having different charges and masses *but* with the same charge-to-mass ratio will undergo the same acceleration and follow the same trajectory.

Motion in a Uniform Field

The motion of a charged particle in a *uniform* electric field is especially important for its basic simplicity and because of its many valuable applications. A uniform field is *constant* at all points—constant in both magnitude and direction—within the region of space where the charged particle is moving. It follows, from Equation 26.30, that **a charged particle in a uniform electric field will move with constant acceleration.** The magnitude of the acceleration is

$$a = \frac{qE}{m} = \text{constant} \tag{26.31}$$

where E is the electric field strength, and the direction of $\vec{a}$ is parallel or antiparallel to $\vec{E}$, depending on the sign of q.

Identifying the motion of a charged particle in a uniform field as being one of constant acceleration brings into play all the kinematic machinery that we developed in Chapter 2 for constant-acceleration motion. The basic trajectory of a charged particle in a uniform field is a *parabola*, analogous to the projectile motion of a mass in the near-earth uniform gravitational field. In the special case of a charged particle moving parallel to the electric field vectors, the motion is one-dimensional, analogous to the one-dimensional vertical motion of a mass tossed straight up or falling straight down.

NOTE ▶ The gravitational acceleration $\vec{a}_{\text{grav}}$ always points straight down. The electric field acceleration $\vec{a}_{\text{elec}}$ can point in *any* direction. You must determine the electric field $\vec{E}$ in order to learn the direction of $\vec{a}$. ◀

EXAMPLE 26.8 An electron moving across a capacitor
Two 6.0-cm-diameter electrodes are spaced 5.0 mm apart. They are charged by transferring 1.0×10^{11} electrons from one electrode to the other. An electron is released from rest at the surface of the negative electrode. How long does it take the electron to cross to the positive electrode? What is its speed as it collides with the positive electrode? Assume the space between the electrodes is a vacuum.

MODEL The electrodes form a parallel-plate capacitor. The electric field inside a parallel-plate capacitor is a uniform field, so the electron will have constant acceleration.

VISUALIZE Figure 26.27 shows the capacitor and the electron. The force on the negative electron is *opposite* the electric field, so the electron is repelled by the negative electrode as it accelerates across the gap of width d.

The capacitor was charged by transferring 10^{11} electrons from the right electrode to the left electrode.

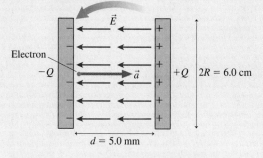

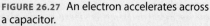

FIGURE 26.27 An electron accelerates across a capacitor.

SOLVE The electrodes are not point charges, so we cannot use Coulomb's law to find the force on the electron. Instead, we must analyze the electron's motion in terms of the electric field inside the capacitor. The field is the agent that exerts the force on the electron, causing it to accelerate. The electric field strength inside a parallel-plate capacitor with charge $Q = Ne$ is

$$E = \frac{\eta}{\epsilon_0} = \frac{Q}{\epsilon_0 A} = \frac{Ne}{\epsilon_0 \pi R^2} = 639{,}000 \text{ N/C}$$

The electron's acceleration in this field is

$$a = \frac{eE}{m} = 1.12 \times 10^{17} \text{ m/s}^2$$

where we used the electron mass $m = 9.11 \times 10^{-31}$ kg. This is an enormous acceleration compared to accelerations we're familiar with for macroscopic objects. We can use one-dimensional kinematics, with $x_i = 0$ and $v_i = 0$, to find the time required for the electron to cross the capacitor:

$$x_f = d = \frac{1}{2}a(\Delta t)^2$$

$$\Delta t = \sqrt{\frac{2d}{a}} = 2.98 \times 10^{-10} \text{ s} = 0.298 \text{ ns}$$

The electron's speed as it reaches the positive electrode is

$$v = a\Delta t = 3.34 \times 10^7 \text{ m/s}.$$

ASSESS We used e rather than $-e$ to find the acceleration because we already knew the direction; we only needed the magnitude. The electron's speed, after traveling a mere 5 mm, is approximately 10% the speed of light.

Parallel electrodes such as those in Example 26.8 are often used to accelerate charged particles. If the positive plate has a small hole in the center, a *beam* of electrons will pass through the hole, after accelerating across the capacitor gap, and emerge with a speed of 3.34×10^7 m/s. This is the basic idea of the *electron gun* used in televisions, oscilloscopes, computer display terminals, and other *cathode-ray tube* (CRT) devices. (A negatively charged electrode is called a *cathode,* so the physicists who first learned to produce electron beams in the late 19th century called them *cathode rays.*) The following example shows that parallel electrodes can also be used to deflect charged particles sideways.

EXAMPLE 26.9 Deflecting an electron beam
An electron gun creates a beam of electrons moving horizontally with a speed of 3.34×10^7 m/s. The electrons enter a 2.0-cm-long gap between two parallel electrodes where the electric field is $\vec{E} = (5.0 \times 10^4 \text{ N/C}, \text{down})$. In which direction, and by what angle, is the electron beam deflected by these electrodes?

MODEL The electric field between the electrodes is uniform. Assume that the electric field outside the electrodes is zero.

VISUALIZE Figure 26.28 shows an electron moving through the electric field. The electric field points down, so the force on the (negative) electrons is upward. The electrons will follow a parabolic trajectory, analogous to that of a ball thrown horizontally, except that the electrons "fall up" rather than down.

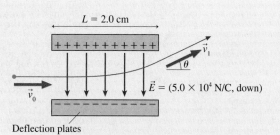

FIGURE 26.28 The deflection of an electron beam in a uniform electric field.

SOLVE This is a two-dimensional motion problem. The electron enters the capacitor with velocity *vector* $\vec{v}_0 = v_{0x}\hat{\imath} = 3.34 \times 10^7\hat{\imath}$ m/s and leaves with velocity $\vec{v}_1 = v_{1x}\hat{\imath} + v_{1y}\hat{\jmath}$. The electron's angle of travel upon leaving the electric field is

$$\theta = \tan^{-1}\left(\frac{v_{1y}}{v_{1x}}\right)$$

This is the *deflection angle*. To find θ we must compute the final velocity vector $\vec{v}_1$.

There is no horizontal force on the electron, so $v_{1x} = v_{0x} = 3.34 \times 10^7$ m/s. The electron's upward acceleration has magnitude

$$a = \frac{eE}{m} = \frac{(1.60 \times 10^{-19}\,\text{C})(5.0 \times 10^4\,\text{N/C})}{9.11 \times 10^{-31}\,\text{kg}}$$

$$= 8.78 \times 10^{15}\,\text{m/s}^2$$

We can use the fact that the horizontal velocity is constant to determine the time interval Δt needed to travel length 2.0 cm:

$$\Delta t = \frac{L}{v_{0x}} = \frac{0.020\,\text{m}}{3.34 \times 10^7\,\text{m/s}} = 5.99 \times 10^{-10}\,\text{s}$$

Vertical acceleration will occur during this time interval, resulting in a final vertical velocity

$$v_{1y} = v_{0y} + a\Delta t = 0 + (8.78 \times 10^{15}\,\text{m/s}^2)(5.99 \times 10^{-10}\,\text{s})$$

$$= 5.26 \times 10^6\,\text{m/s}$$

The electron's velocity as it leaves the capacitor is thus

$$\vec{v}_1 = (3.34 \times 10^7\hat{\imath} + 5.26 \times 10^6\hat{\jmath})\,\text{m/s}$$

and the deflection angle θ is

$$\theta = \tan^{-1}\left(\frac{v_{1y}}{v_{1x}}\right) = 8.95°$$

ASSESS The accelerations of charged particles in electric fields are enormous in comparison to the gravitational acceleration g. Thus it is rarely necessary to include the weight force when calculating the trajectories of charged particles. The only exception might be for a macroscopic charged object, such as a charged plastic bead, in a weak electric field.

Example 26.9 demonstrates how an electron beam is steered to a point on the screen of a cathode-ray tube. First, a high-speed electron beam is created by an electron gun like that of Example 26.8. The beam then passes first through a set of *vertical deflection plates,* as in Example 26.9, then through a second set of *horizontal deflection plates*. After leaving the deflection plates, it travels in a straight line (through vacuum, to eliminate collisions with air molecules) to the screen of the CRT, where it strikes a phosphor coating on the inside surface and makes a dot of light. Properly choosing the electric fields within the deflection plates, which is done by controlling their voltage, steers the electron beam to any point on the screen.

Motion in a Nonuniform Field

The motion of a charged particle in a nonuniform electric field can be quite complicated. Sophisticated mathematical techniques and computers are used to determine the trajectories. However, one type of motion in a nonuniform field is easy to analyze: the circular orbit of a charged particle around a charged sphere or wire.

Figure 26.29 shows a negatively charged particle orbiting a positively charged sphere, much as the moon orbits the earth. You will recall from Chapter 7 that Newton's second law for circular motion is $(F_{\text{net}})_r = mv^2/r$. Here the radial force has magnitude $|q|E$, where E is the electric field strength at distance r. Thus the charge can move in a circular orbit if

$$|q|E = \frac{mv^2}{r} \tag{26.32}$$

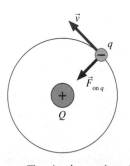

FIGURE 26.29 The circular motion of a charged particle around a charged sphere.

Specific examples of circular orbits will be left for homework problems.

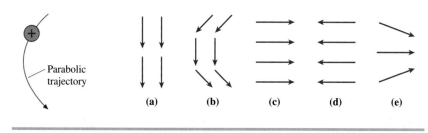

STOP TO THINK 26.6 Which electric field is responsible for the trajectory of the proton?

Parabolic trajectory

(a) (b) (c) (d) (e)

26.7 Motion of a Dipole in an Electric Field

Let us conclude this chapter by returning to one of the more striking puzzles we faced when making the observations at the beginning of Chapter 25. There you found that charged objects of *either* sign exert forces on neutral objects, such as when a comb used to brush your hair picks up pieces of paper. Our qualitative understanding of the *polarization force* was that it required two steps:

■ The charge polarizes the neutral object, creating an induced electric dipole.
■ The charge then exerts an attractive force on the near end of the dipole that is slightly stronger than the repulsive force on the far end.

We are now in a position to make that understanding more quantitative. We will analyze the force on a *permanent* dipole. A homework problem will let you think about *induced* dipoles.

Dipoles in a Uniform Field

Figure 26.30a shows an electric dipole in a *uniform* external electric field $\vec{E}$ that has been created by source charges we do not see. That is, $\vec{E}$ is *not* the field of the dipole but, instead, is a field to which the dipole is responding. In this case, because the field is uniform, the dipole is presumably inside an unseen parallel-plate capacitor.

The net force on the dipole is the sum of the forces on the two charges forming the dipole. Because the charges $\pm q$ are equal in magnitude but opposite in sign, the two forces $\vec{F}_+ = +q\vec{E}$ and $\vec{F}_- = -q\vec{E}$, are also equal but opposite. Thus the net force on the dipole is

$$\vec{F}_{net} = \vec{F}_+ + \vec{F}_- = \vec{0} \qquad (26.33)$$

There is no net force on a dipole in a uniform electric field.

There may be no net force, but the electric field *does* affect the dipole. Because the two forces in Figure 26.30a are in opposite directions but not aligned with each other, the electric field causes the dipole to *rotate*. The "twist" that the field exerts on the dipole is called a *torque*.

The torque causes the dipole to rotate until it is aligned with the electric field, as shown in Figure 26.30b. In this position, the dipole experiences not only no net force but also no torque. Thus Figure 26.30b represents the *equilibrium position* for a dipole in a uniform electric field. Notice that the positive end of the dipole is in the direction in which $\vec{E}$ points.

Figure 26.31 shows a sample of permanent dipoles, such as water molecules, in an external electric field. All the dipoles rotate until they are aligned with the electric field. This is the mechanism by which the sample becomes *polarized*. Once the dipoles are aligned, there is an excess of positive charge at one end of

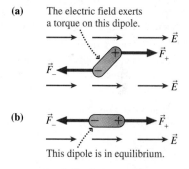

(a) The electric field exerts a torque on this dipole.

(b) This dipole is in equilibrium.

FIGURE 26.30 A dipole in a uniform electric field.

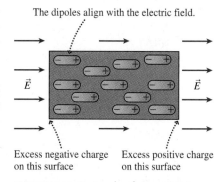

The dipoles align with the electric field.

Excess negative charge on this surface

Excess positive charge on this surface

FIGURE 26.31 A sample of permanent dipoles is *polarized* in an electric field.

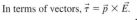

The torque due to a couple is $\tau = lF = pE\sin\theta$.

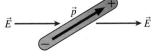

In terms of vectors, $\vec{\tau} = \vec{p} \times \vec{E}$.

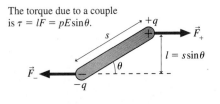

FIGURE 26.32 Calculating the torque on a dipole.

the sample and an excess of negative charge at the other end. The excess charges at the ends of the sample are the basis of the polarization forces we discussed in Section 25.3.

It's not hard to calculate the torque on a dipole. The two forces on the dipole in Figure 26.32 form what we called a *couple* in Chapter 13. There you learned that the torque τ on a couple is the product of the force F with the distance l between the lines along which the forces act. You can see that $l = s\sin\theta$, where θ is the angle the dipole makes with the electric field $\vec{E}$. Thus the torque on the dipole is

$$\tau = lF = (s\sin\theta)(qE) = pE\sin\theta \qquad (26.34)$$

where $p = qs$ was our definition of the dipole moment. The torque is zero when the dipole is aligned with the field, making $\theta = 0$.

If you studied torque in Chapter 13, you will recall that the torque can be written in a compact mathematical form as the cross product between two vectors. The terms p and E in Equation 26.34 are the magnitudes of vectors, and θ is the angle between them. Thus in vector notation, the torque exerted on a dipole moment $\vec{p}$ by an electric field $\vec{E}$ is

$$\vec{\tau} = \vec{p} \times \vec{E} \qquad (26.35)$$

The torque is greatest when $\vec{p}$ is perpendicular to $\vec{E}$, zero when $\vec{p}$ is aligned with or opposite to $\vec{E}$.

EXAMPLE 26.10 The angular acceleration of a dipole dumbbell

Two 1.0 g balls are connected by a 2.0-cm-long insulating rod of negligible mass. One ball has a charge of $+10$ nC, the other a charge of -10 nC. The rod is held in a 1.0×10^4 N/C uniform electric field at an angle of 30° with respect to the field, then released. What is its initial angular acceleration?

MODEL The two oppositely charged balls form an electric dipole. The electric field exerts a torque on the dipole, causing an angular acceleration.

VISUALIZE Figure 26.33 shows the dipole in the electric field. The angle between the dipole moment and the field is $\theta = 30°$.

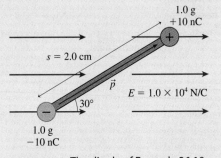

FIGURE 26.33 The dipole of Example 26.10.

SOLVE The dipole moment is $p = qs = (1.0 \times 10^{-8}\,\text{C}) \times (0.020\,\text{m}) = 2.0 \times 10^{-10}\,\text{C}\,\text{m}$. The torque exerted on the dipole moment by the electric field is

$$\tau = pE\sin\theta = (2.0 \times 10^{-10}\,\text{C}\,\text{m})(1.0 \times 10^4\,\text{N/C})\sin 30°$$

$$= 1.0 \times 10^{-6}\,\text{N}\,\text{m}$$

You learned in Chapter 13 that a torque causes an angular acceleration $\alpha = \tau/I$, where I is the moment of inertia. The dipole rotates about its center of mass, which is at the center of the rod, so the moment of inertia is

$$I = m_1 r_1^2 + m_2 r_2^2 = 2m\left(\frac{1}{2}s\right)^2 = \frac{1}{2}ms^2 = 2.0 \times 10^{-7}\,\text{kg}\,\text{m}^2$$

Thus the rod's angular acceleration is

$$\alpha = \frac{\tau}{I} = \frac{1.0 \times 10^{-6}\,\text{N}\,\text{m}}{2.0 \times 10^{-7}\,\text{kg}\,\text{m}^2} = 5.0\,\text{rad/s}^2$$

ASSESS This value of α is the initial angular acceleration, when the rod is first released. The torque and the angular acceleration will decrease as the rod rotates toward alignment with $\vec{E}$.

Dipoles in a Nonuniform Field

Suppose that a dipole is placed in a *nonuniform* electric field, one in which the field strength changes with position. For example, Figure 26.34 shows a dipole in the nonuniform field of a point charge. The first response of the dipole is to rotate until it is aligned with the field, with the dipole's positive end pointing in the same direction as the field. Now, however, there is a slight *difference* between the

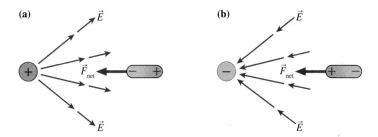

FIGURE 26.34 An aligned dipole is drawn toward a point charge.

forces acting on the two ends of the dipole. This difference occurs because the electric field, which depends on the distance from the point charge, is stronger at the end of the dipole nearest the charge. This causes a *net force* to be exerted on the dipole.

Which way does the force point? Figure 26.34a shows a positive point charge. Once the dipole is aligned, the leftward attractive force on its negative end will be slightly stronger than the rightward repulsive force on its positive end. This causes a net force to the *left,* toward the point charge. The dipole in Figure 26.34b aligns in the opposite orientation in the field of a negative point charge, but the net force is still to the left.

As you can see, **the net force on a dipole is toward the direction of the strongest field.** Because any finite-size charged object, such as a charged rod or a charged disk, has a field strength that increases as you get closer to the object, we can conclude that **a dipole will experience a net force toward any charged object.**

EXAMPLE 26.11 The force on a water molecule

The water molecule H_2O has a permanent dipole moment of magnitude 6.2×10^{-30} Cm. A water molecule is located 10 nm from a Na^+ ion in a saltwater solution. What force does the ion exert on the water molecule?

VISUALIZE Figure 26.35 shows the ion and the dipole. The forces are an action/reaction pair.

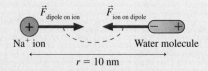

FIGURE 26.35 The interaction between an ion and a permanent dipole.

SOLVE A Na^+ ion has charge $q = +e$. The electric field of the ion aligns the water's dipole moment and exerts a net force on it. We could calculate the net force on the dipole as the small difference between the attractive force on its negative end and the repulsive force on its positive end. Alternatively, we know from Newton's third law that the force $\vec{F}_{\text{dipole on ion}}$ has the

same magnitude as the force $\vec{F}_{\text{ion on dipole}}$ that we are seeking. We calculated the on-axis field of a dipole in Section 26.2. An ion of charge $q = e$ will experience a force of magnitude $F = qE_{\text{dipole}} = eE_{\text{dipole}}$ when placed in that field. The dipole's electric field, which we found in Equation 26.11, is

$$E_{\text{dipole}} = \frac{1}{4\pi\epsilon_0} \frac{2p}{r^3}$$

The force on the ion at distance $r = 1.0 \times 10^{-8}$ m is

$$F_{\text{dipole on ion}} = eE_{\text{dipole}} = \frac{1}{4\pi\epsilon_0} \frac{2ep}{r^3} = 1.79 \times 10^{-14}\,\text{N}$$

Thus the force on the water molecule is $F_{\text{ion on dipole}} = 1.79 \times 10^{-14}$ N.

ASSESS While 1.79×10^{-14} N may seem like a very small force, it is $\approx 10^{11}$ times larger than the size of the earth's gravitational force on these atomic particles. Forces such as these cause water molecules to cluster around any ions that are in solution. This clustering, a phenomenon studied in a physical chemistry course, plays an important role in the microscopic physics of solutions.

SUMMARY

The goal of Chapter 26 has been to learn how to calculate and use the electric field.

GENERAL PRINCIPLES

Sources of $\vec{E}$

Electric fields are created by charges.

Two major tools for calculating $\vec{E}$ are

- The field of a point charge

$$\vec{E} = \frac{1}{4\pi\epsilon_0}\frac{q}{r^2}\hat{r}$$

- The principle of superposition

Multiple point charges

Use superposition: $\vec{E} = \vec{E}_1 + \vec{E}_2 + \vec{E}_3 + \ldots$

Continuous distribution of charge

- Divide the charge into point-like ΔQ
- Find the field of each ΔQ
- Find $\vec{E}$ by summing the fields of all ΔQ

The summation usually becomes an integral. A critical step is replacing ΔQ with an expression involving a **charge density** (λ or η) and an integration coordinate.

Consequences of $\vec{E}$

The electric field exerts a force on a charged particle.

$$\vec{F} = q\vec{E}$$

The force causes acceleration

$$\vec{a} = (q/m)\vec{E}$$

Trajectories of charged particles are calculated with kinematics.

The electric field exerts a torque on a dipole.

$$\tau = pE\sin\theta$$

The torque tends to align the dipoles with the field.

In a nonuniform electric field, a dipole has a net force in the direction of increasing field strength.

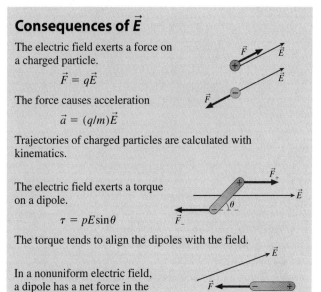

APPLICATIONS

The following fields are important models of the electric field:

Electric dipole

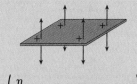

The electric dipole moment is

$\vec{p} = (qs, \text{from negative to positive})$

Field on axis $\vec{E} = \frac{1}{4\pi\epsilon_0}\frac{2\vec{p}}{r^3}$

Field in bisecting plane $\vec{E} = -\frac{1}{4\pi\epsilon_0}\frac{\vec{p}}{r^3}$

Infinite line of charge with linear charge density λ

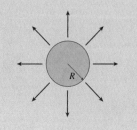

$\vec{E} = \left(\frac{1}{4\pi\epsilon_0}\frac{2\lambda}{r}, \text{perpendicular to line}\right)$

Infinite plane of charge with surface charge density η

$\vec{E} = \left(\frac{\eta}{2\epsilon_0}, \text{perpendicular to plane}\right)$

Sphere of charge

Same as a point charge Q for $r > R$

Parallel-plate capacitor

The electric field inside an ideal capacitor is a **uniform electric field**

$\vec{E} = \left(\frac{\eta}{\epsilon_0}, \text{from positive to negative}\right)$

A real capacitor has a weak **fringe field** around it.

TERMS AND NOTATION

dipole moment, $\vec{p}$	uniformly charged	plane of charge	fringe field
electric field line	line of charge	sphere of charge	uniform electric field
linear charge density, λ	electrode	parallel-plate capacitor	charge-to-mass ratio, q/m
surface charge density, η			

EXERCISES AND PROBLEMS

Exercises

Section 26.2 The Electric Field of Multiple Point Charges

1. What are the strength and direction of the electric field at the position indicated by the dot in Figure Ex26.1? Specify the direction as an angle above or below horizontal.

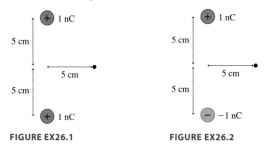

FIGURE EX26.1   FIGURE EX26.2

2. What are the strength and direction of the electric field at the position indicated by the dot in Figure Ex26.2? Specify the direction as an angle above or below horizontal.
3. What are the strength and direction of the electric field at the position indicated by the dot in Figure Ex26.3? Specify the direction as an angle above or below horizontal.

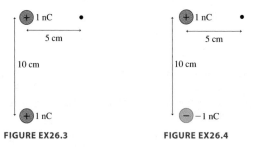

FIGURE EX26.3   FIGURE EX26.4

4. What are the strength and direction of the electric field at the position indicated by the dot in Figure Ex26.4? Specify the direction as an angle above or below horizontal.
5. An electric dipole is formed from ± 1.0 nC charges spaced 2.0 mm apart. The dipole is at the origin, oriented along the y-axis. What is the electric field strength at the points (a) $(x, y) = (10 \text{ cm}, 0 \text{ cm})$ and (b) $(x, y) = (0 \text{ cm}, 10 \text{ cm})$?
6. An electric dipole is formed from two charges, $\pm q$, spaced 1.0 cm apart. The dipole is at the origin, oriented along the y-axis. The electric field strength at the point $(x, y) = (0 \text{ cm}, 10 \text{ cm})$ is 360 N/C.
   a. What is the charge q? Give your answer in nC.
   b. What is the electric field strength at the point $(x, y) = (10 \text{ cm}, 0 \text{ cm})$?

Section 26.3 The Electric Field of a Continuous Charge Distribution

7. A 10-cm-long thin glass rod uniformly charged to $+10$ nC and a 10-cm-long thin plastic rod uniformly charged to -10 nC are placed side by side, 4.0 cm apart. What are the electric field strengths E_1 to E_3 at distances 1.0 cm, 2.0 cm, and 3.0 cm from the glass rod along the line connecting the midpoints of the two rods?
8. Two 10-cm-long thin glass rods uniformly charged to $+10$ nC are placed side by side, 4.0 cm apart. What are the electric field strengths E_1 to E_3 at distances 1.0 cm, 2.0 cm, and 3.0 cm from the rod on the left along the line connecting the midpoints of the two rods?
9. A 10-cm-long thin glass rod is uniformly charged to $+50$ nC. A small plastic bead, charged to -5.0 nC, is 4.0 cm from the center of the rod. What is the force (magnitude and direction) on the bead?
10. The electric field strength 5.0 cm from a very long charged wire is 2000 N/C. What is the electric field strength 10.0 cm from the wire?
11. The electric field 5.0 cm from a very long charged wire is (2000 N/C, toward the wire). What is the charge (in nC) on a 1.0-cm-long segment of the wire?

Section 26.4 The Electric Fields of Rings, Planes, and Spheres

12. Two 10-cm-diameter charged rings face each other, 20 cm apart. The left ring is charged to -20 nC and the right ring is charged to $+20$ nC.
    a. What is the electric field $\vec{E}$, both magnitude and direction, at the midpoint between the two rings?
    b. What is the force $\vec{F}$ on a 1.0 nC charge placed at the midpoint?
13. Two 10-cm-diameter charged rings face each other, 20 cm apart. Both rings are charged to $+20$ nC. What is the electric field strength at (a) the midpoint between the two rings and (b) the center of the left ring?
14. Two 10-cm-diameter charged disks face each other, 20 cm apart. The left disk is charged to -50 nC and the right disk is charged to $+50$ nC.
    a. What is the electric field $\vec{E}$, both magnitude and direction, at the midpoint between the two disks?
    b. What is the force $\vec{F}$ on a -1.0 nC charge placed at the midpoint?

15. Two 10-cm-diameter charged disks face each other, 20 cm apart. Both disks are charged to -50 nC. What is the electric field strength at (a) the midpoint between the two disks and (b) the center of the left disk?

16. A 20 cm × 20 cm metal electrode is uniformly charged to $+80$ nC. What is the electric field strength 2.0 mm above the center of the electrode?

17. The electric field strength 5.0 cm from a very wide charged electrode is 1000 N/C. What is the charge (in nC) on a 1.0-cm-diameter circular segment of the electrode?

18. The electric field strength 2.0 cm from a 10-cm-diameter metal ball is 50,000 N/C. What is the charge (in nC) on the ball?

19. Two 2.0-cm-diameter insulating spheres have a 6.0 cm space between them. One sphere is charged to $+10$ nC, the other to -15 nC. What is the electric field strength at the midpoint between the two spheres?

Section 26.5 The Parallel-Plate Capacitor

20. A parallel-plate capacitor is formed from two 4.0 cm × 4.0 cm electrodes spaced 2.0 mm apart. The electric field strength inside the capacitor is 1.0×10^6 N/C. What is the charge (in nC) on each electrode?

21. Two closely spaced circular disks form a parallel-plate capacitor. Transferring 1.5×10^9 electrons from one disk to the other causes the electric field strength to be 1.0×10^5 N/C. What are the diameters of the disks?

22. Air "breaks down" when the electric field strength reaches 3×10^6 N/C, causing a spark. A parallel-plate capacitor is made from two 4.0-cm-diameter disks. How many electrons must be transferred from one disk to the other to create a spark between the disks?

Section 26.6 Motion of a Charged Particle in an Electric Field

23. A 0.10 g plastic bead is charged by the addition of 1.0×10^{10} excess electrons. What electric field $\vec{E}$ (strength and direction) will cause the bead to hang suspended in the air?

24. Two 2.0-cm-diameter disks face each other, 1.0 mm apart. They are charged to ± 10 nC.
    a. What is the electric field strength between the disks?
    b. A proton is shot from the negative disk toward the positive disk. What launch speed must the proton have to just barely reach the positive disk?

25. The electron gun in a television tube uses a uniform electric field to accelerate electrons from rest to 5.0×10^7 m/s in a distance of 1.2 cm. What is the electric field strength?

26. An electron is released from rest 2.0 cm from an infinite charged plane. It accelerates toward the plane and collides with a speed of 1.0×10^7 m/s. What are (a) the surface charge density of the plane and (b) the time required for the electron to travel the 2.0 cm?

27. The surface charge density on an infinite charged plane is -2.0×10^{-6} C/m². A proton is shot straight away from the plane at 2.0×10^6 m/s. How far does the proton travel before reaching its turning point?

Section 26.7 Motion of a Dipole in an Electric Field

28. A point charge Q is distance r from the center of a dipole consisting of charges $\pm q$ separated by distance s. The charge is located in the plane that bisects the dipole. At this instant, what are (a) the force (magnitude and direction) and (b) the magnitude of the torque on the dipole? You can assume $r \gg s$.

29. An ammonia molecule (NH_3) has a permanent electric dipole moment 5.0×10^{-30} Cm. A proton is 2.0 nm from the molecule in the plane that bisects the dipole. What is the electric force of the molecule on the proton?

30. The permanent electric dipole moment of the water molecule (H_2O) is 6.2×10^{-30} Cm. What is the maximum possible torque on a water molecule in a 5.0×10^8 N/C electric field?

Problems

31. What are the strength and direction of the electric field at the position indicated by the dot? Give your answer in both component form and as a magnitude and direction.

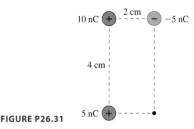

FIGURE P26.31

32. What are the strength and direction of the electric field at the position indicated by the dot? Give your answer in both component form and as a magnitude and direction.

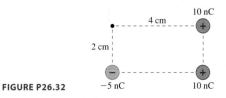

FIGURE P26.32

33. What are the strength and direction of the electric field at the position indicated by the dot in Figure P26.33? Give your answer in both component form and as a magnitude and direction.

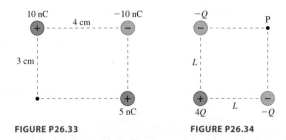

FIGURE P26.33   FIGURE P26.34

34. Figure P26.34 shows three charges at the corners of a square.
    a. Write the electric field at point P in component form.
    b. A particle with positive charge q and mass m is placed at point P and released. What is the initial magnitude of its acceleration?

35. Charges $-q$ and $+2q$ in Figure P26.35 are located at $x = \pm a$.
    a. Determine the electric field at points 1 to 4. Write each field in component form.
    b. Reproduce Figure P26.35, then draw the four electric field vectors on the figure.

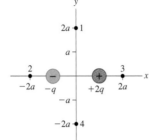

FIGURE P26.35

36. Two positive charges q are distance s apart on the y-axis.
    a. Find an expression for the electric field strength at distance x on the axis that bisects the two charges.
    b. For $q = 1.0$ nC and $s = 6.0$ mm, evaluate E at $x = 0, 2, 4, 6$, and 10 mm.
    c. Draw a graph of E versus x for $0 \le x \le \infty$.
37. Derive Equation 26.12 for the field $\vec{E}_{dipole}$ in the plane that bisects an electric dipole.
38. Three charges are on the y-axis. Charges $-q$ are at $y = \pm d$ and charge $+2q$ is at $y = 0$.
    a. Determine the electric field $\vec{E}$ along the x-axis.
    b. Verify that your answer to part a has the expected behavior as x becomes very small and very large.
    c. Sketch a graph of E_x versus x for $0 \le x \le \infty$.
39. Three 10-cm-long rods form an equilateral triangle. Two of the rods are charged to $+10$ nC, the third to -10 nC. What is the electric field strength at the center of the triangle?
40. Figure P26.40 is a cross section of two infinite lines of charge that extend out of the page. Both have linear charge density λ.
    a. Find an expression for the electric field strength E at height y above the midpoint between the lines.
    b. Draw a graph of E versus y.

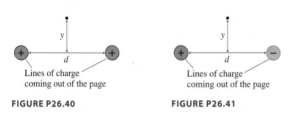

FIGURE P26.40   **FIGURE P26.41**

41. Figure P26.41 is a cross section of two infinite lines of charge that extend out of the page. The linear charge densities are $\pm\lambda$.
    a. Find an expression for the electric field strength E at height y above the midpoint between the lines.
    b. Draw a graph of E versus y.
42. Two infinite lines of charge, each with linear charge density λ, lie along the x- and y-axes, crossing at the origin. What is the electric field strength at position (x, y)?
43. A proton orbits a long charged wire, making 1.0×10^6 revolutions per second. The radius of the orbit is 1.0 cm. What is the wire's linear charge density?

44. Figure P26.44 shows a thin rod of length L with total charge Q.
    a. Find an expression for the electric field strength on the axis of the rod at distance r from the center.
    b. Verify that your expression has the expected behavior if $r \gg L$.
    c. Evaluate E at $r = 3.0$ cm if $L = 5.0$ cm and $Q = 3.0$ nC.

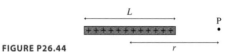

FIGURE P26.44

45. Figure P26.45 shows a thin rod of length L with total charge Q. Find an expression for the electric field $\vec{E}$ at distance x from the end of the rod. Give your answer in component form.

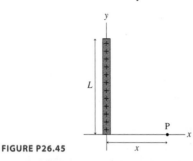

FIGURE P26.45

46. Show that the on-axis electric field of a ring of charge has the expected behavior when $z \ll R$ and when $z \gg R$.
47. a. Show that the maximum electric field strength on the axis of a ring of charge occurs at $z = R/\sqrt{2}$.
    b. What is the electric field strength at this point?
48. Charge Q is uniformly distributed along a thin, flexible rod of length L. The rod is then bent into the semicircle shown in Figure P26.48.
    a. Find an expression for the electric field $\vec{E}$ at the center of the semicircle.
    Hint: A small piece of arc length Δs spans a small angle $\Delta\theta = \Delta s/R$, where R is the radius.
    b. Evaluate the field strength if $L = 10$ cm and $Q = 30$ nC.

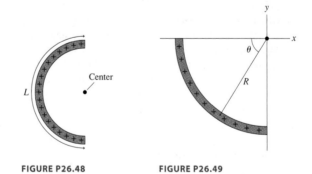

FIGURE P26.48   **FIGURE P26.49**

49. A plastic rod with linear charge density λ is bent into the quarter circle shown in Figure P26.49. We want to find the electric field at the origin.
    a. Write expressions for the x- and y-components of the electric field at the origin due to a small piece of charge at angle θ.
    b. Write, but do not evaluate, definite integrals for the x- and y-components of the net electric field at the origin.
    c. Evaluate the integrals and write $\vec{E}_{net}$ in component form.

50. You've hung two very large sheets of plastic facing each other with distance d between them, as shown in Figure P26.50. By rubbing them with wool and silk, you've managed to give one sheet a uniform surface charge density $\eta_1 = -\eta_0$ and the other a uniform surface charge density $\eta_2 = +3\eta_0$. What is the electric field vector at points 1, 2, and 3?

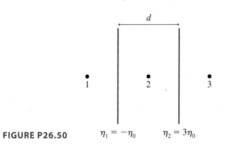

FIGURE P26.50 $\eta_1 = -\eta_0$ $\eta_2 = 3\eta_0$

51. Two parallel plates 1.0 cm apart are equally and oppositely charged. An electron is released from rest at the surface of the negative plate and simultaneously a proton is released from rest at the surface of the positive plate. How far from the negative plate is the point at which the electron and proton pass each other?

52. A proton traveling at a speed of 1.0×10^6 m/s enters the gap between the plates of a 2.0-cm-wide parallel-plate capacitor. The surface charge densities on the plates are $\pm 1.0 \times 10^{-6}$ C/m^2. How far has the proton been deflected sideways when it reaches the far edge of the capacitor? Assume the electric field is uniform inside the capacitor and zero outside.

53. The two parallel plates in Figure P26.53 are 2.0 cm apart and the electric field strength between them is 1.0×10^4 N/C. An electron is launched at a 45° angle from the positive plate. What is the maximum initial speed v_0 the electron can have without hitting the negative plate?

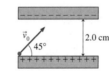

FIGURE P26.53

54. An electron is launched at a 45° angle and a speed of 5.0×10^6 m/s from the positive plate of the parallel-plate capacitor shown in Figure P26.54. The electron lands 4.0 cm away.
    a. What is the electric field strength inside the capacitor?
    b. What is the minimum spacing between the plates?

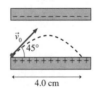

FIGURE P26.54   4.0 cm

55. A problem of practical interest is to make a beam of electrons turn a 90° corner. This can be done with the parallel-plate capacitor shown in Figure P26.55. An electron with kinetic energy 3.0×10^{-17} J moves up through a small hole in the bottom plate of the capacitor.
    a. Should the bottom plate be charged positive or negative relative to the top plate if you want the electron to turn to the right? Explain.

b. What strength electric field is needed if the electron is to emerge from an exit hole 1.0 cm away from the entrance hole, traveling at right angles to its original direction?
   Hint: The difficulty of this problem depends on how you choose your coordinate system.
    c. What minimum separation $d_{\min}$ must the capacitor plates have?

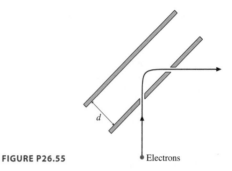

FIGURE P26.55   • Electrons

56. You have a summer intern position at a laboratory that uses a high-speed proton beam. The protons exit the machine at a speed of 2.0×10^6 m/s, and you've been asked to design a device to stop the protons safely. You know that protons will embed themselves in a metal target, but protons traveling faster than 2.0×10^5 m/s emit dangerous x rays when they hit. You decide to slow the protons to an acceptable speed, then let them hit a target. You take two metal plates, space them 2.0 cm apart, then drill a small hole through the center of one plate to let the proton beam enter. The opposite plate is the target in which the protons will embed themselves.
    a. What are the minimum surface charge densities you need to place on each plate? Which plate, positive or negative, faces the incoming proton beam?
    b. What happens if you charge the plates to $\pm 1.0 \times 10^{-5}$ C/m^2? Does your device still work?

57. One type of ink-jet printer, called an electrostatic ink-jet printer, forms the letters by using deflecting electrodes to steer charged ink drops up and down vertically as the ink jet sweeps horizontally across the page. The ink jet forms 30-μm-diameter drops of ink, charges them by spraying 800,000 electrons on the surface, and shoots them toward the page at a speed of 20 m/s. Along the way, the drops pass through two parallel electrodes that are 6.0 mm long, 4.0 mm wide, and spaced 1.0 mm apart. The distance from the center of the plates to the paper is 2.0 cm. To form the letters, which have a maximum height of 6.0 mm, the drops need to be deflected up or down a maximum of 3.0 mm. Ink, which consists of dye particles suspended in alcohol, has a density of 800 kg/m^3.
    a. Estimate the maximum electric field strength needed in the space between the electrodes.
    b. What amount of charge is needed on each electrode to produce this electric field?

58. A 2.0-mm-diameter glass sphere has a charge of $+1.0$ nC. What speed does an electron need to orbit the sphere 1.0 mm above the surface?

59. A proton orbits a 1.0-cm-diameter metal ball 1.0 mm above the surface. The orbital period is 1.0 μs. What is the charge on the ball?

60. In a classical model of the hydrogen atom, the electron orbits the proton in a circular orbit of radius 0.050 nm. What is the orbital frequency? The proton is so much more massive than the electron that you can assume the proton is at rest.

61. In a classical model of the hydrogen atom, the electron orbits a stationary proton in a circular orbit. What is the radius of the orbit for which the orbital frequency is 1.0×10^{12} s^{-1}?

62. A *positron* is an elementary particle identical to an electron except that its charge is $+e$. An electron and a positron can rotate about their center of mass as if they were a dumbbell connected by a massless rod. What is the orbital frequency for an electron and a positron 1.0 nm apart?

63. An electric field can *induce* an electric dipole in a neutral atom or molecule by pushing the positive and negative charge in opposite directions. The dipole moment of an induced dipole is directly proportional to the electric field. That is, $\vec{p} = \alpha \vec{E}$, where α is called the *polarizability* of the molecule. A bigger field stretches the molecule farther and causes a larger dipole moment.
   a. What are the units of α?
   b. An ion with charge q is distance r from a molecule with polarizability α. Find an expression for the force $\vec{F}_{\text{ion on dipole}}$.

64. Show that a line of charge with linear charge density λ exerts an attractive force on an electric dipole with magnitude $F = 2\lambda p/4\pi\epsilon_0 r^2$. Assume that r is much larger than the charge separation in the dipole.

In Problems 65 through 68 you are given the equation(s) used to solve a problem. For each of these
   a. Write a realistic problem for which this is the correct equation(s).
   b. Finish the solution of the problem.

65. $(9.0 \times 10^9 \,\mathrm{N\,m^2/C^2}) \dfrac{(2.0 \times 10^{-9}\,\mathrm{C})\,s}{(0.025\,\mathrm{m})^3} = 1150$ N/C

66. $(9.0 \times 10^9 \,\mathrm{N\,m^2/C^2}) \dfrac{2(2.0 \times 10^{-7}\,\mathrm{C/m})}{r} = 25{,}000$ N/C

67. $\dfrac{\eta}{2\epsilon_0}\left[1 - \dfrac{z}{\sqrt{z^2 + R^2}}\right] = \dfrac{1}{2}\dfrac{\eta}{2\epsilon_0}$

68. $2.0 \times 10^{12}\,\mathrm{m/s^2} = \dfrac{(1.60 \times 10^{-19}\,\mathrm{C})\,E}{(1.67 \times 10^{-27}\,\mathrm{kg})}$

   $E = \dfrac{Q}{(8.85 \times 10^{-12}\,\mathrm{C^2/N\,m^2})(0.020\,\mathrm{m})^2}$

Challenge Problems

69. Your physics assignment is to figure out a way to use electricity to launch a small 6.0-cm-long plastic drink stirrer. You decide that you'll charge the little plastic rod by rubbing it with fur, then hold it near a long, charged wire. When you let go, the electric force of the wire on the plastic rod will shoot it away. Suppose you can charge the plastic stirrer to 10 nC and that the linear charge density of the long wire is $1.0 \times$ 10^{-7} C/m. What is the electric force on the plastic stirrer if the end closest to the wire is 2.0 cm away?

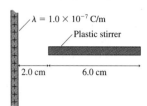

$\lambda = 1.0 \times 10^{-7}$ C/m

Plastic stirrer

2.0 cm   6.0 cm

FIGURE CP26.69

70. A rod of length L lies along the y-axis with its center at the origin. The rod has a nonuniform linear charge density $\lambda = a|y|$, where a is a constant with the units C/m^2.
   a. Draw a graph of λ versus y over the length of the rod.
   b. Determine the constant a in terms of L and the rod's total charge Q.
   Hint: This requires an integration. Think about how to handle the absolute value sign.
   c. Find the electric field strength of the rod at distance x on the x-axis.

71. a. An infinitely long *sheet* of charge of width L lies in the xy-plane between $x = -L/2$ and $x = L/2$. The surface charge density is η. Derive an expression for the electric field $\vec{E}$ at height z above the centerline of the sheet.
   b. Verify that your expression has the expected behavior if $z \ll L$ and if $z \gg L$.
   c. Draw a graph of field strength E versus z.

72. a. An infinitely long *sheet* of charge of width L lies in the xy-plane between $x = -L/2$ and $x = L/2$. The surface charge density is η. Derive an expression for the electric field $\vec{E}$ along the x-axis for points outside the sheet ($x > L/2$).
   b. Verify that your expression has the expected behavior if $x \gg L$.
   Hint: $\ln(1 + u) \approx u$ if $u \ll 1$.
   c. Draw a graph of field strength E versus x for $x > L/2$.

73. You have a summer intern position with a company that designs and builds nanomachines. An engineer with the company is designing a microscopic oscillator to help keep time, and you've been assigned to help him analyze the design. He wants to place a negative charge at the center of a very small, positively charged metal loop. His claim is that the negative charge will undergo simple harmonic motion at a frequency determined by the amount of charge on the loop.
   a. Consider a negative charge near the center of a positively charged ring. Show that there is a restoring force on the charge if it moves along the z-axis but stays close to the center. That is, show there's a force that tries to keep the charge at $z = 0$.
   b. Show that for *small* oscillations, with amplitude $\ll R$, a particle of mass m with charge $-q$ undergoes simple harmonic motion with frequency

   $$f = \frac{1}{2\pi}\sqrt{\frac{qQ}{4\pi\epsilon_0 m R^3}}$$

   R and Q are the radius and charge of the ring.
   c. Evaluate the oscillation frequency for an electron at the center of a 2.0-μm-diameter ring charged to 1.0×10^{-13} C.

74. We want to analyze how a charged object picks up a neutral piece of metal. Figure CP26.74 shows a small circular disk of aluminum foil lying flat on a table. The foil disk has radius R and thickness t. A glass ball with positive charge Q is at height h above the foil. Assume that $R \ll h$ and $t \ll R$. These assumptions imply that the electric field of the ball is very nearly constant over the volume of the foil disk.

    a. What are the magnitude and direction of the ball's electric field at the position of the foil? Your answer will be an expression involving Q and h.

    b. The ball's electric field polarizes the foil. The foil surfaces, with charges $+q$ and $-q$, then act as the plates of a parallel-plate capacitor with separation t. But the foil is a conductor in electrostatic equilibrium, so the electric field E_{in} *inside* the foil must be zero. $E_{in} = 0$ seems to be inconsistent with the surfaces of the foil acting as a parallel-plate capacitor. Use words *and* pictures to explain how $E_{in} = 0$ even though the surfaces of the foil are charged.

    c. Now write the condition that $E_{in} = 0$ as a mathematical statement and use it to find an expression for the charge q on the upper surface of the foil.

    d. Suppose $Q = 50$ nC, $R = 1$ mm, and $t = 0.01$ mm. These are all typical values. The density of aluminum is $\rho = 2700$ kg/m^3. How close must the ball be to lift the foil?

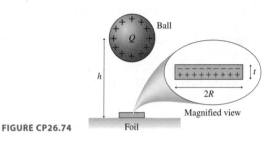

FIGURE CP26.74

Stop to Think 26.1: c. From symmetry, the fields of the positive charges cancel. The net field is that of the negative charge, which is toward the charge.

Stop to Think 26.2: $\eta_3 = \eta_2 = \eta_1$. All pieces of a uniformly charged surface have the same surface charge density.

Stop to Think 26.3: b, e, and **h.** b and e both increase the linear charge density λ.

Stop to Think 26.4: $E_a = E_b = E_c = E_d = E_e$. The field strength of a charged plane is the same at all distances from the plane. An electric field diagram shows the electric field vectors at only a few points; the field exists at all points.

Stop to Think 26.5: $F_a = F_b = F_c = F_d = F_e$. The field strength inside a capacitor is the same at all points, hence the force on a charge is the same at all points. The electric field exists at all points whether or not a vector is shown at that point.

Stop to Think 26.6: c. Parabolic trajectories require *constant* acceleration and thus a *uniform* electric field. The proton has an initial velocity component to the left, but it's being pushed back to the right.

27 Gauss's Law

The nearly spherical shape of the girl's head determines the shape of the electric field that causes her hair to stream outwards.

▶ **Looking Ahead**
The goal of Chapter 27 is to understand and apply Gauss's law. In this chapter you will learn to:

- Recognize and use symmetry to determine the shape of electric fields.
- Calculate the electric flux through a surface.
- Use Gauss's law to calculate the electric field of symmetric charge distributions.
- Use Gauss's law to understand the properties of conductors in electrostatic equilibrium.

◀ **Looking Back**
This chapter builds on basic ideas about electric fields. Please review:

- Sections 25.4–25.6 Coulomb's law and the electric field of a point charge.
- Section 26.2 Electric field vectors and electric field lines.

The electric field of this charged sphere points straight out because that's the only field direction compatible with the *symmetry* of the sphere. Spheres, cylinders, and planes—common shapes for electrodes—all have a high degree of symmetry. As you'll see in this chapter, their symmetry determines the geometry of their electric fields.

You learned in Chapter 26 how to calculate electric fields by starting from Coulomb's law for the electric field of a point charge. This is a foolproof method in principle, but in practice it often requires excessive mathematical gymnastics to carry through the necessary integrals. In this chapter you'll learn how some important electric fields, those with a high degree of symmetry, can be deduced simply from the shape of the charge distribution. The principle underlying this approach to calculating electric fields is called *Gauss's law*.

Gauss's law and Coulomb's law are equivalent in the sense that each can be derived from the other. But Gauss's law gives us a very different perspective on electric fields, much as conservation laws give us a perspective on mechanics different from that of Newton's laws. In practice, Gauss's law will allow us to find some static electric fields that would be difficult to find using Coulomb's law. Ultimately, we'll find that Gauss's law is more general in that it applies not only to electrostatics but also to the electrodynamics of moving charges and fields that change with time.

27.1 Symmetry

Suppose we knew only two things about electric fields:

1. An electric field points away from positive charges, toward negative charges, and
2. An electric field exerts a force on a charged particle.

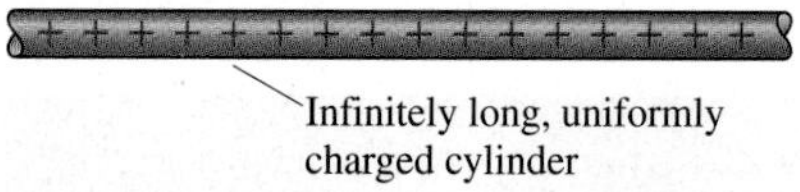

From this information alone, what can we deduce about the electric field of the infinitely long charged cylinder shown in Figure 27.1?

We don't know if the cylinder's diameter is large or small. We don't know if the charge density is the same at the outer edge as along the axis. All we know is that the charge is positive and the charge distribution has *cylindrical symmetry*.

Symmetry is an especially important idea in science and mathematics. We say that a charge distribution is **symmetric** if there are a group of *geometrical transformations* that don't cause any *physical* change. To make this idea concrete, suppose you close your eyes while a friend transforms a charge distribution in one of the following three ways. He or she can

■ *Translate* (that is, displace) the charge parallel to an axis,
■ *Rotate* the charge about an axis, or
■ *Reflect* the charge in a mirror.

When you open your eyes, will you be able to tell if the charge distribution has been changed? You might tell by observing a visual difference in the distribution. Or the results of an experiment with charged particles could reveal that the distribution has changed. If nothing you can see or do reveals any change, then we say that the charge distribution is symmetric under that particular transformation.

Figure 27.2 shows that the charge distribution of Figure 27.1 is symmetrical with respect to

■ Translations parallel to the cylinder axis. Shifting an infinitely long cylinder by 1 mm or 1000 m makes no noticeable or measurable change.
■ Rotation by any angle about the cylinder axis. Turning a cylinder about its axis by 1° or 100° makes no detectable change.
■ Reflections in any plane containing or perpendicular to the cylinder axis. Exchanging top and bottom, front and back, or left and right makes no detectable change.

But this isn't to say that *all* transformations are undetectable. You would notice immediately if your friend rotated the cylinder by 45° about an axis perpendicular to the cylinder axis.

A charge distribution that is symmetrical under these three groups of geometrical transformations is said to be *cylindrically symmetric*. Other charge distributions have other types of symmetries. Some charge distributions have no symmetry at all. Our interest in symmetry can be summed up in a single statement:

The symmetry of the electric field must match the symmetry of the charge distribution.

The proof of this statement is seen by assuming its converse. Suppose the electric field *did not* match the symmetry of the charge distribution. In that case, you could test whether or not the charge distribution had undergone a transformation by observing the motion of charged particles in the field. But the definition of symmetry requires that you not be able to tell. Thus the electric field must have the same symmetry as the charge distribution.

Now we're ready to see what we can learn about the electric field in Figure 27.1. Could the field look like Figure 27.3a? (Imagine this picture rotated about the axis.

Infinitely long, uniformly charged cylinder

FIGURE 27.1 A charge distribution with cylindrical symmetry.

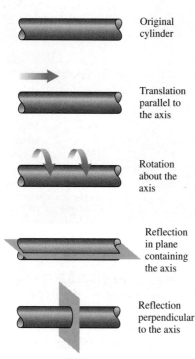

Original cylinder

Translation parallel to the axis

Rotation about the axis

Reflection in plane containing the axis

Reflection perpendicular to the axis

FIGURE 27.2 Transformations that don't change an infinite cylinder of charge.

Field vectors are also coming out of the page and going into the page.) That is, is this a *possible* field? This field looks the same if it's translated parallel to the cylinder axis, if up and down are exchanged by reflecting the field in a plane coming out of the page, or if you rotate the cylinder about its axis.

(a) Is this a possible electric field of an infinitely long charged cylinder? Suppose the charge and the field are reflected in a plane perpendicular to the axis.

(b) The charge distribution is not changed by the reflection, but the field is. This field doesn't match the symmetry of the cylinder, so the cylinder's field can't look like this.

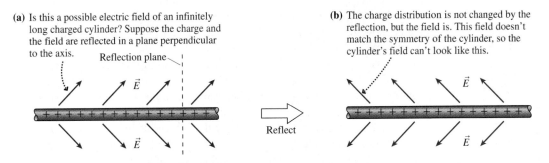

FIGURE 27.3 Could the field of a cylindrical charge distribution look like this?

But the proposed field fails one test. Suppose we reflect the field in a plane perpendicular to the axis, a reflection that exchanges left and right. This reflection, which would *not* make any change in the charge distribution itself, produces the field shown in Figure 27.3b. This change in the field is detectable because a positively charged particle would now have a component of motion to the left instead of to the right.

The field of Figure 27.3a, which makes a distinction between left and right, is not cylindrically symmetric and thus is *not* a possible field. In general, **the electric field of a cylindrically symmetric charge distribution cannot have a component parallel to the cylinder axis.**

Well then, what about the electric field shown in Figure 27.4a? Here we're looking down the axis of the cylinder. The electric field vectors are restricted to planes perpendicular to the cylinder and thus do not have any component parallel to the cylinder axis. This field is symmetric for rotations about the axis, but it's *not* symmetric for a reflection in a plane containing the axis.

The field of Figure 27.4b, after such a reflection, is easily distinguishable from the field of Figure 27.4a. Thus **the electric field of a cylindrically symmetric charge distribution cannot have a component tangent to the circular cross section.**

Figure 27.5 shows the only remaining possible field shape. The electric field is radial, pointing straight out from the cylinder like the bristles on a bottle brush. This is the one electric field shape that matches the symmetry of the charge distribution.

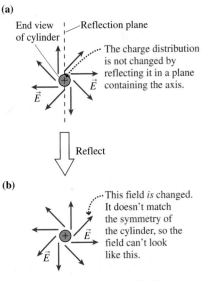

(a)
End view of cylinder — Reflection plane

The charge distribution is not changed by reflecting it in a plane containing the axis.

Reflect

(b)
This field *is* changed. It doesn't match the symmetry of the cylinder, so the field can't look like this.

FIGURE 27.4 Or might the field of a cylindrical charge distribution look like this?

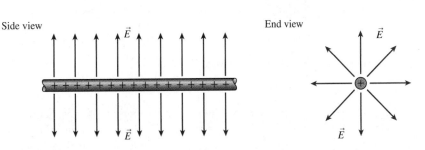

FIGURE 27.5 This is the only shape for the electric field that matches the symmetry of the charge distribution.

What Good Is Symmetry?

Given how little we assumed about Figure 27.1—that the charge distribution is cylindrically symmetric and that electric fields point away from positive charges—we've been able to deduce a great deal about the electric field. In particular, we've deduced the *shape* of the electric field.

Now, shape is not everything. We've learned nothing about the strength of the field or how strength changes with distance. Is E constant? Does it decrease like $1/r$ or $1/r^2$? We don't yet have a complete description of the field, but knowing what shape the field *has* to have will make finding the field strength a much easier task.

That's the good of symmetry. Symmetry arguments allow us to *rule out* many conceivable field shapes as simply being incompatible with the symmetry of the charge distribution. Knowing what doesn't happen, or can't happen, is often as useful as knowing what does happen. By the process of elimination, we're led to the one and only shape the field can possibly have. Reasoning on the basis of symmetry is a sometimes subtle but always powerful means of reasoning.

Three Fundamental Symmetries

Three fundamental symmetries appear frequently in electrostatics. The first row of Figure 27.6 shows the simplest form of each symmetry. The second row shows a more complex, but more realistic, situation with the same symmetry. We may not know the field strength, but the field *shape* in these more complex situations must match the symmetry of the charge distribution.

> **NOTE** ▶ Figures must be finite in extent, but the planes and cylinders in Figure 27.6 are assumed to be infinite. ◀

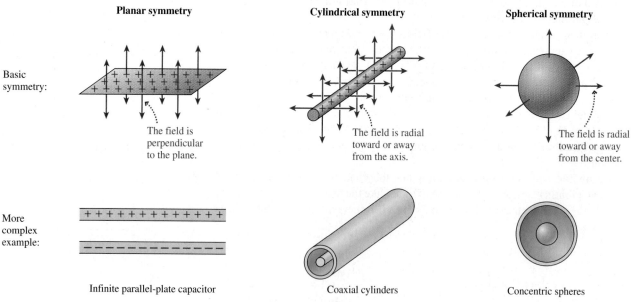

FIGURE 27.6 Three fundamental symmetries.

Objects do exist that are extremely close to being perfect spheres, but no real cylinder or plane can be infinite in extent. Even so, the fields of infinite planes and cylinders are good models for the fields of finite planes and cylinders at points not too close to an edge or an end. Planar and cylindrical electrodes are common in a vast number of practical devices, so the fields that we'll study in this chapter, even if idealized, are not without important applications.

STOP TO THINK 27.1   A uniformly charged rod has a *finite* length *L*. The rod is symmetric under rotations about the axis and under reflection in any plane containing the axis. It is *not* symmetric under translations or under reflections in a plane perpendicular to the axis unless that plane bisects the rod. Which field shape or shapes match the symmetry of the rod?

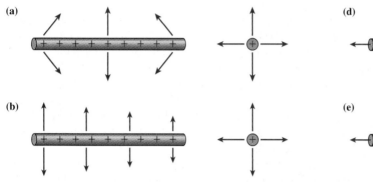

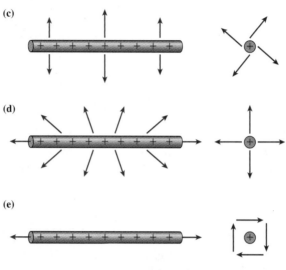

27.2 The Concept of Flux

Figure 27.7a shows an opaque box surrounding a region of space. We can't see what's in the box, but there's an electric field vector coming out of each face of the box. Can you figure out what's in the box?

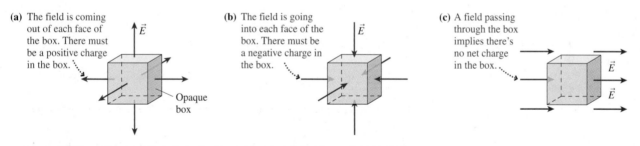

(a) The field is coming out of each face of the box. There must be a positive charge in the box.

Opaque box

(b) The field is going into each face of the box. There must be a negative charge in the box.

(c) A field passing through the box implies there's no net charge in the box.

FIGURE 27.7 Although we can't see into the boxes, the electric fields passing through the faces tell us something about what's in them.

Of course you can. Because electric fields point away from positive charges, and the electric field is coming out of every face of the box, it seems clear that the box contains a positive charge or charges. Similarly, the box in Figure 27.7b certainly contains a negative charge.

What can we tell about the box in Figure 27.7c? The electric field points into the box on the left. An equal electric field points out on the right. This might be the electric field between a large positive electrode somewhere out of sight on the left and a large negative electrode off to the right. An electric field passes through the box, but we see no evidence there's any charge (or at least any net charge) inside the box.

These examples suggest that the electric field as it passes into, out of, or through the box is in some way connected to the charge within the box. However, these simple pictures don't tell us how much charge there is or where within the box the charge is located. Perhaps a better box would be more informative.

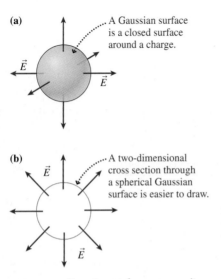

(a)

A Gaussian surface is a closed surface around a charge.

(b)

A two-dimensional cross section through a spherical Gaussian surface is easier to draw.

FIGURE 27.8 Gaussian surface surrounding a charge. A two-dimensional cross section is usually easier to draw.

Suppose we surround a region of space with a *closed surface,* a surface that divides space into distinct inside and outside regions. Within the context of electrostatics, a closed surface through which an electric field passes is called a **Gaussian surface,** named after the 19th-century mathematician Karl Gauss who developed the mathematical foundations of geometry. This is an imaginary, mathematical surface, not a physical surface, although it might coincide with a physical surface. For example, Figure 27.8a shows a spherical Gaussian surface surrounding a charge.

A closed surface must, of necessity, be a surface in three dimensions. But three-dimensional pictures are hard to draw, so we'll often look at two-dimensional cross sections through a Gaussian surface, such as the one shown in Figure 27.8b. Now, a better choice of box makes it more clear what's inside. We can tell from the *spherical symmetry* of the electric field vectors poking through the surface that the positive charge inside must be spherically symmetric and centered at the *center* of the sphere. Notice two features that will soon be important: The electric field is everywhere *perpendicular* to the spherical surface and has the *same magnitude* at each point on the surface.

Figure 27.9 shows another example. An electric field emerges from four sides of the cube in Figure 27.9a but not from the top or bottom. We might be able to guess what's within the box, but we can't be sure. Figure 27.9b uses a different Gaussian surface, a *closed* cylinder (i.e., the cylindrical wall *and* the flat ends), and Figure 27.9c simplifies the drawing by showing two-dimensional end and side views. Now, with a better choice of surface, we can tell that the cylindrical Gaussian surface surrounds some kind of cylindrical charge distribution, such as a charged wire. Again, the electric field is everywhere *perpendicular* to the cylindrical surface and has the *same magnitude* at each point on the surface.

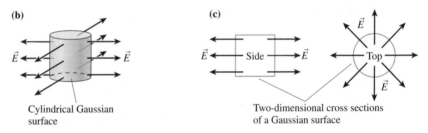

(a)

Cubic Gaussian surface

(b)

Cylindrical Gaussian surface

(c)

Side

Top

Two-dimensional cross sections of a Gaussian surface

FIGURE 27.9 Gaussian surface is most useful when it matches the shape of the field.

For contrast, consider the spherical surface in Figure 27.10a. This is also a Gaussian surface, and the protruding electric field tells us there's a positive charge inside. It might be a point charge located on the left side, but we can't really say. A Gaussian surface that doesn't match the symmetry of the charge distribution isn't terribly useful.

The nonclosed surface of Figure 27.10b doesn't provide much help either. What appears to be a uniform electric field to the right could be due to a large positive plate on the left, a large negative plate on the right, or both. A nonclosed surface doesn't provide enough information.

These examples lead us to two conclusions:

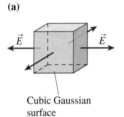

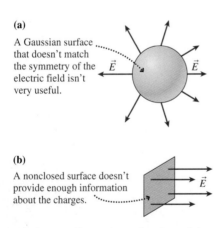

(a)

A Gaussian surface that doesn't match the symmetry of the electric field isn't very useful.

(b)

A nonclosed surface doesn't provide enough information about the charges.

FIGURE 27.10 Not every surface is useful for learning about charge.

1. The electric field, in some sense, "flows" *out of* a closed surface surrounding a region of space containing a net positive charge and *into* a closed surface surrounding a net negative charge. The electric field may flow *through* a closed surface surrounding a region of space in which there is no net charge, but the *net flow* is zero.
2. The electric field pattern through the surface is particularly simple if the closed surface matches the symmetry of the charge distribution inside.

Now, it's true that the electric field doesn't flow like a fluid, but the metaphor is a useful one. The Latin word for flow is *flux,* and the amount of electric field passing through a surface is called the **electric flux.** Our first conclusion, stated in terms of electric flux, is

- There is an outward flux through a closed surface around a net positive charge.
- There is an inward flux through a closed surface around a net negative charge.
- There is no net flux through a closed surface around a region of space in which there is no net charge.

This chapter has been entirely qualitative thus far as we've established pictorially what we mean by symmetry, the idea of flux, and the fact that the electric flux through a closed surface has something to do with the charge inside. Understanding these qualitative ideas is essential, but to go further we need to make these ideas quantitative and precise. In the next section, you'll learn how to calculate the electric flux through a surface. Then, in the section following that, we'll establish a precise relationship between the net flux through a Gaussian surface and the enclosed charge. That relationship, Gauss's law, will allow us to determine the electric fields of some interesting and useful charge distributions.

STOP TO THINK 27.2   This box contains

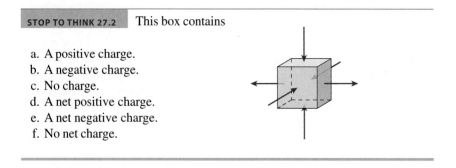

a. A positive charge.
b. A negative charge.
c. No charge.
d. A net positive charge.
e. A net negative charge.
f. No net charge.

27.3 Calculating Electric Flux

Let's start with a brief overview of where this section will take us. We'll begin with a definition of flux that is easy to understand, then we'll turn that simple definition into a formidable-looking integral. We need the integral because the simple definition applies only to uniform electric fields and flat surfaces. Those are a good starting point, but we'll soon need to calculate the flux of nonuniform fields through curved surfaces.

Actiy
Physics 11.7

Mathematically, the flux of a nonuniform field through a curved surface is described by a special kind of integral called a *surface integral.* It's quite possible that you have not yet encountered surface integrals in your calculus course, and the "novelty factor" contributes to making this integral look worse than it really is. We will emphasize over and over the idea that an integral is just a fancy way of doing a sum, in this case the sum of the small amounts of flux through many small pieces of a surface.

The good news is that *every* surface integral we need to evaluate in this chapter, or that you will need to evaluate for the homework problems, is either zero or is so easy that you will be able to do it in your head. This seems like an astounding claim, but you will soon see it is true. The key will be to make effective use of the *symmetry* of the electric field.

Now that you've been warned, you needn't panic at the sight of the mathematical notation that will be introduced. We'll go step by step, and you'll see that, at least as far as electrostatics is concerned, calculating the electric flux is not difficult.

The Basic Definition of Flux

Imagine holding a rectangular wire loop of area A in front of a fan. As Figure 27.11 shows, the volume of air flowing through the loop each second depends on the angle between the loop and the direction of flow. The flow is maximum through a loop that is perpendicular to the air flow; no air goes through the same loop if it lies parallel to the flow.

(a)

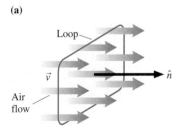

The air flowing through the loop is maximum when $\theta = 0°$.

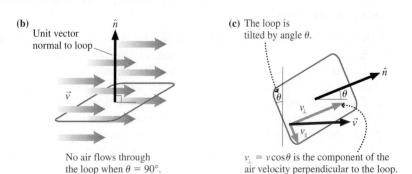

(b)
Unit vector normal to loop

No air flows through the loop when $\theta = 90°$.

(c) The loop is tilted by angle θ.

$v_\perp = v\cos\theta$ is the component of the air velocity perpendicular to the loop.

FIGURE 27.11 The amount of air flowing through a loop depends on the angle between $\vec{v}$ and $\hat{n}$.

The flow direction is identified by the velocity vector $\vec{v}$. We can identify the loop's orientation by defining a unit vector $\hat{n}$ normal to the plane of the loop. Angle θ is then the angle between $\vec{v}$ and $\hat{n}$. The loop perpendicular to the flow in Figure 27.11a has $\theta = 0°$; the loop parallel to the flow in Figure 27.11b has $\theta = 90°$. You can think of θ as the angle by which a loop has been tilted away from perpendicular.

NOTE ▶ A surface has two sides, so $\hat{n}$ could point either way. We'll choose the side that makes $\theta \leq 90°$. ◀

You can see from Figure 27.11c that the velocity vector $\vec{v}$ can be decomposed into components $v_\perp = v\cos\theta$ perpendicular to the loop and $v_\parallel = v\sin\theta$ parallel to the loop. Only the perpendicular component $v_\perp$ carries air *through* the loop. Consequently, the volume of air flowing through the loop each second is

$$\text{volume of air per second (m}^3/\text{s)} = v_\perp A = vA\cos\theta \qquad (27.1)$$

$\theta = 0°$ is the orientation for maximum flow through the loop, as expected, and no air flows through the loop if it is tilted $90°$.

An electric field doesn't flow in a literal sense, but we can apply the same idea to an electric field passing through a surface. Figure 27.12 shows a surface of area A in a uniform electric field $\vec{E}$. Unit vector $\hat{n}$ is normal to the surface and θ is the angle between $\hat{n}$ and $\vec{E}$. Only the component $E_\perp = E\cos\theta$ passes *through* the surface.

With this in mind, and using Equation 27.1 as an analog, let's define the *electric flux* Φ_e as

$$\Phi_e = E_\perp A = EA\cos\theta \qquad (27.2)$$

The electric flux measures the amount of electric field passing through a surface of area A if the normal to the surface is tilted at angle θ from the field.

Equation 27.2 looks very much like a vector dot product: $\vec{E} \cdot \vec{A} = EA\cos\theta$. For this idea to work, let's define an **area vector** $\vec{A} = A\hat{n}$ to be a vector in the direction of $\hat{n}$—that is, *perpendicular* to the surface—with a magnitude A equal to the area of the surface. Vector $\vec{A}$ has units of m^2. Figure 27.13a shows two area vectors.

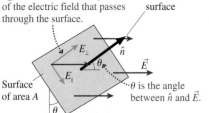

$E_\perp = E\cos\theta$ is the component of the electric field that passes through the surface.

Normal to surface

Surface of area A

θ is the angle between $\hat{n}$ and $\vec{E}$.

FIGURE 27.12 An electric field passing through a surface.

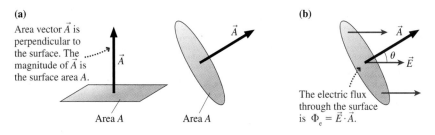

(a) Area vector $\vec{A}$ is perpendicular to the surface. The magnitude of $\vec{A}$ is the surface area A.

$\vec{A}$

Area A

Area A

(b) $\vec{A}$

θ

$\vec{E}$

The electric flux through the surface is $\Phi_e = \vec{E} \cdot \vec{A}$.

FIGURE 27.13 The electric flux can be defined in terms of the area vector $\vec{A}$.

Figure 27.13b shows an electric field passing through a surface of area A. The angle between vectors $\vec{A}$ and $\vec{E}$ is the same angle used in Equation 27.2 to define the electric flux, so Equation 27.2 really is a dot product. We can define the electric flux more concisely as

$$\Phi_e = \vec{E} \cdot \vec{A} \qquad \text{(electric flux of a constant electric field)} \qquad (27.3)$$

Writing the flux as a dot product helps make clear how angle θ is defined: θ is the angle between the electric field and a line *perpendicular* to the plane of the surface.

NOTE ▶ Figure 27.13b shows a circular area, but the shape of the surface is not relevant. However, Equation 27.3 is restricted to a *constant* electric field passing through a *planar* surface. ◀

EXAMPLE 27.1 **The electric flux inside a parallel-plate capacitor**
Two 100 cm² parallel electrodes are spaced 2.0 cm apart. One is charged to +5 nC, the other to −5 nC. A 1.0 cm × 1.0 cm surface between the electrodes is tilted to where its normal makes a 45° angle with the electric field. What is the electric flux through this surface?

MODEL Assume the surface is located near the center of the capacitor where the electric field is uniform. The electric flux doesn't depend on the shape of the surface.

VISUALIZE The surface is square, rather than circular, but otherwise the situation looks like Figure 27.13b.

SOLVE In Chapter 26, we found the electric field inside a parallel-plate capacitor to be

$$E = \frac{Q}{\epsilon_0 A_{\text{plates}}} = \frac{5.0 \times 10^{-9}\,\text{C}}{(8.85 \times 10^{-12}\,\text{C}^2/\text{Nm}^2)(1.0 \times 10^{-2}\,\text{m}^2)}$$

$$= 5.65 \times 10^4\,\text{N/C}$$

A 1.0 cm × 1.0 cm surface has $A = 1.0 \times 10^{-4}\,\text{m}^2$. The electric flux through this surface is

$$\Phi_e = \vec{E} \cdot \vec{A} = EA\cos\theta$$

$$= (5.65 \times 10^4\,\text{N/C})(1.0 \times 10^{-4}\,\text{m}^2)\cos 45°$$

$$= 4.00\,\text{N m}^2/\text{C}$$

ASSESS The units of electric flux are the product of electric field and area units: $\text{N m}^2/\text{C}$.

The Electric Flux of a Nonuniform Electric Field

Our initial definition of the electric flux assumed that the electric field $\vec{E}$ was constant over the surface. How should we calculate the electric flux if $\vec{E}$ varies from point to point on the surface? We can answer this question by returning to the analogy of air flowing through a loop. Suppose the air flow varies from point to point. We can still find the total volume of air passing through the loop each second by dividing the loop into many small areas, finding the flow through each small area, then adding them. Similarly, **the electric flux through a surface can be calculated as the sum of the fluxes through smaller pieces of the surface.** Because flux is a scalar, adding fluxes is easier than adding electric fields.

Figure 27.14 on the next page shows a surface in a nonuniform electric field. Imagine dividing the surface into many small pieces of area δA. Each little area

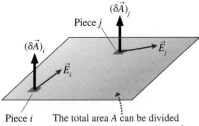

Piece i The total area A can be divided into many small pieces of area δA. $\vec{E}$ may be different at each piece.

FIGURE 27.14 A surface in a nonuniform electric field.

has an area vector $\delta\vec{A}$ perpendicular to the surface. Two of the little pieces, labeled i and j, are shown in the figure. The electric fluxes through these two pieces differ because the electric fields are different.

Consider the small piece i where the electric field is $\vec{E}_i$. The small electric flux $\delta\Phi_i$ through area $(\delta\vec{A})_i$ is

$$\delta\Phi_i = \vec{E}_i \cdot (\delta\vec{A})_i \tag{27.4}$$

The flux through every other little piece of the surface is found the same way. The total electric flux through the entire surface is then the sum of the fluxes through each of the small areas:

$$\Phi_e = \sum_i \delta\Phi_i = \sum_i \vec{E}_i \cdot (\delta\vec{A})_i \tag{27.5}$$

Now let's go to the limit $\delta\vec{A} \to d\vec{A}$. That is, the little areas become infinitesimally small, and there are infinitely many of them. Then the sum becomes an integral, and the electric flux through the surface is

$$\Phi_e = \int_{\text{surface}} \vec{E} \cdot d\vec{A} \tag{27.6}$$

The integral in Equation 27.6 is called a **surface integral.**

Equation 27.6 may look rather frightening if you haven't seen surface integrals before. Despite its appearance, a surface integral is no more complicated than integrals you know from calculus. After all, what does $\int f(x)\,dx$ really mean? This expression is a shorthand way to say "Divide the x-axis into many little segments of length δx, evaluate the function $f(x)$ in each of them, then add up $f(x)\,\delta x$ for all the segments along the line." The integral in Equation 27.6 differs only in that we're dividing a surface into little pieces instead of a line into little segments. In particular, we're summing the fluxes through a vast number of very tiny pieces

You may be thinking, "OK, I understand the idea, but I don't know what to *do*. In calculus, I learned formulas for evaluating integrals such as $\int x^2 dx$. How do I evaluate a surface integral?" This is a good question. We'll deal with evaluation shortly, and it will turn out that the surface integrals in electrostatics are quite easy to evaluate. But don't confuse *evaluating* the integral with understanding what the integral *means*. The surface integral in Equation 27.6 is simply a shorthand notation for the summation of the electric fluxes through a vast number of very tiny pieces of a surface.

The electric field might be different at every point on the surface, but suppose it isn't. That is, suppose the surface is in a uniform electric field $\vec{E}$. A field that is the same at every single point on a surface is a constant as far as the integration of Equation 27.6 is concerned, so we can take it outside the integral. In that case,

$$\Phi_e = \int_{\text{surface}} \vec{E} \cdot d\vec{A} = \int_{\text{surface}} E\cos\theta\, dA = E\cos\theta \int_{\text{surface}} dA \tag{27.7}$$

The integral that remains in Equation 27.7 tells us to add up all the little areas into which the full surface was subdivided. But the sum of all the little areas is simply the area of the surface:

$$\int_{\text{surface}} dA = A \tag{27.8}$$

This idea—that the surface integral of dA is the area of the surface—is one we'll use to evaluate most of the surface integrals of electrostatics. If we substitute Equation 27.8 into Equation 27.7, we find that the electric flux in a uniform electric field is $\Phi_e = EA\cos\theta$. We already knew this, from Equation 27.2, but it was important to see that the surface integral of Equation 27.6 gives the correct result for the case of a uniform electric field.

The Flux Through a Curved Surface

Most of the Gaussian surfaces we considered in the last section were curved surfaces. Figure 27.15 shows an electric field passing through a curved surface. How do we find the electric flux through this surface? Just as we did for a flat surface!

Divide the surface into many small pieces of area δA. For each, define the area vector $\vec{\delta A}$ perpendicular to the surface *at that point*. Compared to Figure 27.14, the only difference that the curvature of the surface makes is that the $\vec{\delta A}$ are no longer parallel to each other. Find the small electric flux $\delta\Phi_i = \vec{E}_i \cdot (\vec{\delta A})_i$ through each little area, then add them all up. The result, once again, is

$$\Phi_e = \int_{\text{surface}} \vec{E} \cdot d\vec{A} \tag{27.9}$$

We *assumed,* in deriving this expression the first time, that the surface was flat and that all the $\vec{\delta A}$ were parallel to each other. But that assumption wasn't necessary. The *meaning* of Equation 27.9—a summation of the fluxes through a vast number of very tiny pieces—is unchanged if the pieces happen to lie on a curved surface.

We seem to be getting more and more complex, using surface integrals first for nonuniform fields and now for curved surfaces. But consider the two situations shown in Figure 27.16. The electric field $\vec{E}$ in Figure 27.16a is everywhere tangent, or parallel, to the curved surface. We don't need to know the magnitude of $\vec{E}$ to recognize that $\vec{E} \cdot d\vec{A}$ is *zero at every point* on the surface because $\vec{E}$ is perpendicular to $d\vec{A}$ at every point. Thus $\Phi_e = 0$. A tangent electric field never pokes through the surface, so it has no flux through the surface.

The electric field in Figure 27.16b is everywhere perpendicular to the surface *and* has the same magnitude E at every point. $\vec{E}$ differs in direction at different points on a curved surface, but at any particular point $\vec{E}$ is parallel to $d\vec{A}$ and $\vec{E} \cdot d\vec{A}$ is simply $E\,dA$. In this case,

$$\Phi_e = \int_{\text{surface}} \vec{E} \cdot d\vec{A} = \int_{\text{surface}} E\,dA = E \int_{\text{surface}} dA = EA \tag{27.10}$$

In evaluating the integral, the fact that E has the same magnitude at every point on the surface allowed us to bring the constant value out of the integral. We then made use of the fact that the integral of dA over the surface is the surface area A.

We can summarize these two situations with a Tactics Box.

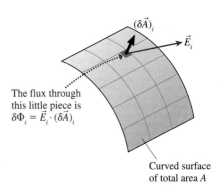

FIGURE 27.15 A curved surface in an electric field.

The flux through this little piece is $\delta\Phi_i = \vec{E}_i \cdot (\vec{\delta A})_i$

Curved surface of total area A

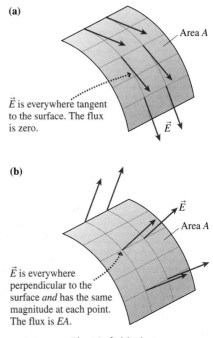

(a)

Area A

$\vec{E}$ is everywhere tangent to the surface. The flux is zero.

(b)

$\vec{E}$

Area A

$\vec{E}$ is everywhere perpendicular to the surface *and* has the same magnitude at each point. The flux is EA.

FIGURE 27.16 Electric fields that are everywhere tangent to or everywhere perpendicular to a curved surface.

TACTICS BOX 27.1 Evaluating surface integrals

❶ If the electric field is everywhere tangent to a surface, the electric flux through the surface is $\Phi_e = 0$.

❷ If the electric field is everywhere perpendicular to a surface *and* has the same magnitude E at every point, the electric flux through the surface is $\Phi_e = EA$.

These two results will be of immeasurable value for using Gauss's law because *every* flux we'll need to calculate will be one of these situations. This is the basis for our earlier claim that the evaluation of surface integrals is not going to be difficult.

The Electric Flux through a Closed Surface

Our final step, to calculate the electric flux through a closed surface such as a box, a cylinder, or a sphere, requires nothing new. We've already learned how to calculate the electric flux through flat and curved surfaces, and a closed surface is nothing more than a surface that happens to be closed.

However, the mathematical notation for the surface integral over a closed surface differs slightly from what we've been using. It is customary to use a little circle on the integral sign to indicate that the surface integral is to be performed over a closed surface. With this notation, the electric flux through a closed surface is

$$\Phi_e = \oint \vec{E} \cdot d\vec{A} \qquad (27.11)$$

Only the notation has changed. The electric flux is still simply the summation of the fluxes through a vast number of very tiny pieces, pieces that now cover a closed surface.

> **NOTE** ▶ A closed surface has a distinct inside and outside. The area vector $d\vec{A}$ is defined to always point *toward the outside*. This removes an ambiguity that was present for a single surface, where $d\vec{A}$ could point to either side. ◀

EXAMPLE 27.2 Calculating the electric flux through a closed cylinder

A cylindrical charge distribution has created the electric field $\vec{E} = E_0(r^2/r_0^2)\hat{r}$, where E_0 and r_0 are constants and where unit vector $\hat{r}$ lies in the xy-plane. Calculate the electric flux through a closed cylinder of length L and radius R that is centered along the z-axis.

MODEL The electric field extends radially outward from the z-axis with cylindrical symmetry. The z-component is $E_z = 0$. The cylinder is a Gaussian surface.

VISUALIZE Figure 27.17a is a view of the electric field looking along the z-axis. The field strength increases with increasing radial distance, and it's symmetric about the z-axis. Figure 27.17b is the closed Gaussian surface for which we need to calculate the electric flux. We can place the cylinder anywhere along the z-axis because the electric field extends forever in that direction.

SOLVE To calculate the flux, divide the closed cylinder into three surfaces: the top, the bottom, and the cylindrical wall. The electric field is tangent to the surface at every point on the top and bottom surfaces. Hence, according to Step 1 in Tactics Box 27.1, the flux through those two surfaces is zero. For the cylindrical wall, the electric field is perpendicular to the surface at every point *and* has the constant magnitude $E = E_0(R^2/r_0^2)$ at every point on the surface. Thus, from Step 2 in Tactics Box 27.1,

$$\Phi_{\text{wall}} = EA_{\text{wall}}$$

If we add up the three pieces, the net flux through the closed surface is

$$\Phi_e = \oint \vec{E} \cdot d\vec{A} = \Phi_{\text{top}} + \Phi_{\text{bottom}} + \Phi_{\text{wall}} = 0 + 0 + EA_{\text{wall}}$$

$$= EA_{\text{wall}}$$

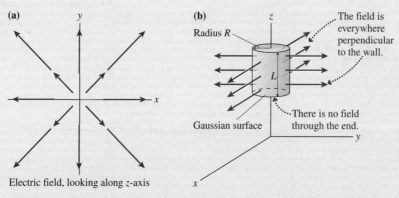

(a)
Electric field, looking along z-axis

(b)
Radius R
The field is everywhere perpendicular to the wall.
L
Gaussian surface
There is no field through the end.

FIGURE 27.17 The electric field and the closed surface through which we will calculate the electric flux.

We've evaluated the surface integral, using the two steps in Tactics Box 27.1, and there was nothing to it! To finish, all we need to recall is that the surface area of a cylindrical wall is circumference × height, or $A_{wall} = 2\pi RL$. Thus

$$\Phi_e = \left(E_0\frac{R^2}{r_0^2}\right)(2\pi RL) = \frac{2\pi LR^3}{r_0^2}E_0$$

ASSESS LR^3/r_0^2 has units of m^2, an area, so this expression for Φ_e has units of Nm^2/C. These are the correct units for electric flux, giving us confidence in our answer. Notice the important role played by symmetry. The electric field was perpendicular to the wall and of constant value at every point on the wall *because* the Gaussian surface had the same symmetry as the charge distribution. We would not have been able to evaluate the surface integral in such an easy way for a surface of any other shape. Symmetry is the key.

Example 27.2 illustrated a two-step approach to performing a flux integral over a closed surface. In summary:

TACTICS BOX 27.2 Finding the flux through a closed surface

❶ Divide the closed surface into pieces that are everywhere tangent to the electric field and everywhere perpendicular to the electric field.
❷ Use Tactics Box 27.1 to evaluate the surface integrals over these surfaces, then add the results.

STOP TO THINK 27.3 The total electric flux through this box is

a. $0\,Nm^2/C$.   b. $1\,Nm^2/C$.   c. $2\,Nm^2/C$.
d. $4\,Nm^2/C$.   e. $6\,Nm^2/C$.   f. $8\,Nm^2/C$.

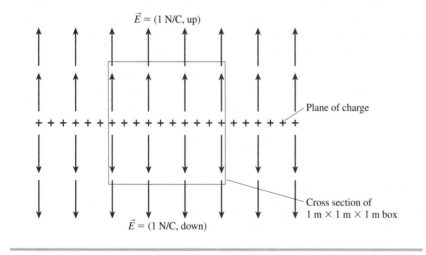

$\vec{E} = (1\,N/C, up)$

Plane of charge

Cross section of $1\,m \times 1\,m \times 1\,m$ box

$\vec{E} = (1\,N/C, down)$

27.4 Gauss's Law

The last section was long, but knowing how to calculate the electric flux through a closed surface is essential for the main topic of this chapter: Gauss's law. Gauss's law is equivalent to Coulomb's law for static charges, although Gauss's law will look very different.

Activ Physics ONLINE 11.8

The purpose of learning Gauss's law is twofold:

- Gauss's law allows the electric fields of some continuous distributions of charge to be found much more easily than does Coulomb's law.
- Gauss's law is valid for *moving* charges, but Coulomb's law is not (although it's a very good approximation for velocities that are much less than the speed of light). Thus Gauss's law is ultimately a more fundamental statement about electric fields than is Coulomb's law.

Let's start with Coulomb's law for the electric field of a point charge. Figure 27.18 shows a spherical Gaussian surface of radius r centered on a positive charge q. Keep in mind that this is an imaginary, mathematical surface, not a physical surface. There is a net flux through this surface because the electric field points outward at every point on the surface. To evaluate the flux, given formally by the surface integral of Equation 27.11, notice that the electric field is perpendicular to the surface at every point on the surface *and,* from Coulomb's law, it has the same magnitude $E = q/4\pi\epsilon_0 r^2$ at every point on the surface. This simple situation arises because **the Gaussian surface has the same symmetry as the electric field.**

Thus we know, without having to do any hard work, that the flux integral is

$$\Phi_e = \oint \vec{E} \cdot d\vec{A} = EA_{\text{sphere}} \qquad (27.12)$$

The surface area of a sphere of radius r is $A_{\text{sphere}} = 4\pi r^2$. If we use A_{sphere} and the Coulomb-law expression for E in Equation 27.12, we find that the electric flux through the spherical surface is

$$\Phi_e = \frac{q}{4\pi\epsilon_0 r^2} 4\pi r^2 = \frac{q}{\epsilon_0} \qquad (27.13)$$

You should examine the logic of this calculation closely. We really did evaluate the surface integral of Equation 27.11, although it may appear, at first, as if we didn't do much. The integral was easily evaluated, we reiterate for emphasis, because the closed surface on which we performed the integration matched the *symmetry* of the charge distribution. In such cases, the surface integral for the flux is simply a field strength multiplied by a surface area.

NOTE ▶ We found Equation 27.13 for a positive charge, but it applies equally to negative charges. According to Equation 27.13, Φ_e is negative if q is negative. And that's what we would expect from the basic definition of flux, $\vec{E} \cdot \vec{A}$. The electric field of a negative charge points inward while the area vector of a closed surface points outward, making the dot product negative. ◀

The Electric Flux Is Independent of the Surface Shape and Radius

Notice something interesting about Equation 27.13. The electric flux depends on the amount of charge but *not* on the radius of the sphere. Although this may seem a bit surprising, it's really a direct consequence of what we *mean* by flux. Think of the fluid analogy with which we introduced the term "flux." If fluid flows outward from a central point, all the fluid crossing a small-radius spherical surface will, at some later time, cross a large-radius spherical surface. No fluid is lost along the way, and no new fluid is created. Similarly, the point charge in Figure 27.19 is the only source of electric field. Every electric field line passing through a small-radius spherical surface also passes through a large-radius spherical surface. Hence the electric flux is independent of r.

NOTE ▶ This argument hinges on the fact that Coulomb's law is an inverse-square force law. The electric field strength, which is proportional to $1/r^2$, decreases with distance. But the surface area, which increases in proportion

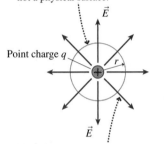

Cross section of a Gaussian sphere of radius r. This is a mathematical surface, not a physical surface.

$\vec{E}$

Point charge q

$\vec{E}$

The electric field is everywhere perpendicular to the surface *and* has the same magnitude at every point.

FIGURE 27.18 A spherical Gaussian surface surrounding a point charge.

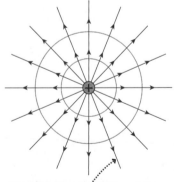

Every field line passing through the smaller sphere also passes through the larger sphere. Hence the flux through the two spheres is the same.

FIGURE 27.19 The electric flux is the same through *every* sphere centered on a point charge.

to r^2, exactly compensates for this decrease. Consequently, the electric flux of a point charge through a spherical surface is independent of the radius of the sphere. ◄

This conclusion about the flux has an extremely important generalization. Figure 27.20a shows a point charge and a closed Gaussian surface of arbitrary shape and dimensions. All we know is that the charge is *inside* the surface. What is the electric flux through this surface?

One way to answer the question is to approximate the surface as a patchwork of spherical and radial pieces. The spherical pieces are centered on the charge and the radial pieces lie along lines extending outward from the charge. The figure, of necessity, has shown fairly large pieces that don't match the actual surface all that well. However, we can make this approximation as good as we want by letting the pieces become sufficiently small.

The electric field is everywhere tangent to the radial pieces. Hence the electric flux through the radial pieces is zero. The spherical pieces, although at varying distances from the charge, form a *complete sphere*. That is, any line drawn radially outward from the charge will pass through exactly one spherical piece, and no radial lines can avoid passing through a spherical piece. You could even imagine, as Figure 27.20b shows, sliding the spherical pieces in and out *without changing the angle they subtend* until they come together to form a complete sphere.

Consequently, the electric flux through these spherical pieces that, when assembled, form a complete sphere must be exactly the same as the flux q/ϵ_0 through a spherical Gaussian surface. In other words, **the flux through** *any* **closed surface surrounding a point charge** q **is**

$$\Phi_e = \oint \vec{E} \cdot d\vec{A} = \frac{q}{\epsilon_0} \qquad (27.14)$$

This surprisingly simple result is a consequence of the fact that Coulomb's law is an inverse-square force law. Even so, the reasoning that got us to Equation 27.14 is rather subtle and well worth reviewing.

Charge Outside the Surface

The closed surface shown in Figure 27.21a has a point charge q outside the surface but no charges inside. Now what can we say about the flux? By approximating this surface with spherical and radial pieces *centered on the charge,* as we did in Figure 27.20, we can reassemble the surface into the equivalent surface of Figure 27.21b. This closed surface consists of sections of two spherical shells, and it is equivalent in the sense that the electric flux through this surface is the same as the electric flux through the original surface of Figure 27.21a.

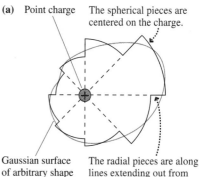

(a) Point charge   The spherical pieces are centered on the charge.

Gaussian surface of arbitrary shape

The radial pieces are along lines extending out from the charge. There's no flux through these.

The approximation with spherical and radial pieces can be as good as desired by letting the pieces become sufficiently small.

(b)

The spherical pieces can slide in or out to form a complete sphere. Hence the flux through the pieces is the same as the flux through a sphere.

FIGURE 27.20 An arbitrary Gaussian surface can be approximated with spherical and radial pieces.

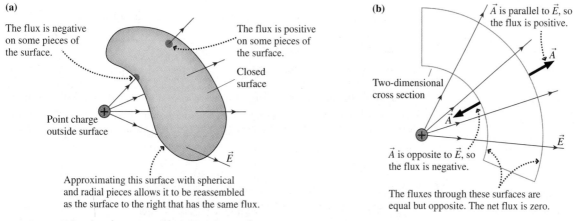

(a)

The flux is negative on some pieces of the surface.

The flux is positive on some pieces of the surface.

Closed surface

Point charge outside surface

$\vec{E}$

Approximating this surface with spherical and radial pieces allows it to be reassembled as the surface to the right that has the same flux.

(b)

$\vec{A}$ is parallel to $\vec{E}$, so the flux is positive.

$\vec{A}$

Two-dimensional cross section

$\vec{A}$

$\vec{E}$

$\vec{A}$ is opposite to $\vec{E}$, so the flux is negative.

The fluxes through these surfaces are equal but opposite. The net flux is zero.

FIGURE 27.21 A point charge outside a Gaussian surface.

If the electric field were a fluid flowing outward from the charge, all the fluid *entering* the closed region through the first spherical surface would later *exit* the region through the second spherical surface. There is no *net* flow into or out of the closed region. Similarly, every electric field line entering this closed volume through one spherical surface exits through the other spherical surface.

Mathematically, the electric fluxes through the two spherical surfaces have the same magnitude because Φ_e is independent of r. But they have *opposite signs* because the outward-pointing area vector $\vec{A}$ is parallel to $\vec{E}$ on one surface but opposite to $\vec{E}$ on the other. The sum of the fluxes through the two surfaces is zero, and we are led to the conclusion that **the net electric flux is zero through a closed surface that does not contain any net charge.** Charges outside the surface do not produce a net flux through the surface.

This isn't to say that the flux through a small piece of the surface is zero. In fact, as Figure 27.21a shows, nearly every piece of the surface has an electric field either entering or leaving and thus has a nonzero flux. But some of these are positive and some are negative. When summed over the *entire* surface, the positive and negative contributions exactly cancel to give no *net* flux.

Multiple Charges

Finally, consider an arbitrary Gaussian surface and a group of charges q_1, q_2, q_3, ... such as those shown in Figure 27.22. Some of these charges are inside the surface, others outside. The charges can be either positive or negative. What is the net electric flux through the closed surface?

By definition, the net flux is

$$\Phi_e = \oint \vec{E} \cdot d\vec{A}$$

From the principle of superposition, the electric field is $\vec{E} = \vec{E}_1 + \vec{E}_2 + \vec{E}_3 + \cdots$, where $\vec{E}_1, \vec{E}_2, \vec{E}_3, \ldots$ are the fields of the individual charges. Thus the flux can be written

$$\Phi_e = \oint \vec{E}_1 \cdot d\vec{A} + \oint \vec{E}_2 \cdot d\vec{A} + \oint \vec{E}_3 \cdot d\vec{A} + \cdots$$
$$= \Phi_1 + \Phi_2 + \Phi_3 + \cdots \tag{27.15}$$

where $\Phi_1, \Phi_2, \Phi_3, \ldots$ are the fluxes through the Gaussian surface due to the individual charges. That is, the net flux is the sum of the fluxes due to individual charges. But we know what those are: q/ϵ_0 for the charges inside the surface and zero for the charges outside. Thus

$$\Phi_e = \left(\frac{q_1}{\epsilon_0} + \frac{q_2}{\epsilon_0} + \cdots + \frac{q_i}{\epsilon_0} \text{ for all charges inside the surface} \right) \tag{27.16}$$
$$+ (0 + 0 + \cdots + 0 \text{ for all charges outside the surface})$$

Define

$$Q_{in} = q_1 + q_2 + \cdots + q_i \text{ for all charges inside the surface} \tag{27.17}$$

as the total charge enclosed *within* the surface. With this definition, we can write our result for the net electric flux in a very neat and compact fashion. For any *closed* surface enclosing total charge Q_{in}, the net electric flux through the surface is

$$\Phi_e = \oint \vec{E} \cdot d\vec{A} = \frac{Q_{in}}{\epsilon_0} \tag{27.18}$$

This result for the electric flux is known as **Gauss's law.**

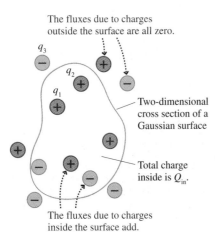

The fluxes due to charges outside the surface are all zero.

q_3
q_2
q_1

Two-dimensional cross section of a Gaussian surface

Total charge inside is Q_{in}.

The fluxes due to charges inside the surface add.

FIGURE 27.22 Charges both inside and outside a Gaussian surface.

What Does Gauss's Law Tell Us?

In one sense, Gauss's law doesn't say anything new or anything that we didn't already know from Coulomb's law. After all, we derived Gauss's law from Coulomb's law. But in another sense, Gauss's law is more important than Coulomb's law. Gauss's law states a very general property of electric fields—namely, that charges create electric fields in just such a way that the net flux of the field is the same through *any* surface surrounding the charges, no matter what its size and shape may be. This fact may have been implied by Coulomb's law, but it was by no means obvious. And Gauss's law will turn out to be particularly useful later when we combine it with other electric and magnetic field equations.

Gauss's law is the mathematical statement of our observations in Section 27.2. There we noticed a net "flow" of electric field out of a closed surface containing charges. Gauss's law quantifies this idea by making a specific connection between the "flow," now measured as electric flux, and the amount of charge.

But is it useful? Although to some extent Gauss's law is a formal statement about electric fields, not a tool for solving practical problems, there are exceptions: Gauss's law will allow us to find the electric fields of some very important and very practical charge distributions much more easily than if we had to rely on Coulomb's law. We'll consider some examples in the next section.

STOP TO THINK 27.4 These are two-dimensional cross sections through three-dimensional closed spheres and a cube. Rank order, from largest to smallest, the electric fluxes Φ_a to Φ_e through surfaces a to e.

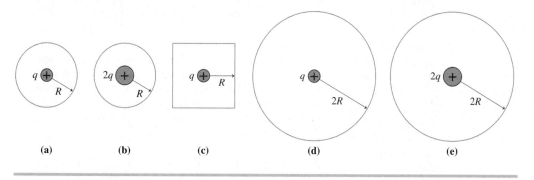

(a) (b) (c) (d) (e)

27.5 Using Gauss's Law

In this section, we'll use Gauss's law to determine the electric fields of several important charge distributions. Some of these you already know, from Chapter 26; others will be new. Several important observations can be made about using Gauss's law:

1. Gauss's law applies only to a *closed* surface, called a Gaussian surface.
2. A Gaussian surface is not a physical surface. It need not coincide with the boundary of any physical object (although it could if we wished). It is an imaginary, mathematical surface in the space surrounding one or more charges.
3. We can't find the electric field from Gauss's law alone. We need to apply Gauss's law in situations where, from symmetry and superposition, we already can guess the *shape* of the field.

These observations and our previous discussion of symmetry and flux lead to the following strategy for solving electric field problems with Gauss's law.

PROBLEM-SOLVING STRATEGY 27.1 **Gauss's law**

MODEL Model the charge distribution as a distribution with symmetry.

VISUALIZE Draw a picture of the charge distribution.

- Determine the symmetry of its electric field.
- Choose and draw a Gaussian surface with the *same symmetry*.
- You need not enclose all the charge within the Gaussian surface.
- Be sure every part of the Gaussian surface is either tangent to or perpendicular to the electric field.

SOLVE The mathematical representation is based on Gauss's law

$$\Phi_e = \oint \vec{E} \cdot d\vec{A} = \frac{Q_{in}}{\epsilon_0}$$

- Use Tactics Boxes 27.1 and 27.2 to evaluate the surface integral.

ASSESS Check that your result has the correct units, is reasonable, and answers the question.

EXAMPLE 27.3 **Outside a sphere of charge**

In Chapter 26 we asserted, without proof, that the electric field outside a sphere of total charge Q is the same as the field of a point charge Q at the center. Use Gauss's law to prove this result.

MODEL The charge distribution within the sphere need not be uniform (i.e., the charge density might increase or decrease with r), but it must have spherical symmetry in order for us to use Gauss's law. We will assume that it does.

VISUALIZE Figure 27.23 shows a sphere of charge Q and radius R. We want to find $\vec{E}$ outside this sphere, for distances $r > R$. The spherical symmetry of the charge distribution tells us that the electric field must point *radially outward* from the sphere. Although Gauss's law is true for any surface surrounding the charged sphere, it is useful only if we choose a Gaussian surface to match the spherical symmetry of the charge distribution and the field. Thus a spherical surface of radius $r > R$ *concentric with* the charged sphere will be our Gaussian surface. Because this surface surrounds the entire sphere of charge, the enclosed charge is simply $Q_{in} = Q$.

SOLVE Gauss's law is

$$\Phi_e = \oint \vec{E} \cdot d\vec{A} = \frac{Q_{in}}{\epsilon_0} = \frac{Q}{\epsilon_0}$$

To calculate the flux, notice that the electric field is everywhere perpendicular to the spherical surface. And although we don't know the electric field magnitude E, spherical symmetry dictates that E must have the same value at all points equally distant from the center of the sphere. Thus we have the simple result that the net flux through the Gaussian surface is

$$\Phi_e = EA_{sphere} = 4\pi r^2 E$$

where we used the fact that the surface area of a sphere is $A_{sphere} = 4\pi r^2$. With this result for the flux, Gauss's law is

$$4\pi r^2 E = \frac{Q}{\epsilon_0}$$

Thus the electric field at distance r outside a sphere of charge is

$$E_{outside} = \frac{1}{4\pi\epsilon_0} \frac{Q}{r^2}$$

Or in vector form, making use of the fact that $\vec{E}$ is radially outward,

$$\vec{E}_{outside} = \frac{1}{4\pi\epsilon_0} \frac{Q}{r^2} \hat{r}$$

where $\hat{r}$ is a radial unit vector.

ASSESS The field is exactly that of a point charge Q, which is what we wanted to show.

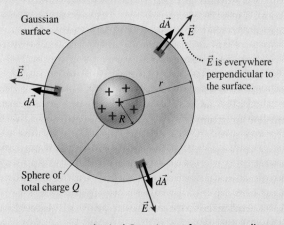

Gaussian surface

$d\vec{A}$ $\vec{E}$

$\vec{E}$ is everywhere perpendicular to the surface.

$\vec{E}$

$d\vec{A}$

r

R

Sphere of total charge Q

$d\vec{A}$

$\vec{E}$

FIGURE 27.23 A spherical Gaussian surface surrounding a sphere of charge.

The derivation of the electric field of a sphere of charge depended crucially on a proper choice of the Gaussian surface. We would not have been able to evaluate the flux integral so simply for any other choice of surface. It's worth noting that the result of Example 27.3 can also be proven by the superposition of point-charge fields, but it requires a difficult three-dimensional integral and about a page of algebra. We obtained the answer using Gauss's law in just a few lines. Where Gauss's law works, it works *extremely* well! However, it works only in situations, such as this, with a very high degree of symmetry.

EXAMPLE 27.4 Inside a sphere of charge

What is the electric field *inside* a uniformly charged sphere?

MODEL We haven't considered a situation like this before. To begin, we don't know if the field strength is increasing or decreasing as we move outward from the center of the sphere. But the field inside must have spherical symmetry. That is, the field must point radially inward or outward, and the field strength can depend only on r. This is sufficient information to solve the problem because it allows us to choose a Gaussian surface.

VISUALIZE Figure 27.24 shows a spherical Gaussian surface with radius $r \leq R$ *inside*, and *concentric with*, the sphere of charge. This surface matches the symmetry of the charge distribution, hence $\vec{E}$ is perpendicular to this surface and the field strength E has the same value at all points on the surface.

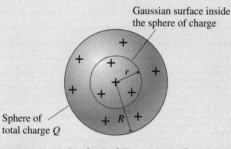

Gaussian surface inside the sphere of charge

Sphere of total charge Q

FIGURE 27.24 A spherical Gaussian surface inside a uniform sphere of charge.

SOLVE The flux integral is identical to that of Example 27.3:

$$\Phi_e = EA_{sphere} = 4\pi r^2 E$$

Consequently, Gauss's law is

$$\Phi_e = 4\pi r^2 E = \frac{Q_{in}}{\epsilon_0}$$

The difference between this example and Example 27.3 is that Q_{in} is no longer the total charge of the sphere. Instead, Q_{in} is the amount of charge *inside* the Gaussian sphere of radius r. Because the charge distribution is *uniform*, the ratio Q_{in}/Q must be the same as the ratio V_r/V_R of the volumes of the spheres of radii r and R. That is,

$$\frac{Q_{in}}{Q} = \frac{V_r}{V_R} = \frac{\frac{4}{3}\pi r^3}{\frac{4}{3}\pi R^3} = \frac{r^3}{R^3}$$

from which we find Q_{in} to be

$$Q_{in} = \frac{r^3}{R^3}Q$$

The amount of enclosed charge increases with the cube of the distance r from the center and, as expected, equals $Q_{in} = Q$ if $r = R$. With this expression for Q_{in}, Gauss's law is

$$4\pi r^2 E = \frac{(r^3/R^3)Q}{\epsilon_0}$$

Thus the electric field at radius r inside a uniformly charged sphere is

$$E_{inside} = \frac{1}{4\pi\epsilon_0}\frac{Q}{R^3}r$$

The electric field strength inside the sphere increases *linearly* with the distance r from the center.

ASSESS The field inside and the field outside a sphere of charge match at the boundary of the sphere, $r = R$, where both give $E = Q/4\pi\epsilon_0 R^2$. In other words, the field strength is *continuous* as we cross the boundary of the sphere. These results are shown graphically in Figure 27.25.

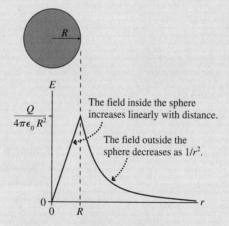

The field inside the sphere increases linearly with distance.

The field outside the sphere decreases as $1/r^2$.

FIGURE 27.25 The electric field strength of a uniform sphere of charge of radius R.

EXAMPLE 27.5 The electric field of a long, charged wire
In Chapter 26, we used superposition to find the electric field of an infinitely long line of charge with linear charge density (C/m) λ. It was not an easy derivation. Find the electric field using Gauss's law.

MODEL A long, charged wire can be modeled as an infinitely long line of charge.

VISUALIZE Figure 27.26 shows an infinitely long line of charge. We can use the symmetry of the situation to see that the only possible shape of the electric field is to point straight into or out from the wire, rather like the bristles on a bottle brush. The shape of the field suggests that we choose our Gaussian surface to be a cylinder of radius r and length L, centered on the wire. Because Gauss's law refers to *closed* surfaces, we must include the ends of the cylinder as part of the surface.

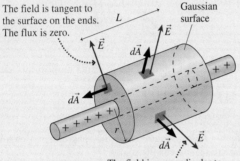

The field is tangent to the surface on the ends. The flux is zero.

Gaussian surface

The field is perpendicular to the surface on the cylinder wall.

FIGURE 27.26 A Gaussian surface around a charged wire.

SOLVE Gauss's law is

$$\Phi_e = \oint \vec{E} \cdot d\vec{A} = \frac{Q_{in}}{\epsilon_0}$$

where Q_{in} is the charge *inside* the closed cylinder. We have two tasks: to evaluate the flux integral, and to determine how much charge is inside the closed surface. The wire has linear charge density λ, so the amount of charge inside a cylinder of length L is simply

$$Q_{in} = \lambda L$$

Finding the net flux is just as straightforward. We can divide the flux through the entire closed surface into the flux through each end plus the flux through the cylindrical wall. The electric field $\vec{E}$, pointing straight out from the wire, is tangent to the end surfaces at every point. Thus the flux through these two surfaces is zero. On the wall, $\vec{E}$ is perpendicular to the surface and has the same strength E at every point. Thus

$$\Phi_e = \Phi_{top} + \Phi_{bottom} + \Phi_{wall} = 0 + 0 + EA_{cyl} = 2\pi rLE$$

where we used $A_{cyl} = 2\pi rL$ as the surface area of a cylindrical wall of radius r and length L. Once again, the proper choice of the Gaussian surface reduces the flux integral merely to finding a surface area. With these expressions for Q_{in} and Φ_e, Gauss's law is

$$\Phi_e = 2\pi rLE = \frac{Q_{in}}{\epsilon_0} = \frac{\lambda L}{\epsilon_0}$$

Thus the electric field at distance r from a long, charged wire is

$$E_{wire} = \frac{\lambda}{2\pi\epsilon_0 r}$$

ASSESS This agrees exactly with the result of the more complex derivation in Chapter 26. Notice that the result does not depend on our choice of L. A Gaussian surface is an imaginary device, not a physical object. We needed a finite-length cylinder to do the flux calculation, but the electric field of an *infinitely* long wire can't depend on the length of an imaginary cylinder.

Example 27.5, for the electric field of a long, charged wire contains a subtle but important idea, one that often occurs when using Gauss's law. The Gaussian cylinder of length L encloses only some of the wire's charge. The pieces of the charged wire outside the cylinder are not enclosed by the Gaussian surface and consequently do not contribute anything to the net flux. Even so, *they are essential* to the use of Gauss's law because it takes the *entire* charged wire to produce an electric field with cylindrical symmetry. In other words, the wire outside the cylinder may not contribute to the flux, but it affects the *shape* of the electric field. Our ability to write $\Phi_e = EA_{cyl}$ depended on knowing that E is the same at every point on the wall of the cylinder. That would not be true for a charged wire of finite length, so we cannot use Gauss's law to find the electric field of a finite-length charged wire.

EXAMPLE 27.6 **The electric field of a plane of charge**

Use Gauss's law to find the electric field of an infinite plane of charge with surface charge density (C/m^2) η.

MODEL A uniformly charged flat electrode can be modeled as an infinite plane of charge.

VISUALIZE Figure 27.27 shows a uniformly charged plane with surface charge density η. We will assume that the plane extends infinitely far in all directions, although we obviously have to show "edges" in our drawing. The planar symmetry allows the electric field to point only straight toward or away from the plane. With this in mind, choose as a Gaussian surface a cylinder with length L and faces of area A centered on the plane of charge. Although we've drawn them as circular, the shape of the faces is not relevant.

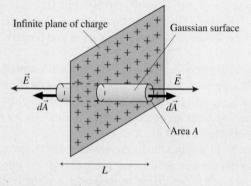

FIGURE 27.27 The Gaussian surface extends to both sides of a plane of charge.

SOLVE The electric field is perpendicular to both faces of the cylinder, so the total flux through both faces is $\Phi_{faces} = 2EA$. (The fluxes add rather than cancel because the area vector $\vec{A}$ points *outward* on each face.) There's *no* flux through the wall of the cylinder because the field vectors are tangent to the wall. Thus the net flux is simply

$$\Phi_e = 2EA$$

The charge inside the cylinder is the charge contained in area A of the plane. This is

$$Q_{in} = \eta A$$

With these expressions for Q_{in} and Φ_e, Gauss's law is

$$\Phi_e = 2EA = \frac{Q_{in}}{\epsilon_0} = \frac{\eta A}{\epsilon_0}$$

Thus the electric field of an infinite charged plane is

$$E_{plane} = \frac{\eta}{2\epsilon_0}$$

This agrees with the result of Chapter 26.

ASSESS Again, the details of how we chose the Gaussian surface don't affect the result. Notice that this is another example of a Gaussian surface enclosing only some of the charge. Most of the plane's charge is outside the Gaussian surface and does not contribute to the flux, but it does affect the shape of the field. We wouldn't have planar symmetry, with the electric field exactly perpendicular to the plane, without all the rest of the charge on the plane.

The plane of charge is an especially good example of how powerful Gauss's law can be. Finding the electric field of a plane of charge via superposition was a difficult and tedious derivation. With Gauss's law, once you see how to apply it, the problem is simple enough to solve in your head!

You might wonder, then, why we bothered with superposition at all. The reason is that Gauss's law, powerful though it may be, is effective only in a very limited number of situations where the field is highly symmetrical. Superposition always works, even if the derivation is messy, because superposition goes directly back to the fields of individual point charges. It's good to use Gauss's law when you can, but superposition is often the only way to attack real-world charge distributions.

STOP TO THINK 27.5 Which Gaussian surface would allow you to use Gauss's law to determine the electric field outside a uniformly charged cube?

a. A sphere whose center coincides with the center of the charged cube.
b. A cube whose center coincides with the center of the charged cube and which has parallel faces.
c. Either a or b.
d. Neither a nor b.

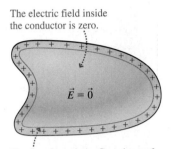

The electric field inside the conductor is zero.

$$\vec{E} = \vec{0}$$

The flux through the Gaussian surface is zero. There's no net charge inside the conductor. Hence all the excess charge is on the surface.

FIGURE 27.28 A Gaussian surface just inside a conductor that's in electrostatic equilibrium.

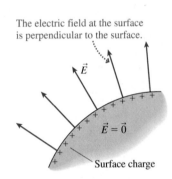

The electric field at the surface is perpendicular to the surface.

$$\vec{E}$$

$$\vec{E} = \vec{0}$$

Surface charge

FIGURE 27.29 The electric field at the surface of a charged conductor.

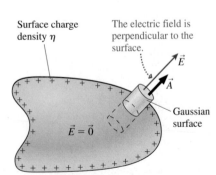

Surface charge density η

The electric field is perpendicular to the surface.

$$\vec{E}$$

$$\vec{A}$$

Gaussian surface

$$\vec{E} = \vec{0}$$

FIGURE 27.30 A Gaussian surface extending through the surface of the conductor has a flux only through the outer face.

27.6 Conductors in Electrostatic Equilibrium

Consider a charged conductor, such as a charged metal electrode, in electrostatic equilibrium. That is, there is no current through the conductor and the charges are all stationary. In Chapter 25, you learned that **the electric field is zero at all points within a conductor in electrostatic equilibrium.** That is, $\vec{E}_{in} = \vec{0}$. If this weren't true, the electric field would cause the charge carriers to move and thus violate the assumption that all the charges are at rest. Let's use Gauss's law to see what else we can learn.

At the Surface of a Conductor

Figure 27.28 shows a Gaussian surface just barely inside the physical surface of a conductor that's in electrostatic equilibrium. The electric field is zero at all points within the conductor, hence the electric flux Φ_e through this Gaussian surface must be zero. But if $\Phi_e = 0$, Gauss's law tells us that $Q_{in} = 0$. That is, there's no net charge within this surface. There are charges—electrons and positive ions—but no *net* charge.

If there's no net charge in the interior of a conductor in electrostatic equilibrium, then **all the excess charge on a charged conductor resides on the exterior surface of the conductor.** Any charges added to a conductor quickly spread across the surface until reaching positions of electrostatic equilibrium, but there is no net charge *within* the conductor.

There may be no electric field within a charged conductor, but the presence of net charge requires an exterior electric field in the space outside the conductor. Figure 27.29 shows that **the electric field right at the surface of the conductor has to be perpendicular to the surface.** To see that this is so, suppose $\vec{E}_{surface}$ had a component tangent to the surface. This component of $\vec{E}_{surface}$ would exert a force on the surface charges and cause a surface current, thus violating the assumption that all charges are at rest. The only exterior electric field consistent with electrostatic equilibrium is one that is perpendicular to the surface.

We can use Gauss's law to relate the field strength at the surface to the charge density on the surface. Figure 27.30 shows a small Gaussian cylinder with faces very slightly above and below the surface of a charged conductor. The charge inside this Gaussian cylinder is ηA, where η is the surface charge density at this point on the conductor. There's a flux $\Phi = AE_{surface}$ through the outside face of this cylinder but, unlike Example 27.6 for the plane of charge, *no* flux through the inside face because $\vec{E}_{in} = \vec{0}$ within the conductor. Furthermore, there's no flux through the wall of the cylinder because $\vec{E}_{surface}$ is perpendicular to the surface. Thus the net flux is $\Phi_e = AE_{surface}$. Gauss's law is

$$\Phi_e = AE_{surface} = \frac{Q_{in}}{\epsilon_0} = \frac{\eta A}{\epsilon_0} \tag{27.19}$$

from which we can conclude that the electric field at the surface of a charged conductor is

$$\vec{E}_{surface} = \left(\frac{\eta}{\epsilon_0}, \text{perpendicular to surface}\right) \tag{27.20}$$

In general, the surface charge density η is *not* constant on the surface of a conductor but varies in a complicated way that depends on the shape of the conductor. If we can determine η, either by calculating it or measuring it, then Equation 27.20 tells us the electric field at that point on the surface. Alternatively, we can use

Equation 27.20 to deduce the charge density on the conductor's surface if we know the electric field just outside the conductor.

Charges and Fields within a Conductor

Figure 27.31 shows a charged conductor with a hole inside. Can there be charge on this interior surface? To find out, place a Gaussian surface around the hole, infinitesimally close but entirely within the conductor. The electric flux Φ_e through this Gaussian surface is zero because the electric field is zero everywhere inside the conductor. Thus we must conclude that $Q_{in} = 0$. There's no net charge inside this Gaussian surface and thus no charge on the surface of the hole. Any excess charge resides on the *exterior* surface of the conductor, not on any interior surfaces.

Furthermore, because there's no electric field inside the conductor and no charge inside the hole, the electric field inside the hole must also be zero. This conclusion has an important practical application. For example, suppose we need to exclude the electric field from the region in Figure 27.32a enclosed within dotted lines. We can do so by surrounding this region with the neutral conducting box of Figure 27.32b.

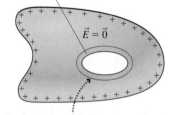

A hollow completely enclosed by the conductor

The flux through the Gaussian surface is zero. There's no net charge inside, hence no charge on this interior surface.

FIGURE 27.31 A Gaussian surface surrounding a hole inside a conductor in electrostatic equilibrium.

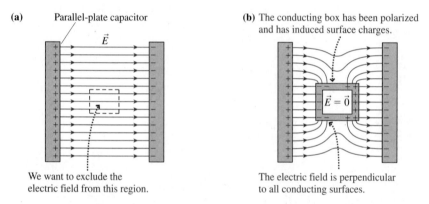

(a) Parallel-plate capacitor

$\vec{E}$

We want to exclude the electric field from this region.

(b) The conducting box has been polarized and has induced surface charges.

$\vec{E} = \vec{0}$

The electric field is perpendicular to all conducting surfaces.

FIGURE 27.32 The electric field can be excluded from a region of space by surrounding it with a conducting box.

This region of space is now a hole inside a conductor, thus the interior electric field is zero. The use of a conducting box to exclude electric fields from a region of space is called **screening.** Solid metal walls are ideal, but in practice wire screen or wire mesh provides sufficient screening for all but the most sensitive applications. The price we pay is that the exterior field is now very complicated.

Finally, Figure 27.33 shows a charge q inside a hole within a neutral conductor. The electric field *within* the conductor is still zero, hence the electric flux through the Gaussian surface is zero. But $\Phi_e = 0$ requires $Q_{in} = 0$. Consequently, the charge inside the hole attracts an equal charge of opposite sign, and charge $-q$ now lines the inner surface of the hole.

The conductor as a whole is neutral, so moving $-q$ to the surface of the hole must leave $+q$ behind somewhere else. Where is it? It can't be in the interior of the conductor, as we've seen, and that leaves only the exterior surface. In essence, an internal charge polarizes the conductor just as an external charge would. Net charge $-q$ moves to the inner surface and net charge $+q$ is left behind on the exterior surface.

In summary, conductors in electrostatic equilibrium have the properties described in Tactics Box 27.3 on the next page.

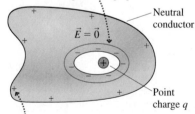

The flux through the Gaussian surface is zero, hence there's no *net* charge inside this surface. There must be charge $-q$ on the inside surface to balance point charge q.

Neutral conductor

$\vec{E} = \vec{0}$

Point charge q

The outer surface must have charge $+q$ in order that the conductor remain neutral.

FIGURE 27.33 A charge in the hole causes a net charge on the interior and exterior surfaces.

EXAMPLE 27.7 **The electric field at the surface of a charged metal sphere**

A 2.0-cm-diameter brass sphere has been given a charge of 2.0 nC. What is the electric field strength at the surface?

MODEL Brass is a conductor. The excess charge resides on the surface.

VISUALIZE The charge distribution has spherical symmetry. The electric field points radially outward from the surface.

SOLVE We can solve this problem two ways. One uses the fact that a sphere is the one shape for which any excess charge will spread out to a *uniform* surface charge density. Thus

$$\eta = \frac{q}{A_{\text{sphere}}} = \frac{q}{4\pi R^2} = \frac{2.0 \times 10^{-9}\,\text{C}}{4\pi (0.010\,\text{m})^2} = 1.59 \times 10^{-6}\,\text{C/m}^2$$

From Equation 27.20, we know the electric field at the surface has strength

$$E_{\text{surface}} = \frac{\eta}{\epsilon_0} = \frac{1.59 \times 10^{-6}\,\text{C/m}^2}{8.85 \times 10^{-12}\,\text{C}^2/\text{N m}^2} = 180{,}000\,\text{N/C}$$

Alternatively, we could have used the result, obtained earlier in the chapter, that the electric field strength outside a sphere of charge Q is $E_{\text{outside}} = Q_{\text{in}}/(4\pi\epsilon_0 r^2)$. But $Q_{\text{in}} = q$ and, at the surface, $r = R$. Thus

$$E_{\text{surface}} = \frac{1}{4\pi\epsilon_0} \frac{q}{R^2} = (9.0 \times 10^9\,\text{N m}^2/\text{C}^2)\frac{2.0 \times 10^{-9}\,\text{C}}{(0.010\,\text{m})^2}$$

$$= 180{,}000\,\text{N/C}$$

As we can see, both methods lead to the conclusion that $E_{\text{surface}} = 180{,}000$ N/C.

SUMMARY

The goal of Chapter 27 has been to understand and apply Gauss's law.

GENERAL PRINCIPLES

Gauss's Law

For any *closed* surface enclosing net charge Q_{in}, the net electric flux through the surface is

$$\Phi_e = \oint \vec{E} \cdot d\vec{A} = \frac{Q_{in}}{\epsilon_0}$$

The electric flux Φ_e is the same for *any* closed surface enclosing charge Q_{in}.

Symmetry

The symmetry of the electric field must match the symmetry of the charge distribution.

In practice, Φ_e is computable only if the symmetry of the Gaussian surface matches the symmetry of the charge distribution.

IMPORTANT CONCEPTS

Charge creates the electric field that is responsible for the electric flux.

Q_{in} is the sum of all enclosed charges. This charge contributes to the flux.

Gaussian surface

Charges outside the surface contribute to the electric field, but they don't contribute to the flux.

Flux is the amount of electric field passing through a surface of area A.

$$\Phi_e = \vec{E} \cdot \vec{A}$$

where $\vec{A}$ is the **area vector.**

For closed surfaces:
A net flux in or out indicates that the surface encloses a net charge. Field lines through but with no *net* flux mean that the surface encloses no *net* charge.

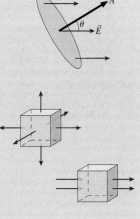

Surface integrals calculate the flux by summing the fluxes through many small pieces of the surface.

$$\Phi_e = \sum \vec{E} \cdot \delta\vec{A}$$
$$\rightarrow \int \vec{E} \cdot d\vec{A}$$

Two important situations:
If the electric field is everywhere tangent to the surface, then

$$\Phi_e = 0$$

If the electric field is everywhere perpendicular to the surface *and* has the same strength E at all points, then

$$\Phi_e = EA$$

APPLICATIONS

Conductors in electrostatic equilibrium

- The electric field is zero at all points within the conductor.
- Any excess charge resides entirely on the exterior surface.
- The external electric field is perpendicular to the surface and of magnitude η/ϵ_0, where η is the surface charge density.
- The electric field is zero inside any hole within a conductor unless there is a charge in the hole.

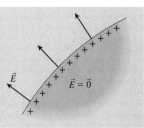

TERMS AND NOTATION

symmetric	electric flux, Φ_e	surface integral	screening
Gaussian surface	area vector, $\vec{A}$	Gauss's law	

EXERCISES AND PROBLEMS

Exercises

Section 27.1 Symmetry

1. Figure Ex27.1 shows two cross sections of two infinitely long coaxial cylinders. The inner cylinder has a positive charge, the outer cylinder has an equal negative charge. Draw this figure on your paper, then draw electric field vectors showing the shape of the electric field.

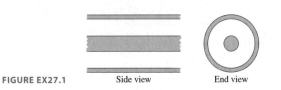

FIGURE EX27.1   Side view   End view

2. Figure Ex27.2 shows a cross section of two concentric spheres. The inner sphere has a negative charge. The outer sphere has a positive charge larger in magnitude than the charge on the inner sphere. Draw this figure on your paper, then draw electric field vectors showing the shape of the electric field.

FIGURE EX27.2

3. Figure Ex27.3 shows a cross section of two infinite parallel planes of charge. Draw this figure on your paper, then draw electric field vectors showing the shape of the electric field.

++++++++++++++++++++

FIGURE EX27.3   ++++++++++++++++++++

Section 27.2 The Concept of Flux

4. The electric field is constant over each face of the cube shown in Figure Ex27.4. Does the box contain positive charge, negative charge, or no charge? Explain.

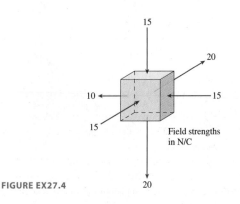

FIGURE EX27.4

5. The electric field is constant over each face of the cube shown in Figure Ex27.5. Does the box contain positive charge, negative charge, or no charge? Explain.

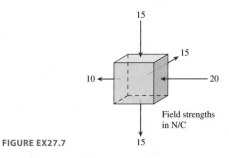

FIGURE EX27.5

6. The cube in Figure Ex27.6 contains negative charge. The electric field is constant over each face of the cube. Does the missing electric field vector on the front face point in or out? What is the minimum possible field strength?

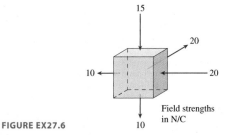

FIGURE EX27.6

7. The cube in Figure Ex27.7 contains no net charge. The electric field is constant over each face of the cube. Does the missing electric field vector on the front face point in or out? What is the field strength?

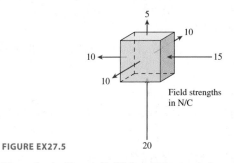

FIGURE EX27.7

Section 27.3 Calculating Electric Flux

8. What is the electric flux through the surface shown in Figure Ex27.8?

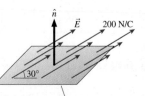

FIGURE EX27.8   10 cm × 10 cm

9. What is the electric flux through the surface shown in Figure Ex27.9?

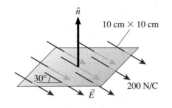

FIGURE EX27.9

10. The electric flux through the surface shown in Figure Ex27.10 is 15.0 Nm^2/C. What is the electric field strength?

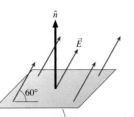

FIGURE EX27.10

10 cm × 10 cm

11. A 2.0 cm × 3.0 cm rectangle lies in the xy-plane. What is the electric flux through the rectangle if
    a. $\vec{E} = (50\hat{i} + 100\hat{k})$ N/C?
    b. $\vec{E} = (50\hat{i} + 100\hat{j})$ N/C?
12. A 2.0 cm × 3.0 cm rectangle lies in the xz-plane. What is the electric flux through the rectangle if
    a. $\vec{E} = (50\hat{i} + 100\hat{k})$ N/C?
    b. $\vec{E} = (50\hat{i} + 100\hat{j})$ N/C?
13. A 4.0-cm-diameter circle lies in the xy-plane in a region where the electric field is $\vec{E} = (1000\hat{i} + 1000\hat{j} + 1000\hat{k})$ N/C. What is the electric flux through the circle?
14. A 1.0 cm × 1.0 cm × 1.0 cm box is between the plates of a parallel-plate capacitor with two faces of the box perpendicular to $\vec{E}$. The electric field strength is 1000 N/C. What is the net electric flux through the box?
15. What is the net electric flux through the two cylinders shown in Figure Ex27.15?

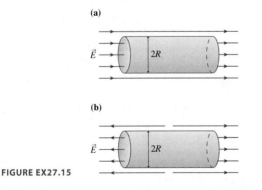

FIGURE EX27.15

Section 27.4 Gauss's Law

Section 27.5 Using Gauss's Law

16. The net electric flux through a closed surface is $-1000 \, Nm^2/C$. How much charge is enclosed within the surface?
17. 55.3 million excess electrons are inside a closed surface. What is the net electric flux through the surface?

18. A 10 nC point charge is at the center of a 2.0 m × 2.0 m × 2.0 m cube. What is the electric flux through the top surface of the cube?
19. The electric flux through each face of a 2.0 m × 2.0 m × 2.0 m cube is 100 Nm^2/C. How much charge is inside the cube?
20. Figure Ex27.20 shows three charges. Draw these charges on your paper four times. Then draw two-dimensional cross sections of three-dimensional closed surfaces through which the electric flux is (a) $2q/\epsilon_0$, (b) $3q/\epsilon_0$, (c) 0, and (d) $-q/\epsilon_0$.

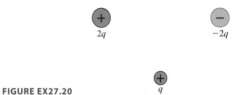

FIGURE EX27.20

21. Figure Ex27.21 shows three charges. Draw these charges on your paper four times. Then draw two-dimensional cross sections of three-dimensional closed surfaces through which the electric flux is (a) $-q/\epsilon_0$, (b) q/ϵ_0, (c) $3q/\epsilon_0$, and (d) $4q/\epsilon_0$.

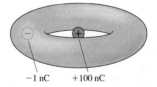

FIGURE EX27.21

22. What is the net electric flux through the torus (i.e., doughnut shape) of Figure Ex27.22?

FIGURE EX27.22   −1 nC   +100 nC

23. What is the net electric flux through the cylinder of Figure Ex27.23?

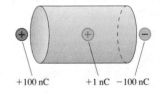

FIGURE EX27.23   +100 nC   +1 nC −100 nC

Section 27.6 Conductors in Electrostatic Equilibrium

24. The electric field strength just above one face of a copper penny is 2000 N/C. What is the surface charge density on this face of the penny?
25. A spark occurs at the tip of a metal needle if the electric field strength exceeds 3×10^6 N/C, the field strength at which air breaks down. What is the minimum surface charge density for producing a spark?

26. The conducting box in Figure Ex27.26 has been given an excess negative charge. The surface density of excess electrons at the center of the top surface is 5.0×10^{10} electrons/m². What are the electric field strengths E_1 to E_3 at points 1 to 3?

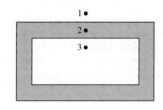

FIGURE EX27.26

Problems

27. Figure P27.27 shows the four sides of a $3.0 \text{ cm} \times 3.0 \text{ cm} \times 3.0 \text{ cm}$ cube.
    a. What are the electric fluxes Φ_1 to Φ_4 through sides 1 to 4?
    b. What is the net flux through these four sides?

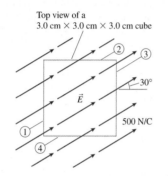

FIGURE P27.27

28. Find the electric fluxes Φ_1 to Φ_5 through surfaces 1 to 5 in Figure P27.28.

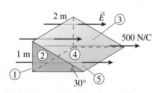

FIGURE P27.28

29. A tetrahedron has an equilateral triangle base with 20-cm-long edges and three equilateral triangle sides. The base is parallel to the ground, and a vertical uniform electric field of strength 200 N/C passes upward through the tetrahedron.
    a. What is the electric flux through the base?
    b. What is the electric flux through each of the three sides?
30. Figure P27.30 shows three Gaussian surfaces and the electric flux through each. What are the three charges q_1, q_2, and q_3?

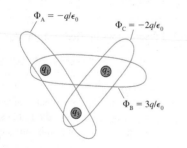

FIGURE P27.30

31. Charges $q_1 = -4Q$ and $q_2 = +2Q$ are located at $x = -a$ and $x = +a$, respectively. What is the net electric flux through a sphere of radius $2a$ centered at the origin?
32. A spherically symmetric charge distribution produces the electric field $\vec{E} = (5000r^2)\hat{r}$ N/C, where r is in m.
    a. What is the electric field strength at $r = 20$ cm?
    b. What is the electric flux through a 40-cm-diameter spherical surface that is concentric with the charge distribution?
    c. How much charge is inside this 40-cm-diameter spherical surface?
33. A spherically symmetric charge distribution produces the electric field $\vec{E} = (200/r)\hat{r}$ N/C, where r is in m.
    a. What is the electric field strength at $r = 10$ cm?
    b. What is the electric flux through a 20-cm-diameter spherical surface that is concentric with the charge distribution?
    c. How much charge is inside this 20-cm-diameter spherical surface?
34. A thin, horizontal 10 cm × 10 cm copper plate is charged with 1.0×10^{10} electrons. If the electrons are uniformly distributed on the surface, what are the strength and direction of the electric field
    a. 0.1 mm above the center of the top surface of the plate?
    b. at the plate's center of mass?
    c. 0.1 mm below the center of the bottom surface of the plate?
35. An initially neutral conductor contains a hollow cavity in which there is a +100 nC point charge. A charged rod transfers −50 nC to the conductor. Afterward, what is the charge (a) on the inner wall of the cavity wall, and (b) on the exterior surface of the conductor?
36. Figure P27.36 shows a hollow cavity within a neutral conductor. A point charge Q is inside the cavity. What is the net electric flux through the closed surface that surrounds the conductor?

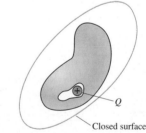

FIGURE P27.36   Closed surface

37. A 20-cm-radius ball is uniformly charged to 80 nC.
    a. What is the ball's charge density (C/m³)?
    b. How much charge is enclosed by spheres of radii 5, 10, and 20 cm?
    c. What is the electric field strength at points 5, 10, and 20 cm from the center?
38. A hollow metal sphere has inner radius a and outer radius b. The hollow sphere has charge $+2Q$. A point charge $+Q$ sits at the center of the hollow sphere.
    a. Determine the electric fields in the three regions $r \leq a$, $a < r < b$, and $r \geq b$.
    b. How much charge is on the inside surface of the hollow sphere? On the exterior surface?
39. Figure P27.39 shows a solid metal sphere at the center of a hollow metal sphere. What is the total charge on (a) the

exterior of the inner sphere, (b) the inside surface of the hollow sphere, and (c) the exterior surface of the hollow sphere?

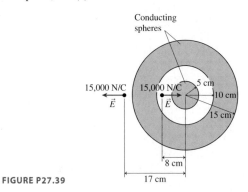

Conducting spheres

15,000 N/C $\vec{E}$ 15,000 N/C $\vec{E}$ 5 cm 10 cm 15 cm 8 cm 17 cm

FIGURE P27.39

40. The earth has a vertical electric field at the surface, pointing down, that averages 100 N/C. This field is maintained by various atmospheric processes, including lightning. What is the excess charge on the surface of the earth?

41. Figure 27.32b showed a conducting box inside a parallel-plate capacitor. The electric field inside the box is $\vec{E} = \vec{0}$. Suppose the surface charge on the exterior of the box could be frozen. Draw a picture of the electric field inside the box after the box, with its frozen charge, is removed from the capacitor.
    Hint: Superposition.

42. A hollow metal sphere has 6 cm and 10 cm inner and outer radii, respectively. The surface charge density on the inside surface is -100 nC/m². The surface charge density on the exterior surface is $+100$ nC/m². What are the strength and direction of the electric field at points 4, 8, and 12 cm from the center?

43. A positive point charge q sits at the center of a hollow spherical shell. The shell, with radius R and negligible thickness, has net charge $-2q$. Find an expression for the electric field strength (a) inside the sphere, $r < R$, and (b) outside the sphere, $r > R$. In what direction does the electric field point in each case?

44. Find the electric field inside and outside a hollow plastic ball of radius R that has charge Q uniformly distributed on its outer surface.

45. A uniformly charged ball of radius a and charge $-Q$ is at the center of a hollow metal shell with inner radius b and outer radius c. The hollow sphere has net charge $+2Q$.
    a. Determine the electric field strength in the four regions $r \le a, a < r < b, b \le r \le c,$ and $r > c$.
    b. Draw a graph of E versus r from $r = 0$ to $r = 2c$.

46. The three parallel planes of charge shown in Figure P27.46 have surface charge densities $-\frac{1}{2}\eta$, η, and $-\frac{1}{2}\eta$. Find the electric fields $\vec{E}_1$ to $\vec{E}_4$ in regions 1 to 4.

$-\frac{1}{2}\eta$ — — — — — — — — 1
η ++++++++++++++++ 2
3
$-\frac{1}{2}\eta$ — — — — — — — — 4

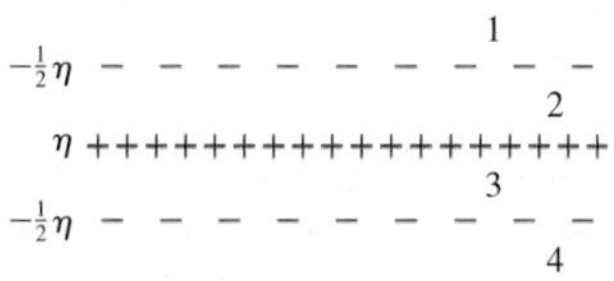

FIGURE P27.46

47. An infinite slab of charge of thickness $2z_0$ lies in the xy-plane between $z = -z_0$ and $z = +z_0$. The charge density ρ_0 (C/m³) is a constant.
    a. Use Gauss's law to find an expression for the electric field strength inside the slab ($-z_0 \le z \le z_0$).
    b. Find an expression for the electric field strength above the slab ($z \ge z_0$).
    c. Sketch a graph of E from $z = 0$ to $z = 3z_0$.

48. Figure P27.48 shows an infinitely wide conductor parallel to and distance d from an infinitely wide plane of charge with surface charge density η. What are the electric fields $\vec{E}_1$ to $\vec{E}_4$ in regions 1 to 4?

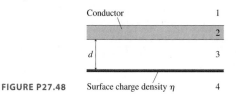

Conductor 1 2 d 3

FIGURE P27.48   Surface charge density η 4

49. Figure P27.49 shows two very large slabs of metal that are parallel and distance d apart. Each slab has a total surface area (top + bottom) A. The thickness of each slab is so small in comparison to its lateral dimensions that the surface area around the sides is negligible. Metal 1 has total charge $Q_1 = Q$ and metal 2 has total charge $Q_2 = 2Q$. Assume Q is positive. In terms of Q and A, determine
    a. The electric field strengths E_1 to E_5 in regions 1 to 5.
    b. The surface charge densities η_a to η_d on the four surfaces a to d.

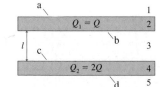

a 1 $Q_1 = Q$ 2 b 3 l c $Q_2 = 2Q$ 4 d 5

FIGURE P27.49

50. A long, thin straight wire with linear charge density λ runs down the center of a thin, hollow metal cylinder of radius R. The cylinder has a net linear charge density 2λ. Assume λ is positive. Find expressions for the electric field strength (a) inside the cylinder, $r < R$, and (b) outside the cylinder, $r > R$. In what direction does the electric field point in each of the cases?

51. A very long, uniformly charged cylinder has radius R and linear charge density λ. Find the cylinder's electric field (a) outside the cylinder, $r \ge R$, and (b) inside the cylinder, $r \le R$. (c) Show that your answers to parts a and b match at the boundary, $r = R$.

52. A spherical shell has inner radius R_{in} and outer radius R_{out}. The shell contains total charge Q, uniformly distributed. The interior of the shell is empty of charge and matter.
    a. Find the electric field outside the shell, $r \ge R_{out}$.
    b. Find the electric field in the interior of the shell, $r \le R_{in}$.
    c. Find the electric field within the shell, $R_{in} \le r \le R_{out}$.
    d. Show that your solutions match at both the inner and outer boundaries.
    e. Draw a graph of E versus r.

53. An early model of the atom, proposed by Rutherford after his discovery of the atomic nucleus, had a positive point charge $+Ze$ (the nucleus) at the center of a sphere of radius R with uniformly distributed negative charge $-Ze$. Z is the atomic number, the number of protons in the nucleus and the number of electrons in the negative sphere.
    a. Show that the electric field inside this atom is

$$E_{in} = \frac{Ze}{4\pi\epsilon_0}\left(\frac{1}{r^2} - \frac{r}{R^3}\right)$$

    b. What is E at the surface of the atom? Is this the expected value? Explain.
    c. A uranium atom has $Z = 92$ and $R = 0.10$ nm. What is the electric field strength at $r = \frac{1}{2}R$?

Challenge Problems

54. All examples of Gauss's law have used highly symmetrical surfaces where the flux integral is either zero or EA. Yet we've claimed that the net $\Phi_e = Q_{in}/\epsilon_0$ is independent of the surface. This is worth checking. Figure CP27.54 shows a cube of edge length L centered on a long thin wire with linear charge density λ. The flux through one face of the cube is *not* simply EA because, in this case, the electric field varies in both strength and direction. But you can calculate the flux by actually doing the flux integral.

    a. Consider the face parallel to the yz-plane. Define area $d\vec{A}$ as a strip of width dy and height L with the vector pointing in the x-direction. One such strip is located at position y. Use the known electric field of a wire to calculate the electric flux $d\Phi$ through this little area. Your expression should be written in terms of y, which is a variable, and various constants. It should not explicitly contain any angles.

    b. Now integrate $d\Phi$ to find the total flux through this face.

    c. Finally, show that the net flux through the cube is $\Phi_e = Q_{in}/\epsilon_0$.

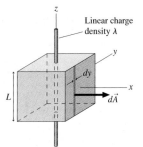

FIGURE CP27.54

Linear charge density λ

55. An infinite cylinder of radius R has a linear charge density λ. The volume charge density (C/m^3) within the cylinder $(r \le R)$ is $\rho(r) = r\rho_0/R$, where ρ_0 is a constant to be determined.

    a. Draw a graph of ρ versus x for an x-axis that crosses the cylinder perpendicular to the cylinder axis. Let x range from $-2R$ to $2R$.

    b. The charge within a small volume dV is $dq = \rho dV$. The integral of ρdV over a cylinder of length L is the total charge $Q = \lambda L$ within the cylinder. Use this fact to show that $\rho_0 = 3\lambda/2\pi R^2$.
    Hint: Let dV be a cylindrical shell of length L, radius r, and thickness dr. What is the volume of such a shell?

    c. Use Gauss's law to find an expression for the electric field E inside the cylinder, $r \le R$.

    d. Does your expression have the expected value at the surface, $r = R$? Explain.

56. A sphere of radius R has total charge Q. The volume charge density (C/m^3) within the sphere is $\rho(r) = C/r^2$, where C is a constant to be determined.

    a. The charge within a small volume dV is $dq = \rho dV$. The integral of ρdV over the entire volume of the sphere is the total charge Q. Use this fact to determine the constant C in terms of Q and R.
    Hint: Let dV be a spherical shell of radius r and thickness dr. What is the volume of such a shell?

    b. Use Gauss's law to find an expression for the electric field E inside the sphere, $r \le R$.

    c. Does your expression have the expected value at the surface, $r = R$? Explain.

57. A sphere of radius R has total charge Q. The volume charge density (C/m^3) within the sphere is

$$\rho = \rho_0\left(1 - \frac{r}{R}\right)$$

    This charge density decreases linearly from ρ_0 at the center to zero at the edge of the sphere.

    a. Show that $\rho_0 = 3Q/\pi R^3$.
    Hint: You'll need to do a volume integral.

    b. Show that the electric field inside the sphere points radially outward with magnitude

$$E = \frac{Qr}{4\pi\epsilon_0 R^3}\left(4 - 3\frac{r}{R}\right)$$

    c. Show that your result of part b has the expected value at $r = R$.

58. A spherical ball of charge has radius R and total charge Q. The electric field strength inside the ball $(r \le R)$ is $E(r) = E_{max}(r^4/R^4)$.

    a. What is E_{max} in terms of Q and R?

    b. Find an expression for the volume charge density (C/m^3) $\rho(r)$ inside the ball as a function of r.

    c. Verify that your charge density gives the total charge Q when integrated over the volume of the ball.

STOP TO THINK ANSWERS

Stop to Think 27.1: a and d. Symmetry requires the electric field to be unchanged if front and back are reversed, if left and right are reversed, or if the field is rotated about the wire's axis. Fields a and d both have the proper symmetry. Other factors would now need to be considered to determine the correct field.

Stop to Think 27.2: e. The net flux is into the box.

Stop to Think 27.3: c. There's no flux through the four sides. The flux is positive $1 \text{ N m}^2/\text{C}$ through both the top and bottom because $\vec{E}$ and $\vec{A}$ both point outward.

Stop to Think 27.4: $\Phi_b = \Phi_e > \Phi_a = \Phi_c = \Phi_d$. The flux through a closed surface depends only on the amount of enclosed charge, not the size or shape of the surface.

Stop to Think 27.5: d. A cube doesn't have enough symmetry to use Gauss's law. The electric field of a charged cube is *not* constant over the face of a cubic Gaussian surface, so we can't evaluate the surface integral for the flux.

28 Current and Conductivity

The transition from candles to lightbulbs was both a technical and a social revolution.

▶ **Looking Ahead**

The goal of Chapter 28 is to learn how and why charge moves through a conductor as what we call a current. In this chapter you will learn to:

- Understand how charge moves through a conductor.
- Understand how an electric field is established inside a current-carrying wire.
- Use a microscopic model of conduction.
- Use the law of conservation of current.
- Relate the current in a wire to the conductivity of the metal.

◀ **Looking Back**

This chapter depends on the basic properties of charges and electric fields and on the parallel-plate capacitor. Please review:

- Sections 25.2–25.3 Charges and conductors.
- Section 26.5 The parallel-plate capacitor.
- Section 26.6 The motion of a charge in an electric field.

Thomas Edison's invention of the lightbulb revolutionized the way people work and play. Can you imagine doing your physics homework by candle light?

Lights, stereos, microwave ovens, and computers are an important part of our contemporary lives. Devices such as these are connected by wires to a battery or an electrical outlet. What is happening *inside* the wire that makes the light come on or the stereo play? And *why* is it happening? We say that "electricity flows through the wire," but what does that statement mean? And equally important, *how do we know?* Simply looking at a wire between a battery and a lightbulb does not reveal whether anything is moving or flowing. As far as visual appearance is concerned, the wire is absolutely the same whether it is "conducting electricity" or not.

The objective of this chapter is to learn about current. We want to understand what moves through a current-carrying wire, and why. We'll also need to establish a connection between an electric current and the charging and electrostatic processes we have studied in the last three chapters.

28.1 The Electron Current

Let's start our exploration of current with a very simple system. Figure 28.1a shows a charged parallel-plate capacitor. We can verify that the plates are charged—one positively and the other negatively—by holding them, one at a time, near charged glass and plastic rods that are hanging from threads. If we now connect the two capacitor plates to each other with a metal wire, as shown in Figure 28.1b, the plates quickly become neutral. We say that the capacitor has been *discharged.*

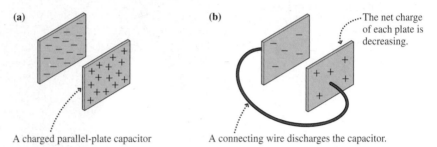

(a)

A charged parallel-plate capacitor

(b)

····The net charge of each plate is decreasing.

A connecting wire discharges the capacitor.

FIGURE 28.1 A capacitor is discharged by a metal wire.

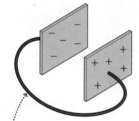

The connecting wire gets warm.

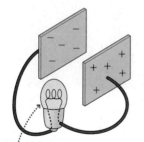

A light bulb glows. The light bulb filament is part of the connecting wire.

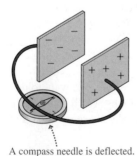

A compass needle is deflected.

FIGURE 28.2 Properties of a current.

The wire is a conductor, a material through which charge easily moves. Apparently the excess charge on one capacitor plate is able to move through the wire to the other plate, neutralizing both plates. In Chapter 25, we defined a **current** as the motion of charge through a conductor. It would seem that the capacitor is discharged by a current in the connecting wire. Our goal in this chapter is to understand how it happens.

First, let's see what else we can observe. Figure 28.2 shows that the connecting wire gets warmer. If the wire is very thin in places, such as the thin filament in a lightbulb, the wire gets hot enough to glow. The current-carrying wire also deflects a compass needle. We will explore the connection between currents and magnetism in Chapter 32. For now, we will use "makes the wire warmer" and "deflects a compass needle" as *indicators* that a current is present in a wire.

Charge Carriers

Opposite charges attract, but the oppositely charged plates of a capacitor don't spontaneously discharge because the charges can't leap from one plate to the other. A connecting wire discharges the capacitor by providing a pathway for charge to move from one side of the capacitor to the other. However, merely observing that a wire discharges a capacitor doesn't answer an important question: Does positive charge move toward the negative plate, or does negative charge move toward the positive plate? *Either* motion would explain the observations we have made.

The charges that move in a current-carrying conductor are called the *charge carriers.* In Chapter 25 we simply *asserted* that the charge carriers in a metal are electrons, but we offered no evidence. One of the first clues was found by J. J. Thompson, the discoverer of the electron. In the 1890s, Thompson found that metals heated until they glow emit electrons. (This *thermal emission* from hot tungsten filaments is now the source of electrons in electron guns.) Thompson's observation suggested that the electrons are moving around inside the metal and can escape if they have sufficient thermal energy.

The *Tolman-Stewart experiment* of 1916 was the first direct evidence that electrons are the charge carriers in metals. Tolman and Stewart caused a metal rod to accelerate very rapidly. As Figure 28.3 shows, the inertia of the charge carriers within the metal (and Newton's first law) causes them to be "thrown" to the rear

When a metal bar accelerates to the right, inertia causes the charge carriers to be displaced to the rear surface. The front surface becomes oppositely charged.

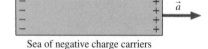

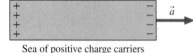
Sea of positive charge carriers   Sea of negative charge carriers

FIGURE 28.3 The Tolman-Stewart experiment to determine the sign of the charge carriers in a metal.

surface of the metal rod as it accelerates away. If the charge carriers are positive, their displacement relative to the metal should cause the rear surface to become positively charged and leave the front surface negatively charged, much as if the metal were polarized by an electric field. Negative charge carriers should give the rear surface a negative charge while leaving the front surface positive.

Tolman and Stewart found that the rear surface of a metal rod becomes negatively charged as it accelerates. The only negatively charged particles are electrons, thus *experimental evidence* tells us that **the charge carriers in metals are electrons.** We noted in Chapter 25 that the electrons act rather like a negatively charged gas or liquid in between the atoms of the lattice. This *model,* called the *sea of electrons,* is illustrated in Figure 28.4. It's not a perfect model, because it overlooks some quantum effects, but it is the basis for a reasonably good description of current in a metal. Notice that the conduction electrons are not attached to any particular atom in the metal.

NOTE ▶ Electrons are the charge carriers in *metals.* Other conductors, such as ionic solutions or semiconductors, have different charge carriers. We will focus on metals, because of their importance to circuits, but don't think that electrons are *always* the charge carrier. ◀

The Electron Current

The conduction electrons in a metal, like molecules in a gas, are moving around quite rapidly, but there is no *net* motion. We can change that by pushing on the sea of electrons, causing the entire sea of electrons to move in one direction like a gas or liquid flowing through a pipe. This net motion, which takes place at what we'll call the **drift speed** v_d, is superimposed on top of the random thermal motions of the individual electrons. The drift speed is quite small, with 10^{-4} m/s being a fairly typical value. We'll develop this model more fully in Section 28.2.

Figure 28.5 shows a cross section through a current-carrying wire in which the entire sea of electrons is moving from left to right at the drift speed. Suppose a microscopic observer could count the electrons that pass through this cross section. Let's define the **electron current** i to be the number of electrons *per second* that pass through a cross section of a wire or other conductor. The units of electron current are s^{-1}. Stated another way, the number N_e of electrons that pass through the cross section during the time interval Δt is

$$N_e = i\Delta t \tag{28.1}$$

It seems likely that the electron current is related to the drift speed. After all, increasing the electron speed should increase the number of electrons that are able to pass through a wire each second. To help us establish a relationship, Figure 28.6 on the next page shows the sea of electrons moving through a wire at the drift speed v_d. This is the *net* speed with which the electrons move, not the speed at which any one electron is bouncing around. The electrons that pass through a particular cross section of the wire during the interval Δt are shaded. How many of them are there?

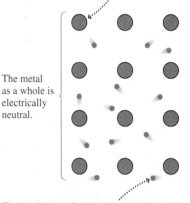

Ions (the metal atoms minus one valence electron) occupy fixed positions in the lattice.

The metal as a whole is electrically neutral.

The conduction electrons (one per atom) are free to move around. They are bound to the solid as a whole, not to any particular atom.

FIGURE 28.4 The sea of electrons is a model of how conduction electrons behave in a metal.

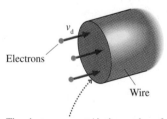

Electrons   Wire

The electron current i is the number of electrons passing through this cross section of the wire per second.

FIGURE 28.5 The electron current.

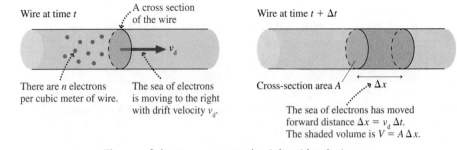

Wire at time t

A cross section of the wire

Wire at time $t + \Delta t$

There are n electrons per cubic meter of wire.

The sea of electrons is moving to the right with drift velocity v_d.

Cross-section area A

Δx

The sea of electrons has moved forward distance $\Delta x = v_\mathrm{d} \Delta t$. The shaded volume is $V = A \Delta x$.

FIGURE 28.6 The sea of electrons moves to the right with velocity v_d.

The electrons travel distance $\Delta x = v_\mathrm{d}\Delta t$ to the right during the interval Δt. This section of wire is a cylinder with volume $V = A \Delta x$. If the *number density* of conduction electrons is n electrons per cubic meter, then the total number of electrons in the cylinder is

$$N_\mathrm{e} = nV = nA\,\Delta x = nAv_\mathrm{d}\Delta t \qquad (28.2)$$

Comparing Equation 28.2 to Equation 28.1, you can see that the electron current in the wire is

$$i = nAv_\mathrm{d} \qquad (28.3)$$

TABLE 28.1 Conduction-electron density in metals

Metal	Electron density (m^{-3})
Aluminum	6.0×10^{28}
Copper	8.5×10^{28}
Iron	8.5×10^{28}
Gold	5.9×10^{28}
Silver	5.8×10^{28}

In most metals, each atom contributes one valence electron to the sea of electrons. Thus the number of conduction electrons per cubic meter is the same as the number of atoms per cubic meter, a quantity that can be determined from the metal's mass density. Table 28.1 gives values of the conduction-electron density n for several common metals. There is not a great deal of variation.

EXAMPLE 28.1 **The size of the electron current**

What is the electron current in a 2.0-mm-diameter copper wire if the electron drift speed is 1.0×10^{-4} m/s?

SOLVE The wire's cross-section area is $A = \pi r^2 = 3.14 \times 10^{-6}$ m^2. Table 28.1 gives the electron density for copper as 8.5×10^{28} m^{-3}. Thus

$i = nAv_\mathrm{d}$

$= (8.5 \times 10^{28} \text{ m}^{-3})(3.14 \times 10^{-6} \text{ m}^2)(1.0 \times 10^{-4} \text{ m/s})$

$= 2.7 \times 10^{19} \text{ s}^{-1}$

ASSESS This is an incredible number of electrons to pass through a section of the wire every second. The number is high not because the sea of electrons moves fast—in fact, it moves at literally a snail's pace—but because the density of electrons is so enormous. This is a fairly typical electron current.

The electron density n and cross-section area A are properties of the wire. We can't change those. Once we choose a particular piece of wire, the only parameter we can vary to control the size of a current is the drift speed v_d. As you'll soon see, the drift speed is determined by the electric field inside the wire.

STOP TO THINK 28.1   These four wires are made of the same metal. Rank in order, from largest to smallest, the electron currents i_a to i_d.

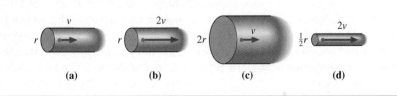

(a)   (b)   (c)   (d)

Conservation of Current

Returning to our capacitor, we can now say with confidence that the wire discharges the capacitor by allowing an electron current to flow from the negative plate, where there is an excess of electrons, to the positive plate. The flow of electrons continues until both plates are electrically neutral.

Figure 28.7 shows a lightbulb in the wire connecting two capacitor plates. The bulb glows while the electron current is discharging the capacitor. How do you think the electron current at point A compares to the electron current at point B? Are the electron currents at these points the same? Or is one larger than the other? Think about this before going on.

You might have predicted that the electron current at B is less than the electron current at A because the bulb, in order to be glowing, must use up some of the current. It's easy to test this prediction. We can compare the electron currents at A and B by comparing how far two compass needles are deflected. Or we could insert two small lightbulbs at A and B, small enough not to significantly influence the main lightbulb in Figure 28.7, and compare their brightness. All methods give the same result: The electron current at point B is *exactly equal* to the electron current at point A. The current leaving a lightbulb is exactly the same as the current entering the lightbulb.

This is an important observation, one that demands an explanation. After all, "something" makes the bulb glow, so why don't we observe a decrease in the electron current? The electron current i is the number of electrons moving through the wire per second. There are only two ways to decrease i: either decrease the number of electrons, or decrease their drift speed through the wire. Electrons are charged particles. The lightbulb can't destroy electrons without violating both the law of conservation of mass and the law of conservation of charge. Thus the *number* of electrons is not changed by the lightbulb.

Can the electrons slow down as they pass through the bulb? This is a little trickier, so first consider the analogous situation shown in Figure 28.8 where we see water flowing through a hose of constant diameter. Suppose the water flows into one end at a speed of 2.0 m/s. Is it possible for the water to flow out the other end at a speed of only 1.5 m/s?

We can't destroy water molecules any more than we can destroy electrons, nor can we increase the density of water by pushing the molecules closer together. Water can go in faster than it flows out only if water molecules are somehow stored inside the hose. If the hose is made of rubber, perhaps it's expanding in diameter as more and more water is stored inside. But this isn't sustainable because eventually the hose will burst. A sustainable flow exists only if the water flows out at the same speed it flows in.

The same is true for electrons in a constant-diameter wire. The only way the drift speed at B can be less than the drift speed at A is if electrons are being stored inside the bulb, making the bulb increasingly negative. But this isn't sustainable, and there are two reasons we know it doesn't happen. First, we can use charged plastic and glass rods to test whether a glowing bulb accumulates a negative charge. It doesn't. Second, if the bulb were to become increasingly negative, there would come a point when the repulsive force would stop the flow of new electrons into the bulb and the light would go out. This doesn't happen. **The rate of electrons going through the wire at B is the same as at A.**

The lightbulb doesn't "use up" current, but it *does* use energy. Imagine pushing a block along a rough surface at a constant speed. You have to do *work* on the block to keep it moving at a steady speed, and the energy you supply as work is dissipated by friction, making the block and the surface warmer. We've not yet identified what it is, but something is doing work by "pushing" the electrons through the wire *at constant speed*. The energy is dissipated by atomic-level friction as the electrons move through the atoms, making the wire hotter until, in the case of the light-bulb filament, it glows.

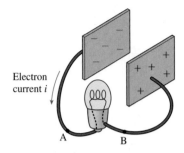

Electron current i

FIGURE 28.7 How does the electron current at A compare to the electron current at B?

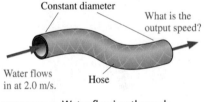

Constant diameter

What is the output speed?

Water flows in at 2.0 m/s.   Hose

FIGURE 28.8 Water flowing through a hose.

There are many issues here that we'll need to look at before we can say that we understand how currents work, and we'll take them one at a time in this and subsequent chapters. For now, we draw a first important conclusion:

> **Law of conservation of current** The electron current is the same at all points in a current-carrying wire.

A Puzzle

Let's end this section with a puzzle. Figure 28.9 shows a capacitor charged to ± 16 nC as it is being discharged by a 2.0-mm-diameter, 20-cm-long copper wire. *How long does it take* to discharge the capacitor?

We've noted that a fairly typical drift speed of the electron current through a wire is 10^{-4} m/s. At this rate, it would take 2000 s, or about a half hour, for an electron to travel 20 cm. We should have time to go for a cup of coffee while we wait for the discharge to occur!

But this isn't what happens. As far as our senses are concerned, the discharge of a capacitor by a copper wire is instantaneous. So what's wrong with our simple calculation?

The important point we overlooked is that the wire is *already full* of electrons. Return to the water-in-a-hose analogy. If the hose is already full of water, adding a drop to one end immediately (or very nearly so) pushes a drop out the other end. Likewise with the wire. As soon as the excess electrons move from the negative capacitor plate into the wire, they immediately (or very nearly so) push an equal number of electrons out the other end of the wire and onto the positive plate, thus neutralizing it. We don't have to wait for electrons to move all the way through the wire from one plate to the other. Instead, we just need to slightly rearrange the charges on the plates *and* in the wire.

Let's do a rough estimate of how much rearrangement is needed and how long the discharge takes. The negative plate in Figure 28.10, with $Q = -16$ nC, has 10^{11} excess electrons. Using the conduction-electron density of copper in Table 28.1, we can calculate that there are 5×10^{22} conduction electrons in the wire, a vastly larger number. The length of copper wire needed to hold 10^{11} electrons is a mere 4×10^{-13} m, only about 1% the diameter of an atom.

The instant the wire joins the capacitor plates, the repulsive forces between the excess 10^{11} electrons on the negative plate cause them to push their way into the wire. As they do so, 10^{11} electrons are squeezed out of the final 4×10^{-13} m of the wire and onto the positive plate. If the electrons all move together, and if they move at the typical drift speed of 10^{-4} m/s, both less than perfect assumptions but OK for the purpose of making an estimate, it takes 4×10^{-9} s, or 4 ns, to move 4×10^{-13} m and to discharge the capacitor. And, indeed, this is the right order of magnitude for how long the electrons take to rearrange themselves so that the capacitor plates are neutral.

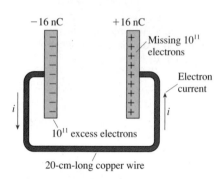

−16 nC   +16 nC

Missing 10^{11} electrons

Electron current

10^{11} excess electrons

20-cm-long copper wire

FIGURE 28.9 How long does it take to discharge a capacitor?

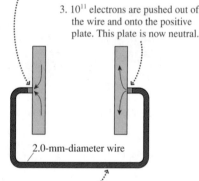

1. The 10^{11} excess electrons on the negative plate move into the wire. The length of wire needed to accommodate these electrons is only 4×10^{-13} m.

3. 10^{11} electrons are pushed out of the wire and onto the positive plate. This plate is now neutral.

2.0-mm-diameter wire

2. The sea of 5×10^{22} electrons in the wire is pushed to the side. It moves only 4×10^{-13} m, taking almost no time.

FIGURE 28.10 The sea of electrons needs only a minuscule rearrangement to discharge the capacitor.

STOP TO THINK 28.2 Why does the light in a room come on instantly when you flip a switch several meters away?

28.2 Creating a Current

Suppose you want to slide a book across the table to your friend. You give it a quick push to start it moving, but it begins slowing down because of friction as soon as you take your hand off. The book's kinetic energy is transformed into thermal energy, leaving the book and the table slightly warmer. The only way to keep the book moving at a *constant* speed is to continue pushing it.

As Figure 28.11 shows, the sea of electrons is similar to the book. If you push the sea of electrons, you create a current of electrons moving through the conductor. But the electrons aren't moving in a vacuum. Collisions between the electrons and the atoms of the metal transform the electrons' kinetic energy into the thermal energy of the metal, making the metal warmer. (Recall that "makes the wire warmer" is one of our indicators of a current.) Consequently, the sea of electrons will quickly slow down and stop *unless you continue pushing.* How do you push on electrons? With an electric field!

One of the important conclusions of Chapter 25 was that $\vec{E} = \vec{0}$ inside a conductor in electrostatic equilibrium. But a conductor with electrons moving through it is *not* in electrostatic equilibrium. **An electron current is a nonequilibrium motion of charges sustained by an internal electric field.**

Thus the quick answer to "What creates a current?" is "An electric field." But why is there an electric field in a current-carrying wire? How does it get established? What is the relationship between the strength of the electric field and the size of the electron current? These are the questions we need to answer.

Establishing the Electric Field in a Wire

Figure 28.12a shows two metal wires attached to the plates of a charged capacitor. The ends of the wires are close together, but not touching. The wires are conductors, so some of the charges on the capacitor plates spread out along the wires as a surface charge. (Remember that all excess charge on a conductor is located on the surface.)

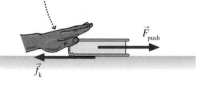

Because of friction, a steady push is needed to move the book at steady speed.

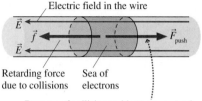

Electric field in the wire

Retarding force due to collisions   Sea of electrons

Because of collisions with atoms, a steady push is needed to move the sea of electrons at steady speed. Electrons are negative, so $\vec{F}_{push}$ is opposite to $\vec{E}$.

FIGURE 28.11 An electron current is sustained by pushing on the sea of electrons with an electric field.

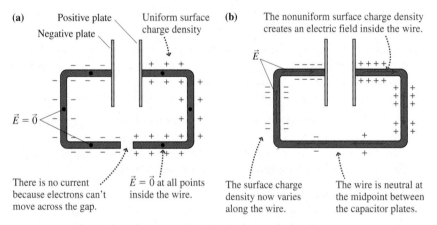

(a) Positive plate   Uniform surface charge density

Negative plate

$\vec{E} = \vec{0}$

There is no current because electrons can't move across the gap.

$\vec{E} = \vec{0}$ at all points inside the wire.

(b) The nonuniform surface charge density creates an electric field inside the wire.

$\vec{E}$

The surface charge density now varies along the wire.

The wire is neutral at the midpoint between the capacitor plates.

FIGURE 28.12 The surface charge on the wires before and after they are connected.

This is an electrostatic situation, with no charges in motion. Consequently, the electric field must be zero at *every* point inside the wire. Symmetry requires that there be equal amounts of charge to either side of a point in order to make $\vec{E} = \vec{0}$, hence the surface charge density must be uniform along each wire except near the ends (where the details need not concern us). We implied this uniform density in Figure 28.12a by drawing equally spaced $+$ and $-$ symbols along the wire. Remember that a positively charged surface is a surface that is *missing* electrons.

Now connect the ends of the wires together. What happens? The excess electrons on the surface of the negative wire suddenly have an opportunity to move onto the positive wire that is missing electrons. And, because opposite charges attract, they do. Within a *very* brief interval of time ($\approx 10^{-9}$ s), the sea of electrons shifts slightly and the surface charge is rearranged into a *nonuniform* distribution like that shown in Figure 28.12b. The surface charge near the positive and negative plates remains strongly positive and negative because of the large amount of charge on the capacitor plates, but the midpoint of the wire, halfway between the positive and negative plates, becomes electrically neutral. The new surface charge density on the wire varies from positive at the positive capacitor plate through zero at the midpoint to negative at the negative plate.

The new distribution of surface charge has an *extremely* important consequence. Figure 28.13 shows a section from a wire on which the surface charge density becomes more positive toward the left and more negative toward the right. Calculating the exact electric field is complicated, but we can understand the basic idea if we *model* this section of wire with four circular rings of charge.

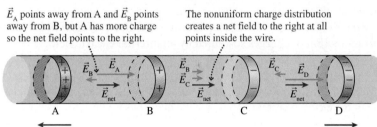

$\vec{E}_A$ points away from A and $\vec{E}_B$ points away from B, but A has more charge so the net field points to the right.

The nonuniform charge distribution creates a net field to the right at all points inside the wire.

The four rings A through D model the nonuniform charge distribution on the wire.

A   B   C   D

More positive   More negative

FIGURE 28.13 A varying surface charge distribution creates an internal electric field inside the wire.

In Chapter 26, we found that the on-axis field of a ring of charge

1. Points away from a positive ring, toward a negative ring;
2. Is proportional to the amount of charge on the ring; and
3. Decreases with distance away from the ring.

Because the field strength decreases with distance from the ring, the field at the midpoint between rings A and B is well approximated as $\vec{E}_{net} \approx \vec{E}_A + \vec{E}_B$. Ring A has more charge than ring B, so $\vec{E}_{net}$ points away from A.

The analysis of Figure 28.13 leads to a very important conclusion:

The *nonuniform* distribution of surface charges along a wire creates a net electric field *inside* the wire that points from the more positive end of the wire toward the more negative end of the wire. This is the internal electric field $\vec{E}$ that pushes the electron current through the wire.

The following example shows that the electric field inside a current-carrying wire can be established with an extremely small amount of surface charge.

EXAMPLE 28.2 The surface charge on a current-carrying wire

Table 26.1 in Chapter 26 gave the typical electric field strength in a current-carrying wire as 0.01 N/C. (We'll verify this value later in this chapter.) Two 2.0-mm-diameter rings are 2.0 mm apart. They are charged to $\pm Q$. What value of Q causes the electric field at the midpoint to be 0.010 N/C?

MODEL Use the on-axis electric field of a ring of charge from Chapter 26.

VISUALIZE Figure 28.14 shows the two rings. Both contribute equally to the field strength, so the electric field strength of the positive ring is $E_+ = 0.0050$ N/C. The distance $z = 1.0$ mm is half the ring spacing.

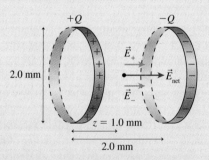

FIGURE 28.14 The electric field of two charged rings.

SOLVE Chapter 26 gives the on-axis electric field of a ring of charge Q as

$$E_+ = \frac{1}{4\pi\epsilon_0} \frac{zQ}{(z^2 + R^2)^{3/2}}$$

Thus the charge needed to produce the desired field is

$$Q = \frac{4\pi\epsilon_0(z^2 + R^2)^{3/2}}{z} E_+$$

$$= \frac{((0.0010 \text{ m})^2 + (0.0010 \text{ m})^2)^{3/2}}{(9.0 \times 10^9 \text{ Nm}^2/\text{C}^2)(0.0010 \text{ m})} (0.0050 \text{ N/C})$$

$$= 1.6 \times 10^{-18} \text{ C}$$

ASSESS The electric field of a ring of charge is largest when $z \approx R$, so these two rings are a simple but reasonable model for estimating the electric field inside a 2.0-mm-diameter wire. We find that the surface charge needed to establish the electric field is *very small*. A mere 10 electrons have to be moved from one ring to the other to charge them to $\pm 1.6 \times 10^{-18}$ C. This is sufficient to drive a sizable electron current through the wire.

Figure 28.13 showed how a varying surface charge density creates an electric field inside a straight wire. You might wonder how an electron current turns a corner. As Figure 28.15 shows, all it takes is just a little extra negative surface charge on the outside edge of a corner. When the electron current first begins, the electrons try to move straight ahead and thus pile up on the outside edge of the corner. This little bit of excess negative charge then exerts a repulsive force on the oncoming sea of electrons, just enough to turn them around the corner. Detailed calculations show that two or three extra electrons on the surface are usually all that are needed to make the current follow the wire around a bend.

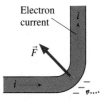

A few extra negative charges on the outside corner exert a repulsive force on the electrons, forcing the current to turn the corner.

FIGURE 28.15 The electron current turning a corner.

STOP TO THINK 28.3 The two charged rings are a model of the surface charge distribution along a wire. Rank in order, from largest to smallest, the electron currents E_a to E_e at the midpoint between the rings.

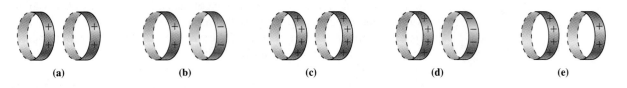

(a)   (b)   (c)   (d)   (e)

A Model of Conduction

An electric field inside a conductor pushes on the sea of electrons to create the electron current. The field has to *keep* pushing because the electrons continuously lose energy in collisions with the positive ions that form the structure of the solid. These collisions provide a drag force, much like friction.

The conduction electrons are analogous to the molecules in a gas. We characterized gases by their macroscopic parameters of temperature and pressure, but we needed an atomic-level perspective of colliding molecules in order to understand what temperature and pressure really are. The result was the kinetic theory of gases. We need a similar micro/macro link to help us understand how metals conduct electricity.

We will treat the conduction electrons—those electrons that make up the sea of electrons—as free particles moving through the crystal lattice of the metal. In the absence of an electric field, the electrons, like the molecules in a gas, move randomly in all directions with a distribution of speeds. If we assume that the average thermal energy of the electrons is given by the same $\frac{3}{2}kT$ that applies to an ideal gas, we can find that the average electron speed at room temperature is

(a) No electric field   Ions in the lattice of the metal

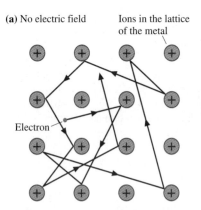

Electron

The electron has frequent collisions with ions, but it undergoes no net displacement.

(b) With an electric field   Parabolic trajectories in the electric field

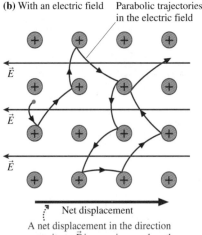

$\vec{E}$

$\vec{E}$

$\vec{E}$

Net displacement

A net displacement in the direction opposite to $\vec{E}$ is superimposed on the random thermal motion.

FIGURE 28.16 A microscopic view of a conduction electron moving through a metal.

$\approx 10^5$ m/s. This estimate turns out, for quantum physics reasons, to be not quite right, but it correctly indicates that the conduction electrons are moving very fast.

However, an individual electron does not travel far before colliding with an ion and being scattered to a new direction. Figure 28.16a shows that an electron moves in straight lines between collisions. Its *average* velocity is zero, as it bounces back and forth between collisions, and it undergoes no *net* displacement. This is similar to molecules in a container of gas.

Suppose we now turn on an electric field. Figure 28.16b shows that the steady electric force causes the electrons to move along *parabolic trajectories* between collisions. Because of the curvature of the trajectories, the negatively charged electrons begin to drift slowly in the direction opposite the electric field. The motion is similar to a ball moving in a pinball machine that has a slight downward tilt. An individual electron ricochets back and forth between the ions at a high rate of speed, but now there is a slow *net* motion in the "downhill" direction. Even so, this net displacement is a *very* small effect superimposed on top of the much larger thermal motion. Figure 28.16b has greatly exaggerated the rate at which the drift would occur.

Suppose an electron just had a collision with an ion and has rebounded with velocity $\vec{v}_i$. The acceleration of the electron between collisions is

$$a_x = \frac{F}{m} = \frac{eE}{m} \tag{28.4}$$

where E is the electric field strength inside the wire and m is the mass of the electron. (We'll assume that $\vec{E}$ points in the negative x-direction.) The field causes the x-component of the electron's velocity to increase linearly with time:

$$v_x = v_{ix} + a_x \Delta t = v_{ix} + \frac{eE}{m}\Delta t \tag{28.5}$$

The electron speeds up, with increasing kinetic energy, until its next collision with an ion. The collision transfers much of the electron's kinetic energy to the ion and thus to the thermal energy of the metal. **This energy transfer is the "friction" that raises the temperature of the wire.** The electron then rebounds, in a random direction, with a new initial velocity $\vec{v}_i$ and starts the process all over.

Figure 28.17a shows how the electron velocity abruptly changes due to a collision. Notice that the acceleration (the slope of the line) is the same before and after the collision. Figure 28.17b then follows an electron through a series of collisions. You can see that each collision "resets" the velocity. The primary observation we can make from Figure 28.17b is that this repeated process of speeding up and colliding gives the electron a nonzero *average* velocity. This average velocity, due to the electric field, is the *drift speed* v_d of the electron.

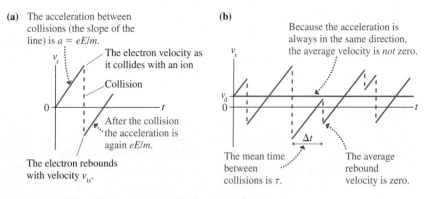

FIGURE 28.17 The electron velocity as a function of time.

If we observe all the electrons in the metal at one instant of time, their average velocity is

$$v_\mathrm{d} = \overline{v_x} = \overline{v_{\mathrm{i}x}} + \frac{eE}{m}\overline{\Delta t} \tag{28.6}$$

where a bar over a quantity indicates an average value. The average value of $v_{\mathrm{i}x}$, the velocity with which an electron rebounds after a collision, is zero. We know this because, in the absence of an electric field, the sea of electrons moves neither right nor left.

At any one instant of time, some electrons will have just recently collided and their acceleration time Δt will be shorter than average. Other electrons will be "overdue" for a collision and have Δt longer than average. When averaged over all electrons, the average value of Δt is the **mean time between collisions,** which we designate τ. The mean time between collisions is analogous to the mean free path between collisions in the kinetic theory of gases.

Thus the average speed at which the electrons are pushed along by the electric field is

$$v_\mathrm{d} = \frac{e\tau}{m}E \tag{28.7}$$

We can complete our model of conduction by using Equation 28.7 for v_d in the electron-current equation $i = nAv_\mathrm{d}$. Upon doing so, we find that an electric field strength E in a wire of cross-section area A causes an electron current

$$i = \frac{ne\tau A}{m}E \tag{28.8}$$

The electron density n and the mean time between collisions τ are properties of the metal.

Equation 28.8 is the main result of this model of conduction. We've found that **the electron current is directly proportional to the electric field strength.** A stronger electric field pushes the electrons faster and thus increases the electron current.

EXAMPLE 28.3 The electron current in a copper wire
The mean time between collisions for electrons in room-temperature copper is 2.5×10^{-14} s. What is the electron current in a 2.0-mm-diameter copper wire where the internal electric field strength is 0.010 N/C?

MODEL Use the model of conduction to relate the drift speed to the field strength.

SOLVE The electron current is $i = nAv_\mathrm{d}$. The electron drift speed in a 0.010 N/C electric field can be found from Equation 28.7:

$$v_\mathrm{d} = \frac{e\tau}{m}E = \frac{(1.60 \times 10^{-19}\ \mathrm{C})(2.5 \times 10^{-14}\ \mathrm{s})(0.010\ \mathrm{N/C})}{9.11 \times 10^{-31}\ \mathrm{kg}}$$
$$= 4.4 \times 10^{-5}\ \mathrm{m/s}$$

Copper has an electron density $n = 8.5 \times 10^{28}\ \mathrm{m}^{-3}$, and a 2.0-mm-diameter wire has a cross-section area $A = \pi r^2 = 3.14 \times 10^{-6}\ \mathrm{m}^2$. Thus the electron current is

$$i = nAv_\mathrm{d} = 1.2 \times 10^{19}\ \mathrm{electrons/s}$$

ASSESS A *lot* of electrons are going past each second!

28.3 Batteries

We've focused our attention on the discharge of a capacitor because we can easily understand where all the charges are and how they move. By contrast, we can't easily see what's happening to the charges inside a battery. Nonetheless, batteries are the primary source of current in circuits and other practical applications, so it's important to understand how our ideas about the electron current apply to a battery.

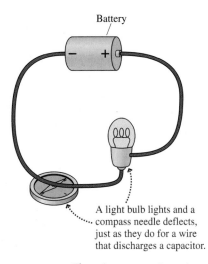

Battery

A light bulb lights and a compass needle deflects, just as they do for a wire that discharges a capacitor.

FIGURE 28.18 There is a current in a wire connecting the terminals of a battery.

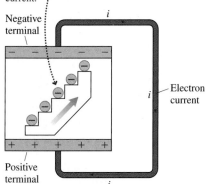

The "charge escalator" inside a battery continuously "lifts" electrons from the positive to the negative terminal. This renewal of charge sustains the electron current.

Negative terminal

i

Electron current

Positive terminal

i

FIGURE 28.19 A battery can be thought of as a charge escalator.

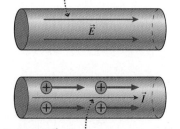

Surface charges have created an electric field inside the wire.

$\vec{E}$

$\vec{I}$

The current $\vec{I}$ is the rate at which the electric field seems to push *positive* charge through the wire. $\vec{I}$ is in the direction of $\vec{E}$.

FIGURE 28.20 The current I.

Figure 28.18 shows a wire connecting the two terminals of a battery, much like the wire that connected the capacitor plates in Figure 28.1. Just like that wire, the wire connecting the battery terminals gets warm, deflects a compass needle, and makes a lightbulb inserted into it glow brightly. These indicators tell us that charges flow through the wire from one terminal to the other. The current *in the wire* is the same whether it is supplied by a capacitor or a battery. Everything you've learned in this chapter about the electron current applies equally well to the current supplied by a battery—with one important difference.

That difference is the duration of the current. The current that discharges a capacitor is transient. It ceases as soon as the excess charge on the capacitor plates is removed; the lightbulb then goes out and the compass needle returns to its initial position. By contrast, the wire connecting the battery terminals *continues* to deflect the compass needle and *continues* to light the lightbulb. There is a *sustained* current in the wire—a sustained motion of charges—when it is connected to the battery. The capacitor quickly runs out of excess charge, but the battery can keep the motion going.

To understand why, we need to peer inside a battery. Figure 28.19 shows that the inner workings of a battery act like a *charge escalator* between two capacitor plates. Electrons are removed from the positive terminal and "lifted" to the negative terminal. It is the charge escalator that sustains the electron current in the wire by providing a continually renewed supply of electrons at the negative terminal.

The charge escalator does, however, have to be powered by some external source of energy. It is, in a very real sense, lifting the electrons "uphill" against the electric field. As you know, the energy to run the charge escalator comes from chemical reactions within the battery. A dead battery is one in which the supply of chemicals has been exhausted.

Our interest is in what happens to the charges, so we can conveniently ignore the details of the chemical reactions. All we care about is that the chemical reactions move electrons from the positive to the negative terminal of the battery. However, we can't overlook the fact that a description of the battery is going to require the ideas of work and energy. We will return to batteries and the charge escalator in Chapter 30. For now we need only to see that the role of a battery is to *maintain a charge separation* and thus to sustain a current.

28.4 Current and Current Density

We have developed the idea of a current as the motion of electrons through metals. But the properties of currents were known and used for a century before the discovery that electrons are the charge carriers in metals. We need to connect our ideas about the electron current to the conventional definition of current.

Because the coulomb is the SI unit of charge, and because currents are charges in motion, it seemed quite natural in the 19th century to define current as the *rate,* in coulombs per second, at which charge moves through a wire. Figure 28.20 shows a wire in which the electric field is $\vec{E}$. We define the current $\vec{I}$ in the wire to be

$$\vec{I} \equiv \left(\frac{dQ}{dt}, \text{ in the direction of } \vec{E} \right) \tag{28.9}$$

Strictly speaking, current is a vector. It has both a magnitude and a direction, and we will draw current arrows on diagrams to indicate the direction. For calculations, however, we are usually interested only in the magnitude $I = dQ/dt$.

The SI unit for current is the coulomb per second, which is called the **ampere** A:

$$1 \text{ ampere} = 1 \text{ A} \equiv 1 \text{ coulomb per second} = 1 \text{ C/s}$$

The current unit is named after the French scientist André Marie Ampère, who made major contributions to the study of electricity and magnetism in the early 19th century. The *amp* is an informal abbreviation of ampere. Household currents are typically ≈ 1 A. For example, the current through a 100 watt lightbulb is 0.85 A. The current in an electric hair dryer is ≈ 10 A. Currents in consumer electronics, such as stereos and computers are much less. They are typically measured in milliamps (1 mA $= 10^{-3}$ A) or microamps (1 μA $= 10^{-6}$ A).

For a *steady current,* which will be our primary focus, the amount of charge delivered by current I during the time interval Δt is

$$Q = I\Delta t \qquad (28.10)$$

Equation 28.10 is closely related to Equation 28.1, which said that the number of electrons delivered during a time interval Δt is $N_e = i\Delta t$. Each electron has charge of magnitude e, hence the total charge of N_e electrons is $Q = eN_e$. Consequently, the conventional current I and the electron current i are related by

$$I = \frac{Q}{\Delta t} = \frac{eN_e}{\Delta t} = ei \qquad (28.11)$$

Because electrons are the charge carriers, the rate at which charge moves is e times the rate at which the electrons move.

EXAMPLE 28.4 The current in a copper wire
The electron current in the copper wire of Example 28.3 was 1.2×10^{19} electrons/s. What is the current I? How much charge flows through a cross section of the wire each hour?

SOLVE The current in the wire is

$$I = ei = (1.60 \times 10^{-19}\,\text{C})(1.2 \times 10^{19}\,\text{s}^{-1}) = 1.9\,\text{A}$$

The amount of charge passing through the wire in 1 hr $= 3600$ s is

$$Q = I\Delta t = (1.9\,\text{A})(3600\,\text{s}) = 6840\,\text{C}$$

In one sense, the current I and the electron current i merely differ by a scale factor. The electron current i is the rate at which electrons move through a wire. The current I is the rate at which the charge of the electrons moves through the wire. The electron current i is more *fundamental* because it looks directly at the charge carriers. The current I is more *practical* because we can measure charge more easily than we can count electrons.

Despite the close connection between i and I, there's one important distinction. Because currents were known and studied before it was known what the charge carriers are, the direction of current was *defined* to be the direction in which positive charges *seem* to move. Thus the direction of the current $\vec{I}$ is the same as that of the internal electric field $\vec{E}$. But because the charge carriers turned out to be negative, at least for a metal, **the direction of the current $\vec{I}$ in a metal is opposite the direction of motion of the electrons.**

The situation shown in Figure 28.21 may seem disturbing, but it makes no real difference. A capacitor is discharged regardless of whether positive charges move toward the negative plate or negative charges move toward the positive plate. The primary application of the current is to the analysis of circuits, and in a circuit—a macroscopic device—we simply can't tell what is moving through the wires. All of our calculations will be correct and all of our circuits will work perfectly well if we choose to think of current as the flow of positive charge. The distinction is important only at the microscopic level.

Because the current I is in the direction of the electric field $\vec{E}$, **the current direction in a wire is from the positive terminal of a battery to the negative terminal.**

The current $\vec{I}$ is defined to point in the direction of $\vec{E}$. It is the direction in which positive charge carriers would move.

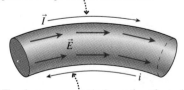

The electron current i is the motion of actual charge carriers. It is opposite to $\vec{E}$ and $\vec{I}$.

FIGURE 28.21 The current $\vec{I}$ is opposite the direction of motion of the electrons in a metal.

The Current Density in a Wire

We found that the electron current in a wire of cross-section area A is $i = nAv_d$. Thus the current I is

$$I = ei = nev_d A \qquad (28.12)$$

The quantity nev_d depends on the charge carriers and on the internal electric field that determines the drift speed, whereas A is simply a physical dimension of the wire. It will be useful to separate these quantities by defining the **current density** J in a wire as the current per square meter of cross section:

$$J = \text{current density} = \frac{I}{A} = nev_d \qquad (28.13)$$

The current density has units of A/m^2.

You learned earlier that the mass density ρ characterizes all pieces of a particular material, such as lead. A *specific* piece of the material, with known dimensions, is then characterized by its mass $m = \rho V$. Similarly, the current density J describes how charge flows through *any* piece of a particular kind of metal in response to an electric field. A *specific* piece of metal, shaped into a wire with cross-section area A, then has the specific current

$$I = JA \qquad (28.14)$$

EXAMPLE 28.5 Finding the electron drift speed

A 1.0 A current passes through a 1.0-mm-diameter aluminum wire. What is the drift speed of the electrons in the wire?

SOLVE We can find the drift speed from the current density. The current density is

$$J = \frac{I}{A} = \frac{I}{\pi r^2} = \frac{1.0 \text{ A}}{\pi (0.00050 \text{ m})^2} = 1.3 \times 10^6 \text{ A/m}^2$$

The electron drift speed is thus

$$v_d = \frac{J}{ne} = 1.3 \times 10^{-4} \text{ m/s} = 0.13 \text{ mm/s}$$

where the conduction-electron density for aluminum was taken from Table 28.1.

ASSESS We earlier used 1.0×10^{-4} m/s as a typical electron drift speed. This example shows where that value comes from.

Conservation of Current Revisited

Figure 28.22a shows a current-carrying wire. We've already seen that the electron current is the same at all points in a current-carrying wire. The conventional current I is measured differently, in terms of charge rather than the number of electrons, but its properties are the same. The law of conservation of current tells us that the current I is the same at all points in a current-carrying wire.

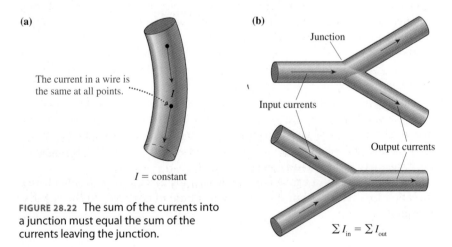

(a)

The current in a wire is the same at all points. ·········· I

I = constant

(b)

Junction

Input currents

Output currents

$\Sigma I_{\text{in}} = \Sigma I_{\text{out}}$

FIGURE 28.22 The sum of the currents into a junction must equal the sum of the currents leaving the junction.

Figure 28.22b shows two wires merging into one and one wire splitting into two. A point where a wire branches is called a **junction.** The presence of a junction doesn't change our basic reasoning. We cannot create or destroy electrons in the wire, and neither can we store them in the junction. The rate at which electrons flow into one *or many* wires must be exactly balanced by the rate at which they flow out of others. For a *junction,* the law of conservation of charge requires that

$$\sum I_{in} = \sum I_{out} \tag{28.15}$$

where, as usual, the $\sum$ symbol means summation.

This basic conservation statement, that the sum of the currents into a junction equals the sum of the currents leaving, is called **Kirchhoff's junction law.** The junction law, which will play an important role in circuit analysis, is a very important consequence of the conservation of charge.

STOP TO THINK 28.4   What are the magnitude and the direction of the current in the fifth wire?

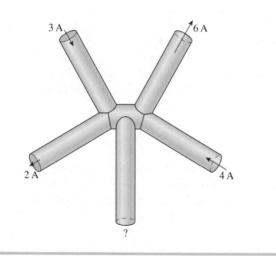

3 A

6 A

2 A

4 A

?

28.5 Conductivity and Resistivity

The current density $J = nev_d$ is directly proportional to the electron drift speed v_d. We used the microscopic model of conduction to find that the drift speed is $v_d = e\tau E/m$, where τ is the mean time between collisions. Thus we can write the current density as

$$J = nev_d = ne\left(\frac{e\tau E}{m}\right) = \frac{ne^2\tau}{m}E \tag{28.16}$$

The quantity $ne^2\tau/m$ in Equation 28.16 depends *only* on the conducting material. It contains the electron density n and the mean time between collisions τ. Materials with larger values of these numbers conduct current better than materials with smaller values. That is, a given electric field strength will generate a larger current density in a material that has larger values of n and τ.

It makes sense, then, to define the **conductivity** σ of a material as

$$\sigma = \text{conductivity} = \frac{ne^2\tau}{m} \tag{28.17}$$

where m is the mass of the electron. Conductivity, like density, characterizes a material as a whole. All pieces of copper (at the same temperature) have the same value of σ, but the conductivity of copper is different from that of aluminum. Notice that the mean time between collisions τ can be inferred from measured values of the conductivity.

With this definition of conductivity, Equation 28.16 becomes

$$J = \sigma E \qquad (28.18)$$

This is a result of fundamental importance. Equation 28.18 tells us three things:

1. Current is caused by an electric field exerting forces on the charge carriers.
2. The current density, and hence the current $I = JA$, depends linearly on the strength of the electric field.
3. The current density also depends on the *conductivity* of the material. Different conducting materials have different conductivities because they have different values of the electron density n and, especially, different values of the mean time between electron collisions with the lattice of atoms.

The value of the conductivity is affected by the crystalline structure, by any impurities in the metal, and by the temperature. As the temperature increases, so do the thermal vibrations of the lattice atoms. This makes them "bigger targets" and causes collisions to be more frequent, thus lowing τ and decreasing the conductivity. Metals conduct better at low temperatures than they do at high temperatures.

For many practical applications of current it will be convenient to use the inverse of the conductivity, $1/\sigma$. This is called the **resistivity** and has the symbol ρ. Thus

$$\rho = \text{resistivity} = \frac{1}{\sigma} = \frac{m}{ne^2\tau} \qquad (28.19)$$

The resistivity of a material tells us how reluctantly the electrons move in response to an electric field. Table 28.2 gives values of the conductivity and resistivity for several metals and for carbon. You can see that they do vary quite a bit, with copper and silver being the best two conductors.

The units of conductivity, from Equation 28.18, are those of J/E, namely $A\,C/N\,m^2$. These are clearly awkward. In Chapter 31 we will introduce a new unit called the *ohm*, symbolized by Ω (uppercase Greek omega). It will then turn out that resistivity has units of $\Omega\,m$ and conductivity has $\Omega^{-1}m^{-1}$.

NOTE ▶ *Resistivity* is not the same as *resistance,* a related quantity that you will meet in Chapter 31. ◀

TABLE 28.2 Resistivity and conductivity of conducting materials

Material	Resistivity $(\Omega\,m)$	Conductivity $(\Omega^{-1}m^{-1})$
Aluminum	2.8×10^{-8}	3.5×10^7
Copper	1.7×10^{-8}	6.0×10^7
Gold	2.4×10^{-8}	4.1×10^7
Iron	9.7×10^{-8}	1.0×10^7
Silver	1.6×10^{-8}	6.2×10^7
Tungsten	5.6×10^{-8}	1.8×10^7
Nichrome*	1.5×10^{-6}	6.7×10^5
Carbon	3.5×10^{-5}	2.9×10^4

*Nickel-chromium alloy used for heating wires

EXAMPLE 28.6 **The electric field in a wire**
A 2.0-mm-diameter aluminum wire carries a current of 800 mA. What is the electric field strength inside the wire?

SOLVE The electric field strength is

$$E = \frac{J}{\sigma} = \frac{I}{\sigma A} = \frac{I}{\sigma \pi r^2} = \frac{0.80\,\text{A}}{(3.5 \times 10^7\,\Omega^{-1}m^{-1})\pi(0.0010\,\text{m})^2}$$

$$= 0.0072\,\text{N/C}$$

where the conductivity of aluminum was taken from Table 28.2.

ASSESS This is a *very* small field in comparison with those we calculated in Chapters 25 and 26 for point charges and charged objects. This calculation justifies the claim in Table 26.1 that a typical electric field strength inside a current-carrying wire is 0.01 N/C.

EXAMPLE 28.7 Mean time between collisions

What is the mean time between collisions for electrons in copper?

SOLVE The mean time between collisions is related to a material's conductivity by

$$\tau = \frac{m\sigma}{ne^2}$$

The electron density of copper is found in Table 28.1 and the measured conductivity is found in Table 28.2. With this information,

$$\tau = \frac{(9.11 \times 10^{-31} \text{ kg})(6.0 \times 10^7 \ \Omega^{-1}\text{m}^{-1})}{(8.5 \times 10^{28} \text{ m}^{-3})(1.60 \times 10^{-19} \text{ C})^2} = 2.5 \times 10^{-14} \text{ s}$$

This is the value that was given in Example 28.1.

The electric field strength found in Example 28.6 is roughly the size of the electric field 1 mm from a *single* electron. The lesson to be learned from this example is that it takes *very few* surface charges on a wire to create the internal electric field necessary for the wire to carry considerable current. Just a few excess electrons every centimeter are sufficient. The reason, once again, is the enormous value of the charge-carrier density n. Even through the electric field is very tiny and the drift speed is agonizingly slow, a wire can carry a substantial current due to the vast number of charge carriers able to move.

Superconductivity

In 1911, the Dutch physicist Kamerlingh Onnes was studying the conductivity of metals at very low temperatures. Scientists had just recently discovered how to liquefy helium, and this opened a whole new field of *low-temperature physics*. As we noted above, metals become better conductors (i.e., they have higher conductivity and lower resistivity) at lower temperatures. But the effect is gradual. Onnes, however, found that mercury suddenly and dramatically loses *all* resistance to current when cooled below a temperature of 4.2 K. This complete loss of resistance at low temperatures is called **superconductivity.**

Later experiments established that the resistivity of a superconducting metal is not just small, it is truly zero. The electrons are moving in a frictionless environment, and charge will continue to move through a superconductor *without an electric field*. Superconductivity was not understood until the 1950s, when it was explained as being a specific quantum effect.

Superconducting wires can carry enormous currents because the wires are not heated by electrons colliding with the atoms. Very strong magnetic fields can be created with superconducting electromagnets, but applications remained limited for many decades because all known superconductors required temperatures less than 20 K. This situation changed dramatically in 1986 with the discovery of *high-temperature superconductors*. These ceramic-like materials are superconductors at temperatures as "high" as 125 K. Although −150°C may not seem like a high temperature to you, the technology for producing such temperatures is simple and inexpensive. Thus many new superconductor applications are likely to appear in coming years.

Superconductors have unusual magnetic properties. Here a small permanent magnet levitates above a disk of the high-temperature superconductor $YBa_2Cu_3O_7$ that has been cooled to liquid-nitrogen temperature.

STOP TO THINK 28.5 Rank in order, from largest to smallest, the current densities J_a to J_d in these four wires.

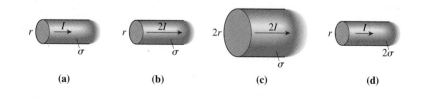

| (a) | (b) | (c) | (d) |

SUMMARY

The goal of Chapter 28 has been to learn how and why charge moves through a conductor as a current.

GENERAL PRINCIPLES

Current is a nonequilibrium motion of charges sustained by an electric field. Nonuniform surface charge density creates an electric field in a wire. The electric field pushes the electron current i in a direction opposite to $\vec{E}$. The conventional current I is in the direction in which positive charge *seems* to move.

Electron current

i = rate of electron flow

$N_e = i\Delta t$

Conventional current

I = rate of charge flow = ei

$Q = I\Delta t$

Current density

$J = I/A$

Conservation of Current

The current is the same at any two points in a wire.

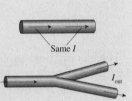

At a junction,

$$\sum I_{in} = \sum I_{out}$$

Batteries

The role of a battery is to maintain a charge separation and thus sustain a current. Chemical reactions power the "charge escalator" that moves electrons from the positive terminal to the negative terminal.

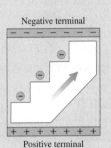

IMPORTANT CONCEPTS

Sea of electrons

Conduction electrons move freely around the positive ions that form the atomic lattice.

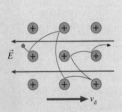

Conduction

An electric field causes a slow drift at speed v_d to be superimposed on the rapid but random thermal motions of the electrons.

Collisions of electrons with the ions transfer energy to the atoms. This makes the wire warm and lightbulbs glow. More collisions mean a higher resistivity ρ and a lower conductivity σ.

The drift speed is $v_d = \dfrac{e\tau}{m}E$ where τ is the mean time between collisions. v_d is related to the electron current by

$$i = nAv_d$$

where n is the electron density.

An electric field E in a conductor causes a current density $J = nev_d = \sigma E$ where the conductivity is

$$\sigma = \frac{ne^2\tau}{m}$$

The resistivity is $\rho = 1/\sigma$.

TERMS AND NOTATION

current, I	mean time between collisions, τ	Kirchhoff's junction law
drift speed, v_d	ampere, A	conductivity, σ
electron current, i	current density, J	resistivity, ρ
law of conservation of current	junction	superconductivity

EXERCISES AND PROBLEMS

Exercises

Section 28.1 The Electron Current

1. 1.0×10^{20} electrons flow through a cross section of a 2.0-mm-diameter iron wire in 5.0 s. What is the electron drift speed?

2. Estimate how long it takes an electron to go from the wall switch to the overhead lightbulb in your bedroom.

3. The electron drift speed in a 1.0-mm-diameter gold wire is 5.0×10^{-5} m/s. How long does it take 1 mole of electrons to flow through a cross section of the wire?

4. 1.0×10^{16} electrons flow through a cross section of silver wire in 320 μs with a drift speed of 8.0×10^{-4} m/s. What is the diameter of the wire?

5. 1.44×10^{14} electrons flow through a cross section of a 2.0 mm $\times$ 2.0 mm square wire in 3.0 μs. The electron drift speed is 2.0×10^{-4} m/s. What metal is the wire made of?

Section 28.2 Creating a Current

6. What is the surface charge density of a 1.0-mm-diameter wire with 1000 excess electrons per centimeter of length?

7. a. How many conduction electrons are there in a 1.0-mm-diameter gold wire that is 10 cm long?
   b. How far must the sea of electrons in the wire move to deliver -32 nC of charge to an electrode?

8. The electron drift speed is 2.0×10^{-4} m/s in a metal with a mean free time between collisions of 5.0×10^{-14} s. What is the electric field strength?

9. The mean free time between collisions in iron is 4.2×10^{-15} s. What electric field strength causes a 5.0×10^{19} s^{-1} electron current in a 1.8-mm-diameter iron wire?

Section 28.3 Batteries

10. Draw Figure Ex28.10 on your paper.
    a. Use plusses and minuses to show how the surface charge is distributed along the wire.
    b. Show the electric field vector $\vec{E}$ at each of the seven points in the wire marked with a dot. The length of each vector should be proportional to E at that point.

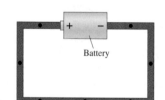

FIGURE EX28.10

11. A battery supplies a steady 1.5 A current to a circuit. How many electrons per second does the charge escalator transport from the positive terminal to the negative terminal?

Section 28.4 Current and Current Density

12. The current in an electric hair dryer is 10.0 A. How much charge and how many electrons flow through the hair dryer in 5.0 min?

13. 2.0×10^{13} electrons flow through a transistor in 1.0 ms. What is the current through the transistor?

14. In an ionic solution, 5.0×10^{15} positive ions with charge $+2e$ pass to the right each second while 6.0×10^{15} negative ions with charge $-e$ pass to the left. What is the current in the solution?

15. The current in a 2.0 mm $\times$ 2.0 mm square aluminum wire is 2.5 A. What are (a) the current density and (b) the electron drift speed?

16. The wires leading to and from a 0.12-mm-diameter lightbulb filament are 1.5 mm in diameter. The wire to the filament carries a current with a current density of 450,000 A/m^2. What are (a) the current and (b) the current density in the filament?

17. The current in a 100 watt lightbulb is 0.85 A. The filament inside the bulb is 0.25 mm in diameter.
    a. What is the current density in the filament?
    b. What is the electron current in the filament?

18. The electron drift speed in a gold wire is 3.0×10^{-4} m/s.
    a. What is the current density in the wire?
    b. What is the current if the wire diameter is 0.50 mm?

19. In an integrated circuit, the current density in a 2.5-μm-thick $\times$ 75-μm-wide gold film is 750,000 A/m^2. What is the current in the film?

20. A hollow copper wire with an inner diameter of 1.0 mm and an outer diameter of 2.0 mm carries a current of 10 A. What is the current density in the wire?

Section 28.5 Conductivity and Resistivity

21. What is the mean free time between collisions for electrons in an aluminum wire and in an iron wire?

22. What is the mean free time between collisions for electrons in silver and in gold?

23. The electric field in a 2.0 mm $\times$ 2.0 mm square aluminum wire is 0.012 N/C. What is the current in the wire?

24. What electric field strength is needed to create a 5.0 A current in a 2.0-mm-diameter iron wire?

25. A 3.0-mm-diameter wire carries a 12 A current when the electric field is 0.085 N/C. What is the wire's resistivity?

26. A 0.0075 N/C electric field creates a 3.9 mA current in a 1.0-mm-diameter wire. What material is the wire made of?

27. A 0.50-mm-diameter silver wire carries a 20 mA current. What are (a) the electric field and (b) the electron drift speed in the wire?

28. A metal cube 1.0 cm on each side is sandwiched between two electrodes. The electrodes create a 0.0050 N/C electric field in the metal. A current of 9.0 A passes through the cube, from the positive electrode to the negative electrode. Identify the metal.

Problems

29. The density of aluminum is 2700 kg/m^3. Verify the conduction-electron density for aluminum given in Table 28.1. The mass of an aluminum atom is 27 u.

30. For what electric field strength would the current in a 2.0-mm-diameter nichrome wire be the same as the current in a 1.0-mm-diameter aluminum wire in which the electric field strength is 0.0080 N/C?

31. The current in a wire is doubled. What happens to
    a. The current density?
    b. The conduction-electron density?
    c. The mean time between collisions?
    d. The electron drift speed?

32. The electric field strength inside a wire is doubled. What happens to
    a. The current?
    b. The conduction-electron density?
    c. The mean time between collisions?
    d. The electron drift speed?

33. The electron beam inside a television picture tube is 0.4 mm in diameter and carries a current of 50 μA. This electron beam impinges on the inside of the picture tube screen.
    a. How many electrons strike the screen each second?
    b. What is the current density in the electron beam?
    c. The electrons move with a velocity of 4.0×10^7 m/s. What electric field strength is needed to accelerate electrons from rest to this velocity in a distance of 5.0 mm?
    d. Each electron transfers its kinetic energy to the picture tube screen upon impact. What is the *power* delivered to the screen by the electron beam?

34. Figure P28.34 shows a 4.0-cm-wide plastic film being wrapped onto a 2.0-cm-diameter roller that turns at 90 rpm. The plastic has a uniform surface charge density -2.0 nC/cm^2.
    a. What is the current of the moving film?
    b. How long does it take the roller to accumulate a charge of -10 μC?

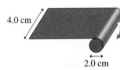

4.0 cm

FIGURE P28.34   2.0 cm

35. A sculptor has asked you to help electroplate gold onto a brass statue. You know that the charge carriers in the ionic solution are gold ions, and you've calculated that you must deposit 0.50 g of gold to reach the necessary thickness. How much current do you need, in mA, to plate the statue in 3.0 hours?

36. The biochemistry that takes place inside cells depends on various elements, such as sodium, potassium, and calcium, that are dissolved in water as ions. These ions enter cells through narrow pores in the cell membrane known as *ion channels*. Each ion channel, which is formed from a specialized protein molecule, is selective for one type of ion. Measurements with microelectrodes have shown that a 0.30-nm-diameter potassium ion (K$^+$) channel carries a current of 1.8 pA.
    a. How many potassium ions pass through if the ion channel opens for 1.0 ms?
    b. What is the current density in the ion channel?

37. The starter motor of a car engine draws a current of 150 A from the battery. The copper wire to the motor is 5.0 mm in diameter and 1.2 m long. The starter motor runs for 0.80 s until the car engine starts.
    a. How much charge passes through the starter motor?
    b. How far does an electron travel along the wire while the starter motor is on?

38. A car battery is rated at 90 A hr, meaning that it can supply a 90 A current for 1 hr before being completely discharged. If you leave your headlights on until the battery is completely dead, how much charge leaves the battery?

39. What fraction of the current in a wire of radius R flows in the part of the wire with radius $r \leq \frac{1}{2}R$?

40. You need to design a 1.0 A fuse that "blows" if the current exceeds 1.0 A. The fuse material in your stockroom melts at a current density of 500 A/cm^2. What diameter wire of this material will do the job?

41. A 62 g hollow copper cylinder is 10 cm long and has an inner diameter of 1.0 cm. The current density along the length of the cylinder is 150,000 A/m^2. What is the current in the cylinder?

42. A hollow metal cylinder has inner radius a, outer radius b, length L, and conductivity σ. The current I is *radially* outward from the inner surface to the outer surface.
    a. Find an expression for the electric field strength inside the metal as a function of the radius r from the cylinder's axis.
    b. Evaluate the electric field strength at the inner and outer surfaces of an iron cylinder if $a = 1.0$ cm, $b = 2.5$ cm, $L = 10$ cm, and $I = 25$ A.

43. A hollow metal sphere has inner radius a, outer radius b, and conductivity σ. The current I is *radially* outward from the inner surface to the outer surface.
    a. Find an expression for the electric field strength inside the metal as a function of the radius r from the center.
    b. Evaluate the electric field strength at the inner and outer surfaces of a copper sphere if $a = 1.0$ cm, $b = 2.5$ cm, and $I = 25$ A.

44. The total amount of charge in coulombs that has entered a wire at time t is given by the expression $Q = 4t - t^2$, where t is in seconds and $t \geq 0$.
    a. Graph Q versus t for the interval $0 \leq t \leq 4$ s.
    b. Find an expression for the current in the wire at time t.
    c. Graph I versus t for the interval $0 \leq t \leq 4$ s.
    d. Explain *why* I has the value at $t = 2.0$ s that you observe.

45. The total amount of charge that has entered a wire at time t is given by the expression $Q = (20 \text{ C})(1 - e^{-t/(2.0 \text{ s})})$, where t is in seconds and $t \geq 0$.
    a. Graph Q versus t for the interval $0 \leq t \leq 10$ s.
    b. Find an expression for the current in the wire at time t.
    c. What is the maximum value of the current?
    d. Graph I versus t for the interval $0 \leq t \leq 10$ s.

46. The current in a wire at time t is given by the expression $I = (2.0 \text{ A})e^{-t/(2.0 \, \mu s)}$, where t is in microseconds and $t \geq 0$.
    a. Graph I versus t for the interval $0 \leq t \leq 10$ μs.
    b. Find an expression for total amount of charge (in coulombs) that has entered the wire at time t. The initial conditions are $Q = 0$ C at $t = 0$ μs.
    c. Graph Q versus t for the interval $0 \leq t \leq 10$ μs.

47. The electric field in a current-carrying wire can be modeled as the electric field at the midpoint between two charged rings. Model a 3.0-mm-diameter aluminum wire as two 3.0-mm-diameter rings 2.0 mm apart. What is the current in the wire after 20 electrons are transferred from one ring to the other?

48. The two wires in Figure P28.48 are made of the same material. What are the current and the electron drift speed in the 2.0-mm-diameter segment of the wire?

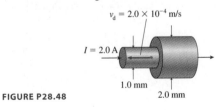

$v_d = 2.0 \times 10^{-4}$ m/s

$I = 2.0$ A

1.0 mm

FIGURE P28.48   2.0 mm

49. What is the electron drift speed at the 3.0-mm-diameter end (the left end) of the wire in Figure P28.49?

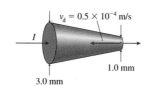

$v_d = 0.5 \times 10^{-4}$ m/s

1.0 mm

FIGURE P28.49 3.0 mm

50. What diameter should the nichrome wire in Figure P28.50 be in order for the electric field strength to be the same in both wires?

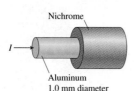

Nichrome

I

FIGURE P28.50

Aluminum
1.0 mm diameter

51. The two segments of the wire in Figure P28.51 have equal diameters but different conductivities σ_1 and σ_2. Current I passes through this wire. If the conductivities have the ratio $\sigma_2/\sigma_1 = 2$, what is the ratio E_2/E_1 of the electric field strengths in the two segments of the wire?

FIGURE P28.51 σ_1 σ_2 $\rightarrow I$

52. An aluminum wire consists of the three segments shown in Figure P28.52. The current in the top segment is 10 A. For each of these three segments, find the
   a. Current I.
   b. Current density J.
   c. Electric field E.
   d. Drift velocity v_d.
   e. Mean time between collisions τ.
   f. Electron current i.
   Place your results in a table for easy viewing.

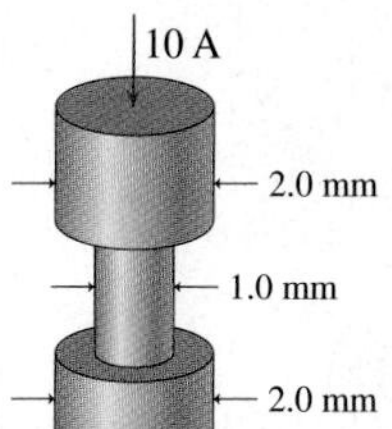

10 A

2.0 mm

1.0 mm

2.0 mm

FIGURE P28.52

Challenge Problems

53. The current supplied by a battery slowly decreases as the battery runs down. Suppose that the current as a function of time is $I = (0.75 \text{ A})e^{-t/(6 \text{ hr})}$. What is the total number of electrons transported from the positive electrode to the negative electrode by the charge escalator from the time the battery is first used until it is completely dead?

54. In a classical model of the hydrogen atom, the electron moves around the proton in a circular orbit of radius 0.053 nm.
   a. What is the electron's orbital frequency?
   b. What is the effective current of the electron?

55. Assume the conduction electrons in a metal can be treated as classical particles in an ideal gas.
   a. What is the rms velocity of electrons in room-temperature copper?
   b. How far, on average, does an electron move between collisions?

56. A 5.0-mm-diameter proton beam carries a total current of 1.5 mA. The current density in the proton beam, which increases with distance from the center, is given by $J = J_{edge}(r/R)$, where R is the radius of the beam and J_{edge} is the current density at the edge.
   a. How many protons per second are delivered by this proton beam?
   b. Determine the value of J_{edge}.

57. Figure CP28.57 shows a wire that is made of two equal-diameter segments with conductivities σ_1 and σ_2. When current I passes through the wire, a thin layer of charge appears at the boundary between the segments.
   a. Find an expression for the surface charge density η on the boundary. Give your result in terms of I, σ_1, σ_2, and the wire's cross-section area A.
   b. A 1.0-mm-diameter wire made of copper and iron segments carries a 5 A current. How much charge accumulates at the boundary between the segments?

FIGURE CP28.57 Surface charge density η

STOP TO THINK ANSWERS

Stop to Think 28.1: $i_c > i_b > i_a > i_d$. The electron current is proportional to $r^2 v_d$. Changing r by a factor of 2 has more influence than changing v_d by a factor of 2.

Stop to Think 28.2: The electrons don't have to move from the switch to the bulb, which could take hours. Because the wire between the switch and the bulb is already full of electrons, a flow of electrons from the switch into the wire immediately causes electrons to flow from the other end of the wire into the lightbulb.

Stop to Think 28.3: $E_d > E_b > E_e > E_a = E_c$. The electric field strength depends on the *difference* in the charge on the two wires.

The electric fields of the rings in a and c are opposed to each other, so the net field is zero. The rings in d have the largest charge *difference*.

Stop to Think 28.4: 1 A into the junction. The total current entering the junction must equal the total current leaving the junction.

Stop to Think 28.5: $J_b > J_a = J_d > J_c$. The current density $J = I/\pi r^2$ is independent of the conductivity σ, so a and d are the same. Changing r by a factor of 2 has more influence than changing I by a factor of 2.

29 The Electric Potential

Millions of lightbulbs transform electric energy into light and thermal energy.

▶ **Looking Ahead**

The goal of Chapter 29 is to calculate and use the electric potential and electric potential energy. In this chapter you will learn to:

- Use electric potential energy and conservation of energy to analyze the motion of charged particles.
- Use the electric potential to find the electric potential energy.
- Calculate the electric potential of useful and important charge distributions.
- Represent the electric potential graphically.

◀ **Looking Back**

This chapter depends heavily on the concepts of work, energy, and conservation of energy. Please review:

- Sections 10.2–10.5 Kinetic, gravitational, and elastic energy.
- Section 10.7 Energy diagrams.
- Sections 11.2–11.5 Work and potential energy.
- Section 26.3 Calculating the electric field of a continuous distribution of charge.

The sparkling lights of a big city are an awesome spectacle. These lights use a tremendous amount of energy. Where does all that energy come from?

Energy has been a theme throughout most of this textbook. Energy allows things to happen. A system without a source of energy is not terribly interesting; it just sits there. You want your lights to light, your computer to compute, and your stereo to keep your neighbors awake. In other words, you want devices that use electricity to *do* something, and that takes energy. It is time to see how the concept of energy helps us to understand and analyze electric phenomena.

In Chapter 25, we introduced the idea of the *electric field* to understand how one set of charges, the source charges, exerts electric forces on other charges. Now, to understand electric energy, we will introduce a new concept called the *electric potential*. In this chapter, you will study the basic properties of the electric potential and learn how it is connected to electric energy. In Chapter 30, we'll explore the relationship between the electric potential and the electric field. These two chapters will lead us directly into electric circuits, which are a practical application of the ideas of the electric potential and electric field.

29.1 Electric Potential Energy

We've alluded to energy many times in the last four chapters. We noted that a parallel-plate capacitor is charged by transferring electrons from one plate to the other, but the transfer doesn't happen by magic. It takes energy. A current flows in a wire only as long as energy is expended to push the sea of electrons forward against the "friction" of collisions with the atoms in the metal.

Our study of electric energy has two practical, and related, goals:

- To understand the motion of charged particles.
- To understand the fundamental ideas of electric circuits.

To meet these goals, we need to find out how electric energy is related to electric charges, forces, and fields.

Mechanical Energy

We will begin our investigation of electric energy by exploiting the close analogy between gravitational forces and electric forces. The gravitational force between two masses depends inversely on the square of the distance between them, as does the electric force between two point charges. Similarly, the uniform gravitational field near the earth's surface looks very much like the uniform electric field inside a parallel-plate capacitor.

You will recall that a system's mechanical energy $E_{mech} = K + U$ is conserved for particles that interact with each other via *conservative forces,* where K and U are the kinetic and potential energy. That is,

$$\Delta E_{mech} = \Delta K + \Delta U = 0 \qquad (29.1)$$

We need to be careful with notation because we are now using E to represent the electric field strength. To avoid confusion, we will represent mechanical energy either as the explicit sum $K + U$ or as E_{mech}, with an explicit subscript.

NOTE ▶ Recall that for any quantity X, the *change* in X is $\Delta X = X_{final} - X_{initial}$. ◀

The kinetic energy $K = \sum K_i$, where $K_i = \frac{1}{2}m_i v_i^2$, is the sum of the kinetic energies of all the particles in the system. The potential energy U is the *interaction energy* of the system. In particular, we defined the *change* in potential energy in terms of the work W done by the forces of interaction as the system moves from an initial position or configuration i to a final position or configuration f:

$$\Delta U = U_f - U_i = -W_{interaction\ forces} \qquad (\text{position i} \rightarrow \text{position f}) \qquad (29.2)$$

This formal definition of ΔU is rather abstract and will make more sense when we see specific applications.

NOTE ▶ The potential energy is an energy *of the system,* not of a particular particle in the system. That is, the familiar $U_{grav} = mgy$ is the energy of the earth + particle system due to their mutual gravitational interaction. Even so, we often speak of the *particle's* gravitational potential energy because the earth remains essentially at rest while the much less massive particle moves. ◀

A *constant* force does work

$$W = \vec{F} \cdot \Delta\vec{r} = F\Delta r \cos\theta \qquad (29.3)$$

on a particle that undergoes a linear displacement $\Delta\vec{r}$, where θ is the angle between the force $\vec{F}$ and $\Delta\vec{r}$. Figure 29.1 reminds you of the three special cases $\theta = 0°, 90°$, and $180°$. It also shows that, in general, the work is done by the force component F_r in the direction of motion.

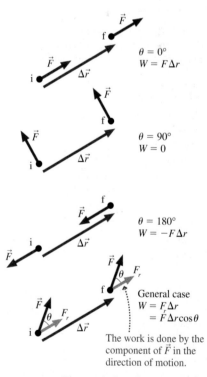

FIGURE 29.1 The work done by a constant force.

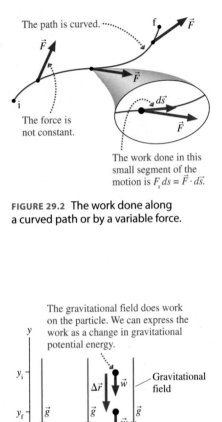

The path is curved.

$\vec{F}$

The force is not constant.

$d\vec{s}$

$\vec{F}$

$\vec{F}$

The work done in this small segment of the motion is $F_s ds = \vec{F} \cdot d\vec{s}$.

FIGURE 29.2 The work done along a curved path or by a variable force.

The gravitational field does work on the particle. We can express the work as a change in gravitational potential energy.

y

y_i

$\Delta \vec{r}$ $\vec{w}$

Gravitational field

y_f $\vec{g}$ $\vec{g}$ $\vec{w}$ $\vec{g}$

0

The net force on the particle is down. It gains kinetic energy (i.e., speeds up) as it loses potential energy.

FIGURE 29.3 Potential energy is transformed into kinetic energy as a particle moves in a gravitational field.

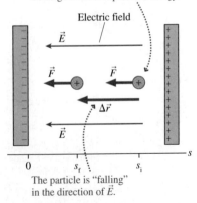

The electric field does work on the particle. We can express the work as a change in electric potential energy.

Electric field

$\vec{E}$

$\vec{F}$ $\vec{F}$

$\Delta \vec{r}$

$\vec{E}$

0 s_f s_i s

The particle is "falling" in the direction of $\vec{E}$.

FIGURE 29.4 The electric field does work on the charged particle.

NOTE ▶ Work is *not* the oft-remembered "force times distance." Work is force times distance only in the one very special case in which the force is both constant *and* parallel to the displacement. ◀

If the force is *not* constant or the displacement is *not* along a linear path, we can calculate the work by dividing the path into many small segments. Figure 29.2 shows how this is done. The work done as the particle moves distance ds is $F_s ds$, where F_s is the force component parallel to ds (i.e., the component in the direction of motion). The total work done on the particle is

$$W = \sum_j (F_s)_j \Delta s_j \to \int_{s_i}^{s_f} F_s \, ds = \int_i^f \vec{F} \cdot d\vec{s} \tag{29.4}$$

The second integral recognizes that $F_s ds = F \cos\theta \, ds$ is equivalent to the dot product $\vec{F} \cdot d\vec{s}$, allowing us to write the work in vector notation. As with Gauss's law, this integral looks more formidable than it really is. We'll look at examples shortly.

Finally, recall that a *conservative force* is one for which the work done as a particle moves from position i to position f is *independent of the path followed*. In other words, the integral in Equation 29.4 gives the same value for *any* path between points i and f. We'll assert for now, and prove later, that **the electric force is a conservative force.**

A Uniform Field

The gravitational field $\vec{g}$ near the earth's surface is a *uniform* field, $\vec{g} = (g, \text{down})$. Figure 29.3 shows a particle of mass m falling in the gravitational field. You learned in Chapter 10 that the particles *loses* potential energy and *gains* kinetic energy as it moves in the direction of the gravitational field. This is a fancy way of saying that the particle speeds up as it falls.

The gravitational force is in the same direction as the particle's displacement, so the gravitational field does a *positive* amount of work on the particle. The gravitational force is constant, so the work done by gravity is

$$W_{grav} = w \Delta r \cos 0° = mg|y_f - y_i| = mgy_i - mgy_f \tag{29.5}$$

We have to be careful with signs because the displacement Δr must be a positive number.

Now we can see how the definition of ΔU in Equation 29.2 makes sense. The *change* in gravitational potential energy is

$$\Delta U_{grav} = U_f - U_i = -W_{grav}(i \to f) = mgy_f - mgy_i \tag{29.6}$$

Comparing the initial and final terms on the two sides of the equation, we see that the gravitational potential energy near the earth is

$$U_{grav} = U_0 + mgy \tag{29.7}$$

where U_0 is the value of U_{grav} at $y = 0$. We usually choose $U_0 = 0$, in which case $U_{grav} = mgy$, but such a choice is not necessary. The zero point of potential energy is an arbitrary choice because we have defined ΔU rather than U.

The uniform electric field between the plates of the parallel-plate capacitor of Figure 29.4 is analogous to the uniform gravitational field near the earth's surface. The one difference is that $\vec{g}$ always points down whereas the electric field inside a capacitor can point in any direction. To deal with this, let's define a coordinate axis s that points *from* the negative plate, which we define to be $s = 0$, *toward* the positive plate. The electric field $\vec{E}$ then points in the negative s-direction, just as the gravitational field $\vec{g}$ points in the negative y-direction. This s-axis, which is valid no matter how the capacitor is oriented, is analogous to the y-axis used for gravitational potential energy.

A positive charge q inside the capacitor speeds up and gains kinetic energy as it "falls" toward the negative plate. Is the charge losing potential energy as it gains kinetic energy? Indeed it is, and the calculation of the potential energy is just like the calculation of gravitational potential energy. The electric field exerts a *constant* force $F = qE$ on the charge in the direction of motion, thus the work done on the charge by the electric field is

$$W_{elec} = F\Delta r\cos 0° = qE|s_f - s_i| = qEs_i - qEs_f \qquad (29.8)$$

where we again have to be careful with the signs because $s_f < s_i$.

The work done by the electric field causes the charge to experience a change in *electric* potential energy given by

$$\Delta U_{elec} = U_f - U_i = -W_{elec}(i \rightarrow f) = qEs_f - qEs_i \qquad (29.9)$$

Comparing the initial and final terms on the two sides of the equation, we see that the **electric potential energy** of charge q in a uniform electric field is

$$U_{elec} = U_0 + qEs$$
(potential energy of charge q in a uniform electric field) $\qquad (29.10)$

where s is the distance from the negative plate and U_0 is the potential energy at the negative plate ($s = 0$). It will often be convenient to choose $U_0 = 0$, but the choice has no physical consequences because it doesn't affect ΔU_{elec}, the *change* in potential energy.

Equation 29.10 was derived with the assumption that q is positive, but it is valid for either sign of q. A negative value for q in Equation 29.10 causes the potential energy U_{elec} to become *more negative* as s increases. As Figure 29.5 shows, a negative charge speeds up and gains kinetic energy as it moves *away from* the negative plate of the capacitor.

NOTE ▶ Equation 29.10 is called "the potential energy of charge q," but we noted above that this is really the potential energy of the charge + capacitor system. However, to the extent that the charges on the capacitor plate stay fixed and only charge q moves, we're justified in thinking of this as the potential energy of just the charge. ◀

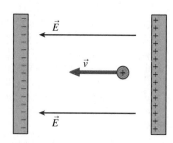

The potential energy of a positive charge decreases in the direction of $\vec{E}$. The charge gains kinetic energy as it moves toward the negative plate.

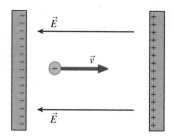

The potential energy of a negative charge decreases in the direction opposite to $\vec{E}$. The charge gains kinetic energy as it moves away from the negative plate.

FIGURE 29.5 A charged particle of either sign gains kinetic energy as it moves in the direction of decreasing potential energy.

EXAMPLE 29.1 Conservation of energy inside a capacitor

A 2.0 cm × 2.0 cm parallel-plate capacitor with a 2.0 mm spacing is charged to ±1.0 nC. First a proton, then an electron are released from rest at the midpoint of the capacitor.

a. What is each particle's change in electric potential energy from its release until it collides with one of the plates?
b. What is each particle's speed as it reaches the plate?

MODEL The mechanical energy of each particle is conserved. A parallel-plate capacitor has a uniform electric field.

VISUALIZE Figure 29.6 is a before-and-after pictorial representation similar to those you learned to draw in Part II. The proton moves toward the negative plate, the electron toward the positive plate.

SOLVE

a. The s-axis was defined to point from the negative toward the positive plate of the capacitor. Both charged particles have $s_i = \frac{1}{2}d$, where $d = 2.0$ mm is the plate separation. The

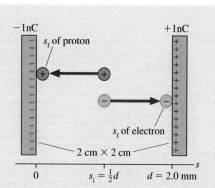

FIGURE 29.6 A proton and an electron in a capacitor.

positive proton loses potential energy and gains kinetic energy as it moves toward the negative plate. For the proton, with $q = +e$ and $s_f = 0$, the change in potential energy is

$$\Delta U_p = U_f - U_i = (U_0 + 0) - \left(U_0 + eE\frac{d}{2}\right) = -\frac{1}{2}eEd$$

where we used the electric potential energy for a charge in a uniform electric field. ΔU_p is negative, as expected. Notice that U_0 cancels when ΔU is calculated, which is why the value of U_0 has no physical consequences.

The electron moves toward the positive plate, which is the direction of decreasing potential energy for a negative charge. The electron has $q = -e$ and ends at $s_f = d$. Thus

$$\Delta U_e = U_f - U_i = (U_0 + (-e)Ed) - \left(U_0 + (-e)E\frac{d}{2}\right)$$

$$= -\frac{1}{2}eEd$$

Both particles have the *same* change in potential energy. The capacitor's electric field is

$$E = \frac{\eta}{\epsilon_0} = \frac{Q}{\epsilon_0 A} = 2.82 \times 10^5 \text{ N/C}$$

Using $d = 0.0020$ m, we find

$$\Delta U_p = \Delta U_e = -4.52 \times 10^{-17} \text{ J}$$

b. The law of conservation of energy is $\Delta K + \Delta U = 0$. Both particles are released from rest, hence $\Delta K = K_f - 0 = \frac{1}{2}mv_f^2$. Thus $\frac{1}{2}mv_f^2 = -\Delta U$, or

$$v_f = \sqrt{\frac{-2\Delta U}{m}} = \begin{cases} 2.33 \times 10^5 \text{ m/s for the proton} \\ 9.96 \times 10^6 \text{ m/s for the electron} \end{cases}$$

where we used the masses of the proton and the electron.

ASSESS Even though both particles have the same ΔU, the electron reaches a much larger final speed due to its much smaller mass.

STOP TO THINK 29.1 The positive charge is the end view of a positively charged glass rod. A negatively charged particle moves in a circular arc around the glass rod. Is the work done on the charged particle by the rod's electric field positive, negative, or zero?

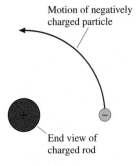

Motion of negatively charged particle

End view of charged rod

29.2 The Potential Energy of Point Charges

Now that we've introduced the idea of electric potential energy, let's look at *the fundamental interaction of electricity*—the force between two point charges. This force, given by Coulomb's law, varies with the distance between the two charges, so we'll begin by recalling how we dealt with variable forces in mechanics.

Figure 29.7 shows an object being pushed to the right by an expanding spring that has spring constant k. The force exerted by the spring decreases as it expands, hence we need to use the integral expression of Equation 29.4 to calculate the work done by the spring. The force, given by Hooke's law, is entirely in the direction of motion, so $F_s ds = F dx = -kx dx$. (The force is positive and to the right because, in this coordinate system, x is negative.) Thus the work done by the spring is

$$W_{sp} = \int_{x_i}^{x_f} F dx = -k \int_{x_i}^{x_f} x dx = -\frac{1}{2}kx_f^2 + \frac{1}{2}kx_i^2 \qquad (29.11)$$

The potential energy of the mass + spring system is related to the work done by

$$\Delta U_{sp} = U_f - U_i = -W_{sp}(i \rightarrow f) = \frac{1}{2}kx_f^2 - \frac{1}{2}kx_i^2 \qquad (29.12)$$

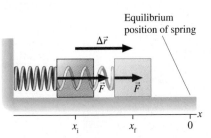

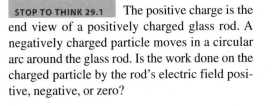

Equilibrium position of spring

$\Delta \vec{r}$

$\vec{F}$ $\vec{F}$

x_i x_f 0 x

FIGURE 29.7 The force exerted by a spring is not a constant force.

By comparing the left and right sides of this equation, we see that the potential energy of the mass + spring system is

$$U_{sp} = \frac{1}{2}kx^2 \qquad (29.13)$$

You should recall this result from Chapter 10. Here, where we can see the spring stretching and compressing, we really do need to think of this as the energy of the mass + spring system, not the energy of either the spring or the mass alone.

Figure 29.8 is the *energy diagram* for a mass + spring system. Recall that an energy diagram is a graphical representation of how the kinetic and potential energy are transformed as a particle moves. The parabola is the potential energy as a function of the object's position x, and the total energy line E_{mech} is the system's total mechanical energy. The particle oscillates between the *turning points* x_L and x_R where the total energy line crosses the potential-energy curve.

Now let's apply the same procedure to the interaction between two point charges. Figure 29.9a shows two charges q_1 and q_2, which we will assume to be like charges, exerting repulsive forces on each other. The potential energy of their interaction can be found by calculating the work done by q_1 on q_2 as q_2 moves from position x_i to position x_f. We'll assume that q_1 has been glued down and is unable to move.

We'll calculate the work just as we did for the spring, only we'll use Coulomb's law rather than Hooke's law for the force:

$$W_{elec} = \int_{x_i}^{x_f} F_{1 \text{ on } 2} \, dx = \int_{x_i}^{x_f} \frac{Kq_1q_2}{x^2} \, dx = Kq_1q_2 \frac{-1}{x}\Big|_{x_i}^{x_f} = -\frac{Kq_1q_2}{x_f} + \frac{Kq_1q_2}{x_i} \quad (29.14)$$

The potential energy of the two charges is related to the work done by

$$\Delta U_{elec} = U_f - U_i = -W_{elec}(i \rightarrow f) = \frac{Kq_1q_2}{x_f} - \frac{Kq_1q_2}{x_i} \qquad (29.15)$$

By comparing the left and right sides of the equation, as we did before, we see that the potential energy of the two-point-charge system is

$$U_{elec} = \frac{Kq_1q_2}{x} \qquad (29.16)$$

We chose to integrate along the x-axis for convenience, but what is really important is the *distance* between the charges. Thus a more general expression for the electric potential energy is

$$U_{elec} = \frac{Kq_1q_2}{r} = \frac{1}{4\pi\epsilon_0}\frac{q_1q_2}{r} \qquad \text{(two point charges)} \qquad (29.17)$$

This is explicitly the energy *of the system*, not the energy of just q_1 or q_2.

NOTE ▶ The electric potential energy of two point charges looks *almost* the same as the force between the charges. The difference is the r in the denominator of the potential energy compared to the r^2 in Coulomb's law. Make sure you remember which is which! ◀

Two important points need to be noted:

■ We derived Equation 29.17 for two like charges, but it is equally valid for two opposite charges. The potential energy of two like charges is *positive* and of two opposite charges is *negative*.

■ Because the electric field outside a *sphere of charge* is the same as that of a point charge at the center, Equation 29.17 is also the electric potential energy of two charged spheres. Distance r is the distance between their centers.

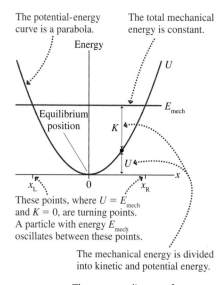

The potential-energy curve is a parabola.

The total mechanical energy is constant.

These points, where $U = E_{mech}$ and $K = 0$, are turning points. A particle with energy E_{mech} oscillates between these points.

The mechanical energy is divided into kinetic and potential energy.

FIGURE 29.8 The energy diagram for a mass + spring system.

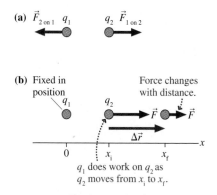

(a) $\vec{F}_{2 \text{ on } 1}$ q_1 q_2 $\vec{F}_{1 \text{ on } 2}$

(b) Fixed in position q_1 q_2 Force changes with distance.

q_1 does work on q_2 as q_2 moves from x_i to x_f.

FIGURE 29.9 The interaction between two point charges.

(a) Like charges

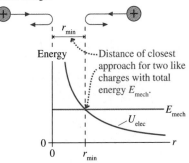

(b) Opposite charges

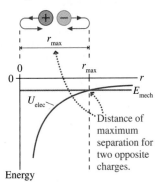

FIGURE 29.10 The potential-energy diagrams for two like charges and two opposite charges.

Figure 29.10 shows the potential-energy curves for two like charges and two opposite charges as a function of the distance r between them. Both curves are hyperbolas. Distance must be a positive number, so these graphs show only $r > 0$.

A charge released from rest must move in the direction of *decreasing* potential energy. Figure 29.10 shows that two like charges will move apart, because U decreases as r increases, and that two opposite charges will move toward one another. This is the expected behavior, based on Coulomb's law, but energy conservation provides an alternative perspective on the process.

More interesting, perhaps, is to consider two like charges that are shot toward each other with total mechanical energy E_{mech}. You can see in Figure 29.10a that the total energy line crosses the potential-energy curve at r_{min}. Two like charges shot toward each other will gradually slow down, because of the repulsive force between them, until the distance between them is r_{min}. At this point the kinetic energy is zero and both charged particles are instantaneously at rest. Both then reverse direction and move apart, speeding up as they go. r_{min} is the *distance of closest approach*. It is a quantity determined by energy conservation, not by analyzing the forces.

Similarly, you can see in Figure 29.10b that two oppositely charged particles shot apart from each other will slow down, losing kinetic energy, until reaching *maximum separation* r_{max}. Both reverse directions at the same instant, then they "fall" back together.

The Electric Force Is a Conservative Force

Potential energy can be defined only if the force is *conservative,* meaning that the work done on the particle as it moves from position i to position f is independent of the path followed between i and f. We *asserted* earlier that the electric force is a conservative force; now it's time to show it.

The work calculation in Equation 29.14 was based on Figure 29.9, where charge q_2 moved *straight out* from position i to position f. Figure 29.11 shows an alternative path between i and f. We can calculate the work done by the electric field as q_2 moves along this curved path by breaking the path into many small segments that are either radially outward from q_1 or circular arcs around q_1. The version shown in the figure is rather crude, with only a few segments, but you can imagine that we can approximate the true path arbitrarily closely by letting the number of segments approach infinity.

An alternative path for q_2 to move from i to f

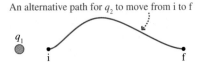

Approximate the path using circular arcs and radial lines centered on q_1.

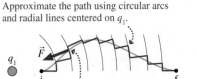

The electric force does zero work as q_2 moves along a circular arc because the force is perpendicular to the displacement.

All the work is done along the radial line segments, which are equivalent to a straight line from i to f.

FIGURE 29.11 Calculating the work done as q_2 moves along a curved path from i to f.

The electric force is a *central force*, directed straight at q_1. The electric force does *zero work* as q_2 moves along any of the circular arcs because the displacement is perpendicular to $\vec{F}$. All the work is done during the motion along the radial segments. The fact that the segments are displaced from each other doesn't affect the amount of work done along each segment. The total work, found by adding the work done along each radial segment, is equal to the total work that we calculated in Equation 29.14. The work is independent of the path, thus the electric force is a conservative force.

The Zero of Potential Energy

You can see both from Equation 29.17 for U_{elec} and from the graphs of Figure 29.10 that the potential energy of two charged particles approaches zero as $r \to \infty$. As you know, we can choose any point we wish as the zero of potential energy. The fact that only *changes* of potential energy ΔU appear in energy equations means that the zero of potential energy has no physical consequences.

Even if the zero point has no physical consequences, it's useful to think that $U = 0$ means "no interaction." A ball on the floor still has a weight, but its "ability to fall" is zero. Hence it makes sense to put the zero of gravitational potential energy at the floor—not essential, but useful.

For the two point charges, the zero point at infinity appeared from the integral in Equation 29.14 without our having made a conscious choice, but it's a good choice because two charged particles cease interacting only if they are infinitely far apart. A zero of potential energy at infinity allows us to think of U_{elec} as "the amount of interaction."

A zero point at infinity presents the one slight difficulty of interpreting negative energies. All a negative energy means is that the system has *less* energy than two charged particles that are infinitely far apart ($U_{\text{elec}} = 0$) *and* at rest ($K = 0$). Figure 29.12 shows that a system with the negative total energy $E_1 < 0$ is a *bound* system. The two charged particles cannot escape from each other. The electron and proton of a hydrogen atom are an example of a bound system.

Two opposite charges with the positive total energy $E_2 > 0$ *can* escape. They'll slow down as they move apart, but eventually the potential energy vanishes and they continue to coast apart with kinetic energy K_∞. The threshold condition is a system with $E = 0$. Two charged particles with $E = 0$ can escape, but it will take infinitely long because the kinetic energy approaches zero as the particles get far apart. The initial velocity that allows a particle to reach $r_f = \infty$ with $v_f = 0$ is called the **escape velocity.**

> **NOTE** ▶ Real particles can't be infinitely far apart, but because U_{elec} decreases with distance, there comes a point when $U_{\text{elec}} = 0$ is an excellent approximation. Two charged particles for which $U_{\text{elec}} \approx 0$ are sometimes described as "far apart" or "far away." ◄

Two particles with total energy $E_2 > 0$ can move apart forever. Their kinetic energy is K_∞ as $r \to \infty$.

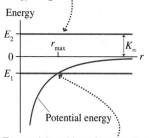

Two particles with total energy $E_1 < 0$ are a bound system. They can't get farther apart than r_{max}.

FIGURE 29.12 A system with $E_{\text{mech}} < 0$ is a *bound system.*

EXAMPLE 29.2 Approaching a charged sphere

A proton is fired from far away at a 1.0-mm-diameter glass sphere that has been charged to $+100$ nC. What initial speed must the proton have to just reach the surface of the glass?

MODEL Energy is conserved. The glass sphere can be treated as a charged particle, so the potential energy is that of two point charges. The proton starts "far away," which we interpret as sufficiently far to make $U_i \approx 0$.

VISUALIZE Figure 29.13 shows the before-and-after pictorial representation. To "just reach" the glass sphere means that the proton comes to rest, $v_f = 0$, as it reaches $r_f = 0.50$ mm, the *radius* of the sphere.

SOLVE Conservation of energy $K_f + U_f = K_i + U_i$ is

$$0 + \frac{K q_p q_{\text{sphere}}}{r_f} = \frac{1}{2} m v_i^2 + 0$$

The proton charge is $q_p = e$. With this, we can solve for the proton's initial speed:

$$v_i = \sqrt{\frac{2 K e q_{\text{sphere}}}{m r_f}} = 1.86 \times 10^7 \text{ m/s}$$

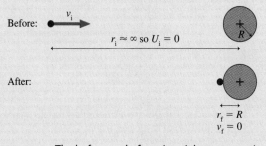

FIGURE 29.13 The before-and-after pictorial representation of a proton approaching a glass sphere.

EXAMPLE 29.3 Escape velocity

An interaction between two elementary particles causes an electron and a positron (a positive electron) to be shot out back-to-back with equal speeds. What minimum speed must each particle have when they are 100 fm apart in order to escape each other?

MODEL Energy is conserved. The particles end "far apart," which we interpret as sufficiently far to make $U_f \approx 0$.

VISUALIZE Figure 29.14 shows the before-and-after pictorial representation. The minimum speed to escape is the speed that allows the particles to reach $r_f = \infty$ with $v_f = 0$.

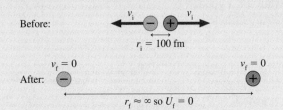

Before:

$r_i = 100$ fm

$v_f = 0$ $v_f = 0$

After:

$r_f \approx \infty$ so $U_f = 0$

FIGURE 29.14 The before-and-after pictorial representation of an electron and a positron flying apart.

SOLVE Here it is essential to interpret U_{elec} as the potential energy of the electron+positron system. Similarly, K is the *total* kinetic energy of the system. The electron and the positron, with equal masses and equal speeds, have equal kinetic energies. Conservation of energy $K_f + U_f = K_i + U_i$ is

$$0 + 0 + 0 = \frac{1}{2}mv_i^2 + \frac{1}{2}mv_i^2 + \frac{Kq_eq_p}{r_i} = mv_i^2 - \frac{Ke^2}{r_i}$$

Using $r_i = 100$ fm $= 1.0 \times 10^{-13}$ m, we can calculate the minimum initial speed to be

$$v_i = \sqrt{\frac{Ke^2}{mr_i}} = 5.0 \times 10^7 \text{ m/s}$$

ASSESS v_i is a little more than 10% the speed of light, just about the limit of what a "classical" calculation can predict. We would need to use the theory of relativity if v_i were any larger.

Multiple Point Charges

If more than two charges are present, the potential energy is the sum of the potential energies due to all pairs of charges:

$$U_{elec} = \sum_{i<j} \frac{Kq_iq_j}{r_{ij}} \tag{29.18}$$

where r_{ij} is the distance between q_i and q_j. The summation contains the $i < j$ restriction to ensure that each pair of charges is counted only once.

> **NOTE** ▶ If two or more charges don't move, their potential energy doesn't change and can be thought of as an additive constant with no physical consequences. It's necessary to calculate only the potential energy for those pairs of charges for which the distance r_{ij} changes. ◀

EXAMPLE 29.4 Launching an electron

Three electrons are spaced 1.0 mm apart along a vertical line. The outer two electrons are fixed in position.

a. Is the center electron at a point of stable or unstable equilibrium?
b. If the center electron is displaced horizontally by an infinitesimal distance, what will be its speed when it is very far away?

MODEL Energy is conserved. The outer two electrons don't move, so we don't need to include the potential energy of their interaction.

VISUALIZE Figure 29.15 shows the before-and-after pictorial representation.

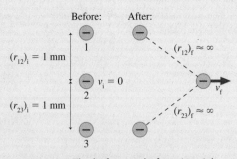

FIGURE 29.15 The before-and-after pictorial representation of three electrons.

SOLVE

a. The center electron is in equilibrium *exactly* in the center because the two electric forces on it balance. But if it moves a little to the right or left, no matter how little, then the horizontal components of the forces from both outer electrons will push the center electron farther away. This is an unstable equilibrium, like being on the top of a hill.

b. An infinitesimal displacement will cause the electron to move away. If the displacement is only infinitesimal, the initial conditions are $(r_{12})_i = (r_{23})_i = 1.0$ mm and $v_i = 0$. "Far away" is interpreted as $r_f \to \infty$, where $U_f \approx 0$. There are

now *two* terms in the potential energy, so conservation of energy $K_f + U_f = K_i + U_i$ gives

$$\frac{1}{2}mv_f^2 + 0 + 0 = 0 + \left[\frac{Kq_1q_2}{(r_{12})_i} + \frac{Kq_2q_3}{(r_{23})_i}\right]$$

$$= \left[\frac{Ke^2}{(r_{12})_i} + \frac{Ke^2}{(r_{23})_i}\right]$$

This is easily solved to give

$$v_f = \sqrt{\frac{2}{m}\left[\frac{Ke^2}{(r_{12})_i} + \frac{Ke^2}{(r_{23})_i}\right]} = 1006 \text{ m/s}$$

STOP TO THINK 29.2 Rank in order, from largest to smallest, the potential energies U_a to U_d of these four pairs of charges. Each + symbol represents the same amount of charge.

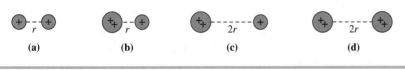

(a)   (b)   (c)   (d)

29.3 The Potential Energy of a Dipole

The electric dipole has been our model for understanding how charged objects interact with neutral objects. In Chapter 26 we found that an electric field exerts a *torque* on a dipole. We can complete the picture by calculating the potential energy of an electric dipole in a uniform electric field.

Figure 29.16a shows a dipole in an electric field $\vec{E}$. Recall that the dipole moment $\vec{p}$ is a vector that points from $-q$ to q with magnitude $p = qd$. The forces $\vec{F}_+$ and $\vec{F}_-$ exert a torque on the dipole, but now we're interested in calculating the *work* done by these forces as the dipole rotates from angle ϕ_i to angle ϕ_f. The work on the positive and negative charges is the same, so we'll analyze the positive charge and then multiply by 2.

Figure 29.16b looks more closely at the force on the positive charge. As the dipole rotates, the charge's small displacement is $ds = r\,d\phi = (\frac{1}{2}d)d\phi$. The small amount of work done by force $\vec{F}_+$ is

$$dW_+ = \vec{F}_+ \cdot d\vec{s} = F_+(ds)\cos\theta = \left(\frac{1}{2}q\,dE\right)\cos\theta\,d\phi = \left(\frac{1}{2}pE\right)\cos\theta\,d\phi \quad (29.19)$$

You can see from the figure that $\theta = \phi + 90°$, hence $\cos\theta = \cos(\phi + 90°) = -\sin\phi$. Thus the work done by the electric field on the dipole as it rotates through the small angle $d\phi$ is

$$dW_{elec} = 2\,dW_+ = -pE\sin\phi\,d\phi \quad (29.20)$$

where the factor of 2 includes the work done on the negative charge.

The total work done by the electric field as the dipole turns from ϕ_i to ϕ_f is

$$W_{elec} = -pE\int_{\phi_i}^{\phi_f} \sin\phi\,d\phi = pE\cos\phi_f - pE\cos\phi_i \quad (29.21)$$

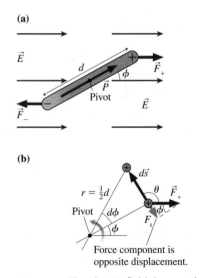

FIGURE 29.16 The electric field does work as a dipole rotates.

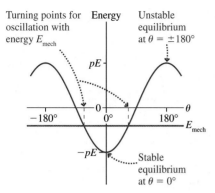

FIGURE 29.17 The energy of a dipole in an electric field.

The potential energy associated with the work done on the dipole is

$$\Delta U_{dipole} = U_f - U_i = -W_{elec}(i \rightarrow f) = -pE\cos\phi_f + pE\cos\phi_i \quad (29.22)$$

By comparing the left and right sides of Equation 29.22, we see that the potential energy of an electric dipole $\vec{p}$ in a uniform electric field $\vec{E}$ is

$$U_{dipole} = -pE\cos\phi = -\vec{p} \cdot \vec{E} \quad (29.23)$$

Figure 29.17 shows the energy diagram of a dipole. The potential energy is minimum at $\theta = 0°$ where the dipole is aligned with the electric field. This is a point of stable equilibrium. A dipole exactly opposite $\vec{E}$, at $\theta = \pm 180°$, is at a point of unstable equilibrium. The slightest disturbance will cause it to flip around. A frictionless dipole with mechanical energy E_{mech} will oscillate back and forth between turning points on either side of $\theta = 0°$.

EXAMPLE 29.5 Rotating a molecule

The water molecule is a permanent electric dipole with dipole moment 6.2×10^{-30} Cm. A water molecule is aligned in an electric field with field strength 1.0×10^7 N/C. How much energy is needed to rotate the molecule 90°?

MODEL The molecule is at the point of minimum energy. It won't spontaneously rotate 90°. However, an external force that supplies energy, such as a collision with another molecule, can cause the water molecule to rotate.

SOLVE The molecule starts at $\theta_i = 0°$ and ends at $\theta_f = 90°$. The increase in potential energy is

$$\Delta U_{dipole} = U_f - U_i = -pE\cos 90° - (-pE\cos 0°)$$
$$= pE = 6.2 \times 10^{-23} \text{ J}$$

This is the energy needed to rotate the molecule 90°.

ASSESS ΔU_{dipole} is significantly less than $k_B T$ at room temperature. Thus collisions with other molecules can easily supply the energy to rotate the water molecules and keep them from staying aligned with the electric field.

29.4 The Electric Potential

11.11 Activ Physics

We introduced the concept of the *electric field* in Chapter 25 because action at a distance raised concerns and difficulties. The field provides an intermediary through which two charges exert forces on each other. Charge q_1 somehow alters the space around it by creating an electric field $\vec{E}_1$. Charge q_2 then responds to the field, experiencing force $\vec{F} = q_2 \vec{E}_1$.

When we try to understand electric potential energy we face the same kinds of difficulties that action at a distance presented to an understanding of electric force. For a mass on a spring, we can *see* how the energy is stored in the stretched or compressed spring. But when we say two charged particles have a potential energy, an energy that can be converted to a tangible kinetic energy of motion, *where is the energy?* It's indisputable that two positive charges fly apart when you release them, gaining kinetic energy, but there's no obvious place that the energy had been stored.

In defining the electric field, we chose to separate the charges that are the *source* of the field from the charge *in* the field. The force on charge q is related to the electric field of the source charges by

force on q = [charge q] × [alteration of space by the source charges]

Let's try a similar procedure for the potential energy. The electric potential energy is due to the interaction of charge q with other charges, so let's divide up the potential energy of the system such that

potential energy of q + sources

= [charge q] × [*potential* for interaction of the source charges]

Figure 29.18 shows this idea schematically.

The potential at this point is V.

The source charges alter the space around them by creating an electric potential.

Source charges

If charge q is in the potential, the electric potential energy is $U_{q+sources} = qV$.

FIGURE 29.18 Source charges alter the space around them by creating an electric potential.

Equations 29.10 and 29.17 show that the charge q can indeed be separated out from the expressions for the potential energy of a charge in a parallel-plate capacitor and for the potential energy of a point charge interacting with another point charge. Thus we will define the **electric potential** V (or, for brevity, just *the potential*) as

$$V \equiv \frac{U_{q+\text{sources}}}{q} \tag{29.24}$$

Charge q is used as a probe to determine the electric potential, but the value of V is *independent of q.* **The electric potential is a property of the source charges.**

In practice, we're usually more interested in knowing the potential energy if a charge q happens to be at a point in space where the electric potential of the source charges is V. Turning Equation 29.24 around, we see that the electric potential energy is

$$U_{q+\text{sources}} = qV \tag{29.25}$$

Once the potential has been determined, it's very easy to find the potential energy.

The unit of electric potential is the joule per coulomb, which is called the **volt** V:

$$1 \text{ volt} = 1 \text{ V} \equiv 1 \text{ J/C}$$

This unit is named for Alessandro Volta, who invented the electric battery in the year 1800. Microvolts (μV), millivolts (mV), and kilovolts (kV) are commonly used units because the electric potentials used in practical applications differ significantly in magnitude.

NOTE ▶ Once again, commonly used symbols and notation are in conflict. The symbol V is widely used to represent *volume,* and now we're introducing the same symbol to represent *potential.* To make matters more confusing, V is the abbreviation for *volts.* In printed text, V for potential is italicized and V for volts is not, but you can't make such a distinction in handwritten work. This is not a pleasant state of affairs, but these are the commonly accepted symbols. It's incumbent upon you to be especially alert to the *context* in which a symbol is used. ◀

What Good Is the Electric Potential?

The electric potential is an abstract idea, and it will take some practice to see just what it means and how it is useful. We'll use multiple representations—words, pictures, graphs, and analogies—to explain and describe the electric potential.

We start with two essential ideas:

- The electric potential depends only on the source charges and their geometry. The potential is the "ability" of the source charges to have an interaction *if* a charge q shows up. This ability, or potential, is present throughout space **regardless of whether or not charge q is there to experience it.**
- If we know the electric potential V throughout a region of space, we'll immediately know the potential energy $U = qV$ of any charge that enters that region of space.

NOTE ▶ It is unfortunate that the terms *potential* and *potential energy* are so much alike. It is easy to confuse the two. Despite the similar names, they are very different concepts and are not interchangeable. Table 29.1 will help you to distinguish between the two. ◀

The source charges exert influence through the electric potential they establish throughout space. Once we know the potential—and the rest of this chapter deals

TABLE 29.1 Distinguishing electric potential and potential energy

The *electric potential* is a property of the source charges and, as you'll soon see, is related to the electric field. The electric potential is present whether or not a charged particle is there to experience it. Potential is measured in J/C, or V.

The *electric potential energy* is the interaction energy of a charged particle with the source charges. Potential energy is measured in J.

with how to calculate the potential—we can ignore the source charges and work just with the potential. The source charges remain hidden offstage.

The potential energy of a charged particle is determined by the electric potential: $U = qV$. Consequently, charged particles speed up or slow down as they move through a region of changing potential. Figure 29.19 illustrates this idea. It's useful to say that a particle moves through a **potential difference,** which is the difference $\Delta V = V_f - V_i$ between the potential at a starting point i and an ending point f. The potential difference between two points is often called the **voltage.** In illustrations, the potential difference between two points is represented by a double-headed green arrow.

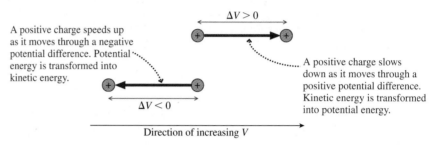

FIGURE 29.19 A charged particle speeds up or slows down as it moves through a potential difference.

If a particle moves through a potential difference ΔV, its electric potential energy is $\Delta U = q\Delta V$. We can write the conservation of energy equation in terms of the electric potential as $\Delta K + \Delta U = \Delta K + q\Delta V = 0$ or, as is often more practical,

$$K_f + qV_f = K_i + qV_i \qquad (29.26)$$

Conservation of energy is the basis of a powerful problem-solving strategy.

(MP) PROBLEM-SOLVING STRATEGY 29.1 **Conservation of energy in charge interactions**

MODEL Check whether there are any dissipative forces that would keep the mechanical energy from being conserved.

VISUALIZE Draw a before-and-after pictorial representation. Define symbols that will be used in the problem, list known values, and identify what you're trying to find.

SOLVE The mathematical representation is based on the law of conservation of mechanical energy:

$$K_f + qV_f = K_i + qV_i$$

■ Is the electric potential given in the problem statement? If not, you'll need to use a known potential, such as that of a point charge, or calculate the potential using the procedure given later, in Problem-Solving Strategy 29.2.
■ K_i and K_f are the sums of the kinetic energies of all moving particles.
■ Some problems may need additional conservation laws, such as conservation of charge or conservation of momentum.

ASSESS Check that your result has the correct units, is reasonable, and answers the question.

EXAMPLE 29.6 Moving through a potential difference

A proton with a speed of 2.0×10^5 m/s enters a region of space in which source charges have created an electric potential. What is the proton's speed after it moves through a potential difference of 100 V? What will be the final speed if the proton is replaced by an electron?

MODEL Energy is conserved. The electric potential determines the potential energy.

VISUALIZE Figure 29.20 is a before-and-after pictorial representation of a charged particle moving through a potential difference. A positive charge *slows down* as it moves into a region of higher potential ($K \rightarrow U$). A negative charge *speeds up* ($U \rightarrow K$).

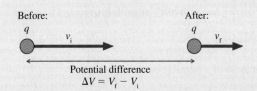

FIGURE 29.20 A charged particle moving through a potential difference.

SOLVE The potential energy of charge q is $U = qV$. Conservation of energy, now expressed in terms of the electric potential V, is $K_f + qV_f = K_i + qV_i$, or

$$K_f = K_i - q\Delta V$$

where $\Delta V = V_f - V_i$ is the potential difference. In terms of the speeds, energy conservation is

$$\frac{1}{2}mv_f^2 = \frac{1}{2}mv_i^2 - q\Delta V$$

We can solve this for the final speed:

$$v_f = \sqrt{v_i^2 - \frac{2q}{m}\Delta V}$$

For a proton, with $q = e$, the final speed is

$$(v_f)_{proton} = \sqrt{(2.0 \times 10^5 \text{ m/s})^2 - \frac{2(1.60 \times 10^{-19} \text{ C})(100 \text{ V})}{(1.67 \times 10^{-27} \text{ kg})}}$$

$$= 1.44 \times 10^5 \text{ m/s}$$

An electron, though, with $q = -e$ and a different mass, speeds up to $(v_f)_{electron} = 5.93 \times 10^6$ m/s.

ASSESS The electric potential *already existed* in space due to other charges that are not explicitly seen in the problem. The electron and proton have nothing to do with creating the potential. Instead, they *respond* to the potential by having potential energy $U = qV$.

STOP TO THINK 29.3   A proton is released from rest at point B, where the potential is 0 V. Afterward, the proton

a. Remains at rest at B.
b. Moves toward A with a steady speed.
c. Moves toward A with an increasing speed.
d. Moves toward C with a steady speed.
e. Moves toward C with an increasing speed.

-100 V   0 V $+100$ V

A•   B•   C•

29.5 The Electric Potential Inside a Parallel-Plate Capacitor

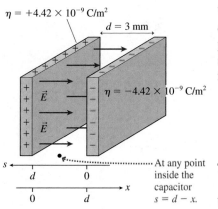

$\eta = +4.42 \times 10^{-9}$ C/m²

$d = 3$ mm

$\vec{E}$

$\vec{E}$

$\eta = -4.42 \times 10^{-9}$ C/m²

At any point inside the capacitor
$s = d - x$.

FIGURE 29.21 A parallel-plate capacitor.

We began this chapter with the potential energy of a charge inside a parallel-plate capacitor. Now let's investigate the electric potential. Figure 29.21 shows two parallel electrodes, separated by distance d, with surface charge density $\pm\eta$. As a specific example, we'll let $d = 3.0$ mm and $\eta = 4.42 \times 10^{-9}$ C/m². The electric field inside the capacitor, as you learned in Chapter 26, is

$$\vec{E} = \left(\frac{\eta}{\epsilon_0}, \text{from positive toward negative} \right) \tag{29.27}$$

$$= (500 \text{ N/C, from left to right})$$

This electric field is due to the *source charges* on the capacitor plates.

In Section 29.1, we found that the electric potential energy of a charge q in the uniform electric field of a parallel-plate capacitor is

$$U_{\text{elec}} = U_{q+\text{sources}} = qEs \tag{29.28}$$

We've set the constant term U_0 to zero. U_{elec} is the energy of q interacting with the source charges on the capacitor plates.

Our new view of the interaction is to separate the role of charge q from the role of the source charges by defining the electric potential $V = U_{q+\text{sources}}/q$. Thus the electric potential inside a parallel-plate capacitor is

$$V = Es \qquad \text{(electric potential inside a parallel-plate capacitor)} \tag{29.29}$$

where s is the distance from the *negative* electrode. The electric potential, like the electric field, exists at *all points* inside the capacitor. The electric potential is created by the source charges on the capacitor plates and exists whether or not charge q is inside the capacitor.

A first point to notice is that the electric potential increases linearly from the negative plate, where $V_- = 0$, to the positive plate, where $V_+ = Ed$. Let's define the *potential difference* ΔV_C between the two capacitor plates to be

$$\Delta V_C = V_+ - V_- = Ed \tag{29.30}$$

In our specific example, $\Delta V_C = (500 \text{ N/C})(0.003 \text{ m}) = 1.5$ V. The units work out because 1.5 (Nm)/C = 1.5 J/C = 1.5 V.

NOTE ▶ People who work with circuits would call ΔV_C "the voltage across the capacitor" or simply "the voltage." ◀

Equation 29.30 has an interesting implication. Thus far, we've determined the electric field inside a capacitor by specifying the surface charge density η on the plates. Alternatively, we could specify the capacitor voltage ΔV_C (i.e., the potential difference between the capacitor plates) and then determine the electric field strength as

$$E = \frac{\Delta V_C}{d} \tag{29.31}$$

In fact, this is how E is determined in practical applications because it's easy to measure ΔV_C with a voltmeter but difficult, in practice, to know the value of η.

Equation 29.31 implies that the units of electric field are volts per meter, or V/m. We have been using electric field units of newtons per coulomb. In fact, as you can show as a homework problem, these units are equivalent to each other. That is,

$$1 \text{ N/C} = 1 \text{ V/m}$$

NOTE ▶ Volts per meter are the electric field units used by scientists and engineers in practice. We will now adopt them as our standard electric field units. ◀

Returning to the electric potential, we can substitute Equation 29.31 for E into Equation 29.29 for V. It will also be useful to measure the distance along the usual left-to-right x-axis, and you can see in Figure 29.21 that $s = d - x$. Thus the electric potential inside the capacitor is

$$V = Es = \frac{\Delta V_C}{d}(d - x) = \left(1 - \frac{x}{d}\right)\Delta V_C \qquad (29.32)$$

The potential *decreases* linearly from $V_+ = \Delta V_C = 1.5$ V at the positive plate ($x = 0$) to $V_- = 0$ V at the negative plate ($x = d = 3.0$ mm).

Let's explore the electric potential inside the capacitor by looking at several different, but related, ways that the potential can be represented graphically.

Graphical representations of the electric potential inside a capacitor

A graph of potential versus x. You can see the potential decreasing from 1.5 V at the positive plate to 0 V at the negative plate.	A three-dimensional view showing **equipotential surfaces.** These are mathematical surfaces, not physical surfaces, that have the same value of V at every point. The equipotential surfaces of a capacitor are planes parallel to the capacitor plates. The capacitor plates are also equipotential surfaces.	A two-dimensional **contour map.** The capacitor plates and the equipotential surfaces are seen edge on, so you need to imagine them extending above and below the plane of the page.	A three-dimensional **elevation graph.** The potential is graphed vertically versus the x-coordinate on one axis and a generalized "yz-coordinate" on the other axis. The face-on view of the elevation graph gives you the potential graph.

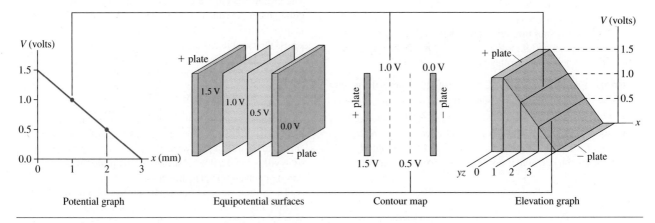

Potential graph   Equipotential surfaces   Contour map   Elevation graph

These four graphical representations show the same information from different perspectives, and the connecting lines help you see how they are related. If you think of the elevation graph as a "mountain," then the contour lines on the contour map are like the lines of a topographic map.

The potential graph and the contour map are the two representations most widely used in practice because they are easy to draw. Their limitation is that they are trying to convey three-dimensional information in a two-dimensional presentation. When you see graphs or contour maps, you need to imagine the three-dimensional equipotential surfaces or the three-dimensional elevation graph.

There's nothing special about showing equipotential surfaces or contour lines every 0.5 V. We chose these intervals because they were convenient. As an alternative, Figure 29.22 shows how the contour map looks if the contour lines are spaced every 0.3 V. Contour lines and equipotential surfaces are *imaginary* lines and surfaces drawn to help us visualize how the potential changes in space. Drawing the

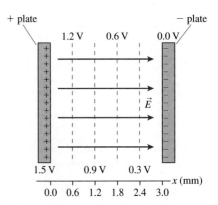
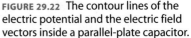

FIGURE 29.22 The contour lines of the electric potential and the electric field vectors inside a parallel-plate capacitor.

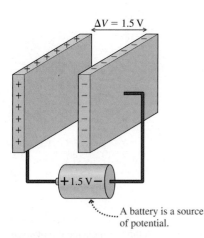

$\Delta V = 1.5$ V

$+1.5$ V$-$

A battery is a source of potential.

FIGURE 29.23 Using a battery to charge a capacitor to a precise value of ΔV_C.

map more than one way reinforces the idea that there is an electric potential at *every* point inside the capacitor, not just at the points where we happened to draw a contour line or an equipotential surface.

Figure 29.22 also shows the electric field vectors. Notice that

- The electric field vectors are perpendicular to the equipotential surfaces.
- The potential decreases along the direction in which the electric field points. In other words, the electric field points "downhill" on a graph or map of the electric potential.

Chapter 30 will present a more in-depth exploration of the connection between the electric field and the electric potential. There you will find that these observations are always true. They are not unique to the parallel-plate capacitor.

Finally, you might wonder how we can arrange a capacitor to have a surface charge density of precisely 4.42×10^{-9} C/m². It turns out to be simple. As Figure 29.23 shows, we merely use wires to attach the capacitor plates to a 1.5 V battery. This is another topic that we'll look at more carefully in Chapter 30, but it's worth noting now that **a battery is a source of potential.** That's why batteries are labeled in volts, and it's a major reason that we need to thoroughly understand the concept of potential.

EXAMPLE 29.7 A proton in a capacitor

A parallel-plate capacitor is constructed of two 2.0-cm-diameter disks spaced 2.0 mm apart. It is charged to a potential difference of 500 V.

a. What is the electric field strength inside?
b. How much charge is on each plate?
c. A proton is shot through a small hole in the negative plate with a speed of 2.0×10^5 m/s. Does it reach the other side? If not, where is the turning point?

MODEL Energy is conserved. The proton's potential energy inside the capacitor can be found from the capacitor's electric potential.

VISUALIZE Figure 29.24 is a before-and-after pictorial representation of the proton in the capacitor. Notice the *terminal symbols* where the potential is applied to the capacitor plates.

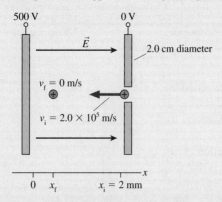

500 V   0 V

$\vec{E}$

2.0 cm diameter

$v_f = 0$ m/s

$v_i = 2.0 \times 10^5$ m/s

$0 \quad x_f \qquad x_i = 2$ mm

FIGURE 29.24 A before-and-after pictorial representation of a proton moving in a capacitor.

SOLVE

a. The electric field strength inside the capacitor is

$$E = \frac{\Delta V_C}{d} = \frac{500 \text{ V}}{0.0020 \text{ m}} = 2.5 \times 10^5 \text{ V/m}$$

b. Because $E = \eta/\epsilon_0$ for a parallel-plate capacitor, with $\eta = Q/A = Q/\pi R^2$, we find

$$Q = \pi R^2 \epsilon_0 E = 6.95 \times 10^{-10} \text{ C} = 0.695 \text{ nC}$$

c. The proton has charge $q = e$, and its potential energy at a point where the capacitor's potential is V is $U = eV$. It will gain potential energy $\Delta U = e\Delta V_C$ if it moves all the way across the capacitor. The increase in potential energy comes at the expense of kinetic energy, so the proton has sufficient kinetic energy to make it all the way across only if

$$K_i \geq e\Delta V_C$$

We can calculate that $K_i = 3.34 \times 10^{-17}$ J and that $e\Delta V_C = 8.00 \times 10^{-17}$ J. The proton does *not* have sufficient kinetic energy to be able to gain 8.00×10^{-17} J of potential energy, so it will not make it across. Instead, the proton will reach a turning point and reverse direction.

The proton starts at the negative plate, where $x_i = d = 2.0$ mm. Let the turning point be at x_f. The potential inside the capacitor is given by $V = \Delta V_C(1 - x/d)$ with $d = 0.0020$ m and $\Delta V_C = 500$ V. Conservation of energy requires $K_f + eV_f = K_i + eV_i$. This is

$$0 + e\Delta V_C\left(1 - \frac{x_f}{d}\right) = \frac{1}{2}mv_i^2 + 0$$

where we used $V_i = 0$ V at the negative plate ($x_i = d$). The solution for the turning point is

$$x_f = d - \frac{mdv_i^2}{2e\Delta V_C} = 1.16 \text{ mm}$$

The proton travels 0.84 mm, less than halfway across, before being turned back.

ASSESS We were able to use the electric potential inside the capacitor to determine the proton's potential energy. Notice that we used V/m as the electric field units.

We've assumed that $V = 0$ V at the negative plate, but that is not the only possible choice. Figure 29.25 shows three views of a parallel-plate capacitor that has a potential *difference* $\Delta V_C = 100$ V. These are contour maps, showing the edges of the equipotential surfaces.

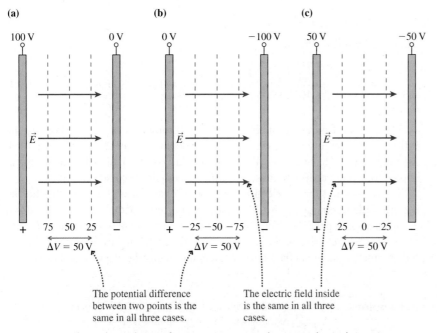

FIGURE 29.25 These three choices for $V = 0$ represent the same physical situation.

In Figure 29.25a we have chosen to let $V_{neg} = 0$ V. In this case, the potential of the positive plate $V_{pos} = \Delta V_C = 100$ V. At a point midway between the plates, at $x = d/2$, $V_{mid} = 50$ V.

Suppose, as Figure 29.25b shows, we choose the positive plate to have a potential of zero: $V_{pos} = 0$ V. After all, we're free to put the zero of potential and potential energy wherever we wish. Now the negative plate has potential $V_{neg} = -100$ V and the potential at the midpoint is $V_{mid} = -50$ V. Figure 29.25c shows yet a third choice, the symmetrical choice with $V = 0$ V in the center.

The important point to be made is that the three contour maps in Figure 29.25 represent the *same physical situation*. The potential difference between any two points is the same in all three maps. No matter which choice of zero we use, a charge q moving from $x = 0.75d$ to $x = 0.25d$ experiences a change of potential energy $\Delta U = 50q$ joules. We may *prefer* one of these figures over the other, but there is no measurable, physical difference between them.

STOP TO THINK 29.4   Rank in order, from largest to smallest, the potentials V_a to V_e at the points a to e.

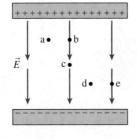

To determine the potential of q at this point . . .

q

q'

r . . . place charge q' at the point as a probe and measure the potential energy $U_{q'+q}$.

q

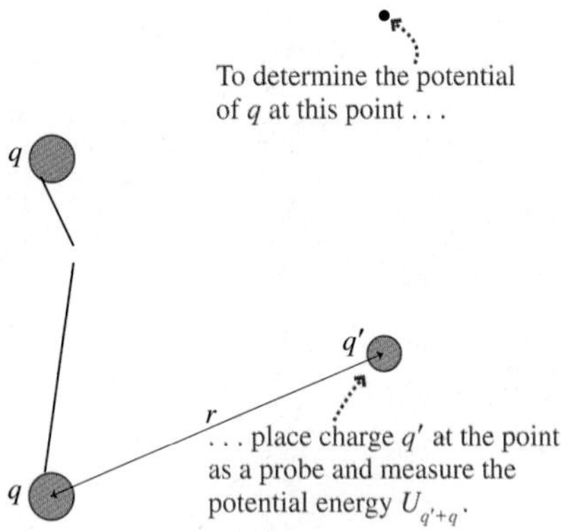

FIGURE 29.26 Measuring the electric potential of charge q.

29.6 The Electric Potential of a Point Charge

Another important electric potential is that of a point charge. Let q in Figure 29.26 be the source charge, and let a second charge q' probe the electric potential of q. The potential energy of the two point charges is

$$U_{q'+q} = \frac{1}{4\pi\epsilon_0} \frac{qq'}{r} \tag{29.33}$$

Thus, by definition, the electric potential of charge q is

$$V = \frac{U_{q'+q}}{q'} = \frac{1}{4\pi\epsilon_0} \frac{q}{r} \qquad \text{(electric potential of a point charge)} \tag{29.34}$$

The potential of Equation 29.34 extends through all of space, showing the influence of charge q, but it weakens with distance as $1/r$. This expression for V assumes that we have chosen $V = 0$ V to be at $r = \infty$. This is the most logical choice for a point charge because the influence of charge q ends at infinity.

The expression for the electric potential of charge q is similar to that for the electric field of charge q. The difference most quickly seen is that V depends on $1/r$ whereas $\vec{E}$ depends on $1/r^2$. But it is also important to notice that **the potential is a scalar** whereas the field is a *vector*. Thus the mathematics of using the potential are much easier than the vector mathematics using the electric field requires.

EXAMPLE 29.8 Calculating the potential of a point charge

What is the electric potential 1.0 cm from a +1.0 nC charge? What is the potential difference between a point 1.0 cm away and a second point 3.0 cm away?

SOLVE The potential at $r = 1.0$ cm is

$$V_{1\,cm} = \frac{1}{4\pi\epsilon_0} \frac{q}{r} = (9.0 \times 10^9 \,\text{N}\,\text{m}^2/\text{C}^2) \frac{1.0 \times 10^{-9}\,\text{C}}{0.010\,\text{m}}$$

$$= 900\,\text{V}$$

We can similarly calculate $V_{3\,cm} = 300$ V. Thus the potential difference between these two points is $\Delta V = V_{1\,cm} - V_{3\,cm} = 600$ V.

ASSESS 1 nC is typical of the electrostatic charge produced by rubbing, and you can see that such a charge creates a fairly large potential nearby. Why are we not shocked and injured when working with the "high voltages" of such charges? The sensation of being shocked is a result of current, not potential. Some high-potential sources simply do not have the ability to generate much current. We will look at this issue in Chapter 31.

Visualizing the Potential of a Point Charge

Figure 29.27 shows four graphical representations of the electric potential of a point charge. These match the four representations of the electric potential inside a capacitor, and a comparison of the two is worthwhile. This figure assumes that q is positive; you may want to think about how the representations would change if q were negative.

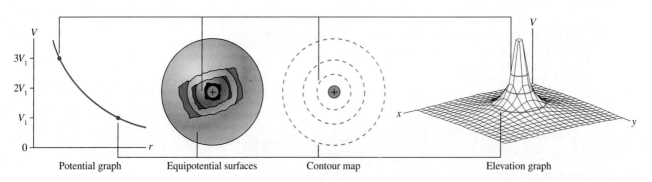

Potential graph   Equipotential surfaces   Contour map   Elevation graph

FIGURE 29.27 Four graphical representations of the electric potential of a point charge.

STOP TO THINK 29.5 Rank in order, from largest to smallest, the potential differences ΔV_{12}, ΔV_{13}, and ΔV_{23} between points 1 and 2, points 1 and 3, and points 2 and 3.

The Electric Potential of a Charged Sphere

In practice, you are more likely to work with a charged sphere, of radius R and total charge Q, than with a point charge. Outside a uniformly charged sphere, the electric potential is identical to that of a point charge Q at the center. That is,

$$V = \frac{1}{4\pi\epsilon_0}\frac{Q}{r} \qquad \text{(sphere of charge, } r \geq R) \qquad (29.35)$$

We can cast this result in a more useful form. It is customary to speak of charging an electrode, such as a sphere, "to" a certain potential, as in "Bob charged the sphere to a potential of 3000 volts." This potential, which we will call V_0, is the potential right on the surface of the sphere. We can see from Equation 29.35 that

$$V_0 = V(\text{at } r = R) = \frac{Q}{4\pi\epsilon_0 R} \qquad (29.36)$$

Consequently, a sphere of radius R that is charged to potential V_0 has total charge

$$Q = 4\pi\epsilon_0 R V_0 \qquad (29.37)$$

If we substitute this expression for Q into Equation 29.35, we can write the potential outside a sphere that is charged to potential V_0 as

$$V = \frac{R}{r}V_0 \qquad \text{(sphere charged to potential } V_0) \qquad (29.38)$$

Equation 29.38 tells us that the potential of a sphere is V_0 on the surface and decreases inversely with the distance. The potential at $r = 3R$ is $\frac{1}{3}V_0$.

EXAMPLE 29.9 A proton and a charged sphere
A proton is released from rest at the surface of a 1.0-cm-diameter sphere that has been charged to $+1000$ V.

a. What is the charge of the sphere?
b. What is the proton's speed when it is 1.0 cm from the sphere?

MODEL Energy is conserved. The potential outside the charged sphere is the same as the potential of a point charge at the center.

VISUALIZE Figure 29.28 is a before-and-after pictorial representation.

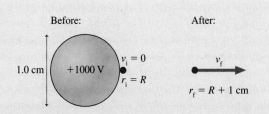

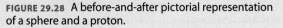

FIGURE 29.28 A before-and-after pictorial representation of a sphere and a proton.

SOLVE

a. We can use the sphere's potential to find that the charge of the sphere is

$$Q = 4\pi\epsilon_0 R V_0 = 0.56 \times 10^{-9}\,\text{C} = 0.56\,\text{nC}$$

b. A sphere charged to $V_0 = +1000\,\text{V}$ is positively charged. The proton will be repelled by this charge and move away from the sphere. The conservation of energy equation $K_f + eV_f = K_i + eV_i$, with Equation 29.38 for the potential of a sphere, is

$$\frac{1}{2}mv_f^2 + \frac{eR}{r_f}V_0 = \frac{1}{2}mv_i^2 + \frac{eR}{r_i}V_0$$

The proton starts from the surface of the sphere, $r_i = R$, with $v_i = 0$. When the proton is 1.0 cm from the *surface* of the sphere, it has $r_f = 1.0\,\text{cm} + R = 1.5\,\text{cm}$. Using these, we can solve for v_f:

$$v_f = \sqrt{\frac{2eV_0}{m}\left(1 - \frac{R}{r_f}\right)} = 3.57 \times 10^5\,\text{m/s}$$

ASSESS This example illustrates how the ideas of electric potential and potential energy work together, yet they are *not* the same thing.

29.7 The Electric Potential of Many Charges

Suppose there are many source charges $q_1, q_2, \ldots$ The electric potential V at a point in space is the sum of the potentials due to each charge:

$$V = \sum_i \frac{1}{4\pi\epsilon_0}\frac{q_i}{r_i} \tag{29.39}$$

where r_i is the distance from charge q_i to the point in space where the potential is being calculated. In other words, **the electric potential, like the electric field, obeys the principle of superposition.**

As an example, the contour map and elevation graph in Figure 29.29 show that the potential of an electric dipole is the sum of the potentials of the positive and negative charges. Potentials such as these have many practical applications. For example, electrical activity within the body can be monitored by measuring equipotential lines on the skin. Figure 29.29c shows that the equipotentials near the heart are a slightly distorted but recognizable electric dipole.

(a) Contour map

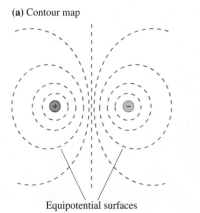

Equipotential surfaces

(b) Elevation graph

(c)

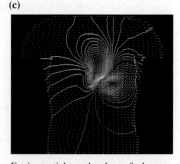

Equipotentials on the chest of a human are a slightly distorted electric dipole.

FIGURE 29.29 The electric potential of an electric dipole.

EXAMPLE 29.10 The potential of two charges

What is the electric potential at the point indicated in Figure 29.30?

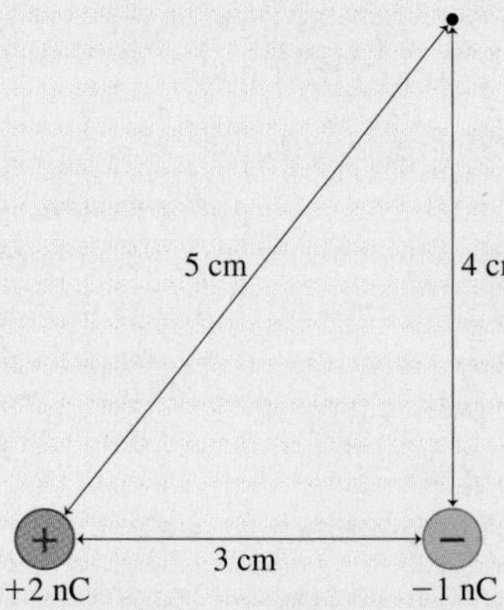

FIGURE 29.30 Finding the potential of two charges.

MODEL The potential is the sum of the potentials due to each charge.

SOLVE The potential at the indicated point is

$$V = \frac{1}{4\pi\epsilon_0}\frac{q_1}{r_1} + \frac{1}{4\pi\epsilon_0}\frac{q_2}{r_2}$$

$$= (9.0 \times 10^9 \, \text{N}\,\text{m}^2/\text{C}^2)\left(\frac{2.0 \times 10^{-9}\,\text{C}}{0.050\,\text{m}} + \frac{-1.0 \times 10^{-9}\,\text{C}}{0.040\,\text{m}}\right)$$

$$= 135\,\text{V}$$

ASSESS The potential is a *scalar*, so we found the net potential by adding two scalars. We don't need any angles or components to calculate the potential.

A Continuous Distribution of Charge

Equation 29.39 is the basis for determining the potential of a continuous distribution of charge, such as a charged rod or a charged disk. The procedure is much like the one you learned in Chapter 26 for calculating the electric field of a continuous distribution of charge, but *easier* because the potential is a scalar. We will continue to assume that the object is *uniformly charged,* meaning that the charges are evenly spaced over the object.

The strategy uses three basic steps:

1. Divide the total charge Q into many small point-like charges ΔQ.
2. Use our knowledge of the potential of a point charge to find the electric potential of each ΔQ.
3. Calculate the net potential by summing the potentials of all the ΔQ.

In practice, as with electric field calculations, we usually will let the sum become an integral.

Before looking at examples, let's flesh out the steps of the problem-solving strategy on the next page. The goal of the strategy is to break a problem down into small steps that are individually manageable.

(MP) **PROBLEM-SOLVING STRATEGY 29.2** **The electric potential of a continuous distribution of charge**

MODEL Model the charges as a simple shape, such as a line or a disk. Assume the charge is uniformly distributed.

VISUALIZE For the pictorial representation:

❶ Draw a picture and establish a coordinate system.
❷ Identify the point P at which you want to calculate the electric potential.
❸ Divide the total charge Q into small pieces of charge ΔQ, using shapes for which you *already know* how to determine V. This division is often, but not always, into point charges.
❹ Identify distances that need to be calculated.

SOLVE The mathematical representation is $V = \sum V_i$.

- Use superposition to form an algebraic expression for the potential at P.
- Let the (x, y, z) coordinates of the point remain as variables.
- Replace the small charge ΔQ with an equivalent expression involving a *charge density* and a *coordinate,* such as dx, that describes the shape of charge ΔQ. **This is the critical step in making the transition from a sum to an integral** because you need a coordinate to serve as the integration variable.
- All distances must be expressed in terms of the coordinates.
- Let the sum become an integral. The integration will be over the coordinate variable that is related to ΔQ. The integration limits for this variable will depend on the coordinate system you have chosen. Carry out the integration and simplify the result.

ASSESS Check that your result is consistent with any limits for which you know what the potential should be.

EXAMPLE 29.11 **The potential of a ring of charge**
Figure 29.31 shows a thin, uniformly charged ring of radius R with total charge Q. Find the potential at a point on the axis of the ring.

MODEL Because the ring is thin, we'll assume the charge lies along a circle of radius R.

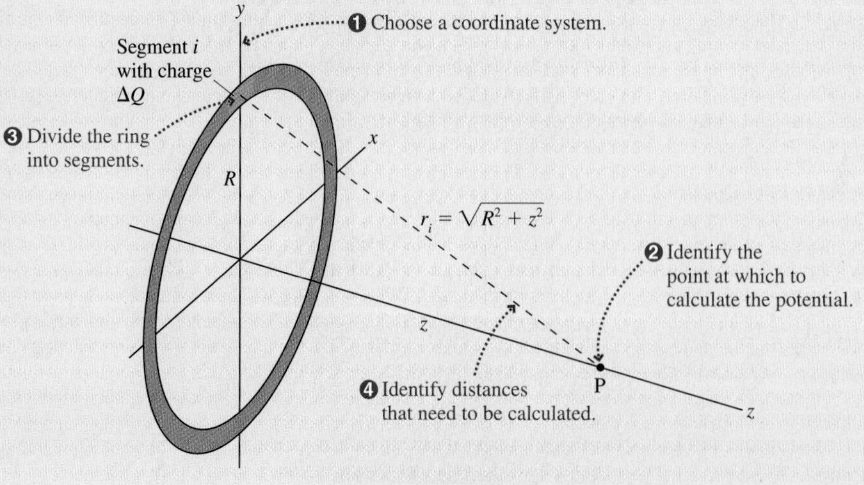

FIGURE 29.31 Finding the potential of a ring of charge.

VISUALIZE Figure 26.31 illustrates the four steps of the problem-solving strategy. We've chosen a coordinate system in which the ring lies in the xy-plane and point P is on the z-axis. We've then divided the ring into N small segments of charge ΔQ, each of which can be modeled as a point charge. The distance r_i between segment i and point P is

$$r_i = \sqrt{R^2 + z^2}$$

Note that r_i is a constant distance, the same for every charge segment. The potential V at P is the sum of the potentials due to each segment of charge:

$$V = \sum_{i=1}^{N} V_i = \sum_{i=1}^{N} \frac{1}{4\pi\epsilon_0} \frac{\Delta Q}{r_i} = \frac{1}{4\pi\epsilon_0} \frac{1}{\sqrt{R^2 + z^2}} \sum_{i=1}^{N} \Delta Q$$

We were able to bring all terms involving z to the front because z is a constant as far as the summation is concerned. Surprisingly, we don't need to convert the sum to an integral to complete this calculation. The sum of all the ΔQ charge segments around the ring is simply the ring's total charge, $\sum(\Delta Q) = Q$, hence the electric potential on the axis of a charged ring is

$$V_{\text{ring on axis}} = \frac{1}{4\pi\epsilon_0} \frac{Q}{\sqrt{R^2 + z^2}}$$

ASSESS From far away, the ring appears as a point charge Q in the distance. Thus we expect the potential of the ring to be that of a point charge when $z \gg R$. You can see that $V_{\text{ring}} \approx Q/4\pi\epsilon_0 z$ when $z \gg R$, which is, indeed, the potential of a point charge Q.

EXAMPLE 29.12 The potential of a disk of charge

Figure 29.32 shows a thin, uniformly charged disk of radius R with total charge Q. Find the potential at a point on the axis of the disk.

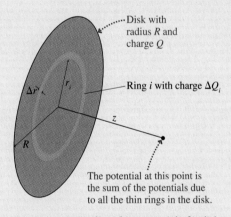

..... Disk with radius R and charge Q

— Ring i with charge ΔQ_i

The potential at this point is the sum of the potentials due to all the thin rings in the disk.

FIGURE 29.32 Finding the potential of a disk of charge.

MODEL The disk has uniform surface charge density $\eta = Q/A = Q/\pi R^2$. We can take advantage of now knowing the on-axis potential of a ring of charge.

VISUALIZE Orient the ring in the xy-plane, as shown in Figure 29.32, with point P at distance z. Then divide the disk into *rings* of equal width Δr. Ring i has radius r_i and charge ΔQ_i. We can use the result of Example 29.11 to write the potential at distance z of ring i as

$$V_i = \frac{1}{4\pi\epsilon_0} \frac{\Delta Q_i}{\sqrt{r_i^2 + z^2}}$$

The potential at P due to all the rings is the sum

$$V = \sum_i V_i = \frac{1}{4\pi\epsilon_0} \sum_{i=1}^{N} \frac{\Delta Q_i}{\sqrt{r_i^2 + z^2}}$$

The critical step is to relate ΔQ_i to a coordinate. Because we now have a surface, rather than a line, the charge in ring i is $\Delta Q_i = \eta \Delta A_i$, where ΔA_i is the area of ring i. We can find ΔA_i, as you've learned to do in calculus, by "unrolling" the ring to form a narrow rectangle of length $2\pi r_i$ and height Δr. Thus the area of ring i is $\Delta A_i = 2\pi r_i \Delta r$ and the charge is

$$\Delta Q_i = \eta \Delta A_i = \frac{Q}{\pi R^2} 2\pi r_i \Delta r = \frac{2Q}{R^2} r_i \Delta r$$

With this substitution, the potential at P is

$$V = \frac{1}{4\pi\epsilon_0} \sum_{i=1}^{N} \frac{2Q}{R^2} \frac{r_i \Delta r_i}{\sqrt{r_i^2 + z^2}} \rightarrow \frac{Q}{2\pi\epsilon_0 R^2} \int_0^R \frac{r\,dr}{\sqrt{r^2 + z^2}}$$

where, in the last step, we let $N \rightarrow \infty$ and the sum become an integral. This integral can be found in a table of integrals, but it's not hard to evaluate with a change of variables. Let $u = r^2 + z^2$, in which case $r\,dr = \frac{1}{2} du$. Changing variables requires that we also change the integration limits. You can see that $u = z^2$ when $r = 0$, and $u = R^2 + z^2$ when $r = R$. With these changes, the on-axis potential of a charged disk is

$$V_{\text{disk on axis}} = \frac{Q}{2\pi\epsilon_0 R^2} \int_{z^2}^{R^2+z^2} \frac{\frac{1}{2}du}{u^{1/2}} = \frac{Q}{2\pi\epsilon_0 R^2} u^{1/2} \Big|_{z^2}^{R^2+z^2}$$

$$= \frac{Q}{2\pi\epsilon_0 R^2} \left(\sqrt{R^2 + z^2} - z \right)$$

ASSESS Although we had to go through a number of steps, this procedure is easier than evaluating the electric field because we do not have to worry about components. Similar procedures can be followed to find the potential for any continuous distribution of charge.

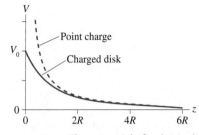

FIGURE 29.33 The potential of a charged disk and a point charge with the same Q.

We can find the potential V_0 of the disk itself by setting $z = 0$, giving $V_0 = Q/2\pi\epsilon_0 R$. In other words, placing charge Q on a disk of radius R charges it to potential V_0. The on-axis potential of the disk can be written in terms of V_0 as

$$V_{\text{disk on axis}} = V_0\left[\frac{\sqrt{R^2 + z^2} - z}{R}\right]$$

$$= V_0\left[\sqrt{1 + (z/R)^2} - (z/R)\right]$$ (29.40)

Figure 29.33 shows a graph of $V_{\text{disk on axis}}$ as a function of distance z along the axis. The potential of a point charge Q is shown for comparison. You can see that the charged disk begins to look like a point charge for $z \gg R$ but differs significantly from a point charge for $z \lesssim R$.

EXAMPLE 29.13 The potential of a dime

A dime, which is 17.5 mm in diameter, is given a charge of +5.0 nC.

a. What is the potential of the dime?
b. What is the potential energy of an electron 1.0 cm above the dime?

MODEL The dime is a thin charged disk.

SOLVE

a. The potential of the dime itself is the potential of a disk at $z = 0$:

$$V_0 = \frac{Q}{2\pi\epsilon_0 R} = 10{,}300 \text{ V}$$

b. To calculate the potential energy $U = qV$ of charge q, we first need to determine the potential of the disk at $z = 1.0$ cm. Using Equation 29.40 for the potential on the axis of the dime, we find

$$V = V_0\left[\sqrt{1 + (z/R)^2} - (z/R)\right] = 3870 \text{ V}$$

The electron's charge is $q = -e = -1.60 \times 10^{-19}$ C, so its potential energy at $z = 1.0$ cm is $U = -6.2 \times 10^{-16}$ J.

SUMMARY

The goal of Chapter 29 has been to calculate and use the electric potential and electric potential energy.

GENERAL PRINCIPLES

Sources of V

The electric potential, like the electric field, is created by charges.

Two major tools for calculating V are

- The potential of a point charge $V = \dfrac{1}{4\pi\epsilon_0}\dfrac{q}{r}$

- The principle of superposition

Multiple point charges

Use superposition: $V = V_1 + V_2 + V_3 + \ldots$

Continuous distribution of charge

- Divide the charge into point-like ΔQ.

- Find the potential of each ΔQ.

- Find V by summing the potentials of all ΔQ.

The summation usually becomes an integral. A critical step is replacing ΔQ with an expression involving a charge density and an integration coordinate. Calculating V is usually easier than calculating $\vec{E}$ because the potential is a scalar.

Consequences of V

A charged particle has potential energy

$$U = qV$$

at a point where source charges have created an electric potential V.

The electric force is a conservative force, so the mechanical energy is conserved for a charged particle in an electric potential:

$$K_f + U_f = K_i + U_i$$

The potential energy of **two point charges** separated by distance r is

$$U_{q_1+q_2} = \frac{Kq_1q_2}{r} = \frac{1}{4\pi\epsilon_0}\frac{q_1q_2}{r}$$

The **zero point** of potential and potential energy is chosen to be convenient. For point charges, we let $U = 0$ when $r \to \infty$.

The potential energy in an electric field of an **electric dipole** with dipole moment $\vec{p}$ is

$$U_{\text{dipole}} = -pE\cos\theta = -\vec{p}\cdot\vec{E}$$

APPLICATIONS

Graphical representations of the potential:

Potential graph **Equipotential surfaces**

Contour map **Elevation graph**

Sphere of charge Q

Same as a point charge if $r \geq R$.

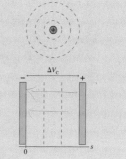

Parallel-plate capacitor

$V = Es$, where s is measured from the negative plate. The electric field inside is

$$E = \Delta V_C/d$$

Units

Electric potential: $1\ \text{V} = 1\ \text{J/C}$

Electric field: $1\ \text{V/m} = 1\ \text{N/C}$

TERMS AND NOTATION

electric potential energy, U	volt, V	equipotential surface
escape velocity	potential difference, ΔV	contour map
electric potential, V	voltage, ΔV	elevation graph

EXERCISES AND PROBLEMS

Exercises

Section 29.1 Electric Potential Energy

1. The electric field strength is 50,000 N/C inside a parallel-plate capacitor with a 2.0 mm spacing. A proton is released from rest at the positive plate. What is the proton's speed when it reaches the negative plate?
2. The electric field strength is 20,000 N/C inside a parallel-plate capacitor with a 1.0 mm spacing. An electron is released from rest at the negative plate. What is the electron's speed when it reaches the positive plate?
3. A proton is released from rest at the positive plate of a parallel-plate capacitor. It crosses the capacitor and reaches the negative plate with a speed of 50,000 m/s. What will be the proton's final speed if the experiment is repeated with double the amount of charge on each capacitor plate?
4. A proton is released from rest at the positive plate of a parallel-plate capacitor. It crosses the capacitor and reaches the negative plate with a speed of 50,000 m/s. The experiment is repeated with a He^+ ion (charge e, mass 4 u). What is the ion's speed at the negative plate?

Section 29.2 The Potential Energy of Point Charges

5. What is the electric potential energy of the group of charges in Figure Ex29.5?

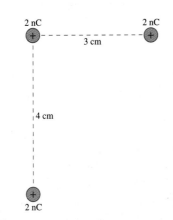

FIGURE EX29.5

6. What is the electric potential energy of the group of charges in Figure Ex29.6?

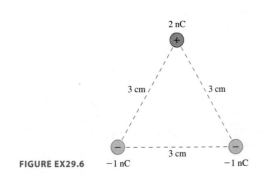

FIGURE EX29.6

7. What is the electric potential energy of the electron in Figure Ex29.7?

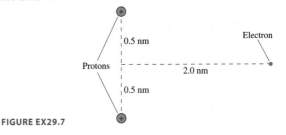

FIGURE EX29.7

8. Three electrons form an equilateral triangle 1.0 nm on each side. A proton is at the center of the triangle. What is the potential energy of this group of charges?

Section 29.3 The Potential Energy of a Dipole

9. A water molecule perpendicular to an electric field has 1.0×10^{-21} J more potential energy than a water molecule aligned with the field. The dipole moment of a water molecule is 6.2×10^{-30} C m. What is the strength of the electric field?
10. The graph shows the potential energy of an electric dipole. Consider a dipole that oscillates between $\pm 60°$.
    a. What is the dipole's mechanical energy?
    b. What is the dipole's kinetic energy when it is aligned with the electric field?

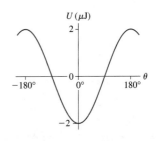

FIGURE EX29.10

Section 29.4 The Electric Potential

11. What is the speed of an electron that has been accelerated from rest through a potential difference of 1000 V?
12. What is the speed of a proton that has been accelerated from rest through a potential difference of -1000 V?
13. What potential difference is needed to accelerate a He^+ ion (charge $+e$, mass 4 u) from rest to a speed of 1.0×10^6 m/s?
14. What potential difference is needed to accelerate an electron from rest to a speed of 1.0×10^6 m/s?
15. An electron with an initial speed of 500,000 m/s is brought to rest by an electric field.
    a. Did the electron move into a region of higher potential or lower potential?
    b. What was the potential difference that stopped the electron?
16. A proton with an initial speed of 800,000 m/s is brought to rest by an electric field.
    a. Did the proton move into a region of higher potential or lower potential?
    b. What was the potential difference that stopped the proton?

Section 29.5 The Electric Potential Inside a Parallel-Plate Capacitor

17. Show that 1 V/m = 1 N/C.
18. a. What is the potential of an ordinary AA or AAA battery? (If you're not sure, find one and look at the label.)
    b. An AA battery is connected to a parallel-plate capacitor having 4.0-cm-diameter plates spaced 2 mm apart. How much charge does the battery supply to each plate?
19. A 2.0 cm × 2.0 cm parallel-plate capacitor has a 2.0 mm spacing. The electric field strength inside the capacitor is 1.0×10^5 V/m.
    a. What is the potential difference across the capacitor?
    b. How much charge is on each plate?
20. Two 2.0 cm × 2.0 cm plates that form a parallel-plate capacitor are charged to ±0.708 nC. What are the electric field strength inside and the potential difference across the capacitor if the spacing between the plates is (a) 1.0 mm and (b) 2.0 mm?
21. a. In Figure Ex29.21, which capacitor plate, left or right, is the positive plate?
    b. What is the electric field strength inside the capacitor?
    c. What is the potential energy of a proton at the midpoint of the capacitor?

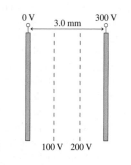

FIGURE EX29.21

Section 29.6 The Electric Potential of a Point Charge

22. A +25 nC charge is at the origin.
    a. What are the radii of the 1000 V, 2000 V, 3000 V, and 4000 V equipotential surfaces?
    b. Draw a contour map in the xy-plane showing the charge and these four surfaces.
23. a. What is the electric potential at points A, B, and C in Figure Ex29.23?
    b. What is the potential energy of an electron at each of these points?
    c. What are the potential differences ΔV_{AB} and ΔV_{BC}?

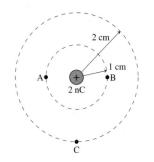

FIGURE EX29.23

24. A 1.0-mm-diameter ball bearing has 2.0×10^9 excess electrons. What is the ball bearing's potential?
25. A 2.0-mm-diameter glass bead is positively charged. The potential difference between a point 2.0 mm from the bead and a point 4.0 mm from the bead is 500 V. What is the charge on the bead?
26. In a semiclassical model of the hydrogen atom, the electron orbits the proton at a distance of 0.053 nm.
    a. What is the electric potential of the proton at the position of the electron?
    b. What is the electron's potential energy?

Section 29.7 The Electric Potential of Many Charges

27. What is the electric potential at the point indicated with the dot in Figure Ex29.27?

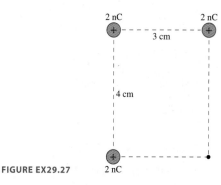

FIGURE EX29.27

28. What is the electric potential at the point indicated with the dot in Figure Ex29.28?

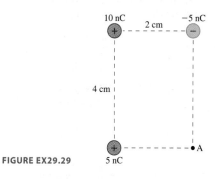

FIGURE EX29.28

29. a. What is the electric potential at point A in Figure Ex29.29?
    b. What is the potential energy of a proton at point A?

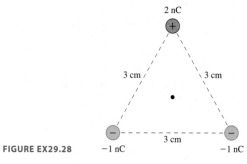

FIGURE EX29.29

30. a. What is the electric potential at point B in Figure Ex29.30?
    b. What is the potential energy of an electron at point B?

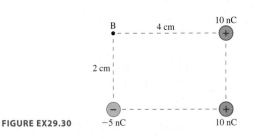

FIGURE EX29.30

31. A +3.0 nC charge is at $x = 0$ cm and a -1.0 nC charge is at $x = 4$ cm. At what point or points on the x-axis is the electric potential zero?

32. A -3.0 nC charge is on the x-axis at $x = -9$ cm and a $+4.0$ nC charge is on the x-axis at $x = 16$ cm. At what point or points on the y-axis is the electric potential zero?

33. Two point charges are located on the x-axis at $x = a$ and $x = b$. Figure Ex29.33 is a graph of E_x, the x-component of the electric field.
    a. What can you conclude about the signs and the relative magnitudes of the two charges?
    b. Draw a graph of the electric potential as a function of x.

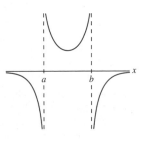

FIGURE EX29.33

34. Two point charges are located on the x-axis at $x = a$ and $x = b$. Figure Ex29.34 is a graph of V, the electric potential.
    a. What can you conclude about the signs and the relative magnitudes of the two charges?
    b. Draw a graph of E_x, the x-component of the electric field, as a function of x.

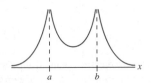

FIGURE EX29.34

35. The two halves of the rod are uniformly charged to $\pm Q$. What is the electric potential at the point indicated by the dot?

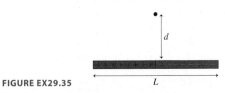

FIGURE EX29.35

Problems

36. Two point charges 2.0 cm apart have an electric potential energy $-180\ \mu$J. The total charge is 30 nC. What are the two charges?

37. Two positive point charges are 5.0 cm apart. If the electric potential energy is 72 μJ, what is the magnitude of the force between the two charges?

38. Bead A has a mass of 15 g and a charge of -5.0 nC. Bead B has a mass of 25 g and a charge of -10.0 nC. The beads are held 12 cm apart (measured between their centers) and released. What maximum speed and maximum acceleration are achieved by each bead?
    Hint: There are *two* conserved quantities. Make use of both.

39. A -2.0 nC charge and a $+2.0$ nC charge are located on the x-axis at $x = -1.0$ cm and $x = +1.0$ cm, respectively.
    a. At what position or positions on the x-axis is the electric field zero?
    b. At what position or positions on the x-axis is the electric potential zero?
    c. Draw graphs of the electric field strength and the electric potential along the x-axis.

40. A -10.0 nC point charge and a $+20.0$ nC point charge are 15 cm apart on the x-axis.
    a. What is the electric potential at the point on the x-axis where the electric field is zero?
    b. What is the magnitude of the electric field at the point on the x-axis, between the charges, where the electric potential is zero?

41. Two small metal spheres with masses 2.0 g and 4.0 g are tied together by a 5.0-cm-long massless string and are at rest on a frictionless surface. Each is charged to $+2.0\ \mu$C.
    a. What is the energy of this system?
    b. What is the tension in the string?
    c. The string is cut. What is the speed of each sphere when they are far apart?
    Hint: There are *two* conserved quantities. Make use of both.

42. A proton and an alpha particle ($q = +2e$, $m = 4$ u) are fired directly toward each other from far away, each with an initial speed of $0.01c$. What is their distance of closest approach, as measured between their centers?
    Hint: There are *two* conserved quantities. Make use of both.

43. Figure P29.43 shows four charged particles at the corners of a rectangle. What is the total kinetic energy a long time after these charges have been released?

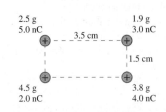

FIGURE P29.43

44. A proton's speed as it passes point A is 50,000 m/s. It follows the trajectory shown in Figure P29.44. What is the proton's speed at point B?

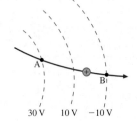

FIGURE P29.44

45. Two 2.0-cm-diameter disks spaced 2.0 mm apart form a parallel-plate capacitor. The electric field between the disks is 5.0×10^5 V/m.
    a. What is the voltage across the capacitor?
    b. How much charge is on each disk?
    c. An electron is launched from the negative plate. It strikes the positive plate at a speed of 2.0×10^7 m/s. What was the electron's speed as it left the negative plate?

46. What is the ratio $\Delta V_p / \Delta V_e$ of the potential differences that will accelerate a proton and an electron from rest to (a) the same final speed and (b) the same final kinetic energy?

47. An arrangement of source charges produces the electric potential $V = 5000x^2$ along the x-axis, where V is in volts and x is in meters.
    a. Graph the potential between $x = -10$ cm and $x = +10$ cm.
    b. Describe the motion of a positively charged particle in this potential.
    c. What is the mechanical energy of a 1.0 g, 10 nC charged particle if its turning points are at ± 8.0 cm?
    d. What is the particle's maximum speed?

48. In Figure P29.48, a proton is fired with a speed of 200,000 m/s from the midpoint of the capacitor toward the positive plate.
    a. Show that this is insufficient speed to reach the positive plate.
    b. What is the proton's speed as it collides with the negative plate?

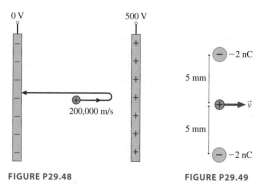

FIGURE P29.48   **FIGURE P29.49**

49. What is the escape speed of the proton in Figure P29.49?

50. What is the escape speed of an electron launched from the surface of a 1.0-cm-diameter plastic sphere that has been charged to 10 nC?

51. Your lab assignment is to use positive charge Q to launch a proton, starting from rest, so that it acquires the maximum possible speed. You can launch the proton from the surface of a sphere of positive charge Q and radius R, or from the center of a ring of charge Q and radius R, or from the center of a disk of charge Q and radius R. Which will you choose?

52. An electric dipole consists of 1.0 g spheres charged to ± 2.0 nC at the ends of a 10-cm-long massless rod. The dipole rotates on a frictionless pivot at its center. The dipole is held perpendicular to a uniform electric field with field strength 1000 V/m, then released. What is the dipole's angular velocity at the instant it is aligned with the electric field?

53. A proton is fired from far away toward the nucleus of an iron atom. Iron is element number 26, and the diameter of the nucleus is 9.0 fm. What initial speed does the proton need to just reach the surface of the nucleus? Assume the nucleus remains at rest.

54. A proton is fired from far away toward the nucleus of a mercury atom. Mercury is element number 80, and the diameter of the nucleus is 14.0 fm. If the proton is fired at a speed of 4.0×10^7 m/s, what is its closest approach to the surface of the nucleus? Assume the nucleus remains at rest.

55. In the form of radioactive decay known as *alpha decay*, an unstable nucleus emits a helium-atom nucleus, which is called an *alpha particle*. An alpha particle contains two protons and two neutrons, thus having mass $m = 4$ u and charge $q = 2e$. Suppose a uranium nucleus with 92 protons decays into thorium, with 90 protons, and an alpha particle. The alpha particle is initially at rest at the surface of the thorium nucleus, which is 15 fm in diameter. What is the speed of the alpha particle when it is detected in the laboratory? Assume the thorium nucleus remains at rest.

56. One form of nuclear radiation, *beta decay*, occurs when a neutron changes into a proton, an electron, and a chargeless particle called a *neutrino*: $n \rightarrow p^+ + e^- + \nu$ where ν is the symbol for a neutrino. When this change happens to a neutron within the nucleus of an atom, the proton remains behind in the nucleus while the electron and neutrino are ejected from the nucleus. The ejected electron is called a *beta particle*. One nucleus that exhibits beta decay is the isotope of hydrogen ^{3}H, called *tritium*, whose nucleus consists of one proton (making it hydrogen) and two neutrons (giving tritium an atomic mass $m = 3$ u). Tritium is radioactive, and it decays to helium: $^3\text{H} \rightarrow \,^3\text{He} + e^- + \nu$.
    a. Is charge conserved in the beta decay process? Explain.
    b. Why is the final product a helium atom? Explain.
    c. The nuclei of both ^{3}H and ^{3}He have radii of 1.5×10^{-15} m. With what minimum speed must the electron be ejected if it is to escape from the nucleus and not fall back?

57. Two 10.0-cm-diameter electrodes 0.5 cm apart form a parallel-plate capacitor. The electrodes are attached by metal wires to the terminals of a 15 V battery. After a long time, the capacitor is disconnected from the battery but is not discharged. What are the charge on each electrode, the electric field strength inside the capacitor, and the potential difference between the electrodes
    a. Right after the battery is disconnected?
    b. After insulating handles are used to pull the electrodes away from each other until they are 1.0 cm apart?
    c. After the original electrodes (not the modified electrodes of part b) are expanded until they are 20.0 cm in diameter?

58. Two 10.0-cm-diameter electrodes 0.5 cm apart form a parallel-plate capacitor. The electrodes are attached by metal wires to the terminals of a 15 V battery. What are the charge on each electrode, the electric field strength inside the capacitor, and the potential difference between the electrodes
    a. While the capacitor is attached to the battery?
    b. After insulating handles are used to pull the electrodes away from each other until they are 1.0 cm apart? The electrodes remain connected to the battery during this process.
    c. After the original electrodes (not the modified electrodes of part b) are expanded until they are 20.0 cm in diameter while remaining connected to the battery?

59. a. Find an algebraic expression for the electric field strength E_0 at the surface of a charged sphere in terms of the sphere's potential V_0 and radius R.
    b. What is the electric field strength at the surface of a 1.0-cm-diameter marble charged to 500 V?

60. Two spherical drops of mercury each have a charge of 0.1 nC and a potential of 300 V at the surface. The two drops merge to form a single drop. What is the potential at the surface of the new drop?

61. A Van de Graaff generator is a device for generating a large electric potential by building up charge on a hollow metal sphere. A typical classroom-demonstration model has a diameter of 30.0 cm.
    a. The generator is charged up by placing charge on the *inside* surface of the metal sphere. What happens to the charge after it is placed there? Explain.
    b. How much charge is needed on the sphere for its potential to be 500,000 V?
    c. What is the electric field strength just *inside* and just *outside* the surface of the sphere when it is charged to 500,000 V?

62. A thin spherical shell of radius R has total charge Q. What is the electric potential at the center of the shell?

63. Figure P29.63 shows two uniformly charged spheres. What is the potential difference between points a and b? Which point is at the higher potential?
    Hint: The potential at any point is the superposition of the potentials due to *all* charges.

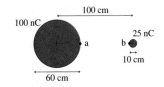

100 nC   100 cm

25 nC

a   b

10 cm

FIGURE P29.63   60 cm

64. An electric dipole with dipole moment p is oriented along the y-axis.
    a. Find an expression for the electric potential on the y-axis at a point where y is much larger than the charge spacing s. Write your expression in terms of the dipole moment p.
    b. The dipole moment of a water molecule is 6.2×10^{-30} C m. What is the electric potential 1.0 nm from a water molecule along the axis of the dipole?

65. Two positive point charges q are located on the x-axis at $x = \pm\frac{1}{2}s$.
    a. Find an expression for the electric potential at a point on the x-axis where $|x| \ll s$.
    b. Describe the motion of a proton that moves along the x-axis near the origin.

66. Two positive point charges q are located on the y-axis at $y = \pm\frac{1}{2}s$.
    a. Find an expression for the potential along the x-axis.
    b. Draw a graph of V versus x for $-\infty < x < \infty$. For comparison, use a dotted line to show the potential of a point charge $2q$ located at the origin.

67. The arrangement of charges shown in Figure P29.67 is called a *linear electric quadrupole*. The positive charges are located at $y = \pm s$. Notice that the net charge is zero. Find an expression for the electric potential on the y-axis at distances $y \gg s$.

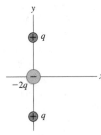

FIGURE P29.67

68. Show that the on-axis potential of a charged disk reduces to the potential of a point charge Q when $z \gg R$.

69. Figure P29.69 shows a thin rod of length L and charge Q. Find an expression for the electric potential a distance x away from the center of the rod on the axis of the rod.

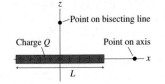

z

Point on bisecting line

Charge Q Point on axis

x

FIGURE P29.69 L

70. Figure P29.69 showed a thin rod of length L and charge Q. Find an expression for the electric potential a distance z away from the center of rod on the line that bisects the rod.

71. Figure P29.71 shows a thin rod with charge Q that has been bent into a semicircle of radius R. Find an expression for the electric potential at the center.

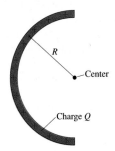

R

Center

Charge Q

FIGURE P29.71

72. A disk with a hole has inner radius R_{in} and outer radius R_{out}. The disk is uniformly charged with total charge Q. Find an expression for the on-axis electric potential at distance z from the center of the disk. Verify that your expression has the correct behavior when $R_{in} \rightarrow 0$.

In Problems 73 through 76 you are given the equation(s) used to solve a problem. For each of these,
    a. Write a realistic problem for which this is the correct equation(s).
    b. Finish the solution of the problem.

73. $\dfrac{(9.0 \times 10^9 \text{ N m}^2/\text{C}^2)q_1 q_2}{0.030 \text{ m}} = 90 \times 10^{-6} \text{ J}$

    $q_1 + q_2 = 40 \text{ nC}$

74. $\frac{1}{2}(1.67 \times 10^{-27} \text{ kg})(2.5 \times 10^6 \text{ m/s})^2 + 0 =$

    $\frac{1}{2}(1.67 \times 10^{-27} \text{ kg})v_i^2 +$

    $\dfrac{(9.0 \times 10^9 \text{ N m}^2/\text{C}^2)(2.0 \times 10^{-9} \text{ C})(1.60 \times 10^{-19} \text{ C})}{0.0010 \text{ m}}$

75. $\dfrac{100 \text{ V}}{0.0010 \text{ m}} = \dfrac{Q/(0.020 \text{ m})(0.020 \text{ m})}{8.85 \times 10^{-12} \text{ C}^2/\text{N m}^2}$

76. $\dfrac{(9.0 \times 10^9 \text{ N m}^2/\text{C}^2)(3.0 \times 10^{-9} \text{ C})}{0.030 \text{ m}} +$

    $\dfrac{(9.0 \times 10^9 \text{ N m}^2/\text{C}^2)(3.0 \times 10^{-9} \text{ C})}{(0.030 \text{ m}) + d} = 1200 \text{ V}$

Challenge Problems

77. A 1.0 nC charge is at the origin. A -3.0 nC charge is on the x-axis at $x = 4.0$ cm. Find *all* the points in the xy-plane at which the potential is zero. Give your answer as a contour map showing the $V = 0$ V equipotential line.

78. The four 1.0 g spheres shown in Figure CP29.78 are released simultaneously and allowed to move away from each other. What is the speed of each sphere when they are very far apart?

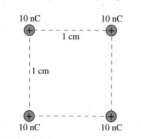

FIGURE CP29.78

79. The 2.0-mm-diameter spheres in Figure CP29.79 are released from rest. What are their speeds v_A and v_B when they are very far apart?

FIGURE CP29.79

80. The 2.0-mm-diameter spheres in Figure CP29.80 are released from rest. What are their speeds v_C and v_D at the instant they collide?

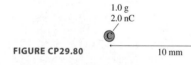

FIGURE CP29.80

81. An electric dipole has dipole moment p. If $r \gg s$, where s is the separation between the charges, show that the electric potential of the dipole can be written

$$V = \frac{1}{4\pi\epsilon_0} \frac{p\cos\theta}{r^2}$$

where r is the distance from the center of the dipole and θ is the angle from the dipole axis.

82. Electrodes of area A are spaced distance d apart to form a parallel-plate capacitor. The electrodes are charged to $\pm q$.
    a. What is the infinitesimal increase in electric potential energy dU if an infinitesimal amount of charge dq is moved from the negative electrode to the positive electrode?
    b. An uncharged capacitor can be charged to $\pm Q$ by transferring charge dq over and over and over. Use your answer to part a to show that the potential energy of a capacitor charged to $\pm Q$ is $U_{cap} = \frac{1}{2} Q\Delta V_C$.

83. A sphere of radius R has charge q.
    a. What is the infinitesimal increase in electric potential energy dU if an infinitesimal amount of charge dq is brought from infinity to the surface of the sphere?
    b. An uncharged sphere can acquire total charge Q by the transfer of charge dq over and over and over. Use your answer to part a to find an expression for the potential energy of a sphere of radius R with total charge Q.
    c. Your answer to part b is the amount of energy needed to assemble a charged sphere. It is often called the *self-energy* of the sphere. What is the self-energy of a proton, assuming it to be a charged sphere with a diameter of 1.0×10^{-15} m?

84. A hollow cylindrical shell of length L and radius R has charge Q uniformly distributed along its length. What is the electric potential at the center of the cylinder?

STOP TO THINK ANSWERS

Stop to Think 29.1: Zero. The motion is always perpendicular to the electric force.

Stop to Think 29.2: $U_b = U_d > U_a = U_c$. The potential energy depends inversely on r. The effects of doubling the charge and doubling the distance cancel each other.

Stop to Think 29.3: c. The proton gains speed by losing potential energy. It loses potential energy by moving in the direction of decreasing electric potential.

Stop to Think 29.4: $V_a = V_b > V_c > V_d = V_e$. The potential decreases steadily from the positive to the negative plate. It depends only on the distance from the positive plate.

Stop to Think 29.5: $\Delta V_{13} = \Delta V_{23} > \Delta V_{12}$. The potential depends only on the *distance* from the charge, not the direction. $\Delta V_{12} = 0$ because these points are at the same distance.

30 Potential and Field

One way to establish a potential difference is with a battery.

▶ **Looking Ahead**
The goal of Chapter 30 is to understand how the electric potential is connected to the electric field. In this chapter you will learn to:

- Calculate the electric potential from the electric field.
- Calculate the electric field from the electric potential.
- Understand the geometry of the potential and the field.
- Understand and use sources of electric potential.
- Understand and use capacitors.

◀ **Looking Back**
This chapter continues our exploration of topics that we began in Chapters 28 and 29. Please review:

- Sections 28.3–28.5 Batteries, current, and resistivity.
- Sections 29.4–29.6 The electric potential and its graphical representation.

Everyone is familiar with batteries. You probably have several near you as you read this, in your watch, your calculator, your computer, and your CD player. Batteries are essential for electronics, but just what does a battery actually *do*?

Batteries are just one of several topics that we'll explore as we continue our investigation of the electric potential. The larger issue that we must first address is the connection between the electric potential and the electric field. The potential and the field are not two independent ideas, merely two different perspectives on how the source charges alter the space around them. Exploring the connection between the potential and the field will strengthen your understanding of both.

Our discussion of the electric potential will lead naturally into important applications, including batteries, capacitors, and the current in wires. These applications will set the stage for Chapter 31, on electric circuits.

30.1 Connecting Potential and Field

Chapter 29 introduced the concept of the *electric potential*. To continue our investigation of this important idea, Figure 30.1 shows schematically the four key ideas of force, field, potential energy, and potential. We explored the connection between the force on a particle and the particle's potential energy in Chapters 10 and 11. In Chapter 25, we dealt with the long-range nature of the electric force by generalizing the idea of a force to that of the electric field, and we've based the idea of electric potential on that of potential energy.

The "missing link" is the connection between the electric potential and the electric field, and that is the focus of this chapter. **The electric potential and electric field are not two distinct entities but, instead, two different perspectives or two different mathematical representations of how source charges alter the space around them.**

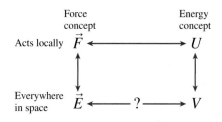

FIGURE 30.1 The four key ideas.

Finding the Potential from the Electric Field

Suppose we know the electric field in some region of space. If the potential and the field are really the same thing seen from two perspectives, we should be able to find the electric potential from the electric field. Chapter 29 introduced all the pieces for doing so, we just need to assemble them into a general statement.

First, we used the potential energy of charge q and the source charges to define the electric potential as

$$V \equiv \frac{U_{q+\text{sources}}}{q} \tag{30.1}$$

Second, we defined the potential energy in terms of the work done by force $\vec{F}$ on charge q as it moves from position i to position f:

$$\Delta U = -W(\text{i} \rightarrow \text{f}) = -\int_{s_i}^{s_f} F_s\,ds \tag{30.2}$$

Third, we found that the force exerted on charge q by the electric field is $\vec{F} = q\vec{E}$. Putting these three pieces together, you can see that the charge q cancels out and the potential difference between two points in space is

$$\Delta V = V(s_f) - V(s_i) = -\int_{s_i}^{s_f} E_s\,ds \tag{30.3}$$

where s is the position along a line from point i to point f.

As a simple example, suppose the electric field is uniform. A constant E_s can be taken outside the integral, and we find that the potential difference between two points is

$$\Delta V = -E_s \Delta s \quad \text{(uniform electric field)} \tag{30.4}$$

The potential difference depends only on the distance between the two points and on the component of $\vec{E}$ along the line between them.

NOTE ▶ Equation 30.3 determines only the potential *difference* ΔV. If you want to assign a specific value of V to a point in space, you must first make a choice of the zero point of the potential. ◀

Equation 30.3 is the basis for a procedure, shown in Tactics Box 30.1 on the next page, that can be used to find the potential difference between two points whenever you already know the electric field.

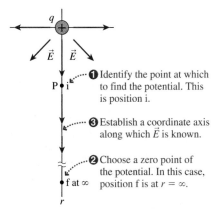

FIGURE 30.2 Finding the potential of a point charge.

① Identify the point at which to find the potential. This is position i.

③ Establish a coordinate axis along which $\vec{E}$ is known.

② Choose a zero point of the potential. In this case, position f is at $r = \infty$.

> **TACTICS BOX 30.1** **Finding the potential difference from the electric field**
>
> ❶ Draw a picture and identify the two points between which you wish to find the potential. Call them positions i and f.
> ❷ If you need to assign a specific value of V to a point in space, choose position f to be the zero point of the potential, usually at infinity.
> ❸ Establish a coordinate axis from i to f along which you already know or can easily determine the electric field component E_s.
> ❹ Carry out the integral of Equation 30.3 to find the potential difference ΔV.

To see how this works, let's use the electric field of a point charge to find its electric potential. Figure 30.2 identifies a point P at which we want to know the potential and calls this position i. In this case, because we want to assign a specific value of V to point P, we've chosen position f to be at $r = \infty$ and identified that as the zero point of the potential. The integration of Equation 30.3 is straight outward along the radial line from i to f:

$$\Delta V = V(\infty) - V(r) = -\int_r^\infty E_r dr \tag{30.5}$$

The electric field is radially outward, parallel to the line of integration, hence

$$E_r = \frac{1}{4\pi\epsilon_0} \frac{q}{r^2}$$

Thus the potential at distance r from a point charge q is

$$V(r) = V(\infty) + \frac{q}{4\pi\epsilon_0} \int_r^\infty \frac{dr}{r^2} = V(\infty) + \frac{q}{4\pi\epsilon_0} \frac{-1}{r}\bigg|_r^\infty = 0 + \frac{1}{4\pi\epsilon_0} \frac{q}{r} \tag{30.6}$$

We've rediscovered the potential of a point charge that you learned in Chapter 29:

$$V_{\text{point charge}} = \frac{1}{4\pi\epsilon_0} \frac{q}{r}$$

EXAMPLE 30.1 **The potential of a disk of charge**

In Chapter 26, the electric field strength at distance z on the axis of a disk of charge Q and radius R was found to be

$$E = \frac{Q}{2\pi R^2 \epsilon_0}\left[1 - \frac{z}{(z^2 + R^2)^{1/2}}\right]$$

Find the electric potential on the axis of the disk.

VISUALIZE Figure 30.3 shows the disk, the z-axis, and the point P where we want to know the potential. We've chosen $z = \infty$ as the zero point of the potential.

SOLVE We'll integrate along the z-axis to point f at infinity. The electric field is parallel to the line of integration, so $E_z = E$. Equation 30.3 is

$$V(z) = V(\infty) + \int_z^\infty E_z(z)\,dz$$

$$= 0 + \frac{Q}{2\pi R^2 \epsilon_0} \int_z^\infty \left[1 - \frac{z}{(z^2 + R^2)^{1/2}}\right]dz$$

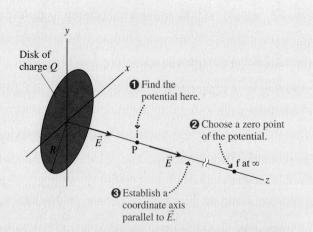

① Find the potential here.

② Choose a zero point of the potential.

③ Establish a coordinate axis parallel to $\vec{E}$.

Disk of charge Q

f at ∞

FIGURE 30.3 Finding the on-axis potential of a charged disk.

The integral requires some care with the upper limit. The integral doesn't diverge, as it might appear, because the integrand becomes zero as $z \to \infty$. The second integration is done by changing variables to $u = z^2 + R^2$. Then, with $z\,dz = \frac{1}{2}du$, we find

$$V(z) = \frac{Q}{2\pi R^2 \epsilon_0}\left[\int dz - \int_z^\infty \frac{\frac{1}{2}du}{u^{1/2}}\right] = \frac{Q}{2\pi R^2 \epsilon_0}[z - u^{1/2}]_z^\infty$$

$$= \frac{Q}{2\pi R^2 \epsilon_0}[z - \sqrt{z^2 + R^2}]_z^\infty$$

The two halves of the integral cancel when $z \to \infty$, leaving

$$V_{\text{disk}} = \frac{Q}{2\pi R^2 \epsilon_0}(\sqrt{z^2 + R^2} - z)$$

ASSESS This agrees with Example 29.12, where we found the on-axis potential of a disk of charge by working directly with the continuous distribution of charge. Here we found the potential by explicitly recognizing the connection between the potential and the field.

30.2 Finding the Electric Field from the Potential

Now let's look at the reverse operation: finding the electric field when we know the potential. Figure 30.4 shows two points i and f separated by a very small distance Δs, so small that the electric field is essentially constant over this very short distance. The work done by the electric field as a charge q moves through this small distance is $W = F_s\Delta s = qE_s\Delta s$. Consequently, the potential difference between these two points is

$$\Delta V = \frac{\Delta U_{q+\text{sources}}}{q} = \frac{-W}{q} = -E_s\Delta s \qquad (30.7)$$

In terms of the potential, the component of the electric field in the s-direction is $E_s = -\Delta V/\Delta s$. In the limit $\Delta s \to 0$,

$$E_s = -\frac{dV}{ds} \qquad (30.8)$$

Now we have reversed Equation 30.3 and have a way to find the electric field from the potential. We'll begin with examples where the field is parallel to a coordinate axis, then we'll look at what Equation 30.8 tells us about the geometry of the field and the potential.

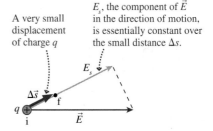

E_s, the component of $\vec{E}$ in the direction of motion, is essentially constant over the small distance Δs.

A very small displacement of charge q

FIGURE 30.4 The electric field does work on charge q.

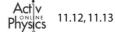

Actlv Physics   11.12, 11.13

Field Parallel to a Coordinate Axis

The derivative of Equation 30.8 gives E_s, the component of the electric field parallel to the displacement $\Delta\vec{s}$. It doesn't tell us about the electric field component perpendicular to $\Delta\vec{s}$. Thus Equation 30.8 is most useful if we can use symmetry to select a coordinate axis that is parallel to $\vec{E}$ and along which the perpendicular component of $\vec{E}$ is known to be zero.

For example, suppose we knew the potential of a point charge to be $V = q/4\pi\epsilon_0 r$ but didn't remember the electric field. Symmetry requires that the field point straight outward from the charge, with only a radial component E_r. If we choose the s-axis to be in the radial direction, parallel to $\vec{E}$, we can use Equation 30.8 to find

$$E = E_r = -\frac{dV}{dr} = -\frac{d}{dr}\left(\frac{q}{4\pi\epsilon_0 r}\right) = \frac{1}{4\pi\epsilon_0}\frac{q}{r^2} \qquad (30.9)$$

This is, indeed, the well-known electric field of a point charge.

Equation 30.8 is especially useful for a continuous distribution of charge because calculating V, which is a scalar, is usually much easier than calculating the vector $\vec{E}$ directly from the charge. Once V is known, $\vec{E}$ is found simply by taking a derivative.

EXAMPLE 30.2 **The electric field of a ring of charge**

In Chapter 29, we found the on-axis potential of a ring of radius R and charge Q to be

$$V_{\text{ring}} = \frac{1}{4\pi\epsilon_0} \frac{Q}{\sqrt{z^2 + R^2}}$$

Find the on-axis electric field of a ring of charge.

SOLVE Symmetry requires the electric field along the axis to point straight outward from the ring with only a z-component E_z. The electric field at distance z is

$$E = E_z = -\frac{dV}{dz} = -\frac{d}{dz}\left(\frac{1}{4\pi\epsilon_0} \frac{Q}{\sqrt{z^2 + R^2}}\right)$$

$$= \frac{1}{4\pi\epsilon_0} \frac{zQ}{(z^2 + R^2)^{3/2}}$$

ASSESS This result is in perfect agreement with the electric field we found in Chapter 26, but this calculation was easier because, unlike in Chapter 26, we didn't have to deal with angles.

A geometric interpretation of Equation 30.8 is that the electric field is the negative of the *slope* of the V-versus-s graph. This interpretation should be familiar. You learned in Chapter 11 that the force on a particle is the negative of the slope of the potential-energy graph: $F = -dU/ds$. In fact, Equation 30.8 is simply $F = -dU/ds$ with both sides divided by q to yield E and V. This geometric interpretation is an important step in developing an understanding of potential.

EXAMPLE 30.3 **Finding E from the slope of V**

Figure 30.5a is a graph of the electric potential in a region of space where $\vec{E}$ is parallel to the x-axis. Draw a graph of E_x versus x.

MODEL The electric field is the *negative* of the slope of the potential graph.

VISUALIZE Figure 30.5a shows the graph of the potential.

SOLVE There are three regions of different slope:

$0 < x < 2$ cm $\begin{cases} \Delta V/\Delta x = (20\ \text{V})/(0.020\ \text{m}) = 1000\ \text{V/m} \\ E_x = 1{,}000\ \text{V/m} \end{cases}$

$2 < x < 4$ cm $\begin{cases} \Delta V/\Delta x = 0\ \text{V/m} \\ E_x = 0\ \text{V/m} \end{cases}$

$4 < x < 8$ cm $\begin{cases} \Delta V/\Delta x = (-20\ \text{V})/(0.040\ \text{m}) = -500\ \text{V/m} \\ E_x = 500\ \text{V/m} \end{cases}$

The results are shown in Figure 30.5b.

ASSESS The electric field $\vec{E}$ points to the left (E_x is negative) for $0 < x < 2$ cm and to the right (E_x is positive) for $4 < x < 8$ cm. Notice that **the electric field is zero in a region of space where the potential is not changing.**

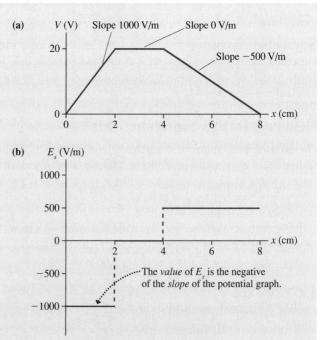

FIGURE 30.5 Graphs of V and E_x versus position x.

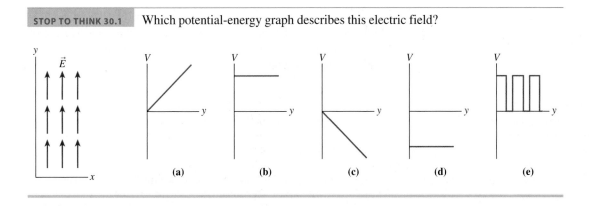

Which potential-energy graph describes this electric field?

(a)   (b)   (c)   (d)   (e)

The Geometry of Potential and Field

Equations 30.3 for V in terms of E_s and 30.8 for E_s in terms of V have profound implications for the geometry of the potential and the field. Figure 30.6 shows two equipotential surfaces, with V_+ positive relative to V_-. To learn about the electric field $\vec{E}$ at point P, allow a charge to move through the two displacements $\Delta\vec{s}_1$ and $\Delta\vec{s}_2$. Displacement $\Delta\vec{s}_1$ is *tangent* to the equipotential surface, hence a charge moving in this direction experiences *no* potential difference. According to Equation 30.8, the electric field component along a direction of *constant* potential is $E_s = -dV/ds = 0$. In other words, the electric field component tangent to the equipotential is $E_\parallel = 0$.

Displacement $\Delta\vec{s}_2$ is *perpendicular* to the equipotential surface. There is a potential difference along $\Delta\vec{s}_2$, hence the electric field component is

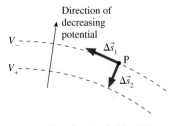

FIGURE 30.6 The electric field at P is related to the shape of the equipotential surfaces.

$$E_\perp = -\frac{dV}{ds} \approx -\frac{\Delta V}{\Delta s} = -\frac{V_+ - V_-}{\Delta s}$$

You can see that the electric field is inversely proportional to Δs_2, the spacing between the equipotential surfaces. Furthermore, because $(V_+ - V_-) > 0$, the minus sign tells us that the electric field is *opposite* in direction to $\Delta\vec{s}_2$. In other words, $\vec{E}$ points in the direction of *decreasing* potential.

These important ideas about the geometry of the potential and the field are summarized in Figure 30.7.

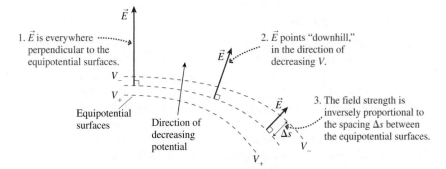

1. $\vec{E}$ is everywhere perpendicular to the equipotential surfaces.

2. $\vec{E}$ points "downhill," in the direction of decreasing V.

3. The field strength is inversely proportional to the spacing Δs between the equipotential surfaces.

Equipotential surfaces

Direction of decreasing potential

FIGURE 30.7 The geometry of the potential and the field.

Mathematically, we can calculate the individual components of $\vec{E}$ at any point by extending Equation 30.8 to three dimensions:

$$E_x = -\frac{\partial V}{\partial x} \qquad E_y = -\frac{\partial V}{\partial y} \qquad E_z = -\frac{\partial V}{\partial z} \qquad (30.10)$$

where $\partial V/\partial x$ is the partial derivative of V with respect to x while y and z are held constant. More advanced treatments of the electric field make extensive use of this mathematical relationship, but for the most part we'll limit our investigations to those we can analyze graphically.

EXAMPLE 30.4 Finding the electric field from the equipotential surfaces

In Figure 30.8 a 1 cm × 1 cm grid is superimposed on a contour map of the potential. Estimate the strength and direction of the electric field at points 1, 2, and 3. Show your results graphically by drawing the electric field vectors on the contour map.

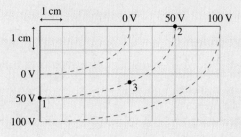

FIGURE 30.8 Equipotential lines.

MODEL The electric field is perpendicular to the equipotential lines, points "downhill," and depends on the slope of the potential hill.

VISUALIZE The potential is highest on the bottom and the right. An elevation graph of the potential would look like the lower-right quarter of a bowl or a football stadium.

SOLVE Some distant but unseen source charges have created an electric field and potential. We do not need to see the source charges to relate the field to the potential. Because $E \approx -\Delta V/\Delta s$, the electric field is stronger where the equipotential lines are closer together and weaker where they are far-

ther apart. If Figure 30.8 were a topographic map, you would interpret the closely spaced contour lines at the bottom of the figure as a steep slope.

Figure 30.9 shows how measurements of Δs from the grid are combined with values of ΔV to determine $\vec{E}$. Point 3 requires an estimate of the spacing between the 0 V and the 100 V surfaces. Notice that we're using the 0 V and 100 V equipotential surfaces to determine $\vec{E}$ at a point on the 50 V equipotential.

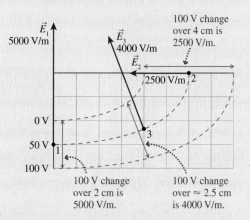

FIGURE 30.9 The electric field at points 1 to 3.

ASSESS The *directions* of $\vec{E}$ are found by drawing downhill vectors perpendicular to the equipotentials. The distances between the equipotential surfaces are needed to determine the field strengths.

Kirchhoff's Loop Law

Chapter 28 introduced *Kirchhoff's junction law*, which said that the sum of the currents flowing into a junction must equal the sum of the currents flowing out. The junction law is really a statement about conservation of charge.

A second law, called *Kirchhoff's loop law*, is a statement about conservation of energy. Figure 30.10 shows two points, 1 and 2, in a region of electric field and potential. You learned in Chapter 29 that the work done in moving a charge between points 1 and 2 is *independent of the path*. Consequently, the potential difference between points 1 and 2 along any two paths that join them is $\Delta V = 20$ V. This must be true in order for the idea of an equipotential surface to make sense.

Now consider the path 1–a–b–c–2–d–1 that ends where it started. What is the potential difference "around" this closed path? The potential increases by 20 V in moving from 1 to 2, but then decreases by 20 V in moving from 2 back to 1. Thus $\Delta V = 0$ V around the closed path.

The numbers are specific to this example, but the idea applies to any loop (i.e., a closed path) through an electric field. The situation is analogous to hiking on the side of a mountain. You may walk uphill during parts of your hike and downhill during other parts, but if you return to your starting point your *net* change of

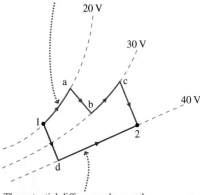

The potential difference along path 1-a-b-c-2 is $\Delta V = 0$ V + 10 V + 0 V + 10 V = 20 V.

The potential difference along path 1-d-2 is $\Delta V = 20$ V + 0 V = 20 V.

FIGURE 30.10 The potential difference between points 1 and 2 is the same along either path.

elevation is zero. So for any path that starts and ends at the same point, we can conclude that

$$\Delta V_{\text{loop}} = \sum_i (\Delta V)_i = 0 \qquad (30.11)$$

Stated in words, **the sum of all the potential differences encountered while moving around a loop or closed path is zero.** This is **Kirchhoff's loop law.**

Kirchhoff's loop law is a statement of energy conservation, because a charge that moves around a loop and returns to its starting point has $\Delta U = q\Delta V = 0$. Kirchhoff's loop law and Kirchhoff's junction law will turn out to be the two fundamental principles of circuit analysis.

STOP TO THINK 30.2   Which set of equipotential surfaces matches this electric field?

$\vec{E}$

<table>
<tr><td>0 V   50 V</td><td>0 V   50 V</td><td>0 V   50 V</td></tr>
<tr><td>(a)</td><td>(b)</td><td>(c)</td></tr>
<tr><td>50 V   0 V</td><td>50 V   0 V</td><td>50 V   0 V</td></tr>
<tr><td>(d)</td><td>(e)</td><td>(f)</td></tr>
</table>

30.3 A Conductor in Electrostatic Equilibrium

The basic relationships between potential and field allow us to draw some interesting and important conclusions about conductors. Consider a conductor, such as a metal, that is in electrostatic equilibrium. The conductor may be charged, but all the charges are at rest.

You learned in Chapter 25 that any excess charges on a conductor in electrostatic equilibrium are always located on the *surface* of the conductor. Using similar reasoning, we can conclude that **the electric field is zero at any interior point of a conductor in electrostatic equilibrium.** Why? If the field were other than zero, then there would be a force $\vec{F} = q\vec{E}$ on the charge carriers and they would move, creating a current. But there are no currents in a conductor in electrostatic equilibrium, so it must be that $\vec{E} = \vec{0}$ at all interior points.

NOTE ▶ $\vec{E} = \vec{0}$ inside a conductor in electrostatic equilibrium is in contrast with a current-carrying conductor, where there most definitely *is* an interior electric field to create the current density $J = \sigma E$. ◀

The two points inside the conductor in Figure 30.11 are connected by a line that remains entirely inside the conductor. We can find the potential difference $\Delta V = V_2 - V_1$ between these points by using Equation 30.3 to integrate E_s along the line from 1 to 2. But $E_s = 0$ at all points along the line, because $\vec{E} = \vec{0}$, thus the value of the integral is zero and $\Delta V = 0$. In other words, **any two points inside a conductor in electrostatic equilibrium are at the same potential.**

When a conductor is in electrostatic equilibrium, the *entire conductor* is at the same potential. If we charge a metal sphere, then the entire sphere is at a single potential. Similarly, a charged metal rod or wire is at a single potential *if* it is in electrostatic equilibrium. (This conclusion does *not* apply to a current-carrying wire because it is not in electrostatic equilibrium.)

A corona discharge, with crackling noises and glimmers of light, occurs at pointed metal tips where the electric field can be very strong.

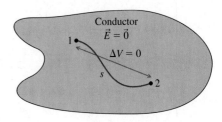

FIGURE 30.11 All points inside a conductor in electrostatic equilibrium are at the same potential.

If $\vec{E} = \vec{0}$ inside a charged conductor but $\vec{E} \neq \vec{0}$ outside, what happens right at the surface? If the entire conductor is at the same potential, then the surface is an equipotential surface. You have seen that the electric field is always perpendicular to an equipotential surface, hence **the exterior electric field $\vec{E}$ of a charged conductor is perpendicular to the surface.**

Figure 30.12 summarizes what we know about conductors in electrostatic equilibrium. Item 6, that the charge density and thus the electric field strength are largest at "sharp points," we'll assert without proof. These are important and practical conclusions because conductors are the primary components of electrical devices.

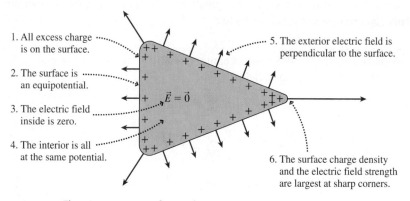

1. All excess charge is on the surface.

2. The surface is an equipotential.

3. The electric field inside is zero.

4. The interior is all at the same potential.

5. The exterior electric field is perpendicular to the surface.

6. The surface charge density and the electric field strength are largest at sharp corners.

$\vec{E} = \vec{0}$

FIGURE 30.12 Electric properties of a conductor in electrostatic equilibrium.

We can use similar reasoning to estimate the electric field and potential in between two charged conductors. As an example, Figure 30.13 shows a positively charged metal sphere near a flat metal plate. The surfaces of the sphere and the flat plate are equipotentials, hence the electric field must be perpendicular to both. Close to a surface, the electric field is still *nearly* perpendicular to the surface. Consequently, **an equipotential surface close to an electrode must roughly match the shape of the electrode.**

In between, the equipotential surfaces *gradually* change as they "morph" from one electrode shape to the other. It's not hard to sketch a contour map showing a plausible set of equipotential surfaces. You can then draw electric field lines (field lines are usually easier to draw than field vectors in these situations) that are perpendicular to the equipotentials, point "downhill," and are closer together where the contour line spacing is smaller.

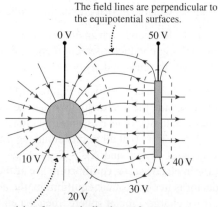

The field lines are perpendicular to the equipotential surfaces.

0 V   50 V

10 V   40 V

20 V   30 V

The equipotential surfaces gradually change from the shape of one electrode to the shape of the other.

FIGURE 30.13 Estimating the electric field and potential between two charged conductors.

STOP TO THINK 30.3   Three charged metal spheres of different radii are connected by a thin metal wire. The potential and electric field at the surface of each sphere are V and E. Which of the following is true?

a. $V_1 = V_2 = V_3$ and $E_1 = E_2 = E_3$ b. $V_1 = V_2 = V_3$ and $E_1 > E_2 > E_3$

c. $V_1 > V_2 > V_3$ and $E_1 = E_2 = E_3$ d. $V_1 > V_2 > V_3$ and $E_1 > E_2 > E_3$

e. $V_3 > V_2 > V_1$ and $E_3 = E_2 = E_1$ f. $V_3 > V_2 > V_1$ and $E_3 > E_2 > E_1$

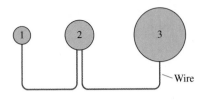

30.4 Sources of Electric Potential

A *separation of charge* creates an electric potential difference. Shuffling your feet on the carpet transfers electrons from the carpet to you, creating a potential difference between you and a doorknob that causes a spark and a shock as you touch it. Charging a capacitor by moving electrons from one plate to the other creates a potential difference across the capacitor.

In fact, as Figure 30.14 shows, *any* separation of charge causes a potential difference. The charge separation between the two electrodes creates an electric field $\vec{E}$ pointing from the positive toward the negative electrode. As a consequence, there is a potential difference between the electrodes that is given by

$$\Delta V = V_{\text{pos}} - V_{\text{neg}} = -\int_{\text{neg}}^{\text{pos}} E_s \, ds$$

where the integral runs from any point on the negative electrode to any point on the positive. The key idea is that **we can create a potential difference by creating a charge separation.**

One source of electric potential is a charged capacitor. However, you learned in Chapter 28 that the charge separation cannot be maintained if we "use" the capacitor. In Figure 30.15, each charge in the discharge current I loses potential energy as it "falls" through the wire, reaching the negative plate with $U_{\text{bottom}} = 0$. A charge at the negative plate has no means to return to the other side.

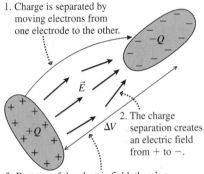

1. Charge is separated by moving electrons from one electrode to the other.

2. The charge separation creates an electric field from + to −.

3. Because of the electric field, there's a potential difference between the electrodes.

FIGURE 30.14 A charge separation creates a potential difference.

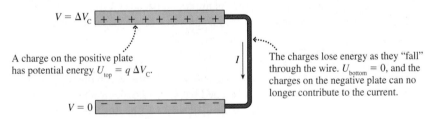

$V = \Delta V_C$

A charge on the positive plate has potential energy $U_{\text{top}} = q \, \Delta V_C$.

$V = 0$

The charges lose energy as they "fall" through the wire. $U_{\text{bottom}} = 0$, and the charges on the negative plate can no longer contribute to the current.

FIGURE 30.15 A charged capacitor is a source of potential difference, but it cannot sustain a current.

To be practical, a source of potential needs a way to "lift" charges from the negative electrode back to the positive electrode where they can be reused. Once there, they could again "fall" through the wire and sustain the current. A device that continuously separates charge, moving it from the negative to the positive electrode, can *sustain* the potential difference and thus sustain a current in the wire.

The **Van de Graaff generator** shown in Figure 30.16a on the next page is a mechanical charge separator—essentially a fancy foot shuffler. A moving plastic or leather belt is charged, then the charge is mechanically transported via the conveyor belt to the spherical electrode at the top of the insulating column. The charging of the belt could be done by friction, but in practice a *corona discharge* due to the strong electric field at the tip of a needle is more efficient and reliable.

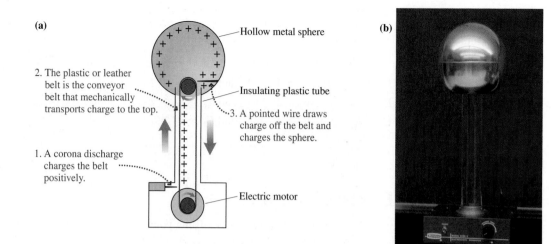

(a)

2. The plastic or leather belt is the conveyor belt that mechanically transports charge to the top.

1. A corona discharge charges the belt positively.

Hollow metal sphere

Insulating plastic tube

3. A pointed wire draws charge off the belt and charges the sphere.

Electric motor

(b)

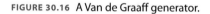

FIGURE 30.16 A Van de Graaff generator.

A Van de Graaff generator has two noteworthy features:

- Charge is mechanically transported from the negative side to the positive side. This charge separation creates *and sustains* a potential difference between the spherical electrode and its surroundings.
- The electric field of the spherical electrode exerts a downward force on the positive charges moving up the belt. Consequently, *work must be done* to "lift" the positive charges. The work is done by the electric motor that runs the belt.

A classroom demonstration Van de Graaff generator like the one shown in Figure 30.16b creates a potential difference of several hundred thousand volts between the upper sphere and its surroundings. The maximum potential is reached when the electric field near the sphere becomes large enough to cause a breakdown of the air. This produces a spark and temporarily discharges the sphere. A large Van de Graaff generator surrounded by vacuum can reach a potential of 20 MV or more. These generators are used to accelerate protons for nuclear physics experiments.

For us, the Van de Graaff generator is important because it shows that a potential difference can be created and sustained by a device that separates charge.

Batteries and EMF

The most common source of electric potential is a **battery.** A battery consists of chemicals, called *electrolytes,* sandwiched between two electrodes made of different metals. Chemical reactions in the electrolytes separate charge by moving positive ions to one electrode and negative ions to the other. In other words, chemical reactions, rather than a mechanical conveyor belt, transport charge from one electrode to the other. The procedure is different, but the outcome is the same: a potential difference.

Figure 30.17 reminds you of the **charge escalator** model of a battery that we introduced in Section 28.3. The escalator "lifts" positive charges from the negative terminal to the positive terminal, where they can be reused. Lifting positive charges to a positive terminal requires that work be done, and the chemical reactions within the battery provide the energy to do this work. When the chemicals are used up, the reactions cease, and the battery is dead.

By separating the charge, the charge escalator establishes a potential difference ΔV_{bat} between the terminals. The value of ΔV_{bat} is determined by the specific chemical reactions employed by the battery. To see how, suppose the chemical

The charge "falls downhill" through the wire, but it can be sustained because of the charge escalator.

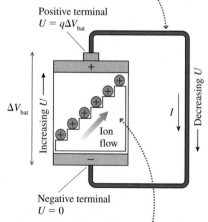

Positive terminal
$U = q\Delta V_{\text{bat}}$

ΔV_{bat}

Increasing U

Decreasing U

I

Ion flow

Negative terminal
$U = 0$

The charge escalator "lifts" charge from the negative side to the positive side. Charge q gains energy $\Delta U = q\Delta V_{\text{bat}}$.

FIGURE 30.17 The charge escalator model of a battery.

reactions do work W_{chem} to move charge q from the negative to the positive terminal. In an **ideal battery,** in which there are no internal energy losses, the charge gains electric potential energy $\Delta U = W_{chem}$. This is analogous to a book gaining gravitational potential energy as you do work to lift it from the floor to a shelf.

The quantity W_{chem}/q, which is the work done *per charge* by the charge escalator, is called the **emf** of the battery, pronounced as the sequence of three letters "e-m-f." The symbol for emf is $\mathcal{E}$, a script E, and the units are those of the electric potential: joules per coulomb, or volts. The *rating* of a battery, such as 1.5 V or 9 V, is the battery's emf. Originally the term emf was an abbreviation of "electromotive force." That is an outdated term (work per charge is not a force!), so today we just call it emf and it is not an abbreviation of anything.

> **NOTE** ▶ The term *emf,* often capitalized as EMF, is widely used in popular science articles in newspapers and magazines to mean "electromagnetic field." You may have seen the abbreviation EMF if you've read about the debate over whether electric transmission lines, which generate electromagnetic fields, are a health hazard. This is *not* how we will use the term *emf.* ◀

By definition, the electric potential is related to the electric potential energy of charge q by $\Delta V = \Delta U/q$. But $\Delta U = W_{chem}$ for the charges in a battery, hence the potential difference between the terminals of an ideal battery is

$$\Delta V_{bat} = \frac{W_{chem}}{q} = \mathcal{E} \qquad \text{(ideal battery)} \qquad (30.12)$$

In other words, a battery constructed to have an emf of 1.5 V (i.e., the chemical reactions do 1.5 J of work to separate 1 C of charge) creates a 1.5 V potential difference. In practice, the measured potential difference ΔV_{bat} between the terminals of a real battery, called the **terminal voltage,** is usually slightly less than $\mathcal{E}$. You will learn the reason for this in the next chapter.

Electric generators, photocells, and other sources of potential difference use different means to separate charges, but otherwise they function exactly the same as a battery. The common feature of all such devices is that they use a *nonelectrical* means to separate charge and, thus, to create a potential difference. The emf $\mathcal{E}$ of any device is the work done per charge to separate the charge.

30.5 Connecting Potential and Current

An important consequence of the charge escalator model is that **a battery is a source of potential difference.** It is true that charges flow through a wire that connects the battery terminals, but this current is a *consequence* of the battery's potential difference. You can think of the battery's emf as being the *cause.* Current, heat, light, sound, and so on are all *effects* that happen when the battery is used in certain ways. Distinguishing between cause and effect will be vitally important for understanding how a battery functions in a circuit.

Our goal in this section is to find the connection between potential and current. Let's start by connecting a wire between the two terminals of a battery, as shown in Figure 30.18. You learned in Section 30.2 that the potential difference between any two points is independent of the path between them. Consequently, the potential difference between the two ends of the wire, along a path through the wire, is equal to the potential difference between the two terminals of the battery:

$$\Delta V_{wire} = \Delta V_{bat} \qquad (30.13)$$

Thus the battery is the *source* of a potential difference between the ends of the wire.

The potential difference between these two points is the same whether you go through the battery or through the wire.

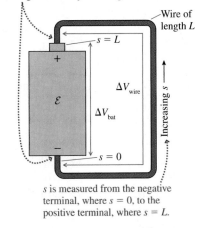

s is measured from the negative terminal, where $s = 0$, to the positive terminal, where $s = L$.

FIGURE 30.18 The potential difference along the wire is the same as the potential difference between the battery terminals.

(a) The potential difference between the ends of the wire establishes an electric field inside the wire.

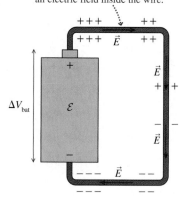

(b) The electric field drives a current through the wire.

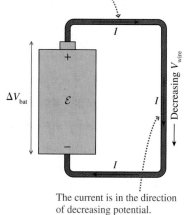

The current is in the direction of decreasing potential.

FIGURE 30.19 The electric field and the current inside the wire.

TABLE 30.1 Resistivity of materials

Material	Resistivity (Ω m)
Aluminum	2.8×10^{-8}
Copper	1.7×10^{-8}
Gold	2.4×10^{-8}
Iron	9.7×10^{-8}
Silver	1.6×10^{-8}
Tungsten	5.6×10^{-8}
Nichrome*	1.5×10^{-6}
Carbon	3.5×10^{-5}

*Nickel-chromium alloy used for heating wires

If there is a potential difference between the ends of the wire, there must be an electric field *inside* the wire. And if there's an internal electric field, a current will flow! The wire's electric field is related to the potential difference ΔV_{wire} by Equation 30.3:

$$\Delta V_{\text{wire}} = -\int_0^L E_s \, ds \tag{30.14}$$

where L is the length of the wire and s is measured along the wire from $s = 0$ at the battery's negative end to $s = L$ at the positive end.

The field vector $\vec{E}_{\text{wire}}$ inside the wire points "downhill," from the positive toward the negative end of the wire. Thus the s-component of $\vec{E}_{\text{wire}}$, along the line of integration, is $E_s = -E_{\text{wire}}$. You learned in Chapter 28 that the electric field inside a constant-diameter wire is constant (a consequence of conservation of current), so E_{wire} can be taken outside the integral. Equation 30.14 is then

$$\Delta V_{\text{wire}} = -\int_0^L (-E_{\text{wire}}) \, ds = E_{\text{wire}} \int_0^L ds = E_{\text{wire}} L \tag{30.15}$$

Hence, the electric field strength inside the wire is

$$E_{\text{wire}} = \frac{\Delta V_{\text{wire}}}{L} \tag{30.16}$$

Equation 30.16 is an important result. The electric field strength inside a constant-diameter wire—the field that drives the current forward—is simply the potential difference between the ends of the wire divided by the length of the wire. And ΔV_{wire} can be established by connecting the wire to a battery.

Figure 30.19a shows the electric field $\vec{E}_{\text{wire}}$ inside the wire. Notice how the surface charges on the wire—charges supplied by the battery—cause $\vec{E}$ to follow the wire, as you learned in Chapter 28. The electric field drives the current I shown in Figure 30.19b. Notice that the field points in the direction of *decreasing* potential, implying that **the current is in the direction of decreasing potential.**

Now we can use E_{wire} to find the current I in the wire. You learned in Chapter 28 that the current density in a wire of conductivity σ is $J = \sigma E_{\text{wire}}$. The current density is $J = I/A$, where A is the cross-section area of the wire. Thus

$$I = AJ = A\sigma E_{\text{wire}} = \frac{A}{\rho} E_{\text{wire}} \tag{30.17}$$

where $\rho = 1/\sigma$ is the resistivity. Values of σ and ρ were tabulated for several metals in Table 28.2; Table 30.1 reproduces only the resistivity values.

The electric field strength in the wire is $E_{\text{wire}} = \Delta V_{\text{wire}}/L$, hence the current is

$$I = \frac{A}{\rho L} \Delta V_{\text{wire}} \tag{30.18}$$

You can see that the **current in the wire is proportional to the potential difference between the ends of the wire.** Equation 30.18 is the connection we were seeking between potential and current, but we can cast it into a more useful form if we define the **resistance** of the wire as

$$R = \frac{\rho L}{A} \tag{30.19}$$

The resistance is a property of a *specific* wire or conductor because it depends on the conductor's length and diameter as well as on the resistivity of the material from which it is made.

The SI unit of resistance is the **ohm,** defined as

$$1 \text{ ohm} = 1 \ \Omega \equiv 1 \text{ V/A}$$

where Ω is an uppercase Greek omega. The ohm is the basic unit of resistance, although kilohms (1 kΩ = 10^3 Ω) and megohms (1 MΩ = 10^6 Ω) are widely used. You can see from Equation 30.19 why the resistivity ρ has units of Ω m.

The resistance of a wire increases as the length increases. This seems reasonable, because it should be harder to push electrons through a long wire than a short one. Decreasing the cross-section area also increases the resistance. This again seems reasonable, because the same electric field can push more electrons through a fat wire than a skinny one.

NOTE ▶ It is important to distinguish between resistivity and resistance. *Resistivity* describes just the *material,* not any particular piece of it. *Resistance* characterizes a specific piece of the conductor having a specific geometry. The relationship between resistivity and resistance is analogous to that between mass density and mass. ◀

The definition of resistance allows us to write the current in a wire as

$$I = \frac{\Delta V_{\text{wire}}}{R} \qquad \text{(current in a wire)} \qquad (30.20)$$

In other words, establishing a potential difference ΔV_{wire} between the ends of a wire of resistance R creates an electric field that, in turn, causes a current $I = \Delta V_{\text{wire}}/R$ in the wire. The smaller the resistance, the larger the current.

EXAMPLE 30.5 **The current in a wire**
A 20-cm-long, 1.0-mm-diameter nichrome wire is connected to the terminals of a 1.5 V battery. What are the electric field and the current in the wire?

MODEL Assume the battery is an ideal battery, with $\Delta V_{\text{bat}} = \mathcal{E} = 1.5$ V.

SOLVE Connecting the wire to the battery makes $\Delta V_{\text{wire}} = \Delta V_{\text{bat}} = 1.5$ V. The electric field inside the wire is

$$E_{\text{wire}} = \frac{\Delta V_{\text{wire}}}{L} = \frac{1.5 \text{ V}}{0.20 \text{ m}} = 7.5 \text{ V/m}$$

This is not a very large field. The wire's resistance is

$$R = \frac{\rho L}{A} = \frac{\rho L}{\pi r^2}$$

From Table 30.1 we find that $\rho = 1.5 \times 10^{-6}$ Ω m for nichrome. After converting L and r to meters, we find $R = 0.382$ Ω. Thus the current in the wire is

$$I = \frac{\Delta V}{R} = \frac{1.5 \text{ V}}{0.382 \ \Omega} = 3.93 \text{ A}$$

The next chapter will look more carefully at currents in circuits. Our purpose here was to establish the connection between a source of potential difference and the current in a wire. The cause-and-effect sequence is the main idea.

1. A battery is a source of potential difference ΔV_{bat}. An ideal battery has $\Delta V_{\text{bat}} = \mathcal{E}$.
2. The battery creates a potential difference $\Delta V_{\text{wire}} = \Delta V_{\text{bat}}$ between the ends of a wire.
3. The potential difference ΔV_{wire} causes an electric field $E = \Delta V_{\text{wire}}/L$ in the wire.
4. The electric field establishes a current $I = JA = \sigma AE$ in the wire.
5. The magnitude of the current is determined *jointly* by the battery and the wire's resistance R to be $I = \Delta V_{\text{wire}}/R$.

STOP TO THINK 30.4   A wire connects the positive and negative terminals of a battery. Two identical wires connect the positive and negative terminals of an identical battery. Rank in order, from largest to smallest, the currents I_a to I_d at points a to d.

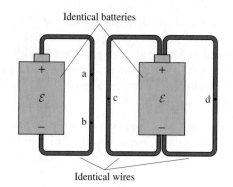

Identical batteries

Identical wires

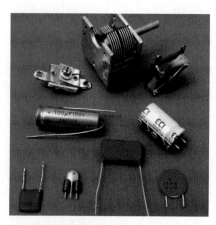

Capacitors are important elements in electric circuits. They come in a variety of sizes and shapes.

30.6 Capacitance and Capacitors

We introduced the parallel-plate capacitor in Chapter 26 and have made frequent use of it since. We've assumed that the capacitor is charged, but we haven't really addressed the issue of *how* it gets charged. Figure 30.20 shows the two plates of a capacitor connected with conducting wires to the two terminals of a battery. What happens? And how is the potential difference ΔV_C across the capacitor related to the battery's potential difference ΔV_{bat}?

Figure 30.20a shows the situation shortly after the capacitor is connected and before it is fully charged. According to Kirchhoff's loop law, Equation 30.11, the sum of the four potential differences around the closed path must be zero. However, we have to be careful with signs. The potential increases (becomes more positive) from the negative to the positive terminal of the battery (ΔV_{bat}) as the charges move "uphill" on the charge escalator. The potential then decreases (becomes more negative) from the positive plate of the battery to the capacitor (ΔV_1), from the positive to the negative plate of the capacitor (ΔV_C), and from

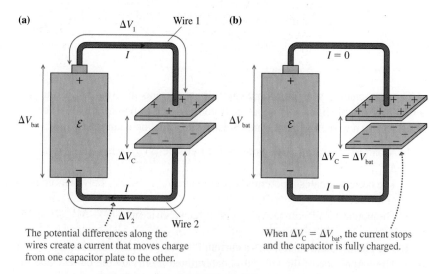

The potential differences along the wires create a current that moves charge from one capacitor plate to the other.

When $\Delta V_C = \Delta V_{bat}$, the current stops and the capacitor is fully charged.

FIGURE 30.20 A parallel-plate capacitor is charged by a battery.

the negative plate of the capacitor back to the battery (ΔV_2). Thus the requirement that the potential return to its starting point after a complete loop is

$$\Delta V_{\text{bat}} - \Delta V_1 - \Delta V_C - \Delta V_2 = 0 \qquad (30.21)$$

Initially, when the capacitor is uncharged, $\Delta V_C = 0$. The potential differences along the wires are $\Delta V_1 = \Delta V_2 = \frac{1}{2}\Delta V_{\text{bat}}$ if we assume the wires are equal. These potential differences establish currents in the wires, and charges begin to flow through them. The current in wire 2 carries positive charge from the lower capacitor plate to the negative terminal of the battery. The charge escalator in the battery lifts this charge to the positive terminal and sends it as a current through wire 1 to the upper capacitor plate. The net effect is that the battery removes charge from one plate of the capacitor and transfers it to the other plate. In other words, the battery "charges the capacitor."

As the charge on the capacitor plates increases, the potential difference ΔV_C increases and the potential differences along the wire decrease. As ΔV_1 and ΔV_2 decrease, so does the current in the wire. When ΔV_1 and ΔV_2 reach zero, the current stops and the capacitor is fully charged. This is the situation in Figure 30.20b, where $\Delta V_C = \Delta V_{\text{bat}}$. The capacitor stops charging when $\Delta V_C = \Delta V_{\text{bat}}$ because there is no longer a potential *difference* along the wires to cause any further current.

You learned in Chapter 29 that a parallel-plate capacitor's potential difference is related to the electric field inside by $\Delta V_C = Ed$, where d is the separation between the plates. And you know from Chapter 26 that a capacitor's electric field is

$$E = \frac{Q}{\epsilon_0 A} \qquad (30.22)$$

where A is the surface area of the plates. Combining these gives

$$Q = \frac{\epsilon_0 A}{d}\Delta V_C \qquad (30.23)$$

In other words, **the charge on the capacitor plates is directly proportional to the potential difference between the plates.**

The ratio of the charge Q to the potential difference ΔV_C is called the **capacitance** C:

$$C = \frac{Q}{\Delta V_C} = \frac{\epsilon_0 A}{d} \qquad \text{(parallel-plate capacitor)} \qquad (30.24)$$

Capacitance is a purely *geometric* property of two electrodes because it depends only on their surface area and spacing. The SI unit of capacitance is the **farad,** named in honor of Michael Faraday. One farad is defined as

$$1 \text{ farad} = 1 \text{ F} \equiv 1 \text{ C/V}$$

One farad turns out to be an enormous amount of capacitance. Practical capacitors are usually measured in units of microfarads (μF) or picofarads (1 pF $= 10^{-12}$ F).

With this definition of capacitance, Equation 30.24 can be written

$$Q = C\,\Delta V_C \qquad \text{(charge on a capacitor)} \qquad (30.25)$$

Equation 30.25 for the charge on a capacitor is analogous to Equation 30.20 for the current in a wire, $I = \Delta V_{\text{wire}}/R$. There we found that the current in a wire is determined jointly by the potential difference supplied by a battery *and* a property of the wire called resistance. Now we see that the charge on a capacitor is determined jointly by the potential difference supplied by a battery *and* a property of the electrodes called capacitance.

EXAMPLE 30.6 Charging a capacitor
The spacing between the plates of a 1.0 μF capacitor is 0.050 mm.

a. What is the surface area of the plates?
b. How much charge is on the plates if this capacitor is attached to a 1.5 V battery?

MODEL Assume the battery is ideal and that the capacitor is a parallel-plate capacitor.

SOLVE

a. From the definition of capacitance,

$$A = \frac{dC}{\epsilon_0} = 5.65 \text{ m}^2$$

b. The charge is $Q = C\Delta V_C = 1.5 \times 10^{-6} \text{ C} = 1.5 \ \mu\text{C}$.

ASSESS The surface area needed to construct a 1.0 μF capacitor (a fairly typical value) is enormous. In practice, two very large sheets of thin metal foil are separated by a thin insulator ($d \approx 0.05$ mm). The sheets are then rolled up to form a cylinder. The insulator and the rolling do have some effect on the capacitance, which you can learn about in more advanced courses.

Forming a Capacitor

The parallel-plate capacitor is important because it is straightforward to analyze and it produces a uniform electric field. But capacitors and capacitance are not limited to flat, parallel electrodes. *Any* two electrodes, regardless of their shape, form a capacitor.

Figure 30.21 shows two arbitrary electrodes charged to $\pm Q$. The net charge, as was the case with a parallel-plate capacitor, is zero. By definition, the capacitance of the two electrodes is

$$C = \frac{Q}{\Delta V_C} \tag{30.26}$$

where ΔV_C is the potential difference between the positive and negative electrode. It might appear that the capacitance depends on the amount of charge, but the potential difference is proportional to Q. Consequently, **the capacitance depends only on the geometry of the electrodes.**

To make use of Equation 30.26, we must be able to determine the potential difference between the electrodes when they are charged to $\pm Q$. The following example shows how this is done.

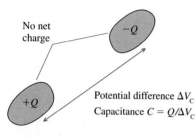

No net charge

Potential difference ΔV_C
Capacitance $C = Q/\Delta V_C$

FIGURE 30.21 Any two electrodes form a capacitor.

EXAMPLE 30.7 A spherical capacitor
A metal sphere of radius R_1 is inside and concentric with a hollow metal sphere of radius R_2. What is the capacitance of this spherical capacitor?

MODEL Assume the inner sphere is negative and the outer is positive.

VISUALIZE Figure 30.22 shows the two spheres. The electric field between them points from the positive outer sphere to the inner negative sphere.

SOLVE You might think we could find the potential difference between the spheres by using the Chapter 29 result for the potential of a charged sphere. However, that was the potential of an *isolated* charged sphere. To find the potential difference between *two* spheres we need to use Equation 30.3:

$$\Delta V = V(s_f) - V(s_i) = -\int_{s_i}^{s_f} E_s \, ds$$

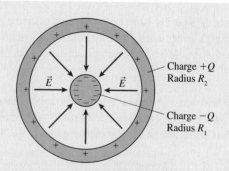

Charge $+Q$
Radius R_2

Charge $-Q$
Radius R_1

FIGURE 30.22 A spherical capacitor.

The electric field between the spheres is the superposition of the fields of the inner and outer sphere. The field of the inner sphere is that of point charge $-Q$, while, from Gauss's law, the interior field of the outer sphere is zero. If we integrate along a radial line from $r_i = R_1$ on the inner sphere to $r_f = R_2$ on the outer sphere, the potential difference between them is

$$\Delta V_C = -\int_{R_1}^{R_2}\left(\frac{-Q}{4\pi\epsilon_0 r^2}\right)dr = \frac{Q}{4\pi\epsilon_0}\int_{R_1}^{R_2}\frac{dr}{r^2} = \frac{Q}{4\pi\epsilon_0}\left(\frac{1}{R_1} - \frac{1}{R_2}\right)$$

Then, from the definition of capacitance,

$$C = \frac{Q}{\Delta V_C} = 4\pi\epsilon_0\left(\frac{1}{R_1} - \frac{1}{R_2}\right)^{-1} = 4\pi\epsilon_0\frac{R_1 R_2}{R_2 - R_1}$$

ASSESS As expected, the capacitance depends on the geometry but not on the charge Q. Note that we did not need to assume a negative inner sphere, but a positive inner sphere would have required us to integrate inward, from R_2 to R_1, to get a positive ΔV_C.

Combinations of Capacitors

In practice, two or more capacitors are sometimes joined together. Figure 30.23 illustrates two basic combinations: **parallel capacitors** and **series capacitors.** Notice that a capacitor, no matter what its actual geometrical shape, is represented in *circuit diagrams* by two parallel lines.

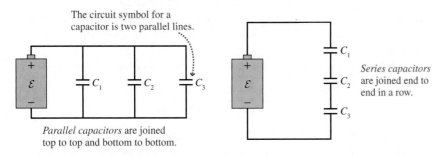

FIGURE 30.23 Parallel and series capacitors.

NOTE ▶ The terms "parallel capacitors" and "parallel-plate capacitor" do not describe the same thing. The former term describes how two or more capacitors are connected to each other, the latter describes how a particular capacitor is constructed. ◀

As we'll show, parallel or series capacitors (or, as is sometimes said, capacitors "in parallel" or "in series") can be represented by a single **equivalent capacitance.** We'll demonstrate this first with the two parallel capacitors C_1 and C_2 of Figure 30.24a. Because the two top electrodes are connected by a conducting wire, they form a single conductor in electrostatic equilibrium. Thus the two top electrodes are at the same potential. Similarly, the two connected bottom electrodes are at the same potential. Consequently, two (or more) capacitors in parallel each have the *same* potential difference ΔV_C between the two electrodes.

The charges on the two capacitors are $Q_1 = C_1\Delta V_C$ and $Q_2 = C_2\Delta V_C$. Altogether, the battery's charge escalator moved total charge $Q = Q_1 + Q_2$ from the negative electrodes to the positive electrodes. Suppose, as in Figure 30.24b, we replaced the two capacitors with a single capacitor having charge $Q = Q_1 + Q_2$ and potential difference ΔV_C. This capacitor is equivalent to the original two in the sense that the battery can't tell the difference. In either case, the battery has to establish the same potential difference and move the same amount of charge.

By definition, the capacitance of this equivalent capacitor is

$$C_{eq} = \frac{Q}{\Delta V_C} = \frac{Q_1 + Q_2}{\Delta V_C} = \frac{Q_1}{\Delta V_C} + \frac{Q_2}{\Delta V_C} = C_1 + C_2 \qquad (30.27)$$

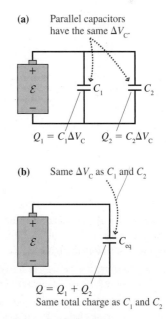

(a) Parallel capacitors have the same ΔV_C.

$Q_1 = C_1\Delta V_C \qquad Q_2 = C_2\Delta V_C$

(b) Same ΔV_C as C_1 and C_2

$Q = Q_1 + Q_2$
Same total charge as C_1 and C_2

FIGURE 30.24 Replacing two parallel capacitors with an equivalent capacitor.

This analysis hinges on the fact that **parallel capacitors each have the same potential difference** ΔV_C. We could easily extend this analysis to more than two capacitors. If capacitors $C_1, C_2, C_3, \ldots$ are in parallel, their equivalent capacitance is

$$C_{eq} = C_1 + C_2 + C_3 + \cdots \qquad \text{(parallel capacitors)} \qquad (30.28)$$

Neither the battery nor any other part of a circuit can tell if the parallel capacitors are replaced by a single capacitor having capacitance C_{eq}.

Now consider the two series capacitors in Figure 30.25a. The center section, consisting of the bottom plate of C_1, the top plate of C_2, and the connecting wire, is electrically isolated. The battery cannot remove charge from or add charge to this section. If it starts out with no net charge, it must end up with no net charge. As a consequence, the two capacitors in series have equal charges $\pm Q$. The battery transfers Q from the bottom of C_2 to the top of C_1. This transfer polarizes the center section, as shown, but it still has $Q_{net} = 0$.

The potential differences across the two capacitors are $\Delta V_1 = Q/C_1$ and $\Delta V_2 = Q/C_2$. The total potential difference across both capacitors is $\Delta V_C = \Delta V_1 + \Delta V_2$. Suppose, as in Figure 30.25b, we replaced the two capacitors with a single capacitor having charge Q and potential difference $\Delta V_C = \Delta V_1 + \Delta V_2$. This capacitor is equivalent to the original two because the battery has to establish the same potential difference and move the same amount of charge in either case.

By definition, the *inverse* of the capacitance of this equivalent capacitor is

$$\frac{1}{C_{eq}} = \frac{\Delta V_C}{Q} = \frac{\Delta V_1 + \Delta V_2}{Q} = \frac{\Delta V_1}{Q} + \frac{\Delta V_2}{Q} = \frac{1}{C_1} + \frac{1}{C_2} \qquad (30.29)$$

This analysis hinges on the fact that **series capacitors each have the same charge** Q. We could easily extend this analysis to more than two capacitors. If capacitors $C_1, C_2, C_3, \ldots$ are in series, their equivalent capacitance is

$$C_{eq} = \left(\frac{1}{C_1} + \frac{1}{C_2} + \frac{1}{C_3} + \cdots \right)^{-1} \qquad \text{(series capacitors)} \qquad (30.30)$$

NOTE ▶ Be careful to avoid the common error of adding the inverses but forgetting to invert the sum. ◀

Let's summarize the key facts before looking at a numerical example:

- Parallel capacitors all have the same potential difference ΔV_C. Series capacitors all have the same amount of charge $\pm Q$.
- The equivalent capacitance of a parallel combination of capacitors is *larger* than any single capacitor in the group. The equivalent capacitance of a series combination of capacitors is *smaller* than any single capacitor in the group.

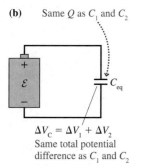

(a) Series capacitors have the same Q.

$\Delta V_1 = Q/C_1$

C_1 $+Q$ / $-Q$

No net charge on this isolated segment

C_2 $+Q$ / $-Q$

$\Delta V_2 = Q/C_2$

(b) Same Q as C_1 and C_2

C_{eq}

$\Delta V_C = \Delta V_1 + \Delta V_2$
Same total potential difference as C_1 and C_2

FIGURE 30.25 Replacing two series capacitors with an equivalent capacitor.

EXAMPLE 30.8 A capacitor circuit

Find the charge on and the potential difference across each of the three capacitors in Figure 30.26.

MODEL Assume the battery is ideal, with $\Delta V_{bat} = \mathcal{E} = 12$ V. Use the results for parallel and series capacitors.

SOLVE The three capacitors are neither in parallel nor in series, but we can break them into smaller groups that are. A useful method of *circuit analysis* is first to combine elements until reaching a single equivalent element, then to reverse the process

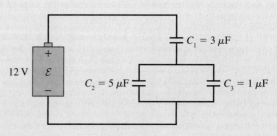

FIGURE 30.26 A capacitor circuit.

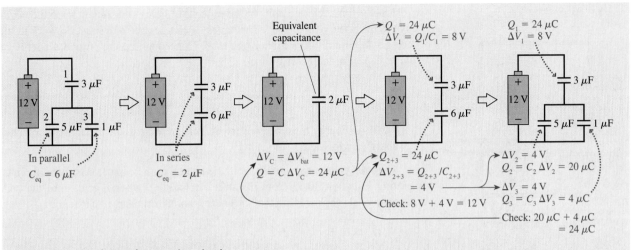

FIGURE 30.27 Analyzing the capacitor circuit.

and calculate values for each element. Figure 30.27 shows the analysis of this circuit. Notice that we redraw the circuit after every step. The equivalent capacitance of the 3 μF and 6 μF capacitors in series is found from

$$C_{eq} = \left(\frac{1}{3\ \mu F} + \frac{1}{6\ \mu F}\right)^{-1} = \left(\frac{2}{6} + \frac{1}{6}\right)^{-1} \mu F$$

$$= \left(\frac{1}{2}\right)^{-1} \mu F = 2\ \mu F$$

Once we get to the single equivalent capacitance, we find that $\Delta V_C = \Delta V_{bat} = 12$ V and $Q = C\Delta V_C = 24\ \mu C$. Now we can reverse direction. Capacitors in series all have the same charge, so the charge on C_1 and on C_{2+3} is $\pm 24\ \mu C$. This is enough to

determine that $\Delta V_1 = 8$ V and $\Delta V_{2+3} = 4$ V. Capacitors in parallel all have the same potential difference, so $\Delta V_2 = \Delta V_3 = 4$ V. This is enough to find that $Q_2 = 20\ \mu C$ and $Q_3 = 4\ \mu C$. The charge on and the potential difference across each of the three capacitors is shown in the final step of Figure 30.27.

ASSESS Notice that we had two important checks of internal consistency. $\Delta V_1 + \Delta V_{2+3} = 8$ V + 4 V add up to the 12 V we had found for the 2 μF equivalent capacitor. Then $Q_2 + Q_3 = 20\ \mu C + 4\ \mu C$ add up to the 24 μC we had found for the 6 μF equivalent capacitor. We'll do much more circuit analysis of this type in the next chapter, but it's worth noting now that circuit analysis becomes nearly foolproof *if* you make use of these checks of internal consistency.

STOP TO THINK 30.5   Rank in order, from largest to smallest, the equivalent capacitance $(C_{eq})_a$ to $(C_{eq})_d$ of circuits a to d.

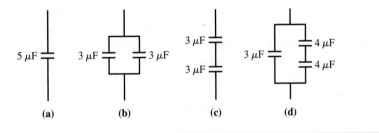

(a)   (b)   (c)   (d)

30.7 The Energy Stored in a Capacitor

Capacitors are important elements in electric circuits because of their ability to store energy. Figure 30.28 on the next page shows a capacitor being charged. The instantaneous value of the charge on the two plates is $\pm q$, and this charge separation has established a potential difference $\Delta V = q/C$ between the two electrodes.

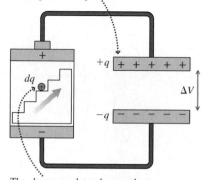

The instantaneous charge on the plates is $\pm q$.

$+q$

dq

$-q$

ΔV

The charge escalator does work $dq\,\Delta V$ to move charge dq from the negative plate to the positive plate.

FIGURE 30.28 The charge escalator does work on charge dq as the capacitor is being charged.

An additional charge dq is in the process of being transferred from the negative to the positive electrode. The battery's charge escalator must do work to lift charge dq "uphill" to a higher potential. Consequently, the potential energy of dq + capacitor increases by

$$dU = dq\,\Delta V = \frac{q\,dq}{C} \tag{30.31}$$

NOTE ▶ Energy must be conserved. This increase in the capacitor's potential energy is provided by the battery. ◀

The total energy transferred from the battery to the capacitor is found by integrating Equation 30.31 from the start of charging, when $q = 0$, until the end, when $q = Q$. Thus we find that the energy stored in a charged capacitor is

$$U_C = \frac{1}{C}\int_0^Q q\,dq = \frac{Q^2}{2C} \tag{30.32}$$

In practice, it is often easier to write the stored energy in terms of the capacitor's potential difference $\Delta V_C = Q/C$. This is

$$U_C = \frac{Q^2}{2C} = \frac{1}{2}C(\Delta V_C)^2 \tag{30.33}$$

The potential energy stored in a capacitor depends on the *square* of the potential difference across it. This result is reminiscent of the potential energy $U = \frac{1}{2}k(\Delta x)^2$ stored in a spring, and a charged capacitor really is analogous to a stretched spring. A stretched spring holds the energy until we release it, then that potential energy is transformed into kinetic energy. Likewise, a charged capacitor holds energy until we discharge it. Then the potential energy is transformed first into the kinetic energy of moving electrons (the current) and ultimately, as the electrons collide with atoms in the metal, into the thermal energy of the wire.

EXAMPLE 30.9 Storing energy in a capacitor
How much energy is stored in a 2.0 μF capacitor that has been charged to 5000 V? What is the average power dissipation if this capacitor is discharged in 10 μs?

SOLVE The energy stored in the charged capacitor is

$$U_C = \frac{1}{2}C(\Delta V_C)^2 = \frac{1}{2}(2.0 \times 10^{-6}\,\text{F})(5000\,\text{V})^2 = 25\,\text{J}$$

If this energy is released in 10 μs, the average power dissipation is

$$P = \frac{\Delta E}{\Delta t} = \frac{25\,\text{J}}{1.0 \times 10^{-5}\,\text{s}} = 2.5 \times 10^6\,\text{W} = 2.5\,\text{MW}$$

ASSESS The stored energy is equivalent to raising a 1 kg mass 2.5 m. This is a rather large amount of energy, which you can see by imagining the damage a 1 kg mass could do after falling 2.5 m. When this energy is released very quickly, which is possible in an electric circuit, it provides an *enormous* amount of power.

The usefulness of a capacitor stems from the fact that it can be charged very slowly, over many seconds, and then can release the energy very quickly. A mechanical analogy would be using a crank to slowly stretch the spring of a catapult, then quickly releasing the energy to launch a massive rock.

The capacitor described in Example 30.9 is typical of the capacitors used in high-power pulsed lasers. The capacitor is charged relatively slowly, in about 0.1 s, then quickly discharged into the laser tube to generate a high-power laser pulse.

Exactly the same thing occurs, only on a smaller scale, in the flash unit of a camera. The camera batteries charge a capacitor, then the energy stored in the capacitor is quickly discharged into a *flashlamp*. The charging process in a camera takes several seconds, which is why you can't fire a camera flash twice in quick succession.

An important medical application of capacitors is the *defibrillator*. A heart attack or a serious injury can cause the heart to enter a state known as *fibrillation* in which the heart muscles twitch randomly and cannot pump blood. A strong electric shock through the chest can sometimes restore the proper heart rhythm. A defibrillator has a large capacitor that can store up to 360 J of energy. This energy is released in about 2 ms through two "paddles" pressed against the patient's chest. It takes several seconds to charge the capacitor, which is why, on television medical shows, you hear an emergency room doctor or nurse shout, "Charging!"

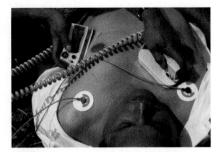

A defibrillator, which can restart the heart, discharges a capacitor through the patient's chest.

The Energy in the Electric Field

We can "see" the potential energy of a stretched spring in the tension of the coils. If a charged capacitor is analogous to a stretched spring, where is the stored energy? It's in the electric field!

Figure 30.29 shows a parallel-plate capacitor in which the plates have area A and are separated by distance d. The potential difference across the capacitor is related to the electric field inside the capacitor by $\Delta V_C = Ed$. The capacitance, which we found in Equation 30.24, is $C = \epsilon_0 A/d$. Substituting these into Equation 30.33, we find that the energy stored in the capacitor is

$$U_C = \frac{1}{2}C(\Delta V_C)^2 = \frac{1}{2}\frac{\epsilon_0 A}{d}(Ed)^2 = \frac{\epsilon_0}{2}(Ad)E^2 \qquad (30.34)$$

The quantity Ad is the volume *inside* the capacitor, the region in which the capacitor's electric field exists. (Recall that an ideal capacitor has $\vec{E} = \vec{0}$ everywhere except between the plates.) Although we talk about "the energy stored in the capacitor," Equation 30.34 suggests that, strictly speaking, **the energy is stored in the capacitor's electric field.**

Because Ad is the volume in which the energy is stored, we can define an **energy density** u_E of the electric field:

$$u_E = \frac{\text{energy stored}}{\text{volume in which it is stored}} = \frac{U_C}{Ad} = \frac{\epsilon_0}{2}E^2 \qquad (30.35)$$

The energy density has units J/m^3. We've derived Equation 30.35 for a parallel-plate capacitor, but it turns out to be the correct expression for any electric field.

From this perspective, charging a capacitor stores energy in the capacitor's electric field as the field grows in strength. Later, when the capacitor is discharged, the energy is released as the field collapses.

We first introduced the electric field as a way to visualize how a long-range force operates. But if the field can store energy, the field must be real, not merely a pictorial device. We'll explore this idea further in Chapter 34, where we'll find that the energy transported by a light wave—the very real energy of warm sunshine—is the energy of electric and magnetic fields.

Capacitor plate with area A

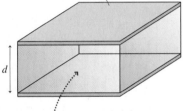

The capacitor's energy is stored in the electric field in volume Ad between the plates.

FIGURE 30.29 A capacitor's energy is stored in the electric field.

EXAMPLE 30.10 The energy density of the electric field
The plates of a parallel-plate capacitor are separated by 1.0 mm. What is the energy density in the capacitor's electric field if the capacitor is charged to 500 V?

SOLVE The electric field inside the capacitor is

$$E = \frac{\Delta V_C}{d} = \frac{500 \text{ V}}{0.0010 \text{ m}} = 5.0 \times 10^5 \text{ V/m}$$

Consequently, the energy density in the electric field is

$$u_E = \frac{\epsilon_0}{2}E^2 = \frac{1}{2}(8.85 \times 10^{-12} \text{ C}^2/\text{Nm}^2)(5.0 \times 10^5 \text{ V/m})^2$$

$$= 1.1 \text{ J/m}^3$$

SUMMARY

The goal of Chapter 30 has been to understand how the electric potential is connected to the electric field.

GENERAL PRINCIPLES

Connecting V and $\vec{E}$

The electric potential and the electric field are two different perspectives of how source charges alter the space around them. V and $\vec{E}$ are related by

$$\Delta V = V(s_f) - V(s_i) = -\int_{s_i}^{s_f} E_s \, ds$$

where s is measured from point i to point f and E_s is the component of $\vec{E}$ parallel to the line of integration.

Graphically

ΔV = the negative of the area under the E_s graph

and

$$E_s = -\frac{dV}{ds}$$

= the negative of the slope of the potential graph.

The Geometry of Potential and Field

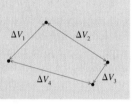

The electric field

- Is perpendicular to the equipotential surfaces.
- Points "downhill" in the direction of decreasing V.
- Is inversely proportional to the spacing Δs between the equipotential surfaces.

Conservation of Energy

The sum of all potential differences around a closed path is zero. $\sum (\Delta V)_i = 0$.

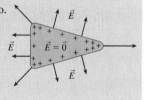

IMPORTANT CONCEPTS

A **battery** is a source of potential. The charge escalator in a battery uses chemical reactions to move charges from the negative terminal to the positive terminal.

$$\Delta V_{bat} = \mathcal{E}$$

where the emf $\mathcal{E}$ is the work per charge done by the charge escalator.

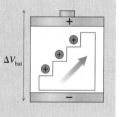

For a conductor in electrostatic equilibrium

- The interior electric field is zero.
- The exterior electric field is perpendicular to the surface.
- The surface is an equipotential.
- The interior is at the same potential as the surface.

APPLICATIONS

Resistors

A potential difference ΔV_{wire} between the ends of a wire creates an electric field inside the wire

$$E_{wire} = \frac{\Delta V_{wire}}{L}$$

The electric field causes a current

$$I = \frac{\Delta V_{wire}}{R}$$

where $R = \frac{\rho L}{A}$ is the wire's **resistance.**

Capacitors

The **capacitance** of two conductors charged to $\pm Q$ is

$$C = \frac{Q}{\Delta V_C}$$

The energy stored in a capacitor is

$$U_C = \tfrac{1}{2} C (\Delta V_C)^2$$

Series capacitors

$$C_{eq} = \left(\frac{1}{C_1} + \frac{1}{C_2} + \frac{1}{C_3} + \cdots \right)^{-1}$$

Parallel capacitors

$$C_{eq} = C_1 + C_2 + C_3 + \cdots$$

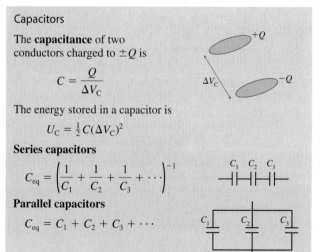

TERMS AND NOTATION

Kirchhoff's loop law	emf, $\mathcal{E}$	farad, F
Van de Graaff generator	terminal voltage, ΔV_{bat}	parallel capacitors
battery	resistance, R	series capacitors
charge escalator	ohm, Ω	equivalent capacitance, C_{eq}
ideal battery	capacitance, C	energy density, u_E

EXERCISES AND PROBLEMS

Exercises

Section 30.1 Connecting Potential and Field

1. What is the potential difference between $x_i = 10$ cm and $x_f = 30$ cm in the uniform electric field $E_x = 1000$ V/m?

2. What is the potential difference between $y_i = -5$ cm and $y_f = 5$ cm in the uniform electric field $\vec{E} = (20{,}000\hat{i} - 50{,}000\hat{j})$ V/m?

Section 30.2 Finding the Electric Field from the Potential

3. The electric potential in a region of uniform electric field is -1000 V at $x = -1.0$ m and $+1000$ V at $x = +1.0$ m. What is E_x?

4. The electric potential along the x-axis is $V = 100x^2$ V, where x is in meters. What is E_x at $x = 0$ m? At $x = 1$ m?

5. Figure Ex30.5 is a graph of V versus x. Draw the corresponding graph of E_x versus x.

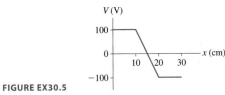

FIGURE EX30.5

6. Figure Ex30.6 is a graph of V versus x. Draw the corresponding graph of E_x versus x.

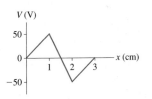

FIGURE EX30.6

7. What are the magnitude and direction of the electric field at the dot in Figure Ex30.7?

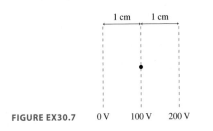

FIGURE EX30.7

8. What are the magnitude and direction of the electric field at the dot in Figure Ex30.8?

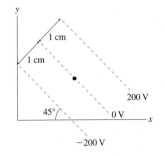

FIGURE EX30.8

9. What is the potential difference ΔV_{34} in Figure Ex30.9?

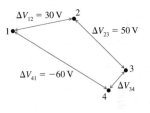

FIGURE EX30.9

Section 30.3 A Conductor in Electrostatic Equilibrium

10. The excess charge on a 20-cm-diameter metal sphere is 10 nC. Draw a graph of (a) E versus r and (b) V versus r over the range $0 < r < 40$ cm.

Section 30.4 Sources of Electric Potential

11. How much work does the charge escalator do to move 1.0 μC of charge from the negative terminal to the positive terminal of a 1.5 V battery?

12. How much work does the electric motor of a Van de Graaff generator do to lift a positive ion ($q = e$) if the potential of the spherical electrode is 1.0 MV?

13. What is the emf of a battery that does 0.60 J of work to transfer 0.050 C of charge from the negative to the positive terminal?

Section 30.5 Connecting Potential and Current

14. Wires 1 and 2 are made of the same metal. Wire 2 has twice the length and twice the diameter of wire 1. What are the ratios (a) ρ_2/ρ_1 of the resistivities and (b) R_2/R_1 of the resistances of the two wires?

15. What is the resistance of
    a. A 1.0-m-long copper wire that is 0.50 mm in diameter?
    b. A 10-cm-long piece of carbon with a 1.0 mm × 1.0 mm square cross section?
16. A 10-m-long wire with a diameter of 0.80 mm has a resistance of 1.1 Ω. Of what material is the wire made?
17. The electric field inside a 30-cm-long copper wire is 10 V/m. What is the potential difference between the ends of the wire?
18. a. How long must a 0.60-mm-diameter aluminum wire be to have a 0.50 A current when connected to the terminals of a 1.5 V flashlight battery?
    b. What is the current if the wire is half this length?
19. The terminals of a 0.70 V watch battery are connected by a 100-m-long gold wire with a diameter of 0.10 mm. What is the current in the wire?

Section 30.6 Capacitance and Capacitors

20. Two 2.0 cm × 2.0 cm square aluminum electrodes are spaced 0.50 mm apart. The electrodes are connected to a 100 V battery.
    a. What is the capacitance?
    b. What is the charge on each electrode?
21. You need to construct a 100 pF capacitor for a science project. You plan to cut two $L \times L$ metal squares and place spacers between them. The thinnest spacers you have are 0.20 mm thick. What is the proper value of L?
22. A switch that connects a battery to a 10 μF capacitor is closed. Several seconds later you find that the capacitor plates are charged to ± 30 μC. What is the emf of the battery?
23. What is the emf of a battery that will charge a 2.0 μF capacitor to ± 48 μC?
24. Two electrodes connected to a 9.0 V battery are charged to ± 45 nC. What is the capacitance of the electrodes?
25. A 6 μF capacitor, a 10 μF capacitor, and a 16 μF capacitor are connected in parallel. What is their equivalent capacitance?
26. A 6 μF capacitor, a 10 μF capacitor, and a 16 μF capacitor are connected in series. What is their equivalent capacitance?
27. What is the capacitance of the two metal spheres shown in Figure Ex30.27?

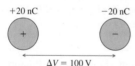

+20 nC   −20 nC

FIGURE EX30.27 $\Delta V = 100$ V

28. Initially, the switch in Figure Ex30.28 is open and the capacitor is uncharged. How much charge flows through the switch after the switch is closed?

Switch

1.5 V   10 μF

FIGURE EX30.28

Section 30.7 The Energy Stored in a Capacitor

29. To what potential should you charge a 1.0 μF capacitor to store 1.0 J of energy?

30. Figure Ex30.30 shows Q versus t for a 2.0 μF capacitor. Draw a graph showing U_C versus t.

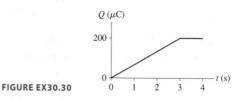

FIGURE EX30.30

31. Capacitor 2 has half the capacitance and twice the potential difference as capacitor 1. What is the ratio U_{C1}/U_{C2}?
32. 50 pJ of energy is stored in a 2.0 cm × 2.0 cm × 2.0 cm region of uniform electric field. What is the electric field strength?
33. A 2.0-cm-diameter parallel-plate capacitor with a spacing of 0.50 mm is charged to 200 V. What are (a) the total energy stored in the electric field and (b) the energy density?

Problems

34. a. Which point, A or B, has a larger electric potential?
    b. What is the potential difference between A and B?

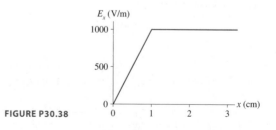

FIGURE P30.34   7 cm

35. The electric field in a region of space is $E_x = -1000x$ V/m, where x is in meters.
    a. Graph E_x versus x over the region -1 m $\leq x \leq 1$ m.
    b. What is the potential difference between $x_i = -20$ cm and $x_f = 30$ cm?
36. The electric field in a region of space is $E_x = 5000x$ V/m, where x is in meters.
    a. Graph E_x versus x over the region -1 m $\leq x \leq 1$ m.
    b. Find an expression for the potential V at position x. As a reference, let $V = 0$ V at the origin.
    c. Graph V versus x over the region -1 m $\leq x \leq 1$ m.
37. An infinitely long cylinder of radius R has linear charge density λ. The potential on the surface of the cylinder is V_0, and the electric field outside the cylinder is $E_r = \lambda/2\pi\epsilon_0 r$. Find the potential relative to the surface at a point that is distance r from the axis, assuming $r > R$.
38. Figure P30.38 shows E_x, the x-component of the electric field, as a function of position along the x-axis. Find and graph V versus x over the region 0 cm $\leq x \leq 3$ cm. As a reference, let $V = 0$ V at $x = 3$ cm.

FIGURE P30.38

39. The three metal electrodes in Figure P30.39 are charged as shown. Draw a graph of (a) E_x versus x and (b) V versus x over the region $0 \le x \le 3$ cm.

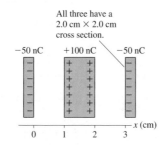

All three have a 2.0 cm × 2.0 cm cross section.

−50 nC   +100 nC   −50 nC

FIGURE P30.39   0   1   2   3   x (cm)

40. Figure P30.40 shows a graph of V versus x in a region of space. The potential is independent of y and z.
    a. Draw a graph of E_x versus x.
    b. Draw a contour map of the potential in the xy-plane in the square-shaped region -3 m $\le x \le 3$ m and -3 m $\le y \le 3$ m. Show and label the -10 V, -5 V, 0 V, $+5$ V, and $+10$ V equipotential surfaces.
    c. Draw electric field vectors on your contour map of part b.

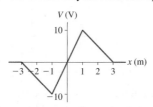

FIGURE P30.40

41. Use the on-axis potential of a charged disk from Chapter 29 to find the on-axis electric field of a charged disk.
42. a. Use the methods of Chapter 29 to find the potential at distance x on the axis of the charged rod shown in Figure P30.42.
    b. Use the result of part a to find the electric field at distance x on the axis of a rod.

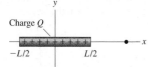

Charge Q

$-L/2$ $L/2$

FIGURE P30.42

43. Determine the magnitude and direction of the electric field at points 1 and 2 in Figure P30.43.

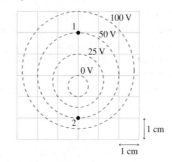

100 V
50 V
25 V
0 V

FIGURE P30.43   1 cm   1 cm

44. Figure P30.44 shows a set of equipotential lines and five labeled points.
    a. From measurements made on this figure with a ruler, using the scale on the figure, estimate the electric field strength E at the five points indicated.

b. Trace the figure on your paper, then show the electric field vectors $\vec{E}$ at the five points.

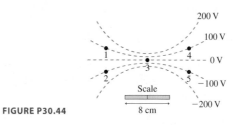

200 V
100 V
0 V
−100 V
Scale
−200 V

FIGURE P30.44   8 cm

45. The electric potential in a region of space is $V = (150x^2 - 200y^2)$ V, where x and y are in meters. What are the strength and the direction of the electric field at $(2.0$ m, 2.0 m$)$?

46. Figure P30.46 shows the electric potential at points on a 5.0 cm × 5.0 cm grid.
    a. Reproduce this figure on your paper, then draw the 50 V, 75 V, and 100 V equipotential surfaces.
    b. Determine the electric field (strength and direction) at the points A, B, C, and D.
    c. Draw the electric field vectors at points A, B, C, and D on your diagram.

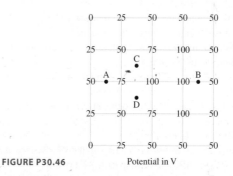

FIGURE P30.46   Potential in V

47. Metal sphere 1 has a positive charge of 6.0 nC. Metal sphere 2, which is twice the diameter of sphere 1, is initially uncharged. The spheres are then connected together by a long, thin metal wire. What are the final charges on each sphere?

48. The metal spheres in Figure P30.48 are charged to ± 300 V. Draw this figure on your paper, then draw a plausible contour map of the potential, showing and labeling the -300 V, -200 V, -100 V, $\ldots$, 300 V equipotential surfaces.

−300 V   +300 V

FIGURE P30.48

49. The potential at the center of a 4.0-cm-diameter copper sphere is 500 V, relative to $V = 0$ V at infinity. How much excess charge is on the sphere?

50. A 15-cm-long nichrome wire is connected across the terminals of a 1.5 V battery.
    a. What is the electric field inside the wire?
    b. What is the current density inside the wire?
    c. If the current in the wire is 2.0 A, what is the wire's diameter?

51. A 20-cm-long hollow nichrome tube of inner diameter 2.8 mm, outer diameter 3.0 mm is connected to a 3.0 V battery. What is the current in the tube?

52. A 1.5 V battery provides 0.50 A of current.
    a. At what rate (C/s) is charge lifted by the charge escalator?
    b. How much work does the charge escalator do to lift 1.0 C of charge?
    c. What is the power output of the charge escalator?
53. A 1.5 V flashlight battery is connected to a wire with a resistance of 3.0 Ω. Figure P30.53 shows the battery's potential difference as a function of time. What is the total charge lifted by the charge escalator?

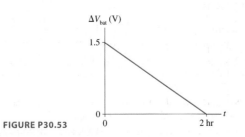

FIGURE P30.53

54. Two 10-cm-diameter metal plates are 1.0 cm apart. They are charged to ± 12.5 nC. They are suddenly connected together by a 0.224-mm-diameter copper wire stretched taut from the center of one plate to the center of the other.
    a. What is the maximum current in the wire?
    b. What is the largest electric field in the wire?
    c. Does the current increase with time, decrease with time, or remain steady? Explain.
    d. What is the total amount of energy dissipated in the wire?
55. Two 2.0 cm $\times$ 2.0 cm metal electrodes are spaced 1.0 mm apart and connected by wires to the terminals of a 9.0 V battery.
    a. What are the charge on each electrode and the potential difference between them?
    The wires are disconnected, and insulated handles are used to pull the plates apart to a new spacing of 2.0 mm.
    b. What are the charge on each electrode and the potential difference between them?
56. Two 2.0 cm $\times$ 2.0 cm metal electrodes are spaced 1.0 mm apart and connected by wires to the terminals of a 9.0 V battery.
    a. What are the charge on each electrode and the potential difference between them?
    While the plates are still connected to the battery, insulated handles are used to pull them apart to a new spacing of 2.0 mm.
    b. What are the charge on each electrode and the potential difference between them?
57. A spherical capacitor with a 1.0 mm gap between the spheres has a capacitance of 100 pF. What are the diameters of the two spheres?
58. You need a capacitance of 50 μF, but you don't happen to have a 50 μF capacitor. You do have a 30 μF capacitor. What additional capacitor do you need to produce a total capacitance of 50 μF? Should you join the two capacitors in parallel or in series?
59. You need a capacitance of 50 μF, but you don't happen to have a 50 μF capacitor. You do have a 75 μF capacitor. What additional capacitor do you need to produce a total capacitance of 50 μF? Should you join the two capacitors in parallel or in series?
60. Find expressions for the equivalent capacitance of (a) N identical capacitors C in parallel and (b) N identical capacitors C in series.

61. What is the equivalent capacitance of the three capacitors in Figure P30.61?

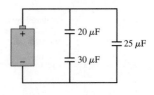

FIGURE P30.61

62. What is the equivalent capacitance of the three capacitors in Figure P30.62?

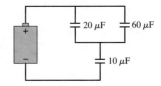

FIGURE P30.62

63. What are the charge on and the potential difference across each capacitor in Figure P30.63?

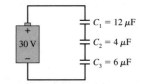

FIGURE P30.63

64. What are the charge on and the potential difference across each capacitor in Figure P30.64?

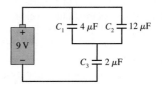

FIGURE P30.64

65. What are the charge on and the potential difference across each capacitor in Figure P30.65?

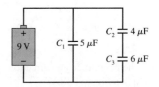

FIGURE P30.65

66. You have three 12 μF capacitors. Draw diagrams showing how you could arrange all three so that their equivalent capacitance is (a) 4.0 μF, (b) 8.0 μF, (c) 18 μF, and (d) 36 μF.
67. What is the capacitance of the three concentric metal spherical shells in Figure P30.67?
    Hint: Can you think of this as a combination of capacitors?

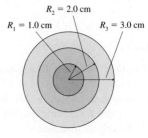

FIGURE P30.67

68. Six identical capacitors with capacitance C are connected as shown in Figure P30.68.
    a. What is the equivalent capacitance of these six capacitors?
    b. What is the potential difference between points a and b?

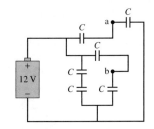

FIGURE P30.68

69. What is the capacitance of the two electrodes in Figure P30.69? **Hint:** Can you think of this as a combination of capacitors?

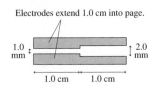

FIGURE P30.69

70. Initially, the switch in Figure P30.70 is in position A and capacitors C_2 and C_3 are uncharged. Then the switch is flipped to position B. Afterward, what are the charge on and the potential difference across each capacitor?

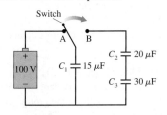

FIGURE P30.70

71. A battery with an emf of 60 V is connected to the two capacitors shown in Figure P30.71. Afterward, the charge on capacitor 2 is 450 μC. What is the capacitance of capacitor 2?

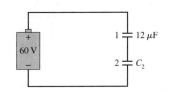

FIGURE P30.71

72. Capacitors $C_1 = 10\ \mu$F and $C_2 = 20\ \mu$F are each charged to 10 V, then disconnected from the battery without changing the charge on the capacitor plates. The two capacitors are then connected in parallel, with the positive plate of C_1 connected to the negative plate of C_2 and vice versa. Afterward, what are the charge on and the potential difference across each capacitor?

73. An isolated 5.0 μF parallel-plate capacitor has 4.0 mC of charge. An external force changes the distance between the electrodes until the capacitance is 2.0 μF. How much work is done by the external force?

74. A parallel-plate capacitor is constructed from two 10 cm × 10 cm electrodes spaced 1.0 mm apart. The capacitor plates are charged to ±10 nC, then disconnected from the battery.
    a. How much energy is stored in the capacitor?
    b. Insulating handles are used to pull the capacitor plates apart until the spacing is 2.0 mm. Now how much energy is stored in the capacitor?

c. Energy must be conserved. How do you account for the difference between a and b?

75. What is the energy density in the electric field at the surface of a 1.0-cm-diameter sphere charged to a potential of 1000 V?

76. The flash unit in a camera uses a 3.0 V battery to charge a capacitor. The capacitor is then discharged through a flashlamp. The discharge takes 10 μs, and the average power dissipated in the flashlamp is 10 W. What is the capacitance of the capacitor?

77. You need to melt a 0.50 kg block of ice at −10°C in a hurry. The stove isn't working, but you do have a 50 V battery. It occurs to you that you could build a capacitor from a couple of pieces of sheet metal that are nearby, charge the capacitor with the battery, then discharge it through the block of ice. If you use square sheets spaced 2.0 mm apart, what must the dimensions of the sheets be to accomplish your goal? Is this feasible?

In Problems 78 through 81 you are given the equation(s) used to solve a problem. For each of these, you are to
    a. Write a realistic problem for which this is the correct equation(s).
    b. Finish the solution of the problem.

78. $2z$ V/m $= -\dfrac{dV}{dz}$
    $V(z = 0) = 10$ V

79. $\dfrac{3.0\ \text{V}}{R} = 5.0$ A
    $R = \dfrac{(9.7 \times 10^{-8}\ \Omega\,\text{m})L}{\pi(0.00050\ \text{m})^2}$

80. $400\ \text{nC} = (100\ \text{V})\,C$
    $C = \dfrac{(8.85 \times 10^{-12}\ \text{C}^2/\text{Nm}^2)(0.10\ \text{m} \times 0.10\ \text{m})}{d}$

81. $\left(\dfrac{1}{3\ \mu\text{F}} + \dfrac{1}{6\ \mu\text{F}}\right)^{-1} + C = 4\ \mu$F

Challenge Problems

82. The electric potential in a region of space is $V = 100(x^2 - y^2)$ V, where x and y are in meters.
    a. Draw a contour map of the potential, showing and labeling the −400 V, −100 V, 0 V, +100 V, and +400 V equipotential surfaces.
    b. Find an expression for the electric field $\vec{E}$ at position (x, y).
    c. Draw the electric field lines on your diagram of part a.

83. Charge is uniformly distributed with charge density ρ inside a very long cylinder of radius R. Find the potential difference between the surface and the axis of the cylinder.

84. An electric dipole at the origin consists of two charges $\pm q$ spaced distance s apart along the y-axis.
    a. Find an expression for the potential $V(x, y)$ at an arbitrary point in the xy-plane. Your answer will be in terms of q, s, x, and y.
    b. Use the binomial approximation to simplify your result of part a when $s \ll x$ and $s \ll y$.
    c. Assuming $s \ll x$ and y, find expressions for E_x and E_y, the components of $\vec{E}$ for a dipole.
    d. What is the on-axis field $\vec{E}$? Does your result agree with Equation 26.11?
    e. What is the field $\vec{E}$ on the bisecting axis? Does your result agree with Equation 26.12?

85. Consider a uniformly charged sphere of radius R and total charge Q. The electric field E_{out} *outside* the sphere ($r \geq R$) is simply that of a point charge Q. In Chapter 27, we used Gauss's law to find that the electric field E_{in} *inside* the sphere ($r \leq R$) is radially outward with field strength

$$E_{in} = \frac{1}{4\pi\epsilon_0}\frac{Q}{R^3}r$$

a. Graph E versus R for $0 \leq r \leq 3R$.
b. The electric potential V_{out} *outside* the sphere is that of a point charge Q. Find an expression for the electric potential V_{in} at position r *inside* the sphere. As a reference, let $V_{in} = V_{out}$ at the surface of the sphere.
c. What is the ratio $V_{center}/V_{surface}$?
d. Graph V versus R for $0 \leq r \leq 3R$.

86. High-frequency signals are often transmitted along a *coaxial cable,* such as the one shown in Figure CP30.86. For example, the cable TV hookup coming into your home is a coaxial cable. The signal is carried on a wire of radius R_1 while the outer conductor of radius R_2 is grounded (i.e., at $V = 0$ V). An insulating material fills the space between them, and an insulating plastic coating goes around the outside.

a. Find an expression for the capacitance per meter of a coaxial cable. Assume that the insulating material between the cylinders is air.
b. Evaluate the capacitance per meter of a cable having $R_1 = 0.50$ mm and $R_2 = 3.0$ mm.

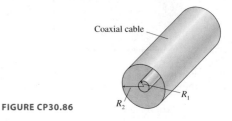

FIGURE CP30.86

STOP TO THINK ANSWERS

Stop to Think 30.1: c. E_y is the negative of the slope of the V-versus-y graph. E_y is positive, because $\vec{E}$ points up, so the graph has a negative slope. E_y has constant magnitude, so the slope has a constant value.

Stop to Think 30.2: c. $\vec{E}$ points "downhill," so V must decrease from right to left. E is larger on the left than on the right, so the contour lines must be closer together on the left.

Stop to Think 30.3: b. Because of the connecting wire, the three spheres form a single conductor in electrostatic equilibrium. Thus all points are at the same potential. The electric field of a sphere is related to the sphere's potential by $E = V/R$, so a smaller-radius sphere has a larger E.

Stop to Think 30.4: $I_a = I_b = I_c = I_d$. Conservation of current requires $I_a = I_b$. The current in each wire is $I = \Delta V/R$. All the wires have the same resistance, because they are identical, and they all have the same potential difference, because the battery is a *source of potential.*

Stop to Think 30.5: $(C_{eq})_b > (C_{eq})_a = (C_{eq})_d > (C_{eq})_c$. $(C_{eq})_b = 3\ \mu F + 3\ \mu F = 6\ \mu F$. The equivalent capacitance of series capacitors is less than any capacitor in the group, so $(C_{eq})_c < 3\ \mu F$. Only d requires any real calculation. The two $4\ \mu F$ capacitors are in series and are equivalent to a single $2\ \mu F$ capacitor. The $2\ \mu F$ equivalent capacitor is in parallel with $3\ \mu F$, so $(C_{eq})_d = 5\ \mu F$.

31 Fundamentals of Circuits

This microprocessor, the heart of a computer, is an extraordinarily complex electric circuit. Even so, its operations can be understood on the basis of a few fundamental physical principles.

▶ **Looking Ahead**
The goal of Chapter 31 is to understand the fundamental physical principles that govern electric circuits. In this chapter you will learn to:

- Understand the conducting and insulating materials used in circuits.
- Draw and use basic circuit diagrams.
- Analyze circuits containing resistors in series and in parallel.
- Calculate power dissipation in circuit elements.
- Understand the growth and decay of current in *RC* circuits.

◀ **Looking Back**
This chapter is based on our earlier development of the ideas of current and potential. Please review:

- Sections 28.4–28.5 Current and conductivity.
- Sections 30.4–30.5 Batteries, emf, and current.
- Section 30.6 Capacitors.

A computer is an incredible device. Surprising as it may seem, the power of a computer is achieved simply by the controlled flow of charges through tiny wires and circuit elements. The most powerful supercomputer is a direct descendant of the charged rods with which we began Part VI.

This chapter will bring together many of the ideas you have learned about the electric field and potential and apply them to the analysis of electric circuits. This single chapter will not pretend to be a full course on circuit analysis. Instead, as the title implies, our more modest goal is to describe the *fundamental physical principles* by which circuits operate. An understanding of these basic principles will prepare you to undertake a course in circuit analysis and design at a later time.

Our primary interest is with circuits in which the battery's potential difference is unchanging and all the currents in the circuit are constant. Such circuits are called **direct current,** or DC, circuits. We'll consider alternating-current (AC) circuits, in which the potential difference oscillates sinusoidally, in Chapter 35.

31.1 Resistors and Ohm's Law

In Part VI we have developed a theoretical understanding of charges and their interactions. Now we are ready to see how this information can be put to practical use. Figure 31.1 summarizes several key ideas that are of particular relevance to electric circuits.

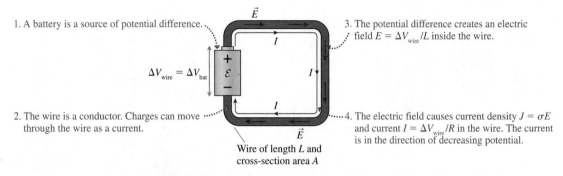

1. A battery is a source of potential difference.

$\Delta V_{wire} = \Delta V_{bat}$

2. The wire is a conductor. Charges can move through the wire as a current.

3. The potential difference creates an electric field $E = \Delta V_{wire}/L$ inside the wire.

4. The electric field causes current density $J = \sigma E$ and current $I = \Delta V_{wire}/R$ in the wire. The current is in the direction of decreasing potential.

Wire of length L and cross-section area A

FIGURE 31.1 A summary of the properties of charges, fields, and potentials that are relevant to electric circuits.

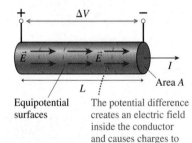

Equipotential surfaces

The potential difference creates an electric field inside the conductor and causes charges to flow through it.

FIGURE 31.2 The current I is related to the potential difference ΔV.

You learned in Chapter 30 that there is a simple relationship between the current I in a conducting element and ΔV, the potential difference between its ends. Figure 31.2 shows a conductor of length L and cross-section area A. The current density in the conductor is

$$J = \sigma E = \frac{E}{\rho} \tag{31.1}$$

where $E = \Delta V/L$ is the field strength created by the potential difference and $\rho = 1/\sigma$ is the material's resistivity. The actual current in the conductor, $I = JA$, is

$$I = JA = \frac{E}{\rho}A = \frac{\Delta V/L}{\rho}A = \frac{\Delta V}{(\rho L/A)} \tag{31.2}$$

The quantity

$$R = \frac{\rho L}{A} \tag{31.3}$$

is the *resistance* of the conductor. The unit of resistance is the *ohm*, where $1 \text{ ohm} = 1\ \Omega = 1 \text{ V/A}$.

In terms of resistance, Equation 31.2 is

$$I = \frac{\Delta V}{R} \tag{31.4}$$

This relationship between I and ΔV is known as **Ohm's law.**

Circuit textbooks often write Ohm's law as $V = IR$ rather than $I = \Delta V/R$. This can be misleading until you have sufficient experience with circuit analysis. First, Ohm's law relates the current to the potential *difference* between the ends of the conductor. Engineers and circuit designers *mean* "potential difference" when they use the symbol V, but this use of the symbol is easily overlooked by beginners who think that it means "the potential."

Second, $V = IR$ or even $\Delta V = IR$ suggests that a current I causes a potential difference ΔV. As you have seen, the proper cause-and-effect sequence is the other way around. Current is a *consequence* of a potential difference, hence $I = \Delta V/R$ is a better description of cause and effect.

EXAMPLE 31.1 The current in a wire

What is the current in a 1.0-mm-diameter, 10-cm-long copper wire that is attached to the terminals of a 1.5 V battery?

SOLVE The resistivity of copper, from Table 30.1, is $\rho = 1.7 \times 10^{-8}\ \Omega\,\text{m}$. Thus the resistance of the copper wire is

$$R = \frac{\rho L}{A} = \frac{\rho L}{\pi r^2} = 2.2 \times 10^{-3}\ \Omega$$

The current in the wire is then found from Ohm's law:

$$I = \frac{\Delta V}{R} = 680\ \text{A}$$

ASSESS This is an enormous current because the resistance of a copper wire is extremely small. In fact, as you will learn later in this chapter, no real 1.5 V battery can deliver a current anywhere near this large.

Because the resistance of metals is so small, a circuit made exclusively of copper wires would have enormous currents and would quickly deplete the battery. It is useful to limit the current by using circuit elements that, although they are conductors, have a resistance significantly larger than the wires. Such devices are called **resistors.** Resistors are made either by using poorly conducting materials, such as carbon or nichrome, or by depositing very thin metal films on an insulating substrate. The cross-section area A of a thin film is so small that a film's resistance can be much larger than that of a solid wire. The resistors used in circuits typically range from $10\ \Omega$ to $1\ \text{M}\Omega$. The resistance of the connecting metal wires, by comparison, is essentially zero.

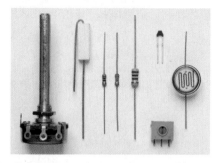

The resistors used in circuits range from a few ohms to millions of ohms of resistance.

Ohmic and Nonohmic Materials

Despite its name, Ohm's law is *not* a law of nature. It is limited to those materials whose resistance R remains constant—or very nearly so—during use. The materials to which Ohm's law applies are called **ohmic.** Figure 31.3a shows that the current through an ohmic material is directly proportional to the potential difference. Doubling the potential difference results in a doubling of the current. Resistors and metal wires are ohmic devices.

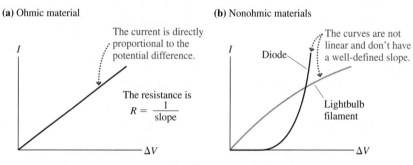

(a) Ohmic material

The current is directly proportional to the potential difference.

The resistance is $R = \dfrac{1}{\text{slope}}$

(b) Nonohmic materials

Diode

The curves are not linear and don't have a well-defined slope.

Lightbulb filament

FIGURE 31.3 Current-versus-potential-difference graphs for ohmic and nonohmic materials.

Many materials and devices are **nonohmic,** meaning that the current through the device is *not* directly proportional to the potential difference. Three important examples of nonohmic devices are

1. Batteries, where $\Delta V = \mathcal{E}$ is determined by chemical reactions, independent of I.
2. Semiconductors, where the I-versus-ΔV curve is very nonlinear.
3. Capacitors, where, as you'll learn later in this chapter, the relationship between I and ΔV differs from that of a resistor.

Figure 31.3b shows the I-versus-ΔV graphs of a semiconductor device called a *diode* and of a lightbulb filament. You can see that the inverse of the slope does not have a unique value that we can call "resistance."

NOTE ▶ Ohm's law is an important part of circuit analysis because resistors are essential components of almost any circuit. However, it is important that you apply Ohm's law *only* to the resistors and not to anything else. ◀

We can identify three basic classes of ohmic circuit materials:

1. *Wires* are metals with very small resistivities ρ and thus very small resistances ($R \ll 1\ \Omega$). An **ideal wire** has $R = 0\ \Omega$, hence the potential difference between the ends of an ideal wire is $\Delta V = 0$ V *even if there is a current in it*. We will usually adopt the *ideal-wire model* of assuming that any connecting wires in a circuit are ideal.
2. *Resistors* are poor conductors that have resistances in the range 10^1 to $10^6\ \Omega$. They are used to control the current in a circuit. Most resistors in a circuit have a specified value of R, such as 500 Ω. The filament in a light-bulb (a tungsten wire with a high resistance due to an extremely small cross-section area A) functions as a resistor as long as it is glowing, but the filament is slightly nonohmic because the value of its resistance when hot is larger than its room-temperature value.
3. *Insulators* are materials such as glass, plastic, or air. An **ideal insulator** has $R = \infty\ \Omega$, hence there is no current in an insulator even if there is a potential difference across it ($I = \Delta V/R = 0$ A). This is why insulators can be used to hold apart two conductors that are at different potentials. All practical insulators have $R \gg 10^9\ \Omega$ and can be treated, for our purposes, as ideal.

The Ideal-Wire Model

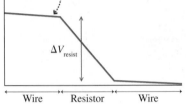

(a) The current is constant along the wire-resistor-wire combination. ⋯

Wire   Resistor   Wire

$R_{\text{resist}} \gg R_{\text{wire}}$

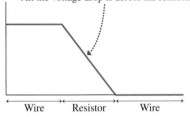

(b) The voltage drop along the wires is much less than across the resistor because the wires have much less resistance.

V

ΔV_{resist}

Wire   Resistor   Wire

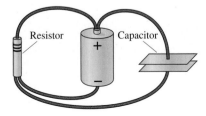

(c) In the ideal wire model, with $R_{\text{wire}} = 0\ \Omega$, there is no voltage drop along the wires. All the voltage drop is across the resistor.

V

Wire   Resistor   Wire

FIGURE 31.4 The potential along a wire-resistor-wire combination.

Figure 31.4a illustrates the ideal-wire model. Current must be conserved, hence the current I in the resistor is the same as the current in each wire. However, the resistor's resistance is *much* larger than the resistance of the wires: $R_{\text{resist}} \gg R_{\text{wire}}$. Consequently, the potential difference across the resistor $\Delta V_{\text{resist}} = IR_{\text{resist}}$ is *much* larger than the potential difference $\Delta V_{\text{wire}} = IR_{\text{wire}}$ between the ends of each wire.

Figure 31.4b shows the potential along the wire-resistor-wire combination. You can see a large *voltage drop,* or potential difference, across the resistor. The voltage drops across the two wires are extremely small. A very reasonable approximation is to assume that *all* the voltage drop occurs along the resistor, none along the wires. This is the approximation made by the ideal-wire model, which assumes that $R_{\text{wire}} = 0\ \Omega$ and $\Delta V_{\text{wire}} = 0$ V. With this approximation, shown in Figure 31.4c, the segments of the graph corresponding to the wires are horizontal.

STOP TO THINK 31.1 Conductors a to d are all made of the same material. Rank in order, from largest to smallest, the resistances R_a to R_d.

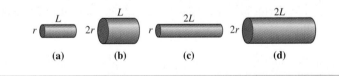

L L $2L$ $2L$

r $2r$ r $2r$

(a)   **(b)**   **(c)**   **(d)**

31.2 Circuit Elements and Diagrams

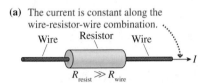

Resistor   Capacitor

FIGURE 31.5 An electric circuit.

Figure 31.5 shows an electric circuit in which a resistor and a capacitor are connected by wires to a battery. To understand the functioning of this circuit, we do not need to know whether the wires are bent or straight, or whether the battery is to the right or to the left of the resistor. The literal picture of Figure 31.5 provides many irrelevant details. It is customary when describing or analyzing circuits to use a more abstract picture called a **circuit diagram.** A circuit diagram is a *logical* picture of what is connected to what. The actual circuit, once it is built, may *look* quite different from the circuit diagram, but it will have the same logic and connections.

A circuit diagram also replaces pictures of the circuit elements with symbols. Figure 31.6 shows the basic symbols that we will need. Notice that the *longer* line at one end of the battery symbol represents the positive terminal of the battery.

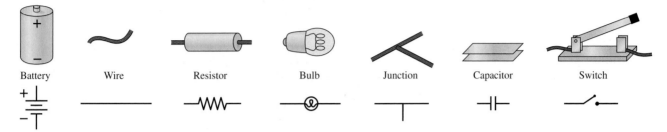

FIGURE 31.6 A library of basic symbols used for electric circuit drawings.

Figure 31.7 is a circuit diagram of the circuit shown in Figure 31.5. Notice how circuit elements are labeled. The battery's emf $\mathcal{E}$ is shown beside the battery, and $+$ and $-$ symbols, even though somewhat redundant, are shown beside the terminals. The resistance R of the resistor and capacitance C of the capacitor are written beside them. We would use numerical values for $\mathcal{E}$ and R if we knew them. The wires, which in practice may bend and curve, are shown as straight-line connections between the circuit elements. You should get into the habit of drawing your own circuit diagrams in a similar fashion.

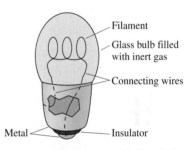

FIGURE 31.7 A circuit diagram for the circuit of Figure 31.5.

Lightbulbs are important circuit elements, and Figure 31.8 gives more information about the anatomy of a light bulb. The important point to understand is that a lightbulb, like a wire or a resistor, has two "ends," and that current passes *through* the bulb. It is often useful to think of a lightbulb as a resistor that happens to give off light when a current is present. Even though, as Figure 31.3b showed, a lightbulb filament is not a perfectly ohmic material, the resistance of a *glowing* lightbulb remains reasonably constant if you don't change ΔV by much. A lightbulb's resistance is typically in the range from 10 Ω to 500 Ω.

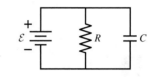

FIGURE 31.8 The anatomy of a lightbulb.

STOP TO THINK 31.2 Which of these diagrams represent the same circuit?

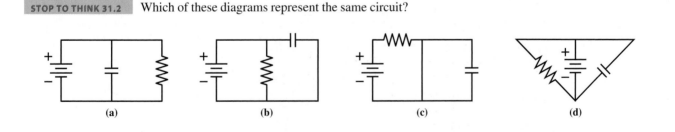

31.3 Kirchhoff's Laws and the Basic Circuit

We are now ready to begin analyzing circuits. To analyze a circuit means to find:

1. The potential difference across each circuit component.
2. The current in each circuit component.

Circuit analysis is based on *Kirchhoff's laws,* which we introduced in Chapters 28 and 30.

(a)

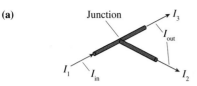

Junction law: $I_1 = I_2 + I_3$

(b)

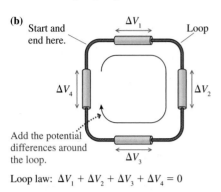

Loop law: $\Delta V_1 + \Delta V_2 + \Delta V_3 + \Delta V_4 = 0$

FIGURE 31.9 Kirchhoff's laws apply to junctions and loops.

You learned in Chapter 28 that charge and current are conserved. Consequently, the total current into the junction of Figure 31.9a must equal the total current leaving the junction. That is,

$$\sum I_{\text{in}} = \sum I_{\text{out}} \tag{31.5}$$

This statement is **Kirchhoff's junction law.**

An important property of the electric potential that you learned in Chapter 30 is that the sum of the potential differences around any loop or closed path is zero. This is a statement of energy conservation, because a charge that moves around a closed path and returns to its starting point has $\Delta U = 0$. We apply this idea to the circuit of Figure 31.9b by adding all of the potential differences *around* the loop formed by the circuit. Doing so gives

$$\Delta V_{\text{loop}} = \sum_i (\Delta V)_i = 0 \tag{31.6}$$

where $(\Delta V)_i$ is the potential difference of the ith component in the loop. This statement is **Kirchhoff's loop law.**

Kirchhoff's loop law can be true only if at least one of the $(\Delta V)_i$ is negative. To apply the loop law, we need to explicitly identify which potential differences are positive and which are negative.

TACTICS BOX 31.1 Using Kirchhoff's loop law

❶ **Draw a circuit diagram.** Label all known and unknown quantities.

❷ **Assign a direction to the current.** Draw and label a current arrow I to show your choice.
- If you know the actual current direction, choose that direction.
- If you don't know the actual current direction, make an arbitrary choice. All that will happen if you choose wrong is that your value for I will end up negative.

❸ **"Travel" around the loop.** Start at any point in the circuit, then go all the way around the loop in the direction you assigned to the current in step 2. As you go through each circuit element, ΔV is interpreted to mean

$$\Delta V = V_{\text{downstream}} - V_{\text{upstream}}$$

- For an ideal battery in the negative-to-positive direction:
$$\Delta V_{\text{bat}} = +\mathcal{E}$$

Potential increases

- For an ideal battery in the positive-to-negative direction:
$$\Delta V_{\text{bat}} = -\mathcal{E}$$

Potential decreases

- For a resistor:
$$\Delta V_R = -IR$$

Potential decreases

❹ **Apply the loop law:**
$$\sum (\Delta V)_i = 0$$

NOTE ▶ ΔV_R for a resistor seems to be opposite Ohm's law, but Ohm's law was only concerned with the *magnitude* of the potential difference. Kirchhoff's law requires us to recognize that the electric potential inside a resistor *decreases* in the direction of the current. Thus $\Delta V = V_{\text{downstream}} - V_{\text{upstream}} < 0$. ◀

The Basic Circuit

The most basic electric circuit is that of a single resistor connected to the two terminals of a battery. Figure 31.10a shows a literal picture of the circuit elements and the connecting wires; Figure 31.10b is the circuit diagram. Notice that this is a **complete circuit,** forming a continuous path between the battery terminals.

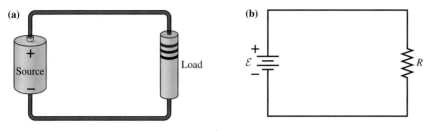

FIGURE 31.10 The basic circuit of a resistor connected to a battery.

The resistor might be a known resistor, such as "a 10 Ω resistor," or it might be some other resistive device, such as a lightbulb. Regardless of what the resistor is, it is called the **load.** The battery is called the **source.**

Figure 31.11 shows the use of Kirchhoff's loop law to analyze this circuit. Two things are worth noting:

1. This circuit has no junctions, so the current I is the same in all four sides of the circuit. Kirchhoff's junction law is not needed.
2. We're assuming the ideal-wire model, in which there are *no* potential differences along the connecting wires.

Kirchhoff's loop law, with two circuit elements, is

$$\Delta V_{\text{loop}} = \sum_i (\Delta V)_i = \Delta V_{\text{bat}} + \Delta V_{\text{R}} = 0 \qquad (31.7)$$

Let's look at each of the two terms in Equation 31.7:

1. The potential *increases* as we travel through the battery on our clockwise journey around the loop. We enter the negative terminal and, farther downstream, exit the positive terminal after having gained potential $\mathcal{E}$. Thus

$$\Delta V_{\text{bat}} = +\mathcal{E}$$

2. The *magnitude* of the potential difference across the resistor is $\Delta V = IR$, but Ohm's law does not tell us whether this should be positive or negative—and the difference is crucial. The potential of a conductor *decreases* in the direction of the current, which we've indicated with the + and − signs in Figure 31.11. Thus

$$\Delta V_{\text{R}} = V_{\text{downstream}} - V_{\text{upstream}} = -IR$$

NOTE ▶ Determining which potential differences are positive and which negative is perhaps *the* most important step in circuit analysis. ◀

With this information about ΔV_{bat} and ΔV_{R}, the loop equation becomes

$$\mathcal{E} - IR = 0 \qquad (31.8)$$

We can solve the loop equation to find that the current in the circuit is

$$I = \frac{\mathcal{E}}{R} \qquad (31.9)$$

We can then use the current to find that the resistor's potential difference is

$$\Delta V_{\text{R}} = -IR = -\mathcal{E} \qquad (31.10)$$

❶ Draw circuit diagram.

❷ The orientation of the battery indicates a clockwise current, so assign a clockwise direction to I.

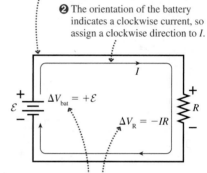

❸ Determine ΔV for each circuit element.

FIGURE 31.11 Analysis of the basic circuit using Kirchhoff's loop law.

This result should come as no surprise. The potential energy that the charges gain in the battery is subsequently lost as they "fall" through the resistor.

Equations 31.9 and 31.10 are the primary results of our circuit analysis. Notice that the current depends on the size of the resistance. The emf of a battery is a fixed quantity; the current that the battery delivers depends jointly on the emf and on the load.

EXAMPLE 31.2 A single-resistor circuit

A 15 Ω resistor is connected to the terminals of a 1.5 V battery.

a. What is the current in the circuit?
b. Draw a graph showing the potential as a function of distance traveled through the circuit, starting from $V = 0$ V at the negative terminal of the battery.

MODEL Assume ideal connecting wires and an ideal battery for which $\Delta V_{bat} = \mathcal{E}$.

VISUALIZE Figure 31.12 shows the circuit. We'll choose a clockwise (cw) direction for I.

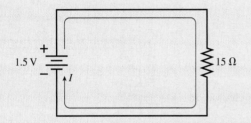

FIGURE 31.12 The circuit of Example 31.2.

SOLVE

a. This is the basic circuit of a single resistor connected to a single battery. The current is given by Equation 31.9:

$$I = \frac{\mathcal{E}}{R} = \frac{1.5\ \text{V}}{15\ \Omega} = 0.10\ \text{A}$$

b. The battery's potential difference is $\Delta V_{bat} = \mathcal{E} = 1.5$ V. The resistor's potential difference is $\Delta V_R = -\mathcal{E} = -1.5$ V. Based on this, Figure 31.13 shows the potential experienced by charges flowing around the circuit. The distance s is measured from the battery's negative terminal, and we have chosen to let $V = 0$ V at that point. The potential ends at the value from which it started.

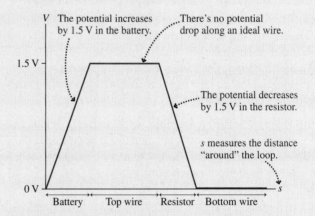

FIGURE 31.13 A graphical presentation of how the potential changes around the loop of the circuit in Figure 31.12.

ASSESS The value of I is positive. This tells us that the *actual* current direction is cw.

EXAMPLE 31.3 A more complex circuit

Analyze the circuit shown in Figure 31.14a.

a. Find the current in and the potential difference across each resistor.
b. Draw a graph showing how the potential changes around the circuit, starting from $V = 0$ V at the negative terminal of the 6 V battery.

MODEL Assume ideal connecting wires and ideal batteries, for which $\Delta V_{bat} = \mathcal{E}$.

VISUALIZE In Figure 31.14b the circuit has been redrawn; $\mathcal{E}_1$, $\mathcal{E}_2$, R_1, and R_2 defined; and the cw direction chosen for the current. This direction is an arbitrary choice because, with two batteries, we may not be sure of the actual current direction.

(a)

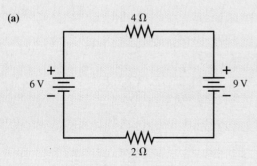

(b)

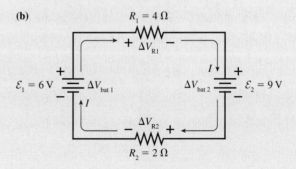

FIGURE 31.14 Circuit for Example 31.3.

SOLVE

a. How do we deal with *two* batteries? Can charge flow "backward" through a battery, from positive to negative? Consider the charge-escalator analogy. Left to itself, a charge escalator lifts charge from lower to higher potential. But it *is* possible to run down an up escalator, as many of you have probably done. If two escalators are placed "head to head," whichever is stronger will, indeed, force the charge to run down the up escalator of the other battery. The current in a battery *can* be from positive to negative if driven in that direction by a larger emf from a second battery. Indeed, this is how rechargeable batteries are recharged.

Because there are no junctions, the current is the same through *each* component in the circuit. With some thought, we might deduce whether the current is cw or ccw, but we do not need to know in advance of our analysis. We will choose a clockwise direction for the current and solve for the value of I. If our solution is positive, then the current really is cw. If the solution should turn out to be negative, we will know that the current is ccw. Kirchhoff's loop law, going clockwise from the negative terminal of battery 1, is

$$\Delta V_{\text{closed loop}} = \sum_i (\Delta V)_i = \Delta V_{\text{bat 1}} + \Delta V_{\text{R1}}$$
$$+ \Delta V_{\text{bat 2}} + \Delta V_{\text{R2}} = 0$$

All the signs are $+$ because this is a formal statement of *adding* potential differences around the loop. Next we can evaluate each ΔV. As we go cw, the charges *gain* potential in battery 1 but *lose* potential in battery 2. Thus $\Delta V_{\text{bat 1}} = +\mathcal{E}_1$ and $\Delta V_{\text{bat 2}} = -\mathcal{E}_2$. There is a *loss* of potential in traveling through each resistor, because we're traversing them in the direction we assigned to the current, so $\Delta V_{\text{R1}} = -IR_1$ and $\Delta V_{\text{R2}} = -IR_2$. Thus Kirchhoff's loop law becomes

$$\sum (\Delta V)_i = \mathcal{E}_1 - IR_1 - \mathcal{E}_2 - IR_2$$
$$= \mathcal{E}_1 - \mathcal{E}_2 - I(R_1 + R_2) = 0$$

We can solve this equation to find the current in the loop:

$$I = \frac{\mathcal{E}_1 - \mathcal{E}_2}{R_1 + R_2} = \frac{6\text{ V} - 9\text{ V}}{4\text{ }\Omega + 2\text{ }\Omega} = -0.50\text{ A}$$

The value of I is negative, hence the actual current in this circuit is 0.50 A *counterclockwise*. You perhaps anticipated this from the orientation of the larger 9 V battery.

b. The potential difference across the 4 Ω resistor is

$$\Delta V_{\text{R1}} = -IR_1 = -(-0.50\text{ A})(4\text{ }\Omega) = +2.0\text{ V}$$

Because the current is actually ccw, the resistor's potential *increases* in the cw direction of our travel around the loop. Similarly, the potential difference across the 2 Ω resistor is $\Delta V_{\text{R2}} = 1.0$ V. Figure 31.15 is a graph of potential versus position, following a cw path around the loop starting from $V = 0$ V at the negative terminal of the 6 V battery.

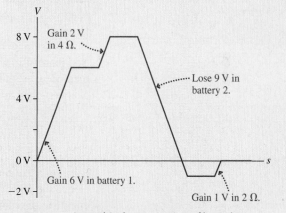

FIGURE 31.15 A graphical presentation of how the potential changes around the loop.

ASSESS Notice how the potential *drops* 9 V upon passing through battery 2 in the cw direction. It then gains 2 V upon passing through R_2 to end at the starting potential.

STOP TO THINK 31.3 What is ΔV across the unspecified circuit element? Does the potential increase or decrease when traveling through this element in the direction assigned to I?

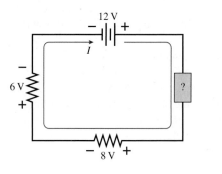

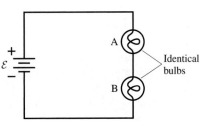

FIGURE 31.16 Which lightbulb is brighter?

31.4 Energy and Power

The circuit of Figure 31.16 has two identical lightbulbs, A and B. Which is brighter? Or are they equally bright? Think about this before going on.

You might have been tempted to say that A is brighter. After all, the current gets to A first, so A might "use up" some of the current and leave less for B. But this would violate the laws of conservation of charge and conservation of current. There are no junctions between A and B, so the current through the two bulbs must be the same. Hence the bulbs are equally bright.

It's not current that the bulbs use up, it's *energy*. A battery not only supplies a potential difference, it also supplies the energy to a circuit. The charge escalator is an energy-transfer process, transferring chemical energy E_{chem} stored in the battery to the potential energy U of the charges. That energy is then dissipated as the charges move through the wires and resistors, increasing their thermal energy until, in the case of the lightbulb filaments, they glow.

A charge gains potential energy $\Delta U = q\Delta V_{bat}$ as it moves up the charge escalator in the battery. For an ideal battery, with $\Delta V_{bat} = \mathcal{E}$, the battery supplies energy $\Delta U = q\mathcal{E}$ to charge q as it lifts the charge from the negative to the positive terminal.

It is useful to know the *rate* at which the battery supplies energy to the charges. You learned in Chapter 11 that the rate at which energy is transferred is *power,* measured in joules per second or *watts*. If energy $\Delta U = q\mathcal{E}$ is transferred to charge q, then the *rate* at which energy is transferred from the battery to the moving charges is

$$P_{bat} = \text{rate of energy transfer} = \frac{dU}{dt} = \frac{dq}{dt}\mathcal{E} \tag{31.11}$$

But dq/dt, the rate at which charge moves through the battery, is the current I. Hence the power supplied by a battery, or the rate at which the battery transfers energy to the charges passing through it, is

$$P_{bat} = I\mathcal{E} \quad \text{(power delivered by an emf)} \tag{31.12}$$

$I\mathcal{E}$ has units of J/s, or W.

EXAMPLE 31.4 Delivering power
A 90 Ω load is connected to a 120 V battery. How much power is delivered by the battery?

SOLVE This is our basic battery-and-resistor circuit, which we analyzed earlier. In this case

$$I = \frac{\mathcal{E}}{R} = \frac{120 \text{ V}}{90 \text{ }\Omega} = 1.33 \text{ A}$$

Thus the power delivered by the battery is

$$P_{bat} = I\mathcal{E} = (1.33 \text{ A})(120 \text{ V}) = 160 \text{ W}$$

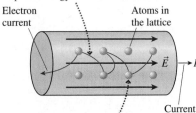

The electric field causes electrons to speed up. The energy transformation is $U \rightarrow K$.

Electron current

Atoms in the lattice

$\vec{E}$ → I

Current

Collisions transfer energy to the lattice. The energy transformation is $K \rightarrow E_{th}$.

FIGURE 31.17 A current-carrying resistor dissipates power because the electric force does work on the charges.

P_{bat} is the energy transferred per second from the battery's store of chemicals to the moving charges that make up the current. But what happens to this energy? Where does it end up? Figure 31.17, a section of a current-carrying resistor, reminds you of our microscopic model of conduction. The electrons accelerate in the electric field, then collide with atoms in the lattice. The acceleration phase is a transformation of potential to kinetic energy. The collisions then transfer the electron's kinetic energy to the *thermal* energy of the lattice. The potential energy was acquired in the battery, from the conversion of chemical energy, so the entire energy transfer process looks like

$$E_{chem} \rightarrow U \rightarrow K \rightarrow E_{th}$$

The net result is that **the battery's chemical energy is transferred to the thermal energy of the resistors,** raising their temperature.

Let's look more closely at the energy transfer in the resistor. Suppose the average distance between collisions is d. The electric force $\vec{F} = q\vec{E}$ exerted on charge q does work as it pushes the charge through distance d. The field is constant inside the resistor, so the work is simply

$$W = F\Delta s = qEd \tag{31.13}$$

According to the work-kinetic energy theorem, this work increases the kinetic energy of charge q by $\Delta K = W = qEd$. This kinetic energy is transferred to the lattice when charge q collides with a lattice atom, causing the energy of the lattice to increase by

$$\Delta E_{\text{per collision}} = \Delta K = qEd$$

Collisions occur over and over as the charge makes its way through a resistor of length L. The total energy transferred while traveling distance L is

$$\Delta E_{\text{th}} = qEL \tag{31.14}$$

But EL is the potential difference ΔV_R between the two ends of the resistor. Thus *each* charge q, as it travels the length of the resistor, transfers energy to the atomic lattice in the amount

$$\Delta E_{\text{th}} = q\Delta V_R \tag{31.15}$$

The *rate* at which energy is transferred from the current to the resistor is thus

$$P_R = \frac{dE_{\text{th}}}{dt} = \frac{dq}{dt}\Delta V_R = I\Delta V_R \tag{31.16}$$

We say that this power—so many joules per second—is *dissipated* by the resistor as charge flows through it. The resistor, in turn, transfers this energy to the air and to the circuit board on which it is mounted, causing the circuit and all its surroundings to heat up.

From our analysis of the basic circuit, in which a single resistor is connected to a battery, we learned that $\Delta V_R = \mathcal{E}$. That is, the potential difference across the resistor is exactly the emf supplied by the battery. But then Equations 31.12 and 31.16, for P_{bat} and P_R, are numerically equal, and we find that

$$P_R = P_{\text{bat}} \tag{31.17}$$

The answer to the question "What happens to the energy supplied by the battery?" is "The battery's chemical energy is transformed into the thermal energy of the resistor." The *rate* at which the battery supplies energy is exactly equal to the *rate* at which the resistor dissipates energy. This is, of course, exactly what we would have expected from energy conservation.

EXAMPLE 31.5 **The power of light**
How much current is "drawn" by a 100 W lightbulb connected to a 120 V outlet?

MODEL Most household appliances, such as a 100 W lightbulb or a 1500 W hair dryer, have a power rating. The rating does *not* mean that these appliances *always* dissipate that much power. These appliances are intended for use at a standard household voltage of 120 V, and their rating is the power they will dissipate *if* operated with a potential difference of 120 V. Their power consumption will differ from the rating if they are operated at any other potential difference.

SOLVE Because the lightbulb is operating as intended, it will dissipate 100 W of power. Thus

$$I = \frac{P}{\Delta V_R} = \frac{100 \text{ W}}{120 \text{ V}} = 0.833 \text{ A}$$

ASSESS A current of 0.833 A in this lightbulb transfers 100 J/s to the thermal energy of the filament which, in turn, dissipates 100 J/s as heat and light to its surroundings.

A resistor obeys Ohm's law, $\Delta V_R = IR$. This gives us two alternative ways of writing the power dissipated by a resistor. We can either substitute IR for ΔV_R or substitute $\Delta V_R/R$ for I. Thus

$$P_R = I\Delta V_R = I^2 R = \frac{(\Delta V_R)^2}{R} \qquad \text{(power dissipated by a resistor)} \qquad (31.18)$$

If the same current I passes through several resistors in series, then $P_R = I^2 R$ tells us that most of the power will be dissipated by the largest resistance. This is why a lightbulb filament glows but the connecting wires do not. Essentially *all* of the power supplied by the battery is dissipated by the high-resistance lightbulb filament and essentially no power is dissipated by the low-resistance wires. The filament gets very hot, but the wires do not.

EXAMPLE 31.6 The power of sound

Most loudspeakers are designed to have a resistance of 8 Ω. If an 8 Ω loudspeaker is connected to a stereo amplifier with a rating of 100 W, what is the maximum possible current to the loudspeaker?

MODEL The rating of an amplifier is the *maximum* power it can deliver. Most of the time it delivers far less, but the maximum might be reached for brief, intense sounds like cymbal clashes.

SOLVE The loudspeaker is a resistive load. The maximum current to the loudspeaker occurs when the amplifier delivers maximum power $P_{max} = (I_{max})^2 R$. Thus

$$I_{max} = \sqrt{\frac{P_{max}}{R}} = \sqrt{\frac{100 \text{ W}}{8 \text{ }\Omega}} = 3.5 \text{ A}$$

EXAMPLE 31.7 A dim bulb

How much power is dissipated by a 60 W (120 V) lightbulb when operated, using a dimmer switch, at 100 V?

MODEL The 60 W rating is for operation at 120 V.

SOLVE The lightbulb dissipates 60 W at $\Delta V_R = 120$ V. Thus the filament's resistance is

$$R = \frac{(\Delta V_R)^2}{P_R} = \frac{(120 \text{ V})^2}{60 \text{ W}} = 240 \text{ }\Omega$$

The power dissipation when operated at $\Delta V_R = 100$ V is

$$P_R = \frac{(\Delta V_R)^2}{R} = \frac{(100 \text{ V})^2}{240 \text{ }\Omega} = 42 \text{ W}$$

ASSESS Actually, this result is not quite correct. As noted previously, a filament is not a true ohmic material because the resistance changes somewhat with temperature. The filament's resistance at 100 V, where it glows less brightly, is not quite the same as its resistance at 120 V. The voltage in this example is still near 120 V, so the temperature of the filament will decrease only slightly, and our answer should be fairly accurate. However, this calculation would not give a good result for the power dissipation at 20 V, where the filament's temperature would be much less than at 120 V.

The electric meter on the side of your house or apartment records the kilowatt hours of electric energy that you use.

It is the stored chemical energy of the battery that is used up by a lightbulb. The energy is used by being converted first to the energy of the charges, then to the thermal energy of the filament, thus heating it, and lastly to the heat and light energy that we feel and see coming from the bulb. Conservation of energy is of paramount importance for understanding electric circuits.

Kilowatt Hours

The energy dissipated (i.e., transformed into thermal energy) by a resistor during time Δt is $E_{th} = P_R \Delta t$. The product of watts and seconds is joules, the SI unit of energy. However, your local electric company prefers to use a different unit, called *kilowatt hours,* to measure the energy you use each month.

A load that consumes P_R kW of electricity in Δt hours has used $P_R \Delta t$ **kilowatt hours** of energy, abbreviated kWh. For example, a 4000 W electric water heater uses 40 kWh of energy in 10 hours. A 1500 W hair dryer uses 0.25 kWh of energy in 10 minutes. Despite the rather unusual name, a kilowatt hour is a unit of energy. A homework problem will let you find the conversion factor from kilowatt hours to joules.

Your monthly electric bill specifies the number of kilowatt hours you used last month. This is the amount of energy that the electric company delivered to you, via an electric current, and that you transformed into light and thermal energy inside your home. The cost of electricity varies throughout the country, but the average cost of electricity in the United States is approximately 10¢ per kWh ($0.10/kWh). Thus it costs about $4.00 to run your water heater for 10 hours, about 2.5¢ to dry your hair.

STOP TO THINK 31.4   Rank in order, from largest to smallest, the powers P_a to P_d dissipated in resistors a to d.

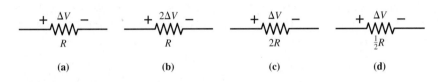

     (a)   (b)   (c)   (d)

31.5 Series Resistors

Many circuits contain two or more resistors connected to each other in various ways. Thus much of circuit analysis consists of analyzing different *combinations* of resistors. As an example, consider the three lightbulbs in Figure 31.18. The batteries are identical and the bulbs are identical. You learned in the previous section that B and C are equally bright, because of conservation of current, but how does the brightness of B compare to that of A? Think about this before going on.

Figure 31.19a shows two resistors placed end to end between points a and b. Resistors that are aligned end to end, *with no junctions between them,* are called **series resistors** or, sometimes, resistors "in series." Because there are no junctions, and because current is conserved, the current I must be the same through each of these resistors. That is, the current out of the last resistor in a series is equal to the current into the first resistor.

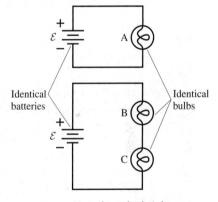

FIGURE 31.18 How does the brightness of bulb B compare to that of A?

(a) Two resistors in series   **(b)** An equivalent resistor

FIGURE 31.19 Replacing two series resistors with an equivalent resistor.

The potential differences across the two resistors are $\Delta V_1 = IR_1$ and $\Delta V_2 = IR_2$. The total potential difference ΔV_{ab} between points a and b is the sum of the individual potential differences:

$$\Delta V_{ab} = \Delta V_1 + \Delta V_2 = IR_1 + IR_2 = I(R_1 + R_2) \qquad (31.19)$$

Suppose, as in Figure 31.19b, we replaced the two resistors with a single resistor having current I and potential difference $\Delta V_{ab} = \Delta V_1 + \Delta V_2$. We can then use Ohm's law to find that the resistance R_{ab} between points a and b is

$$R_{ab} = \frac{\Delta V_{ab}}{I} = \frac{I(R_1 + R_2)}{I} = R_1 + R_2 \qquad (31.20)$$

Because the battery has to establish the same potential difference and provide the same current in both cases, the two resistors R_1 and R_2 act exactly the same as a *single* resistor of value $R_1 + R_2$. We can say that the single resistor R_{ab} is *equivalent* to the two resistors in series.

There was nothing special about having only two resistors. If we have N resistors in series, their **equivalent resistance** is

$$R_{eq} = R_1 + R_2 + \cdots + R_N \qquad \text{(series resistors)} \qquad (31.21)$$

The behavior of the circuit will be unchanged if the N series resistors are replaced by the single resistor R_{eq}. The key idea in this analysis is the fact that **resistors in series all have the same current.**

EXAMPLE 31.8 **A series resistor circuit**

a. What is the current in the circuit of Figure 31.20a?
b. Draw a graph of potential versus position in the circuit, going cw from $V = 0$ V at the battery's negative terminal.

MODEL The three resistors are end to end, with no junctions between them, and thus are in series. Assume ideal connecting wires and an ideal battery.

SOLVE

a. Nothing about the circuit's behavior will change if we replace the three series resistors by their equivalent resistance

$$R_{eq} = 15\ \Omega + 4\ \Omega + 8\ \Omega = 27\ \Omega$$

This is shown as an equivalent circuit in Figure 31.20b. Now we have a circuit with a single battery and a single resistor, for which we know the current to be

$$I = \frac{\mathcal{E}}{R_{eq}} = \frac{9\ \text{V}}{27\ \Omega} = 0.333\ \text{A}$$

b. $I = 0.333$ A is the current in each of the three resistors in the original circuit. Thus the potential differences across the resistors are $\Delta V_{R1} = -IR_1 = -5.00$ V, $\Delta V_{R2} = -IR_2 = -1.33$ V, and $\Delta V_{R3} = -IR_3 = -2.67$ V for the 15 Ω, the 4 Ω, and the 8 Ω resistors, respectively. Figure 31.20c shows that the potential increases by 9 V due to the battery's emf, then decreases by 9 V in three steps.

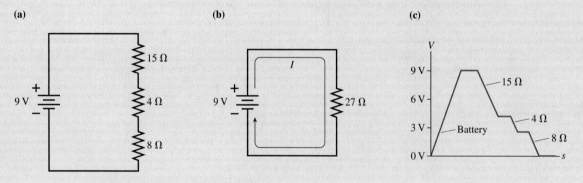

FIGURE 31.20 Analyzing a circuit with series resistors.

Now we can answer the lightbulb question posed at the beginning of this section. Suppose the resistance of each lightbulb is R. The battery drives current $I_A = \mathcal{E}/R$ through bulb A. Bulbs B and C are in series, with an equivalent resistance $R_{eq} = 2R$, but the battery has the same emf $\mathcal{E}$. Thus the current through bulbs B and C is $I_{B+C} = \mathcal{E}/R_{eq} = \mathcal{E}/2R = \frac{1}{2}I_A$. Bulb B has only half the current of bulb A, so B is dimmer.

Many people predict that A and B should be equally bright. It's the same battery, so shouldn't it provide the same current to both circuits? No! A battery is a source of emf, *not* a source of current. In other words, the battery's emf is the same no matter how the battery is used. When you buy a 1.5 V battery you're buying a device that provides a specified amount of potential difference, not a specified amount of current. The battery does provide the current to the circuit, but the *amount* of current depends on the resistance of the load. Your 1.5 V battery causes 1 A to pass through a 1.5 Ω load but only 0.1 A to pass through a 15 Ω load. As an analogy, think about a water faucet. The pressure in the water main underneath the street is a fixed and unvarying quantity set by the water company. It's like the emf of a battery. But the amount of water coming out of a faucet depends on how far you open it. A faucet opened slightly has a "high resistance," so only a little water flows. A wide-open faucet has a "low resistance," and the water flow is large.

We're spending a lot of time on this property of a battery because it's a critical idea for understanding circuits. In summary, **a battery provides a fixed and unvarying emf (potential difference). It does *not* provide a fixed and unvarying current. The amount of current depends jointly on the battery's emf *and* the resistance of the circuit attached to the battery.**

Ammeters

A device that measures the current in a circuit element is called an **ammeter.** Because charge flows *through* circuit elements, an ammeter must be placed *in series* with the circuit element whose current is to be measured.

 12.1

Figure 31.21a shows a simple one-resistor circuit with an unknown emf $\mathcal{E}$. We can measure the current in the circuit by inserting the ammeter as shown in Figure 31.21b. Notice that we have to *break the connection* between the battery and the resistor in order to insert the ammeter. Now the current in the resistor has to first pass through the ammeter.

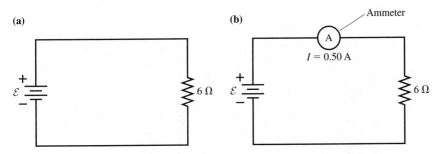

FIGURE 31.21 An ammeter measures the current in a circuit element.

Because the ammeter is now in series with the resistor, the total resistance seen by the battery is $R_{eq} = 6\ \Omega + R_{ammeter}$. In order that the ammeter measure the current without changing the current, the ammeter's resistance must, in this case, be $\ll 6\ \Omega$. Indeed, an ideal ammeter has $R_{ammeter} = 0\ \Omega$ and thus has no effect on the current. Real ammeters come very close to this ideal.

The ammeter in Figure 31.21b reads 0.50 A, meaning that the current in the 6 Ω resistor is $I = 0.50$ A. Thus the resistor's potential difference is $\Delta V_R = -IR = -3.0$ V. If the ammeter is ideal, as we will assume, then, from Kirchhoff's loop law, the battery's emf is $\mathcal{E} = -\Delta V_R = 3.0$ V.

STOP TO THINK 31.5   What are the current and the potential at points a to e?

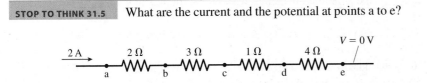

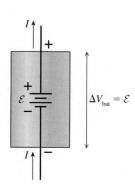

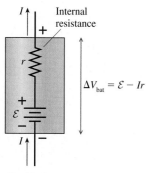

FIGURE 31.22 An ideal battery and a real battery.

31.6 Real Batteries

Let's look at how real batteries differ from the ideal battery we have been assuming. Real batteries, like ideal batteries, separate charge and create a potential difference. However, real batteries also provide a slight resistance to the charges on the charge escalator. They have what is called an **internal resistance,** which is symbolized by r. Figure 31.22 shows both an ideal and a real battery.

From our vantage point outside a battery, we cannot see $\mathcal{E}$ and r separately. To the user, the battery provides a potential difference ΔV_{bat} called the **terminal voltage.** $\Delta V_{bat} = \mathcal{E}$ for an ideal battery, but the presence of the internal resistance affects ΔV_{bat}. Suppose the current in the battery is I. As charges travel from the negative to the positive terminal, they gain potential $\mathcal{E}$ but *lose* potential $\Delta V_{int} = -Ir$ due to the internal resistance. Thus the terminal voltage of the battery is

$$\Delta V_{bat} = \mathcal{E} - Ir \le \mathcal{E} \tag{31.22}$$

Only when $I = 0$, meaning that the battery is not being used, is $\Delta V_{bat} = \mathcal{E}$.

Figure 31.23 shows a single resistor R connected to the terminals of a battery having emf $\mathcal{E}$ and internal resistance r. Resistances R and r are in series, so we can replace them, for the purpose of circuit analysis, with a single equivalent resistor $R_{eq} = R + r$. Hence the current in the circuit is

$$I = \frac{\mathcal{E}}{R_{eq}} = \frac{\mathcal{E}}{R + r} \tag{31.23}$$

If $r \ll R$, so that the internal resistance of the battery is negligible, then $I = \mathcal{E}/R$, exactly the result we found before. But the current decreases significantly if $r \approx R$.

Although physically separated, the internal resistance r is electrically in series with R.

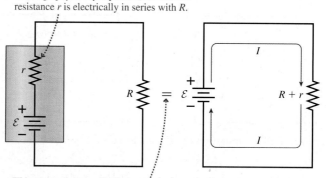

This means the two circuits are equivalent.

FIGURE 31.23 A single resistor connected to a real battery is in series with the battery's internal resistance, giving $R_{eq} = R + r$.

We can use Ohm's law to find that the potential difference across the load resistor R is

$$\Delta V_R = IR = \frac{R}{R + r}\mathcal{E} \tag{31.24}$$

Similarly, the potential difference across the terminals of the battery is

$$\Delta V_{bat} = \mathcal{E} - Ir = \mathcal{E} - \frac{r}{R + r}\mathcal{E} = \frac{R}{R + r}\mathcal{E} \tag{31.25}$$

The potential difference across the resistor is equal to the potential difference between the *terminals* of the battery, where the resistor is attached, *not* equal to the battery's emf.

EXAMPLE 31.9 Lighting up a flashlight

A 6 Ω flashlight bulb is powered by a 3 V battery having an internal resistance of 1 Ω. What are the power dissipation of the bulb and the terminal voltage of the battery?

MODEL Assume ideal connecting wires but not an ideal battery.

VISUALIZE The circuit diagram looks like Figure 31.23. R is the resistance of the bulb's filament.

SOLVE Equation 31.23 gives us the current:

$$I = \frac{\mathcal{E}}{R + r} = \frac{3\text{ V}}{6\text{ }\Omega + 1\text{ }\Omega} = 0.43\text{ A}$$

This is 15% less than the 0.5 A an ideal battery would supply. The potential difference across the resistor is $\Delta V_R = IR = 2.57$ V, thus the power dissipation is

$$P_R = I\Delta V = 1.1\text{ W}$$

The battery's terminal voltage is

$$\Delta V_{bat} = \frac{R}{R + r}\mathcal{E} = \frac{6\text{ }\Omega}{6\text{ }\Omega + 1\text{ }\Omega}\,3\text{ V} = 2.6\text{ V}$$

ASSESS 1 Ω is a typical internal resistance for a flashlight battery. The internal resistance causes the battery's terminal voltage to be 0.4 V less than its emf in this circuit.

A Short Circuit

In Figure 31.24 we've replaced the resistor with an ideal wire having $R_{wire} = 0$ Ω. When a connection of very low or zero resistance is made between two points in a circuit that are normally separated by a higher resistance, then we have what is called a **short circuit.** The wire in Figure 31.24 is *shorting out* the battery.

If the battery were ideal, shorting it with an ideal wire ($R = 0$ Ω) would cause the current to be $I = \mathcal{E}/0 = \infty$. The current, of course, cannot really become infinite. Instead, the battery's internal resistance r becomes the only resistance in the circuit. If we use $R = 0$ Ω in Equation 31.23, we find that the *short-circuit current* is

$$I_{short} = \frac{\mathcal{E}}{r} \qquad (31.26)$$

A 3 V battery with 1 Ω internal resistance generates a short circuit current of 3 A. This is the *maximum possible current* that this battery can produce. Adding any external resistance R will decrease the current to a value less than 3 A.

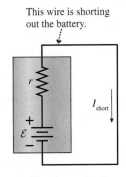

This wire is shorting out the battery.

FIGURE 31.24 The short-circuit current of a battery.

EXAMPLE 31.10 A short-circuited battery

What is the short-circuit current of a 12 V car battery with an internal resistance of 0.020 Ω? What happens to the power supplied by the battery?

SOLVE The short-circuit current is

$$I_{short} = \frac{\mathcal{E}}{r} = \frac{12\text{ V}}{0.02\text{ }\Omega} = 600\text{ A}$$

Power is generated by chemical reactions in the battery and dissipated by the load resistance. But with a short-circuited battery, the load resistance is *inside* the battery! The "shorted" battery has to dissipate power $P = I^2r = 7200$ W *internally.*

ASSESS This value is realistic. Car batteries are designed to drive the starter motor, which has a very small resistance and can draw a current of a few hundred amps. That is why the battery cables are so thick. A shorted car battery can produce an *enormous* amount of current. The normal response of a shorted car battery is to explode; it simply cannot dissipate this much power. Shorting a flashlight battery can make it rather hot, but your life is not in danger. Although the voltage of a car battery is relatively small, a car battery can be dangerous and should be treated with great respect.

Most of the time a battery is used under conditions in which $r \ll R$ and the internal resistance is negligible. The ideal battery model is fully justified in that case. Thus we will assume that batteries are ideal *unless stated otherwise*. But keep in mind that batteries (like all other sources of emf) do have an internal resistance, and that this internal resistance limits the maximum possible current of the battery.

31.7 Parallel Resistors

Figure 31.25 is another lightbulb puzzle. Initially the switch is open. The current is the same through bulbs A and B, because of conservation of current, and they are equally bright. Bulb C is not glowing. What happens to the brightness of A and B when the switch is closed? And how does the brightness of C then compare to that of A and B? Think about this before going on.

Figure 31.26a shows two resistors aligned side by side with their ends connected at junctions c and d. Resistors that are connected *at both ends* are called **parallel resistors** or, sometimes, resistors "in parallel." The left ends of both resistors are at the same potential V_c. Likewise, the right ends are at the same potential V_d. Thus the potential *differences* ΔV_1 and ΔV_2 are the *same* and are simply ΔV_{cd}.

Kirchhoff's junction law applies at the junctions. The input current I splits into currents I_1 and I_2 at the left junction. On the right, the two currents are recombined into current I. According to the junction law,

$$I = I_1 + I_2 \tag{31.27}$$

We can apply Ohm's law to each resistor, along with $\Delta V_1 = \Delta V_2 = \Delta V_{cd}$, to find that the current is

$$I = \frac{\Delta V_1}{R_1} + \frac{\Delta V_2}{R_2} = \frac{\Delta V_{cd}}{R_1} + \frac{\Delta V_{cd}}{R_2} = \Delta V_{cd}\left[\frac{1}{R_1} + \frac{1}{R_2}\right] \tag{31.28}$$

Suppose, as in Figure 31.26b, we replaced the two resistors with a single resistor having current I and potential difference ΔV_{cd}. This resistor is equivalent to the original two because the battery has to establish the same potential difference and provide the same current in either case. A second application of Ohm's law shows that the resistance between points c and d is

$$R_{cd} = \frac{\Delta V_{cd}}{I} = \left[\frac{1}{R_1} + \frac{1}{R_2}\right]^{-1} \tag{31.29}$$

The two resistors R_1 and R_2 act exactly the same as the single resistor R_{cd}. Resistor R_{cd} is *equivalent* to the two resistors in parallel.

There is nothing special about having chosen two resistors to be in parallel. If we have N resistors in parallel, the *equivalent resistance* is

$$R_{eq} = \left(\frac{1}{R_1} + \frac{1}{R_2} + \cdots + \frac{1}{R_N}\right)^{-1} \quad \text{(parallel resistors)} \tag{31.30}$$

The behavior of the circuit will be unchanged if the N parallel resistors are replaced by the single resistor R_{eq}. The key idea of this analysis is the fact that **resistors in parallel all have the same potential difference.**

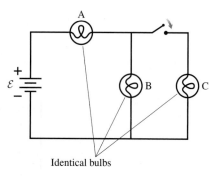

FIGURE 31.25 What happens to the brightness of the bulbs when the switch is closed?

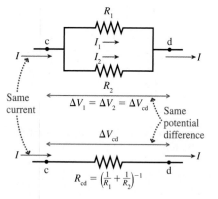

(a) Two resistors in parallel

(b) An equivalent resistor

FIGURE 31.26 Replacing two parallel resistors with an equivalent resistor.

12.2 Activ Physics

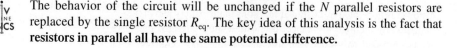

Two useful results of our analysis are the equivalent resistances of two identical resistors $R_1 = R_2 = R$ in series and in parallel:

Two identical resistors in series $\quad R_{eq} = 2R$

Two identical resistors in parallel $\quad R_{eq} = \dfrac{R}{2}$ $\qquad$ (31.31)

EXAMPLE 31.11 A parallel resistor circuit

The three resistors of Figure 31.27 are connected to a 9 V battery. Find the potential difference across and current through each resistor.

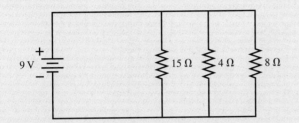

FIGURE 31.27 Parallel resistor circuit of Example 31.11.

MODEL The resistors are in parallel. Assume an ideal battery and ideal connecting wires.

SOLVE The three parallel resistors can be replaced by a single equivalent resistor

$$R_{eq} = \left(\frac{1}{15\ \Omega} + \frac{1}{4\ \Omega} + \frac{1}{8\ \Omega} \right)^{-1} = (0.4417\ \Omega^{-1})^{-1} = 2.26\ \Omega$$

The equivalent circuit is shown in Figure 31.28a, from which we find the current to be

$$I = \frac{\mathcal{E}}{R_{eq}} = \frac{9\ \text{V}}{2.26\ \Omega} = 3.98\ \text{A}$$

The potential difference across R_{eq} is $\Delta V_{eq} = \mathcal{E} = 9.0$ V. Now we have to be careful. Current I divides at the junction into the smaller currents I_1, I_2 and I_3 shown in Figure 31.28b. However, the division is *not* into three equal currents. According to Ohm's law, resistor i has current $I_i = \Delta V_i / R_i$. Because the resistors are in parallel, their potential differences are equal:

$$\Delta V_1 = \Delta V_2 = \Delta V_3 = \Delta V_{eq} = 9.0\ \text{V}$$

Thus the currents are

$$I_1 = \frac{9\ \text{V}}{15\ \Omega} = 0.60\ \text{A} \qquad I_2 = \frac{9\ \text{V}}{4\ \Omega} = 2.25\ \text{A}$$

$$I_3 = \frac{9\ \text{V}}{8\ \Omega} = 1.13\ \text{A}$$

ASSESS The *sum* of the three currents is 3.98 A, as required by Kirchhoff's junction law.

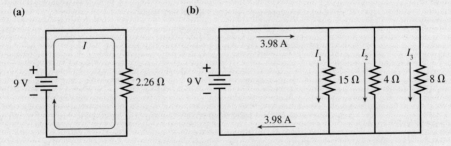

FIGURE 31.28 The parallel resistors can be replaced by a single equivalent resistor.

The result of Example 31.11 seems surprising. The equivalent of a parallel combination of 15 Ω, 4 Ω, and 8 Ω was found to be 2.26 Ω. How can the equivalent of a group of resistors be *less* than any single resistance in the group? Should not more resistors imply more resistance? The answer is yes for resistors in series but not for resistors in parallel. Even though a resistor is an obstacle to the flow of charge, parallel resistors provide more pathways for charge to get through. Consequently, the equivalent of several resistors in parallel is always *less* than any single resistor in the group.

Complex combinations of resistors can often be reduced to a single equivalent resistance through a step-by-step application of the series and parallel rules. The final example in this section illustrates this idea.

EXAMPLE 31.12 A combination of resistors

What is the equivalent resistance of the group of resistors shown in Figure 31.29a?

MODEL This circuit contains both series and parallel resistors.

SOLVE Reduction to a single equivalent resistance is best done in a series of steps, with the circuit being redrawn after each step. The procedure is shown in Figure 31.29b. Note that the

10 Ω and 25 Ω resistors are *not* in parallel. They are connected at their top ends but not at their bottom ends. Resistors must be connected at *both* ends to be in parallel. Similarly, the 10 Ω and 45 Ω resistors are *not* in series because of the junction between them. If the original group of four resistors occurred within a larger circuit, they could be replaced with a single 15.4 Ω resistor without having any effect on the rest of the circuit.

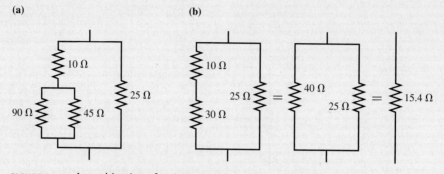

FIGURE 31.29 A combination of resistors is reduced to a single equivalent resistor.

Returning to the lightbulb question at the beginning of this section, suppose the resistance of each bulb in Figure 31.25 is R. Initially, before the switch is closed, bulbs A and B are in series with equivalent resistance $2R$. The current from the battery is

$$I_{before} = \frac{\mathcal{E}}{2R} = \frac{1}{2}\frac{\mathcal{E}}{R}$$

This is the current in both bulbs.

Closing the switch places bulbs B and C in parallel with each other. The equivalent resistance of two identical resistors in parallel is $R_{eq} = \frac{1}{2}R$. This equivalent resistance of B and C is in series with bulb A, hence the total resistance of the circuit is $\frac{3}{2}R$ and the current leaving the battery is

$$I_{after} = \frac{\mathcal{E}}{(3R/2)} = \frac{2}{3}\frac{\mathcal{E}}{R} > I_{before}$$

Closing the switch *decreases* the circuit resistance and thus *increases* the current leaving the battery.

All the charge flows through A, so A *increases* in brightness when the switch is closed. The current I_{after} then splits at the junction. Bulbs B and C have equal resistance, so the current splits equally. The current in B is $\frac{1}{3}(\mathcal{E}/R)$, which is *less* than I_{before}. Thus B *decreases* in brightness when the switch is closed. Bulb C has the same brightness as bulb B.

Voltmeters

A device that measures the potential difference across a circuit element is called a **voltmeter.** Because potential difference is measured *across* a circuit element, from one side to the other, a voltmeter is placed in *parallel* with the circuit element whose potential difference is to be measured.

Figure 31.30a shows a simple circuit in which a 17 Ω resistor is connected across a 9 V battery with an unknown internal resistance. By connecting a voltmeter across the resistor, as shown in Figure 31.30b, we can measure the potential difference across the resistor. Unlike an ammeter, using a voltmeter does *not* require that we break the connections.

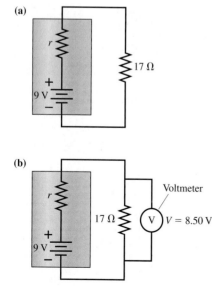

FIGURE 31.30 A voltmeter measures the potential difference across a circuit element.

Because the voltmeter is now in parallel with the resistor, the total resistance seen by the battery is $R_{eq} = (1/17\ \Omega + 1/R_{voltmeter})^{-1}$. In order that the voltmeter measure the voltage without changing the voltage, the voltmeter's resistance must, in this case, be $\gg 17\ \Omega$. Indeed, an *ideal voltmeter* has $R_{voltmeter} = \infty\ \Omega$, and thus has no effect on the voltage. Real voltmeters come very close to this ideal, and we will always assume them to be so.

The voltmeter in Figure 31.30b reads 8.50 V. This is less than $\mathcal{E}$ because of the battery's internal resistance. Equation 31.24 found an expression for the resistor's potential difference ΔV_R. That equation is easily solved for the internal resistance r:

$$r = \frac{\mathcal{E} - \Delta V_R}{\Delta V_R} R = \frac{0.5\ \text{V}}{8.5\ \text{V}}\, 17\ \Omega = 1.0\ \Omega$$

Here a voltmeter reading was the one piece of experimental data we needed in order to determine the battery's internal resistance.

STOP TO THINK 31.6   Rank in order, from brightest to dimmest, the identical bulbs A to D.

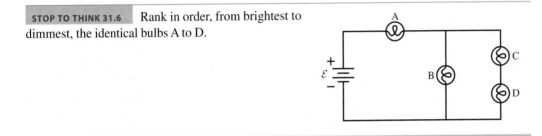

31.8 Resistor Circuits

We can use the information in this chapter to analyze a variety of more complex but more realistic circuits. We will thus have a chance to bring together the many ideas of this chapter and to see how they are used in practice.

Activ
Physics   ONLINE   12.3–12.5

(MP)   **PROBLEM-SOLVING STRATEGY 31.1**   **Resistor circuits**

MODEL Assume that wires are ideal and, where appropriate, that batteries are ideal.

VISUALIZE Draw a circuit diagram. Label all known and unknown quantities.

SOLVE Base your mathematical analysis on Kirchhoff's laws and on the rules for series and parallel resistors.

- Step by step, reduce the circuit to the smallest possible number of equivalent resistors.
- Determine the current through and potential difference across the equivalent resistors.
- Rebuild the circuit, using the facts that the current is the same through all resistors in series and the potential difference is the same for all parallel resistors.

ASSESS Use two important checks as you rebuild the circuit.

- Verify that the sum of the potential differences across series resistors matches ΔV for the equivalent resistor.
- Verify that the sum of the currents through parallel resistors matches I for the equivalent resistor.

EXAMPLE 31.13 Analyzing a complex circuit

Find the current through and the potential difference across each of the four resistors in the circuit shown in Figure 31.31.

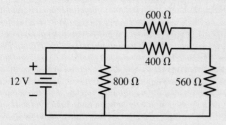

FIGURE 31.31 A complex resistor circuit for analysis.

MODEL Assume an ideal battery, with no internal resistance, and ideal connecting wires.

VISUALIZE Figure 31.31 shows the circuit diagram. We'll keep redrawing the diagram as we analyze the circuit.

SOLVE First, break the circuit down, step-by-step, into one with a single resistor. Figure 31.32a shows this done in four steps, using series and parallel resistors. The final circuit battery-and-resistor circuit is our basic circuit, which we know how to analyze. The current is

$$I = \frac{\mathcal{E}}{R} = \frac{12\ \text{V}}{400\ \Omega} = 0.030\ \text{A} = 30\ \text{mA}$$

The potential difference across the 400 Ω resistor is $\Delta V_{400} = \Delta V_{\text{bat}} = \mathcal{E} = 12$ V.

Second, rebuild the circuit, step-by-step, finding the currents and potential differences at each step. Figure 31.32b repeats the steps of Figure 31.32a exactly, but in reverse order. The 400 Ω resistor came from two 800 Ω resistors in parallel. Because $\Delta V_{400} = 12$ V, it must be true that each $\Delta V_{800} = 12$ V. The current through each 800 Ω is then $I = \Delta V/R = 15$ mA. The checkpoint is to note that 15 mA + 15 mA = 30 mA.

The right 800 Ω resistor was formed by 240 Ω and 560 Ω in series. Because $I_{800} = 15$ mA, it must be true that $I_{240} = I_{560} = 15$ mA. The potential difference across each is $\Delta V = IR$, so $\Delta V_{240} = 3.6$ V and $\Delta V_{560} = 8.4$ V. Here the checkpoint is to note that 3.6 V + 8.4 V = 12 V = ΔV_{800}, so the potential differences add as they should.

Finally, the 240 Ω resistor came from 600 Ω and 400 Ω in parallel, so they each have the same 3.6 V potential difference as their 240 Ω equivalent. The currents are $I_{600} = 6$ mA and $I_{400} = 9$ mA. Note that 6 mA + 9 mA = 15 mA, which is our third checkpoint. We now know all currents and potential differences.

ASSESS We *checked our work* at each step of the rebuilding process by verifying that currents summed properly at junctions and that potential differences summed properly along a series of resistances. This "check as you go" procedure is extremely important. It provides you, the problem solver, with a built-in error finder that will immediately inform you if a mistake has been made.

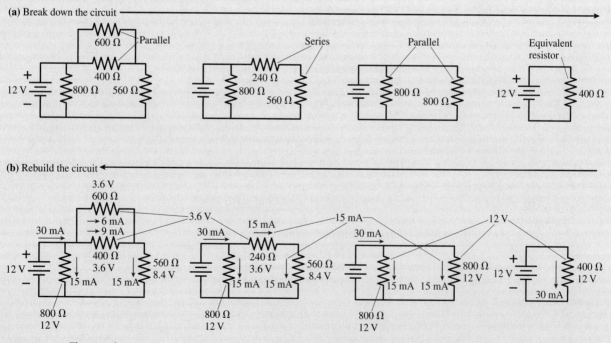

FIGURE 31.32 The step-by-step circuit analysis.

31.9 Getting Grounded

People who work with electronics are often heard to talk about things being "grounded." It always sounds quite serious, perhaps somewhat mysterious. What is it? Why do it?

The circuit analysis procedures we have discussed so far deal only with potential *differences*. Although we are free to choose the zero point of potential anywhere that is convenient, our analysis of circuits has not revealed any need to establish a zero point. Potential differences are all we have needed.

Difficulties can begin to arise, however, if you want to connect two *different* circuits together. Perhaps you would like to connect your CD player to your amplifier or your computer monitor to the computer itself. Incompatibilities can arise unless all the circuits to be connected have a *common* reference point for the potential.

You learned previously that the earth itself is a conductor. Suppose we have two circuits. If we connect *one* point of each circuit to the earth by an ideal wire, and we also agree to call the potential of the earth $V_{earth} = 0$ V, then both circuits have a common reference point. But notice something very important: *one* wire connects the circuit to the earth, but there is not a second wire returning to the circuit. That is, the wire connecting the circuit to the earth is not part of a complete circuit, so there is *no current* in this wire! Because the wire is an equipotential, it gives one point in the circuit the same potential as the earth, but it does *not* in any way change how the circuit functions. A circuit connected to the earth in this way is said to be **grounded,** and the wire is called the *ground wire.*

Figure 31.33a shows a fairly simple circuit with a 10 V battery and two resistors in series. The symbol beneath the circuit is the *ground symbol.* In this circuit, the symbol indicates that a wire has been connected between the negative battery terminal and the earth. This ground wire does not make a complete circuit, so there is no current in it. Consequently, the presence of the ground wire does not affect the circuit's behavior. The total resistance is $8\ \Omega + 12\ \Omega = 20\ \Omega$, so the current in the loop is $I = (10\text{ V})/(20\ \Omega) = 0.5$ A. The potential differences across the two resistors are found, using Ohm's law, to be $\Delta V_8 = 4$ V and $\Delta V_{12} = 6$ V. These are the same values of the current and the potential differences that we would find if the ground wire were *not* present. So what has grounding the circuit accomplished?

Figure 31.33b shows the potential at several points in the circuit. By definition, $V_{earth} = 0$ V. The negative battery terminal and the bottom of the 12 Ω resistor are connected by ideal wires to the earth, so *the* potential at these two points must also be zero. The positive terminal of the battery is 10 V more positive than the negative terminal, so $V_{neg} = 0$ V implies $V_{pos} = +10$ V. Similarly, the fact that the potential *decreases* by 6 V as charge flows through the 12 Ω resistor now implies that *the* potential at the junction of the resistors must be $+6$ V. The potential difference across the 8 Ω resistor is 4 V, so the top has to be at $+10$ V. This agrees with the potential at the positive battery terminal, as it must because these two points are connected by an ideal wire.

Grounding the circuit has not changed the current or any of the potential differences. All that grounding the circuit does is allow us to have *specific values* for the potential at each point in the circuit. Now we can say "The voltage at the resistor junction is 6 V," whereas before all we could say was that "there is a 6 V potential difference across the 12 Ω resistor."

There is one important lesson from this: Nothing happens in a circuit "because" it is grounded. You cannot use "because it is grounded" to *explain* anything about a circuit's behavior. **Being grounded does not affect the circuit's behavior under normal conditions.**

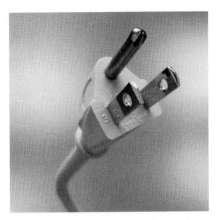

The circular prong of a three-prong plug is a connection to ground.

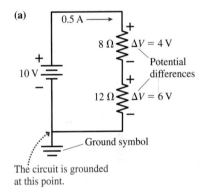

The circuit is grounded at this point.

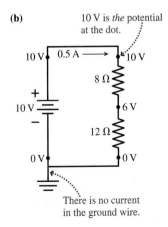

There is no current in the ground wire.

FIGURE 31.33 A circuit that is grounded at one point.

We added "under normal conditions" because there is one exception. Most circuits are enclosed in a case of some sort that is held away from the circuit with insulators. Sometimes, however, a circuit breaks or malfunctions in some way such that the case comes into electrical contact with the circuit. If the circuit uses high voltage, or even ordinary 120 V household voltage, anyone touching the case could be injured or killed by electrocution. To prevent this, many appliances or electrical instruments have the case itself grounded. Grounding ensures that the potential of the case will always remain at 0 V and be safe. If a malfunction occurs that connects the case to the circuit, a large current will pass through the ground wire to the earth and cause a fuse to blow. This is the *only* time the ground wire would ever have a current, and it is *not* a normal operation of the circuit.

Thus grounding a circuit serves two functions. First, it provides a common reference potential so that different circuits or instruments can be correctly interconnected. Second, it is an important safety feature to prevent injury or death from a defective circuit. For this reason you should *never* tamper with or try to defeat the ground connection (the third prong) on an electrical instrument's plug. If it has a ground connection, then it *needs* a ground connection and you should not try to plug it into a two-prong ungrounded outlet. Grounding the instrument does not affect its operation *under normal conditions,* but the abnormal and the unexpected are always with us. Play it safe.

EXAMPLE 31.14 A grounded circuit

Suppose the circuit of Figure 31.33 were grounded at the junction between the two resistors instead of at the bottom. Find the potential at each corner of the circuit.

VISUALIZE Figure 31.34 shows the new circuit. (It is customary to draw the ground symbol so that its "point" is always down.)

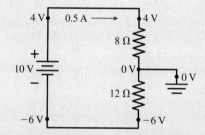

FIGURE 31.34 Circuit of Figure 31.33 grounded at the point between the resistors.

SOLVE Changing the ground point does not affect the circuit's behavior. The current is still 0.5 A, and the potential differences across the two resistors are still 4 V and 6 V. All that has happened is that we have moved the $V = 0$ V reference point. Because the earth has $V_{earth} = 0$ V, the junction itself now has a potential of 0 V. The potential decreases by 4 V as charge flows through the 8 Ω resistor. Because it *ends* at 0 V, the potential at the top of the 8 Ω must be +4 V. Similarly, the potential decreases by 6 V through the 12 Ω resistor. Because it *starts* at 0 V, the bottom of the resistor must be at −6 V. The negative battery terminal is at the same potential as the bottom of the 12 Ω resistor, because they are connected by a wire, so $V_{neg} = -6$ V. Finally, the potential increases by 10 V as the charge flows through the battery, so $V_{pos} = +4$ V, in agreement, as it should be, with the potential at the top of the 8 Ω.

You may wonder about the negative voltages. A negative voltage means only that the potential at that point is less than the potential at some other point that we chose to call $V = 0$ V. Only potential *differences* are physically meaningful, and only potential differences enter into Ohm's law: $I = \Delta V/R$. The potential difference across the 12 Ω resistor in this example is 6 V, decreasing from top to bottom, regardless of which point we choose to call $V = 0$ V.

31.10 *RC* Circuits

Activ Physics ONLINE 12.6–12.8

Thus far we've considered only circuits in which the current is steady and continuous. There are many circuits in which the time dependence of the current is a crucial feature. Charging and discharging a capacitor is an important example.

Figure 31.35a shows a charged capacitor, a switch, and a resistor. The capacitor has charge Q_0 and potential difference $\Delta V_C = Q_0/C$. There is no current, so the potential difference across the resistor is zero. Then, at $t = 0$, the switch closes and the capacitor begins to discharge through the resistor. A circuit such as this, with resistors and capacitors, is called an *RC* **circuit.**

How long does the capacitor take to discharge? How does the current through the resistor vary as a function of time? To answer these questions, Figure 31.35b shows the circuit after the switch has closed. Now the potential difference across the resistor is $\Delta V_R = -IR$, where I is the current discharging the capacitor.

Kirchhoff's loop law is valid for any circuit, not just circuits with batteries. The loop law applied to the circuit of Figure 31.35b, going around the loop cw, is

$$\Delta V_C + \Delta V_R = \frac{Q}{C} - IR = 0 \qquad (31.32)$$

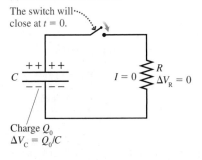
(a) Before the switch closes

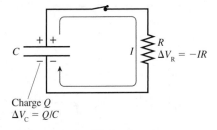
(b) After the switch closes

FIGURE 31.35 An *RC* circuit.

Q and I in this equation are the *instantaneous* values of the capacitor charge and the resistor current.

The current I is the rate at which charge flows through the resistor: $I = dq/dt$. But the charge flowing through the resistor is charge that was *removed* from the capacitor. That is, an infinitesimal charge dq flows through the resistor when the capacitor charge *decreases* by dQ. Thus $dq = -dQ$, and the resistor current is related to the instantaneous capacitor charge by

$$I = -\frac{dQ}{dt} \qquad (31.33)$$

Now I is positive when Q is decreasing, as we would expect. The reasoning that has led to Equation 31.33 is rather subtle but very important. You'll see the same reasoning later in other contexts.

If we substitute Equation 31.33 into Equation 31.32 and then divide by R, the loop law for the *RC* circuit becomes

$$\frac{dQ}{dt} + \frac{Q}{RC} = 0 \qquad (31.34)$$

Equation 31.34 is a first-order differential equation for the capacitor charge Q, but one that we can solve by direct integration. First, rearrange Equation 31.34 to get all the charge terms on one side of the equation:

$$\frac{dQ}{Q} = -\frac{1}{RC}dt$$

The product RC is a constant for any particular circuit.

We know the capacitor charge was Q_0 at $t = 0$ when the switch was closed. We want to integrate from these starting conditions to charge Q at the unspecified time t. That is,

$$\int_{Q_0}^{Q} \frac{dQ}{Q} = -\frac{1}{RC}\int_{0}^{t} dt \qquad (31.35)$$

Both are well-known integrals, giving

$$\ln Q \Big|_{Q_0}^{Q} = \ln Q - \ln Q_0 = \ln\left(\frac{Q}{Q_0}\right) = -\frac{t}{RC}$$

We can solve for the capacitor charge Q by taking the exponential of both sides, then multiplying by Q_0. Doing so gives

$$Q = Q_0 e^{-t/RC} \tag{31.36}$$

Notice that $Q = Q_0$ at $t = 0$, as expected.

The argument of an exponential function must be dimensionless, so the quantity RC must have dimensions of time. It is useful to define the **time constant** τ of the RC circuit to be

$$\tau = RC \tag{31.37}$$

We can then write Equation 31.36 as

$$Q = Q_0 e^{-t/\tau} \tag{31.38}$$

The meaning of Equation 31.38 is easier to understand if we portray it graphically. Figure 31.36a shows the capacitor charge as a function of time. The charge decays exponentially, starting from Q_0 at $t = 0$ and asymptotically approaching zero as $t \rightarrow \infty$. The time constant τ is the time at which the charge has decreased to e^{-1} (about 37%) of its initial value. At time $t = 2\tau$, the charge has decreased to e^{-2} (about 13%) of its initial value.

NOTE ▶ The *shape* of the graph of Q is always the same, regardless of the specific value of the time constant τ. ◀

We find the resistor current by using Equation 31.33:

$$I = -\frac{dQ}{dt} = \frac{Q_0}{\tau} e^{-t/\tau} = \frac{Q_0}{RC} e^{-t/\tau} = \frac{\Delta V_0}{R} e^{-t/\tau} = I_0 e^{-t/\tau} \tag{31.39}$$

where I_0 is the initial current, immediately after the switch closes. Figure 31.36b is a graph of the resistor current versus t. You can see that the current undergoes the same exponential decay, with the same time constant, as the capacitor charge.

NOTE ▶ There's no specific time at which the capacitor has been discharged, because Q approaches zero asymptotically, but the charge and current have dropped to less than 1% of their initial values at $t = 5\tau$. Thus 5τ is a reasonable answer to the question, "How long does it take to discharge a capacitor?" ◀

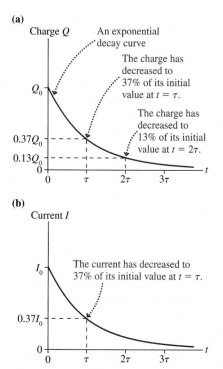

(a)

Charge Q

An exponential decay curve

The charge has decreased to 37% of its initial value at $t = \tau$.

Q_0

The charge has decreased to 13% of its initial value at $t = 2\tau$.

$0.37Q_0$

$0.13Q_0$

0

$0 \quad \tau \quad 2\tau \quad 3\tau$

(b)

Current I

I_0

The current has decreased to 37% of its initial value at $t = \tau$.

$0.37I_0$

0

$0 \quad \tau \quad 2\tau \quad 3\tau$

FIGURE 31.36 The decay curves of the capacitor charge and the resistor current.

EXAMPLE 31.15 Exponential decay in an *RC* circuit

The switch in Figure 31.37 has been in position a for a long time. It is changed to position b at $t = 0$ s. What are the charge on the capacitor and the current through the resistor at $t = 5.0\ \mu s$?

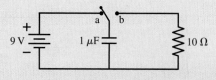

FIGURE 31.37 An *RC* circuit.

MODEL The battery charges the capacitor to 9.0 V. Then, when the switch is changed to position b, the capacitor discharges through the 10 Ω resistor. Assume ideal wires.

SOLVE The time constant of the *RC* circuit is

$$\tau = RC = (10\ \Omega)(1.0 \times 10^{-6}\ \text{F}) = 10 \times 10^{-6}\ \text{s} = 10\ \mu s$$

The capacitor is initially charged to 9.0 V, giving $Q_0 = C\Delta V_C = 9.0\ \mu C$. The capacitor charge at $t = 5.0\ \mu s$ is

$$Q = Q_0 e^{-t/RC} = (9.0\ \mu C)e^{-(5.0\ \mu s)/(10\ \mu s)}$$
$$= (9.0\ \mu C)e^{-0.5} = 5.5\ \mu C$$

The initial current, immediately after the switch is closed, is $I_0 = Q_0/\tau = 0.90$ A. The resistor current at $t = 5.0\ \mu s$ is

$$I = I_0 e^{-t/RC} = (0.90\ \text{A})e^{-0.5} = 0.55\ \text{A}$$

ASSESS This capacitor will be almost entirely discharged $5\tau = 50\ \mu s$ after the switch is closed.

Charging a Capacitor

Figure 31.38a shows a circuit that charges a capacitor. After the switch is closed, the battery's charge escalator moves charge from the bottom electrode of the capacitor to the top electrode. The resistor, by limiting the current, slows the process but doesn't stop it. The capacitor charges until $\Delta V_C = \mathcal{E}$, then the charging current ceases. The full charge of the capacitor is $Q_{max} = C(\Delta V_C)_{max} = C\mathcal{E}$.

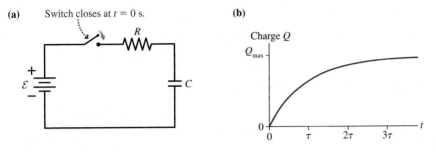

(a) Switch closes at $t = 0$ s. **(b)**

FIGURE 31.38 A circuit for charging a capacitor.

As a homework problem, you can show that the capacitor charge at time t is

$$Q = Q_{max}(1 - e^{-t/\tau}) \tag{31.40}$$

where again $\tau = RC$. This "upside-down decay" to Q_{max} is shown graphically in Figure 31.38b. *RC* circuits that alternately charge and discharge a capacitor are at the heart of time-keeping circuits in computers and other digital electronics.

STOP TO THINK 31.7   The time constant for the discharge of this capacitor is

a. 5 s.
b. 4 s.
c. 2 s.
d. 1 s.
e. The capacitor doesn't discharge because the resistors cancel each other.

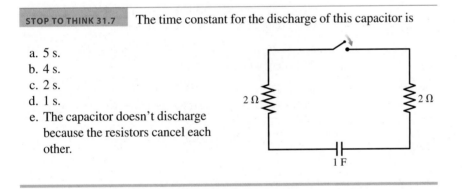

SUMMARY

The goal of Chapter 31 has been to understand the fundamental physical principles that govern electric circuits.

GENERAL STRATEGY

MODEL Assume that wires and, where appropriate, batteries are ideal.

VISUALIZE Draw a circuit diagram. Label known and unknown quantities.

SOLVE The solution is based on Kirchhoff's laws.

- Reduce the circuit to the smallest possible number of equivalent resistors.
- Find the current and the potential difference.
- "Rebuild" the circuit to find I and ΔV for each resistor.

ASSESS Verify that

- The sum of potential differences across series resistors matches ΔV for the equivalent resistor.
- The sum of the currents through parallel resistors matches I for the equivalent resistor.

Kirchhoff's loop law

For a closed loop:

- Assign a direction to the current I.
- $\sum_i (\Delta V)_i = 0$

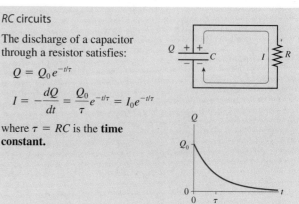

Kirchhoff's junction law

For a junction:

- $\sum I_{in} = \sum I_{out}$

IMPORTANT CONCEPTS

Ohm's law

A potential difference ΔV between the ends of a conductor with resistance R creates a current

$$I = \frac{\Delta V}{R}$$

Signs of ΔV

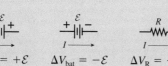

$$\Delta V_{bat} = +\mathcal{E} \qquad \Delta V_{bat} = -\mathcal{E} \qquad \Delta V_R = -IR$$

The energy used by a circuit is supplied by the emf $\mathcal{E}$ of the battery through the energy transformations

$$E_{chem} \rightarrow U \rightarrow K \rightarrow E_{th}$$

The battery *supplies* energy at the rate

$$P_{bat} = I\mathcal{E}$$

The resistors *dissipate* energy at the rate

$$P_R = I\Delta V_R = I^2 R = \frac{(\Delta V_R)^2}{R}$$

APPLICATIONS

Series resistors

$$R_{eq} = R_1 + R_2 + R_3 + \ldots$$

Parallel resistors

$$R_{eq} = \left(\frac{1}{R_1} + \frac{1}{R_2} + \frac{1}{R_3} + \ldots \right)^{-1}$$

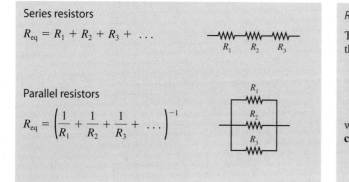

RC circuits

The discharge of a capacitor through a resistor satisfies:

$$Q = Q_0 e^{-t/\tau}$$

$$I = -\frac{dQ}{dt} = \frac{Q_0}{\tau} e^{-t/\tau} = I_0 e^{-t/\tau}$$

where $\tau = RC$ is the **time constant.**

TERMS AND NOTATION

direct current	circuit diagram	kilowatt hour, kWh	short circuit
Ohm's law	Kirchhoff's junction law	series resistors	parallel resistors
resistor	Kirchhoff's loop law	equivalent resistance, R_{eq}	voltmeter
ohmic	complete circuit	ammeter	grounded
nonohmic	load	internal resistance, r	RC circuit
ideal wire	source	terminal voltage, ΔV_{bat}	time constant, τ
ideal insulator			

EXERCISES AND PROBLEMS

Exercises

Section 31.1 Resistors and Ohm's Law

1. Pencil "lead" is actually carbon. What is the resistance of the 0.70-mm-diameter, 6.0-cm-long lead from a mechanical pencil?

2. The resistance of a very fine aluminum wire with a 10 μm $\times$ 10 μm square cross section is 1000 Ω.
   a. How long is the wire?
   b. A 1000 Ω resistor is made by wrapping this wire in a spiral around a 3.0-mm-diameter glass core. How many turns of wire are needed?

3. A 3.0 V potential difference is applied between the ends of a 0.80-mm-diameter, 50-cm-long nichrome wire. What is the current in the wire?

4. A 1.0-mm-diameter, 20-cm-long copper wire carries a 3.0 A current. What is the potential difference between the ends of the wire?

5. A circuit calls for a 0.50-mm-diameter copper wire to be stretched between two points. You don't have any copper wire, but you do have aluminum wire in a wide variety of diameters. What diameter aluminum wire will provide the same resistance?

6. Figure Ex31.6 is a current-versus-potential-difference graph for a material. What is the material's resistance?

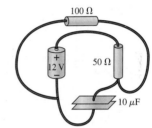

FIGURE EX31.6

Section 31.2 Circuit Elements and Diagrams

7. Draw a circuit diagram for the circuit of Figure Ex31.7.

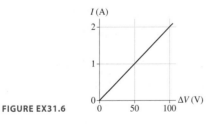

FIGURE EX31.7

8. Draw a circuit diagram for the circuit of Figure Ex31.8.

FIGURE EX31.8

Section 31.3 Kirchhoff's Laws and the Basic Circuit

9. In Figure Ex31.9, what is the current in the wire above the junction? Does charge flow toward or away from the junction?

FIGURE EX31.9

10. a. What are the magnitude and direction of the current in the 30 Ω resistor in Figure Ex31.10?
    b. Draw a graph of the potential as a function of the distance traveled through the circuit, traveling cw from $V = 0$ V at the lower left corner.

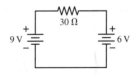

FIGURE EX31.10

11. a. What are the magnitude and direction of the current in the 18 Ω resistor in Figure Ex31.11?
    b. Draw a graph of the potential as a function of the distance traveled through the circuit, traveling cw from $V = 0$ V at the lower left corner.

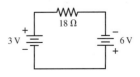

FIGURE EX31.11

12. a. What is the potential difference across each resistor in Figure Ex31.12?
    b. Draw a graph of the potential as a function of the distance traveled through the circuit, traveling cw from $V = 0$ V at the lower left corner.

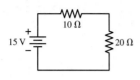

FIGURE EX31.12

Section 31.4 Energy and Power

13. What is the resistance of a 1500 W (120 V) hair dryer? What is the current in the hair dryer when it is used?

14. How much power is dissipated by each resistor in Figure Ex31.14?

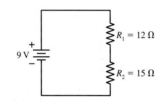

FIGURE EX31.14

15. A standard 100 W (120 V) lightbulb contains a 7.0-cm-long tungsten filament. The high-temperature resistivity of tungsten is 9.0×10^{-7} Ω m. What is the diameter of the filament?

16. How many joules are in 1 kWh?

17. A typical American family uses 1000 kWh of electricity a month.
    a. What is the average current in the 120 V power line to the house?
    b. On average, what is the resistance of a household?

18. A 60 W (120 V) night light is turned on for an average 12 hr a day year round. What is the annual cost of electricity at a billing rate of $0.10/kWh?

Section 31.5 Series Resistors

19. An 80-cm-long wire is made by welding a 1.0-mm-diameter, 20-cm-long copper wire to a 1.0-mm-diameter, 60-cm-long iron wire. What is the resistance of the composite wire?

20. Two of the three resistors in Figure Ex31.20 are unknown but equal. Is the total resistance between points a and b less than, greater than, or equal to 50 Ω? Explain.

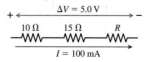

FIGURE EX31.20

21. What is the value of resistor R in Figure Ex31.21?

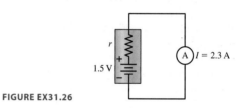

FIGURE EX31.21

22. Two 75 W (120 V) lightbulbs are wired in series, then the combination is connected to a 120 V supply. How much power is dissipated by each bulb?

23. The corroded contacts in a lightbulb socket have 5.0 Ω resistance. How much actual power is dissipated by a 100 W (120 V) lightbulb screwed into this socket?

Section 31.6 Real Batteries

24. The voltage across the terminals of a 9.0 V battery is 8.5 V when the battery is connected to a 20 Ω load. What is the battery's internal resistance?

25. Compared to an ideal battery, by what percentage does the battery's internal resistance reduce the potential difference across the 20 Ω resistor in Figure Ex31.25?

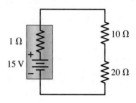

FIGURE EX31.25

26. What is the internal resistance of the battery in Figure Ex31.26? How much power is dissipated inside the battery?

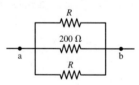

FIGURE EX31.26

Section 31.7 Parallel Resistors

27. A metal wire of resistance R is cut into two pieces of equal length. The two pieces are connected together side by side. What is the resistance of the two connected wires?

28. Two of the three resistors in Figure Ex31.28 are unknown but equal. Is the total resistance between points a and b less than, greater than, or equal to 200 Ω? Explain.

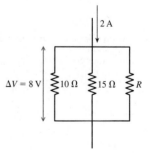

FIGURE EX31.28

29. What is the value of resistor R in Figure Ex31.29?

FIGURE EX31.29

30. What is the equivalent resistance between points a and b in Figure Ex31.30?

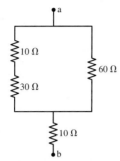

FIGURE EX31.30

31. What is the equivalent resistance between points a and b in Figure Ex31.31?

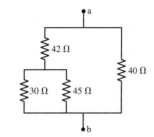

FIGURE EX31.31

32. What is the equivalent resistance between points a and b in Figure Ex31.32?

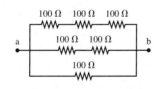

FIGURE EX31.32

33. What is the equivalent resistance between points a and b in Figure Ex31.33?

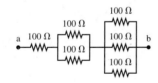

FIGURE EX31.33

Section 31.9 Getting Grounded

34. Determine the value of the potential at points a to d in Figure Ex31.34.

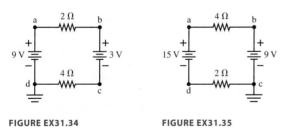

FIGURE EX31.34   FIGURE EX31.35

35. Determine the value of the potential at points a to d in Figure Ex31.35.

Section 31.10 *RC* Circuits

36. Show that the product RC has units of s.
37. What is the time constant for the discharge of the capacitors in Figure Ex31.37?

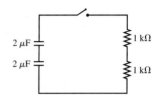

FIGURE EX31.37

38. What is the time constant for the discharge of the capacitors in Figure Ex31.38?

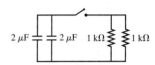

FIGURE EX31.38

39. A 10 μF capacitor initially charged to 20 μC is discharged through a 1.0 kΩ resistor. How long does it take to reduce the capacitor's charge to 10 μC?
40. The switch in Figure Ex31.40 has been in position a for a long time. It is changed to position b at $t = 0$ s. What are the charge Q on the capacitor and the current I through the resistor (a) immediately after the switch is closed? (b) at $t = 50$ μs? (c) at $t = 200$ μs?

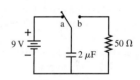

FIGURE EX31.40

41. What value resistor will discharge a 1.0 μF capacitor to 10% of its initial charge in 2.0 ms?
42. A capacitor is discharged through a 100 Ω resistor. The discharge current decreases to 25% of its initial value in 2.5 ms. What is the value of the capacitor?

Problems

43. Figure P31.43 shows five identical bulbs connected to an ideal battery. All the bulbs are glowing. Rank in order, from brightest to dimmest, the brightness of bulbs A to E. Explain.

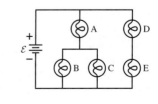

FIGURE P31.43

44. Figure P31.44 shows six identical bulbs connected to an ideal battery. All the bulbs are glowing. Rank in order, from brightest to dimmest, the brightness of bulbs A to F. Explain.

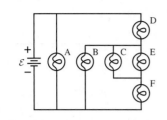

FIGURE P31.44

45. The battery in Figure P31.45 is ideal. Initially, bulbs A and B are both glowing. Bulb B is then removed from its socket. Does removing bulb B cause the potential difference ΔV_{12} between points 1 and 2 to increase, decrease, or become zero? Explain.

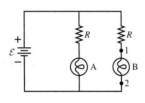

FIGURE P31.45

46. You've made the finals of the Science Olympics! As one of your tasks, you're given 1.0 g of aluminum and asked to make a wire, using all the aluminum, with a resistance of 1.0 Ω. What length and diameter will you choose for your wire?

47. Not too long ago houses were protected from excessive currents by fuses rather than circuit breakers. Sometimes a fuse blew out and a replacement wasn't at hand. Because a copper penny happens to have almost the same diameter as a fuse, some people replaced the fuse with a penny. Unfortunately, a penny never blows out, no matter how large the current, and the use of pennies in fuse boxes caused many house fires. Make the appropriate measurements on a penny, then calculate the resistance between the two faces of a copper penny.

48. You have three 12 Ω resistors. Draw diagrams showing how you could arrange all three so that their equivalent resistance is (a) 4.0 Ω, (b) 8.0 Ω, (c) 18 Ω, and (d) 36 Ω.

49. What is the equivalent resistance between points a and b in Figure P31.49?

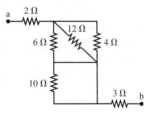

FIGURE P31.49

50. There is a current of 0.25 A in the circuit of Figure P31.50.
    a. What is the direction of the current? Explain.
    b. What is the value of the resistance R?
    c. What is the power dissipated by R?
    d. Make a graph of potential versus position, starting from V = 0 V in the lower left corner and proceeding cw.

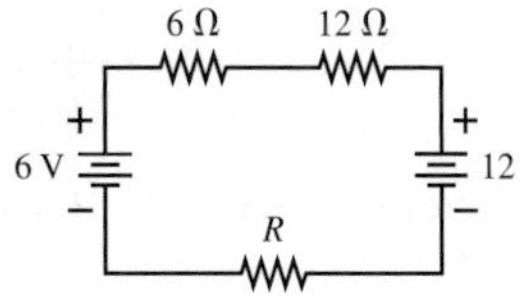

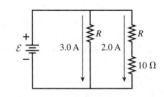

FIGURE P31.50

51. A variable resistor R is connected across the terminals of a battery. Figure P31.51 shows the current in the circuit as R is varied. What are the emf and internal resistance of the battery?

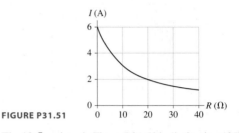

FIGURE P31.51

52. The 10 Ω resistor in Figure P31.52 is dissipating 40 W of power. How much power are the other two resistors dissipating?

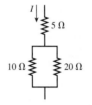

FIGURE P31.52

53. What are the emf and internal resistance of the battery in Figure P31.53?

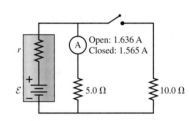

FIGURE P31.53

54. What are the resistance R and the emf of the battery in Figure P31.54?

FIGURE P31.54

55. A 2.5 V battery and a 1.5 V battery, each with an internal resistance of 1 Ω, are connected in parallel. That is, their positive terminals are connected by a wire and their negative terminals are connected by a wire. What is the terminal voltage of each battery in this configuration?

56. a. Load resistor R is attached to a battery of emf $\mathcal{E}$ and internal resistance r. For what value of the resistance R, in terms of $\mathcal{E}$ and r, will the power dissipated by the load resistor be a maximum?
    b. What is the maximum power that the load can dissipate if the battery has $\mathcal{E} = 9.0$ V and $r = 1.0$ Ω?
    c. *Why* should the power dissipated by the load have a maximum value? Explain.
    Hint: What happens to the power dissipation when R is either very small or very large?

57. The ammeter in Figure P31.57 reads 3.0 A. Find I_1, I_2, and $\mathcal{E}$.

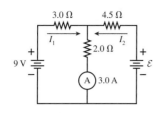

FIGURE P31.57

58. a. Suppose the circuit in Figure P31.58 is grounded at point d. Find the potential at each of the four points a, b, c, and d.
   b. Make a graph of potential versus position, starting from point d and proceeding cw.
   c. Repeat parts a and b for the same circuit grounded at point a instead of d.

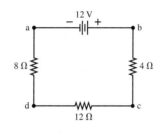

FIGURE P31.58

59. What is the current in the 2 Ω resistor in Figure P31.59?

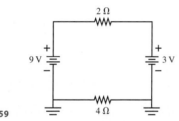

FIGURE P31.59

60. Energy experts tell us to replace regular incandescent light-bulbs with compact fluorescent bulbs, but it seems hard to justify spending $15 on a lightbulb. A 60 W incandescent bulb costs 50¢ and has a lifetime of 1000 hours. A 15 W compact fluorescent bulb produces the same amount of light as a 60 W incandescent bulb and is intended as a replacement. It costs $15 and has a lifetime of 10,000 hours. Compare the *life-cycle costs* of 60 W incandescent bulbs to 15 W compact fluorescent bulbs. The life-cycle cost of an object is the cost of purchasing it plus the cost of fueling and maintaining it over its useful life. Which is the cheaper source of light and which the more expensive? Assume that electricity costs $0.10/kWh.
   Hint: Be sure to compare the two over equal time spans.

61. A refrigerator has a 1000 W compressor, but the compressor runs only 20% of the time.
   a. If electricity costs $0.10/kWh, what is the monthly (30 day) cost of running the refrigerator?
   b. A more energy efficient refrigerator with an 800 W compressor costs $100 more. If you buy the more expensive refrigerator, how many months will it take to recover your additional cost?

62. For an ideal battery ($r = 0\ \Omega$), closing the switch in Figure P31.62 does not affect the brightness of bulb A. In practice, bulb A dims *just a little* when the switch closes. To see why, assume that the 1.5 V battery has an internal resistance

$r = 0.50\ \Omega$ and that the resistance of a glowing bulb is $R = 6\ \Omega$.
   a. What is the current through bulb A when the switch is open?
   b. What is the current through bulb A after the switch has closed?
   c. By what percent does the current through A change when the switch is closed?
   d. Would the current through A change if $r = 0\ \Omega$?

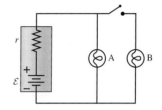

FIGURE P31.62

63. What are the battery current I_{bat} and the potential difference ΔV_{ab} between points a and b when the switch in Figure P31.63 is (a) open and (b) closed?

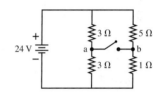

FIGURE P31.63

64. The circuit in Figure P31.64 is called a *voltage divider*. What value of R will make $V_{out} = V_{in}/10$?

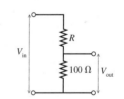

FIGURE P31.64

65. A circuit you're building needs an ammeter that goes from 0 mA to a full-scale reading of 50 mA. Unfortunately, the only ammeter in the storeroom goes from 0 μA to a full-scale reading of only 500 μA. Fortunately, you've just finished a physics class, and you realize that you can make this ammeter work by putting a resistor in parallel with it, as shown in Figure P31.65. You've measured that the resistance of the ammeter is 50 Ω, not the 0 Ω of an ideal ammeter.
   a. What value of R must you use so that the meter will go to full scale when the current I is 50 mA?
   b. What is the effective resistance of your ammeter?

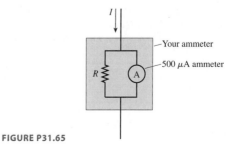

FIGURE P31.65

66. A circuit you're building needs a voltmeter that goes from 0 V to a full-scale reading of 5.0 V. Unfortunately, the only meter in the storeroom is an *ammeter* that goes from 0 μA to a full-scale reading of 500 μA. Fortunately, you've just finished a physics class, and you realize that you can convert this meter to a voltmeter by putting a resistor in series with it, as shown in Figure P31.66. You've measured that the resistance of the ammeter is 50 Ω, not the 0 Ω of an ideal ammeter. What value of R must you use so that the meter will go to full scale when the potential difference across the resistor is 5.0 V?

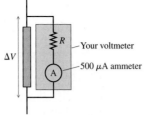

FIGURE P31.66

67. For the circuit shown in Figure P31.67, find the current through and the potential difference across each resistor. Place your results in a table for ease of reading.

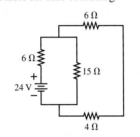

FIGURE P31.67

68. For the circuit shown in Figure P31.68, find the current through and the potential difference across each resistor. Place your results in a table for ease of reading.

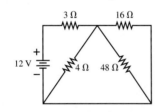

FIGURE P31.68

69. For the circuit shown in Figure P31.69, find the current through and the potential difference across each resistor. Place your results in a table for ease of reading.

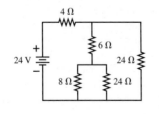

FIGURE P31.69

70. For the circuit shown in Figure P31.70, find the current through and the potential difference across each resistor. Place your results in a table for ease of reading.

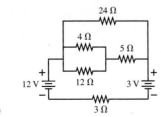

FIGURE P31.70

71. For the circuit in Figure P31.71, what are (a) the current through the 2 Ω resistor, (b) the power dissipated by the 20 Ω resistor, and (c) the potential at point a?

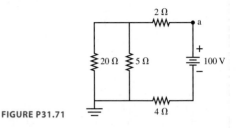

FIGURE P31.71

72. The capacitor in an *RC* circuit is discharged with a time constant of 10 ms. At what time after the discharge begins are (a) the charge on the capacitor reduced to half its initial value and (b) the energy stored in the capacitor reduced to half its initial value?

73. A 50 μF capacitor that had been charged to 30 V is discharged through a resistor. Figure P31.73 shows the capacitor voltage as a function of time. What is the value of the resistance?

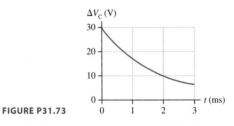

FIGURE P31.73

74. A 0.25 μF capacitor is charged to 50 V. It is then connected in series with a 25 Ω resistor and a 100 Ω resistor and allowed to discharge completely. How much energy is dissipated by the 25 Ω resistor?

75. The capacitors in Figure P31.75 are charged and the switch closes at $t = 0$ s. At what time has the current in the 8 Ω resistor decayed to half the value it had immediately after the switch was closed?

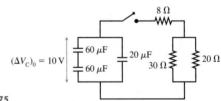

FIGURE P31.75

76. The capacitor in Figure P31.76 begins to charge after the switch closes at $t = 0$ s.
    a. What is ΔV_C a very long time after the switch has closed? Explain.
    b. What is Q_{max} in terms of $\mathcal{E}$, R, and C?
    c. In this circuit, does $I = +dQ/dt$ or $-dQ/dt$? Explain.
    d. Find an expression for the current I at time t. Graph I from $t = 0$ to $t = 5\tau$.

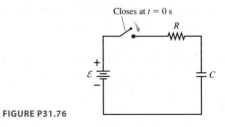

FIGURE P31.76

77. The switch in Figure P31.77 has been closed for a very long time.
    a. What is the charge on the capacitor?
    b. The switch is opened at $t = 0$ s. At what time has the charge on the capacitor decreased to 10% of its initial value?

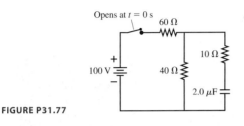

FIGURE P31.77

Challenge Problems

78. The switch in Figure CP31.78 has been in position a for a very long time. It is suddenly flipped to position b for 1.25 ms, then back to a. How much energy is dissipated by the 50 Ω resistor?

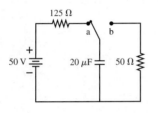

FIGURE CP31.78

79. The capacitor in Figure 31.38a begins to charge after the switch closes at $t = 0$ s. Analyze this circuit and show that $Q = Q_{max}(1 - e^{-t/\tau})$, where $Q_{max} = C\mathcal{E}$.

80. The switch in Figure 31.38a closes at $t = 0$ s and, after a very long time, the capacitor is fully charged. Find expressions for (a) the total energy supplied by the battery as the capacitor is being charged, (b) total energy dissipated by the resistor as the capacitor is being charged, and (c) the energy stored in the capacitor when it is fully charged. Your expressions will be in terms of $\mathcal{E}$, R, and C. (d) Do your results for parts a to c show that energy is conserved? Explain.

81. An *oscillator circuit* is important to many applications. A simple oscillator circuit can be built by adding a neon gas tube to an *RC* circuit, as shown in Figure CP31.81. Gas is normally a good insulator, and the resistance of the gas tube is essentially infinite when the light is off. This allows the capacitor to charge. When the capacitor voltage reaches a value V_{on}, the electric field inside the tube becomes strong enough to ionize the neon gas. Visually, the tube lights with an orange glow. Electrically, the ionization of the gas provides a very-low-resistance path through the tube. The capacitor very rapidly (we can think of it as instantaneously) discharges through the tube and the capacitor voltage drops. When the capacitor voltage has dropped to a value V_{off}, the electric field inside the tube becomes too weak to sustain the ionization and the neon light turns off. The capacitor then starts to charge again. The capacitor voltage oscillates between V_{off}, when it starts charging, and V_{on}, when the light comes on to discharge it.
    a. Show that the oscillation period is

$$T = RC\ln\left(\frac{\mathcal{E} - V_{off}}{\mathcal{E} - V_{on}}\right)$$

    b. A neon gas tube has $V_{on} = 80$ V and $V_{off} = 20$ V. What resistor value should you choose to go with a 10 μF capacitor and a 90 V battery to make a 10 Hz oscillator?

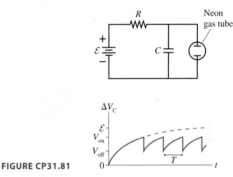

FIGURE CP31.81

STOP TO THINK ANSWERS

Stop to Think 31.1: $R_c > R_a > R_d > R_b$. The resistance is proportional to L/r^2. Increasing r decreases R; increasing L increases R. But the radius has a larger effect because R depends on the square of r.

Stop to Think 31.2: a, b, and **d.** These three are the same circuit because the logic of the connections is the same. In c, the functioning of the circuit is changed by the extra wire connecting the two sides of the capacitor.

Stop to Think 31.3: ΔV increases by 2 V in the direction of I. Kirchhoff's loop law, starting on the left side of the battery, is then $+12$ V $+ 2$ V $- 8$ V $- 6$ V $= 0$ V.

Stop to Think 31.4: $P_b > P_d > P_a > P_c$. The power dissipated by a resistor is $P_R = (\Delta V_R)^2/R$. Increasing R decreases P_R; increasing ΔV_R increases P_R. But the potential has a larger effect because P_R depends on the square of ΔV_R.

Stop to Think 31.5: $I = 2$ A for all. $V_a = 20$ V, $V_b = 16$ V, $V_c = 10$ V, $V_d = 8$ V, $V_e = 0$ V. Current is conserved. The potential is 0 V on the right and increases by IR for each resistor going to the left.

Stop to Think 31.6: $A > B > C = D$. All the current from the battery goes through A, so it is brightest. The current divides at the junction, but not equally. Because B is in parallel with $C + D$ but has half the resistance, twice as much current travels through B as through $C + D$. So B is dimmer than A but brighter than C and D. C and D are equal because of conservation of current.

Stop to Think 31.7: b. The two 2 Ω resistors are in series and equivalent to a 4 Ω resistor. Thus $\tau = RC = 4$ s.

32 The Magnetic Field

The beautiful aurora borealis, the northern lights, is due to the earth's magnetic field.

▶ **Looking Ahead**
The goal of Chapter 32 is to learn how to calculate and use the magnetic field. In this chapter you will learn to:

- Recognize basic magnetic phenomena.
- Calculate the magnetic field of moving charged particles and currents.
- Use the right-hand rule to find the directions of magnetic forces and fields.
- Understand the motion of a charged particle in a magnetic field.
- Calculate magnetic forces and torques on wires and current loops.
- Understand the magnetic properties of materials.

◀ **Looking Back**
This chapter uses what you have learned about circular motion, rotation, and dipoles to understand motion in a magnetic field. Please review:

- Sections 7.1–7.2 Uniform circular motion.
- Sections 13.3 and 13.9 Torque and the cross product of two vectors.
- Sections 25.5–25.6 Basic properties of fields.
- Sections 26.2 and 26.7 The properties of an electric dipole.

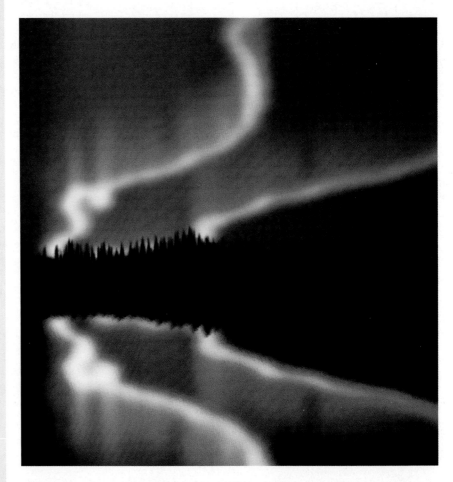

The shimmering aurora is one of the most beautiful of earth's natural displays. These lights, which mystified travelers and the inhabitants of northern realms for centuries, are a consequence of the earth's magnetism.

Magnetism, like electricity, has been known since antiquity. The ancient Greeks knew that certain minerals called *lodestones* could attract iron objects. Chinese navigators were using lodestone compasses by the year 1000, but compasses were not known in the West until nearly 1200. Later, in about 1600, William Gilbert recognized that compasses work because the earth itself is a magnet. The same forces that align compass needles are also responsible for the aurora.

Our task for this chapter is to investigate magnets and magnetism. Magnets are all around you. In addition to holding shopping lists and cartoons on refrigerators, magnets allow you to run electric motors, produce a picture on your television screen, store information on computer disks, cook food in a microwave oven, and listen to music using loudspeakers. Magnets are used in magnetic-resonance imaging to produce images of the interior of the human body, in high-energy physics experiments to identify subatomic particles, and in magnetic levitation trains.

Just what is magnetism? How are magnetic fields created? What are their properties? How are they used? These are the questions we will address.

32.1 Magnetism

We began our investigation of electricity in Chapter 25 by looking at the results of simple experiments with charged rods. Let's follow a similar approach with magnetism.

Discovering magnetism

Experiment 1

If a bar magnet is taped to a piece of cork and allowed to float in a dish of water, it always turns to align itself in an approximate north-south direction. The end of a magnet that points north is called the *north-seeking pole,* or simply, the **north pole.** The other end is the **south pole.**

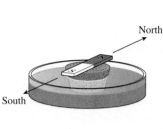

Experiment 2

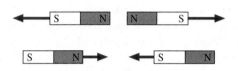

If the north pole of one magnet is brought near the north pole of another magnet, they exert repulsive forces on each other. Two south poles also repel each other, but the north pole of one magnet exerts an attractive force on the south pole of another magnet.

Experiment 3

The north pole of a bar magnet attracts one end of a compass needle and repels the other. Apparently the compass needle itself is a little bar magnet with a north pole and a south pole.

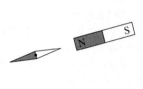

Experiment 4

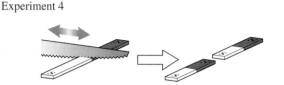

Cutting a bar magnet in half produces two weaker but still complete magnets, each with a north pole and a south pole. No matter how small the magnets are cut, even down to microscopic sizes, each piece remains a complete magnet with two poles.

Experiment 5

Magnets can pick up some objects, such as paper clips, but not all. If an object is attracted to one end of a magnet, it is also attracted to the other end. Most materials, including copper, aluminum, glass, and plastic, experience no force from a magnet.

Experiment 6

A magnet does not affect an electroscope. A charged rod exerts a weak *attractive* force on *both* ends of a magnet. However, the force is the same as the force on a metal bar that isn't a magnet, so it is simply a polarization force like the ones we studied in Chapter 25. Other than polarization forces, charges have *no effects* on magnets.

No effect

What do these experiments tell us?

1. Experiment 6 reveals that **magnetism is not the same as electricity.** Magnetic poles and electric charges share some similar behavior, but they are not the same. The magnetic force is a force of nature that we have not previously encountered.
2. Magnetism is a long-range force. Paper clips leap up to a magnet. You can feel the pull as you bring a refrigerator magnet close to the refrigerator.
3. Magnets have two poles, called north and south poles. The names are merely descriptive; they tell us nothing about how magnetism works. Two like poles exert repulsive forces on each other; two opposite poles exert attractive forces on each other. The behavior is *analogous* to electric charges, but, as noted, magnetic poles and electric charges are *not* the same.
4. The poles of a bar magnet can be identified by using it as a compass. Other magnets, such as flat refrigerator magnets or horseshoe magnets, aren't so easily made into a compass, but their poles can be identified by testing them against a bar magnet. A pole that attracts a known north pole and repels a known south pole must be a south magnetic pole.

5. Materials that are attracted to a magnet, or that a magnet sticks to, are called **magnetic materials.** The most common magnetic material is iron. Others include nickel and cobalt. Magnetic materials are attracted to *both* poles of a magnet. This attraction is analogous to how neutral objects are attracted to both positively and negatively charged rods by the polarization force. The difference is that *all* neutral objects are attracted to a charged rod whereas only a few materials are attracted to a magnet.

Our goal is to develop a theory of magnetism that will enable us to explain these observations.

Monopoles and Dipoles

It is a strange observation that cutting a magnet in half yields two weaker but still complete magnets, each with a north pole and a south pole. Every magnet that has ever been observed has both a north pole and south pole, thus forming a permanent **magnetic dipole.** A magnetic dipole is analogous to an electric dipole, but the two charges in an electric dipole can be separated and used individually. This appears *not* to be true for a magnetic dipole.

An isolated magnetic pole, such as a north pole in the absence of a south pole, would be called a **magnetic monopole.** No one has ever observed a magnetic monopole. On the other hand, no one has ever given a convincing reason why isolated magnetic poles should not exist, and some theories of subatomic particles say they should. Whether or not magnetic monopoles exist in nature remains an unanswered question at the most fundamental level of physics.

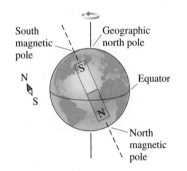

FIGURE 32.1 The earth is a large magnet.

Compasses and Geomagnetism

The north pole of a compass needle is attracted toward the geographic north pole of the earth and repelled by the earth's geographic south pole. Apparently the earth itself is a large magnet, as shown in Figure 32.1. The reasons for the earth's magnetism are complex, but geophysicists generally agree that the earth's magnetic poles arise from currents in its molten iron core. Two interesting facts about the earth's magnetic field are one, that the magnetic poles are offset slightly from the geographic poles of the earth's rotation axis, and two, that the geographic north pole is actually a *south* magnetic pole! You should be able to use what you have learned thus far to convince yourself that this is the case.

STOP TO THINK 32.1 Does the compass needle rotate clockwise (cw), counterclockwise (ccw), or not at all?

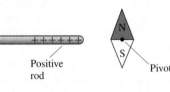

32.2 The Discovery of the Magnetic Field

As electricity began to be seriously studied in the 18th century, some scientists speculated that there might be a connection between electricity and magnetism. Interestingly, the link between electricity and magnetism was discovered *in the midst of a classroom lecture demonstration* in 1819 by the Danish scientist Hans Christian Oersted. Oersted was using a battery to produce a large current in a wire. By chance, a compass was sitting next to the wire, and Oersted noticed that the current caused the compass needle to turn. In other words, the compass responded as if a magnet had been brought near.

Oersted had long been interested in a possible connection between electricity and magnetism, so the significance of this serendipitous observation was immediately apparent to him. Oersted's discovery that **magnetism is caused by an electric current** will be our starting point for developing a theory of magnetism.

The Effect of a Current on a Compass

Let us use compasses to probe the magnetism created when a current passes through a long, straight wire. In Figure 32.2a, before the current is turned on, the compasses are aligned along a north-south line. You can see in Figure 32.2b that a strong current in the wire causes the compass needles to pivot until they are *tangent* to a circle around the wire. Figure 32.2c illustrates a **right-hand rule** that relates the orientation of the compass needles to the direction of the current.

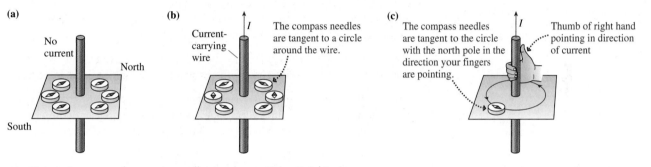

FIGURE 32.2 Response of compass needles to a current in a straight wire.

Magnetism is more demanding than electricity in requiring a three-dimensional perspective, of the sort shown in Figure 32.2. But since two-dimensional figures are easier to draw, we will make as much use of them as we can. Consequentially, we will often need to indicate field vectors or currents that are perpendicular to the page. Figure 32.3a shows the notation we will use. Figure 32.3b demonstrates this notation by showing the compasses around a current that is directed into the page. To use the right-hand rule with this drawing, point your right thumb into the page. Your fingers will curl cw, and that is the direction in which the north poles of the compass needles point.

The Magnetic Field

We introduced the idea of a *field* as a way to understand the long-range electric force. A charge alters the space around it by creating an electric field. A second charge then experiences a force due to the presence of the electric field. The electric field is the *means* by which charges interact with each other. Although this idea appeared rather far-fetched, it turned out to be very useful. We need a similar idea to understand the long-range force exerted by a current on a compass needle.

Let us define the **magnetic field** $\vec{B}$ as having the following properties:

1. A magnetic field is created at *all* points in space surrounding a current-carrying wire.
2. The magnetic field at each point is a vector. It has both a magnitude, which we call the *magnetic field strength B*, and a direction.
3. The magnetic field exerts forces on magnetic poles. The force on a north magnetic pole is parallel to $\vec{B}$; the force on a south magnetic pole is opposite to $\vec{B}$.

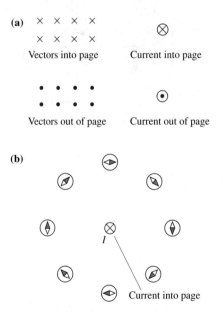

FIGURE 32.3 The notation for vectors and currents that are perpendicular to the page.

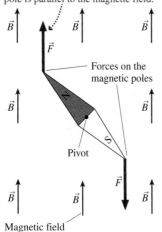

The magnetic force on the north pole is parallel to the magnetic field.

Forces on the magnetic poles

Pivot

Magnetic field

FIGURE 32.4 The magnetic field exerts forces on the poles of a compass, causing the needle to align with the field.

Figure 32.4 shows a compass needle in a magnetic field. The field vectors are shown at a few points, but keep in mind that the field is present at *all* points in space. A magnetic force is exerted on each of the two poles of the compass, parallel to $\vec{B}$ for the north pole and opposite to $\vec{B}$ for the south pole. This pair of opposite forces exerts a torque on the needle, rotating the needle until it is parallel to the magnetic field at that point.

Notice that the north pole of the compass needle, when it reaches the equilibrium position, is in the direction of the magnetic field. Thus a compass needle can be used as a probe of the magnetic field, just as a charge was a probe of the electric field. **Magnetic forces cause a compass needle to become aligned parallel to a magnetic field, with the north pole of the compass showing the direction of the magnetic field at that point.**

Look back at the compass alignments around the current-carrying wire in Figure 32.3b. Because compass needles align with the magnetic field, the magnetic field at each point must be tangent to a circle around the wire. Figure 32.5a shows the magnetic field by drawing field vectors. Notice that the field is weaker (shorter vectors) at greater distances from the wire.

Another way to picture the field is with the use of **magnetic field lines.** These are imaginary lines drawn through a region of space so that

- A tangent to a field line is in the direction of the magnetic field, and
- The field lines are closer together where the magnetic field strength is larger.

Figure 32.5b shows the magnetic field lines around a current-carrying wire. Notice that magnetic field lines form loops, with no beginning or ending point. This is in contrast to electric field lines, which stop and start on charges.

(a)

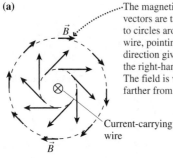

The magnetic field vectors are tangent to circles around the wire, pointing in the direction given by the right-hand rule. The field is weaker farther from the wire.

Current-carrying wire

(b)

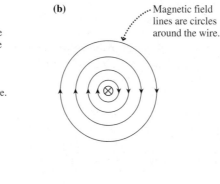

Magnetic field lines are circles around the wire.

FIGURE 32.5 The magnetic field around a current-carrying wire.

(c)

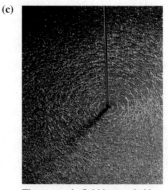

The magnetic field is revealed by the pattern of iron filings around the current-carrying wire.

In the photograph of Figure 32.5c, iron filings that have been sprinkled around a current-carrying wire allow us to visualize the circular magnetic field pattern. It was patterns such as these that first suggested the field concept to Faraday.

NOTE ▶ The magnetic field of a current-carrying wire is very different from the electric field of a charged wire. The electric field of a charged wire points radially outward (positive wire) or inward (negative wire). ◀

Two Kinds of Magnetism?

You might be concerned that we have introduced two kinds of magnetism. We opened this chapter discussing permanent magnets and their forces. Then, without warning, we switched to the magnetic forces caused by a current. It is not at all obvious that these forces are the same kind of magnetism as that exhibited by stationary chucks of metal called "magnets." Perhaps there are two different types of magnetic forces, one having to do with currents and the other being responsible for permanent magnets. One of the major goals for our study of magnetism is to see that these two quite different ways of producing magnetic effects are really just two different aspects of a *single* magnetic force.

STOP TO THINK 32.2   The magnetic field at position P points

a. Up.
b. Down.
c. Into the page.
d. Out of the page.

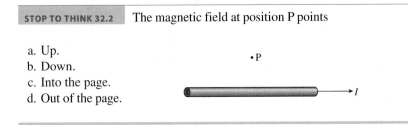

32.3 The Source of the Magnetic Field: Moving Charges

Figure 32.5 is a qualitative picture of the wire's magnetic field. Our first task is to turn that picture into a quantitative description. Because current in a wire generates a magnetic field, and a current is a collection of moving charges, it's natural to wonder if *any* moving charge would do the same. Oersted's discovery encouraged the general assumption among scientists that this was the case, although confirmation was not to come until 1875, 55 years later, when a rapidly spinning charged disk was shown to produce the same magnetic effects as the current in a circular loop of wire.

Thus our starting point is the idea that **moving charges are the source of the magnetic field.** Figure 32.6 shows a charged particle q moving with velocity $\vec{v}$. The magnetic field of this moving charge is found to be

$$\vec{B} = \left(\frac{\mu_0}{4\pi} \frac{qv\sin\theta}{r^2}, \text{ direction given by the right-hand rule} \right) \quad (32.1)$$

where r is the distance from the charge and θ is the angle between $\vec{v}$ and $\vec{r}$.

Equation 32.1 is called the **Biot-Savart law** for a point charge (rhymes with *Leo* and *bazaar*), named for two French scientists whose investigations were motivated by Oersted's observations. It is analogous to Coulomb's law for the electric field of a point charge. Notice that the Biot-Savart law, like Coulomb's law, is an inverse-square law. However, the Biot-Savart law is somewhat more complex than Coulomb's law because the magnetic field depends on the angle θ between the charge's velocity and the line to the point where the field is evaluated.

NOTE ▶ The magnetic field of a moving charge is *in addition* to the charge's electric field. The charge has an electric field whether it is moving or not. ◀

The SI unit of magnetic field strength is the **tesla,** abbreviated as T. The tesla is defined as

$$1 \text{ tesla} = 1 \text{ T} \equiv 1 \text{ N/A m}$$

You will see later in the chapter that this definition is based on the magnetic force on a current-carrying wire. One tesla is quite a large field. Table 32.1 shows some typical magnetic field strengths. Most magnetic fields are a small fraction of a tesla.

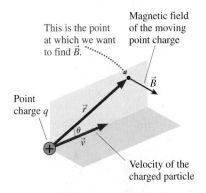

FIGURE 32.6 The magnetic field of a moving point charge.

TABLE 32.1 Typical magnetic field strengths

Field location	Field strength (T)
Surface of the earth	5×10^{-5}
Refrigerator magnet	5×10^{-3}
Laboratory magnet	0.1 to 1
Superconducting magnet	10

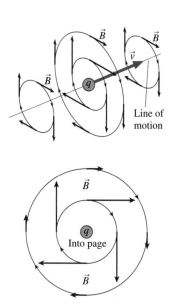

FIGURE 32.7 Two views of the magnetic field of a positive charge moving with velocity $\vec{v}$.

The constant μ_0 in Equation 32.1 is called the **permeability constant.** Its value is

$$\mu_0 = 4\pi \times 10^{-7} \,\text{Tm/A} = 1.257 \times 10^{-6} \,\text{Tm/A}$$

This constant plays a role in magnetism similar to that of the permittivity constant ϵ_0 in electricity.

The right-hand rule for finding the direction of $\vec{B}$ is similar to that used for a current-carrying wire: Point your right thumb in the direction of $\vec{v}$. The magnetic field vector $\vec{B}$ is perpendicular to the plane of $\vec{r}$ and $\vec{v}$, pointing in the direction in which your fingers curl. In other words, the $\vec{B}$ vectors are tangent to circles drawn about the charge's line of motion. Figure 32.7 shows a more complete view than Figure 32.6 of the magnetic field of a positive moving charge. Notice that $\vec{B}$ is zero along the line of motion, where $\theta = 0°$ or $180°$, due to the $\sin\theta$ term in Equation 32.1.

NOTE ▶ The vector arrows in Figure 32.7 would have the same lengths but be reversed in direction for a negative charge. ◀

The requirement that a charge be moving to generate a magnetic field is explicit in Equation 32.1. If the speed v of the particle is zero, the magnetic field (but not the electric field!) is zero. This helps to emphasize a fundamental distinction between electric and magnetic fields: **Charges create electric fields, but only *moving* charges create magnetic fields.**

EXAMPLE 32.1 **The magnetic field of a proton**
A proton moves along the x-axis with velocity $v_x = 1.0 \times 10^7$ m/s. As it passes the origin, what is the magnetic field at the (x, y, z) positions (1 mm, 0 mm, 0 mm), (0 mm, 1 mm, 0 mm), and (1 mm, 1 mm, 0 mm)?

MODEL The magnetic field is that of a moving charged particle.

VISUALIZE Figure 32.8 shows the geometry. The first point is on the x-axis, directly in front of the proton, with $\theta_1 = 0°$. The second point is on the y-axis, with $\theta_2 = 90°$, and the third is in the xy-plane.

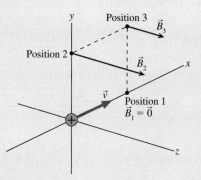

FIGURE 32.8 The magnetic field of Example 32.1.

SOLVE Position 1, which is along the line of motion, has $\theta = 0°$. Thus $\vec{B}_1 = \vec{0}$. Position 2 (at 0 mm, 1 mm, 0 mm) is at distance $r_2 = 1$ mm $= 0.001$ m. Equation 32.1, the Biot-Savart law, gives us the magnetic field strength at this point as

$$B = \frac{\mu_0}{4\pi} \frac{qv\sin\theta_2}{r_2^2}$$

$$= \frac{4\pi \times 10^{-7}\,\text{Tm/A}}{4\pi} \frac{(1.60 \times 10^{-19}\,\text{C})(1.0 \times 10^7\,\text{m/s})\sin 90°}{(0.0010\,\text{m})^2}$$

$$= 1.60 \times 10^{-13}\,\text{T}$$

According to the right-hand rule, the field points in the positive z-direction. Thus

$$\vec{B}_2 = 1.60 \times 10^{-13}\,\hat{k}\,\text{T}$$

where $\hat{k}$ is the unit vector in the positive z-direction. The field at position 3, at (1 mm, 1 mm, 0 mm), also points in the z-direction, but it is weaker than at position 2 both because r is larger *and* because θ is smaller. From geometry we know $r_3 = \sqrt{2}$ mm $= 0.00141$ m and $\theta_3 = 45°$. Another calculation using Equation 32.1 gives

$$\vec{B}_3 = 0.57 \times 10^{-13}\,\hat{k}\,\text{T}$$

ASSESS The magnetic field of a single moving charge is *very* small.

Superposition

The Biot-Savart law is the starting point for generating all magnetic fields, just as our earlier expression for the electric field of a point charge was the starting point for generating all electric fields. You learned in Chapter 26 that the total electric field caused by several charges $q_1, q_2, \ldots, q_n$ is the superposition of the electric fields of each separate charge.

Magnetic fields have been found experimentally to also obey the principle of superposition. If there are n moving point charges, the net magnetic field is given by the vector sum

$$\vec{B}_{total} = \vec{B}_1 + \vec{B}_2 + \cdots + \vec{B}_n \tag{32.2}$$

where each individual $\vec{B}$ is calculated with Equation 32.1. The principle of superposition will be the basis for calculating the magnetic fields of several important current distributions.

The Vector Cross Product

In Chapter 25, we found that the electric field of a point charge could be written concisely and accurately as

$$\vec{E} = \frac{1}{4\pi\epsilon_0} \frac{q}{r^2} \hat{r}$$

where $\hat{r}$ is a *unit vector* that points from the charge to the point at which we wish to calculate the field. Unit vector $\hat{r}$ expresses the idea "away from q."

The unit vector $\hat{r}$ also allows us to write the Biot-Savart law more concisely and more accurately, but we'll need to use the form of vector multiplication called the *cross product*. To remind you, Figure 32.9 shows two vectors, $\vec{C}$ and $\vec{D}$, with angle α between them. The **cross product** of $\vec{C}$ and $\vec{D}$ is defined to be the vector

$$\vec{C} \times \vec{D} = (CD\sin\alpha, \text{ direction given by the right-hand rule}) \tag{32.3}$$

The symbol $\times$ between the vectors is *required* to indicate a cross product.

NOTE ▶ The cross product of two vectors and the right-hand rule used to determine the direction of the cross product were introduced in Section 13.9 to describe torque and angular momentum. If you omitted that section, you will want to turn to it now to read about the cross product. A review would be worthwhile even if you did learn about the cross product earlier. Study the examples in that section carefully if cross products are new to you. ◀

The Biot-Savart law, Equation 32.1, can be written in terms of the cross product as

$$\vec{B} = \frac{\mu_0}{4\pi} \frac{q\vec{v} \times \hat{r}}{r^2} \quad \text{(magnetic field of a moving point charge)} \tag{32.4}$$

where unit vector $\hat{r}$, shown in Figure 32.10, points from charge q to the point at which we want to evaluate the field. This expression for $\vec{B}$ has magnitude $(\mu_0/4\pi)(qv\sin\theta/r^2)$ (because the magnitude of unit vector $\hat{r}$ is 1) and points in the correct direction (given by the right-hand rule), so it agrees completely with Equation 32.1.

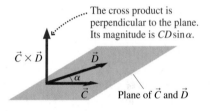

FIGURE 32.9 The cross product $\vec{C} \times \vec{D}$ is a vector perpendicular to the plane of vectors $\vec{C}$ and $\vec{D}$.

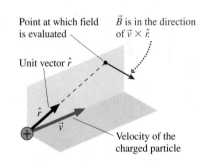

FIGURE 32.10 Unit vector $\hat{r}$ defines the direction from the moving charge to the point at which we want to evaluate the magnetic field.

EXAMPLE 32.2 **The magnetic field direction of an electron**

The electron in Figure 32.11 is moving to the right. What is the direction of the electron's magnetic field at the position indicated with a dot?

FIGURE 32.11 What is the direction of the electron's magnetic field at the dot?

VISUALIZE Because the charge is negative, the magnetic field points in the direction of $-(\vec{v} \times \hat{r})$, or opposite the direction of $\vec{v} \times \hat{r}$. Unit vector $\hat{r}$ points from the charge toward the dot. We can use the right-hand rule to find that $\vec{v} \times \hat{r}$ points *into* the page. Thus the electron's magnetic field at the dot points *out* of the page.

STOP TO THINK 32.3 The positive charge is moving straight out of the page. What is the direction of the magnetic field at the position of the dot?

$\vec{v}$ out of page

a. Up   b. Down   c. Left   d. Right

32.4 The Magnetic Field of a Current

13.1 Activ Physics ONLINE

Moving charges are the source of the magnetic field, but in practice we're more interested in the magnetic field of a current—a collection of moving charges—than in the very small magnetic fields of individual charges. The Biot-Savart law and the principle of superposition will be our primary tools for calculating magnetic fields. First, however, it will be useful to rewrite the Biot-Savart law in terms of current.

Figure 32.12a shows a current-carrying wire. The wire as a whole is electrically neutral, but current I represents the motion of positive charge carriers through the wire. Suppose the small amount of moving charge ΔQ spans the small length Δs. The charge has velocity $\vec{v} = \Delta\vec{s}/\Delta t$, where the vector $\Delta\vec{s}$, which is parallel to $\vec{v}$, is the charge's displacement vector. If ΔQ is small enough to treat as a point charge, the magnetic field it creates at a point in space is proportional to $(\Delta Q)\vec{v}$. We can write $(\Delta Q)\vec{v}$ in terms of the wire's current I as

$$(\Delta Q)\vec{v} = \Delta Q \frac{\Delta\vec{s}}{\Delta t} = \frac{\Delta Q}{\Delta t}\Delta\vec{s} = I\Delta\vec{s} \tag{32.5}$$

where we used the definition of current, $I = \Delta Q/\Delta t$.

(a)

Charge ΔQ in a small length Δs of a current-carrying wire

(b)

The magnetic field of the short segment of current is in the direction of $\Delta\vec{s} \times \hat{r}$.

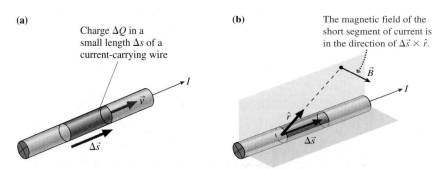

FIGURE 32.12 Relating the magnetic field of wire to the current I.

If we replace $q\vec{v}$ in the Biot-Savart law with $I\Delta\vec{s}$, we find that the magnetic field of a very short segment of wire carrying current I is

$$\vec{B} = \frac{\mu_0}{4\pi}\frac{I\Delta\vec{s} \times \hat{r}}{r^2} \tag{32.6}$$

(magnetic field of a very short segment of current)

Equation 32.6 is still the Biot-Savart law, only now written in terms of current rather than the motion of an individual charge. Figure 32.12b shows the direction of the current segment's magnetic field as determined by using the right-hand rule.

Equation 32.6 is the basis of a strategy for calculating the magnetic field of a current-carrying wire. You will recognize that it is the same basic strategy you learned for calculating the electric field of a continuous distribution of charge. The goal is to break a problem down into small steps that are individually manageable.

(MP) PROBLEM-SOLVING STRATEGY 32.1 The magnetic field of a current

MODEL Model the wire as a simple shape, such as a straight line or a loop.

VISUALIZE For the pictorial representation:

❶ Draw a picture and establish a coordinate system.

❷ Identify the point P at which you want to calculate the magnetic field.

❸ Divide the current-carrying wire into segments for which you *already know* how to determine $\vec{B}$. This is usually, though not always, a division into very short segments of length Δs.

❹ Draw the magnetic field vector for one or two segments. This will help you identify distances and angles that need to be calculated.

❺ Look for symmetries that simplify the field. You may conclude that some components of $\vec{B}$ are zero.

SOLVE The mathematical representation is $\vec{B}_{net} = \sum \vec{B}_k$.

- Use superposition to form an algebraic expression for *each* of the three components of $\vec{B}$ (unless you are sure one or more is zero) at point P.
- Let the (x, y, z) coordinates of the point remain as variables.
- Express all angles and distances in terms of the coordinates.
- Let $\Delta s \to ds$ and the sum become an integral. Think carefully about the integration limits for this variable; they will depend on the boundaries of the wire and on the coordinate system you have chosen to use. Carry out the integration and simplify the results as much as possible.

ASSESS Check that your result is consistent with any limits for which you know what the field should be.

EXAMPLE 32.3 The magnetic field of a long, straight wire
A long, straight wire carries current I in the positive x-direction. Find the magnetic field at a point that is distance d from the wire.

MODEL Because the wire is "long," let's model it as being infinitely long.

VISUALIZE Figure 32.13 illustrates the steps in the problem-solving strategy. We've chosen a coordinate system with point P on the y-axis. We've then divided the rod into small segments, each containing a small amount ΔQ of *moving charge*. Unit

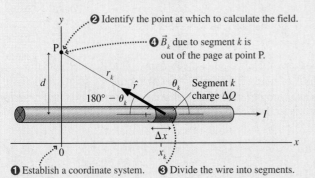

❷ Identify the point at which to calculate the field.

❹ $\vec{B}_k$ due to segment k is out of the page at point P.

Segment k charge ΔQ

❶ Establish a coordinate system. ❸ Divide the wire into segments.

FIGURE 32.13 Calculating the magnetic field of a long, straight wire carrying current I.

vector $\hat{r}$ and angle θ_k are shown for segment k. You should use the right-hand rule to convince yourself that $\vec{B}_k$ points *out of the page*, in the positive z-direction. This is the direction no matter where segment k happens to be along the x-axis. Consequently, B_x (the component of $\vec{B}$ parallel to the wire) and B_y (the component of $\vec{B}$ straight away from the wire) are zero. The only component of $\vec{B}$ we need to evaluate is B_z, the component tangent to a circle around the wire.

SOLVE We can use the Biot-Savart law to find the field $(B_k)_z$ of segment k. The cross product $\Delta \vec{s}_k \times \hat{r}$ has magnitude $(\Delta x)(1) \sin\theta_k$, hence

$$(B_k)_z = \frac{\mu_0}{4\pi} \frac{I\Delta x \sin\theta_k}{r_k^2} = \frac{\mu_0}{4\pi} \frac{I\sin\theta_k}{r_k^2}\Delta x = \frac{\mu_0}{4\pi} \frac{I\sin\theta_k}{x_k^2 + d^2}\Delta x$$

where we wrote the distance r_k in terms of x_k and d. We also need to express θ_k in terms of x_k and d. Because $\sin(180° - \theta) = \sin\theta$, this is

$$\sin\theta_k = \sin(180° - \theta_k) = \frac{d}{r_k} = \frac{d}{\sqrt{x_k^2 + d^2}}$$

With this expression for $\sin\theta_k$, the magnetic field of segment k is

$$(B_k)_z = \frac{\mu_0}{4\pi} \frac{Id}{(x_k^2 + d^2)^{3/2}}\Delta x$$

Now we're ready to sum the magnetic field of all the segments. The superposition is a vector sum, but in this case only the z-components are nonzero. Thus

$$B_{\text{wire}} = \sum_k (B_k)_z$$

$$= \frac{\mu_0 Id}{4\pi} \sum_k \frac{\Delta x}{(x_k^2 + d^2)^{3/2}} \rightarrow \frac{\mu_0 Id}{4\pi} \int_{-\infty}^{\infty} \frac{dx}{(x^2 + d^2)^{3/2}}$$

Only at the very last step did we convert the sum to an integral. Then our model of the wire as being infinitely long sets the integration limits at $\pm\infty$. This is a standard integral that can be found in integral tables. Evaluation gives

$$B_{\text{wire}} = \frac{\mu_0 Id}{4\pi} \frac{x}{d^2(x^2 + d^2)^{1/2}} \Big|_{-\infty}^{\infty} = \frac{\mu_0}{2\pi} \frac{I}{d}$$

This is the magnitude of the field. The field direction is determined by using the right-hand rule. We can combine these two pieces of information to write

$$\vec{B}_{\text{wire}} = \left(\frac{\mu_0}{2\pi} \frac{I}{d}, \begin{array}{l} \text{tangent to a circle around the wire} \\ \text{in the right-hand direction} \end{array} \right)$$

ASSESS Figure 32.14 shows the magnetic field of a current-carrying wire. Compare this to Figure 32.2 and convince yourself that the direction shown is that of the right-hand rule.

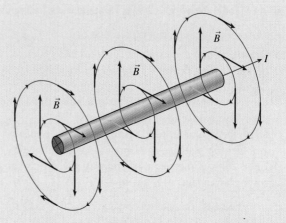

FIGURE 32.14 The magnetic field of a long, straight wire carrying current I.

NOTE ▶ The difficulty magnetic field calculations present is not doing the integration itself, which is the last step, but setting up the calculation and knowing *what* to integrate. The purpose of the problem-solving strategy is to guide you through the process of setting up the integral. ◀

EXAMPLE 32.4 **The magnetic field strength near a heater wire**
A 1.0-m-long, 1.0-mm-diameter nichrome heater wire is connected to a 12 V battery. What is the magnetic field strength 1.0 cm away from the wire?

MODEL 1 cm is much less than the 1 m length of the wire, so model the wire as infinitely long.

SOLVE The current through the wire is $I = \Delta V_{\text{bat}}/R$, where the wire's resistance R is

$$R = \frac{\rho L}{A} = \frac{\rho L}{\pi r^2} = 1.91 \, \Omega$$

The nichrome resistivity $\rho = 1.50 \times 10^{-6} \, \Omega\text{m}$ was taken from Table 28.2. Thus the current is $I = (12 \text{ V})/(1.91 \, \Omega) = 6.28 \text{ A}$. The magnetic field strength at distance $d = 1.0 \text{ cm} = 0.010 \text{ m}$ from the wire is

$$B_{\text{wire}} = \frac{\mu_0}{2\pi} \frac{I}{d} = (2.0 \times 10^{-7} \text{ Tm/A}) \frac{6.28 \text{ A}}{0.010 \text{ m}}$$

$$= 1.26 \times 10^{-4} \text{ T}$$

ASSESS The magnetic field of the wire is slightly more than twice the strength of the earth's magnetic field.

13.2 Activ ONLINE Physics

Motors, loudspeakers, metal detectors, and many other devices generate magnetic fields with *coils* of wire. The simplest possible coil is a single-turn circular loop of wire. A circular loop of wire with a circulating current is called a **current loop.**

EXAMPLE 32.5 **The magnetic field of a current loop**
Figure 32.15a shows a current loop, a circular loop of wire with radius R that carries current I. Find the magnetic field of the current loop at distance z on the axis of the loop.

MODEL Real coils need wires to bring the current in and out, but we'll model the coil as a current moving around the full circle shown in Figure 32.15b.

(a) A practical current loop   **(b) An ideal current loop**

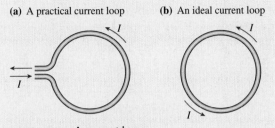

FIGURE 32.15 A current loop.

VISUALIZE Figure 32.16 shows a loop for which we've assumed that the current is circulating ccw. We've chosen a coordinate system in which the loop lies at $z = 0$ in the xy-plane. Let segment k be the segment at the top of the loop. Vector $\Delta\vec{s}_k$ is parallel to the x-axis and unit vector $\hat{r}$ is in the yz-plane, thus angle θ_k, the angle between $\Delta\vec{s}_k$ and $\hat{r}$, is 90°.

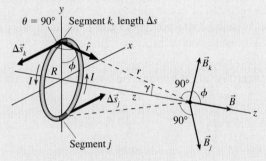

FIGURE 32.16 Calculating the magnetic field of a current loop.

The direction of $\vec{B}_k$, the magnetic field due to the current in segment k, is given by the cross product $\Delta\vec{s}_k \times \hat{r}$. $\vec{B}_k$ must be perpendicular to $\Delta\vec{s}_k$ *and* perpendicular to $\hat{r}$. You should convince yourself that $\vec{B}_k$ in Figure 32.16 points in the correct

direction. Notice that the xy-component of $\vec{B}_k$ is canceled by the xy-component of magnetic field $\vec{B}_j$ due to the current segment at the bottom of the loop, 180° away. In fact, *every* current segment on the loop can be paired with a segment 180° away, on the opposite side of the loop, such that the xy-components of $\vec{B}$ cancel and the components of $\vec{B}$ parallel to the z-axis add. The symmetry of the loop requires the on-axis magnetic field to point along the z-axis. Knowing that we need to sum only the z-components will simplify our calculation.

SOLVE We can use the Biot-Savart law to find the z-component $(B_k)_z = B_k\cos\phi$ of the magnetic field of segment k. The cross product $\Delta\vec{s}_k \times \hat{r}$ has magnitude $(\Delta s)(1)\sin 90° = \Delta s$, thus

$$(B_k)_z = \frac{\mu_0}{4\pi}\frac{I\Delta s}{r^2}\cos\phi = \frac{\mu_0 I\cos\phi}{4\pi(z^2 + R^2)}\Delta s$$

where we wrote distance r in terms of z and R. You can see, because $\phi + \gamma = 90°$, that angle ϕ is also the angle between $\hat{r}$ and the radius of the loop. Hence $\cos\phi = R/r$, and $(B_k)_z$ is

$$(B_k)_z = \frac{\mu_0 IR}{4\pi(z^2 + R^2)^{3/2}}\Delta s$$

The final step is to sum the magnetic fields due to all the segments:

$$B_{\text{loop}} = \sum_k (B_k)_z = \frac{\mu_0 IR}{4\pi(z^2 + R^2)^{3/2}}\sum_k \Delta s$$

In this case, unlike the straight wire, none of the terms multiplying Δs depends on the position of segment k, so all these terms can be factored out of the summation. We're left with a summation that adds up the lengths of all the small segments. But this is just the total length of the wire, which is the circumference $2\pi R$. Thus the on-axis magnetic field of a current loop is

$$B_{\text{loop}} = \frac{\mu_0 IR}{4\pi(z^2 + R^2)^{3/2}}2\pi R = \frac{\mu_0}{2}\frac{IR^2}{(z^2 + R^2)^{3/2}}$$

In practice, a coil often has N *turns* of wire. If the turns are all very close together, so that the magnetic field of each is essentially the same, then the magnetic field of a coil is N times the magnetic field of a current loop. The magnetic field at the center ($z = 0$) of an N-turn coil is

$$B_{\text{coil center}} = \frac{\mu_0}{2}\frac{NI}{R} \tag{32.7}$$

EXAMPLE 32.6 Matching the earth's magnetic field
What current is needed in a 5-turn, 10-cm-diameter coil to cancel the earth's magnetic field at the center of the coil?

MODEL Scientists sometimes need to carry out measurements in zero magnetic field. One way to create a field-free region of space is to generate a magnetic field equal to the earth's field but pointing in the opposite direction. The vector sum of the two fields is zero.

VISUALIZE Figure 32.17 shows a five-turn coil of wire. The magnetic field is five times that of a single current loop.

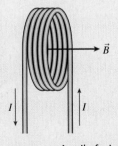

FIGURE 32.17 A coil of wire.

SOLVE The earth's magnetic field, from Table 32.1, is 5×10^{-5} T. We can use Equation 32.7 to find that the current needed to generate a 5×10^{-5} T field is

ASSESS A 0.80 A current is easily produced. Although there are better ways to cancel the earth's field than using a simple current loop, this illustrates the idea.

$$I = \frac{2RB}{\mu_0 N} = \frac{2(0.050 \text{ m})(5.0 \times 10^{-5} \text{ T})}{5(4\pi \times 10^{-7} \text{ Tm/A})} = 0.80 \text{ A}$$

32.5 Magnetic Dipoles

We were able to calculate the on-axis magnetic field of a current loop, but determining the field at other points requires either numerical integrations or an experimental mapping of the field. Figure 32.18 shows the full magnetic field of a current loop. This is a field with *rotational symmetry,* so to picture the full three-dimensional field, imagine Figure 32.18a rotated about the axis of the loop. Figure 32.18b shows the magnetic field in the plane of the loop as seen from the right. There is a clear sense that the magnetic field leaves the loop on one side, "flows" around the outside, then returns to the loop.

(a) Cross section through the current loop

(b) The current loop seen from the right

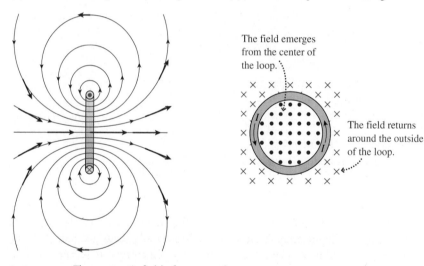

The field emerges from the center of the loop.

The field returns around the outside of the loop.

FIGURE 32.18 The magnetic field of a current loop.

There are two versions of the right-hand rule that you can use to determine which way a loop's field points. Try these in Figure 32.18. Being able to quickly ascertain the field direction of a current loop is an important skill.

TACTICS BOX 32.1 Finding the magnetic field direction of a current loop

Use either of the following methods to find the magnetic field direction:

❶ Point your right thumb in the direction of the current at any point on the loop and let your fingers curl through the center of the loop. Your fingers are then pointing in the direction in which $\vec{B}$ leaves the loop.

❷ Curl the fingers of your right hand around the loop in the direction of the current. Your thumb is then pointing in the direction in which $\vec{B}$ leaves the loop.

A Current Loop Is a Magnetic Dipole

A current loop has two distinct sides. Bar magnets and flat refrigerator magnets also have two distinct sides or ends, so you might wonder if current loops are related to these permanent magnets. Consider the following experiments with a current loop. Notice that we're using a simplified picture that shows the magnetic field only in the plane of the loop.

Investigating current loops

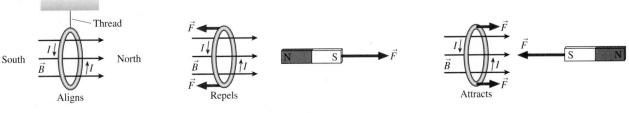

A current loop hung by a thread aligns itself along a north-south axis with the magnetic field pointing north.

The north pole of a permanent magnet repels the side of a current loop from which the magnetic field is emerging.

The south pole of a permanent magnet attracts the side of a current loop from which the magnetic field is emerging.

These investigations show that **a current loop is a magnet,** just like a permanent magnet. A magnet created by a current in a coil of wire is called an **electromagnet.** An electromagnet picks up small pieces of iron, influences a compass needle, and acts in every way like a permanent magnet.

In fact, Figure 32.19 shows that **a flat permanent magnet and a current loop generate the same magnetic field.** It is the field of a magnetic dipole, irrespective of how the dipole was produced. For both, you can identify the north pole as the face or end *from which* the magnetic field emerges. The magnetic field of both point *into* the south pole.

> **NOTE** ▶ The magnetic field *inside* a permanent magnet differs from the magnetic field at the center of a current loop. Only the exterior field of a magnet matches the field of a current loop. ◀

One of the goals of this chapter is to show that magnetic forces exerted by currents and magnetic forces exerted by permanent magnets are just two different aspects of a single magnetism. We've now found a strong connection between permanent magnets and current loops, and this connection will turn out to be a big piece of the puzzle.

The Magnetic Dipole Moment

The expression for the electric field of an electric dipole was considerably simplified when we considered the field at distances significantly larger than the size of the charge separation s. The on-axis field of an electric dipole when $z \gg s$ is

$$\vec{E}_{dipole} = \frac{1}{4\pi\epsilon_0} \frac{2\vec{p}}{z^3}$$

where the electric dipole moment $\vec{p}$ is the vector $\vec{p} = (qs,$ from the negative to the positive charge).

The on-axis magnetic field of a current loop, which we calculated in Example 32.5, is

$$B_{loop} = \frac{\mu_0}{2} \frac{IR^2}{(z^2 + R^2)^{3/2}}$$

(a) Current loop

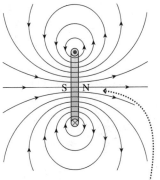

Whether it's a current loop or a permanent magnet, the magnetic field emerges from the north pole.

(b) Permanent magnet

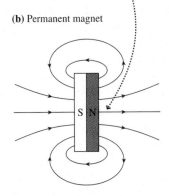

FIGURE 32.19 A current loop has magnetic poles and generates the same magnetic field as a flat permanent magnet.

If z is much larger than the diameter of the current loop, $z \gg R$, we can make the approximation $(z^2 + R^2)^{3/2} \rightarrow z^3$. Then the loop's field is

$$B_{\text{loop}} \approx \frac{\mu_0}{2} \frac{IR^2}{z^3} = \frac{\mu_0}{4\pi} \frac{2(\pi R^2)I}{z^3} = \frac{\mu_0}{4\pi} \frac{2AI}{z^3} \qquad (32.8)$$

where $A = \pi R^2$ is the area of the loop.

A more advanced treatment of current loops shows that, if z is much larger than the size of the loop, Equation 32.8 is the on-axis magnetic field of a current loop of *any* shape, not just a circular loop. The shape of the loop affects the nearby field, but the distant field depends only on the current I and the area A enclosed within the loop. With this in mind, let's define the **magnetic dipole moment** $\vec{\mu}$ of a current loop enclosing area A to be

$$\vec{\mu} = (AI, \text{ from the south pole to the north pole})$$

The SI units of the magnetic dipole moment are $A\,m^2$.

> **NOTE** ▶ Don't confuse the magnetic dipole moment $\vec{\mu}$ with the constant μ_0 in the Biot-Savart law. ◀

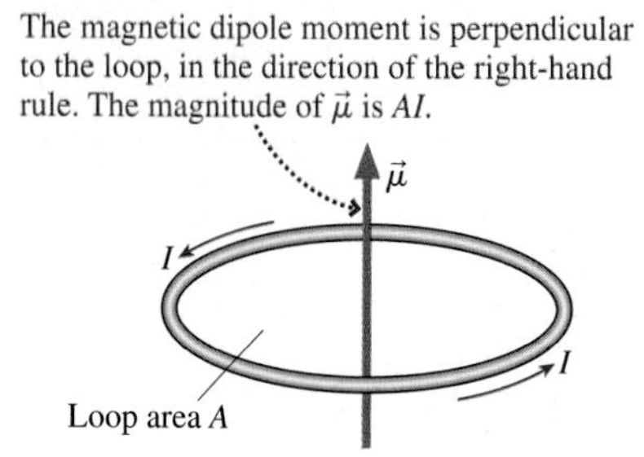

The magnetic dipole moment, like the electric dipole moment, is a vector. It has the same direction as the on-axis magnetic field. Thus the right-hand rule used for determining the direction of $\vec{B}$ also shows the direction of $\vec{\mu}$. Figure 32.20 shows the magnetic dipole moment of a circular current loop.

Because the on-axis magnetic field of a current loop points in the same direction as $\vec{\mu}$, we can combine Equation 32.8 and the definition of $\vec{\mu}$ to write the on-axis field of a magnetic dipole as

$$\vec{B}_{\text{dipole}} = \frac{\mu_0}{4\pi} \frac{2\vec{\mu}}{z^3} \qquad \text{(on the axis of a magnetic dipole)} \qquad (32.9)$$

If you compare $\vec{B}_{\text{dipole}}$ to $\vec{E}_{\text{dipole}}$, you can see that the magnetic field of a magnetic dipole has the same basic shape as the electric field of an electric dipole.

Because a permanent bar magnet is also a magnetic dipole, a permanent magnet also has a magnetic dipole moment. Its on-axis magnetic field is given by Equation 32.9 when z is much larger than the size of the magnet. Equation 32.9 and laboratory measurements of the on-axis magnetic field can be used to determine a permanent magnet's dipole moment.

The magnetic dipole moment is perpendicular to the loop, in the direction of the right-hand rule. The magnitude of $\vec{\mu}$ is AI.

FIGURE 32.20 The magnetic dipole moment of a circular current loop.

EXAMPLE 32.7 The field of a magnetic dipole

a. The on-axis magnetic field strength 10 cm from a magnetic dipole is 1.0×10^{-5} T. What is the size of the magnetic dipole moment?

b. If the magnetic dipole is created by a 4.0-mm-diameter current loop, what is the current?

MODEL Assume that the distance 10 cm is much larger than the size of the dipole.

SOLVE

a. If $z \gg R$, we can use Equation 32.9 to find the magnetic dipole moment:

$$\mu = \frac{4\pi}{\mu_0} \frac{z^3 B}{2}$$

$$= \frac{4\pi}{4\pi \times 10^{-7}\,\text{Tm/A}} \frac{(0.10\,\text{m})^3 (1.0 \times 10^{-5}\,\text{T})}{2}$$

$$= 0.050\,A\,m^2$$

b. The magnetic dipole moment of a current loop is $\mu = AI$, so the necessary current is

$$I = \frac{\mu}{\pi R^2}$$

$$= \frac{(0.050\,A\,m^2)}{\pi (0.0020\,\text{m})^2} = 4000\,\text{A}$$

ASSESS Only a superconducting ring could carry a 4000 A current, so producing this magnetic field with a current loop is not very feasible. But $0.050\,A\,m^2$ is a quite modest dipole moment for a bar magnet, so this field could be produced with a permanent magnet.

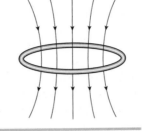

STOP TO THINK 32.4 What is the current direction in this loop? And which side of the loop is the north pole?

   a. Current cw; north pole on top
   b. Current cw; north pole on bottom
   c. Current ccw; north pole on top
   d. Current ccw; north pole on bottom

32.6 Ampère's Law and Solenoids

In principle, the Biot-Savart law can be used to calculate the magnetic field of any current distribution. In practice, the integrals are very difficult to evaluate for anything other than very simple situations. We faced a similar situation for calculating electric fields, but we discovered an alternative method—Gauss's law—for calculating the electric field of charge distributions with a high degree of symmetry. Gauss's law doesn't work in every situation, but it is simple and elegant where it does.

Likewise, there's an alternative method, called *Ampère's law,* for calculating the magnetic fields of current distributions with a high degree of symmetry. Ampère's law, like Gauss's law, doesn't work in all situations, but it is simple and elegant where it does. Whereas Gauss's law is written in terms of a surface integral, Ampère's law is based on the mathematical procedure called a *line integral.*

Line Integrals

We've flirted with the idea of a line integral ever since introducing the concept of work in Chapter 11, but now we need to take a more serious look at what a line integral represents and how it is used. Figure 32.21a shows a curved line that goes from an initial point i to a final point f.

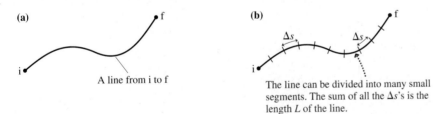

(a)

A line from i to f

(b)

The line can be divided into many small segments. The sum of all the Δs's is the length L of the line.

FIGURE 32.21 Integrating along a line from i to f.

Suppose, as shown in Figure 32.21b, we divide the line up into many small segments of length Δs. The first segment is Δs_1, the second is Δs_2, and so on. The sum of all the Δs's is just the length L of the line between i and f. We can write this mathematically as

$$L = \sum_k \Delta s_k \rightarrow \int_i^f ds \qquad (32.10)$$

where, in the last step, we let $\Delta s \rightarrow ds$ and the sum become an integral.

This integral is called a **line integral.** All we've done is to subdivide a line into infinitely many infinitesimal pieces, then add them up. This is exactly what you do in calculus when you evaluate an integral such as $\int x\,dx$. In fact, an integration along the *x*-axis *is* a line integral, one that happens to be along a straight line. Figure 32.21 differs only in that the line is curved. The underlying idea in both cases is that an integral is just a fancy way of doing a sum.

(a)

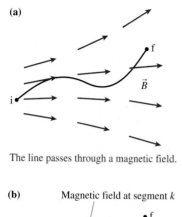

The line passes through a magnetic field.

(b)

Magnetic field at segment k

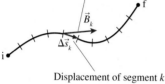

Displacement of segment k

FIGURE 32.22 Integrating $\vec{B}$ along a line from i to f.

The line integral of Equation 32.10 is not a terribly exciting one. Figure 32.22a makes things more interesting by allowing the line to pass through a magnetic field. Figure 32.22b again divides the line into small segments, but this time $\Delta \vec{s}_k$ is the displacement vector of segment k. The magnetic field at this point in space is $\vec{B}_k$.

Suppose we were to evaluate the dot product $\vec{B}_k \cdot \Delta \vec{s}_k$ at each segment, then add the values of $\vec{B}_k \cdot \Delta \vec{s}_k$ due to every segment. Doing so, and again letting the sum become an integral, we have

$$\sum_k \vec{B}_k \cdot \Delta \vec{s}_k \rightarrow \int_i^f \vec{B} \cdot d\vec{s} = \text{the line integral of } \vec{B} \text{ from i to f}$$

Once again, the integral is just a shorthand way to say "Divide the line into lots of little pieces, evaluate $\vec{B}_k \cdot \Delta \vec{s}_k$ for each piece, then add them up."

Although this process of evaluating the integral could be difficult, the only line integrals we'll need to deal with fall into two simple cases. If the magnetic field is *everywhere perpendicular* to the line, then $\vec{B} \cdot d\vec{s} = 0$ at every point along the line and the integral is zero. If the magnetic field is *everywhere tangent* to the line *and* has the same magnitude B at every point, then $\vec{B} \cdot d\vec{s} = B\,ds$ at every point and

$$\int_i^f \vec{B} \cdot d\vec{s} = \int_i^f B\,ds = B\int_i^f ds = BL \tag{32.11}$$

We used Equation 32.10 in the last step to integrate ds along the line.

Tactics Box 32.2 summarizes these two situations.

TACTICS BOX 32.2 Evaluating line integrals

❶ If $\vec{B}$ is everywhere perpendicular to a line, the line integral of $\vec{B}$ is

$$\int_i^f \vec{B} \cdot d\vec{s} = 0$$

❷ If $\vec{B}$ is everywhere tangent to a line of length L *and* has the same magnitude B at every point, the line integral of $\vec{B}$ is

$$\int_i^f \vec{B} \cdot d\vec{s} = BL$$

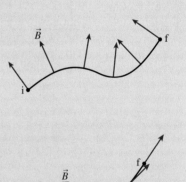

Ampère's Law

The French scientist André-Marie Ampère, for whom the SI unit of current is named, studied the properties of magnetism in the 1820s. Ampère noted that the magnetic field of a current-carrying wire is everywhere tangent to a circle around the wire *and* has the same magnitude $\mu_0 I/2\pi d$ at all points on the circle. According to Tactics Box 32.2, these conditions allow us to easily evaluate the line integral of $\vec{B}$ along a circular path around the wire.

Figure 32.23 shows a wire carrying current I into the page and the magnetic field at distance d. Suppose we were to integrate the magnetic field *all the way around* the circle. That is, the initial point i of the integration path and the final point f will be the same point. This would be a line integral around a *closed curve*, which is denoted

$$\oint \vec{B} \cdot d\vec{s}$$

The little circle on the integral sign indicates that the integration is performed around a closed curve. The notation has changed, but the meaning has not.

Because $\vec{B}$ is tangent to the circle *and* of constant magnitude at every point on the circle, we can use Option 2 from Tactics Box 32.2 to write

$$\oint \vec{B} \cdot d\vec{s} = BL = B(2\pi d) \qquad (32.12)$$

where, in this case, the path length L is the circumference $2\pi d$ of the circle. We know that the magnetic field strength is $B = \mu_0 I / 2\pi d$, thus

$$\oint \vec{B} \cdot d\vec{s} = \mu_0 I \qquad (32.13)$$

The interesting result is that the line integral of $\vec{B}$ around the current-carrying wire is independent of the radius of the circle. Any circle, from one touching the wire to one far away, would give the same result. The integral depends only on the amount of current passing *through* the circle that we integrated around.

This is reminiscent of Gauss's law. In our investigation of Gauss's law, we started with the observation that electric flux Φ_e through a sphere surrounding a point charge depends only on the amount of charge inside, not on the radius of the sphere. After examining several cases, we concluded that the shape of the surface wasn't relevant. The electric flux through *any* closed surface enclosing total charge Q_{in} turned out to be $\Phi_e = Q_{in}/\epsilon_0$.

Although we'll skip the details, the same type of reasoning that we used to prove Gauss's law shows that the result of Equation 32.13

- Is independent of the shape of the curve around the current.
- Is independent of where the current passes through the curve.
- Depends only on the total amount of current through the area enclosed by the integration path.

Thus whenever total current $I_{through}$ passes through an area bounded by a *closed curve*, the line integral of the magnetic field around the curve is

$$\oint \vec{B} \cdot d\vec{s} = \mu_0 I_{through} \qquad (32.14)$$

This result for the magnetic field is known as **Ampère's law.**

To make practical use of Ampère's law, we need to determine which currents are positive and which are negative. The right-hand rule is once again the proper tool. If you curl your right fingers around the closed path in the direction in which you are going to integrate, then any current passing though the bounded area in the direction of your thumb is a positive current. Any current in the opposite direction is a negative current. In Figure 32.24, for example, currents I_1 and I_2 are positive, I_3 is negative. Thus $I_{through} = I_1 + I_2 - I_3$.

> **NOTE** ▶ The integration path of Ampère's law is a mathematical curve through space. It does not have to match a physical surface or boundary, although it could if we want it to. ◀

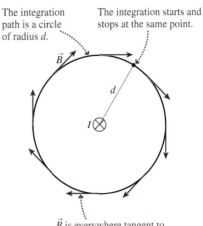

The integration path is a circle of radius d.

The integration starts and stops at the same point.

$\vec{B}$ is everywhere tangent to the integration path.

FIGURE 32.23 Integrating the magnetic field around a wire.

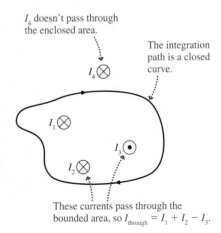

I_4 doesn't pass through the enclosed area.

The integration path is a closed curve.

These currents pass through the bounded area, so $I_{through} = I_1 + I_2 - I_3$.

FIGURE 32.24 Using Ampère's law.

In one sense, Ampère's law doesn't tell us anything new. After all, we derived Ampère's law from the Biot-Savart law for the magnetic field of a current. But in another sense, Ampère's law is more important than the Biot-Savart law because it states a very general property about magnetic fields. Ampère's law will turn out to be especially useful in Chapter 34 when we combine it with other electric and magnetic equations to form Maxwell's equations of the electromagnetic field. In the meantime, Ampère's law will allow us to find the magnetic fields of some important current distributions that have a high degree of symmetry.

EXAMPLE 32.8 **The magnetic field inside a current-carrying wire**

A wire of radius R carries current I. Find the magnetic field inside the wire at distance $r < R$ from the axis.

MODEL Assume the current density is uniform over the cross section of the wire.

VISUALIZE Figure 32.25 shows a cross section through the wire. The wire has perfect cylindrical symmetry, with all the charges moving parallel to the wire, so the magnetic field *must* be tangent to circles that are concentric with the wire. We don't know how the strength of the magnetic field depends on the distance from the center of the wire—that's what we're going to find—but the symmetry of the situation dictates the *shape* of the magnetic field.

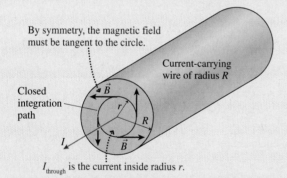

By symmetry, the magnetic field must be tangent to the circle.

Current-carrying wire of radius R

Closed integration path

$\vec{B}$

r

R

I

$\vec{B}$

I_{through} is the current inside radius r.

FIGURE 32.25 Using Ampère's law inside a current-carrying wire.

SOLVE To find the field strength at radius r, draw a circle of radius r. The amount of current passing through this circle is

$$I_{\text{through}} = JA_{\text{circle}} = \pi r^2 J$$

where J is the current density. Our assumption of a uniform current density allows us to use the full current I passing though a wire of radius R to find that

$$J = \frac{I}{A} = \frac{I}{\pi R^2}$$

Thus the current through the circle of radius r is related to the total current I by

$$I_{\text{through}} = \frac{r^2}{R^2} I$$

Let's integrate $\vec{B}$ around the circumference of this circle. According to Ampère's law,

$$\oint \vec{B} \cdot d\vec{s} = \mu_0 I_{\text{through}} = \frac{\mu_0 r^2}{R^2} I$$

We know from the symmetry of the wire that $\vec{B}$ is everywhere tangent to the circle *and* has the same magnitude at all points on the circle. Consequently, the line integral of $\vec{B}$ around the circle can be evaluated using Option 2 of Tactics Box 32.2:

$$\oint \vec{B} \cdot d\vec{s} = BL = 2\pi r B$$

where $L = 2\pi r$ is the path length. If we substitute this expression into Ampère's law, we find that

$$2\pi r B = \frac{\mu_0 r^2}{R^2} I$$

Solving for B, we find that the magnetic field strength at radius r *inside* a current-carrying wire is

$$B = \frac{\mu_0 I}{2\pi R^2} r$$

ASSESS The magnetic field strength increases linearly with distance from the center of the wire until, at the surface of the wire, $B = \mu_0 I/2\pi R$ matches our earlier solution for the magnetic field outside a current-carrying wire. This agreement at $r = R$ gives us confidence in our result. The magnetic field strength both inside and outside the wire is shown graphically in Figure 32.26.

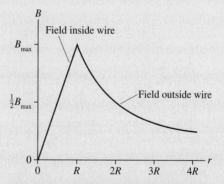

FIGURE 32.26 Graphical representation of the magnetic field inside and outside a current-carrying wire.

The Magnetic Field of a Solenoid

In our study of electricity, we made extensive use of the idea of a uniform electric field: a field that is the same at every point in space. We found that two closely spaced, parallel charged plates generate a uniform electric field between them, and this uniform field was one reason why we focused so much attention on learning about the parallel-plate capacitor.

Similarly, there are many applications of magnetism for which we would like to generate a **uniform magnetic field,** a field that has the same magnitude and the same direction at every point within some region of space. None of the sources we have looked at thus far produces a uniform magnetic field.

In practice, a uniform magnetic field is generated with a **solenoid.** A solenoid, shown in Figure 32.27, is a helical coil of wire with the same current I passing through each loop in the coil. Solenoids may have hundreds or thousands of coils, often called *turns,* sometimes wrapped in several layers.

FIGURE 32.27 A solenoid.

We can understand a solenoid by thinking of it as a stack of current loops. Figure 32.28a shows the magnetic field of a single current loop at three points on the axis and three points equally distant from the axis. The field directly above the loop is opposite in direction to the field inside the loop. Figure 32.28b then shows three parallel loops. We can use information from Figure 32.28a to draw the magnetic fields of each loop at the center of loop 2 and at a point above loop 2.

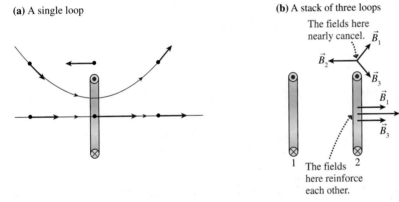

(a) A single loop

(b) A stack of three loops

The fields here nearly cancel. $\vec{B}_1$

$\vec{B}_2$

$\vec{B}_3$

$\vec{B}_1$

$\vec{B}_2$

$\vec{B}_3$

1   2   3

The fields here reinforce each other.

FIGURE 32.28 Using superposition to find the magnetic field of a stack of current loops.

The superposition of the fields at the center of loop 2 produces a *stronger* field than that of loop 2 alone. But the fields at the point above loop 2 tend to cancel, producing a net magnetic field that is either zero or very much weaker than the field at the center of the loop. We've used only three current loops to illustrate the idea, but these tendencies are reinforced by including more loops. With many current loops along the same axis, **the field in the center is strong and roughly parallel to the axis, whereas the field outside the loops is very weak.**

Figure 32.29 is a numerical calculation of the magnetic field of a 15-turn solenoid. You can see that the magnetic field inside the coils is nearly uniform (i.e., the field lines are nearly parallel and equally spaced) but the field outside is much weaker. Our goal of producing a uniform magnetic field can be achieved by increasing the number of coils until we have an *ideal solenoid* that is infinitely long and in which the coils are as close together as possible. **The magnetic field inside an ideal solenoid is uniform and parallel to the axis; the magnetic field outside is zero.** No real solenoid is infinitely long (just as no real capacitor is infinitely wide), but a very uniform magnetic field can be produced near the center of a tightly wound solenoid whose length is much larger than its diameter.

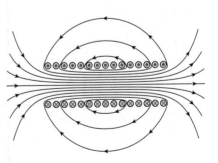

FIGURE 32.29 The magnetic field of a 15-turn solenoid.

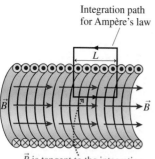

Integration path for Ampère's law

L

$\vec{B}$ $\vec{B}$

$\vec{B}$ is tangent to the integration path along the bottom edge.

FIGURE 32.30 A closed path inside and outside an ideal solenoid.

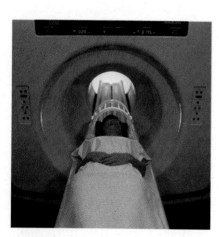

This patient is undergoing magnetic resonance imaging. The large cylinder surrounding the patient contains a solenoid to generate a uniform magnetic field.

We can use Ampère's law to calculate the field of an ideal solenoid if we choose the integration path carefully. Figure 32.30 shows a cross section through an infinitely long solenoid. The integration path that we'll use is a rectangle of width L, enclosing N turns of the solenoid coil. Because this is a mathematical curve, not a physical boundary, there's no difficulty with letting it protrude through the wall of the solenoid wherever we wish. The solenoid's magnetic field direction, given by the right-hand rule, is left to right, so we'll integrate around this path in the ccw direction.

Each of the N wires enclosed by the integration path carries current I, so the total current passing through the rectangle is $I_{\text{through}} = NI$. Ampère's law is thus

$$\oint \vec{B} \cdot d\vec{s} = \mu_0 I_{\text{through}} = \mu_0 NI \qquad (32.15)$$

The line integral around this path is the sum of the line integrals along each side. Along the bottom, where $\vec{B}$ is parallel to $d\vec{s}$ and of constant value B, the integral is simply BL. The integral along the top is zero because the magnetic field outside an ideal solenoid is zero.

The left and right sides sample the magnetic field both inside and outside the solenoid. The magnetic field outside is zero, but the magnetic field inside is not. However, the interior magnetic field is everywhere *perpendicular* to the line of integration. Consequently, as we recognized in Option 1 of Tactics Box 32.2, the line integral is zero.

Only the integral along the bottom path is nonzero, leading to

$$\oint \vec{B} \cdot d\vec{s} = BL = \mu_0 NI$$

Thus the strength of the uniform magnetic field inside a solenoid is

$$B_{\text{solenoid}} = \frac{\mu_0 NI}{L} = \mu_0 nI \qquad (32.16)$$

where $n = N/L$ is the number of turns per unit length.

Objects inserted into the center of a solenoid are in a uniform magnetic field. Measurements that need a uniform magnetic field are often conducted inside a solenoid, which can be built quite large. The cylinder that surrounds a patient undergoing magnetic resonance imaging (MRI) contains a large solenoid made of superconducting wire, allowing it to carry the very large currents needed to generate a strong uniform magnetic field.

EXAMPLE 32.9 Generating a uniform magnetic field
We wish to generate a 0.10 T uniform magnetic field near the center of a 10-cm-long solenoid. How many turns are needed if the wire can carry a maximum current of 10 A?

MODEL Assume that the solenoid is an ideal solenoid.

SOLVE Generating a magnetic field with a solenoid is a trade-off between current and turns of wire. A larger current requires fewer turns. However, wires have resistance, and too large a current can overheat the solenoid. Maximum safe currents are established on the basis of the wire's cross-section area. For a wire that

can carry 10 A, we can use Equation 32.16 to find the required number of turns:

$$N = \frac{LB}{\mu_0 I} = \frac{(0.10\ \text{m})(0.10\ \text{T})}{(4\pi \times 10^{-7}\ \text{T m/A})(10\ \text{A})} = 800\ \text{turns}$$

ASSESS A wire that can carry 10 A without overheating is about 1 mm in diameter, so only 100 turns can be placed in a 10 cm length. Thus it takes eight layers to reach the required number of turns.

For a real solenoid, of finite length, the magnetic field is approximately uniform *inside* the solenoid and weak, but not zero, outside. As Figure 32.31 shows, the magnetic field outside the solenoid looks like that of a bar magnet. Thus a

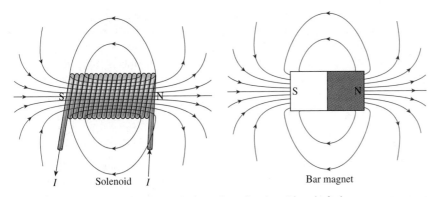

FIGURE 32.31 The magnetic fields of a finite-length solenoid and of a bar magnet.

solenoid is an electromagnet, with one end of the solenoid being a north magnetic pole and the other a south magnetic pole. You can use the right-hand rule to identify the north-pole end. A solenoid with many turns and a large current can be a very powerful magnet.

32.7 The Magnetic Force on a Moving Charge

It's time to switch our attention from how magnetic fields are generated to how magnetic fields exert forces and torques. Oersted discovered that a current passing through a wire causes a magnetic torque to be exerted on a nearby compass needle. Upon hearing of Oersted's discovery, Ampère reasoned that the current was acting like a magnet and, if this were true, that two current-carrying wires should exert magnetic forces on each other.

To find out, Ampère set up two parallel wires that could carry large currents in either the same direction or in opposite (or "antiparallel") directions. Figure 32.32 shows the outcome of his experiment. Notice that, for currents, "likes" attract and "opposites" repel. This is the opposite of what would have happened had the wires been charged and thus exerting electric forces on each other. Ampère's experiment showed that **a magnetic field exerts a force on a current,** but before we can analyze Ampère's experiment we must first investigate the magnetic force on a moving charge.

Magnetic Force

A current consists of moving charges. Ampère's experiment implied that a magnetic field exerts a force on a *moving* charge. This is true, although the exact form of the force law was not discovered until later in the 19th century. The magnetic force turns out to depend not only on the charge and the charge's velocity, but also on how the velocity vector is oriented relative to the magnetic field. Figure 32.33 shows the outcome of three experiments to observe the magnetic force on a charged particle.

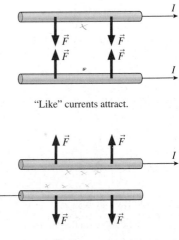

"Like" currents attract.

"Opposite" currents repel.

FIGURE 32.32 Ampère's experiment to observe the forces between parallel current-carrying wires.

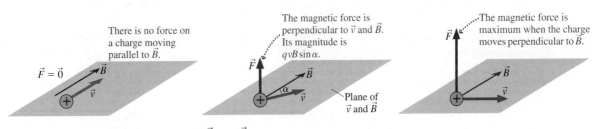

There is no force on a charge moving parallel to $\vec{B}$.

The magnetic force is perpendicular to $\vec{v}$ and $\vec{B}$. Its magnitude is $qvB\sin\alpha$.

Plane of $\vec{v}$ and $\vec{B}$

The magnetic force is maximum when the charge moves perpendicular to $\vec{B}$.

FIGURE 32.33 The relationship between $\vec{v}$, $\vec{B}$, and $\vec{F}$.

If you compare the experiment on the right in Figure 32.33 to Figure 32.9, you'll see that the relationship between $\vec{v}$, $\vec{B}$, and $\vec{F}$ is exactly the same as the geometric relationship between $\vec{C}$, $\vec{D}$, and $\vec{C} \times \vec{D}$. The magnetic force on a charge q as it moves through a magnetic field $\vec{B}$ with velocity $\vec{v}$ depends on the cross product between $\vec{v}$ and $\vec{B}$. The magnetic force on a moving charged particle can be written

$$\vec{F}_{\text{on } q} = q\vec{v} \times \vec{B} = (qvB\sin\alpha, \text{ direction of right-hand rule}) \quad (32.17)$$

where α is the angle between $\vec{v}$ and $\vec{B}$.

13.4 **Activ Physics**

The right-hand rule is that of the cross product. Spread your right thumb and index finger apart by angle α, then bend your middle finger so that it is perpendicular to your thumb and index finger. Orient your hand so that your thumb points in the direction of $\vec{v}$ and your index finger in the direction of $\vec{B}$. Your middle finger is then pointing in the direction of the force $\vec{F}$. **The magnetic force on a moving charged particle is perpendicular to both $\vec{v}$ and $\vec{B}$.**

The magnetic force has several important properties:

1. Only a *moving* charge experiences a magnetic force. There is no magnetic force on a charge at rest ($v = 0$) in a magnetic field.
2. There is no force on a charge moving parallel ($\alpha = 0°$) or antiparallel ($\alpha = 180°$) to a magnetic field.
3. When there is a force, the force is perpendicular to *both* $\vec{v}$ and $\vec{B}$.
4. The force on a negative charge is in the direction *opposite* to $\vec{v} \times \vec{B}$.
5. For a charge moving perpendicular to $\vec{B}$ ($\alpha = 90°$), the magnitude of the magnetic force is $F = |q|vB$.

Figure 32.34 shows the relationship between $\vec{v}$, $\vec{B}$, and $\vec{F}$ for several moving charges. (The *source* of the magnetic field isn't shown, only the field itself.) You can see the inherent three-dimensionality of magnetism, with the force perpendicular to both $\vec{v}$ and $\vec{B}$. Thus the magnetic force is very different from the electric force, which is parallel to the electric field.

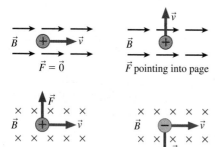

FIGURE 32.34 Magnetic forces on moving charges.

EXAMPLE 32.10 The magnetic force on an electron

A long wire carries a 10 A current from left to right. An electron 1.0 cm above the wire is traveling to the right at a speed of 1.0×10^7 m/s. What are the magnitude and the direction of the magnetic force on the electron?

MODEL The magnetic field is that of a long, straight wire.

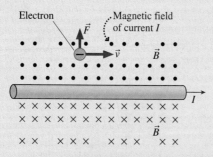

FIGURE 32.35 An electron moving parallel to a current-carrying wire.

VISUALIZE Figure 32.35 shows the current and an electron moving to the right. The right-hand rule tells us that the wire's magnetic field above the wire is out of the page, so the electron is moving perpendicular to the field.

SOLVE The electron charge is negative, thus the direction of the force is opposite the direction of $\vec{v} \times \vec{B}$. The right-hand rule shows that $\vec{v} \times \vec{B}$ points down, toward the wire, so $\vec{F}$ points up, away from the wire. The magnitude of the force is $|q|vB = evB$. The field is that of a long, straight wire,

$$B = \frac{\mu_0 I}{2\pi d} = 2.0 \times 10^{-4} \text{ T}$$

Thus the magnitude of the force on the electron is

$$F = evB = (1.60 \times 10^{-19} \text{ C})(1.0 \times 10^7 \text{ m/s})(2.0 \times 10^{-4} \text{ T})$$

$$= 3.2 \times 10^{-16} \text{ N}$$

The force on the electron is $\vec{F} = (3.2 \times 10^{-16} \text{ N, up})$.

ASSESS This force will cause the electron to curve away from the wire.

We can draw an interesting and important conclusion at this point. You have seen that the magnetic field is *created by* moving charges. Now you also see that magnetic forces are *exerted on* moving charges. Thus it appears that **magnetism** is an interaction between moving charges. Any two charges, whether moving or stationary, interact with each other through the electric field. In addition, two *moving* charges also interact with each other through the magnetic field. This fundamental observation is easy to lose sight of when we talk about currents, magnets, torques, and all the other phenomena of magnetism. But the most basic feature underlying of all these phenomena is an interaction between moving charges.

Cyclotron Motion

Many important applications of magnetism involve the motion of charged particles in a magnetic field. Your television picture tube functions by using magnetic fields to steer electrons as they move through a vacuum from the electron gun to the screen. Microwave generators, which are used in applications ranging from ovens to radar, use a device called a *magnetron* in which electrons oscillate rapidly in a magnetic field.

You've just seen that there is no force on a charge that has velocity $\vec{v}$ parallel or antiparallel to a magnetic field. Consequently, **a magnetic field has no effect on a charge moving parallel or antiparallel to the field.** To understand the motion of charged particles in magnetic fields, we need to consider only motion *perpendicular* to the field.

Figure 32.36 shows a positive charge q moving with a velocity $\vec{v}$ in a plane that is perpendicular to a *uniform* magnetic field $\vec{B}$. According to the right-hand rule, the magnetic force on this particle is *perpendicular* to the velocity $\vec{v}$. A force that is always perpendicular to $\vec{v}$ changes the *direction* of motion, by deflecting the particle sideways, but it cannot change the particle's speed. Thus **a particle moving perpendicular to a uniform magnetic field undergoes uniform circular motion at constant speed.** This motion is called the **cyclotron motion** of a charged particle in a magnetic field.

NOTE ▶ A negative charge will orbit in the opposite direction from that shown in Figure 32.36 for a positive charge. ◀

You've seen analogies to cyclotron motion in Parts I and II of this text. For a mass moving in a circle at the end of a string, the tension force is always perpendicular to $\vec{v}$. For a satellite moving in a circular orbit, the gravitational force is always perpendicular to $\vec{v}$. Now, for a charged particle moving in a magnetic field, it is the magnetic force of strength $F = qvB$ that points toward the center of the circle and causes the particle to have a centripetal acceleration.

Newton's second law for circular motion, which you learned in Chapter 7, is

$$F = qvB = ma_r = \frac{mv^2}{r} \qquad (32.18)$$

Thus the radius of the cyclotron orbit is

$$r_{\text{cyc}} = \frac{mv}{qB} \qquad (32.19)$$

The inverse dependence on B indicates that the size of the orbit can be decreased by increasing the magnetic field strength.

We can also determine the frequency of the cyclotron motion. Recall from your earlier study of circular motion that the frequency of revolution f is related to the speed and radius by $f = v/2\pi r$. A rearrangement of Equation 32.19 gives the **cyclotron frequency:**

$$f_{\text{cyc}} = \frac{qB}{2\pi m} \qquad (32.20)$$

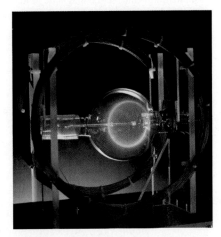

An electron beam undergoing circular motion in a magnetic field.

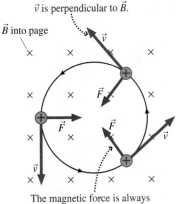

$\vec{v}$ is perpendicular to $\vec{B}$.

The magnetic force is always perpendicular to $\vec{v}$, causing the particle to move in a circle.

FIGURE 32.36 Cyclotron motion of a charged particle moving in a magnetic field.

13.7, 13.8

where the ratio q/m is the particle's *charge-to-mass ratio*. Notice that the cyclotron frequency depends on the charge-to-mass ratio and the magnetic field strength but *not* on the charge's velocity.

EXAMPLE 32.11 The radius of cyclotron motion

In Figure 32.37, an electron is accelerated from rest through a potential difference of 500 V, then injected into a uniform magnetic field. Once in the magnetic field, it completes half a revolution in 2.0 ns. What is the radius of its orbit?

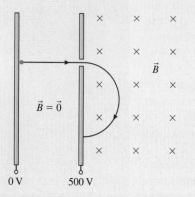

FIGURE 32.37 An electron is accelerated, then injected into a magnetic field.

MODEL Energy is conserved as the electron is accelerated by the potential difference. The electron then undergoes cyclotron motion in the magnetic field, although it completes only half a revolution before hitting the back of the acceleration electrode.

SOLVE The electron accelerates from rest ($v_i = 0$ m/s) at $V_i = 0$ V to speed v_f at $V_f = 500$ V. We can use conservation of energy $K_f + qV_f = K_i + qV_i$ to find the speed v_f with which it enters the magnetic field:

$$\frac{1}{2}mv_f^2 + (-e)V_f = 0 + 0$$

$$v_f = \sqrt{\frac{2eV_f}{m}} = \sqrt{\frac{2(1.60 \times 10^{-19}\,\text{C})(500\,\text{V})}{9.11 \times 10^{-31}\,\text{kg}}}$$

$$= 1.33 \times 10^7 \text{ m/s}$$

The cyclotron radius in the magnetic field is $r_{\text{cyc}} = mv/eB$, but we first need to determine the field strength. Were it not for the electrode, the electron would undergo circular motion with period $T = 4.0$ ns. Hence the cyclotron frequency is $f = 1/T = 2.5 \times 10^8$ Hz. We can use the cyclotron frequency to determine that the magnetic field strength is

$$B = \frac{2\pi m f_{\text{cyc}}}{e} = \frac{2\pi(9.11 \times 10^{-31}\,\text{kg})(2.50 \times 10^8\,\text{Hz})}{1.60 \times 10^{-19}\,\text{C}}$$

$$= 8.94 \times 10^{-3} \text{ T}$$

Thus the radius of the electron's orbit is

$$r = \frac{mv}{qB} = 8.47 \times 10^{-3} \text{ m} = 8.47 \text{ mm}$$

Figure 32.38a shows a more general situation in which the charged particle's velocity $\vec{v}$ is neither parallel to nor perpendicular to $\vec{B}$. The component of $\vec{v}$ parallel to $\vec{B}$ is not affected by the field, so the charged particle spirals around the magnetic field vectors in a helical trajectory. The radius of the helix is determined by $\vec{v}_\perp$, the component of $\vec{v}$ perpendicular to $\vec{B}$.

(a) Charged particles spiral around the magnetic field lines.

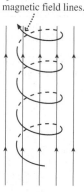

(b) The earth's magnetic field leads particles into the atmosphere near the poles, causing the aurora.

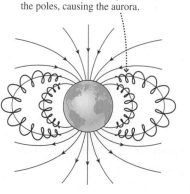

(c) The aurora seen from space

FIGURE 32.38 In general, charged particles spiral along helical trajectories around the magnetic field lines. This motion is responsible for the earth's aurora.

The motion of charged particles in a magnetic field is responsible for the earth's aurora, which you saw in the photograph at the beginning of the chapter. High-energy particles and radiation streaming out from the sun, called the *solar wind,* create ions and electrons as they strike molecules high in the atmosphere. Some of these charged particles become trapped in the earth's magnetic field, creating what is known as the *Van Allen radiation belt.*

As Figure 32.38b shows, the electrons spiral along the magnetic field lines until the field leads them into the atmosphere. The shape of the earth's magnetic field is such that most electrons enter the atmosphere in a circular region around the north magnetic pole and another around the south magnetic pole. There they collide with oxygen and nitrogen atoms, exciting the atoms and causing them to emit auroral light. Figure 32.38c shows a false-color image from space of the ultraviolet light emitted by the aurora.

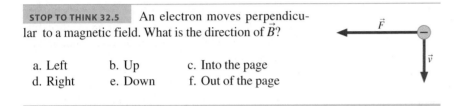

STOP TO THINK 32.5   An electron moves perpendicular to a magnetic field. What is the direction of $\vec{B}$?

a. Left   b. Up   c. Into the page
d. Right   e. Down   f. Out of the page

The Cyclotron

Physicists studying the structure of the atomic nucleus and of elementary particles usually use a device called a *particle accelerator.* Charged particles, typically protons or electrons, are accelerated to very high speeds, close to the speed of light, and then collide with a target. The very large impact energies are sufficient to disrupt the nuclear forces, ejecting elementary particles that can be tracked and studied. The first practical particle accelerator, invented in the 1930s, was the **cyclotron.** Although cyclotrons have been superseded by newer accelerators for studying elementary particles, they remain important for many applications of nuclear physics.

A cyclotron, shown in Figure 32.39, consists of an evacuated chamber within a large, uniform magnetic field. Inside the chamber are two hollow conductors shaped like the letter D and hence called "dees." The dees are made of copper, which doesn't affect the magnetic field; are open along the straight sides; and are separated by a small gap. A charged particle, typically a proton, is injected into the magnetic field from a source near the center of the cyclotron, and it begins to move in and out of the dees in a circular cyclotron orbit.

The cyclotron operates by taking advantage of the fact that the cyclotron frequency f_{cyc} of a charged particle is independent of the particle's speed. An *oscillating* potential difference ΔV is connected across the dees and adjusted until its frequency is exactly the cyclotron frequency. There is almost no electric field inside the dees (you learned in Chapter 27 that the electric field inside a hollow conductor is zero), but a strong electric field points from the positive to the negative dee in the gap between them.

Suppose the proton emerges into the gap from the positive dee. The electric field in the gap *accelerates* the proton across the gap into the negative dee, and it gains kinetic energy $e\Delta V$. A half cycle later, when it next emerges into the gap, the potential of the dees (whose potential difference is oscillating at f_{cyc}) will have changed sign. The proton will *again* be emerging from the positive dee and will *again* accelerate across the gap and gain kinetic energy $e\Delta V$.

Because the dees change potential in time with the proton's orbit, and the proton's cyclotron frequency doesn't change as it speeds up, this pattern will continue orbit after orbit. The proton's kinetic energy increases by $2e\Delta V$ every orbit,

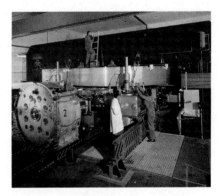

Cyclotrons are used in applied nuclear physics to accelerate ions to very high speeds.

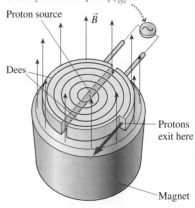

The potential ΔV oscillates at the cyclotron frequency f_{cyc}.

Proton source

$\vec{B}$

Dees

Protons exit here

Magnet

FIGURE 32.39 A cyclotron.

FIGURE 32.40 An electron and a positron moving in a bubble chamber. The magnetic field is perpendicular to the page.

so after N orbits its kinetic energy will be $K = 2Ne\,\Delta V$ (assuming that its initial kinetic energy was near zero). The radius of its orbit increases as it speeds up, hence the proton follows the *spiral* path shown in Figure 32.38 until it finally reaches the outer edge of the dee. It is then directed out of the cyclotron and aimed at a target. Although ΔV is modest, usually a few hundred volts, the fact that the proton can undergo many thousands of orbits before reaching the outer edge allows it to acquire a very large kinetic energy.

Magnetic fields are also important in the analysis of the elementary particles produced in these high-energy collisions. Notice that Equation 32.19 for r_{cyc} can be written $mv = p = r_{cyc}qB$. In other words, the momentum of a charged particle can be determined by measuring the radius of its orbit in a known magnetic field. This is done inside a device called a *bubble chamber*, where the particles leave a map of their trajectories in the form of a string of tiny bubbles in liquid hydrogen. This bubble pattern is photographed, and the radius of the particle's orbit is measured from the photograph. Figure 32.40 is a photograph of a collision in which an electron and a positron (an *antielectron*, having the mass of an electron but charge $+e$) were created. Notice that they spiral in opposite directions, because of their opposite charges, with slowly decreasing radii as they lose energy in collisions with the hydrogen atoms.

The Hall Effect

A charged particle moving through a vacuum is deflected sideways, perpendicular to $\vec{v}$, by a magnetic field. In 1879, a graduate student named Edwin Hall showed that the same is true for the charges moving through a conductor as part of a current. This phenomenon—now called the **Hall effect**—is used to gain information about the charge carriers in a conductor. It is also the basis of a widely used technique for measuring magnetic field strengths.

Figure 32.41a shows a magnetic field perpendicular to a flat, current-carrying conductor. You learned in Chapter 28 that the charge carriers move through a conductor at the drift speed v_d. Their motion is perpendicular to $\vec{B}$, so each charge carrier experiences a magnetic force $F_m = ev_d B$ perpendicular to both $\vec{B}$ and the current I. However, for the first time we have a situation in which it *does* matter whether the charge carriers are positive or negative.

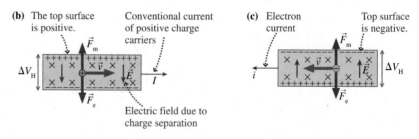

FIGURE 32.41 The charge carriers in a current are deflected to one surface of a conductor, creating the Hall voltage ΔV_H.

Figure 32.41b, which is looking along the magnetic field direction, shows that positive charge carriers moving in the direction of I are pushed toward the top surface of the conductor. This will create an excess positive charge on the top surface and leave an excess negative charge on the bottom. Figure 32.41c, where the electrons in an electron current i move opposite the direction of I, shows that electrons would be pushed toward the top surface. (Be sure to use the right-hand rule and the sign of the electron charge to confirm the deflections shown in these figures.) Thus the sign of the excess charge on the top surface is the same as the sign of the charge carriers. Experimentally, the top surface is negative when the conductor is a metal, and this is one more piece of evidence that the charge carriers in metals are electrons.

Electrons are deflected toward the top surface once the current starts flowing, but the process can't continue indefinitely. The excess charges on the surfaces, like the charges on the plates of a capacitor, create a potential difference ΔV between the two surfaces and an electric field $E = \Delta V/w$ *inside* the conductor. Charge builds up on the surface until the downward electric force $\vec{F}_e$ on the charge carriers exactly balances the upward magnetic force $\vec{F}_m$. Once the forces are balanced, a steady state is reached in which the charge carriers move in the direction of the current and no additional charge is deflected to the surface.

The steady state condition, in which $F_m = F_e$, is

$$F_m = ev_d B = F_e = eE = e\frac{\Delta V}{w} \qquad (32.21)$$

Thus the steady-state potential difference between the two surfaces of the conductor, which is called the **Hall voltage** ΔV_H, is

$$\Delta V_H = wv_d B \qquad (32.22)$$

You learned in Chapter 28 that the drift speed is related to the current density J by $J = nev_d$, where n is the charge-carrier density (charge carriers per m^3). Thus

$$v_d = \frac{J}{ne} = \frac{I/A}{ne} = \frac{I}{wtne} \qquad (32.23)$$

where $A = wt$ is the cross-section area of the conductor. If we use this expression for v_d in Equation 32.23, we find that the Hall voltage is

$$\Delta V_H = \frac{IB}{tne} \qquad (32.24)$$

The Hall voltage is very small for metals in laboratory-sized magnetic fields, typically in the microvolt range. Even so, measurements of the Hall voltage in a known magnetic field are used to determine the charge-carrier density n. Interestingly, the Hall voltage is larger for *poor* conductors that have smaller charge-carrier densities. A laboratory probe for measuring magnetic field strengths, called a *Hall probe*, measures ΔV_H for a poor conductor whose charge-carrier density is known. The magnetic field is then determined from Equation 32.24.

EXAMPLE 32.12 Measuring the magnetic field

A Hall probe consists of a strip of the metal bismuth that is 0.15 mm thick and 5.0 mm wide. Bismuth is a poor conductor with a charge-carrier density 1.35×10^{25} m^{-3}. The Hall voltage on the probe is 2.5 mV when the current through it is 1.50 A. What is the strength of the magnetic field, and what is the electric field strength inside the bismuth?

MODEL Assume the magnetic field is uniform over the Hall probe.

VISUALIZE The bismuth strip looks like Figure 32.41a. The thickness is $t = 1.5 \times 10^{-4}$ m and the width is $w = 5.0 \times 10^{-3}$ m.

SOLVE Equation 32.24 gives the Hall voltage. We can rearrange the equation to find that the magnetic field is

$$B = \frac{tne}{I}\Delta V_H$$

$$= \frac{(1.5 \times 10^{-4}\,\text{m})(1.35 \times 10^{25}\,\text{m}^{-3})(1.60 \times 10^{-19}\,\text{C})}{1.5\,\text{A}}\,0.0025\,\text{V}$$

$$= 0.54\,\text{T}$$

The electric field created inside the bismuth by the excess charge on the surface is

$$E = \frac{\Delta V_H}{w} = \frac{0.0025\,\text{V}}{5.0 \times 10^{-3}\,\text{m}} = 0.50\,\text{V/m}$$

ASSESS 0.54 T is a fairly typical strength for a laboratory magnet.

32.8 Magnetic Forces on Current-Carrying Wires

13.5 Activ Physics ONLINE

We were motivated to look at the magnetic force on moving charges by the experiment in which Ampère observed magnetic forces between current-carrying wires. We're now ready to analyze Ampère's experiment. As a first step, let us find the force that a uniform magnetic field exerts on a long, straight wire that carries current I through the field. If a current-carrying wire is *parallel* to a magnetic field, the force on it is zero. This follows from the fact that there is no force on a charged particle moving parallel to $\vec{B}$.

It's more interesting to consider the wire in Figure 32.42 that is *perpendicular* to the magnetic field. Note that the field is an external magnetic field, created by a permanent magnet or by other currents; it is *not* the field of the current I. The direction of the force on the current is found by considering the force on each charge in the current. By the right-hand rule, each charge has a force of magnitude qvB directed to the left. Consequently, the entire length of wire within the magnetic field experiences a force to the left, perpendicular to both the current direction and the field direction.

NOTE ▶ The familiar right-hand rule can be applied directly to a current-carrying wire. Point your right thumb in the direction of the current (parallel to $\vec{v}$) and your index finger in the direction of $\vec{B}$. Your middle finger is then pointing in the direction of the force $\vec{F}$ on the wire. ◀

To find the magnitude of the force, we must relate qv of the charges to the current I in the wire. Consider a section of wire of length L in which charge carriers with total charge q move with speed v. The current I, by definition, is the charge q divided by the time Δt it takes the charge to flow out of this section of the wire: $I = q/\Delta t$. The time required is $\Delta t = L/v$, giving

$$I = \frac{q}{\Delta t} = \frac{q}{L/v} = \frac{qv}{L}$$

Thus $qv = IL$. If we substitute IL for qv in the force equation $F = qvB$, we find that the magnetic force on length L of a current-carrying wire is

$$F_{\text{wire}} = ILB \quad \text{(force on a current perpendicular to the field)} \qquad (32.25)$$

Equation 32.25 is a simple result, but remember the two assumptions behind it: The wire must be perpendicular to the field, and the field must be constant over the length L of the wire. As an aside, you can see from Equation 32.25 that the magnetic field B must have units of N/A m. This is why we defined 1 T = 1 N/A m in Section 32.3.

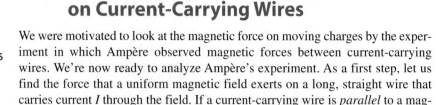

A wire is perpendicular to an externally created magnetic field.

A current through a wire that is fixed at the ends causes the wire to be bent sideways.

FIGURE 32.42 Magnetic force on a current-carrying wire.

EXAMPLE 32.13 Magnetic levitation

The 0.10 T uniform magnetic field of Figure 32.43 is horizontal, parallel to the floor. A straight segment of 1.0-mm-diameter copper wire, also parallel to the floor, is perpendicular to the magnetic field. What current through the wire, and in which direction, will allow the wire to "float" in the magnetic field?

MODEL The wire will float in the magnetic field if the magnetic force on the wire points upward and has magnitude mg, allowing it to balance the downward gravitational force.

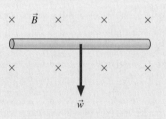

FIGURE 32.43 Magnetic levitation.

SOLVE We can use the right-hand rule to determine which current direction experiences an upward force. With $\vec{B}$ pointing away from us, the direction of the current needs to be from left to right. The forces will balance when

$$F = ILB = mg = \rho(\pi r^2 L)g$$

where $\rho = 8920 \text{ kg/m}^3$ is the density of copper. The length of the wire cancels, leading to

$$I = \frac{\rho \pi r^2 g}{B} = \frac{(8920 \text{ kg/m}^3)\pi(0.00050 \text{ m})^2(9.80 \text{ m/s}^2)}{0.10 \text{ T}}$$

$$= 0.687 \text{ A}$$

A 0.687 A current from left to right will levitate the wire in the magnetic field.

ASSESS A 0.687 A current is quite reasonable, but this idea is useful only if we can get the current into and out of this segment of wire. In practice, we could do so with wires that come in from below the page. These input and output wires would be parallel to $\vec{B}$ and not experience a magnetic force. Although this example is very simple, it is the basis for applications such as magnetic levitation trains.

Force Between Two Parallel Wires

Now consider Ampère's experimental arrangement of two parallel wires of length L, distance d apart. Figure 32.44a shows the currents I_1 and I_2 in the same directions; Figure 32.44b shows the currents in opposite directions. We will assume that the wires are sufficiently long to allow us to use the earlier result for the magnetic field of a long straight wire: $B = \mu_0 I / 2\pi d$.

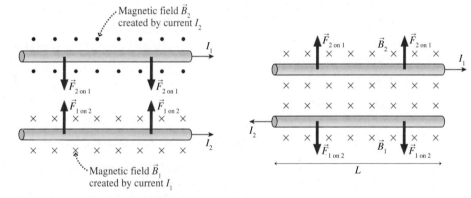

(a) Currents in same direction

(b) Currents in opposite directions

FIGURE 32.44 Magnetic forces between parallel current-carrying wires.

As Figure 32.44a shows, the current I_2 in the lower wire creates a magnetic field $\vec{B}_2$ at the position of the upper wire. $\vec{B}_2$ points out of the page, perpendicular to current I_1. **It is field $\vec{B}_2$, due to the lower wire, that exerts a magnetic force on the upper wire.** Using the right-hand rule, you can see that the force on the upper wire is downward, thus attracting it toward the lower wire. The field of the lower current is not a uniform field, but it is the *same* at all points along the upper wire because the two wires are parallel. Consequently, we can use the field of a long, straight wire to determine the magnetic force exerted by the lower wire on the upper wire when they are separated by distance d:

$$F_{\text{parallel wires}} = I_1 L B_2 = I_1 L \frac{\mu_0 I_2}{2\pi d} = \frac{\mu_0 L I_1 I_2}{2\pi d} \qquad (32.26)$$

(force between two parallel wires)

As an exercise, you should convince yourself that the current in the upper wire exerts an upward-directed magnetic force on the lower wire with exactly the same magnitude. You should also convince yourself, using the right-hand rule, that the forces are repulsive and tend to push the wires apart if the two currents are in opposite directions.

Thus two parallel wires exert equal but opposite forces on each other, as required by Newton's third law. **Parallel wires carrying currents in the same direction attract each other; parallel wires carrying currents in opposite directions repel each other.**

EXAMPLE 32.14 A current balance

Two stiff, 50-cm-long parallel wires are connected at the ends by metal springs. Each spring has an unstretched length of 5.0 cm and a spring constant of 0.020 N/m. The wires push each other apart when a current travels around the loop. How much current is required to stretch the springs to lengths of 6.0 cm?

MODEL Two parallel wires carrying currents in opposite directions exert repulsive magnetic forces on each other.

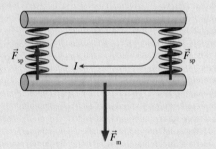

FIGURE 32.45 The current-carrying wires of Example 32.14.

VISUALIZE Figure 32.45 shows the "circuit." The springs are conductors, allowing a current to travel around the loop. In equilibrium, the repulsive magnetic forces between the wires are balanced by the restoring forces $F_{sp} = k\Delta y$ of the springs.

SOLVE Figure 32.45 shows the forces on the lower wire. The net force is zero, hence $F_m = 2F_{sp}$. The repulsive force between the wires is given by Equation 32.26 with $I_1 = I_2 = I$:

$$F_m = \frac{\mu_0 L I^2}{2\pi d} = 2F_{sp} = 2k\Delta y$$

where k is the spring constant and $\Delta y = 1.0$ cm is the amount by which each spring stretches. Solving for the current, we find

$$I = \sqrt{\frac{4\pi k d \Delta y}{\mu_0 L}} = 15.5 \text{ A}$$

ASSESS Devices similar to the one described, in which a magnetic force balances a mechanical force, are called *current balances.* They can be used to make very accurate current measurements.

32.9 Forces and Torques on Current Loops

13.6 **Activ Physics** ONLINE

You have seen that a current loop is a magnetic dipole, much like a permanent magnet. We will now look at some important features of how current loops behave in magnetic fields. This discussion will be largely qualitative, but it will highlight some of the important properties of magnets and magnetic fields. We will then use these ideas in the next section to make the connection between electromagnets and permanent magnets.

Figure 32.46a shows two current loops. Using what we've learned about the forces between parallel and antiparallel currents, you can see that **parallel cur-**

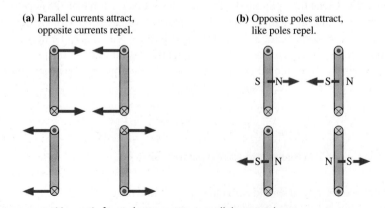

(a) Parallel currents attract, opposite currents repel.

(b) Opposite poles attract, like poles repel.

FIGURE 32.46 Magnetic forces between two parallel current loops.

rent loops exert attractive magnetic forces on each other if the currents circulate in the same direction; they repel each other when the currents circulate in opposite directions.

It is convenient to think of these forces in terms of magnetic poles. Figure 32.46b shows the north and south magnetic poles of the current loops. If the currents circulate in the same direction, a north and a south pole face each other and exert attractive forces on each other. If the currents circulate in opposite directions, the two like poles repel each other.

Here, at last, we have a real connection to the behavior of magnets that opened our discussion of magnetism—namely, that like poles repel and opposite poles attract. Now we have an *explanation* for this behavior, at least for electromagnets. **Magnetic poles attract or repel because the moving charges in one current exert attractive or repulsive magnetic forces on the moving charges in the other current.** Our tour through interacting moving charges is finally starting to show some practical results!

Now let's consider the forces on a current loop in a *uniform* magnetic field. Figure 32.47 shows a square current loop in a uniform magnetic field. The current in each of the four sides experiences a magnetic force due to the field $\vec{B}$. The forces $\vec{F}_{\text{front}}$ and $\vec{F}_{\text{back}}$ are opposite to each other and cancel. Forces $\vec{F}_{\text{top}}$ and $\vec{F}_{\text{bottom}}$ also add to give no net force, but because $\vec{F}_{\text{top}}$ and $\vec{F}_{\text{bottom}}$ don't act along the same line they will *rotate* the loop by exerting a torque on it.

The forces on the top and bottom segments form what we called a *couple* in Chapter 13. The torque due to a couple is the magnitude of the force multiplied by the distance d between the two lines of action. You can see that $d = L\sin\theta$, hence the torque on the loop—a torque exerted by the magnetic field—is

$$\tau = Fd = (ILB)(L\sin\theta) = (IL^2)B\sin\theta = \mu B\sin\theta \qquad (32.27)$$

where $\mu = IL^2 = IA$ is the size of the loop's magnetic dipole moment.

Although we derived Equation 32.27 for a square loop, the result is valid for a current loop of any shape. Notice that Equation 32.27 looks like another example of a cross product. We earlier defined the magnetic dipole moment vector $\vec{\mu}$ to be a vector perpendicular to the current loop in a direction given by the right-hand rule. Figure 32.47 shows that θ is the angle between $\vec{B}$ and $\vec{\mu}$, hence the torque on a magnetic dipole is

$$\vec{\tau} = \vec{\mu} \times \vec{B} \qquad (32.28)$$

The torque is zero when the magnetic dipole moment $\vec{\mu}$ is aligned parallel or antiparallel to the magnetic field, and is maximum when $\vec{\mu}$ is perpendicular to the field. It is this magnetic torque that causes a compass needle—a magnetic moment—to rotate until it is aligned with the magnetic field.

An Electric Motor

The torque on a current loop in a magnetic field is the basis for how an electric motor works. As Figure 32.48 on the next page shows, the *armature* of a motor is a coil of wire wound on an axle that is free to rotate. When a current passes through the coil, the magnetic field exerts a torque on the armature and causes it to rotate. If the current were steady, the armature would oscillate back and forth around the equilibrium position until (assuming there's some friction or damping) it stops with the plane of the coil perpendicular to the field. To keep the motor turning, a device called a *commutator* reverses the current direction in the coils every 180°. (Notice that the commutator is split, so the positive terminal of the battery sends current into whichever wire touches the bottom half of the commutator.) The current reversal prevents the armature from ever reaching an equilibrium position, so the magnetic torque keeps the motor spinning as long as there is a current.

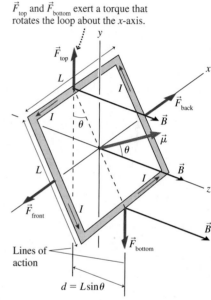

$\vec{F}_{\text{top}}$ and $\vec{F}_{\text{bottom}}$ exert a torque that rotates the loop about the x-axis.

FIGURE 32.47 A uniform magnetic field exerts a torque on a current loop.

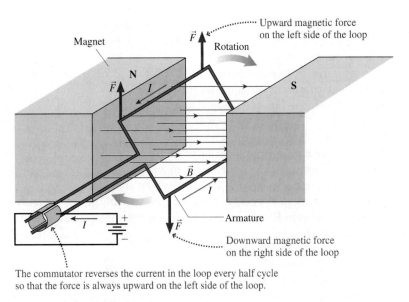

FIGURE 32.48 A simple electric motor.

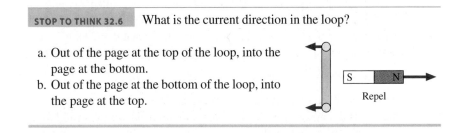

STOP TO THINK 32.6   What is the current direction in the loop?

a. Out of the page at the top of the loop, into the page at the bottom.
b. Out of the page at the bottom of the loop, into the page at the top.

32.10 Magnetic Properties of Matter

Our theory has focused mostly on the magnetic properties of currents, yet our everyday experience is mostly with permanent magnets. We have seen that current loops and solenoids have magnetic poles and exhibit behaviors like those of permanent magnets, but we still lack a specific connection between electromagnets and permanent magnets. The goal of this section is to complete our understanding by developing an atomic-level view of the magnetic properties of matter.

Atomic Magnets

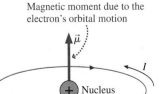

FIGURE 32.49 A classical orbiting electron is a tiny magnetic dipole.

A plausible explanation for the magnetic properties of materials is the orbital motion of the atomic electrons. Figure 32.49 shows a simple, classical model of an atom in which a negative electron orbits a positive nucleus. In this picture of the atom, the electron's motion is that of a current loop! It is a microscopic current loop, to be sure, but a current loop nonetheless. Consequently, an orbiting electron acts as a tiny magnetic dipole, with a north pole and a south pole. You can think of the magnetic dipole as an atomic-size magnet. Experiments with *individual* hydrogen atoms verify that they are, indeed, tiny magnets.

However, the atoms of most elements contain many electrons. Unlike the solar system, where all of the planets orbit in the same direction, electron orbits are arranged to oppose each other: one electron moves counterclockwise for every electron that moves clockwise. Thus the magnetic moments of individual orbits tend to cancel each other and the *net* magnetic moment is either zero or very small.

The cancellation continues as atoms are joined into molecules and the molecules into solids. When all is said and done, the net magnetic moment of any bulk matter due to the orbiting electrons is so small as to be negligible. There are various subtle magnetic effects that can be observed under laboratory conditions, but orbiting electrons cannot explain the very strong magnetic effects of a piece of iron.

The Electron Spin

The key to understanding atomic magnetism was the 1922 discovery that electrons have an *inherent magnetic moment*. Perhaps this shouldn't be surprising. An electron has a *mass,* which allows it to interact with gravitational fields, and a *charge,* which allows it to interact with electric fields. There's no reason an electron shouldn't also interact with magnetic fields, and to do so it comes with a magnetic moment.

An electron's inherent magnetic moment is often called the electron *spin* because, in a classical picture, a spinning ball of charge would have a magnetic moment. This classical picture is not a realistic portrayal of how the electron really behaves, but its inherent magnetic moment makes it seem *as if* the electron were spinning. While it may not be spinning in a literal sense, each electron really is a microscopic bar magnet.

We must appeal to the results of quantum physics to find out what happens in an atom with many electrons. The spin magnetic moments, like the orbital magnetic moments, tend to oppose each other as the electrons are placed into their shells, causing the net magnetic moment of a *filled* shell to be zero. However, atoms containing an odd number of electrons must have at least one valence electron with an unpaired spin. These atoms have net magnetic moment due to the electron's spin.

But atoms with magnetic moments don't necessarily form a solid with magnetic properties. For most elements, the magnetic moments of the atoms are randomly arranged when the atoms join together to form a solid. As Figure 32.50 shows, this random arrangement produces a solid whose net magnetic moment is very close to zero. This agrees with our common experience that most materials are not magnetic; you cannot pick them up with a magnet or make a magnet from them. On the other hand, there are those materials such as iron that do exhibit strong magnetic properties, so we need to discover why these magnetic materials are different.

Ferromagnetism

It happens that in iron, and a few other substances, the spins interact with each other in such a way that atomic magnetic moments tend to all line up in the *same* direction. Materials that behave in this fashion are called **ferromagnetic,** with the prefix *ferro* meaning "iron-like." Figure 32.51 shows how the spin magnetic moments are aligned for the atoms making up a ferromagnetic solid.

In ferromagnetic materials, the individual magnetic moments add together to create a *macroscopic* magnetic dipole. The material has a north and a south magnetic pole, generates a magnetic field, and aligns parallel to an external magnetic field. In other words, it is a magnet!

Although iron is a magnetic material, a typical piece of iron is not a strong permanent magnet. You need not worry that a steel nail, which is mostly iron and is easily lifted with a magnet, will leap from your hands and pin itself against the hammer because of its own magnetism. It turns out, as shown in Figure 32.52, that a piece of iron is divided into small regions called **magnetic domains.** A typical domain size is roughly 0.1 mm—small, but not unreasonably so. The magnetic moments of all of the iron atoms within each domain are perfectly aligned, so each individual domain, like Figure 32.51, is a strong magnet.

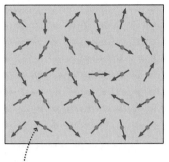

The atomic magnetic moments due to unpaired spins point in random directions. The sample has no net magnetic moment.

FIGURE 32.50 The random magnetic moments of the atoms in a typical solid produce no net magnetic moment.

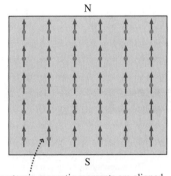

The atomic magnetic moments are aligned. The sample has a north and south magnetic pole.

FIGURE 32.51 The aligned atomic magnetic moments in a ferromagnetic material create a macroscopic magnetic dipole.

Magnetic domains

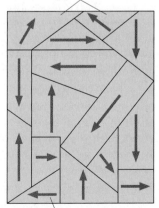

Magnetic moment of the domain

FIGURE 32.52 Magnetic domains in a ferromagnetic material. The net magnetic dipole is nearly zero.

However, the various magnetic domains that form a larger solid, such as you might hold in your hand, are randomly arranged. Their magnetic dipoles largely cancel, much like the cancellation that occurs on the atomic scale for nonferromagnetic substances, so the solid as a whole has only a small magnetic moment. That is why the nail is not a strong permanent magnet.

Induced Magnetic Dipoles

If a ferromagnetic substance is subjected to an *external* magnetic field, the external field exerts a torque on the magnetic dipole of each domain. The torque causes many of the domains to rotate and become aligned with the external field, just as a compass needle aligns with a magnetic field, although internal forces between the domains generally prevent the alignment from being perfect. In addition, atomic-level forces between the spins can cause the *domain boundaries* to move. Domains that are aligned along the external field become larger at the expense of domains that are opposed to the field. These changes in the size and orientation of the domains cause the material to develop a *net magnetic dipole* that is aligned with the external field. This magnetic dipole has been *induced* by the external field, so it is called an **induced magnetic dipole.**

NOTE ▶ The induced magnetic dipole is analogous to the polarization forces and induced electric dipoles that you studied in Chapter 26. ◀

Figure 32.53 shows a ferromagnetic material near the end of a solenoid. The magnetic moments of the domains align with the solenoid's field, creating an induced magnetic dipole whose south pole faces the solenoid's north pole. Consequently, the magnetic force between the poles pulls the ferromagnetic object to the electromagnet.

The fact that a magnet attracts and picks up ferromagnetic objects was one of the basic observations about magnetism with which we started the chapter. Now we have an *explanation* of how it works, based on three ideas:

1. Electrons are microscopic magnets due to their spin.
2. A ferromagnetic material in which the spins are aligned is organized into magnetic domains.
3. The individual domains align with an external magnetic field to produce an induced magnetic dipole moment for the entire object.

The object's magnetic dipole may not return to zero when the external field is removed because some domains remain "frozen" in the alignment they had in the external field. Thus a ferromagnetic object that has been in an external field may be left with a net magnetic dipole moment after the field is removed. In other words, the object has become a **permanent magnet.** A permanent magnet is simply a ferromagnetic material in which a majority of the magnetic domains are aligned with each other to produce a net magnetic dipole moment.

Whether or not a ferromagnetic material can be made into a permanent magnet depends on the internal crystalline structure of the material. *Steel* is an alloy of iron with other elements. An alloy of mostly iron with the right percentages of chromium and nickel produces *stainless steel,* which has virtually no magnetic properties at all because its particular crystalline structure is not conducive to the formation of domains. A very different steel alloy called Alnico V is made with 51% iron, 24% cobalt, 14% nickel, 8% aluminum, and 3% copper. It has extremely prominent magnetic properties and is used to make high-quality permanent magnets. You can see from the complex formula that developing good magnetic materials requires a lot of engineering skill as well as a lot of patience!

So we've come full circle. One of our initial observations about magnetism was that a permanent magnet can exert forces on some materials but not others.

Ferromagnetic material

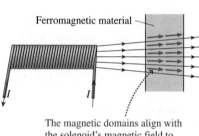

The magnetic domains align with the solenoid's magnetic field to produce an induced magnetic dipole.

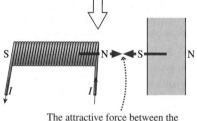

The attractive force between the opposite poles pulls the ferromagnetic material toward the solenoid.

FIGURE 32.53 The magnetic field of the solenoid creates an induced magnetic dipole in the iron.

The *theory* of magnetism that we then proceeded to develop was a theory about the interactions between moving charges. What moving charges had to do with permanent magnets was not obvious. But finally, by considering magnetic effects at the atomic level, we find that properties of permanent magnets and magnetic materials can be traced to the interactions of vast numbers of electron spins.

STOP TO THINK 32.7 Which magnet or magnets produced this induced magnetic dipole?

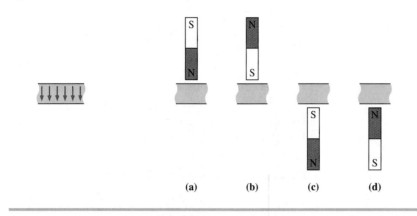

SUMMARY

The goal of Chapter 32 has been to learn how to calculate and use the magnetic field.

GENERAL PRINCIPLES

At its most fundamental level, magnetism is an interaction between moving charges. The magnetic field of one moving charge exerts a force on another moving charge.

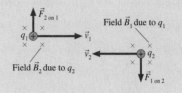

Magnetic Fields

The Biot-Savart law

- A **point charge,** $\vec{B} = \dfrac{\mu_0}{4\pi}\dfrac{q\vec{v}\times\hat{r}}{r^2}$

- A **short current element,** $\vec{B} = \dfrac{\mu_0}{4\pi}\dfrac{I\Delta\vec{s}\times\hat{r}}{r^2}$

To find the magnetic field of a current

- Divide the wire into many short segments.
- Find the field of each segment Δs.
- Find $\vec{B}$ by summing the fields of all Δs, usually as an integral.

An alternative method for fields with a high degree of symmetry is Ampère's law

$$\oint \vec{B}\cdot d\vec{s} = \mu_0 I_{\text{through}}$$

where I_{through} is the current through the area bounded by the integration path.

Magnetic Forces

The magnetic force on a moving charge is

$$\vec{F} = q\vec{v}\times\vec{B}$$

The force is perpendicular to $\vec{v}$ and $\vec{B}$.

The magnetic force on a current-carrying wire perpendicular to a magnetic field is $F = ILB$.

$\vec{F} = \vec{0}$ for a charge or current moving parallel to $\vec{B}$.

The magnetic torque on a magnetic dipole is

$$\vec{\tau} = \vec{\mu}\times\vec{B}$$

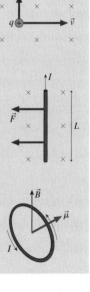

APPLICATIONS

Wire

$$B = \frac{\mu_0}{2\pi}\frac{I}{d}$$

Loop

Solenoid

$$B = \frac{\mu_0 NI}{L}$$

Flat magnet

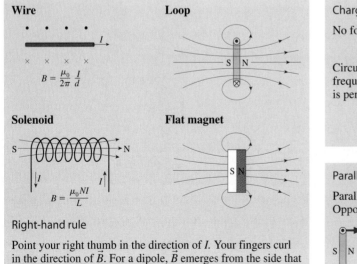

Right-hand rule

Point your right thumb in the direction of I. Your fingers curl in the direction of $\vec{B}$. For a dipole, $\vec{B}$ emerges from the side that is the north pole.

Charged-particle motion

No force if $\vec{v}$ is parallel to $\vec{B}$.

Circular motion at the cyclotron frequency $f_{\text{cyc}} = qB/2\pi m$ if $\vec{v}$ is perpendicular to $\vec{B}$.

Parallel wires and current loops

Parallel currents attract.
Opposite currents repel.

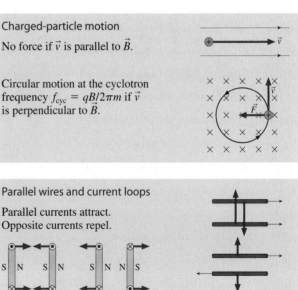

TERMS AND NOTATION

north pole	magnetic field lines	magnetic dipole moment, $\vec{\mu}$	cyclotron
south pole	Biot-Savart law	line integral	Hall effect
magnetic material	tesla, T	Ampère's law	Hall voltage, ΔV_{H}
magnetic dipole	permeability constant, μ_0	uniform magnetic field	ferromagnetic
magnetic monopole	cross product	solenoid	magnetic domain
right-hand rule	current loop	cyclotron motion	induced magnetic dipole
magnetic field, $\vec{B}$	electromagnet	cyclotron frequency, f_{cyc}	permanent magnet

EXERCISES AND PROBLEMS

Exercises

Section 32.2 The Magnetic Field

1. What is the current direction in the wire of Figure Ex32.1? Explain.

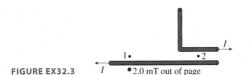

FIGURE EX32.1

FIGURE EX32.2

2. What is the current direction in the wire of Figure Ex32.2? Explain.

Section 32.3 The Source of the Magnetic Field: Moving Charges

3. Points 1 and 2 in Figure Ex32.3 are the same distance from the wires as the point where $B = 2.0$ mT. What are the strength and direction of $\vec{B}$ at points 1 and 2?

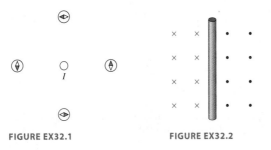

FIGURE EX32.3

4. What is the magnetic field strength at points 2 to 4 in Figure Ex32.4? Assume that the wires overlap closely and that points 1 to 4 are equally distant from the wires.

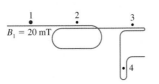

FIGURE EX32.4

5. A proton moves along the y-axis with $v_y = -1.0 \times 10^7$ m/s. As it passes the origin, what are the strength and direction of the magnetic field at the (x, y) positions (a) (1 cm, 0 cm), (b) (0 cm, 1 cm), and (c) (0 cm, −2 cm)?

6. An electron moves along the y-axis with $v_y = 1.0 \times 10^7$ m/s. As it passes the origin, what are the strength and direction of the magnetic field at the (x, y) positions (a) (0 cm, 1 cm), (b) (1 cm, 0 cm), and (c) (1 cm, 1 cm)?

7. What are the magnetic field strength and direction at the dot in Figure Ex32.7?

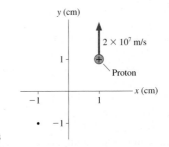

FIGURE EX32.7

8. What are the magnetic field strength and direction at the dot in Figure Ex32.8?

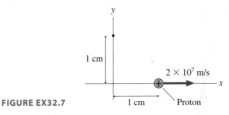

FIGURE EX32.8

9. A proton is passing the origin. The magnetic field at the (x, y, z) position (1 mm, 0 mm, 0 mm) is $1.0 \times 10^{-13} \, \hat{\jmath}$ T. The field at (0 mm, 1 mm, 0 mm) is $-1.0 \times 10^{-13} \, \hat{\imath}$ T. What are the speed and direction of the proton?

Section 32.4 The Magnetic Field of a Current

10. What currents are needed to generate the magnetic field strengths of Table 32.1 at a point 1.0 cm from a long, straight wire?

11. At what distances from a very thin, straight wire carrying a 10 A current would the magnetic field strengths of Table 32.1 be generated?

12. For a current loop, what is the ratio of the magnetic field strength at $z = R$ to the magnetic field strength at the center of the loop?

13. The magnetic field at the center of a 1.0-cm-diameter loop is 2.5 mT.
    a. What is the current in the loop?
    b. A long straight wire carries the same current you found in part a. At what distance from the wire is the magnetic field 2.5 mT?

14. What are the magnetic field strength and direction at points 1 to 3 in Figure Ex32.14?

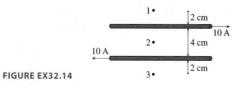

FIGURE EX32.14

15. What is the magnetic field $\vec{B}$ at points 1 to 3 in Figure Ex32.15?

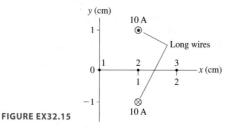

FIGURE EX32.15

Section 32.5 Magnetic Dipoles

16. A 100 A current circulates around a 2.0-mm-diameter superconducting ring.
    a. What is the ring's magnetic dipole moment?
    b. What is the on-axis magnetic field strength 5.0 cm from the ring?

17. The on-axis magnetic field strength 10 cm from a small bar magnet is 5.0 μT.
    a. What is the bar magnet's magnetic dipole moment?
    b. What will be the magnetic field strength if the bar magnet is rotated end-over-end by 180°?

18. The earth's magnetic dipole moment is 8.0×10^{22} A m^2.
    a. What is the magnetic field strength on the surface of the earth at the earth's north magnetic pole? How does this compare to the value in Table 32.1? You can assume that the current loop is deep inside the earth.
    b. Astronauts discover an earth-size planet without a magnetic field. To create a magnetic field, so that compasses will work, they propose running a current through a wire around the equator. What size current would be needed?

19. A current loop with a 25 A current produces an on-axis magnetic field strength of 3.5 nT at a point 50 cm from the loop. What is the area of the loop? You can assume that the loop's radius is much less than 50 cm.

Section 32.6 Ampère's Law and Solenoids

20. What is the line integral of $\vec{B}$ between points i and f in Figure Ex32.20?

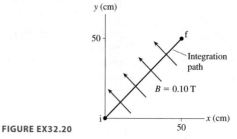

FIGURE EX32.20

21. What is the line integral of $\vec{B}$ between points i and f in Figure Ex32.21?

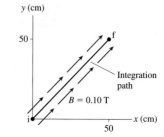

FIGURE EX32.21

22. What is the line integral of $\vec{B}$ between points i and f in Figure Ex32.22?

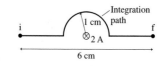

FIGURE EX32.22

23. The value of the line integral of $\vec{B}$ around the closed path in Figure Ex32.23 is 3.77×10^{-6} T m. What is I_3?

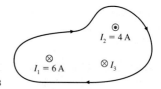

FIGURE EX32.23

24. A 2.0-cm-diameter, 15-cm-long solenoid is tightly wound from 1.0-mm-diameter wire. What current is needed to generate a 3.0 mT field inside the solenoid?

25. Magnetic resonance imaging needs a magnetic field strength of 1.5 T. The solenoid is 1.8 m long and 75 cm in diameter. It is tightly wound with a single layer of 2.0-mm-diameter superconducting wire. What size current is needed?

Section 32.7 The Magnetic Force on a Moving Charge

26. What is the *initial* direction of deflection for the charged particles entering the magnetic fields shown in Figure Ex32.26?

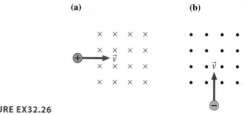

FIGURE EX32.26

27. What is the *initial* direction of deflection for the charged particles entering the magnetic fields shown in Figure Ex32.27?

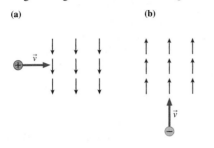

(a) **(b)**

FIGURE EX32.27

28. Determine the magnetic field direction that causes the charged particles shown in Figure Ex32.28 to experience the indicated magnetic force.

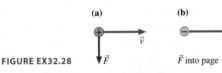

(a) **(b)**

FIGURE EX32.28 $\vec{F}$ $\vec{F}$ into page

29. Determine the magnetic field direction that causes the charged particles shown in Figure Ex32.29 to experience the indicated magnetic force.

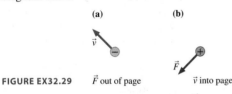

(a) **(b)**

FIGURE EX32.29 $\vec{F}$ out of page $\vec{v}$ into page

30. A proton moves in the magnetic field $\vec{B} = 0.50\,\hat{i}$ T with a speed of 1.0×10^7 m/s in the directions shown in Figure Ex32.30. For each, what is magnetic force $\vec{F}$ on the proton? Give your answers in component form.

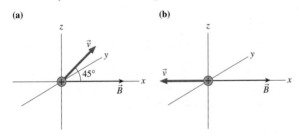

(a) **(b)**

FIGURE EX32.30

31. An electron moves in the magnetic field $\vec{B} = 0.50\,\hat{i}$ T with a speed of 1.0×10^7 m/s in the directions shown in Figure Ex32.31. For each, what is magnetic force $\vec{F}$ on the electron? Give your answers in component form.

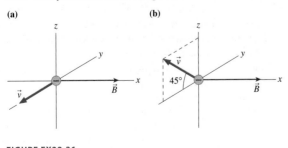

(a) **(b)**

FIGURE EX32.31

32. What is the cyclotron frequency in a 3.00 T magnetic field of the ions (a) N_2^+, (b) O_2^+, and (c) CO^+? Give your answers in MHz. The masses of the atoms are shown in the table. The accuracy of your answers should reflect the accuracy of the data. (For this problem, assume that all the data you need are good to six significant figures. Although N_2^+ and CO^+ both have a *nominal* molecular mass of 28, they are easily distinguished by virtue of their different cyclotron resonance frequencies.)

Atomic masses

^{12}C	12.0000 u
^{14}N	14.0031 u
^{16}O	15.9949 u

33. Radio astronomers detect electromagnetic radiation at 45 MHz from an interstellar gas cloud. They suspect this radiation is emitted by electrons spiraling in a magnetic field. What is the magnetic field strength inside the gas cloud?

34. The aurora is caused when electrons and protons, moving in the earth's magnetic field of $\approx 5 \times 10^{-5}$ T, collide with molecules of the atmosphere and cause them to glow. What is the radius of the cyclotron orbit for
   a. An electron with speed 1.0×10^6 m/s?
   b. A proton with speed 5.0×10^4 m/s?

35. The Hall voltage across a 1.0-mm-thick conductor in a 1.0 T magnetic field is 3.2 μV when the current is 15 A. What is the charge-carrier density in this conductor?

Section 32.8 Magnetic Forces on Current-Carrying Wires

36. What magnetic field strength and direction will levitate the 2.0 g wire in Figure Ex32.36?

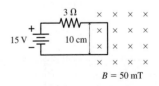

FIGURE EX32.36   10 cm

37. The right edge of the circuit in Figure Ex32.37 extends into a 50 mT uniform magnetic field. What are the magnitude and direction of the net force on the circuit?

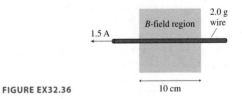

FIGURE EX32.37 $B = 50$ mT

38. What is the net force (magnitude and direction) on each wire in Figure Ex32.38?

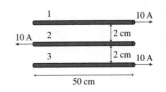

FIGURE EX32.38   50 cm

39. The two 10-cm-long parallel wires in Figure Ex32.39 are separated by 5.0 mm. For what value of the resistor R will the force between the two wires be 5.4×10^{-5} N?

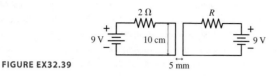

FIGURE EX32.39

Section 32.9 Forces and Torques on Current Loops

40. Figure Ex32.40 shows two square current loops. The loops are far apart and do not interact with each other.
    a. Use force diagrams to show that both loops are in equilibrium, having a net force of zero and no torque.
    b. One of the loop positions is stable. That is, the forces will return it to equilibrium if it is rotated slightly. The other position is unstable, like an upside-down pendulum. Which is which? Explain.

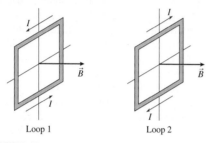

Loop 1   Loop 2

FIGURE EX32.40

41. A square current loop 5.0 cm on each side carries a 500 mA current. The loop is in a 1.2 T uniform magnetic field. The axis of the loop, perpendicular to the plane of the loop, is 30° away from the field direction. What is the magnitude of the torque on the current loop?

42. A small bar magnet experiences a 0.020 N m torque when the axis of the magnet is at 45° to a 0.10 T magnetic field. What is the magnitude of its magnetic dipole moment?

43. a. What is the magnitude of the torque on the current loop in Figure Ex32.43?
    b. What is the loop's equilibrium position?

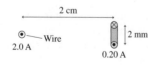

FIGURE EX32.43

Problems

44. You have a horizontal cathode ray tube (CRT) for which the controls have been adjusted such that the electron beam *should* make a single spot of light exactly in the center of the screen. You observe, however, that the spot is deflected to the right. It is possible that the CRT is broken. But as a clever scientist, you realize that your laboratory might be in either an electric or a magnetic field. Assuming that you do not have a compass, any magnets, or any charged rods, how can you use the CRT itself to determine whether the CRT is broken, is in an electric field, or is in a magnetic field? You cannot remove the CRT from the room.

45. Although the evidence is weak, there has been concern in recent years over possible health effects from the magnetic fields generated by transmission lines, the wiring in your house, and electrical appliances.
    a. The current carried by the wiring in the walls of your house rarely exceeds 10 A. What is the magnetic field strength 2 m from a long, straight wire carrying a current of 10 A?
    b. What percentage of the earth's magnetic field is your answer to (a)?
    Because the percentage is small, and because we live in the earth's field with no harmful effects, it is assumed that any possible health effects are due not to the field strength alone but to the 60 Hz oscillatory nature of fields from power lines.
    c. High-voltage transmission lines, on tall towers, typically carry 200 A at voltages of up to 500,000 V. Although this is much larger than household currents, the lines are roughly 20 m overhead. Estimate the magnetic field strength on the ground underneath such lines.
    d. A common electrical appliance is an electric blanket. Some consumer groups urge pregnant women not to use electric blankets, just in case there is a health risk. The current through the heater wires is approximately 1 A. Estimate, stating any assumptions you make, the magnetic field strength a fetus might experience. How does this compare to your answer to part a?

46. A wire carries current I into the junction shown in Figure P32.46. What is the magnetic field at the dot?

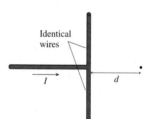

FIGURE P32.46

47. The two insulated wires in Figure P32.47 cross at a 30° angle but do not make electrical contact. Each wire carries a 5.0 A current. Points 1 and 2 are each 4.0 cm from the intersection and equally distant from both wires. What are the magnitude and direction of the magnetic fields at points 1 and 2?

FIGURE P32.47

48. A long wire carrying a 5.0 A current perpendicular to the xy-plane intersects the x-axis at $x = -2.0$ cm. A second, parallel wire carrying a 3.0 A current intersects the x-axis at $x = +2.0$ cm. At what point or points on the x-axis is the magnetic field zero if (a) the two currents are in the same direction and (b) the two currents are in opposite directions?

49. The capacitor in Figure P32.49 is charged to 50 V. The switch closes at $t = 0$ s. Draw a graph showing the magnetic field strength as a function of time at the position of the dot. On your graph indicate the maximum field strength, and provide an appropriate numerical scale on the horizontal axis.

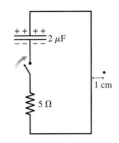

FIGURE P32.49

50. a. Find an expression for the magnetic field at the center (point P) of the circular arc in Figure P32.50.
    b. Does your result agree with the magnetic field of a current loop when $\theta = 2\pi$?

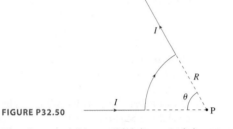

FIGURE P32.50

51. The element niobium, which is a metal, is a superconductor (i.e., no electrical resistance) at temperatures below 9 K. However, the superconductivity is destroyed if the magnetic field at the surface of the metal reaches or exceeds 0.10 T. What is the maximum current in a straight, 3.0-mm-diameter superconducting niobium wire?

52. What are the strength and direction of the magnetic field at point P in Figure P32.52?

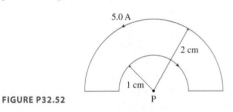

FIGURE P32.52

53. The earth's magnetic field, with a magnetic dipole moment of 8.0×10^{22} A m^2, is generated by currents within the molten iron of the earth's outer core. (The inner core is solid iron.) As a simple model, consider the outer core to be a current loop made of a 1000-km-diameter "wire" of molten iron. The loop diameter, measured between the centers of the "wires," is 3000 km.
    a. What is the current in the current loop?
    b. What is the current density J in the current loop?
    c. To decide whether this is a large or a small current density, compare it to the current density of a 1.0 A current in a 1.0-mm-diameter wire.

54. What is the magnetic field at the center of the loop in Figure P32.54?

FIGURE P32.54

55. Your employer asks you to build a 20-cm-long solenoid with an interior field of 5.0 mT. The specifications call for a single layer of wire, wound with the coils as close together as possible. You have two spools of wire available. Wire with a #18

gauge has a diameter of 1.02 mm and has a maximum current rating of 6 A. Wire with a #26 gauge is 0.41 mm in diameter and can carry up to 1 A. Which wire should you use, and what current will you need?

56. The magnetic field strength at the north pole of a 2.0-cm-diameter, 8-cm-long Alnico magnet is 0.10 T. To produce the same field with a solenoid of the same size, carrying a current of 2.0 A, how many turns of wire would you need? Does this seem feasible? (See Problem 55 for information about wire sizes and maximum current.)

57. Two identical coils are parallel to each other on the same axis. They are separated by a distance equal to their radius. They each have N turns and carry equal currents I in the same direction.
    a. Find an expression for the magnetic field strength at the midpoint between the loops.
    b. Calculate the field strength if the loops are 10 cm in diameter, have 10 turns, and carry a 1.0 A current.

58. You have a 1.0-m-long copper wire. You want to make an N-turn current loop that generates a 1.0 mT magnetic field at the center when the current is 1.0 A. You must use the entire wire. What will be the diameter of your coil?

59. Use the Biot-Savart law to find the magnetic field strength at the center of the semicircle in Figure P32.59.

FIGURE P32.59

60. The *toroid* of Figure P32.60 is a coil of wire wrapped around a doughnut-shaped ring (a *torus*) made of nonconducting material. Toroidal magnetic fields are used to confine fusion plasmas.
    a. From symmetry, what must be the *shape* of the magnetic field in this toroid? Explain.
    b. Use Ampère's law to find an expression for the magnetic field strength at a distance r from the axis of a toroid with N closely spaced turns carrying current I.
    c. Is a toroidal magnetic field a uniform field? Explain.

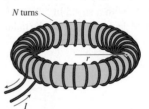

FIGURE P32.60

61. A long, hollow wire has inner radius R_1 and outer radius R_2. The wire carries current I uniformly distributed across the area of the wire. Use Ampère's law to find an expression for the magnetic field strength in the three regions $0 < r < R_1$, $R_1 < r < R_2$, and $R_2 < r$.

62. An electron travels with speed 1.0×10^7 m/s between the two parallel charged plates shown in Figure P32.62. The plates are separated by 1.0 cm and are charged by a 200 V battery. What magnetic field strength and direction will allow the electron to pass between the plates without being deflected?

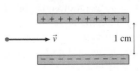

FIGURE P32.62

63. An electron in a cathode-ray tube is accelerated through a potential difference of 10 kV, then passes through the 2.0-cm-wide region of uniform magnetic field in Figure P32.63. What field strength will deflect the electron by 10°?

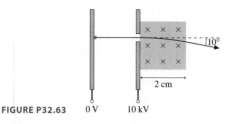

FIGURE P32.63   0 V   10 kV

64. The microwaves in a microwave oven are produced in a special tube called a *magnetron*. The electrons orbit the magnetic field at 2.4 GHz, and as they do so they emit 2.4 GHz electromagnetic waves.
    a. What is the magnetic field strength?
    b. If the maximum diameter of the electron orbit before the electron hits the wall of the tube is 2.5 cm, what is the maximum electron kinetic energy?

65. An antiproton (same properties as a proton except that $q = -e$) is moving in the combined electric and magnetic fields of Figure P32.65.
    a. What are the magnitude and direction of the antiproton's acceleration at this instant?
    b. What would be the magnitude and direction of the acceleration if $\vec{v}$ were reversed?

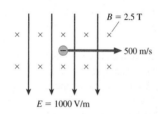

FIGURE P32.65 $E = 1000$ V/m

66. The uniform 30 mT magnetic field in Figure P32.66 points in the positive z-direction. An electron enters the region of magnetic field with a speed of 5.0×10^6 m/s and at an angle of 30° above the xy-plane. Find the radius r and the pitch p of the electron's spiral trajectory.

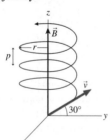

FIGURE P32.66 x

67. a. A 65-cm-diameter cyclotron uses a 500 V oscillating potential difference between the dees. What is the maximum kinetic energy of a proton if the magnetic field strength is 0.75 T?
    b. How many revolutions does the proton make before leaving the cyclotron?

68. For your senior project, you would like to build a cyclotron that will accelerate protons to 10% of the speed of light. The largest vacuum chamber you can find is 50 cm in diameter. What magnetic field strength will you need?

69. A Hall-effect probe to measure magnetic field strengths needs to be calibrated in a known magnetic field. Although it is not easy to do, magnetic fields can be precisely measured by measuring the cyclotron frequency of protons. A testing laboratory adjusts a magnetic field until the proton's cyclotron frequency is 10.0 MHz. At this field strength, the Hall voltage on the probe is 0.543 mV when the current through the probe is 0.150 mA. Later, when an unknown magnetic field is measured, the Hall voltage at the same current is 1.735 mV. What is the strength of this magnetic field?

70. Figure P32.70 shows a *mass spectrometer,* an analytical instrument used to identify the various molecules in a sample by measuring their charge-to-mass ratio e/m. The sample is ionized, the positive ions are accelerated (starting from rest) through a potential difference ΔV, and they then enter a region of uniform magnetic field. The field bends the ions into circular trajectories, but after just half a circle they either strike the wall or pass through a small opening to a detector. As the accelerating voltage is slowly increased, different ions reach the detector and are measured. Typical design values are a magnetic field strength $B = 0.200$ T and a spacing between the entrance and exit holes $d = 8.00$ cm. What accelerating potential difference ΔV is required to detect (a) N_2^+, (b) O_2^+, and (c) CO^+? See Exercise 32 for atomic data, and note there the comment about accuracy and significant figures.

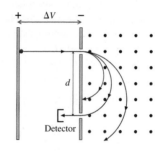

FIGURE P32.70

71. The two springs in Figure P32.71 each have a spring constant of 10 N/m. They are stretched by 1.0 cm when a current passes through the wire. How big is the current?

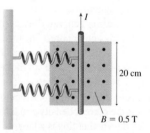

FIGURE P32.71

72. Figure P32.72 is a cross section through three long wires with linear mass density 50 g/m. They each carry equal currents in the directions shown. The lower two wires are 4.0 cm apart and are attached to a table. What current I will allow the upper wire to "float" so as to form an equilateral triangle with the lower wires?

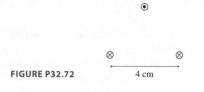

FIGURE P32.72   4 cm

73. A long, straight wire with linear mass density of 50 g/m is suspended by threads, as shown in Figure P32.73. A 10 A current in the wire experiences a horizontal magnetic force that deflects it to an equilibrium angle of 10°. What are the strength and direction of the magnetic field $\vec{B}$?

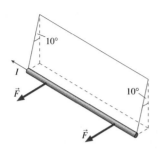

FIGURE P32.73

74. A bar magnet experiences a torque of magnitude 0.075 N m when it is perpendicular to a 0.50 T external magnetic field. What is the strength of the bar magnet's on-axis magnetic field at a point 20 cm from the center of the magnet?

75. In the semiclassical Bohr model of the hydrogen atom, the electron moves in a circular orbit of radius 5.3×10^{-11} m with speed 2.2×10^6 m/s. According to this model, what is the magnetic field at the center of a hydrogen atom?
    Hint: Determine the *average* current of the orbiting electron.

76. A *nonuniform* magnetic field exerts a net force on a current loop of radius R. Figure P32.76 shows a magnetic field that is diverging from the end of a bar magnet. The magnetic field at the position of the current loop makes an angle θ with respect to the vertical.
    a. Find an expression for the net magnetic force on the current.
    b. Calculate the force if $R = 2.0$ cm, $I = 0.50$ A, $B = 200$ mT, and $\theta = 20°$.

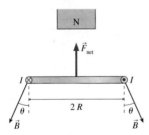

FIGURE P32.76

77. A computer diskette is a plastic disk coated with a ferromagnetic paint. A single magnetic domain can have its magnetic moment oriented to point either up or down, and these two orientations can be interpreted as a binary 0 (up) or 1 (down). Each 0 or 1 is called a *bit* of information. A diskette stores roughly 500,000 *bytes* of data on one side, and each byte contains eight bits. Estimate the width of a magnetic domain, and compare your answer to the typical domain size given in the text. List any assumptions you use in your estimate.

78. The ends of two permanent bar magnets can either attract or repel each other. Give a step-by-step description, using both words and picture, of how these magnetic forces result from the interaction between the electron spins. Consider both the attractive and the repulsive situations.

79. A permanent magnet can pick up a piece of nonmagnetized iron. Give a step-by-step description, using both words and picture, of how the magnetic force on the iron results from the interaction between the electron spins.

Challenge Problems

80. The 10-turn loop of wire shown in Figure CP32.80 lies in a horizontal plane, parallel to a uniform horizontal magnetic field, and carries a 2.0 A current. The loop is free to rotate about a nonmagnetic axle through the center. A 50 g mass hangs from one edge of the loop. What magnetic field strength will prevent the loop from rotating about the axle?

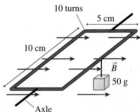

FIGURE CP32.80

81. a. Derive an expression for the magnetic field strength at distance d from the center of a straight wire of finite length L that carries current I.
    b. Determine the field strength at the center of a current-carrying *square* loop having sides of length $2R$.
    c. Compare your answer to part b to the field at the center of a *circular* loop of diameter $2R$. Do so by computing the ratio $B_{\text{square}}/B_{\text{circle}}$.

82. A long, straight conducting wire of radius R has a nonuniform current density $J = J_0 r/R$, where J_0 is a constant. The wire carries total current I.
    a. Find an expression for J_0 in terms of I and R.
    b. Find an expression for the magnetic field strength inside the wire at radius r.
    c. At the boundary, $r = R$, does your solution match the known field outside a long, straight current-carrying wire?

83. The coaxial cable shown in Figure CP32.83 consists of a solid inner conductor of radius R_1 surrounded by a hollow, very thin outer conductor of radius R_2. The two carry equal currents I, but in *opposite* directions. The current density is uniformly distributed over each conductor.
    a. Find expressions for three magnetic fields: within the inner conductor, in the space between the conductors, and outside the outer conductor.
    b. Draw a graph of B versus r from $r = 0$ to $r = 2R_2$ if $R_1 = \frac{1}{3}R_2$.

FIGURE CP32.83

84. An infinitely wide flat sheet of charge flows out of the page in Figure CP32.84. The current per unit width along the sheet (amps per meter) is given by the linear current density J_s.
    a. What is the *shape* of the magnetic field? To answer this question, you may find it helpful to approximate the current sheet as many parallel, closely spaced current-carrying wires. Give your answer as a picture showing magnetic field vectors.

b. Find the magnetic field strength at distance d above or below the current sheet.

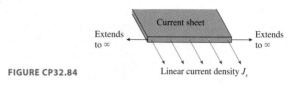

FIGURE CP32.84

STOP TO THINK ANSWERS

Stop to Think 32.1: Not at all. The charge exerts weak, attractive polarization forces on both ends of the compass needle, but in this configuration the forces will balance and have no net effect.

Stop to Think 32.2: d. Point your right thumb in the direction of the current and curl your fingers around the wire.

Stop to Think 32.3: b. Point your right thumb out of the page, in the direction of $\vec{v}$. Your fingers are pointing down as they curl around the left side.

Stop to Think 32.4: b. The right-hand rule gives a downward $\vec{B}$ for a clockwise current. The north pole is on the side from which the field emerges.

Stop to Think 32.5: c. For a field pointing into the page, $\vec{v} \times \vec{B}$ is to the right. But the electron is negative, so the force is in the direction of $-(\vec{v} \times \vec{B})$.

Stop to Think 32.6: b. Repulsion indicates that the south pole of the loop is on the right, facing the bar magnet; the north pole is on the left. Then the right-hand rule gives the current direction.

Stop to Think 32.7: a or c. Any downward magnetic field will align the magnetic domains as shown.

33 Electromagnetic Induction

Electromagnetic induction is the scientific principle that underlies many modern technologies, from the generation of electricity to communications and data storage.

▶ **Looking Ahead**
The goal of Chapter 33 is to understand and apply electromagnetic induction. In this chapter you will learn to:

- Calculate induced current.
- Calculate magnetic flux.
- Use Lenz's law and Faraday's law to determine the direction and size of induced currents.
- Understand how induced electric and magnetic fields lead to electromagnetic waves.
- Analyze circuits with inductors.

◀ **Looking Back**
This chapter will join together ideas about magnetic fields and electric potential. Please review:

- Section 11.3 The vector dot product.
- Section 30.4 Sources of electric potential.
- Sections 32.4–32.8 Magnetic fields and magnetic forces.

What do electric generators, metal detectors, video recorders, computer hard disks, and cell phones have in common? Surprisingly, these diverse technologies all stem from a single scientific principle, electromagnetic induction. **Electromagnetic induction** is the process of generating an electric current by varying the magnetic field that passes through a circuit.

The many applications of electromagnetic induction make it an important topic for study. But more fundamentally, electromagnetic induction establishes an important link between electricity and magnetism. We've been studying electric and magnetic fields as if they were separate, independent fields. Electromagnetic induction forms a link between $\vec{E}$ and $\vec{B}$, a link with important implications for understanding light as an electromagnetic wave.

Electromagnetic induction is a subtle topic, so we will build up to it gradually. We'll first examine different aspects of induction and become familiar with its basic characteristics. Section 33.5 will then introduce Faraday's law, a new law of physics not derivable from any previous laws you have studied. The remainder of the chapter will explore its implications and applications.

33.1 Induced Currents

Oersted's 1820 discovery that a current creates a magnetic field generated enormous excitement. Dozens of scientists immediately began to explore the implications of this discovery. One question they hoped to answer was whether the converse of Oersted's discovery was true. That is, can a magnet be used to create

FIGURE 33.1 Michael Faraday.

a current? There was not yet a good understanding of the origins or properties of electricity and magnetism, so scientists hoping to generate a current from magnetism had little to guide them. Many experiments were reported in which wires and coils were placed in or around magnets of various sizes and shapes, but no one was able to generate a current.

On the other side of the Atlantic, the American scientist Joseph Henry read of these new discoveries with great interest. American professors at the time were expected to devote all of their time to teaching, so Henry had little opportunity for research. It was during a one-month vacation in 1831 that Henry became the first to discover how to produce a current from magnetism, a process that we now call *electromagnetic induction.* But Henry had no time for follow-up studies, and he was not able to publish his discovery until the following year.

At about the same time, in England, Michael Faraday (Figure 33.1) made the same discovery and immediately published his findings. You met Faraday in Chapter 25 as the inventor of the concept of a *field.* The idea came to him as he observed that a compass needle stays tangent to a circle around a current-carrying wire. Faraday ascribed the needle's behavior to "circular lines of force," an idea that soon came to be known as the *magnetic field.* This pictorial representation played a crucial role in Faraday's discovery of the law of electromagnetic induction.

Credit in science usually goes to the first to publish, so today we study *Faraday's law* rather than *Henry's law.* The situation, however, is not entirely unjust. Henry had discovered an *effect,* but he was not able to do the research needed to understand the implications of his discovery. Even if Faraday did not have priority of discovery, it was Faraday who studied the new phenomenon of electromagnetic induction, established its properties, and realized that he had discovered a new law of nature.

Faraday's Discovery

Faraday's 1831 discovery, like Oersted's, was a happy combination of an unplanned event and a mind that was prepared to immediately recognize its significance. Faraday was experimenting with two coils of wire wrapped around an iron ring, as shown in Figure 33.2. He had hoped that the magnetic field generated by a current in the coil on the left would induce a magnetic field in the iron, and that the magnetic field in the iron might then somehow create a current in the circuit on the right.

Like all his previous attempts, this technique failed to generate a current. But Faraday happened to notice that the needle of the current meter jumped ever so slightly at the instant when he closed the switch in the circuit on the left. After the switch was closed, the needle immediately returned to zero. The needle again jumped when he later opened the switch, but this time in the opposite direction. Faraday recognized that the motion of the needle indicated a very slight current in the circuit on the right. But the effect happened only during the very brief interval when the current on the left was starting or stopping, not while it was steady.

Faraday applied his mental picture of lines of force to this discovery. The current on the left first magnetizes the iron ring, then the magnetic field of the iron ring passes through the coil on the right. Faraday's observation that the current-meter needle jumped only when the switch was opened and closed suggested to him that a current was generated only if the magnetic field was *changing* as it passed through the coil. This would explain why all the previous attempts to generate a current were unsuccessful: they had used only steady, unchanging magnetic fields.

Faraday set out to test this hypothesis. If the critical issue was *changing* the magnetic field through the loop, then the iron ring should not be necessary. That is, any method that changes the magnetic field should work. Faraday began a series of experiments to find out if this was true.

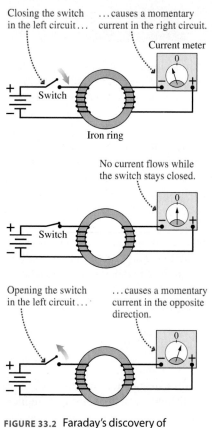

Closing the switch in the left circuit... ...causes a momentary current in the right circuit.

No current flows while the switch stays closed.

Opening the switch in the left circuit... ...causes a momentary current in the opposite direction.

FIGURE 33.2 Faraday's discovery of electromagnetic induction.

Faraday investigates electromagnetic induction

Faraday placed one coil directly above the other, without the iron ring. There was no current in the lower circuit while the switch was in the closed position, but a momentary current appeared whenever the switch was opened or closed.

He pushed a bar magnet into a coil of wire. This action caused a momentary deflection of the current-meter needle, although *holding* the magnet inside the coil had no effect. A quick withdrawal of the magnet deflected the needle in the other direction.

Must the magnet move? Faraday created a momentary current by rapidly pulling a coil of wire out of a magnetic field, although there was no current if the coil was stationary in the magnetic field. Pushing the coil *into* the magnet caused the needle to deflect in the opposite direction.

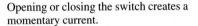

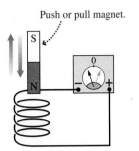

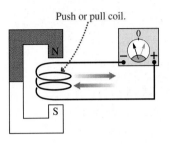

Opening or closing the switch creates a momentary current.

Pushing the magnet into the coil or pulling it out creates a momentary current.

Pushing the coil into the magnet or pulling it out creates a momentary current.

To summarize

Faraday found that there is a current in a coil of wire if and only if the magnetic field passing through the coil is *changing*. This is an informal statement of what we'll soon call *Faraday's law*.

It makes no difference what causes the magnetic field to change: current stopping or starting in a nearby circuit, moving a magnet through the coil, or moving the coil in and out of a magnet. The effect is the same in all cases. There is no current if the field through the coil is not changing, so it's not the magnetic field itself that is responsible for the current but, instead, it is the *changing of the magnetic field.*

The current in a circuit due to a changing magnetic field is called an **induced current.** Opening the switch or moving the magnet *induces* a current in a nearby circuit. An induced current is not caused by a battery. It is a completely new way to generate a current, and we will have to discover how it is similar to and how it is different from currents we have studied previously.

The first induced currents were small, barely noticeable effects. Neither Faraday nor Henry could have answered the question, "What good is it?" Yet electromagnetic induction has became the basis of commercial electricity generation, of radio and television broadcasting, of computer memories and data storage, of cell phones, and much more.

33.2 Motional emf

An induced current can be created two different ways:

1. By changing the size or orientation of a circuit in a stationary magnetic field, or
2. By changing the magnetic field through a stationary circuit.

Although the effects are the same, the causes turn out to be different. We'll start our investigation of electromagnetic induction by looking at situations in which the magnetic field is fixed while the circuit moves or changes.

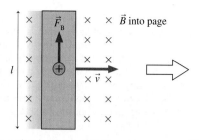

Charge carriers in the wire experience an upward force of magnitude $F_B = qvB$. Being free to move, positive charges flow upward (or, if you prefer, negative charges downward).

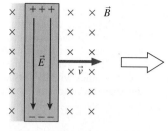

The charge separation creates an electric field in the conductor. $\vec{E}$ increases as more charge flows.

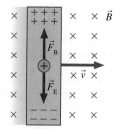

The charge flow continues until the downward electric force $\vec{F}_E$ is large enough to balance the upward magnetic force $\vec{F}_B$. Then the net force on a charge is zero and the current ceases.

FIGURE 33.3 The magnetic force on the charge carriers in a moving conductor creates an electric field inside the conductor.

(a) Magnetic forces separate the charges and cause a potential difference between the ends. This is a motional emf.

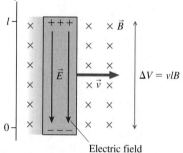

Electric field inside the moving conductor

(b) Chemical reactions separate the charges and cause a potential difference between the ends. This is a chemical emf.

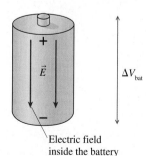

Electric field inside the battery

FIGURE 33.4 Two different ways to generate an emf.

To begin, consider a conductor of length l that moves with velocity $\vec{v}$ through a uniform magnetic field $\vec{B}$, as shown in Figure 33.3. The charge carriers inside the wire also move with velocity $\vec{v}$, so they each experience a magnetic force $\vec{F}_B = q\vec{v} \times \vec{B}$. For simplicity, we will assume that $\vec{v}$ is perpendicular to $\vec{B}$, in which case the strength of the force is $F_B = qvB$. This force causes the charge carriers to move, separating the positive and negative charges and thus creating an electric field inside the conductor.

The charge carriers continue to move until the electric force $F_E = qE$ exactly balances the magnetic force. This balance happens when the electric field strength is

$$E = vB \tag{33.1}$$

In other words, the magnetic force on the charge carriers in a moving conductor creates an electric field $E = vB$ inside the conductor.

The electric field, in turn, creates an electric potential difference between the two ends of the moving conductor. Figure 33.4a defines a coordinate system in which $\vec{E} = -vB\hat{j}$. Using the connection between the electric field and the electric potential that we found in Chapter 30,

$$\Delta V = V_{\text{top}} - V_{\text{bottom}} = -\int_0^l E_y \, dy = -\int_0^l (-vB) dy = vlB \tag{33.2}$$

Thus the motion of the wire through a magnetic field *induces* a potential difference vlB between the ends of the conductor. The potential difference depends on the strength of the magnetic field and on the wire's speed through the field.

There's an important analogy between this potential difference and the potential difference of a battery. Figure 33.4b reminds you that a battery uses a non-electric force—the charge escalator—to separate positive and negative charges. The emf $\mathcal{E}$ of the battery was defined as the work performed per charge (W/q) to separate the charges. An isolated battery, with no current, has a potential difference $\Delta V_{\text{bat}} = \mathcal{E}$. We could refer to a battery, where the charges are separated by chemical reactions, as a source of *chemical emf*.

The moving conductor develops a potential difference because of the work done by magnetic forces to separate the charges. You can think of the moving conductor as a "battery" that stays charged only as long as it keeps moving but "runs down" if it stops. The emf of the conductor is due to its motion, rather than to chemical reactions inside, so we can define the **motional emf** of a conductor moving with velocity $\vec{v}$ to be

$$\mathcal{E} = vlB \tag{33.3}$$

STOP TO THINK 33.1   A square conductor moves through a uniform magnetic field. Which of the figures shows the correct charge distribution on the conductor?

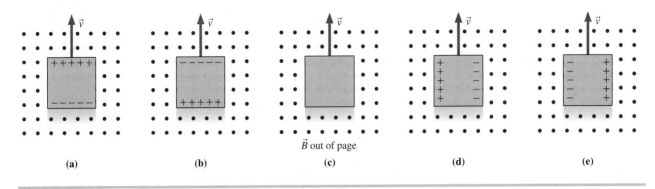

$\vec{B}$ out of page

(a)   (b)   (c)   (d)   (e)

EXAMPLE 33.1   A battery substitute
A 6.0-cm-long flashlight battery has an emf of 1.5 V. With what speed must a 6.0-cm-long wire move through a 0.10 T magnetic field to create a motional emf of 1.5 V?

SOLVE 1.5V is the motional emf. We can use equation 33.3 to find

$$v = \frac{\mathcal{E}}{lB} = \frac{1.5\ \text{V}}{(0.060\ \text{m})(0.10\ \text{T})} = 250\ \text{m/s}$$

ASSESS 250 m/s ≈ 500 mph. This might not be a very practical substitute for a battery, but it would work as long as the wire continued to move through the field with this speed.

EXAMPLE 33.2   Potential difference along a rotating bar
A metal bar of length l rotates with angular velocity ω about a pivot at one end of the bar. A uniform magnetic field $\vec{B}$ is perpendicular to the plane of rotation. What is the potential difference between the ends of the bar?

VISUALIZE Figure 33.5 is a pictorial representation of the bar. The magnetic forces on the charge carriers will cause the outer end to be positive with respect to the pivot.

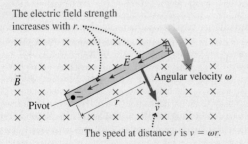

The electric field strength increases with r.

$\vec{B}$

Pivot

r

$\vec{v}$

$\vec{E}$

Angular velocity ω

The speed at distance r is $v = \omega r$.

FIGURE 33.5 Pictorial representation of a metal bar rotating in a magnetic field.

SOLVE Even though the bar is rotating, rather than moving in a straight line, the velocity of each charge carrier is perpendicular to $\vec{B}$. Consequently, the electric field created inside the bar is exactly that given in Equation 33.1, $E = vB$. But v, the speed of the charge carrier, now depends on its distance from the pivot. Recall that in rotational motion the tangential speed at radius r from the center of rotation is $v = \omega r$. Thus the electric field at distance r from the pivot is $E = \omega rB$. The electric field increases in strength as you move outward along the bar.

The electric field $\vec{E}$ points toward the pivot, so its radial component is $E_r = -\omega rB$. If we integrate outward from the center, the potential difference between the ends of the bar is

$$\Delta V = V_{\text{tip}} - V_{\text{pivot}} = -\int_0^l E_r\, dr$$

$$= -\int_0^l (-\omega rB)\, dr = \omega B \int_0^l r\, dr = \frac{1}{2}\omega l^2 B$$

ASSESS $\frac{1}{2}\omega l$ is the speed at the midpoint of the bar. Thus ΔV is $v_{\text{mid}}lB$, which seems reasonable.

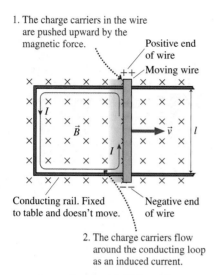

1. The charge carriers in the wire are pushed upward by the magnetic force.

Positive end of wire

Moving wire

Conducting rail. Fixed to table and doesn't move.

Negative end of wire

2. The charge carriers flow around the conducting loop as an induced current.

FIGURE 33.6 A current is induced in the circuit as the wire moves through a magnetic field.

Induced Current in a Circuit

The moving conductor of Figure 33.3 had an emf, but it couldn't sustain a current because the charges had nowhere to go. It's like a battery that is disconnected from a circuit. We can change this by including the moving conductor in a circuit.

Figure 33.6 shows a conducting wire sliding with speed v along a U-shaped conducting rail. We'll assume that the rail is attached to a table and cannot move. The wire and the rail together form a closed conducting loop—a circuit.

Suppose a magnetic field $\vec{B}$ is perpendicular to the plane of the circuit. Charges in the moving wire will be pushed to the ends of the wire by the magnetic force, just as they were in Figure 33.3, but now the charges can continue to flow around the circuit. That is, the moving wire acts like a battery in a circuit.

The current in the circuit is an *induced current.* In this example, the induced current is counterclockwise (ccw). If the total resistance of the circuit is R, the induced current is given by Ohm's law as

$$I = \frac{\mathcal{E}}{R} = \frac{vlB}{R} \tag{33.4}$$

In this situation, the induced current is due to magnetic forces on moving charges.

STOP TO THINK 33.2 Is there an induced current in this circuit? If so, what is its direction?

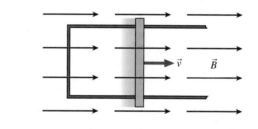

We've assumed that the wire is moving along the rail at constant speed. It turns out that we must apply a continuous pulling force $\vec{F}_{pull}$ to make this happen. Figure 33.7 shows why. The moving wire, which now carries induced current I, is in a magnetic field. You learned in Chapter 32 that a magnetic field exerts a force on a current-carrying wire. According to the right-hand rule, the magnetic force $\vec{F}_{mag}$ on the moving wire points to the left. This "magnetic drag" will cause the wire to slow down and stop *unless* we exert an equal but opposite pulling force $\vec{F}_{pull}$ to keep the wire moving.

NOTE ▶ Think about this carefully. As the wire moves to the right, the magnetic force $\vec{F}_B$ pushes the charge carriers *parallel* to the wire. Their motion, as they continue around the circuit, is the induced current I. Now, because we have a current, a second magnetic force $\vec{F}_{mag}$ enters the picture. This force on the current is *perpendicular* to the wire and acts to slow the wire's motion. ◀

The magnitude of the magnetic force on a current-carrying wire was found in Chapter 32 to be $F_{mag} = IlB$. Using that result, along with Equation 33.4 for the induced current, we find that the force required to pull the wire with a constant speed v is

$$F_{pull} = F_{mag} = IlB = \left(\frac{vlB}{R}\right)lB = \frac{vl^2B^2}{R} \tag{33.5}$$

A pulling force to the right must balance the magnetic force to keep the wire moving at constant speed. This force does work on the wire.

The induced current flows through the moving wire.

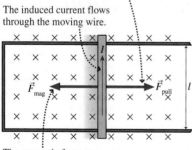

The magnetic force on the current-carrying wire is opposite the motion.

FIGURE 33.7 A pulling force is needed to move the wire to the right.

Energy Considerations

The environment must do work on the wire to pull it. What happens to the energy transferred to the wire by this work? Is energy conserved as the wire moves along the rail? It will be easier to answer this question if we think about power rather than work. Power is the *rate* at which work is done on the wire. You learned in Chapter 11 that the power exerted by a force pushing or pulling an object with velocity v is $P = Fv$. The power provided to the circuit by pulling on the wire is

$$P_{input} = F_{pull}v = \frac{v^2 l^2 B^2}{R} \qquad (33.6)$$

This is the rate at which energy is added to the circuit by the pulling force.

But the circuit also dissipates energy by transforming electric energy into the thermal energy of the wires and components, heating them up. You learned in Chapter 31 that the power dissipated by current I as it passes through resistance R is $P = I^2R$. Equation 33.4 for the induced current I gives us the power dissipated by the circuit of Figure 33.6:

$$P_{dissipated} = I^2 R = \frac{v^2 l^2 B^2}{R} \qquad (33.7)$$

You can see that Equations 33.6 and 33.7 are identical. **The rate at which work is done on the circuit exactly balances the rate at which energy is dissipated.** Thus *energy is conserved.*

If you have to *pull* on the wire to get it to move to the right, you might think that it would spring back to the left on its own. Figure 33.8 shows the same circuit with the wire moving to the left. In this case, you must *push* the wire to the left to keep it moving. The magnetic force is always opposite to the wire's direction of motion.

In both Figure 33.7, where the wire is pulled, and Figure 33.8, where it is pushed, a mechanical force is used to create a current. In other words, we have a conversion of *mechanical* energy to *electric* energy. A device that converts mechanical energy to electric energy is called a **generator.** The slide-wire circuits of Figure 33.7 and 33.8 are simple examples of a generator. We will look at more practical examples of generators later in the chapter.

We can summarize our analysis as follows:

1. Pulling or pushing the wire through the magnetic field at speed v creates a motional emf $\mathcal{E}$ in the wire and induces a current $I = \mathcal{E}/R$ in the circuit.
2. To keep the wire moving at constant speed, a pulling or pushing force must balance the magnetic force on the wire. This force does work on the circuit.
3. The work done by the pulling or pushing force exactly balances the energy dissipated by the current as it passes through the resistance of the circuit.

1. The magnetic force on the charge carriers is down, so the induced current flows clockwise.

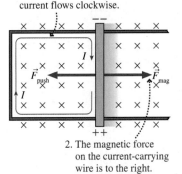

2. The magnetic force on the current-carrying wire is to the right.

FIGURE 33.8 A pushing force is needed to move the wire to the left.

EXAMPLE 33.3 Lighting a bulb

Figure 33.9 shows a circuit consisting of a flashlight bulb, rated 3.0 V/1.5 W, and ideal wires with no resistance. The right wire of the circuit, which is 10 cm long, is pulled at constant speed v through a perpendicular magnetic field of strength 0.10 T.

a. What speed must the wire have to light the bulb to full brightness?
b. What force is needed to keep the wire moving?

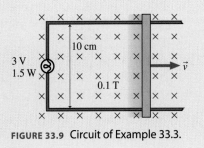

FIGURE 33.9 Circuit of Example 33.3.

MODEL Treat the moving wire as a source of motional emf.

VISUALIZE The direction of the magnetic force on the charge carriers, $\vec{F}_B = q\vec{v} \times \vec{B}$, will cause a counterclockwise (ccw) induced current.

SOLVE

a. The bulb's rating of 3.0 V/1.5 W means that at full brightness it will dissipate 1.5 W at a potential difference of 3.0 V. Because the power is related to the voltage and current by $P = I\Delta V$, the current causing full brightness is

$$I = \frac{P}{\Delta V} = \frac{1.5\ \text{W}}{3.0\ \text{V}} = 0.50\ \text{A}$$

The bulb's resistance, which is the total resistance of the circuit, is

$$R = \frac{\Delta V}{I} = \frac{3.0\ \text{V}}{0.50\ \text{A}} = 6.0\ \Omega$$

Equation 33.5 gives the speed needed to induce this current:

$$v = \frac{IR}{lB} = \frac{(0.50\ \text{A})(6.0\ \Omega)}{(0.10\ \text{m})(0.10\ \text{T})} = 300\ \text{m/s}$$

You can confirm from Equation 33.6 that the input power at this speed is 1.5 W.

b. From Equation 33.5, the pulling force must be

$$F_{\text{pull}} = \frac{vl^2B^2}{R} = 5.0 \times 10^{-3}\ \text{N}$$

You can also obtain this result from $F_{\text{pull}} = P/v$.

ASSESS Example 33.1 showed that high speeds are needed to produce significant potential difference. Thus 300 m/s is not surprising. The pulling force is not very large, but even a small force can deliver large amounts of power $P = Fv$ when v is large.

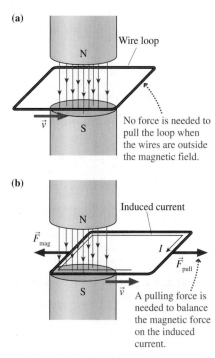

(a)

Wire loop

N

$\vec{v}$ S

No force is needed to pull the loop when the wires are outside the magnetic field.

(b)

Induced current

N

$\vec{F}_{\text{mag}}$

I

$\vec{F}_{\text{pull}}$

S $\vec{v}$ A pulling force is needed to balance the magnetic force on the induced current.

FIGURE 33.10 Pulling a loop of wire out of a magnetic field.

Eddy Currents

Figure 33.10 shows a *rigid* square loop of wire between the poles of a magnet. The upper pole is a north pole, so the magnetic field points downward and is confined to the region between the poles. The magnetic field in Figure 33.10a passes through the loop, but the wires are not in the field. None of the charge carriers in the wire experience a magnetic force, so there is no induced current and it takes no force to pull the loop to the right.

But when the left edge of the loop enters the field, as shown in Figure 33.10b, the magnetic force on the charge carriers induces a current in the loop. The magnetic field then exerts a retarding magnetic force on this current, so **a pulling force must be exerted to pull the loop out of the magnetic field.** Note that the wire, typically copper, is *not* a magnetic material. A piece of the wire held near the magnet would feel no force. Nor would a force be required to pull the wire out if there were a gap in the loop, breaking the circuit and preventing a current. It is the *induced current* in the complete loop that causes the wire to experience a retarding force.

Figure 33.11 is an alternative way of viewing the situation. Pulling the loop out of the field is like pulling a magnet off the refrigerator door. Regardless of which way you look at it, a force is required to pull the loop out of the magnetic field.

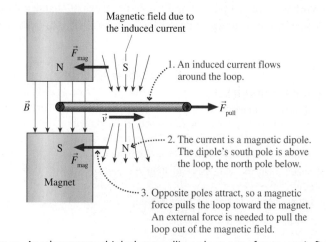

Magnetic field due to the induced current

$\vec{F}_{\text{mag}}$

N   S

$\vec{B}$

$\vec{v}$ $\vec{F}_{\text{pull}}$

S   N $\vec{F}_{\text{mag}}$

Magnet

1. An induced current flows around the loop.

2. The current is a magnetic dipole. The dipole's south pole is above the loop, the north pole below.

3. Opposite poles attract, so a magnetic force pulls the loop toward the magnet. An external force is needed to pull the loop out of the magnetic field.

FIGURE 33.11 Another way to think about pulling a loop out of a magnetic field.

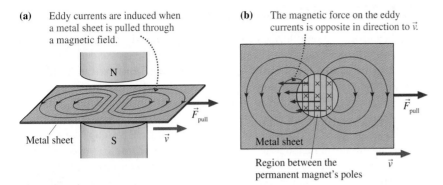

(a) Eddy currents are induced when a metal sheet is pulled through a magnetic field.

(b) The magnetic force on the eddy currents is opposite in direction to $\vec{v}$.

N

S

Metal sheet

$\vec{F}_{pull}$

$\vec{v}$

Metal sheet

Region between the permanent magnet's poles

$\vec{F}_{pull}$

$\vec{v}$

FIGURE 33.12 Eddy currents.

These ideas have interesting implications. Consider pulling a *sheet* of metal through a magnetic field, as shown in Figure 33.12a. The metal, we will assume, is not a magnetic material, so it experiences no magnetic force if it is at rest. The charge carriers in the metal experience a magnetic force as the sheet is dragged between the pole tips of the magnet. A current is induced, just as in the loop of wire, but here the currents do not have wires to define their path. As a consequence, two "whirlpools" of current begin to circulate in the metal. These spread-out current whirlpools in a solid metal are called **eddy currents.**

Figure 33.12b shows the situation if we look down from the north pole of the magnet toward the south pole. There is a magnetic force on the eddy current as it passes between the pole tips. This force is to the left, acting as a retarding force. Thus **an external force is required to pull a metal through a magnetic field.** If the pulling force ceases, the retarding magnetic force quickly causes the metal to decelerate until it stops.

Eddy currents are often undesirable. The power dissipation of eddy currents can cause unwanted heating, and the magnetic forces on eddy currents means that extra energy must be expended to move metals in magnetic fields. But eddy currents also have important useful applications. A good example is magnetic braking, which is used in some trains and transit-system vehicles.

The moving train car has an electromagnet that straddles the rail, as shown in Figure 33.13. During normal travel, there is no current through the electromagnet and no magnetic field. To stop the car, a current is switched into the electromagnet. The current creates a strong magnetic field that passes *through* the rail, and the motion of the rail relative to the magnet induces eddy currents in the rail. The magnetic force between the electromagnet and the eddy currents acts as a braking force on the magnet and, thus, on the car. Magnetic braking systems are very efficient, and they have the added advantage that they heat the rail rather than the brakes.

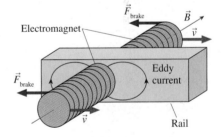

FIGURE 33.13 Magnetic braking systems are an application of eddy currents.

STOP TO THINK 33.3   A square loop of copper wire is pulled through a region of magnetic field. Rank in order, from strongest to weakest, the pulling forces $\vec{F}_1$, $\vec{F}_2$, $\vec{F}_3$, and $\vec{F}_4$ that must be applied to keep the loop moving at constant speed.

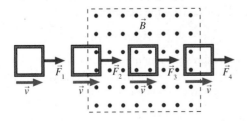

33.3 Magnetic Flux

We've begun our exploration of electromagnetic induction by analyzing a circuit in which one wire moves through a magnetic field. You might be wondering what this has to do with Faraday's discovery. Faraday found that a current is induced when the amount of magnetic field passing through a coil or a loop of wire changes. But that's exactly what happens as the slide wire moves down the rail in Figure 33.6! As the circuit expands, more magnetic field passes through. It's time to define more clearly what we mean by "the amount of field passing through a loop."

Imagine holding a rectangular loop of wire in front of a fan, as shown in Figure 33.14. The amount of air that flows through the loop depends on the effective area of the loop as seen along the direction of flow. You can see from the figure that the effective area (i.e., as seen facing the fan) is

$$A_{\text{eff}} = ab \cos \theta = A \cos \theta \tag{33.8}$$

where A is the area of the loop and θ is the tilt angle of the loop. A loop perpendicular to the flow, with $\theta = 0°$, has $A_{\text{eff}} = A$, the full area of the loop. This is the orientation for maximum flow through the loop. No air at all flows through the loop if it is tilted 90°, and you can see that $A_{\text{eff}} = 0$ in this case.

FIGURE 33.14 The amount of air flowing through a loop depends on the effective area of the loop.

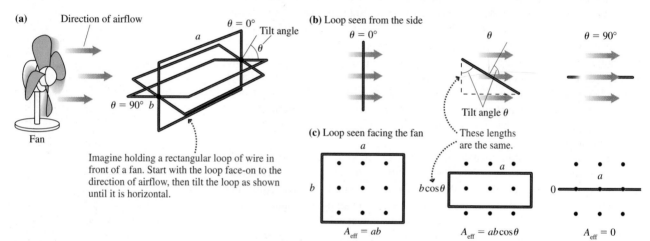

(a) Direction of airflow $\theta = 0°$ Tilt angle a θ $\theta = 90°$ b Fan

Imagine holding a rectangular loop of wire in front of a fan. Start with the loop face-on to the direction of airflow, then tilt the loop as shown until it is horizontal.

(b) Loop seen from the side $\theta = 0°$ θ $\theta = 90°$ Tilt angle θ These lengths are the same.

(c) Loop seen facing the fan a b $A_{\text{eff}} = ab$ $b\cos\theta$ a $A_{\text{eff}} = ab\cos\theta$ 0 a $A_{\text{eff}} = 0$

We can apply this idea to a magnetic field passing through a loop. Figure 33.15 shows a loop of area $A = ab$ in a uniform magnetic field. Think of these field vectors, seen here from behind, as if they were arrows shot into the page. The density of arrows (arrows per m²) is proportional to the strength B of the magnetic field; a stronger field would be represented by arrows spaced closer together. The number of arrows passing through a loop of wire depends on two factors:

1. The density of arrows, which is proportional to B, and
2. The effective area $A_{\text{eff}} = A \cos \theta$ of the loop.

The angle θ is the angle between the magnetic field and the axis of the loop. The maximum number of arrows passes through the loop when it is perpendicular to the magnetic field ($\theta = 0°$). No arrows pass through the loop if it is tilted 90°.

With this in mind, let's define the **magnetic flux** Φ_{m} as

$$\Phi_{\text{m}} = A_{\text{eff}} B = AB \cos \theta \tag{33.9}$$

The magnetic flux measures the amount of magnetic field passing through a loop of area A if the loop is tilted at angle θ from the field. The SI unit of magnetic flux is the **weber**. From Equation 33.9 you can see that

$$1 \text{ weber} = 1 \text{ Wb} = 1 \text{ T}\,\text{m}^2$$

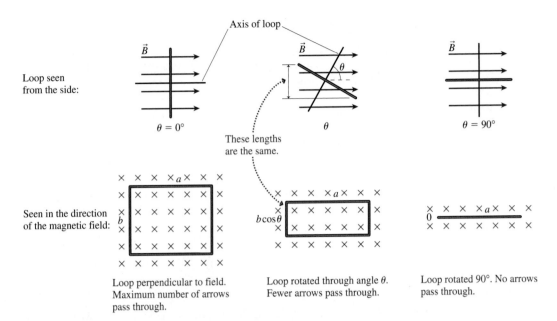

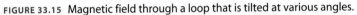

FIGURE 33.15 Magnetic field through a loop that is tilted at various angles.

Equation 33.9 is reminiscent of the vector dot product: $\vec{A} \cdot \vec{B} = AB \cos \theta$. With that in mind, let's define an **area vector** $\vec{A}$ to be a vector that is *perpendicular* to a loop and whose magnitude is equal to the area A of the loop. Vector $\vec{A}$ has units of m^2. Figure 33.16a shows the area vector $\vec{A}$ for a circular loop of area A.

Figure 33.16b shows a magnetic field passing through a loop. The angle between vectors $\vec{A}$ and $\vec{B}$ is the same angle used in Equations 33.8 and 33.9 to define the effective area and the magnetic flux. So Equation 33.9 really is a dot product, and we can define the magnetic flux more concisely as

$$\Phi_m = \vec{A} \cdot \vec{B} \tag{33.10}$$

Writing the flux as a dot product helps make clear how angle θ is defined: θ is the angle between the magnetic field and a line *perpendicular* to the plane of the loop.

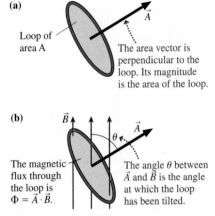

(a)

Loop of area A

The area vector is perpendicular to the loop. Its magnitude is the area of the loop.

(b)

The magnetic flux through the loop is $\Phi = \vec{A} \cdot \vec{B}$.

The angle θ between $\vec{A}$ and $\vec{B}$ is the angle at which the loop has been tilted.

FIGURE 33.16 Magnetic flux can be defined in terms of an area vector $\vec{A}$.

EXAMPLE 33.4 A circular loop rotating in a magnetic field

Figure 33.17 is an edge view of a 10-cm-diameter circular loop rotating in a uniform 0.050 T magnetic field. What is the magnetic flux through the loop when θ is 0°, 30°, 60°, and 90°?

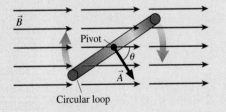

FIGURE 33.17 A circular loop in a magnetic field.

SOLVE Angle θ is the angle between the loop's area vector $\vec{A}$, which is perpendicular to the plane of the loop, and the magnetic field $\vec{B}$. Vector $\vec{A}$ has magnitude $A = \pi r^2 = 7.85 \times 10^{-3}\, m^2$. Thus the magnetic flux is

$$\Phi_m = \vec{A} \cdot \vec{B} = AB \cos \theta = \begin{cases} 3.93 \times 10^{-4}\,\text{Wb} & \theta = 0° \\ 3.40 \times 10^{-4}\,\text{Wb} & \theta = 30° \\ 1.96 \times 10^{-4}\,\text{Wb} & \theta = 60° \\ 0\,\text{Wb} & \theta = 90° \end{cases}$$

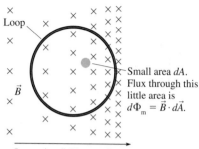

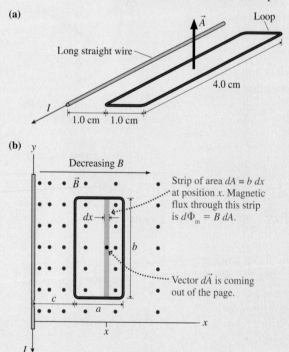

FIGURE 33.18 A loop in a nonuniform magnetic field.

Magnetic Flux in a Nonuniform Field

Equation 33.10 for the magnetic flux assumes that the field is uniform over the area of the loop. We can calculate the flux in a nonuniform field, one where the field strength changes from one edge of the loop to the other, but we'll need to use calculus.

Figure 33.18 shows a loop in a nonuniform magnetic field. Imagine dividing the loop into many small pieces of area dA. The infinitesimal flux $d\Phi_m$ through one such area, where the magnetic field is $\vec{B}$, is

$$d\Phi_m = \vec{B} \cdot d\vec{A} \qquad (33.11)$$

The total magnetic flux through the loop is the sum of the fluxes through each of the small areas. We find that sum by integrating. Thus the total magnetic flux through the loop is

$$\Phi_m = \int_{\text{area of loop}} \vec{B} \cdot d\vec{A} \qquad (33.12)$$

Equation 33.12 is a more general definition of magnetic flux. It may look rather formidable, so we'll illustrate its use with an example.

EXAMPLE 33.5 Magnetic flux from the current in a long straight wire

The 1.0 cm × 4.0 cm rectangular loop of Figure 33.19a is 1.0 cm away from a long straight wire. The wire carries a current of 1.0 A. What is the magnetic flux through the loop?

MODEL We'll treat the wire as if it were infinitely long. The magnetic field strength of a wire decreases with distance from the wire, so the field is *not* uniform over the area of the loop.

(a)

Long straight wire

Loop

4.0 cm

1.0 cm   1.0 cm

I

$\vec{A}$

(b)

y

Decreasing B

$\vec{B}$

Strip of area $dA = b\,dx$ at position x. Magnetic flux through this strip is $d\Phi_m = B\,dA$.

dx

b

c

a

Vector $d\vec{A}$ is coming out of the page.

x

x

I

FIGURE 33.19 Magnetic flux through a loop due to the magnetic field of a long straight wire.

VISUALIZE Using the right-hand rule, we see that the field, as it circles the wire, is perpendicular to the plane of the loop. Figure 33.19b redraws the loop with the field coming out of the page and establishes a coordinate system.

SOLVE Let the loop have dimensions a and b, as shown, with the near edge at distance c from the wire. The magnetic field varies with distance x from the wire, but the field is constant along a line parallel to the wire. This suggests dividing the loop into many narrow rectangular strips of length b and width dx, each forming a small area $dA = b\,dx$. The magnetic field has the same strength at all points within this small area. One such strip is shown in the figure at position x.

The area vector $d\vec{A}$ is perpendicular to the strip (coming out of the page), which makes it parallel to $\vec{B}$ ($\theta = 0°$). Thus the infinitesimal flux through this little area is

$$d\Phi_m = \vec{B} \cdot d\vec{A} = B\,dA = Bb\,dx = \frac{\mu_0 I b}{2\pi x}\,dx$$

where, from Chapter 32, we've used $B = \mu_0 I / 2\pi x$ as the magnetic field at distance x from a long straight wire. Integrating "over the area of the loop" means to integrate from the near edge of the loop at $x = c$ to the far edge at $x = c + a$. Thus

$$\Phi_m = \frac{\mu_0 I b}{2\pi} \int_c^{c+a} \frac{dx}{x} = \frac{\mu_0 I b}{2\pi} \ln x \Big|_c^{c+a} = \frac{\mu_0 I b}{2\pi} \ln\!\left(\frac{c+a}{c}\right)$$

Evaluating for $a = c = 0.010$ m, $b = 0.040$ m, and $I = 1.0$ A gives

$$\Phi_m = 5.55 \times 10^{-9}\ \text{Wb}$$

ASSESS The flux measures how much of the wire's magnetic field passes through the loop, but we had to integrate, rather than simply using Equation 33.10, because the field is stronger at the near edge of the loop than at the far edge.

33.4 Lenz's Law

We started out by looking at a situation in which a moving wire caused a loop to expand in a magnetic field. This is one way to change the magnetic flux through the loop. But Faraday found that a current can be induced by any change in the magnetic flux, no matter how it's accomplished.

For example, a momentary current is induced in the loop of Figure 33.20 as the bar magnet is pushed toward the loop, increasing the flux through the loop. Pulling the magnet back out of the loop causes the current meter to deflect in the opposite direction. The conducting wires aren't moving, so this is not a motional emf. Nonetheless, the induced current is very real.

The German physicist Heinrich Lenz began to study electromagnetic induction after learning of Faraday's discovery. Three years later, in 1834, Lenz announced a rule for determining the direction of the induced current. We now call his rule **Lenz's law,** and it can be stated as follows:

> **Lenz's law** There is an induced current in a closed, conducting loop if and only if the magnetic flux through the loop is changing. The direction of the induced current is such that the induced magnetic field opposes the *change* in the flux.

Lenz's law is rather subtle, and it takes some practice to see how to apply it.

NOTE ▶ One difficulty with Lenz's law is the term *flux*. In everyday language, the word *flux* already implies that something is changing. Think of the phrase, "The situation is in flux." Not so in physics, where *flux* means "passes through." A steady magnetic field through a loop creates a steady, *un*changing magnetic flux. ◀

Lenz's law tells us to look for situations where the flux is *changing*. This can happen in three ways.

1. The magnetic field through the loop changes (increases or decreases),
2. The loop changes in area or angle, or
3. The loop moves into or out of a magnetic field.

Lenz's law depends on an idea that we hinted at in our discussion of eddy currents. If a current is induced in a loop, that current generates its own magnetic field $\vec{B}_{induced}$. This is the *induced magnetic field* of Lenz's law. You learned in Chapter 32 how to use the right-hand rule to determine the direction of this induced magnetic field.

In Figure 33.20, pushing the bar magnet into the loop causes the magnetic flux to *increase* in the downward direction. To oppose the *change* in flux, which is what Lenz's law requires, the loop itself needs to generate the *upward*-pointing magnetic field of Figure 33.21. The induced magnetic field at the center of the loop will point upward if the current is ccw. Thus pushing the north end of a bar magnet toward the loop induces a ccw current around the loop. The induced current ceases as soon as the magnet stops moving.

Now suppose the bar magnet is pulled back away from the loop, as shown in Figure 33.22a on the next page. There is a downward magnetic flux through the loop, but the flux *decreases* as the magnet moves away. According to Lenz's law, the induced magnetic field of the loop will *oppose this decrease*. To do so, the induced field needs to point in the *downward* direction, as shown in Figure 33.22b. Thus as the magnet is withdrawn, the induced current is clockwise (cw), opposite to the induced current of Figure 33.21.

A bar magnet pushed into a loop increases the flux through the loop and induces a current to flow.

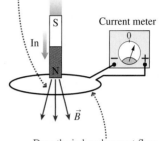

Does the induced current flow clockwise or counterclockwise?

FIGURE 33.20 Pushing a bar magnet toward the loop induces a current in the loop.

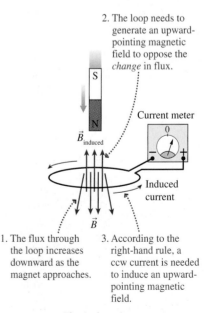

2. The loop needs to generate an upward-pointing magnetic field to oppose the *change* in flux.

1. The flux through the loop increases downward as the magnet approaches.

3. According to the right-hand rule, a ccw current is needed to induce an upward-pointing magnetic field.

FIGURE 33.21 The induced current is ccw.

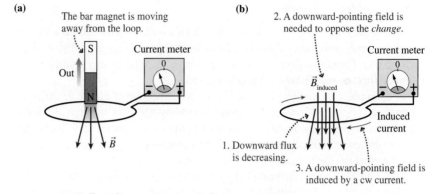

FIGURE 33.22 Pulling the magnet away induces a cw current.

NOTE ▶ Notice that the magnetic field of the bar magnet is pointing downward in both Figures 33.21 and 33.22. It is not the *flux* due to the magnet that the induced current opposes, but the *change* in the flux. This is a subtle but critical distinction. If the induced current opposed the flux itself, the current in both Figures 33.21 and 33.22 would be ccw to generate an upward magnetic field. But that's not what happens. When the field of the magnet points down and is increasing, the induced current opposes the increase by generating an upward field. When the field of the magnet points down but is decreasing, the induced current opposes the decrease by generating a downward field. **◀**

Figure 33.23 shows six basic situations. The magnetic field can point either up or down through the loop. For each, the flux can either increase, hold steady, or decrease in strength. These observations form the basis for a set of rules about using Lenz's law.

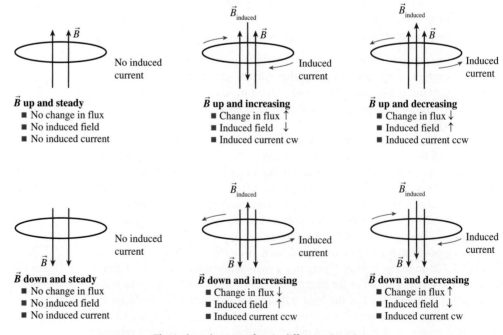

FIGURE 33.23 The induced current for six different situations.

TACTICS BOX 33.1 Using Lenz's law

❶ **Determine the direction of the applied magnetic field.** The field must pass through the loop.

❷ **Determine how the flux is changing.** Is it increasing, decreasing, or staying the same?

❸ **Determine the direction of an induced magnetic field that will oppose the *change* in the flux.**

  ■ Increasing flux: the induced magnetic field points opposite the applied magnetic field.

  ■ Decreasing flux: the induced magnetic field points in the same direction as the applied magnetic field.

  ■ Steady flux: there is no induced magnetic field.

❹ **Determine the direction of the induced current.** Use the right-hand rule to determine the current direction in the loop that generates the induced magnetic field you found in step 3.

Let's look at some examples.

EXAMPLE 33.6 Lenz's law 1

The switch in the circuit of Figure 33.24a has been closed for a long time. What happens in the lower loop when the switch is opened?

MODEL We'll use the right-hand rule to find the magnetic fields of current loops.

SOLVE Figure 33.24b shows the four steps of using Lenz's law. Opening the switch induces a ccw current in the lower loop. This is a momentary current, lasting only until the magnetic field of the upper loop drops to zero.

ASSESS The conclusion is consistent with Figure 33.23.

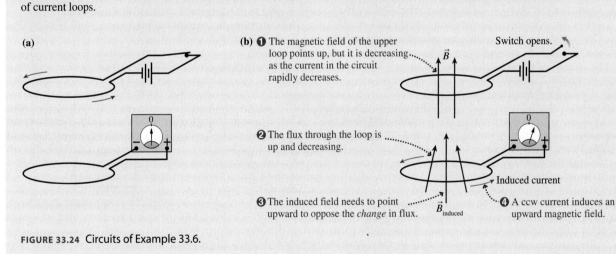

FIGURE 33.24 Circuits of Example 33.6.

EXAMPLE 33.7 Lenz's law 2

Figure 33.25a on the next page shows two solenoids facing each other. When the switch for coil 1 is closed, does the induced current in coil 2 pass from right to left or from left to right through the current meter?

MODEL We'll use the right-hand rule to find the magnetic fields of solenoids.

VISUALIZE It is very important to look at the *direction* in which a solenoid is wound around the cylinder. Notice that the two solenoids in Figure 33.25a are wound in opposite directions.

SOLVE Figure 33.25b shows the four steps of using Lenz's law. Closing the switch induces a current that passes from right to left through the current meter. The induced current is only momentary. It lasts only until the field from coil 1 reaches full strength and is no longer changing.

ASSESS The conclusion is consistent with Figure 33.23.

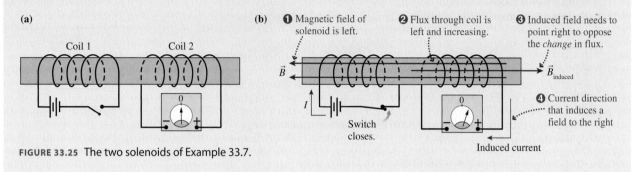

FIGURE 33.25 The two solenoids of Example 33.7.

EXAMPLE 33.8 A rotating loop

The loop of wire in Figure 33.26 was initially in the *xy*-plane, parallel to the magnetic field. It is suddenly rotated 90° about the *y*-axis until it is in the *yz*-plane, perpendicular to the magnetic field. In what direction is the induced current as the loop rotates?

SOLVE Unlike the magnetic fields in the previous examples, this magnetic field is constant and unchanging. Nonetheless, the *flux* through the loop changes as it rotates. To use Lenz's law,

1. The applied magnetic field points to the right.
2. Initially the flux is $\Phi = 0$, but after rotating the flux is $\Phi = AB$ toward the right, where A is the loop area. This is an increasing flux through the loop to the right.
3. To oppose this increase in the flux, the induced magnetic field of the loop must have an *x*-component toward the left.
4. This will be the case if an induced current is in a cw direction, as seen from the perspective of Figure 33.26.

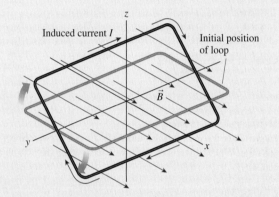

FIGURE 33.26 A current is induced in a loop as the loop rotates in a constant magnetic field.

STOP TO THINK 33.4 A current-carrying wire is pulled away from a conducting loop in the direction shown. As the wire is moving, is there a cw current around the loop, a ccw current, or no current?

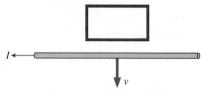

33.5 Faraday's Law

13.9, 13.10 Active Physics ONLINE

Faraday discovered that a current is induced when the magnetic flux through a conducting loop changes. Lenz's law allows us to find the direction of the induced current. To put electromagnetic induction to practical use, we also need to know the *size* of the induced current.

Charges don't start moving spontaneously. A current requires an emf to provide the energy. We started our analysis of induced currents with circuits in which there is a *motional emf*. The motional emf can be understood in terms of magnetic forces on moving charges. But we've also seen that a current can be induced by changing the magnetic field through a stationary circuit, a circuit in which there is no motion. There *must* be an emf in this circuit, even though the mechanism for this emf is not yet clear.

The emf associated with a changing magnetic flux, regardless of what causes the change, is called an **induced emf** $\mathcal{E}$. Then, if there is a complete circuit having resistance R, a current

$$I_{\text{induced}} = \frac{\mathcal{E}}{R} \tag{33.13}$$

is established in the wire as a *consequence* of the induced emf. The direction of the current is given by Lenz's law. The last piece of information we need is the size of the induced emf $\mathcal{E}$.

The research of Faraday and others eventually led to the discovery of the basic law of electromagnetic induction, which we now call **Faraday's law.** Faraday's law is a new law of physics, not derivable from any previous laws you have studied. It states:

> **Faraday's law** An emf $\mathcal{E}$ is induced in a conducting loop if the magnetic flux through the loop changes. The magnitude of the emf is
>
> $$\mathcal{E} = \left| \frac{d\Phi_{\text{m}}}{dt} \right| \tag{33.14}$$
>
> and the direction of the emf is such as to drive an induced current in the direction given by Lenz's law.

In other words, the induced emf is the *rate of change* of the magnetic flux through the loop.

As a corollary to Faraday's law, a coil of wire consisting of N turns in a changing magnetic field acts like N batteries in series. The induced emf of each of the coils adds, so the induced emf of the entire coil is

$$\mathcal{E}_{\text{coil}} = N \left| \frac{d\Phi_{\text{per coil}}}{dt} \right| \quad \text{(Faraday's law for an N-turn coil)} \tag{33.15}$$

As a first example of using Faraday's law, return to the situation of Figure 33.6, where a wire moves through a magnetic field by sliding on a U-shaped conducting rail. Figure 33.27 shows the circuit again. The magnetic field $\vec{B}$ is perpendicular to the plane of the conducting loop, so $\theta = 0°$ and the magnetic flux is $\Phi = AB$, where A is the area of the loop. If the slide wire is distance x from the end, the area is $A = xl$ and the flux at that instant of time is

$$\Phi_{\text{m}} = AB = xlB \tag{33.16}$$

The flux through the loop increases as the wire moves. According to Faraday's law, the induced emf is

$$\mathcal{E} = \left| \frac{d\Phi_{\text{m}}}{dt} \right| = \frac{d}{dt}(xlB) = \frac{dx}{dt}lB = vlB \tag{33.17}$$

where the wire's velocity is $v = dx/dt$. We can now use Equation 33.13 to find that the induced current is

$$I = \frac{\mathcal{E}}{R} = \frac{vlB}{R} \tag{33.18}$$

The flux is increasing into the loop, so the induced magnetic field will oppose this increase by pointing out of the loop. This requires a ccw induced current in the loop. Faraday's law leads us to the conclusion that the loop will have a ccw induced

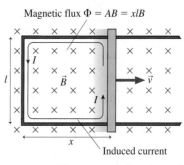

Magnetic flux $\Phi = AB = xlB$

FIGURE 33.27 The magnetic flux through the loop increases as the slide wire moves.

current $I = vlB/R$. This is exactly the conclusion we reached in Section 33.2, where we analyzed the situation from the perspective of magnetic forces on moving charge carriers. Faraday's law confirms what we already knew but, at least in this case, doesn't seem to offer anything new.

Using Faraday's Law

Most electromagnetic induction problems can be solved with a four-step strategy.

(MP) PROBLEM-SOLVING STRATEGY 33.1 **Electromagnetic induction**

MODEL Make simplifying assumptions about wires and magnetic fields.

VISUALIZE Draw a picture or a circuit diagram. Use Lenz's law to determine the direction of the induced current.

SOLVE The mathematical representation is based on Faraday's law

$$\mathcal{E} = \left| \frac{d\Phi_m}{dt} \right|$$

For an N-turn coil, multiply by N. The size of the induced current is $I = \mathcal{E}/R$.

ASSESS Check that your result has the correct units, is reasonable, and answers the question.

EXAMPLE 33.9 Electromagnetic induction in a circular loop

The magnetic field of Figure 33.28 decreases from 1.0 T to 0.4 T in 1.2 s. A 6.0-cm-diameter conducting loop with a resistance of 0.010 Ω is perpendicular to $\vec{B}$. What are the size and direction of the current induced in the loop?

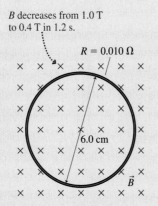

B decreases from 1.0 T to 0.4 T in 1.2 s.

$R = 0.010 \, \Omega$

6.0 cm

$\vec{B}$

FIGURE 33.28 A circular conducting loop in a decreasing magnetic field.

MODEL Assume that B decreases linearly with time.

VISUALIZE The magnetic flux is into the page and decreasing. To oppose the *change* in the flux, the induced field needs to point into the page. This will be true if the induced current in the loop is cw.

SOLVE The magnetic field is perpendicular to the plane of the loop, hence $\theta = 0°$ and the magnetic flux is $\Phi_m = AB = \pi r^2 B$. The radius doesn't change with time, but B does. According to Faraday's law, the induced emf is

$$\mathcal{E} = \left| \frac{d\Phi_m}{dt} \right| = \left| \frac{d(\pi r^2 B)}{dt} \right| = \pi r^2 \left| \frac{dB}{dt} \right|$$

The *rate* at which the magnetic field changes is

$$\frac{dB}{dt} = \frac{\Delta B}{\Delta t} = \frac{-0.60 \, \text{T}}{1.2 \, \text{s}} = -0.50 \, \text{T/s}$$

dB/dt is negative because the field is decreasing, but all we need for Faraday's law is the absolute value. Thus

$$\mathcal{E} = \pi r^2 \left| \frac{dB}{dt} \right| = \pi (0.030 \, \text{m})^2 (0.50 \, \text{T/s}) = 0.00141 \, \text{V}$$

The current induced by this emf is

$$I = \frac{\mathcal{E}}{R} = \frac{0.00141 \, \text{V}}{0.010 \, \Omega} = 0.141 \, \text{A}$$

The decreasing magnetic field causes a 0.141 A cw current that lasts for 1.2 s.

ASSESS We don't have much to go on for assessing the result. The emf is quite small, but, because the resistance of metal wires is also very small, the current is respectable. We know that electromagnetic induction produces currents large enough for practical applications, so this result seems plausible.

EXAMPLE 33.10 Electromagnetic induction in a solenoid

A 2.0-cm-diameter loop of wire with a resistance of 0.010 Ω is placed in the center of a solenoid. The solenoid, shown in Figure 33.29a, is 4.0 cm in diameter, 20 cm long, and wrapped with 1000 turns of wire. Figure 33.29b shows the current through the solenoid as a function of time as the solenoid is "powered up." A positive current is defined to be cw when seen from the left. Find the current in the loop as a function of time and show the result as a graph.

(a)

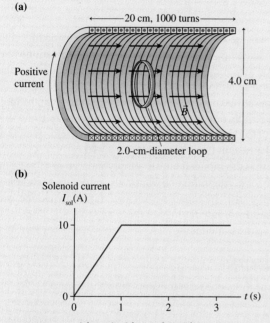

(b)

Solenoid current

FIGURE 33.29 A loop inside a solenoid.

MODEL The solenoid's length is much greater than its diameter, so the field near the center should be nearly uniform.

VISUALIZE The magnetic field of the solenoid creates a magnetic flux through the loop of wire. The solenoid current is always positive, meaning that it is cw as seen from the left. Consequently, from the right-hand rule, the magnetic field inside the solenoid always points to the right. During the first second, while the solenoid current is increasing, the flux through the loop is to the right and increasing. To oppose the *change* in the flux, the loop's induced magnetic field must point to the left. Thus, again using the right-hand rule, the induced current must flow ccw as seen from the left. This is a *negative* current. There's no *change* in the flux for $t > 1$ s, so the induced current is zero.

SOLVE Now we're ready to use Faraday's law to find the magnitude of the current. Because the field is uniform inside the solenoid and perpendicular to the loop ($\theta = 0°$), the flux is $\Phi_m = AB$, where $A = \pi r^2 = 3.14 \times 10^{-4}$ m^2 is the area of the loop (*not* the area of the solenoid). The field of a long solenoid of length l was found in Chapter 32 to be

$$B = \frac{\mu_0 N I_{sol}}{l}$$

The flux through the loop when the solenoid current is I_{sol} is thus

$$\Phi_m = \frac{\mu_0 A N I_{sol}}{l}$$

The changing flux creates an induced emf $\mathcal{E}$ that is given by Faraday's law:

$$\mathcal{E} = \left|\frac{d\Phi_m}{dt}\right| = \frac{\mu_0 A N}{l}\left|\frac{dI_{sol}}{dt}\right| = 1.97 \times 10^{-6}\left|\frac{dI_{sol}}{dt}\right|$$

From the slope of the graph, we find

$$\left|\frac{dI_{sol}}{dt}\right| = \begin{cases} 10 \text{ A/s} & 0.0 \text{ s} < t < 1.0 \text{ s} \\ 0 & 1.0 \text{ s} < t < 3.0 \text{ s} \end{cases}$$

Thus the induced emf is

$$\mathcal{E} = \begin{cases} 1.97 \times 10^{-5} \text{ V} & 0.0 \text{ s} < t < 1.0 \text{ s} \\ 0 \text{ V} & 1.0 \text{ s} < t < 3.0 \text{ s} \end{cases}$$

Finally, the current induced in the loop is

$$I_{loop} = \frac{\mathcal{E}}{R} = \begin{cases} -1.97 \text{ mA} & 0.0 \text{ s} < t < 1.0 \text{ s} \\ 0 \text{ mA} & 1.0 \text{ s} < t < 3.0 \text{ s} \end{cases}$$

where the negative sign comes from Lenz's law. This result is shown in Figure 33.30.

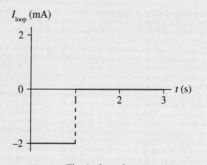

FIGURE 33.30 The induced current in the loop.

What Does Faraday's Law Tell Us?

The induced current in the slide-wire circuit of Figure 33.27 can be understood as a motional emf due to magnetic forces on moving charges. We had not anticipated this kind of current in Chapter 32, but it takes no new laws of physics to understand it.

The induced currents in Examples 33.9 and 33.10 are different. We cannot explain or predict these induced currents on the basis of previous laws or principles. This is new physics.

Faraday recognized that all induced currents are associated with a changing magnetic flux. There are two fundamentally different ways to change the magnetic flux through a conducting loop:

1. The loop can move or expand or rotate, creating a motional emf.
2. The magnetic field can change.

We can see both of these if we write Faraday's law as

$$\mathcal{E} = \left| \frac{d\Phi_{\mathrm{m}}}{dt} \right| = \left| \vec{B} \cdot \frac{d\vec{A}}{dt} + \vec{A} \cdot \frac{d\vec{B}}{dt} \right| \tag{33.19}$$

The first term on the right side represents a motional emf. The magnetic flux changes because the loop itself is changing. This term includes not only situations like the slide-wire circuit, where the area A changes, but also loops that rotate in a magnetic field. The physical area of a rotating loop does not change, but the area *vector* $\vec{A}$ does. The loop's motion causes magnetic forces on the charge carriers in the loop.

The second term on the right side is the new physics in Faraday's law. It says that an emf can also be created simply by changing a magnetic field, even if nothing is moving. This was the case in Examples 33.9 and 33.10.

Faraday's law tells us that the induced emf is simply the rate of change of the magnetic flux through the loop, *regardless* of what causes the flux to change. The "old physics" of motional emf is included within Faraday's law as one way of changing the flux, but Faraday's law then goes on to say that any other way of changing the flux will have the same result.

An Unanswered Question

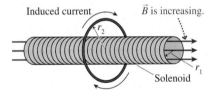

Induced current $\vec{B}$ is increasing.

r_2

r_1

Solenoid

FIGURE 33.31 A changing current in the solenoid induces a current in the loop.

As a final example in this section, consider the loop shown in Figure 33.31. A long, tightly wound solenoid of radius r_1 passes through the center of a conducting loop having a larger radius r_2. What happens to the loop if the solenoid current changes?

You learned in Chapter 32 that the magnetic field is strong inside the solenoid but, if the solenoid is sufficiently long, essentially zero outside the solenoid. Even so, changing the current through the solenoid *does* cause an induced current in the loop. The solenoid's magnetic field establishes a flux through the loop, and changing the solenoid current causes the magnetic flux to change. This is the essence of Faraday's law.

But the loop is completely outside the solenoid. How can the charge carriers in the conducting loop possibly know that the magnetic field inside the solenoid is changing? How do they know which way to move?

In the case of a motional emf, the *mechanism* that causes an induced current is the magnetic force on the moving charges. But here, where there's no motion, what is the mechanism that creates a current when the magnetic flux changes? This is an important question, one that we will answer in the next section.

STOP TO THINK 33.5   A conducting loop is halfway into a magnetic field. Suppose the magnetic field begins to increase rapidly in strength. What happens to the loop?

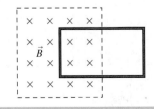

a. The loop is pushed upward, toward the top of the page.
b. The loop is pushed downward, toward the bottom of the page.
c. The loop is pulled to the left, into the magnetic field.
d. The loop is pushed to the right, out of the magnetic field.
e. The tension in the wires increases but the loop does not move.

33.6 Induced Fields and Electromagnetic Waves

Faraday's law is a tool for calculating the strength of an induced current, but one important piece of the puzzle is still missing. What *causes* the current? That is, what *force* pushes the charges around the loop against the resistive forces of the metal?

The only agents that exert forces on charges are electric fields and magnetic fields. Magnetic forces are responsible for motional emfs, but magnetic forces cannot explain the current induced in a *stationary* loop by a changing magnetic field.

Figure 33.32a shows a conducting loop in an increasing magnetic field. According to Lenz's law, there is an induced current in the ccw direction. But something has to act on the charge carriers to make them move, so we infer that there must be an *electric* field tangent to the loop at all points. This electric field is *caused* by the changing magnetic field and is called an **induced electric field.** The induced electric field is the *mechanism* we were seeking that creates a current when there's a changing magnetic field inside a stationary loop.

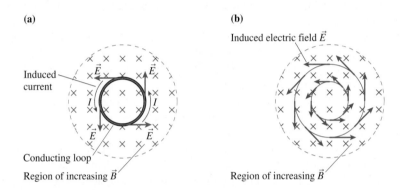

FIGURE 33.32 An induced electric field creates a current in the loop.

The conducting loop isn't necessary. The space in which the magnetic field is changing is filled with the pinwheel pattern of induced electric fields shown in Figure 33.32b. Charges will move if a conducting path is present, but the induced electric field is there as a direct consequence of the changing magnetic field.

But this is a rather peculiar electric field. All the electric fields we have examined until now have been created by charges. Electric field vectors pointed away from positive charges and toward negative charges. An electric field created by charges is called a **Coulomb electric field.** The induced electric field of Figure 33.32b is caused not by charges but by a changing magnetic field. It is called a **non-Coulomb electric field.**

So it appears that there are two different ways to create an electric field:

1. A Coulomb electric field is created by positive and negative charges.
2. A non-Coulomb electric field is created by a changing magnetic field.

Both exert a force $\vec{F} = q\vec{E}$ on a charge, and both create a current in a conductor. However, the origins of the fields are very different. Figure 33.33 is a quick summary of the two ways to create an electric field.

We first introduced the idea of a field as a way of thinking about how two charges exert long-range forces on each other through the emptiness of space. The field may have seemed like a useful pictorial representation of charge interactions, but we had little evidence that fields are *real*, that they actually exist. Now we do. The electric field has shown up in a completely different context, independent of charges, as the explanation of the very real existence of induced currents.

The electric field is not just a pictorial representation; it is real.

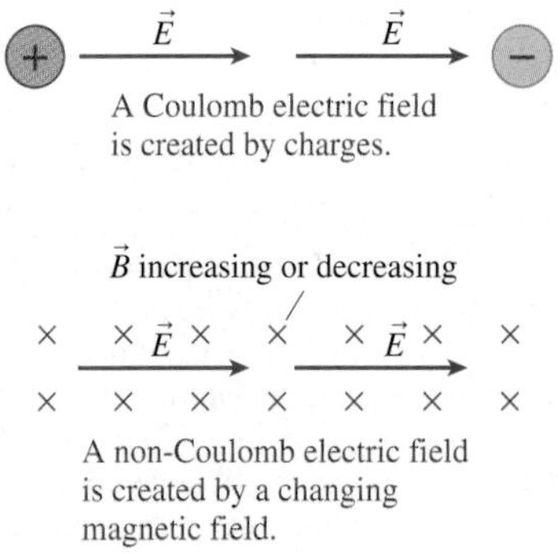

A Coulomb electric field is created by charges.

$\vec{B}$ increasing or decreasing

A non-Coulomb electric field is created by a changing magnetic field.

FIGURE 33.33 Two ways to create an electric field.

Maxwell's Theory

Faraday's field concept was capable of explaining the phenomena of electricity and magnetism as they were known in the 1830s and 1840s. But Faraday, despite his intuitive genius, lacked the mathematical skills to develop a true *theory* of electric and magnetic fields. It was not easy to predict new phenomena or develop applications without a theory.

In 1855, less than two years after receiving his undergraduate degree, the English physicist James Clerk Maxwell presented a paper titled "On Faraday's Lines of Force." In this paper, he began to sketch out how Faraday's pictorial ideas about fields could be given a rigorous mathematical basis. Maxwell then spent the next 10 years developing the mathematical theory of electromagnetism.

Maxwell was troubled by a certain lack of symmetry. Faraday had found that a changing magnetic field creates an induced electric field, a non-Coulomb electric field not tied to charges. But what, Maxwell began to wonder, about a changing *electric* field?

To complete the symmetry, Maxwell proposed that a changing electric field creates an **induced magnetic field,** a new kind of magnetic field not tied to the existence of currents. Figure 33.34 shows a region of space where the *electric* field is increasing. This region of space, according to Maxwell, is filled with a pinwheel pattern of induced magnetic fields. The induced magnetic field looks like the induced electric field, with $\vec{E}$ and $\vec{B}$ interchanged, except that—for technical reasons—the induced $\vec{B}$ points the opposite way from the induced $\vec{E}$. Although there was no experimental evidence that induced magnetic fields existed, Maxwell went ahead and included them in his electromagnetic field theory. This was an inspired hunch, soon to be vindicated.

Maxwell soon realized that it might be possible to establish self-sustaining electric and magnetic fields that would be entirely independent of any charges or currents. That is, a changing electric field $\vec{E}$ creates a magnetic field $\vec{B}$, which then changes in just the right way to recreate the electric field, which then changes in just the right way to again recreate the magnetic field, and so on. The fields are continually recreated through electromagnetic induction without any reliance on charges or currents.

The mathematics of Maxwell's theory is difficult, but eventually Maxwell was able to predict that electric and magnetic fields would be able to sustain themselves, free from charges and currents, if they took the form of an **electromagnetic wave.** The wave would have to have a very specific geometry, shown in Figure 33.35, in which $\vec{E}$ and $\vec{B}$ are perpendicular to each other as well as perpendicular to the direction of travel. That is, an electromagnetic wave would be a *transverse* wave.

Furthermore, Maxwell's theory predicted that the wave would travel with speed

$$v_{\text{em wave}} = \frac{1}{\sqrt{\epsilon_0 \mu_0}} \tag{33.20}$$

where ϵ_0 is the permittivity constant from Coulomb's law and μ_0 is the permeability constant from the law of Biot and Savart. Maxwell computed that an electromagnetic wave, if it existed, would travel with speed $v_{\text{em wave}} = 3.00 \times 10^8$ m/s.

We don't know Maxwell's immediate reaction, but it must have been both shock and excitement. His predicted speed for electromagnetic waves, a prediction that came directly from his theory, was none other than the speed of light! This agreement could be just a coincidence, but Maxwell didn't think so. Making a bold leap of imagination, Maxwell concluded that **light is an electromagnetic wave.**

It took 25 more years for Maxwell's predictions to be tested. In 1886, the German physicist Heinrich Hertz discovered how to generate and transmit radio

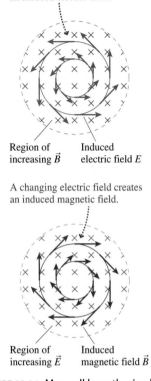

A changing magnetic field creates an induced electric field.

Region of increasing $\vec{B}$ — Induced electric field E

A changing electric field creates an induced magnetic field.

Region of increasing $\vec{E}$ — Induced magnetic field $\vec{B}$

FIGURE 33.34 Maxwell hypothesized the existence of induced magnetic fields.

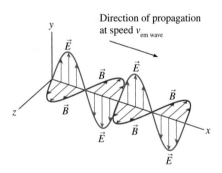

Direction of propagation at speed $v_{\text{em wave}}$

FIGURE 33.35 A self-sustaining electromagnetic wave.

The velocity of transverse undulations in our hypothetical medium, calculated from the electromagnetic experiments of Kohlrausch and Weber [who had measured ϵ_0 and μ_0], agrees so exactly with the velocity of light calculated from the optical experiments of Fizeau that we can scarcely avoid the inference that light consists of the transverse undulations of the same medium which is the cause of electric and magnetic phenomena.

James Clerk Maxwell

waves. Two years later, in 1888, he was able to show that radio waves travel at the speed of light. Maxwell, unfortunately, did not live to see his triumph. He had died in 1879, at the age of 48.

Chapter 34 will develop some of the mathematical details of Maxwell's theory and show how the ideas contained in Faraday's law lead to electromagnetic waves.

33.7 Induced Currents: Three Applications

There are many applications of Faraday's law and induced currents in modern technology. In this section we will look at three: generators, transformers, and metal detectors.

Generators

We noted in Section 33.2 that a slide wire pulled through a magnetic field on a U-shaped track is a simple generator because it transforms mechanical energy into electric energy. Figure 33.36 shows a more practical generator. Here a coil of wire rotates in a magnetic field. Both the field and the area of the loop are constant, but the magnetic flux through the loop changes continuously as the loop rotates. The induced current is removed from the rotating loop by *brushes* that press up against rotating *slip rings*.

The flux through the coil is

$$\Phi_m = \vec{A} \cdot \vec{B} = AB\cos\theta = AB\cos\omega t \qquad (33.21)$$

where ω is the angular frequency ($\omega = 2\pi f$) with which the coil rotates. The induced emf is given by Faraday's law,

$$\mathcal{E}_{coil} = N\frac{d\Phi_m}{dt} = ABN\frac{d}{dt}(\cos\omega t) = -\omega ABN\sin\omega t \qquad (33.22)$$

where N is the number of turns on the coil. We've dropped the absolute value signs to demonstrate that the sign of $\mathcal{E}_{coil}$ alternates between positive and negative.

Because the emf alternates in sign, the current through resistor R alternates back and forth in direction. Hence the generator of Figure 33.36 is an alternating-current generator, producing what we call an *AC voltage*.

A generator inside a hydroelectric dam uses electromagnetic induction to convert the mechanical energy of a spinning turbine into electric energy.

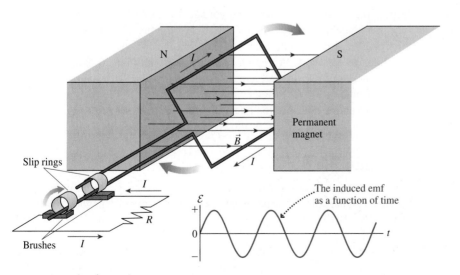

FIGURE 33.36 An alternating-current generator.

EXAMPLE 33.11 An AC generator

A coil with area 2.0 m^2 rotates in a 0.010 T magnetic field at a frequency of 60 Hz. How many turns are needed to generate a peak voltage of 160 V?

SOLVE The coil's maximum voltage is found from Equation 33.22:

$$\mathcal{E}_{max} = \omega ABN = 2\pi fABN$$

The number of turns needed to generate $\mathcal{E}_{max} = 160$ V is

$$N = \frac{\mathcal{E}_{max}}{2\pi fAB} = \frac{160 \text{ V}}{2\pi(60 \text{ Hz})(2.0 \text{ m}^2)(0.010 \text{ T})}$$

$$= 21 \text{ turns}$$

ASSESS A 0.010 T field is modest, so you can see that generating large voltages is not difficult with large (2 m^2) coils. Commercial generators use water flowing through a dam or turbines spun by expanding steam to rotate the generator coils. Work is required to rotate the coil, just as work was required to pull the slide wire in Section 33.2, because the magnetic field exerts retarding forces on the currents in the coil. Thus a generator is a device that turns motion (mechanical energy) into a current (electric energy). A generator is the opposite of a motor, which turns a current into motion.

Transformers

Figure 33.37 shows two coils wrapped on an iron core. The left coil is called the **primary coil.** It has N_1 turns and is driven by an oscillating voltage $V_1 \cos \omega t$. The magnetic field of the primary follows the iron core and passes through the right coil, which has N_2 turns and is called the **secondary coil.** The alternating current through the primary coil causes an oscillating magnetic flux through the secondary coil and, hence, an induced emf. The induced emf of the secondary coil is delivered to resistance R as the oscillating voltage $V_2 \cos \omega t$.

The current through the primary coil is inversely proportional to the number of turns: $I_{prim} \propto 1/N_1$. (This relation is a consequence of the coil's inductance, an idea discussed later in the chapter. But, roughly speaking, the coil's resistance is proportional to the number of turns, and the current, according to Ohm's law, is inversely proportional to the resistance.) According to Faraday's law, the emf induced in the secondary coil is directly proportional to the number of turns: $\mathcal{E}_{sec} \propto N_2$. Combining these two proportionalities, the secondary voltage of an ideal transformer is related to the primary voltage by

$$V_2 = \frac{N_2}{N_1} V_1 \tag{33.23}$$

Depending on the ratio N_2/N_1, the voltage V_2 across the load can be *transformed* to a higher or a lower voltage than V_1. Consequently, this device is called a **transformer.** Transformers are widely used in the commercial generation and transmission of electricity. A *step-up transformer*, with $N_2 \gg N_1$, boosts the voltage of a generator up to several hundred thousand volts. Delivering power with smaller currents at higher voltages reduces losses due to the resistance of the wires. High-voltage transmission lines carry electric power to urban areas, where *step-down transformers* ($N_2 \ll N_1$) lower the voltage to 120 V.

Primary coil
N_1 turns

Iron core

Secondary coil
N_2 turns

$V_1 \cos \omega t$

$V_2 \cos \omega t$

Load

The magnetic field follows the iron core.

FIGURE 33.37 A transformer.

Transformers are essential for transporting electric energy from the power plant to cities and homes.

Metal Detectors

Metal detectors, such as those used in airports for security, seem fairly mysterious. How can they detect the presence of *any* metal—not just magnetic materials such as iron—but not detect plastic or other materials? Metal detectors work because of induced currents.

A metal detector, shown in Figure 33.38, consist of two coils: a *transmitter coil* and a *receiver coil.* A high-frequency alternating current in the transmitter coil generates an alternating magnetic field along the axis. This magnetic field creates a changing flux through the receiver coil and causes an alternating induced current. The transmitter and receiver are similar to a transformer.

Suppose a piece of metal is placed between the transmitter and the receiver. The alternating magnetic field through the metal induces eddy currents in a plane parallel to the transmitter and receiver coils. The receiver coil then responds to the *superposition* of the transmitter's magnetic field and the magnetic field of the eddy currents. Because the eddy currents attempt to prevent the flux from changing, in accordance with Lenz's law, the net field at the receiver *decreases* when a piece of metal is inserted between the coils. Electronic circuits detect the current decrease in the receiver coil and set off an alarm. Eddy currents can't flow in an insulator, so this device detects only metals.

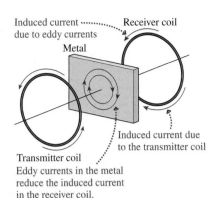

Eddy currents in the metal reduce the induced current in the receiver coil.

FIGURE 33.38 A metal detector.

33.8 Inductors

Capacitors were first introduced as devices that produce a uniform electric field. The capacitance (i.e., the *capacity* to store charge) was defined as the charge-to-voltage ratio $C = Q/\Delta V$. We later found that a capacitor stores potential energy $U_C = \frac{1}{2}C(\Delta V)^2$ and that this energy is released when the capacitor is discharged.

A coil of wire in the form of a solenoid is a device that produces a uniform magnetic field. Do solenoids in circuits have practical uses, as capacitors do? As a starting point to answering this question, notice that the charge on a capacitor is analogous to the magnetic flux through a solenoid. That is, a larger diameter capacitor plate holds more charge just as a larger diameter solenoid contains more flux. Using the definition of capacitance as an analog, let's define the **inductance** L of a magnetic-field device as its flux-to-current ratio

$$L = \frac{\Phi_m}{I} \tag{33.24}$$

Strictly speaking, this is called *self-inductance* because the flux we're considering is the magnetic flux the device creates in itself when there is a current.

The units of inductance are Wb/A. Recalling that $1\ \text{Wb} = 1\ \text{T m}^2$, this is equivalent to $\text{T m}^2/\text{A}$. It's convenient to define an SI unit of inductance called the **henry,** in honor of Joseph Henry, as

$$1\ \text{henry} = 1\ \text{H} \equiv 1\ \text{T m}^2/\text{A}$$

Practical inductances are usually in the range of millihenries (mH) or microhenries (μH).

Any circuit element can have an inductance by virtue of the fact that currents produce magnetic fields. In practice, however, inductance is usually negligible unless the magnetic field is concentrated, as it is in a solenoid. Consequently, a solenoid or coil of wire is our prototype of inductance. A coil of wire used in a circuit for the purpose of providing inductance is called an **inductor.** An *ideal inductor* is one for which the wire forming the coil has no electric resistance. The circuit symbol for an inductor is �⎺ .

It's not hard to find the inductance of a solenoid. In Chapter 32 we found that the magnetic field inside a solenoid having N turns and length l is

$$B = \frac{\mu_0 N I}{l} \tag{33.25}$$

The magnetic flux through *each* coil is $\Phi_{\text{per coil}} = AB$, where A is the cross-section area of the solenoid. The total flux through all N coils is

$$\Phi_m = N\Phi_{\text{per coil}} = \frac{\mu_0 N^2 A}{l} I \tag{33.26}$$

Thus the inductance of the solenoid, using the definition of Equation 33.24, is

$$L_{\text{solenoid}} = \frac{\Phi_{\text{m}}}{I} = \frac{\mu_0 N^2 A}{l} \tag{33.27}$$

The inductance of a solenoid depends only on its geometry, not at all on the current. You may recall that the capacitance of two parallel plates depends only on their geometry, not at all on their potential difference.

EXAMPLE 33.12 The length of an inductor

An inductor is made by tightly wrapping 0.30-mm-diameter wire around a 4.0-mm-diameter cylinder. What length cylinder has an inductance of 10 μH?

SOLVE The cross-section area of the solenoid is $A = \pi r^2$. If the wire diameter is d, the number of turns of wire on a cylinder of length l is $N = l/d$. Thus the inductance is

$$L = \frac{\mu_0 N^2 A}{l} = \frac{\mu_0 (l/d)^2 \pi r^2}{l} = \frac{\mu_0 \pi r^2 l}{d^2}$$

The length needed to give inductance $L = 10^{-5}$ H is

$$l = \frac{d^2 L}{\mu_0 \pi r^2} = \frac{(0.00030 \text{ m})^2 (1.0 \times 10^{-5} \text{ H})}{(4\pi \times 10^{-7} \text{ Tm/A})\pi (0.0020 \text{ m})^2}$$

$$= 0.057 \text{ m} = 5.7 \text{ cm}$$

The Potential Difference Across an Inductor

(a)

Inductor coil

$\vec{B}$

Solenoid magnetic field

Current I

(b)

The induced current is opposite the solenoid current.

The induced magnetic field opposes the change in flux.

$+$ $-$

ΔV_{L}

Increasing current

The induced current carries positive charge carriers to the left and establishes a potential difference across the inductor.

FIGURE 33.39 Increasing the current through an inductor.

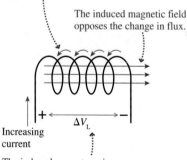

An inductor is not very interesting when the current through it is steady. If the inductor is ideal, with $R = 0 \, \Omega$, the potential difference due to a steady current is zero. Inductors become important circuit elements when currents are changing. Figure 33.39a shows a steady current into the left side of an inductor. The solenoid's magnetic field passes through the coils of the solenoid, establishing a flux.

In Figure 33.39b, the current into the solenoid is increasing. This creates an increasing flux to the left. According to Lenz's law, an induced current in the coils will oppose this increase by creating an induced magnetic field pointing to the right. This requires the induced current to be *opposite* the current into the solenoid. This induced current will carry positive charge carriers to the left until a potential difference ΔV is established across the solenoid.

You saw a similar situation in Section 33.2. The induced current in a conductor moving through a magnetic field carried positive charge carriers to the top of the wire and established a potential difference across the conductor. The induced current in the moving wire was due to magnetic forces on the moving charges. Now, in Figure 33.39b, the induced current is due to the non-Coulomb electric field induced by the changing magnetic field. Nonetheless, the outcome is the same: a potential difference across the conductor.

We can use Faraday's law to find the potential difference. The emf induced in a coil is

$$\mathcal{E}_{\text{coil}} = N \left| \frac{d\Phi_{\text{per coil}}}{dt} \right| = \left| \frac{d\Phi_{\text{m}}}{dt} \right| \tag{33.28}$$

where $\Phi = N\Phi_{\text{per coil}}$ is the total flux through all the coils. The inductance was defined such that $\Phi_{\text{m}} = LI$, so Equation 33.28 becomes

$$\mathcal{E}_{\text{coil}} = L \left| \frac{dI}{dt} \right| \tag{33.29}$$

The induced emf is directly proportional to the *rate of change* of current through the coil. We'll consider the appropriate sign in a moment, but Equation 33.29 gives us the size of the potential difference that is developed across a coil as the current through the coil changes. Note that $\mathcal{E}_{\text{coil}} = 0$ for a steady, unchanging current.

Figure 33.40 shows the same inductor, but now the current (still *in* to the left side) is decreasing. To oppose the decrease in flux, the induced current is in the *same* direction as the input current. The induced current carries charge to the right and establishes a potential difference opposite that in Figure 33.39b.

NOTE ▶ Notice that the induced current does not oppose the current through the inductor, which is from left to right in both Figures 33.39 and 33.40. Instead, in accordance with Lenz's law, the induced current opposes the *change* in the current in the solenoid. The practical result is that it is hard to change the current through an inductor. Any effort to increase or decrease the current is met with opposition in the form of an opposing induced current. You can think of the current in an inductor as having inertia, trying to continue what it was doing without change. ◀

Before we can use inductors in a circuit we need to establish a rule about signs that is consistent with our earlier circuit analysis. Figure 33.41 first shows current *I* passing through a resistor. You learned in Chapter 31 that the potential difference across a resistor is $\Delta V_R = -IR$, where the minus sign indicates that the potential *decreases* in the direction of the current.

We'll use the same convention for an inductor. The potential difference across an inductor, measured along the direction of the current, is

$$\Delta V_L = -L\frac{dI}{dt} \qquad (33.30)$$

If the current is increasing ($dI/dt > 0$), the input side of the inductor is more positive than the output side and the potential decreases in the direction of the current ($\Delta V_L < 0$). This was the situation in Figure 33.39b. If the current is decreasing ($dI/dt < 0$), the input side is more negative and the potential increases in the direction of the current ($\Delta V_L > 0$). This was the situation in Figure 33.40.

The potential difference across an inductor can be very large if the current changes very abruptly (large *dI/dt*). Figure 33.42 shows an inductor connected across a battery. There is a large current through the inductor, limited only by the internal resistance of the battery. Suppose the switch is suddenly opened. A very large induced voltage is created across the inductor as the current rapidly drops to zero. This potential difference (plus ΔV_{bat}) appears across the gap of the switch as it is opened. A large potential difference across a small gap often creates a spark.

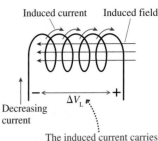

Induced current   Induced field

Decreasing current
ΔV_L
− +

The induced current carries positive charge carriers to the right. The potential difference is opposite that of Figure 33.39.

FIGURE 33.40 Decreasing the current through an inductor.

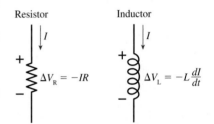

Resistor   Inductor

$\Delta V_R = -IR$ $\Delta V_L = -L\frac{dI}{dt}$

The potential always decreases.

The potential decreases if the current is increasing.

The potential increases if the current is decreasing.

FIGURE 33.41 The potential difference across a resistor and an inductor.

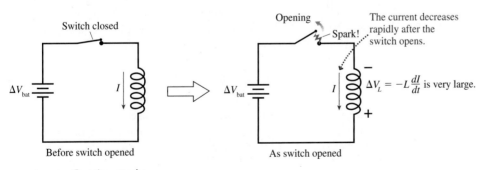

Switch closed

Opening

Spark!

The current decreases rapidly after the switch opens.

ΔV_{bat} I

ΔV_{bat} I $\Delta V_L = -L\frac{dI}{dt}$ is very large.

Before switch opened   As switch opened

FIGURE 33.42 Creating sparks.

Indeed, this is exactly how the spark plugs in your car work. The car's generator sends a large current through the *coil,* which is a big inductor. A switch in the *distributor* is suddenly opened, breaking the current. The induced voltage, typically a few thousand volts, appears across the terminals of the spark plug, creating the spark that ignites the gasoline. A similar phenomenon happens if you unplug appliances such as toaster ovens or hair dryers while they are running. The heating coils in these devices have quite a bit of inductance. Suddenly pulling the plug is like opening a switch. The large induced voltage often causes a spark between the plug and the electric outlet.

EXAMPLE 33.13 Large voltage across an inductor
A 1.0 A current passes through a 10 mH inductor coil. What potential difference is induced across the coil if the current drops to zero in 5 μs?

MODEL Assume this is an ideal inductor, with $R = 0\ \Omega$, and that the current decrease is linear with time.

SOLVE The rate of current decrease is

$$\frac{dI}{dt} \approx \frac{\Delta I}{\Delta t} = \frac{-1.0\ \text{A}}{5.0 \times 10^{-6}\ \text{s}} = -2.0 \times 10^5\ \text{A/s}$$

The induced voltage is

$$\Delta V_\text{L} = -L\frac{dI}{dt} \approx -(0.010\ \text{H})(-2.0 \times 10^5\ \text{A/s})$$
$$= 2000\ \text{V}$$

ASSESS Inductors may be physically small, but they can pack a punch if you try to change the current through them too quickly.

STOP TO THINK 33.6 The potential at a is higher than the potential at b. Which of the following statements about the inductor current I could be true?

a. I is from a to b and steady.
b. I is from a to b and increasing.
c. I is from a to b and decreasing.
d. I is from b to a and steady.
e. I is from b to a and increasing.
f. I is from b to a and decreasing.

Energy in Inductors and Magnetic Fields

An inductor, like a capacitor, stores energy that can later be released. It is energy released from the coil in your car that becomes the spark of the spark plug. You learned in Chapter 31 that electric power is $P_\text{elec} = I\Delta V$. As current passes through an inductor, for which $\Delta V_\text{L} = -L(dI/dt)$, the electric power is

$$P_\text{elec} = I\Delta V_\text{L} = -LI\frac{dI}{dt} \tag{33.31}$$

P_elec is negative because the current is *losing* electric energy. That energy is being transferred to the inductor, which is *storing* energy U_L at the rate

$$\frac{dU_\text{L}}{dt} = +LI\frac{dI}{dt} \tag{33.32}$$

where we've noted that power is the rate of change of energy.

We can find the total energy stored in an inductor by integrating Equation 33.32 from $I = 0$, where $U_\text{L} = 0$, to a final current I. Doing so gives

$$U_\text{L} = L\int_0^I I\,dI = \frac{1}{2}LI^2 \tag{33.33}$$

The potential energy stored in an inductor depends on the square of the current through it. Notice the analogy with the energy $U_\text{C} = \frac{1}{2}C(\Delta V)^2$ stored in a capacitor.

In working with circuits we say that the energy is "stored in the inductor." Strictly speaking, the energy is stored in the inductor's magnetic field, analogous to how a capacitor stores energy in the electric field. We can use the inductance of a solenoid, Equation 33.27, to relate the inductor's energy to the magnetic field strength:

$$U_L = \frac{1}{2}LI^2 = \frac{\mu_0 N^2 A}{2l}I^2 = \frac{1}{2\mu_0}Al\left(\frac{\mu_0 NI}{l}\right)^2 \qquad (33.34)$$

We made the last rearrangement in Equation 33.34 because $\mu_0 NI/l$ is the magnetic field inside the solenoid. Thus

$$U_L = \frac{1}{2\mu_0}AlB^2 \qquad (33.35)$$

But Al is the volume inside the solenoid. Dividing by Al, the magnetic field *energy density* inside the solenoid (energy per m³) is

$$u_B = \frac{1}{2\mu_0}B^2 \qquad (33.36)$$

We've derived this expression for energy density based on the properties of a solenoid, but it turns out to be the correct expression for the energy density anywhere there's a magnetic field. Compare this to the energy density of an electric field $u_E = \frac{1}{2}\epsilon_0 E^2$ that we found in Chapter 30.

Energy in electric and magnetic fields

Electric fields	Magnetic fields
A capacitor stores energy	An inductor stores energy
$U_C = \frac{1}{2}C(\Delta V)^2$	$U_L = \frac{1}{2}LI^2$
Energy density in the field is	Energy density in the field is
$u_E = \frac{\epsilon_0}{2}E^2$	$u_B = \frac{1}{2\mu_0}B^2$

EXAMPLE 33.14 Energy stored in an inductor

The 10 μH inductor of Example 33.12 was 5.7 cm long and 4.0 mm in diameter. Suppose it carries a 100 mA current. What are the energy stored in the inductor, the magnetic energy density, and the magnetic field strength?

SOLVE The stored energy is

$$U_L = \frac{1}{2}LI^2 = \frac{1}{2}(10^{-5}\,\text{H})(0.10\,\text{A})^2 = 5.0 \times 10^{-8}\,\text{J}$$

The solenoid volume is $(\pi r^2)l = 7.16 \times 10^{-7}\,\text{m}^3$. Using this gives the energy density of the magnetic field:

$$u_B = \frac{5.0 \times 10^{-8}\,\text{J}}{7.16 \times 10^{-7}\,\text{m}^3} = 0.070\,\text{J/m}^3$$

From Equation 33.36, the magnetic field with this energy density is

$$B = \sqrt{2\mu_0 u_B} = 4.2 \times 10^{-4}\,\text{T}$$

33.9 *LC* Circuits

Telecommunication—radios, televisions, cell phones—is based on electromagnetic signals that *oscillate* at a well-defined frequency. These oscillations are generated and detected by a simple circuit consisting of an inductor and a capacitor in parallel. This is called an **LC circuit.** In this section we will learn why an *LC* circuit oscillates and determine the oscillation frequency.

Figure 33.43 shows a capacitor with initial charge Q_0, an inductor, and a switch. The switch has been open for a long time, so there is no current in the circuit. Then, at $t = 0$, the switch is closed. How does the circuit respond? Let's think it through qualitatively before getting into the mathematics.

As Figure 33.44 on the next page shows, the inductor provides a conducting path for discharging the capacitor. However, the discharge current has to pass through the inductor, and, as we've seen, an inductor resists changes in current. Consequently, the current doesn't stop when the capacitor charge reaches zero.

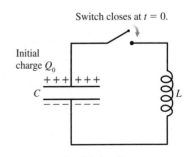

FIGURE 33.43 An *LC* circuit.

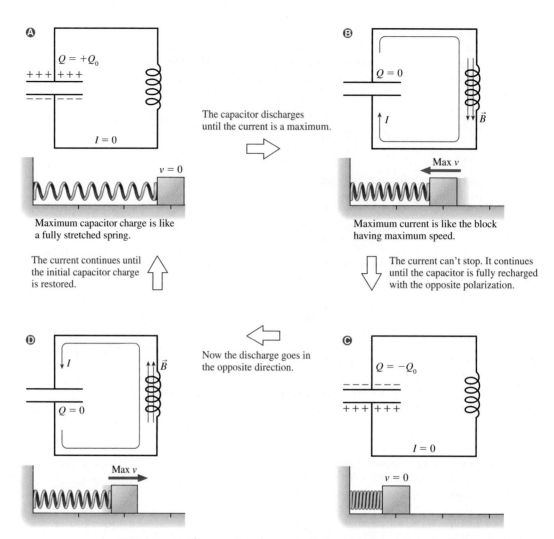

FIGURE 33.44 The capacitor charge oscillates much like a block attached to a spring.

A block attached to a stretched spring is a useful mechanical analogy. Closing the switch to discharge the capacitor is like releasing the block. But the block doesn't stop when it reaches the origin. Its momentum keeps it going until the spring is fully compressed. Likewise, the current continues until it has recharged the capacitor with the opposite polarization. This process repeats over and over, charging the capacitor first one way, then the other. That is, the charge and current *oscillate*.

The goal of our circuit analysis will be to find expressions showing how the capacitor charge Q and the inductor current I change with time. As always, our starting point for circuit analysis is Kirchhoff's voltage law, which says that all the potential differences around a closed loop must sum to zero. Choosing a cw direction for I, Kirchhoff's law is

$$\Delta V_C + \Delta V_L = 0 \qquad (33.37)$$

You learned in Chapter 30 that the potential difference across a capacitor is $\Delta V_C = Q/C$, where Q is the charge on the top plate of the capacitor, and we found the potential difference across an inductor in Equation 33.30 above. Using these, Kirchhoff's law becomes

$$\frac{Q}{C} - L\frac{dI}{dt} = 0 \qquad (33.38)$$

Equation 33.38 has two unknowns, Q and I. We need to eliminate one of the unknowns, and we can do so by finding another relation between Q and I. Current is the rate at which charge moves: $I = dq/dt$. But the charge flowing through the inductor is charge that was *removed* from the capacitor. That is, an infinitesimal charge dq flows through the inductor when the capacitor charge changes by $dQ = -dq$. Thus the current through the inductor is related to the charge on the capacitor by

$$I = -\frac{dQ}{dt} \tag{33.39}$$

Now I is positive when Q is decreasing, as we would expect. This is a subtle but important step in the reasoning, one worth thinking about because it appears in other contexts.

Equations 33.38 and 33.39 are two equations in two unknowns. To solve them, we'll first take the time derivative of Equation 33.39:

$$\frac{dI}{dt} = \frac{d}{dt}\left(-\frac{dQ}{dt}\right) = -\frac{d^2Q}{dt^2} \tag{33.40}$$

We can substitute this result into Equation 33.38:

$$\frac{Q}{C} + L\frac{d^2Q}{dt^2} = 0 \tag{33.41}$$

Now we have an equation for the capacitor charge Q.

Equation 33.41 is a second-order differential equation for Q. Fortunately, it is an equation we've seen before and already know how to solve. To see this, rewrite Equation 33.41 as

$$\frac{d^2Q}{dt^2} = -\frac{1}{LC}Q \tag{33.42}$$

Recall, from Chapter 14, that the equation of motion for an undamped mass on a spring is

$$\frac{d^2x}{dt^2} = -\frac{k}{m}x \tag{33.43}$$

Equation 33.42 is *exactly the same equation,* with x replaced by Q and k/m replaced by $1/LC$. This should be no surprise because we've already seen that a mass on a spring is a mechanical analog of the *LC* circuit.

We know the solution to Equation 33.43. It is simple harmonic motion $x(t) = x_0 \cos \omega t$ with angular frequency $\omega = \sqrt{k/m}$. Thus the solution to Equation 33.42 must be

$$Q(t) = Q_0 \cos \omega t \tag{33.44}$$

where Q_0 is the initial charge, at $t = 0$, and the angular frequency is

$$\omega = \sqrt{\frac{1}{LC}} \tag{33.45}$$

The charge on the upper plate of the capacitor oscillates back and forth between $+Q_0$ (as shown in Figure 33.43) and $-Q_0$ (the opposite polarization) with period $T = 2\pi/\omega$.

As the capacitor charge oscillates, so does the current through the inductor. Using Equation 33.39 gives the current through the inductor:

$$I = -\frac{dQ}{dt} = \omega Q_0 \sin \omega t = I_{max} \sin \omega t \tag{33.46}$$

where $I_{max} = \omega Q_0$ is the maximum current.

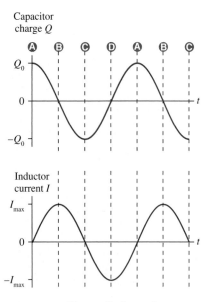

Capacitor charge Q

FIGURE 33.45 The oscillations of an *LC* circuit.

An *LC* circuit is an *electric oscillator*, oscillating at frequency $f = \omega/2\pi$. Figure 33.45 shows graphs of the capacitor charge Q and the inductor current I as functions of time. The letters over the graph match the labels in Figure 33.44, and you should compare the two. Notice that Q and I are 90° out of phase. The current is zero when the capacitor is fully charged, as expected, and the charge is zero when the current is maximum.

EXAMPLE 33.15 An AM radio oscillator
You have a 1.0 mH inductor. What capacitor should you choose to make an oscillator with a frequency of 920 KHz? (This frequency is near the center of the AM radio band.)

SOLVE The angular frequency is $\omega = 2\pi f = 5.78 \times 10^6$ rad/s. Using Equation 33.45 for ω gives the required capacitor:

$$C = \frac{1}{\omega^2 L} = \frac{1}{(5.78 \times 10^6 \text{ rad/s})^2 (0.0010 \text{ H})}$$

$$= 3.0 \times 10^{-11} \text{ F} = 30 \text{ pF}$$

An *LC* circuit, like a mass on a spring, wants to respond only at its natural oscillation frequency $\omega = 1/\sqrt{LC}$. In Chapter 14 we defined a strong response at the natural frequency as a *resonance,* and resonance is the basis for all telecommunications. The input circuit in radios, televisions, and cell phones is an *LC* circuit driven by the signal picked up by the antenna. This signal is the superposition of hundreds of sinusoidal waves at different frequencies, one from each transmitter in the area, but the circuit responds only to the *one* signal that matches the circuit's natural frequency. That particular signal generates a large-amplitude current that can be further amplified and decoded to become the output that you hear.

Turning the dial on your radio or television changes a *variable capacitor,* thus changing the resonance frequency so that you pick up a different station. Cell phones are a bit more complicated. You don't change the capacitance yourself, but a "smart" circuit inside can change its capacitance in response to command signals it receives from the transmitter. The result is the same. Your cell phone responds to the one signal being broadcast to you and ignores the hundreds of other signals that are being broadcast simultaneously at different frequencies.

33.10 *LR* Circuits

14.1 Activ Physics ONLINE

A circuit consisting of an inductor, a resistor, and (perhaps) a battery is called an **LR circuit.** Figure 33.46a is an example of an *LR* circuit. We'll assume that the switch has been in position a for such a long time that the current is steady and unchanging. There's no potential difference across the inductor, because $dI/dt = 0$, so it simply acts like a piece of wire. The current flowing around the circuit is determined entirely by the battery and the resistor: $I_0 = \Delta V_{bat}/R$.

What happens if, at $t = 0$, the switch is suddenly moved to position b? With the battery no longer in the circuit, you might expect the current to stop immediately. But the inductor won't let that happen. The current will continue for some period of time as the inductor's magnetic field drops to zero. In essence, the energy stored in the inductor allows it to act like a battery for a short period of time. Our goal is to determine how the current decays after the switch is moved.

Figure 33.46b shows the circuit after the switch is changed. Our starting point, once again, is Kirchhoff's voltage law. The potential differences around a closed loop must sum to zero. For this circuit, Kirchhoff's law is

$$\Delta V_R + \Delta V_L = 0 \tag{33.47}$$

The potential differences in the direction of the current are $\Delta V_R = -IR$ for the resistor and $\Delta V_L = -L(dI/dt)$ for the inductor. Substituting these into Equation 33.47 gives

$$-RI - L\frac{dI}{dt} = 0 \tag{33.48}$$

We're going to need to integrate to find the current I as a function of time. Before doing so, we need to get all the current terms on one side of the equation and all the time terms on the other. A simple rearrangement of Equation 33.48 gives

$$\frac{dI}{I} = -\frac{R}{L}dt = -\frac{dt}{(L/R)} \tag{33.49}$$

We know that the current at $t = 0$, when the switch was moved, was I_0. We want to integrate from these starting conditions to current I at the unspecified time t. That is,

$$\int_{I_0}^{I} \frac{dI}{I} = -\frac{1}{(L/R)} \int_{0}^{t} dt \tag{33.50}$$

Both are common integrals, giving

$$\ln I \Big|_{I_0}^{I} = \ln I - \ln I_0 = \ln\left(\frac{I}{I_0}\right) = -\frac{t}{(L/R)} \tag{33.51}$$

We can solve for the current I by taking the exponential of both sides, then multiplying by I_0. Doing so gives I, the current as a function of time:

$$I = I_0 e^{-t/(L/R)} \tag{33.52}$$

Notice that $I = I_0$ at $t = 0$, as expected.

The argument of the exponential function must be dimensionless, so L/R must have dimensions of time. If we define the **time constant** τ of the *LR* circuit to be

$$\boxed{\tau = \frac{L}{R}} \tag{33.53}$$

then we can write Equation 33.52 as

$$I = I_0 e^{-t/\tau} \tag{33.54}$$

The time constant is the time at which the current has decreased to e^{-1} (about 37%) of its initial value. We can see this by computing the current at the time $t = \tau$.

$$I(\text{at } t = \tau) = I_0 e^{-\tau/\tau} = e^{-1} I_0 = 0.37 I_0 \tag{33.55}$$

Thus the time constant for an *LR* circuit functions in exactly the same way as the time constant for the *RC* circuit we analyzed in Chapter 31. At time $t = 2\tau$, the current has decreased to $e^{-2}I_0$, or about 13% of its initial value.

The current is graphed in Figure 33.47. You can see that the current decays exponentially. The *shape* of the graph is always the same, regardless of the specific value of the time constant τ.

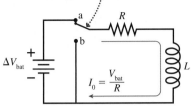

(a) The switch has been in this position for a long time. At $t = 0$ it is moved to position b.

(b) This is the circuit with the switch in position b. The inductor prevents the current from stopping instantly.

FIGURE 33.46 An *LR* circuit.

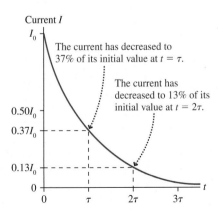

Current I

The current has decreased to 37% of its initial value at $t = \tau$.

The current has decreased to 13% of its initial value at $t = 2\tau$.

FIGURE 33.47 The current decay in an *LR* circuit.

EXAMPLE 33.16 **Exponential decay in an *LR* circuit**
The switch in Figure 33.48 has been in position a for a long time. It is changed to position b at $t = 0$ s.

a. What is the current in the circuit at $t = 5.0 \ \mu s$?
b. At what time has the current decayed to 1% of its initial value?

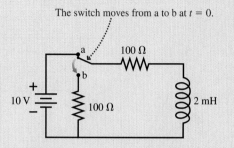

The switch moves from a to b at $t = 0$.

FIGURE 33.48 The *LR* circuit of Example 33.16.

MODEL This is an *LR* circuit. We'll assume ideal wires and an ideal inductor.

VISUALIZE The two resistors will be in series after the switch is thrown.

SOLVE Before the switch is thrown, while $\Delta V_L = 0$, the current is $I_0 = (10 \text{ V})/(100 \ \Omega) = 0.10 \text{ A} = 100 \text{ mA}$. This will be the initial current after the switch is thrown because the current through an inductor can't change instantaneously. The circuit resistance after the switch is thrown is $R = 200 \ \Omega$, so the time constant is

$$\tau = \frac{L}{R} = \frac{2.0 \times 10^{-3} \text{ H}}{200 \ \Omega} = 1.0 \times 10^{-5} \text{ s} = 10 \ \mu s$$

a. The current at $t = 5.0 \ \mu s$ is

$$I = I_0 e^{-t/\tau} = (100 \text{ mA})e^{-(5 \ \mu s)/(10 \ \mu s)} = 61 \text{ mA}$$

b. To find the time at which a particular current is reached we need to go back to Equation 33.52 and solve for t:

$$t = -\frac{L}{R} \ln\left(\frac{I}{I_0}\right) = -\tau \ln\left(\frac{I}{I_0}\right)$$

The time at which the current has decayed to 1 mA (1% of I_0) is

$$t = -(10 \ \mu s) \ln\left(\frac{1 \text{ mA}}{100 \text{ mA}}\right) = 46 \ \mu s$$

ASSESS For all practical purposes, the current has decayed away in $\approx 50 \ \mu s$. The inductance in this circuit is not large, so a short decay time is not surprising.

STOP TO THINK 33.7 Rank in order, from largest to smallest, the time constants $\tau_1, \tau_2,$ and τ_3 of these three circuits.

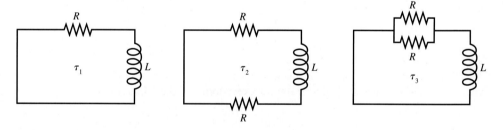

SUMMARY

The goal of Chapter 33 has been to understand and apply electromagnetic induction.

GENERAL PRINCIPLES

Faraday's Law

MODEL Make simplifying assumptions.

VISUALIZE Use Lenz's law to determine the direction of the **induced current.**

SOLVE The **induced emf** is

$$\mathcal{E} = \left| \frac{d\Phi}{dt} \right|$$

Multiply by N for an N-turn coil.
The size of the induced current is $I = \mathcal{E}/R$.

ASSESS Is the result reasonable?

Lenz's Law

There is an induced current in a closed conducting loop if and only if the magnetic flux through the loop is changing. The direction of the induced current is such that the induced magnetic field opposes the *change* in the flux.

Magnetic flux

Magnetic flux measures the amount of magnetic field passing through a surface.

$$\Phi = \vec{A} \cdot \vec{B} = AB \cos\theta$$

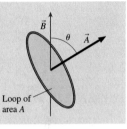

Loop of area A

IMPORTANT CONCEPTS

Three ways to change the flux

1. A loop moves into or out of a magnetic field.

2. The loop changes area or rotates.

3. The magnetic field through the loop increases or decreases.

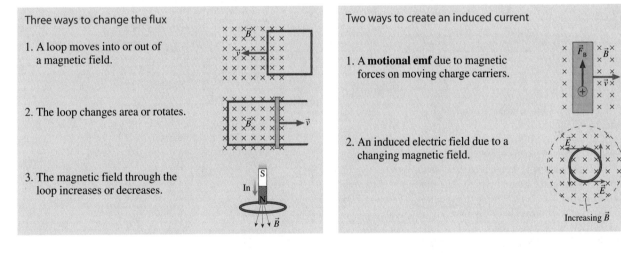

Two ways to create an induced current

1. A **motional emf** due to magnetic forces on moving charge carriers.

2. An induced electric field due to a changing magnetic field.

Increasing $\vec{B}$

APPLICATIONS

Inductors

Solenoid inductance $L_{\text{solenoid}} = \dfrac{\mu_0 N^2 A}{l}$

Potential difference $\Delta V_L = -L\dfrac{dI}{dt}$

Energy stored $U_L = \frac{1}{2}LI^2$

Magnetic energy density $u_B = \dfrac{1}{2\mu_0}B^2$

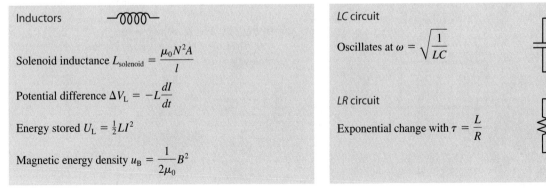

LC circuit

Oscillates at $\omega = \sqrt{\dfrac{1}{LC}}$

LR circuit

Exponential change with $\tau = \dfrac{L}{R}$

TERMS AND NOTATION

electromagnetic induction	induced emf, $\mathcal{E}$	transformer
induced current	Faraday's law	inductance, L
motional emf	induced electric field	henry, H
generator	Coulomb electric field	inductor
eddy current	non-Coulomb electric field	LC circuit
magnetic flux, Φ_m	induced magnetic field	LR circuit
weber, Wb	electromagnetic wave	time constant, τ
area vector, $\vec{A}$	primary coil	
Lenz's law	secondary coil	

EXERCISES AND PROBLEMS

Exercises

Section 33.2 Motional emf

1. A potential difference of 0.050 V is developed across a 10-cm-long wire as it moves through a magnetic field at 5.0 m/s. The magnetic field is perpendicular to the axis of the wire. What are the direction and strength of the magnetic field?

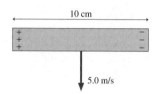

FIGURE EX33.1

2. The earth's magnetic field strength is 5×10^{-5} T. How fast would you have to drive your car to create a 1.0 V motional emf along your 1.0-m-long radio antenna? Assume that the motion of the antenna is perpendicular to $\vec{B}$.

3. A 10-cm-long wire is pulled along a U-shaped conducting rail in a perpendicular magnetic field. The total resistance of the wire and rail is 0.20 Ω. Pulling the wire with a force of 1.0 N causes 4.0 W of power to be dissipated in the circuit.
   a. What is the speed of the wire when pulled with 1.0 N?
   b. What is the strength of the magnetic field?

Section 33.3 Magnetic Flux

4. What is the magnetic flux through the loop shown in Figure Ex33.4?

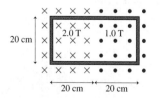

FIGURE EX33.4

5. A 2.0-cm-diameter solenoid passes through the center of a 6.0-cm-diameter loop. The magnetic field inside the solenoid is 0.20 T. What is the magnetic flux through the loop when it is perpendicular to the solenoid and when it is tilted at a 60° angle?

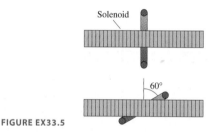

FIGURE EX33.5

Section 33.4 Lenz's Law

6. There is a ccw induced current in the conducting loop shown in Figure Ex33.6. Is the magnetic field inside the loop increasing in strength, decreasing in strength, or steady?

FIGURE EX33.6

7. A solenoid is wound as shown in Figure Ex33.7.
   a. Is there an induced current as magnet 1 is moved away from the solenoid? If so, what is the current direction through resistor R?
   b. Is there an induced current as magnet 2 is moved away from the solenoid? If so, what is the current direction through resistor R?

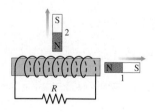

FIGURE EX33.7

8. The metal equilateral triangle in Figure Ex33.8, 20 cm on each side, is halfway into a 0.10 T magnetic field.
   a. What is the magnetic flux through the triangle?
   b. If the magnetic field strength decreases, what is the direction of the induced current in the triangle?

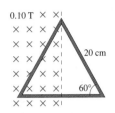

FIGURE EX33.8

9. The current in the solenoid of Figure Ex33.9 is decreasing. The solenoid is surrounded by a conducting loop. Is there a current in the loop? If so, is the loop current cw or ccw?

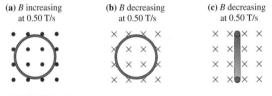

FIGURE EX33.9

Section 33.5 Faraday's Law

10. Figure Ex33.10 shows a 10-cm-diameter loop in three different magnetic fields. The loop's resistance is 0.10 Ω. For each case, determine the induced emf, the induced current, and the direction of the current.

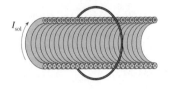

FIGURE EX33.10

11. A 1000-turn coil of wire that is 2.0 cm in diameter is in a magnetic field that drops from 0.10 T to 0 T in 10 ms. The axis of the coil is parallel to the field. What is the emf of the coil?

12. The loop in Figure Ex33.12 is being pushed into the 0.20 T magnetic field at 50 m/s. The resistance of the loop is 0.10 Ω. What are the direction and the magnitude of the current in the loop?

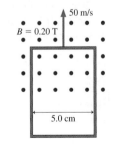

FIGURE EX33.12

13. The resistance of the loop in Figure Ex33.13 is 0.10 Ω. Is the magnetic field strength increasing or decreasing? At what rate (T/s)?

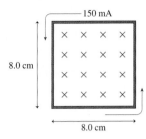

FIGURE EX33.13

Section 33.8 Inductors

14. You need to make a 100 μH inductor on a cylinder that is 5.0 cm long and 1.0 cm in diameter. You plan to wrap four layers of wire around the cylinder. What diameter wire should you use if the coils are tightly wound with no space between them? The wire diameter will be small enough that you don't need to consider the slight change in the coil's diameter for the outer layers.

15. What is the potential difference across a 10 mH inductor if the current through the inductor drops from 150 mA to 50 mA in 10 μs? What is the direction of this potential difference? That is, does the potential increase or decrease along the direction of the current?

16. The maximum allowable potential difference across a 200 mH inductor is 400 V. You need to raise the current through the inductor from 1.0 A to 3.0 A. What is the minimum time you should allow for changing the current?

17. How much energy is stored in a 3.0-cm-diameter, 12-cm-long solenoid that has 200 turns of wire and carries a current of 0.80 A?

Section 33.9 *LC* Circuits

18. A 2.0 mH inductor is connected in parallel with a variable capacitor. The capacitor can be varied from 100 pF to 200 pF. What is the range of oscillation frequencies for this circuit?

19. An FM radio station broadcasts at a frequency of 100 MHz. What inductance should be paired with a 10 pF capacitor to build a receiver circuit for this station?

20. An electric oscillator is made with a 0.10 μF capacitor and a 1.0 mH inductor. The capacitor is initially charged to 5.0 V. What is the maximum current through the inductor as the circuit oscillates?

Section 33.10 *LR* Circuits

21. What value of resistor R gives the circuit in Figure Ex33.21 a time constant of 10 μs?

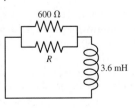

FIGURE EX33.21

22. At $t = 0$ s, the current in the circuit in Figure Ex33.22 is I_0. At what time is the current $\frac{1}{2}I_0$?

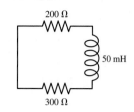

FIGURE EX33.22

Problems

23. A 10 cm × 10 cm square is bent at a 90° angle. A uniform 0.050 T magnetic field points downward at a 45° angle. What is the magnetic flux through the loop?

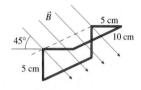

FIGURE P33.23

24. What is the magnetic flux through the loop shown in Figure P33.24?

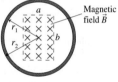

FIGURE P33.24

25. A 20 cm × 20 cm square loop has a resistance of 0.10 Ω. A magnetic field perpendicular to the loop is $B = 4t - 2t^2$, where B is in tesla and t is in seconds.
    a. Determine B, $\mathcal{E}$, and I at half-second intervals from 0 s to 2 s.
    b. Use your results of part a to draw graphs of B and I versus time.

26. A 5.0-cm-diameter coil has 20 turns and a resistance of 0.50 Ω. A magnetic field perpendicular to the coil is $B = 0.020t + 0.010t^2$, where B is in tesla and t is in seconds.
    a. Draw a graph of B as a function of time from $t = 0$ s to $t = 10$ s.
    b. Find an expression for the induced current I as a function of time.
    c. Evaluate I at $t = 5$ s and $t = 10$ s.

27. A 50-turn, 4.0-cm-diameter coil has a resistance of 1.0 Ω. A magnetic field perpendicular to the coil is $B = t - \frac{1}{4}t^2$, where B is in tesla and t is in seconds.
    a. Draw a graph of B as a function of time from $t = 0$ s to $t = 4$ s.
    b. Find an expression for the induced current $I(t)$ as a function of time.
    c. Evaluate I at $t = 1$, 2, and 3 s.

28. A 100-turn, 2.0-cm-diameter coil is at rest in a horizontal plane. A uniform magnetic field 60° away from vertical increases from 0.50 T to 1.50 T in 0.60 s. What is the induced emf in the coil?

29. A 25-turn, 10.0-cm-diameter coil is oriented in a vertical plane with its axis aligned east-west. A magnetic field pointing to the northeast decreases from 0.80 T to 0.20 T in 2.0 s. What is the emf induced in the coil?

30. A 100-turn, 8.0-cm-diameter coil is made of 0.50-mm-diameter copper wire. A magnetic field is perpendicular to the coil. At what rate must B increase to induce a 2.0 A current in the coil?

31. A 2.0 cm × 2.0 cm square loop of wire with resistance 0.010 Ω is parallel to a long straight wire. The near edge of the loop is 1.0 cm from the wire. The current in the wire is increasing at the rate of 100 A/s. What is the current in the loop?

32. A 4.0-cm-diameter loop with resistance 0.10 Ω surrounds a 2.0-cm-diameter solenoid. The solenoid is 10 cm long, has 100 turns, and carries the current shown in the graph. A positive current is cw when seen from the left. Determine the current in the loop as a function of time. Give your answer as a current-versus-time graph from $t = 0$ s to $t = 3$ s.

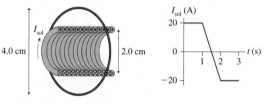

FIGURE P33. 32

33. A 20 cm × 20 cm square loop of wire lies in the xy-plane with its bottom edge on the x-axis. The resistance of the loop is 0.50 Ω. A magnetic field parallel to the z-axis is given by $B = 0.80y^2t$, where B is in tesla, y in meters, and t in seconds. What is the size of the induced current in the loop at $t = 0.50$ s?

34. Figure P33.34 shows a five-turn, 1.0-cm-diameter coil with $R = 0.10$ Ω inside a 2.0-cm-diameter solenoid. The solenoid is 8.0 cm long, has 120 turns, and carries the current shown in the graph. A positive current is cw when seen from the left. Determine the current in the coil as a function of time. Give your answer as a current-versus-time graph from $t = 0$ s to $t = 0.02$ s.

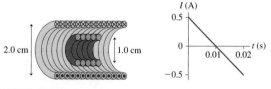

FIGURE P33.34

35. Two 20-turn coils are tightly wrapped on the same 2.0-cm-diameter cylinder with 1.0-mm-diameter wire. The current through coil 1 is shown in the graph. A positive current is into the page at the top of a loop. Determine the current in coil 2 as a function of time. Give your answer as a current-versus-time graph from $t = 0$ s to $t = 0.4$ s. Assume that the magnetic field of coil 1 passes entirely through coil 2.

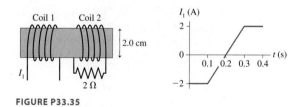

FIGURE P33.35

36. A loop antenna, ᵘch as is used on a television to pick up UHF broadcasts, is 25 ᵐm in diameter. The plane of the loop is perpendicular to th oscillating magnetic field of a 150 MHz electromagnetic ʳave. The magnetic field through the loop is $B = (20 \text{ nT}) \sin$ t.
   a. What is the n ximum emf induced in the antenna?
   b. What is the ᵢ aximum emf if the loop is turned 90° to be perpendicula ᵗo the oscillating electric field?

37. A 50-turn, 4.0-cᵣ -diameter coil with $R = 0.50 \, \Omega$ surrounds a 2.0-cm-diameteʳ olenoid. The solenoid is 20 cm long and has 200 turns. The 60 Hz current through the solenoid is $I_{\text{sol}} = (0.50 \text{ A}) \text{s}$ $(2\pi ft)$. Find an expression for I_{coil}, the induced current ᵢ the coil as a function of time.

38. A 40-turn, 4.0-cᵣ -diameter coil with $R = 0.40 \, \Omega$ surrounds a 3.0-cm-diameteʳ solenoid. The solenoid is 20 cm long and has 200 turns. ᵀ ʰe 60 Hz current through the solenoid is $I = I_0 \sin(2\pi ft)$ ᵂhat is I_0 if the maximum current in the coil is 0.20 A?

39. Electricity is disᵗ ibuted from electrical substations to neighborhoods at 15,( )0 V. This is a 60 Hz oscillating (AC) voltage. Neighborhᵒ d transformers, such as those seen on utility poles, step this ᵥ ltage down to the 120 V that is delivered to your house.
   a. How many tᵤ ns does the primary coil on the transformer have if the seᶜ ondary coil has 100 turns?
   b. No energy iⁿ lost in an ideal transformer, so the output power P_{out} frᵒ n the secondary coil equals the input power P_{in} to the prⁱ ary coil. Suppose a neighborhood transformer delivᵉ s 250 A at 120 V. What is the current in the 15,000 V lineᶠ rom the substation?

40. The square loop ˢhown in Figure P33.40 moves into a 0.80 T magnetic field ᵃ a constant speed of 10 m/s. The loop is 10.0 cm on eachˢ ide and has a resistance of 0.10 Ω. It enters the field at $t = 0$ ˢ.
   a. Find the induᶜ ed current in the loop as a function of time. Give your anˢ wer as a graph of I versus t from $t = 0$ s to $t = 0.020$ s.
   b. What is the mᵃ ximum current? What is the position of the loop when thᵉ current is maximum?

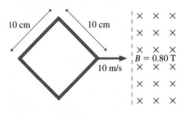

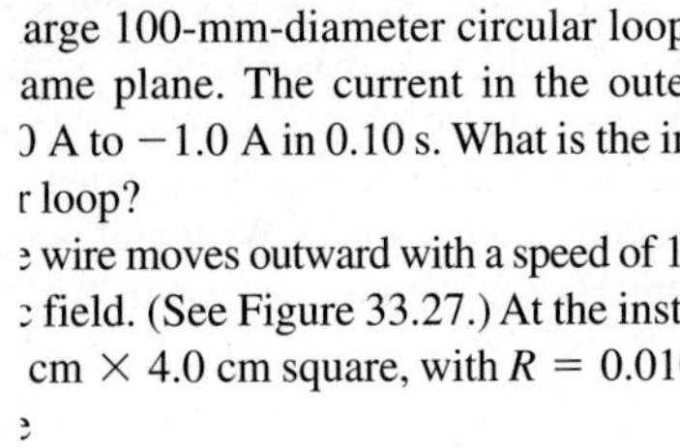

FIGURE P33.40

41. A small, 2.0-mⁿ diameter circular loop with $R = 0.020 \, \Omega$ is at the center of ᵃ arge 100-mm-diameter circular loop. Both loops lie in theˢ ame plane. The current in the outer loop changes from +ᵒ A to -1.0 A in 0.10 s. What is the induced current in the innᵉ r loop?

42. A 4.0-cm-long sliᵈ e wire moves outward with a speed of 100 m/s in a 1.0 T magneᵗ c field. (See Figure 33.27.) At the instant the circuit forms a 4.ᵒ cm × 4.0 cm square, with $R = 0.010 \, \Omega$ on each side, what aᵣ e?
   a. The inducedᵉ nf?
   b. The inducedᶜ urrent?
   c. The potentialᵈ ifference between the two ends of the moving wire?

43. A 20-cm-long, zero-resistance slide wire moves outward, on zero-resistance rails, at a steady speed of 10 m/s in a 0.10 T magnetic field. (See Figure 33.27.) On the opposite side, a 1.0 Ω carbon resistor completes the circuit by connecting the two rails. The mass of the resistor is 50 mg.
   a. What is the induced current in the circuit?
   b. How much force is needed to pull the wire at this speed?
   c. If the wire is pulled for 10 s, what is the temperature increase of the carbon? The specific heat of carbon is 710 J/kgC°.

44. The 10-cm-wide, zero-resistance slide wire shown in Figure P33.44 is pushed toward the 2.0 Ω resistor at a steady speed of 0.50 m/s. The magnetic field strength is 0.50 T.
   a. How big is the pushing force?
   b. How much power does the pushing force supply to the wire?
   c. What are the direction and magnitude of the induced current?
   d. How much power is dissipated in the resistor?

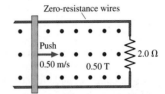

FIGURE P33.44

45. Your camping buddy has an idea for a light to go inside your tent. He happens to have a powerful (and heavy!) horseshoe magnet that he bought at a surplus store. This magnet creates a 0.20 T field between two pole tips 10 cm apart. His idea is to build a hand-cranked generator with a rotating 5.0-cm-radius semicircle between the pole tips. He thinks you can make enough current to fully light a 1.0 Ω lightbulb rated at 4.0 W. That's not super bright, but it should be plenty of light for routine activities in the tent.

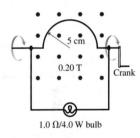

FIGURE P33.45

   a. Find an expression for the induced current as a function of time if you turn the crank at frequency f. Assume that the semicircle is at its highest point at $t = 0$.
   b. With what frequency will you have to turn the crank for the maximum current to fully light the bulb? Is this feasible?

46. You've decided to make a magnetic projectile launcher for your science project. An aluminum bar of length l slides along metal rails through a magnetic field B. The switch closes at $t = 0$ s, while the bar is at rest, and a battery of emf $\mathcal{E}_{\text{bat}}$ starts a current flowing around the loop. The battery has internal resistance r. The resistance of the rails and the bar are effectively zero.
   a. Show that the bar reaches a terminal velocity v_{term}, and find an expression for v_{term}.
   b. Evaluate v_{term} for $\mathcal{E}_{\text{bat}} = 1.0$ V, $r = 0.10 \, \Omega$, $l = 6.0$ cm, and $B = 0.50$ T.

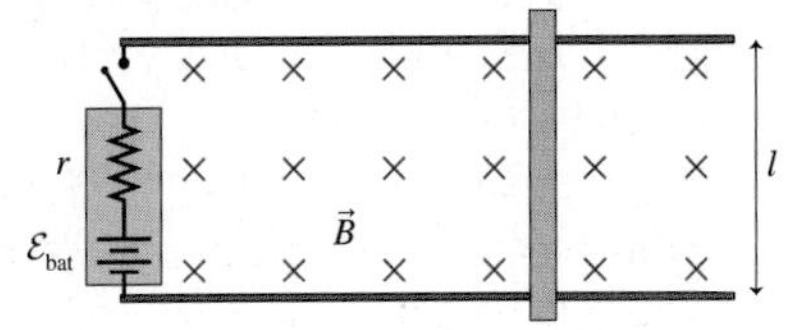

FIGURE P33.46

47. A slide wire of length l, mass m, and resistance R slides down a U-shaped metal track that is tilted upward at angle θ. The track has zero resistance and no friction. A vertical magnetic field B fills the loop formed by the track and the slide wire.
    a. Find an expression for the induced current I when the slide wire moves at speed v.
    b. Show that the slide wire reaches a terminal velocity v_{term}, and find an expression for v_{term}.

48. The plane of a 20 cm × 20 cm metal loop with a mass of 10 g and a resistance of $0.010 \ \Omega$ is oriented vertically. A 1.0 T horizontal magnetic field, perpendicular to the loop, fills the top half of the loop. There is no magnetic field through the bottom half of the loop. The loop is released from rest and allowed to fall.
    a. Show that the loop reaches a terminal velocity v_{term}, and find a value for v_{term}.
    b. How long will it take the loop to leave the field? Assume that the time needed to reach v_{term} is negligible. How does this compare to the time it would take the loop to fall the same distance in the absence of a field?

49. Figure P33.49 shows a U-shaped conducting rail that is oriented vertically in a horizontal magnetic field. The rail has no electric resistance and does not move. A slide wire with mass m and resistance R can slide up and down without friction while maintaining electrical contact with the rail. The slide wire is released from rest.
    a. Show that the slide wire reaches a terminal velocity v_{term}, and find an expression for v_{term}.
    b. Determine the value of v_{term} if $l = 20$ cm, $m = 10$ g, $R = 0.10 \ \Omega$, and $B = 0.50$ T.

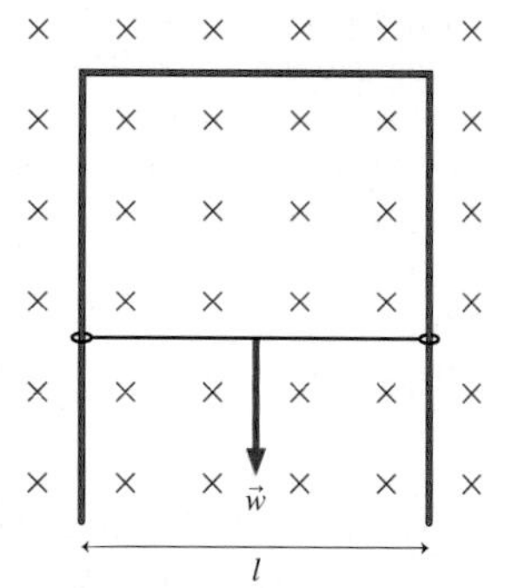

FIGURE P33.49

50. A 10-turn coil of wire having a diameter of 1.0 cm and a resistance of $0.20 \ \Omega$ is in a 1.0 mT magnetic field, with the coil oriented for maximum flux. The coil is connected to an uncharged $1.0 \ \mu F$ capacitor rather than to a current meter. The coil is quickly pulled out of the magnetic field. Afterward, what is the voltage across the capacitor?
    Hint: Use $I = dq/dt$ to relate the *net* change of flux to the amount of charge that flows to the capacitor.

51. The magnetic field at one place on the earth's surface is $55 \ \mu T$ in strength and tilted 60° down from horizontal. A 200-turn coil having a diameter of 4.0 cm and a resistance of $2.0 \ \Omega$ is connected to a $1.0 \ \mu F$ capacitor rather than to a current meter. The coil is held in a horizontal plane and the capacitor is discharged. Then the coil is quickly rotated 180° so that the side that had been facing up is now facing down. Afterward, what is the voltage across the capacitor? See the Hint in Problem 50.

52. A 100 mH inductor whose windings have a resistance of $4.0 \ \Omega$ is connected across a 12 V battery having an internal resistance of $2.0 \ \Omega$. How much energy is stored in the inductor?

53. A solenoid inductor carries a current of 200 mA. It has a magnetic flux of $20 \ \mu Wb$ per turn and stores 1.0 mJ of energy. How many turns does the inductor have?

54. A solenoid inductor has an emf of 0.20 V when the current through it changes at the rate 10.0 A/s. A steady current of 0.10 A produces a flux of $5.0 \ \mu Wb$ per turn. How many turns does the inductor have?

55. a. What is the magnetic energy density at the surface of a 1.0-mm-diameter wire carrying a current of 1.0 A?
    b. A 2.0-cm-diameter tightly wound solenoid is made with 1.0-mm-diameter wire. What is the magnetic energy density inside the solenoid if the current is 1.0 A?

56. a. What is the magnetic energy density at the center of a 4.0-cm-diameter loop carrying a current of 1.0 A?
    b. What current in a straight wire gives the magnetic energy density you found in part a at a point 2.0 cm from the wire?

57. The earth's magnetic field at the earth's surface is approximately $50 \ \mu T$. The earth's atmosphere is approximately 20 km thick.
    a. What is the total energy of the earth's magnetic field in the atmosphere? You can assume that the field strength within the atmosphere is constant.
    b. The world's total energy use is approximately 4×10^{18} J per year. If the magnetic energy in the atmosphere could be "harvested," what percentage of the world's annual energy use could it supply?

58. MRI (magnetic resonance imaging) is a medical technique that produces detailed "pictures" of the interior of the body. The patient is placed into a solenoid that is 40 cm in diameter and 1.0 m long. A 100 A current creates a 5.0 T magnetic field inside the solenoid. To carry such a large current, the solenoid wires are cooled with liquid helium until they become superconducting (no electric resistance).
    a. How much magnetic energy is stored in the solenoid? Assume that the magnetic field is uniform within the solenoid and quickly drops to zero outside the solenoid.
    b. How many turns of wire does the solenoid have?

59. One possible concern with MRI (see Problem 58) is turning the magnetic field on or off too quickly. Bodily fluids are conductors, and a changing magnetic field could cause electric currents to flow through the patient. Suppose a typical patient has a maximum cross-section area of $0.060 \ m^2$. What is the smallest time interval in which a 5.0 T magnetic field can be turned on or off if the induced emf around the patient's body must be kept to less than 0.10 V?

60. Experiments to study vision often need to track the movements of a subject's eye. One way of doing so is to have the subject sit in a magnetic field while wearing special contact lenses that have a coil of very fine wire circling the edge. A current is induced in the coil each time the subject rotates his eye. Consider an experiment in which a 20-turn, 6.0-mm-diameter coil of wire circles the subject's cornea while a 1.0 T magnetic field is directed as shown. The subject begins by looking straight ahead. What emf is induced in the coil if the subject shifts his gaze by 5° in 0.20 s?

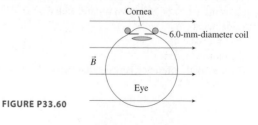

FIGURE P33.60

61. Figure P33.61 shows the current through a 10 mH inductor. Draw a graph showing the potential difference ΔV_L across the inductor for these 6 ms.

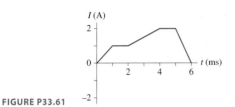

FIGURE P33.61

62. Figure P33.62 shows the current through a 10 mH inductor. Draw a graph showing the potential difference ΔV_L across the inductor for these 6 ms.

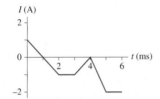

FIGURE P33.62

63. Figure P33.63 shows the potential difference across a 50 mH inductor. The current through the inductor at $t = 0$ s is 0.20 A. Draw a graph showing the current through the inductor from $t = 0$ s to $t = 40$ ms.

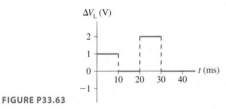

FIGURE P33.63

64. Figure P33.64 shows the potential difference across a 100 mH inductor. The current through the inductor at $t = 0$ s is 0.10 A. Draw a graph showing the current through the inductor from $t = 0$ s to $t = 40$ ms.

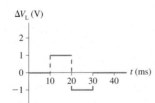

FIGURE P33.64

65. The current through inductance L is given by $I = I_0 e^{-t/\tau}$.
    a. Find an expression for the potential difference ΔV_L across the inductor.
    b. Evaluate ΔV_L at $t = 0$ s, 1, 2, and 3 ms if $L = 20$ mH, $I_0 = 50$ mA, and $\tau = 1.0$ ms.
    c. Draw a graph of ΔV_L versus time from $t = 0$ s to $t = 3$ ms.
66. The current through inductance L is given by $I = I_0 \sin \omega t$.
    a. Find an expression for the potential difference ΔV_L across the inductor.
    b. The maximum voltage across the inductor is 0.20 V when $L = 50 \, \mu$H and $f = 500$ kHz. What is I_0?

67. An *LC* circuit is built with a 20 mH inductor and an 8.0 μF capacitor. The current has its maximum value of 0.50 A at $t = 0$ s.
    a. How long is it until the capacitor is fully charged?
    b. What is the voltage across the capacitor at that time?
68. An *LC* circuit has a 10 mH inductor. The current has its maximum value of 0.60 A at $t = 0$ s. A short time later the capacitor reaches its maximum potential difference of 60 V. What is the value of the capacitance?
69. The maximum charge on the capacitor in an oscillating *LC* circuit is Q_0. What is the capacitor charge, in terms of Q_0, when the energy in the capacitor's electric field equals the energy in the inductor's magnetic field?
70. In recent years it has been possible to buy a 1.0 F capacitor. This is an enormously large amount of capacitance. Suppose you want to build a 1.0 Hz oscillator with a 1.0 F capacitor. You have a spool of 0.25-mm-diameter wire and a long 4.0-cm-diameter plastic cylinder. Design an inductor that will accomplish your goal.
71. The switch in Figure P33.71 has been in position 1 for a long time. It is changed to position 2 at $t = 0$ s.
    a. What is the maximum current through the inductor?
    b. What is the first time at which the current is maximum?

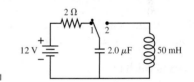

FIGURE P33.71

72. The 300 μF capacitor in Figure P33.72 is initially charged to 100 V, the 1200 μF capacitor is uncharged, and the switches are both open.
    a. What is the maximum voltage to which you can charge the 1200 μF capacitor by the proper closing and opening of the two switches?
    b. How would you do it? Describe the sequence in which you would close and open switches and the times at which you would do so. The first switch is closed at $t = 0$ s.

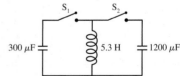

FIGURE P33.72

73. The switch in Figure P33.73 has been open for a long time. It is closed at $t = 0$ s.
    a. What is the current through the battery immediately after the switch is closed?
    b. What is the current through the battery after the switch has been closed a long time?

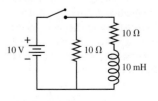

FIGURE P33.73

74. The switch in Figure P33.74 has been open for a long time. It is closed at $t = 0$ s. What is the current through the 20 Ω resistor
    a. immediately after the switch is closed?
    b. after the switch has been closed a long time?
    c. immediately after the switch is reopened?

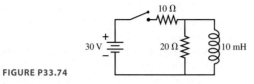

FIGURE P33.74

75. The switch in Figure P33.75 has been open for a long time. It is closed at $t = 0$ s.
    a. After the switch has been closed for a long time, what is the current in the circuit? Call this current I_0.
    b. Find an expression for the current I as a function of time. Write your expression in terms of I_0, R, and L.
    c. Sketch a current-versus-time graph from $t = 0$ s until the current is no longer changing.

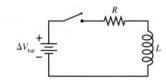

FIGURE P33.75

Challenge Problems

76. The L-shaped conductor in Figure CP33.76 moves at 10 m/s across a stationary L-shaped conductor in a 0.10 T magnetic field. The two vertices overlap, so that the enclosed area is zero, at $t = 0$ s. The conductor has a resistance of 0.010 ohms *per meter.*
    a. What is the direction of the induced current?
    b. Find expressions for the induced emf and the induced current as functions of time.
    c. Evaluate $\mathcal{E}$ and I at $t = 0.10$ s.

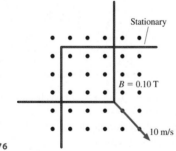

FIGURE CP33.76

77. The metal wire in Figure CP33.77 moves with speed v parallel to a straight wire that is carrying current I. The distance between the two wires is d. Find an expression for the potential difference between the two ends of the moving wire.

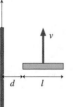

FIGURE CP33.77

78. A closed, square loop is formed with 40 cm of wire having $R = 0.10$ Ω, as shown in Figure CP33.78. A 0.50 T magnetic field is perpendicular to the loop. At $t = 0$ s, two diagonally opposite corners of the loop begin to move apart at 0.293 m/s.
    a. How long does it take the loop to collapse to a straight line?
    b. Find an expression for the induced current I as a function of time while the loop is collapsing. Assume that the sides remain straight lines during the collapse.
    c. Evaluate I at four or five times during the collapse, then draw a graph of I versus t.

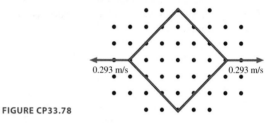

FIGURE CP33.78

79. Let's look at the details of eddy-current braking. A square loop, length l on each side, is shot with velocity v_0 into a uniform magnetic field B. The field is perpendicular to the plane of the loop. The loop has mass m and resistance R, and it enters the field at $t = 0$. Assume that the loop is moving to the right along the x-axis and that the field begins at $x = 0$.
    a. Find an expression for the loop's velocity as a function of time as it enters the magnetic field. You can ignore gravity, and you can assume that the back edge of the loop has not entered the field.
    b. Calculate and draw a graph of v over the interval $0 \text{ s} \le t \le 0.04$ s for the case that $v_0 = 10$ m/s, $l = 10$ cm, $m = 1.0$ g, $R = 0.0010$ Ω, and $B = 0.10$ T. The back edge of the loop does not reach the field during this time interval.

80. An 8.0 cm × 8.0 cm square loop is halfway into a magnetic field that is perpendicular to the plane of the loop. The loop's mass is 10 g and its resistance is 0.010 Ω. A switch is closed at $t = 0$ s, causing the magnetic field to increase from 0 to 1.0 T in 0.010 s.
    a. What is the induced current in the square loop?
    b. What is the force on the loop when the magnetic field is 0.50 T? Is the force directed into the magnetic field or away from the magnetic field?
    c. What is the loop's acceleration at $t = 0.005$ s, when the field strength is 0.50 T? If this acceleration stayed constant, how far would the loop move in 0.010 s?
    d. Because 0.50 T is the average field strength, your answer to c is an estimate of how far the loop moves during the 0.010 s in which the field increases to 1.0 T. If your answer is ≪8 cm, then it is reasonable to neglect the movement of the loop during the 0.010 s that the field ramps up. Is neglecting the movement reasonable?
    e. With what speed is the loop "kicked" away from the magnetic field?
    Hint: What is the impulse on the loop?

81. High-frequency signals are often transmitted along a *coaxial cable,* such as the one shown in Figure CP33.81. For example, the cable TV hookup coming into your home is a coaxial cable. The signal is carried on a wire of radius r_1 while the outer conductor of radius r_2 is grounded. A soft, flexible insulating material fills the space between them, and an insulating plastic coating goes around the outside.

a. Find an expression for the inductance per meter of a coaxial cable. To do so, consider the flux through a rectangle of length l that spans the gap between the inner and outer conductor.

b. Evaluate the inductance per meter of a cable having $r_1 = 0.50$ mm and $r_2 = 3.0$ mm.

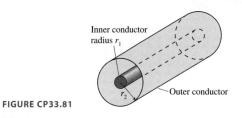

FIGURE CP33.81

STOP TO THINK ANSWERS

Stop to Think 33.1: e. According to the right-hand rule, the magnetic force on a positive charge carrier is to the right.

Stop to Think 33.2: No. The charge carriers in the wire move parallel to $\vec{B}$. There's no magnetic force on a charge moving parallel to a magnetic field.

Stop to Think 33.3: $F_2 = F_4 > F_1 = F_3$. $\vec{F}_1$ is zero because there's no field. $\vec{F}_3$ is also zero because there's no current around the loop. The charge carriers in both the right and left edges are pushed to the bottom of the loop, creating a motional emf but no current. The currents at 2 and 4 are in opposite directions, but the forces on the segments in the field are both to the left and of equal magnitude.

Stop to Think 33.4: Clockwise. The wire's magnetic field as it passes through the loop is into the page. The flux through the loop decreases into the page as the wire moves away. To oppose this decrease, the induced magnetic field needs to point into the page.

Stop to Think 33.5: d. The flux is increasing into the loop. To oppose this increase, the induced magnetic field needs to point out of the page. This requires a ccw induced current. Using the right-hand rule, the magnetic force on the current in the left edge of the loop is to the right, away from the field. The magnetic forces on the top and bottom segments of the loop are in opposite directions and cancel each other.

Stop to Think 33.6: b or f. The potential decreases in the direction of increasing current and increases in the direction of decreasing current.

Stop to Think 33.7: $\tau_3 > \tau_1 > \tau_2$. $\tau = L/R$, so smaller total resistance gives a larger time constant. The parallel resistors have total resistance $R/2$. The series resistors have total resistance $2R$.

34 Electromagnetic Fields and Waves

A laser beam is a subtle interplay of oscillating electric and magnetic fields.

▶ Looking Ahead

The goal of Chapter 34 is to study the properties of electromagnetic fields and waves. In this chapter you will learn that:

- The electric field $\vec{E}$ and the magnetic field $\vec{B}$ are *real,* not just convenient fictions.
- The electric field and magnetic field are interdependent. Furthermore, the fields can exist independently of charges and currents.
- The fields obey four general laws, called Maxwell's equations.
- Maxwell's equations predict the existence of electromagnetic waves that travel at speed c, the speed of light.

◀ Looking Back

This chapter will synthesize many ideas about fields and motion. Please review:

- Section 6.4 Relative motion.
- Section 20.3 Sinusoidal traveling waves.
- Sections 27.3–27.4 The electric flux and Gauss's law.
- Sections 32.3 and 32.6 Magnetic fields and Ampère's law.
- Sections 33.5–33.6 Faraday's law and induced electric fields.

We've now spent nine chapters on electricity and magnetism. You might wonder what more there could be to learn. Surprisingly, there's much more. Our study of the basic properties of charges and currents has been limited mostly to *static* electric and magnetic fields, fields that don't change with time. To understand a laser beam, we need to know how electric and magnetic fields change with time. Other important examples of time-dependent electromagnetic phenomena include high-speed circuits, transmission lines, radar, and optical communications.

Faraday's law has been our one example thus far of time-dependent fields. In Chapter 33, you learned that a *changing* magnetic field creates an electric field. Our goal in this chapter is to explore further the dynamic relationships of electromagnetic fields. We'll also investigate how fields appear to someone moving through them. In the end, we'll find that it makes more sense to think of a single *electromagnetic field* rather than independent electric and magnetic fields.

Our study of electromagnetic fields will culminate in Maxwell's equations for the electromagnetic field. These equations play a role in electricity and magnetism analogous to Newton's laws in mechanics. Maxwell's realization that light is an electromagnetic wave was perhaps the most important discovery of the 19th century.

34.1 Electromagnetic Fields and Forces

We've looked at many examples of electric and magnetic phenomena, but now we want to return to the most basic ideas about electromagnetic fields and forces. The central idea of electricity and magnetism is that a charge alters the space around it by creating in that space an electric field and also, if the charge is moving, a magnetic field. These fields exert electric and magnetic forces on other charges. In other words, fields are the means by which charges interact with each other.

The electric and magnetic fields of a single point charge, we know from Chapters 25 and 32, are

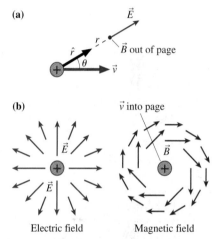

$$\vec{E} = \frac{1}{4\pi\epsilon_0}\frac{q}{r^2}\hat{r} = \left(\frac{1}{4\pi\epsilon_0}\frac{q}{r^2}, \text{ away from } q\right)$$

$$\vec{B} = \frac{\mu_0}{4\pi}\frac{q\vec{v}\times\hat{r}}{r^2} = \left(\frac{\mu_0}{4\pi}\frac{qv\sin\theta}{r^2}, \text{ direction given by right-hand rule}\right)$$

(34.1)

where $\hat{r}$ is a unit vector pointing from the charge to the point at which the field is calculated. Figure 34.1a is the basic geometry, and Figure 34.1b reminds you of the fields of a positive charge.

Electric and magnetic fields can be portrayed using either field vectors or field lines. Both are useful visual aides, but neither is a perfect representation of a field. We'll find it most convenient in this chapter to use the field-line representation. Tactics Box 34.1 reminds you how to draw and use field lines.

FIGURE 34.1 The electric and magnetic fields of a positive point charge.

TACTICS BOX 34.1 Drawing and using field lines

❶ The lines are continuous curves drawn tangent to the field vectors. Conversely, the field vector at any point is tangent to the field line at that point.

❷ Field lines never cross. This conclusion follows from the fact that the field vector has a unique direction at every point in space.

❸ The density of the lines indicates the field strength. Closely spaced lines indicate a large field strength, widely spaced lines a small field strength.

❹ Coulomb electric field lines start or stop only on charges. The lines start from positive charges and end on negative charges.

❺ Magnetic field lines form continuous, noncrossing loops. They do so because there are no magnetic monopoles on which lines could start or stop.

NOTE ▶ Rule 4 refers to *Coulomb electric fields,* which are those electric fields created by source charges. We'll modify rule 4 in Section 34.3 when we discuss the *induced electric fields* associated with Faraday's law. ◀

As examples, Figure 34.2 shows the electric field lines of a point charge and the magnetic field lines of a magnetic dipole. Notice how the magnetic field lines form continuous, noncrossing loops, although some of the field lines exceed the size of the picture. In both examples, the field lines are closer together in regions where the field is stronger.

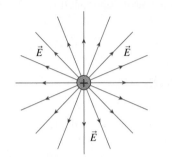
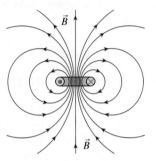

Electric field of a point charge   Magnetic field of a dipole

FIGURE 34.2 Electric and magnetic field-line diagrams.

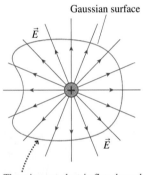

Gaussian surface

There is a net electric flux through this surface that encloses a charge.

FIGURE 34.3 A Gaussian surface enclosing a charge.

A *uniform* electric or magnetic field, which would be represented by parallel, equally spaced field lines, is an idealization that can only approximately be realized in practice. After all, magnetic field lines must eventually curve around and reconnect, although we might be able to make the curvature extremely small in a limited region of space. And electric field lines must start and stop on charges, although the lines might be very close to parallel if the charges are carefully arranged. The uniform field is an important model, one that we will continue to use in this chapter, but it is a good approximation only over a limited region of space.

Gauss's Law Revisited

Gauss's law, which you studied in Chapter 27, states a very general property of the electric field. It says that charges create electric fields in just such a way that the electric flux of the field is the same through *any* closed surface surrounding the charges. Figure 34.3 illustrates this idea by showing the field lines passing through a Gaussian surface that encloses a charge.

The mathematical statement of Gauss's law for the electric field says that for any *closed* surface enclosing total charge Q_{in}, the net electric flux through the surface is

$$\Phi_e = \oint \vec{E} \cdot d\vec{A} = \frac{Q_{in}}{\epsilon_0} \tag{34.2}$$

The circle on the integral sign indicates that the integration is over a closed surface.

We introduced the idea of magnetic flux Φ_m in Chapter 33, but for Faraday's law we considered the flux only through flat surfaces. What is the net magnetic flux over a closed surface? That is, what is the magnetic version of Gauss's law, analogous to Equation 34.2?

Figure 34.4 shows a Gaussian surface around a magnetic dipole. Magnetic field lines form continuous curves, without starting or stopping, so every field line leaving the surface at some point reenters it at another. Consequently, the net magnetic flux over a *closed* surface is zero.

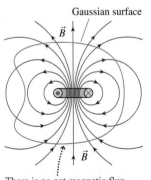

Gaussian surface

There is no net magnetic flux through this closed surface.

FIGURE 34.4 There is no net flux through a Gaussian surface around a magnetic dipole.

We've shown only one surface and one magnetic field, but this conclusion turns out to be a general property of magnetic fields. Because every north pole is always accompanied by a south pole, we can't enclose a "net pole" within a surface. Thus Gauss's law for magnetic fields is

$$\Phi_{\text{m}} = \oint \vec{B} \cdot d\vec{A} = 0 \tag{34.3}$$

Equation 34.2 is the mathematical statement that Coulomb electric field lines start and stop on charges. Equation 34.3 is the mathematical statement that magnetic field lines form closed loops; they don't start or stop (i.e., there are no magnetic monopoles). These two versions of Gauss's law are important statements about what types of fields can and cannot exist. They will become two of Maxwell's equations.

The Lorentz Force

A charge experiences a force in a region of electric or magnetic fields. The force is an interaction with the source charges and currents that created the fields, but one of the wonderful and important aspects of fields is that we don't need to know anything about the sources. Once we've determined their fields, we can forget about the sources and focus on what happens to a charge in the field.

You've learned that the force of an electric field on a charge is

$$\vec{F}_{\text{E}} = q\vec{E}$$

and that the force of a magnetic field on a moving charge is

$$\vec{F}_{\text{B}} = q\vec{v} \times \vec{B}$$

The magnitude of the magnetic force varies from 0 when $\vec{v}$ and $\vec{B}$ are parallel to qvB when the charge moves perpendicular to $\vec{B}$.

If a charge moves through a region of space in which there are both an electric *and* a magnetic field, the net force on the charge is

$$\vec{F} = q(\vec{E} + \vec{v} \times \vec{B}) \tag{34.4}$$

Equation 34.4 is called the **Lorentz force law.** It is the most general statement of the electric and magnetic forces on a charge.

EXAMPLE 34.1 **The motion of a proton**

A proton is launched with velocity $\vec{v}_0 = v_0 \hat{j}$ into a region of space in which an electric field $\vec{E} = E_0 \hat{i}$ and a magnetic field $\vec{B} = B_0 \hat{i}$ are parallel. How many cyclotron orbits will the proton make while traveling distance L along the x-axis? Find an algebraic expression, then evaluate your answer if $E_0 = 10,000$ V/m, $B_0 = 0.10$ T, $v_0 = 1.0 \times 10^5$ m/s, and $L = 10$ cm.

MODEL Assume that the electric and magnetic fields are uniform fields.

VISUALIZE Figure 34.5 shows the proton moving in the par-allel fields. The component of $\vec{v}$ perpendicular to $\vec{B}$ causes the proton to undergo cyclotron motion around the magnetic field. Simultaneously, the electric field causes the proton to accelerate along the x-axis. This component of $\vec{v}$ is parallel to $\vec{B}$ and not affected by the magnetic field. The combined motion is a *helix* whose loops become increasingly stretched out as the proton gains speed.

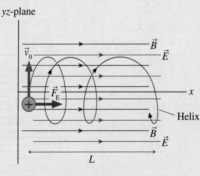

FIGURE 34.5 A proton moving in parallel electric and magnetic fields.

SOLVE The force on the proton is $\vec{F} = q(\vec{E} + \vec{v} \times \vec{B})$. The electric force points along the x-axis and causes the proton to accelerate, starting from $v_{0x} = 0$ m/s, with acceleration

$$a_x = \frac{F_x}{m} = \frac{eE_0}{m}$$

where we used $q = e$ for the charge of a proton. The kinematic equation is

$$\Delta x = L = \frac{1}{2}a_x(\Delta t)^2 = \frac{eE_0}{2m}(\Delta t)^2$$

Thus the time required to travel distance L is

$$\Delta t = \sqrt{\frac{2mL}{eE_0}}$$

The magnetic force is perpendicular to both $\vec{v}$ and $\vec{B}$, causing cyclotron motion in the yz-plane with cyclotron frequency

$$f_{cyc} = \frac{eB_0}{2\pi m}$$

The cyclotron frequency is independent of v_0. The number of cyclotron orbits during the time Δt that it takes the proton to move distance L is

$$N_{orbits} = f_{cyc}\Delta t = \frac{eB_0}{2\pi m}\sqrt{\frac{2mL}{eE_0}} = \frac{B_0}{2\pi}\sqrt{\frac{2eL}{mE_0}}$$

With the values given, $N_{orbits} = 15.6$.

STOP TO THINK 34.1   What is the direction of the net force on the moving charge?

a. Left
b. Right
c. Into the page
d. Out of the page
e. Up and left at 45°
f. Down and left at 45°

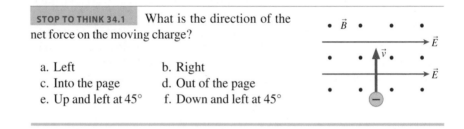

34.2 *E* or *B*? It Depends on Your Perspective

It seems clear, after the last nine chapters, that charges create an electric field $\vec{E}$ and that moving charges, or currents, create a magnetic field $\vec{B}$. Charges other than the source charges always respond to $\vec{E}$, but only moving charges respond to $\vec{B}$. Consider the following, however.

Figure 34.6a shows Sharon running past Bill with velocity $\vec{v}$ while carrying charge q. Bill sees a moving charge, and he knows that this charge creates a magnetic field given by Equation 34.1. But from Sharon's perspective, the charge is at rest. Stationary charges don't create magnetic fields, so Sharon claims that the magnetic field is zero. Is there or is there not a magnetic field?

Or what about the situation in Figure 34.6b? This time Sharon runs through an external magnetic field $\vec{B}$ that Bill has created. Bill sees a charge moving through a magnetic field, so he knows there's a force $\vec{F} = q\vec{v} \times \vec{B}$ on the charge. Using the right-hand rule, Bill determines that the force points straight up. But for Sharon the charge is still at rest. Stationary charges don't experience magnetic forces, so Sharon claims that $\vec{F} = 0$.

Now, we may be a bit uncertain about magnetic fields, because they are an abstract concept, but surely there can be no disagreement over forces. After all, the charge is either going to accelerate upward or it isn't, and Bill and Sharon should be able to agree on the outcome.

Here we have a genuine paradox, not merely faulty reasoning. This paradox has arisen because we have fields and forces that depend on velocity. The difficulty is that we haven't looked at the issue of velocity *with respect to what* or velocity *as measured by whom*. A closer look at how electromagnetic fields are viewed by two experimenters moving relative to each other will lead us to conclude that $\vec{E}$ and $\vec{B}$ are not, as we've been assuming, separate and independent entities. They are closely intertwined.

(a)

Charge q moves with velocity $\vec{v}$ relative to Bill.

(b)

Charge q moves through a magnetic field $\vec{B}$ established by Bill.

FIGURE 34.6 Sharon carries a charge past Bill.

Galilean Relativity

We introduced reference frames and relative motion in Chapter 6, and a review of Section 6.4 is highly recommended. Figure 34.7 shows two reference frames that we'll call frame S and frame S′. Frame S′ moves with velocity $\vec{V}$ with respect to frame S. That is, an experimenter at rest in S sees the origin of S′ go past with velocity $\vec{V}$. Of course, an experimenter at rest in S′ would say that frame S has velocity $-\vec{V}$. We'll use an uppercase *V* for the velocity of reference frames, reserving lowercase *v* for the velocity of objects moving in the reference frames. There's no implication that either frame is "at rest." All we know is that the two frames move relative to each other.

> **NOTE** ▶ We will consider only reference frames that move with respect to each other at *constant* velocity—unchanging speed in a straight line. You learned in Chapter 6 that these are called *inertial reference frames,* and they are the reference frames in which Newton's laws are valid. ◀

Figure 34.8 shows a physical object, such as a charged particle. Experimenters in frame S measure the motion of the particle and find that its velocity *relative to frame S* is $\vec{v}$. At the same time, experimenters in S′ find that the particle's velocity *relative to frame S′* is $\vec{v}'$. In Chapter 6, we found that $\vec{v}$ and $\vec{v}'$ are related by

$$\vec{v}' = \vec{v} - \vec{V} \quad \text{or} \quad \vec{v} = \vec{v}' + \vec{V} \tag{34.5}$$

Equation 34.5, the *Galilean transformation of velocity,* allows us to transform a velocity measured in one reference frame into the velocity that would be measured by an experimenter in a different reference frame.

Suppose the particle in Figure 34.8 is accelerating. How does its acceleration $\vec{a}$, as measured by experimenters in frame S, compare to the acceleration $\vec{a}'$ measured in frame S′? We can answer this question by taking the time derivative of Equation 34.5:

$$\frac{d\vec{v}'}{dt} = \frac{d\vec{v}}{dt} - \frac{d\vec{V}}{dt}$$

The derivatives of $\vec{v}$ and $\vec{v}'$ are the particle's accelerations $\vec{a}$ and $\vec{a}'$ in frames S and S′. But $\vec{V}$ is a *constant* velocity, so $d\vec{V}/dt = 0$. Thus the Galilean transformation of acceleration is simply

$$\vec{a}' = \vec{a} \tag{34.6}$$

Sharon and Bill may measure different positions and velocities for a particle, but they *agree* on its acceleration. This agreement is important because acceleration is directly related to force. An experimenter in frame S would find that a force $\vec{F} = m\vec{a}$ is acting on the particle. Similarly, the force measured in frame S′ is $\vec{F}' = m\vec{a}'$. But $\vec{a}' = \vec{a}$, hence

$$\vec{F}' = \vec{F} \tag{34.7}$$

Experimenters in all inertial reference frames agree about the force acting on a particle. This conclusion is the key to understanding how different experimenters see electric and magnetic fields.

The Transformation of Electric and Magnetic Fields

Now we're ready to return to the paradox that opened this section. Imagine that Bill has measured the electric field $\vec{E}$ and the magnetic field $\vec{B}$ in frame S. Our investigations thus far give us no reason to think that Sharon's measurements of the fields will differ from Bill's. After all, it seems like the fields are just "there," waiting to be measured. Thus our expectation is that Sharon, in frame S′, will measure $\vec{E}' = \vec{E}$ and $\vec{B}' = \vec{B}$.

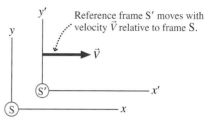

FIGURE 34.7 Reference frames S and S′.

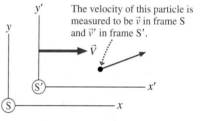

FIGURE 34.8 The particle's velocity is measured in both frame S and frame S′.

In S, the force on q is due to a magnetic field.

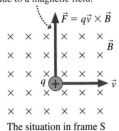

FIGURE 34.9 A charge in frame S moves through a magnetic field and experiences a magnetic force.

In S', the force on q is due to an electric field.

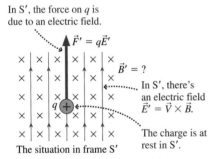

In S', there's an electric field $\vec{E}' = \vec{V} \times \vec{B}$.

The charge is at rest in S'.

FIGURE 34.10 In frame S' the charge experiences an electric force.

To find out if this is true, Bill establishes a region of space in which there is a uniform magnetic field $\vec{B}$ but where $\vec{E} = \vec{0}$. Then, as shown in Figure 34.9, he shoots a positive charge q through the magnetic field. At an instant when q is moving horizontally with velocity $\vec{v}$, the Lorentz force $\vec{F} = q(\vec{E} + \vec{v} \times \vec{B}) = q\vec{v} \times \vec{B}$ is straight up.

Suppose that Sharon, in frame S', moves alongside the charge with velocity $\vec{V} = \vec{v}$. In other words, the charge is at rest in S'. We've just seen that experimenters in S and S' agree about forces, so if Bill finds an upward force in S then Sharon *must* observe an upward force in S'. But there is *no* magnetic force on a stationary charge, so how can this be?

Because Sharon in S' sees a stationary charge with an upward force that depends on the size of q, her only possible conclusion is that there is an upward-pointing *electric field*. After all, the electric field was initially defined in terms of the force experienced by a stationary charge. If the electric field in frame S' is $\vec{E}'$, then the force on the charge is $\vec{F}' = q\vec{E}'$. But we know that $\vec{F}' = \vec{F}$, and Bill has already measured $\vec{F} = q\vec{v} \times \vec{B} = q\vec{V} \times \vec{B}$. Thus we're led to the conclusion that

$$\vec{E}' = \vec{V} \times \vec{B} \qquad (34.8)$$

As Sharon runs past Bill, she finds that at least part of Bill's magnetic field has become an electric field! **Whether a field is seen as "electric" or "magnetic" depends on the motion of the reference frame relative to the sources of the field.**

Figure 34.10 shows the situation from Sharon's perspective. The force on charge q is the same as that measured by Bill in Figure 34.9, but Sharon attributes this force to an electric field rather than a magnetic field. (Sharon needs a moving charge to measure magnetic forces, so we can't determine from this experiment whether or not Sharon experiences a magnetic field $\vec{B}'$. We'll return to this issue.)

More generally, suppose that an experimenter in S creates both an electric field $\vec{E}$ and a magnetic field $\vec{B}$. A charge moving with velocity $\vec{v}$ experiences the Lorentz force $\vec{F} = q(\vec{E} + \vec{v} \times \vec{B})$ shown in Figure 34.11a. The charge is at rest in frame S' that moves with velocity $\vec{V} = \vec{v}$, so the force in S' can be due only to an electric field, $\vec{F}' = q\vec{E}'$. Equating $\vec{F}$ and $\vec{F}'$, because experimenters in all inertial reference frames agree about forces, we find that

$$\vec{E}' = \vec{E} + \vec{V} \times \vec{B} \qquad (34.9)$$

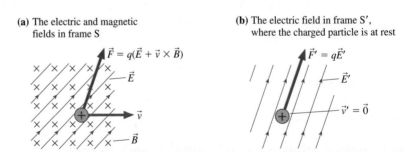

(a) The electric and magnetic fields in frame S

(b) The electric field in frame S', where the charged particle is at rest

FIGURE 34.11 A charge in frame S experiences electric and magnetic forces. The charge experiences the same force in frame S', but it is due only to an electric field.

Equation 34.9 transforms the electric and magnetic fields in S into the electric field measured in frame S'. Figure 34.11b shows the outcome.

EXAMPLE 34.2 Transforming the electric field

Earlier, in Example 34.1, we considered a proton moving in the fields $\vec{E} = 10{,}000\hat{i}$ V/m and $\vec{B} = 0.10\hat{i}$ T. These are the fields in the laboratory. What is the electric field in a reference frame moving through the laboratory with velocity $\vec{V} = 1.0 \times 10^5 \hat{j}$ m/s?

VISUALIZE Figure 34.12 shows the geometry. $\vec{E}$ and $\vec{B}$ are parallel to each other, along the *x*-axis, while velocity $\vec{V}$ of frame S′ is in the *y*-direction. Thus $\vec{V} \times \vec{B}$ points in the negative *z*-direction.

SOLVE $\vec{V}$ and $\vec{B}$ are perpendicular, so the magnitude of $\vec{V} \times \vec{B}$ is $VB = (1.0 \times 10^5 \text{ m/s})(0.10 \text{ T}) = 10{,}000$ V/m. Thus the electric field in frame S′ is

$$\vec{E}' = \vec{E} + \vec{V} \times \vec{B} = (10{,}000\hat{i} - 10{,}000\hat{k}) \text{ V/m}$$

$$= (14{,}100 \text{ V/m}, 45° \text{ below the } x\text{-axis})$$

ASSESS A stationary positive charge in frame S′ experiences an electric force directed 45° below the *x*-axis. The force in frame S is the same, but, because the charge is moving in S, the force is attributed to a combination of electric and magnetic forces.

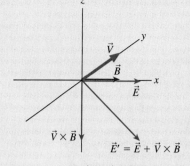

FIGURE 34.12 Finding the direction of field $\vec{E}'$.

Equation 34.9 transforms the fields $\vec{E}$ and $\vec{B}$ of frame S into the electric field $\vec{E}'$ of frame S′. In order to find a transformation equation for $\vec{B}'$, Figure 34.13a shows charge *q* at rest in frame S. Bill measures the fields of a stationary point charge, which we know are

$$\vec{E} = \frac{1}{4\pi\epsilon_0}\frac{q}{r^2}\hat{r} \qquad \vec{B} = \vec{0}$$

What are the fields at this point in space as measured by Sharon in frame S′? We can use Equation 34.9 to find $\vec{E}'$. Because $\vec{B} = \vec{0}$, the electric field in frame S′ is

$$\vec{E}' = \vec{E} = \frac{1}{4\pi\epsilon_0}\frac{q}{r^2}\hat{r} \qquad (34.10)$$

In other words, Coulomb's law is still valid in a frame in which the point charge is moving. We needed to confirm that this is so, rather than just assuming it, because Coulomb's law was introduced in a frame in which the charges were at rest.

But Sharon also measures a magnetic field $\vec{B}'$ because, as seen in Figure 34.13b, charge *q* is moving away from her with velocity $\vec{v}' = -\vec{V}$. The magnetic field of a moving point charge is given by the Biot-Savart law, thus

$$\vec{B}' = \frac{\mu_0}{4\pi}\frac{q}{r^2}\vec{v}' \times \hat{r} = -\frac{\mu_0}{4\pi}\frac{q}{r^2}\vec{V} \times \hat{r} \qquad (34.11)$$

where we used the fact that the charge's velocity in frame S′ is $\vec{v}' = -\vec{V}$.

It will be useful to rewrite Equation 34.11 as

$$\vec{B}' = -\frac{\mu_0}{4\pi}\frac{q}{r^2}\vec{V} \times \hat{r} = -\epsilon_0\mu_0\vec{V} \times \left(\frac{1}{4\pi\epsilon_0}\frac{q}{r^2}\hat{r}\right)$$

The expression in parentheses is simply $\vec{E}$, the electric field in frame S, so we find

$$\vec{B}' = -\epsilon_0\mu_0\vec{V} \times \vec{E} \qquad (34.12)$$

Equation 34.12 expresses the remarkable idea that **the Biot-Savart law for the magnetic field of a moving point charge is nothing other than the Coulomb electric field of a stationary point charge transformed into a moving reference frame.**

(a) In frame S, the static charge creates an electric field but no magnetic field.

(b) In frame S′, the moving charge creates both an electric and a magnetic field.

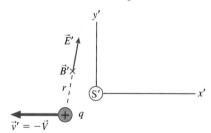

FIGURE 34.13 A charge at rest in frame S is moving in frame S′ and creates a magnetic field $\vec{B}'$.

We will assert without proof that if the experimenters in frame S create a magnetic field $\vec{B}$ in addition to the electric field $\vec{E}$, then the field $\vec{B}'$ measured in frame S' is

$$\vec{B}' = \vec{B} - \epsilon_0 \mu_0 \vec{V} \times \vec{E} \tag{34.13}$$

This is a more general transformation that matches Equation 34.9 for the electric field $\vec{E}'$.

Notice something interesting. The constant μ_0 has units of $T\,m/A$; those of ϵ_0 are $C^2/N\,m^2$. By definition, 1 T = 1 N/A m and 1 A = 1 C/s. Consequently, the units of $\epsilon_0 \mu_0$ turn out to be s^2/m^2. In other words, the quantity $1/\sqrt{\epsilon_0 \mu_0}$, with units of m/s, is a velocity. But what velocity? The constants are well known from measurements of static electric and magnetic fields, so it is straightforward to compute

$$\frac{1}{\sqrt{\epsilon_0 \mu_0}} = \frac{1}{\sqrt{(1.26 \times 10^{-6}\,T\,m/A)(8.85 \times 10^{-12}\,C^2/N\,m^2)}} = 3.00 \times 10^8\ \text{m/s}$$

Can this be a coincidence? Of all the possible values you might get from evaluating $1/\sqrt{\epsilon_0 \mu_0}$, what are the chances it would come out to equal c, the speed of light? Maxwell was the first to discover this unexpected connection between the speed of light and the constants that govern the sizes of electric and magnetic forces, and he knew at once that this couldn't be a random coincidence. In Section 34.6 we'll show that electric and magnetic fields can exist as a *traveling wave*, and that the wave speed is predicted by the theory to be none other than

$$v_{\text{em}} = c = \frac{1}{\sqrt{\epsilon_0 \mu_0}} \tag{34.14}$$

For now, we'll go ahead and write $\epsilon_0 \mu_0 = 1/c^2$. With this, our **Galilean field transformation equations** are

$$
\begin{array}{ccc}
\vec{E}' = \vec{E} + \vec{V} \times \vec{B} & & \vec{E} = \vec{E}' - \vec{V} \times \vec{B}' \\
 & \text{or} & \\
\vec{B}' = \vec{B} - \dfrac{1}{c^2}\vec{V} \times \vec{E} & & \vec{B} = \vec{B}' + \dfrac{1}{c^2}\vec{V} \times \vec{E}'
\end{array}
\tag{34.15}
$$

where $\vec{V}$ is the velocity of frame S' relative to frame S and where, to reiterate, the fields are measured *at the same point in space* by experimenters *at rest* in each reference frame.

NOTE ▶ We'll see shortly that these equations are valid only if $V \ll c$. ◀

We can no longer believe that electric and magnetic fields have a separate, independent existence. Changing from one reference frame to another mixes and rearranges the fields. Different experimenters watching an event will agree on the outcome, such as the deflection of a charged particle, but they will ascribe it to different combinations of fields. Our conclusion is that **there is just a single electromagnetic field that presents different faces, in terms of $\vec{E}$ and $\vec{B}$, to different viewers.** The whole concept of fields is beginning to look more complex, but also more interesting, than we first would have guessed!

EXAMPLE 34.3 **Two views of a magnetic field**
The 1.0 T magnetic field of a laboratory magnet points upward. A rocket flies past the laboratory, parallel to the ground, at 1000 m/s. What are the fields between the magnet's pole tips as measured by a scientist on board the rocket?

MODEL Assume that the laboratory and rocket reference frames are inertial reference frames.

VISUALIZE Figure 34.14 shows the magnet and establishes the reference frames.

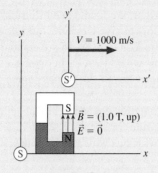

FIGURE 34.14 A rocket flies past a laboratory magnet.

SOLVE The fields in the laboratory frame are $\vec{B} = 1.0\hat{j}$ T and $\vec{E} = \vec{0}$. Frame S′, the frame of the rocket, moves with velocity $\vec{V} = 1000\hat{i}$ m/s. We can use Equations 34.15 to transform the fields measured in the laboratory to the rocket frame S′. We find

$$\vec{E}' = \vec{E} + \vec{V} \times \vec{B} = \vec{V} \times \vec{B}$$

$$\vec{B}' = \vec{B} - \frac{1}{c^2}\vec{V} \times \vec{E} = \vec{B} = 1.0\hat{j}\text{ T}$$

From the right-hand rule, $\vec{V} \times \vec{B}$ is out of the page, or in the $\hat{k}$ direction. $\vec{V}$ and $\vec{B}$ are perpendicular, so

$$\vec{E}' = VB\hat{k} = 1000\hat{k}\text{ V/m} = (1000\text{ V/m, out of page})$$

Thus the rocket scientist measures

$$\vec{B}' = 1.0\hat{j}\text{ T} \quad\text{and}\quad \vec{E}' = 1000\hat{k}\text{ V/m}$$

ASSESS The transformation equations apply only to fields measured at the *same* point in space. Thus these results apply to measurements made between the magnet's pole tips, where $\vec{B}$ is known, but not to other points in the laboratory.

Almost Relativity

Figure 34.15 shows two positive charges moving side by side through frame S with velocity $\vec{v}$. Charge q_1 creates an electric field and a magnetic field at the position of charge q_2. These are

$$\vec{E}_1 = \frac{1}{4\pi\epsilon_0}\frac{q_1}{r^2}\hat{j} \quad\text{and}\quad \vec{B}_1 = \frac{\mu_0}{4\pi}\frac{q_1 v}{r^2}\hat{k}$$

where r is the distance between the charges and we've used $\hat{r} = \hat{j}$ and $\vec{v} \times \hat{r} = v\hat{k}$.

How are the fields seen in frame S′, which moves with $\vec{V} = \vec{v}$ and in which the charges are at rest? From the field transformation equations,

$$\vec{B}_1' = \vec{B}_1 - \frac{1}{c^2}\vec{V} \times \vec{E}_1 = \frac{\mu_0}{4\pi}\frac{q_1 v}{r^2}\hat{k} - \frac{1}{c^2}\left(v\hat{i} \times \frac{1}{4\pi\epsilon_0}\frac{q_1}{r^2}\hat{j}\right)$$

$$= \frac{\mu_0}{4\pi}\frac{q_1 v}{r^2}\left(1 - \frac{1}{\epsilon_0\mu_0 c^2}\right)\hat{k} \tag{34.16}$$

where we used $\hat{i} \times \hat{j} = \hat{k}$. But $\epsilon_0\mu_0 = 1/c^2$, so the term in parentheses is zero and $\vec{B}' = \vec{0}$. This result was expected because q_1 is at rest in S′ and shouldn't create a magnetic field.

The transformation of the electric field is

$$\vec{E}_1' = \vec{E}_1 + \vec{V} \times \vec{B}_1 = \frac{1}{4\pi\epsilon_0}\frac{q_1}{r^2}\hat{j} + v\hat{i} \times \frac{\mu_0}{4\pi}\frac{q_1 v}{r^2}\hat{k}$$

$$= \frac{1}{4\pi\epsilon_0}\frac{q_1}{r^2}(1 - \epsilon_0\mu_0 v^2)\hat{j} = \frac{1}{4\pi\epsilon_0}\frac{q_1}{r^2}\left(1 - \frac{v^2}{c^2}\right)\hat{j} \tag{34.17}$$

where we used $\hat{i} \times \hat{k} = -\hat{j}$ and $\epsilon_0\mu_0 = 1/c^2$.

But now we have a problem. In frame S′, where the two charges are at rest and separated by distance r, the electric field due to charge q_1 should be simply

$$\vec{E}_1' = \frac{1}{4\pi\epsilon_0}\frac{q_1}{r^2}\hat{j}$$

The field transformation equations have given a "wrong" result for the electric field $\vec{E}'$.

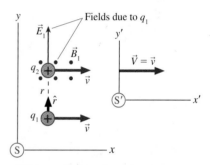

Fields seen in frame S

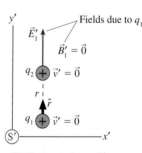

Fields seen in frame S′

FIGURE 34.15 Two charges moving parallel to each other.

It turns out that the field transformations of Equation 34.15, which are based on Galilean relativity, aren't quite right. We would need Einstein's relativity—a topic that we'll take up in Chapter 36—to give the correct transformations. However, the *Galilean* transformations in Equation 34.15 are equivalent to the relativistically correct transformations when $v \ll c$, in which case $v^2/c^2 \ll 1$. You can see that the two expressions for $\vec{E}'_1$ do, in fact, agree if v^2/c^2 can be neglected.

Thus our use of the field transformation equations has an additional rule: Set v^2/c^2 to zero. This is an acceptable rule for speeds $v < 10^7$ m/s. Even with this limitation, our investigation has provided us with a deeper understanding of electric and magnetic fields.

STOP TO THINK 34.2 Which diagram shows the fields in frame S'?

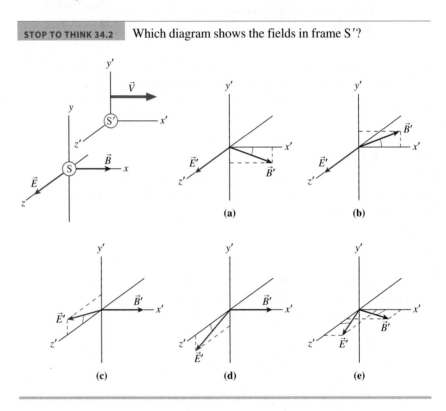

(a)   (b)

(c)   (d)   (e)

34.3 Faraday's Law Revisited

The transformation of electric and magnetic fields can give us new insight into Faraday's law. Figure 34.16a shows a reference frame S, which we can call the laboratory frame, in which a conducting loop is moving with velocity $\vec{v}$ into a magnetic field. You learned in Chapter 33 that the magnetic field exerts an upward force $\vec{F}_B = q\vec{v} \times \vec{B} = (qvB,$ upward) on the charges in the leading edge of the wire, creating an emf $\mathcal{E} = vLB$ and an induced current in the loop. We called this a *motional emf*.

How do things appear to an experimenter who is in frame S' that moves with the loop at velocity $\vec{V} = \vec{v}$ and for whom the loop is at rest? An important lesson of the previous section was that experimenters in different inertial reference frames agree about the outcome of any experiment, hence an experimenter in S' agrees that there is an induced current in the loop. But the charges are at rest in frame S', so there cannot be any magnetic force on them. How is the emf established in frame S'?

We couldn't have answered this question in Chapter 33, but now we've learned that the experimenter in frame S′ doesn't see the same fields as the laboratory experimenter in S. In fact, we can use the field transformations to determine that the fields in S′ are

$$\vec{E}' = \vec{E} + \vec{v} \times \vec{B} = \vec{v} \times \vec{B}$$

$$\vec{B}' = \vec{B} - \frac{1}{c^2}\vec{v} \times \vec{E} = \vec{B} \tag{34.18}$$

where we used the fact that $\vec{E} = \vec{0}$ in the laboratory frame.

An experimenter in the loop's frame sees not only a magnetic field but also the electric field $\vec{E}'$ shown in Figure 34.16b. The magnetic field exerts no force on the charges, because they're at rest in this frame, but the electric field does. The force on charge q is $\vec{F} = q\vec{E}' = q\vec{v} \times \vec{B} = (qvB,\ \text{upward})$. This is the same force as was measured in the laboratory frame, so it will cause the same emf and the same current. The outcome is identical, as we knew it had to be, but the experimenter in S′ attributes the emf to an electric field whereas the experimenter in S attributes it to a magnetic field.

Field $\vec{E}'$ is, in fact, the *induced electric field.* Faraday's law, fundamentally, is a statement that **a changing magnetic field creates an electric field.** But only in frame S′, the frame of the loop, is the magnetic field changing. Thus the induced electric field is seen in the loop's frame but not in the laboratory frame. The induced electric field is a *non-Coulomb* field because it is not created by static charges. It is a field that has been created in a new way.

Calculating the emf

The emf was defined in Chapter 30 as the work required per unit charge to separate the charge and thereby establish a potential difference. That is,

$$\mathcal{E} = \frac{W}{q} \tag{34.19}$$

In batteries, a familiar source of emf, this work is done by chemical forces. But the emf that appears in Faraday's law arises when work is done by the forces of an induced electric field.

Figure 34.17 shows a charged particle moving through an electric field from a to b. We can calculate the work that the electric field does on the charge by dividing the path into many small displacement vectors $d\vec{s}$. The small amount of work done by the electric field as the charge moves through $d\vec{s}$ is $dW = \vec{F} \cdot d\vec{s} = q\vec{E} \cdot d\vec{s}$. The total work done on the particle as it moves from a to b is the sum of all the dW, or

$$W = q\int_a^b \vec{E} \cdot d\vec{s} \tag{34.20}$$

This is a *line integral,* just like the line integral in Ampère's law except that it is an integral of $\vec{E} \cdot d\vec{s}$ rather than $\vec{B} \cdot d\vec{s}$. In Chapter 32 you learned that there are two situations in which evaluating the line integral is very simple:

- If $\vec{E}$ is everywhere perpendicular to the integration path, then $W = 0$.
- If $\vec{E}$ is everywhere tangent to an integration path of length L *and* has the same strength at all points, then $W = qEL$.

These two rules are sufficient for all the calculations we'll need to do *if* we choose the integration path carefully.

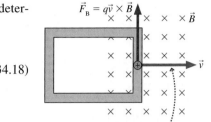

(a) Laboratory frame S

The loop is moving to the right.

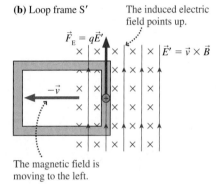

(b) Loop frame S′   The induced electric field points up.

The magnetic field is moving to the left.

FIGURE 34.16 A motional emf as seen in two different reference frames.

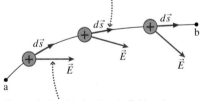

Divide the path into many small displacements.

The work done by the electric field as the charge moves through $d\vec{s}$ is $dW = q\vec{E} \cdot d\vec{s}$.

FIGURE 34.17 The electric field does work on a charge as it moves from a to b.

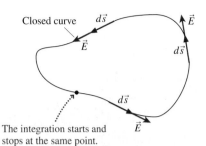

FIGURE 34.18 This electric field can do work as a charged particle moves around the closed curve.

The emf of Faraday's law is an emf around a *closed curve,* such as the one in Figure 34.18, through which the magnetic flux Φ_m is changing. We can calculate the work done by the electric field as a charge moves around a closed curve if we let the end points a and b in Figure 34.17 merge to a single point. The work done is

$$W_{\text{closed curve}} = q \oint \vec{E} \cdot d\vec{s} \qquad (34.21)$$

where the integration symbol with the circle is the same as the one we used in Ampère's law to indicate an integral around a closed curve. If we use this expression for W in Equation 34.19, we find that the emf around a closed loop is

$$\mathcal{E} = \frac{W_{\text{closed curve}}}{q} = \oint \vec{E} \cdot d\vec{s} \qquad (34.22)$$

Equation 34.22 relates the induced electric field to the induced emf of Faraday's law.

> **NOTE** ▶ For a Coulomb electric field, created by charges, $W_{\text{closed curve}} = 0$ because a charge that moves around a closed path experiences no net change in potential energy. That's Kirchhoff's loop law. By contrast, an induced electric field is *not* a conservative field. We cannot associate an electric potential with an induced electric field, and a charge *does* gain energy by traversing a closed path. ◀

Establishing the Sign

Suppose the magnetic flux Φ_m through a loop is changing. According to Faraday's law, the emf around the loop is

$$\mathcal{E} = \left| \frac{d\Phi_m}{dt} \right|$$

in a direction given by Lenz's law. We've related the emf $\mathcal{E}$ to $\vec{E}$ and the flux Φ_m to $\vec{B}$, but we still need to incorporate Lenz's law into the mathematics. The difficulty is that the sign of Φ_m is ambiguous. If you have a surface, such as a circle enclosed by a current loop, how do you know if the flux is going "into" or "out of" of the surface?

We can deal with this ambiguity by establishing a *sign convention*. This convention works for both the magnetic flux Φ_m and the electric flux Φ_e.

TACTICS BOX 34.2 Determining the sign of the flux

❶ For a surface S bounded by a closed curve C, choose either the clockwise (cw) or counterclockwise (ccw) direction around C.

❷ Curl the fingers of your *right* hand around the curve in the chosen direction with your thumb perpendicular to the surface. Your thumb defines the positive direction. The loop's area vector $\vec{A}$ points in the direction of your thumb. The flux Φ through the surface is positive if the field is in the same direction as your thumb, negative if the field is in the opposite direction.

❸ A positive emf creates an induced current in the direction of your fingers; a negative emf creates a current in the opposite direction.

Figure 34.19 applies these rules to a loop moving into a magnetic field. You can see that $\mathcal{E}$ and $d\Phi_m/dt$ have opposite signs. In fact, if you consider all the possible combinations of increasing and decreasing fields, you find that $\mathcal{E}$ **and**

$d\Phi_m/dt$ **always have opposite signs.** This is Lenz's law at work, saying that the induced current (proportional to $\mathcal{E}$) *opposes* the change in the flux ($d\Phi_m/dt$). We can capture this idea mathematically by writing Faraday's law as

$$\mathcal{E} = -\frac{d\Phi_m}{dt} \qquad (34.23)$$

where the minus sign is the mathematical statement of Lenz's law.

Now we can complete our task by using the Equation 34.22 expression for $\mathcal{E}$ in Equation 34.23:

$$\oint \vec{E} \cdot d\vec{s} = -\frac{d\Phi_m}{dt} = -\frac{d}{dt}\left[\int \vec{B} \cdot d\vec{A}\right] \qquad (34.24)$$

where the line integral of $\vec{E}$ is around the closed curve that bounds the surface through which the magnetic flux is calculated. Equation 34.24 is Faraday's law written in terms of the fields $\vec{E}$ and $\vec{B}$.

But $\vec{E}$ and $\vec{B}$ according to which experimenter? **These are the fields in the reference frame of the loop,** where the loop is at rest. There is an induced electric field in the loop's reference frame if the magnetic flux through the loop is changing. The induced electric field is responsible for the emf that drives the induced current around the loop.

The Induced Electric Field

The solenoid in Figure 34.20a, whose upward magnetic field is increasing as the current increases, provides a good example of the connection between $\vec{E}$ and $\vec{B}$. You learned in Chapter 33 that the changing flux induces a current in a conducting loop inside the solenoid, and we could use Lenz's law to determine that the direction of the induced current would be clockwise. But Faraday's law, in the form of Equation 34.24, tells us that **an induced electric field is present whether there's a conducting loop or not.** The electric field is induced simply due to the fact that $\vec{B}$ is changing.

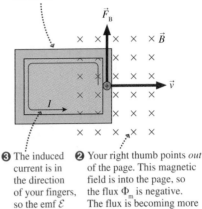

❶ The loop is closed curve C. The shaded area inside the loop is surface S. Choose a ccw direction.

❸ The induced current is in the direction of your fingers, so the emf $\mathcal{E}$ is positive.

❷ Your right thumb points *out* of the page. This magnetic field is into the page, so the flux Φ_m is negative. The flux is becoming more negative as the loop moves into the field, so $d\Phi_m/dt$ is also negative.

FIGURE 34.19 The emf is positive and the flux is negative.

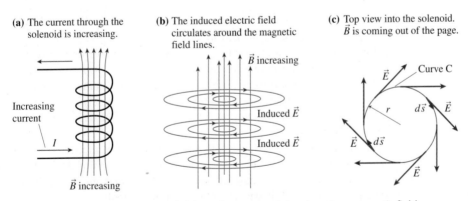

(a) The current through the solenoid is increasing.

Increasing current

$\vec{B}$ increasing

(b) The induced electric field circulates around the magnetic field lines.

$\vec{B}$ increasing

Induced $\vec{E}$

Induced $\vec{E}$

(c) Top view into the solenoid. $\vec{B}$ is coming out of the page.

$\vec{E}$ Curve C

r $d\vec{s}$ $\vec{E}$

$\vec{E}$ $d\vec{s}$

$\vec{E}$

FIGURE 34.20 The induced electric field circulates around the changing magnetic field inside a solenoid.

The shape of the induced electric field has to be such that it *could* drive a current around a conducting loop, if one were present, and it has to be consistent with the cylindrical symmetry of the solenoid. The only possible choice, shown in Figure 34.20b, is an electric field that circulates around the magnetic field lines.

NOTE ▶ Circular electric field lines violate rule 4 of Tactics Box 34.1 for drawing field lines, which said that electric field lines have to start and stop on charges. But we noted there that rule 4 applies only to Coulomb fields that are created by source charges. An induced electric field is a non-Coulomb field created not by source charges but by a changing magnetic field. Without source charges, induced electric field lines *must* form closed loops. ◄

To use Faraday's law, choose a *clockwise* direction around a circle of radius r as the closed curve for evaluating the emf and the flux. Figure 34.20c shows that the electric field vectors are everywhere tangent to the curve. In this case, the line integral of $\vec{E}$ is

$$\oint \vec{E} \cdot d\vec{s} = EL = 2\pi r E \tag{34.25}$$

where $L = 2\pi r$ is the length of the closed curve.

What about the flux? The closed curve surrounds area $A = \pi r^2$. The magnetic field inside a solenoid is uniform, so the flux has magnitude $|\Phi_m| = AB$. To determine the sign, curl your right fingers around the circle in the cw direction. Your right thumb points *into* the page, but $\vec{B}$ is coming out of the page. Thus the flux is a *negative*

$$\Phi_m = -AB = -\pi r^2 B$$

and the rate of change of the flux is

$$\frac{d\Phi_m}{dt} = -\pi r^2 \frac{dB}{dt} \tag{34.26}$$

If we substitute Equations 34.25 and 34.26 into Faraday's law, Equation 34.24, we find

$$\oint \vec{E} \cdot d\vec{s} = 2\pi r E = -\frac{d\Phi_m}{dt} = \pi r^2 \frac{dB}{dt}$$

Thus the induced electric field inside the solenoid ($r < R$) is

$$E = \frac{r}{2} \frac{dB}{dt} \tag{34.27}$$

This result shows very directly that the induced electric field is created by a *changing* magnetic field. A constant $\vec{B}$, with $dB/dt = 0$, would give $E = 0$.

We defined the positive direction to be clockwise. If the solenoid current is increasing, then dB/dt is positive as the magnetic field grows stronger, and E is positive. Thus the induced electric field lines circulate cw for an increasing solenoid current, which is the case shown in Figure 34.20. Conversely, a decreasing solenoid current would make dB/dt negative, hence E would be negative and the field lines would circulate ccw.

We have now extended Faraday's law to make an explicit connection between electric and magnetic fields. Faraday's law, in the form of Equation 34.24, joins Gauss's law and Gauss's law for magnetism as one of the fundamental equations of electromagnetic fields. We have one more equation to go.

EXAMPLE 34.4 An induced electric field

A 4.0-cm-diameter solenoid is wound with 2000 turns per meter. The current through the solenoid oscillates at 60 Hz with an amplitude of 2.0 A. What is the maximum strength of the induced electric field inside the solenoid?

MODEL Assume that the magnetic field inside the solenoid is uniform.

VISUALIZE The electric field lines are concentric circles around the magnetic field lines, as was shown in Figure 34.20. They will reverse direction twice every period as the current oscillates.

SOLVE You learned in Chapter 32 that the magnetic field strength inside a solenoid with n turns per meter is $B = \mu_0 n I$. In this case, the current through the solenoid is $I = I_0 \sin \omega t$, where $I_0 = 2.0$ A is the peak current and $\omega = 2\pi(60\text{ Hz}) = 377$ rad/s. Thus the induced electric field strength at radius r is

$$E = \frac{r}{2}\frac{dB}{dt} = \frac{r}{2}\frac{d}{dt}(\mu_0 n I_0 \sin \omega t) = \frac{1}{2}\mu_0 n r \omega I_0 \cos \omega t$$

The field strength is maximum at maximum radius ($r = R$) *and* at the instant when $\cos\omega t = 1$ (*I* changing at the maximum rate). That is,

$$E_{max} = \frac{1}{2}\mu_0 n R \omega I_0 = 0.019 \text{ V/m}$$

ASSESS This electric field strength, although not large, is similar to the field strength that the emf of a battery creates in a wire. Hence this induced electric field has the ability to drive a substantial induced current through a conducting loop *if* a loop is present. But the induced electric field exists inside the solenoid whether or not there is a conducting loop.

34.4 The Displacement Current

We introduced Ampère's law in Chapter 32 as an alternative method for calculating the magnetic field of a current. Whenever total current $I_{through}$ passes through an area bounded by a closed curve, the line integral of the magnetic field around the curve is

$$\oint \vec{B} \cdot d\vec{s} = \mu_0 I_{through} \tag{34.28}$$

Figure 34.21 illustrates the geometry of Ampère's law. The sign of each current can be determined by using Tactics Box 34.2. In this case, $I_{through} = I_1 - I_2$.

Ampère's law, which is equivalent to the Biot-Savart law for the magnetic field of a moving charge, is the formal statement that **currents create magnetic fields.** Although Ampère's law can be used to calculate magnetic fields in situations with a high degree of symmetry, it is more important as a statement about what types of magnetic field can and cannot exist.

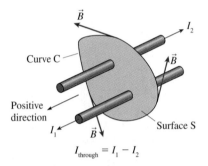

FIGURE 34.21 Ampère's law relates the line integral of $\vec{B}$ around curve C to the current passing through surface S.

Something Is Missing

Nothing restricts the bounded surface of Ampère's law to being flat. It's not hard to see that any current passing through surface S_1 in Figure 34.22 must also pass through the curved surface S_2. To interpret Ampère's law properly, we have to say that the current $I_{through}$ is the net current passing through *any* surface S that is bounded by curve C.

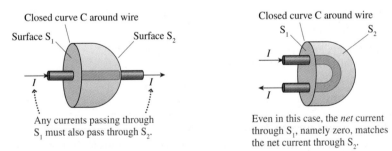

FIGURE 34.22 The *net* current passing through the flat surface S_1 also passes through the curved surface S_2.

But this leads to an interesting puzzle. Figure 34.23a on the next page shows a capacitor being charged. Current *I*, from the left, brings positive charge to the left capacitor plate. The same current carries charges away from the right capacitor plate, leaving the right plate negatively charged. This is a perfectly ordinary current in a conducting wire, and you can use the right-hand rule to verify that its magnetic field is as shown.

(a)   Cross section through a closed
curve C around the wire

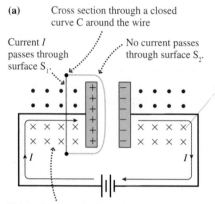

Current I
passes through
surface S_1.

No current passes
through surface S_2.

This is the magnetic field of the
current I that is charging the capacitor.

(b)

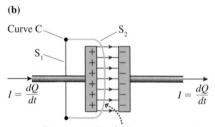

Curve C

S_2

S_1

$I = \dfrac{dQ}{dt}$ $I = \dfrac{dQ}{dt}$

The electric flux Φ_e through surface S_2
increases as the capacitor charges.

FIGURE 34.23 There is no current through
surface S_2 as the capacitor charges, but
there is a changing electric flux.

Curve C is a closed curve encircling the wire on the left. The current passes through surface S_1, a flat surface across C, and we could use Ampère's law to find that the magnetic field is that of a straight wire. But what happens if we try to use surface S_2 to determine $I_{through}$? Ampère's law says that we can consider *any* surface bounded by curve C, and surface S_2 certainly qualifies. But *no* current passes through S_2. Charges are brought to the left plate of the capacitor and charges are removed from the right plate, but *no* charge moves across the gap between the plates. Surface S_1 has $I_{through} = I$, but surface S_2 has $I_{through} = 0$. Another dilemma!

It would appear that Ampère's law is either wrong or incomplete. Maxwell was the first to recognize the seriousness of this problem. He noted that there may be no current passing through S_2, but, as Figure 34.23b shows, there is an electric flux Φ_e through S_2 due to the electric field inside the capacitor. Furthermore, this flux is *changing* with time as the capacitor charges and the electric field strength grows. Faraday had discovered the significance of a changing magnetic flux, but no one had considered a changing electric flux.

The current I passes through S_1, so Ampère's law applied to S_1 gives

$$\oint \vec{B} \cdot d\vec{s} = \mu_0 I_{through} = \mu_0 I$$

We believe this result because it gives the correct magnetic field for a current-carrying wire. Now the line integral depends only on the magnetic field at points on curve C, so its value won't change if we choose a different surface S to evaluate the current. The problem is with the right side of Ampère's law, which would incorrectly give zero if applied to surface S_2. We need to modify the right side of Ampère's law to recognize that an electric flux rather than a current passes through S_2.

The electric flux between two capacitor plates of surface area A is

$$\Phi_e = EA$$

The capacitor's electric field is $E = Q/\epsilon_0 A$, hence the flux is actually independent of the plate size:

$$\Phi_e = \frac{Q}{\epsilon_0 A} A = \frac{Q}{\epsilon_0} \tag{34.29}$$

The *rate* at which the electric flux is changing is

$$\frac{d\Phi_e}{dt} = \frac{1}{\epsilon_0} \frac{dQ}{dt} = \frac{I}{\epsilon_0} \tag{34.30}$$

where we used $I = dQ/dt$. The flux is changing with time at a rate directly proportional to the charging current I.

Equation 34.30 suggests that the quantity $\epsilon_0(d\Phi_e/dt)$ is in some sense "equivalent" to current I. Maxwell called the quantity

$$I_{disp} = \epsilon_0 \frac{d\Phi_e}{dt} \tag{34.31}$$

the **displacement current.** He had started with a fluid-like model of electric and magnetic fields, and the displacement current was analogous to the displacement of a fluid. The fluid model has since been abandoned, but the name lives on despite the fact that nothing is actually being displaced.

Maxwell hypothesized that the displacement current was the "missing" piece of Ampère's law, so he modified Ampère's law to read

$$\oint \vec{B} \cdot d\vec{s} = \mu_0(I_{through} + I_{disp}) = \mu_0\left(I_{through} + \epsilon_0 \frac{d\Phi_e}{dt}\right) \tag{34.32}$$

Equation 34.32 is now known as the Ampère-Maxwell law. When applied to Figure 34.23b, the Ampère-Maxwell law gives

$$S_1: \quad \oint \vec{B} \cdot d\vec{s} = \mu_0 \left(I_{\text{through}} + \epsilon_0 \frac{d\Phi_e}{dt} \right) = \mu_0 (I + 0) = \mu_0 I$$

$$S_2: \quad \oint \vec{B} \cdot d\vec{s} = \mu_0 \left(I_{\text{through}} + \epsilon_0 \frac{d\Phi_e}{dt} \right) = \mu_0 (0 + I) = \mu_0 I$$

where, for surface S_2, we used Equation 34.30 for $d\Phi_e/dt$. Surfaces S_1 and S_2 now both give the same result for the line integral of $\vec{B} \cdot d\vec{s}$ around the closed curve C.

NOTE ▶ The displacement current I_{disp} between the capacitor plates is numerically equal to the current I in the wires leading to and from the capacitor, so in some sense it allows "current" to be conserved all the way through the capacitor. Nonetheless, the displacement current is *not* a flow of charge. The displacement current is equivalent to a real current in the sense that it creates the same magnetic field, but it does so with a changing electric flux rather than a flow of charge. ◀

The Induced Magnetic Field

Ordinary Coulomb electric fields are created by charges, but a second way to create an electric field is by having a changing magnetic field. That's Faraday's law. Ordinary magnetic fields are created by currents, but now we see that a second way to create a magnetic field is by having a changing electric field. Just as the electric field created by a changing $\vec{B}$ is called an induced electric field, the magnetic field created by a changing $\vec{E}$ is called an *induced magnetic field*.

Figure 34.24 shows the close analogy between induced electric fields, governed by Faraday's law, and induced magnetic fields, governed by the second term in the Ampère-Maxwell law. An increasing solenoid current causes an increasing magnetic field. The changing magnetic field, in turn, induces a circular electric field. The negative sign in Faraday's law dictates that the induced electric field direction is ccw when seen looking along the magnetic field direction.

An increasing capacitor charge causes an increasing electric field. The changing electric field, in turn, induces a circular magnetic field. But the sign of the Ampère-Maxwell law is positive, the opposite of the sign of Faraday's law, so the induced magnetic field direction is cw when you're looking along the electric field direction.

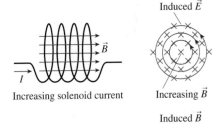

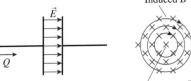

FIGURE 34.24 The close analogy between an induced electric field and an induced magnetic field.

EXAMPLE 34.5 The fields inside a charging capacitor

A 2.0-cm-diameter parallel-plate capacitor with a 1.0 mm spacing is being charged at the rate 0.50 C/s. What is the magnetic field strength inside the capacitor at a point 0.50 cm from the axis?

MODEL The electric field inside a parallel-plate capacitor is uniform. As the capacitor is charged, the changing electric field induces a magnetic field.

VISUALIZE Figure 34.25 shows the fields. The induced magnetic field lines are circles concentric with the capacitor.

SOLVE The electric field of a parallel-plate capacitor is $E = Q/\epsilon_0 A = Q/\epsilon_0 \pi R^2$. The electric flux through the circle of radius r (not the full flux of the capacitor) is

$$\Phi_e = \pi r^2 E = \pi r^2 \frac{Q}{\epsilon_0 \pi R^2} = \frac{r^2}{R^2} \frac{Q}{\epsilon_0}$$

Thus the Ampère-Maxwell law is

$$\oint \vec{B} \cdot d\vec{s} = \epsilon_0 \mu_0 \frac{d\Phi_e}{dt} = \epsilon_0 \mu_0 \frac{d}{dt} \left(\frac{r^2}{R^2} \frac{Q}{\epsilon_0} \right) = \mu_0 \frac{r^2}{R^2} \frac{dQ}{dt}$$

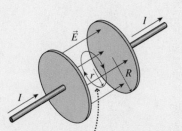

The magnetic field line is a circle concentric with the capacitor. The electric flux through this circle is $\pi r^2 E$.

FIGURE 34.25 The magnetic field strength is found by integrating around a closed curve of radius r.

The magnetic field is everywhere tangent to the circle of radius r, so the integral of $\vec{B} \cdot d\vec{s}$ around the circle is simply $BL = 2\pi rB$. With this value for the line integral, the Ampère-Maxwell law becomes

$$2\pi rB = \mu_0 \frac{r^2}{R^2} \frac{dQ}{dt}$$

and thus

$$B = \frac{\mu_0}{2\pi} \frac{r}{R^2} \frac{dQ}{dt} = (2.0 \times 10^{-7}\,\text{Tm/A}) \frac{0.0050\,\text{m}}{(0.010\,\text{m})^2}(0.50\,\text{C/s})$$

$$= 5.0 \times 10^{-6}\,\text{T}$$

If a changing magnetic field can induce an electric field and a changing electric field can induce a magnetic field, what happens when both fields change simultaneously? That is the question that Maxwell was finally able to answer after he modified Ampère's law to include the displacement current, and it is the subject to which we turn next.

> **STOP TO THINK 34.3** The electric field in four identical capacitors is shown as a function of time. Rank in order, from largest to smallest, the magnetic field strength at the outer edge of the capacitor at time T.

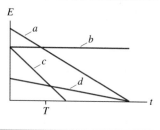

34.5 Maxwell's Equations

James Clerk Maxwell was a young, mathematically brilliant Scottish physicist. In 1855, barely 24 years old and having graduated from Cambridge University just two years earlier, he presented a paper to the Cambridge Philosophical Society entitled "On Faraday's Lines of Force." It had been 30 years and more since the major discoveries of Oersted, Ampère, Faraday, and others, but electromagnetism remained a loose collection of facts and "rules of thumb" without a consistent theory to link these ideas together.

Maxwell's goal, first enunciated in his paper of 1855, was to synthesize this body of knowledge and to place it in a proper mathematical framework. His desire was nothing less than to form a complete *theory* of electromagnetic fields. It took 10 years, until papers published in 1865 and 1868 laid out the theory in a form that looks familiar to us today. The critical step along the way was his recognition of the need to include a displacement current term in Ampère's law.

Maxwell's theory of electromagnetism is embodied in four equations that we today call **Maxwell's equations.** These are

$$\oint \vec{E} \cdot d\vec{A} = \frac{Q_{\text{in}}}{\epsilon_0} \qquad \text{Gauss's law}$$

$$\oint \vec{B} \cdot d\vec{A} = 0 \qquad \text{Gauss's law for magnetism}$$

$$\oint \vec{E} \cdot d\vec{s} = -\frac{d\Phi_m}{dt} \qquad \text{Faraday's law}$$

$$\oint \vec{B} \cdot d\vec{s} = \mu_0 I_{\text{through}} + \epsilon_0 \mu_0 \frac{d\Phi_e}{dt} \qquad \text{Ampère-Maxwell law}$$

You've seen all of these equations earlier in this chapter. It was Maxwell who first wrote them in a consistent mathematical form similar to this. (Not exactly the same, because our present-day vector notation wasn't developed until the 1890s, but Maxwell's versions were mathematically equivalent.) Neither Gauss nor Faraday nor Ampère would recognize these equations, but Maxwell had succeeded in capturing their physical ideas in a concise mathematical form.

Maxwell's claim is that these four equations are a *complete* description of electric and magnetic fields. They tell us how fields are created by charges and currents, and also how fields can be induced by the changing of other fields. We need one more equation for total completeness, an equation that tells us how matter responds to electromagnetic fields. But that's the Lorentz force law, another equation that we already have:

$$\vec{F} = q(\vec{E} + \vec{v} \times \vec{B}) \qquad \text{(Lorentz force law)}$$

Maxwell's equations for the fields, together with the Lorentz force law to tell us how matter responds to the fields, form the complete theory of electromagnetism.

Maxwell's equations bring us to the pinnacle of classical physics. Except at the quantum level of photons, these equations describe everything that is known about electromagnetic phenomena. In fact, they predict many new phenomena not known to Maxwell or his contemporaries, and they're the basis for all of modern circuit theory, electrical engineering, and other electromagnetic technology. When combined with Newton's three laws of motion, his law of gravity, and the first and second laws of thermodynamics, we have all of classical physics—a total of just 11 equations.

While some physicists might quibble over whether all 11 are truly fundamental, the important point is not the exact number but how few equations we need to describe the overwhelming majority of our experience of the physical world. It seems as if we could have written them all on page one of this book and been finished, but it doesn't work that way. Each of these equations is the synthesis of a tremendous number of physical phenomena and conceptual developments. To know physics isn't just to know the equations, but to know what the equations *mean* and how they're used. That's why it's taken us so many chapters and so much effort to get to this point. Each equation is simply a shorthand way to summarize a book's worth of information!

Let's summarize the physical meaning embodied in the five electromagnetic equations:

Classical physics

Newton's first law
Newton's second law
Newton's third law
Newton's law of gravity
Gauss's law
Gauss's law for magnetism
Faraday's law
Ampère-Maxwell law
Lorentz force law
First law of thermodynamics
Second law of thermodynamics

- **Gauss's law:** Charged particles create an electric field.
- **Faraday's law:** An electric field can also be created by a changing magnetic field.
- **Gauss's law for magnetism:** There are no magnetic monopoles.
- **Ampère-Maxwell law, first half:** Currents create a magnetic field.
- **Ampère-Maxwell law, second half:** A magnetic field can also be created by a changing electric field.
- **Lorentz force law, first half:** An electric force is exerted on a charged particle in an electric field.
- **Lorentz force law, second half:** A magnetic force is exerted on a moving charge in a magnetic field.

These are the *fundamental ideas* of electromagnetism. Other important ideas, such as Ohm's law, Kirchhoff's laws, and Lenz's law, despite their practical importance, are not fundamental ideas. They can be derived from Maxwell's equations, sometimes with the addition of empirically based concepts such as that of resistance.

Maxwell's equations can be used to understand motors, generators, antennas and receivers, the transmission of electrical signals through circuits, power lines, microwaves, the electromagnetic properties of materials, and much more. It's true that Maxwell's equations are mathematically more complex than Newton's laws and that their solution, for many problems of practical interest, requires advanced mathematics. Fortunately, we have the mathematical tools to get just far enough into Maxwell's equations to discover their most startling and revolutionary implication—the prediction of electromagnetic waves.

34.6 Electromagnetic Waves

It had been known since the early 19th century, from experiments on interference and diffraction, that light is a wave. We studied the wave properties of light in Part V, but at that time we were not able to determine just what is "waving."

Faraday speculated that light was somehow connected with electricity and magnetism, but Maxwell, using his equations of the electromagnetic field, was the first to understand that light is an oscillation of the electromagnetic field. Maxwell was able to predict that

- Electromagnetic waves can exist at any frequency, not just at the frequencies of visible light. This prediction was the harbinger of radio waves.
- All electromagnetic waves travel in a vacuum with the same speed, a speed that we now call the *speed of light.*

A general wave equation can be derived from Maxwell's equations, but the necessary mathematical techniques are beyond the level of this textbook. We'll adopt a simpler approach in which we *assume* an electromagnetic wave of a certain form and then show that it's consistent with Maxwell's equations. After all, the wave can't exist *unless* it's consistent with Maxwell's equations.

To begin, we're going to assume that electric and magnetic fields can exist independent of charges and currents in a *source-free* region of space. This is a very important assumption because it makes the statement that **fields are real entities.** They're not just cute pictures that tell us about charges and currents, but are real things that can exist all by themselves. Our assertion is that the fields can exist in a self-sustaining mode in which a changing magnetic field creates an electric field (Faraday's law) that in turn changes in just the right way to re-create the original magnetic field (the Ampère-Maxwell law).

The source-free Maxwell's equations, with no charges or currents, are

$$\oint \vec{E} \cdot d\vec{A} = 0 \qquad \oint \vec{E} \cdot d\vec{s} = -\frac{d\Phi_m}{dt}$$

$$\oint \vec{B} \cdot d\vec{A} = 0 \qquad \oint \vec{B} \cdot d\vec{s} = \epsilon_0 \mu_0 \frac{d\Phi_e}{dt}$$

(34.33)

Any electromagnetic wave traveling in empty space must be consistent with these equations.

Let's postulate that an electromagnetic plane wave traveling with speed v_{em} has the characteristics shown in Figure 34.26. It's a useful picture, and one that you'll see in any textbook, but a picture that can be very misleading if you don't think about it carefully. $\vec{E}$ and $\vec{B}$ are *not* spatial vectors. That is, they don't stretch spatially in the y- or z-direction for a certain distance. Instead, these vectors are showing the values of the electric and magnetic fields at *points* along a single line, the x-axis. An $\vec{E}$ vector pointing in the y-direction says that *at that point* on the x-axis, where the vector's tail is, the electric field points in the y-direction and has a certain strength. Nothing is "reaching" to a point in space above the x-axis. In fact, this picture contains no information about any points in space other than those right on the x-axis.

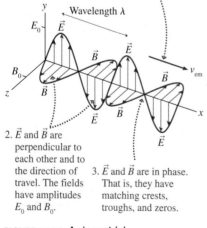

1. A sinusoidal wave with frequency *f* and wavelength λ travels with wave speed v_{em}.

2. $\vec{E}$ and $\vec{B}$ are perpendicular to each other and to the direction of travel. The fields have amplitudes E_0 and B_0.

3. $\vec{E}$ and $\vec{B}$ are in phase. That is, they have matching crests, troughs, and zeros.

FIGURE 34.26 A sinusoidal electromagnetic wave.

However, we are assuming that this is a *plane wave,* which, you'll recall from Chapter 20, is a wave for which the fields are the same at *all points* in any *yz*-plane, perpendicular to the *x*-axis. Figure 34.27a shows a small section of the *xy*-plane where, at this instant of time, $\vec{E}$ is pointing up and $\vec{B}$ is pointing toward you. The field strengths vary with *x*, the direction of travel, but not with *y*. As the wave moves forward, the fields that are now in the x_1-plane will soon arrive in the x_2-plane, and those now in the x_2-plane will move to x_3.

Figure 34.27b shows a section of the *yz*-plane that slices the *x*-axis at x_2. These fields are moving out of the page, coming toward you. The fields are the same at *every point* in this plane, which is what we mean by a plane wave. If you watched a movie of the event, you would see the $\vec{E}$ and $\vec{B}$ fields at each point in this plane *oscillating* in time, but always synchronized with all the other points in the plane. Thus you have to use your imagination to see that the $\vec{E}$ and $\vec{B}$ fields in Figure 34.26 are also the $\vec{E}$ and $\vec{B}$ fields *everywhere* in any *yz*-plane.

Gauss's Laws

Now that we understand the shape of the electromagnetic field, we can check its consistency with Maxwell's equations. This field is a sinusoidal wave, so the components of the fields are

$$E_x = 0 \quad E_y = E_0\sin(2\pi(x/\lambda - ft)) \quad E_z = 0$$
$$B_x = 0 \quad B_y = 0 \qquad\qquad\qquad B_z = B_0\sin(2\pi(x/\lambda - ft)) \tag{34.34}$$

where E_0 and B_0 are the amplitudes of the oscillating electric and magnetic fields.

Figure 34.28 shows an imaginary box—a Gaussian surface—centered on the *x*-axis. Both electric and magnetic field vectors exist at each point in space, but the figure shows them separately for clarity. $\vec{E}$ oscillates along the *y*-axis, so all electric field lines enter and leave the box through the top and bottom surfaces; no electric field lines pass through the sides of the box.

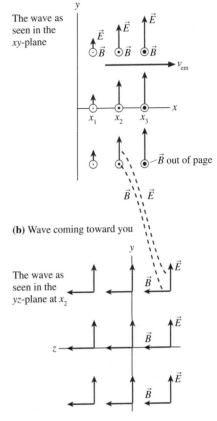

(a) Wave traveling to the right

The wave as seen in the *xy*-plane

(b) Wave coming toward you

The wave as seen in the *yz*-plane at x_2

FIGURE 34.27 Interpreting the electromagnetic wave of Figure 34.26.

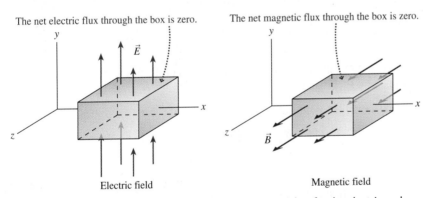

The net electric flux through the box is zero.

The net magnetic flux through the box is zero.

Electric field

Magnetic field

FIGURE 34.28 A closed surface can be used to check Gauss's law for the electric and magnetic fields.

Because this is a plane wave, the magnitude of each electric field vector entering the bottom of the box is exactly matched by the electric field vector leaving the top. The electric flux through the top of the box is equal in magnitude but opposite in sign to the flux through the bottom, and the flux through the sides is zero. Thus the *net* electric flux is $\Phi_e = 0$. There is no charge inside the box, because there are no sources in this region of space, so we also have $Q_{in} = 0$. Hence the electric field of a plane wave is consistent with the first of the source-free Maxwell's equations, Gauss's law.

The exact same argument applies to the magnetic field. The net magnetic flux is $\Phi_m = 0$, thus the magnetic field is consistent with the second of Maxwell's equations.

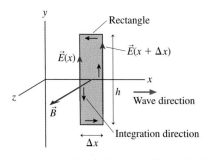

FIGURE 34.29 Faraday's law can be applied to a narrow rectangle in the xy-plane.

Faraday's Law

Faraday's law is concerned with the changing magnetic flux through a closed curve. We'll apply Faraday's law to a narrow rectangle in the xy-plane, shown in Figure 34.29, with height h and width Δx. We'll assume Δx to be so small that $\vec{B}$ is essentially constant over the width of the rectangle.

The magnetic field $\vec{B}$ points in the z-direction, perpendicular to the rectangle. The magnetic flux through the rectangle is $\Phi_m = B_z A_{\text{rectangle}} = B_z h \Delta x$, hence the flux *changes* at the rate

$$\frac{d\Phi_m}{dt} = \frac{d}{dt}(B_z h \Delta x) = \frac{\partial B_z}{\partial t} h \Delta x \tag{34.35}$$

The ordinary derivative dB_z/dt, which is the full rate of change of B from all possible causes, becomes a partial derivative $\partial B_z/\partial t$ in this situation because the change in magnetic flux is due entirely to the change of B with time and not at all to the spatial variation of B.

According to our sign convention, we have to go around the rectangle in a ccw direction to make the flux positive. Thus we must also use a ccw direction to evaluate the line integral

$$\oint \vec{E} \cdot d\vec{s} = \int_{\text{right}} \vec{E} \cdot d\vec{s} + \int_{\text{top}} \vec{E} \cdot d\vec{s} + \int_{\text{left}} \vec{E} \cdot d\vec{s} + \int_{\text{bottom}} \vec{E} \cdot d\vec{s} \tag{34.36}$$

The electric field $\vec{E}$ points in the y-direction, hence $\vec{E} \cdot d\vec{s} = 0$ at all points on the top and bottom edges, and these two integrals are zero.

Along the left edge of the loop, at position x, $\vec{E}$ has the same value at every point. Figure 34.27a shows that the direction of $\vec{E}$ is *opposite* to $d\vec{s}$, thus $\vec{E} \cdot d\vec{s} = -E_y(x)ds$. On the right edge of the loop, at position $x + \Delta x$, $\vec{E}$ is *parallel* to $d\vec{s}$ and $\vec{E} \cdot d\vec{s} = E_y(x + \Delta x)ds$. Thus the line integral of $\vec{E} \cdot d\vec{s}$ around the rectangle is

$$\oint \vec{E} \cdot d\vec{s} = -E_y(x)h + E_y(x + \Delta x)h = \left[E_y(x + \Delta x) - E_y(x)\right]h \tag{34.37}$$

NOTE ▶ $E_y(x)$ indicates that E_y is a function of the position x. It is *not* E_y multiplied by x. ◀

You learned in calculus that the derivative of the function $f(x)$ is

$$\frac{df}{dx} = \lim_{\Delta x \to 0}\left[\frac{f(x + \Delta x) - f(x)}{\Delta x}\right]$$

We've assumed that Δx is very small. If we now let the width of the rectangle go to zero, $\Delta x \to 0$, Equation 34.37 becomes

$$\oint \vec{E} \cdot d\vec{s} = \frac{\partial E_y}{\partial x} h \Delta x \tag{34.38}$$

We've used a partial derivative because E_y is a function of both position x and time t.

Now, using Equations 34.35 and 34.38, we can write Faraday's law as

$$\oint \vec{E} \cdot d\vec{s} = \frac{\partial E_y}{\partial x} h \Delta x = -\frac{d\Phi_m}{dt} = -\frac{\partial B_z}{\partial t} h \Delta x$$

The area $h \Delta x$ of the rectangle cancels, and we're left with

$$\frac{\partial E_y}{\partial x} = -\frac{\partial B_z}{\partial t} \tag{34.39}$$

Equation 34.39, which compares the rate at which E_y varies with position to the rate at which B_z varies with time, is a *required condition* that an electromagnetic wave must satisfy to be consistent with Maxwell's equations. We can use Equation 34.34 for E_y and B_z to evaluate the partial derivatives:

$$\frac{\partial E_y}{\partial x} = \frac{2\pi E_0}{\lambda}\cos\left(2\pi(x/\lambda - ft)\right)$$

$$\frac{\partial B_z}{\partial t} = -2\pi f B_0 \cos\left(2\pi(x/\lambda - ft)\right)$$

Thus the required condition of Equation 34.39 is

$$\frac{\partial E_x}{\partial x} = \frac{2\pi E_0}{\lambda}\cos\left(2\pi(x/\lambda - ft)\right) = -\frac{\partial B_z}{\partial t} = 2\pi f B_0 \cos\left(2\pi(x/\lambda - ft)\right)$$

Canceling the many common factors, and multiplying by λ, we're left with

$$E_0 = (\lambda f)B_0 = v_{em}B_0 \qquad (34.40)$$

where we used the fact that $\lambda f = v$ for any sinusoidal wave.

Equation 34.40, which came from applying Faraday's law, tells us that the field amplitudes E_0 and B_0 of an electromagnetic wave are not arbitrary. **Once the amplitude B_0 of the magnetic field wave is specified, the electric field amplitude E_0 must be $E_0 = v_{em}B_0$.** Otherwise the fields won't satisfy Maxwell's equations.

The Ampère-Maxwell Law

We have one equation to go, but this one will now be easier. The Ampère-Maxwell law is concerned with the changing electric flux through a closed curve. Figure 34.30 shows a very narrow rectangle of width Δx and length l in the xz-plane. The electric field is perpendicular to this rectangle, hence the electric flux through it is $\Phi_e = E_y A_{rectangle} = E_y l \Delta x$. This flux is changing at the rate

$$\frac{d\Phi_e}{dt} = \frac{d}{dt}(E_y l \Delta x) = \frac{\partial E_y}{\partial t}l\Delta x \qquad (34.41)$$

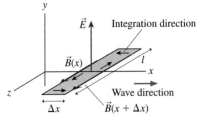

FIGURE 34.30 The Ampère-Maxwell law can be applied to a narrow rectangle in the xz-plane.

The line integral of $\vec{B} \cdot d\vec{s}$ around this closed rectangle is calculated just like the line integral of $\vec{E} \cdot d\vec{s}$ in Figure 34.29. $\vec{B}$ is perpendicular to $d\vec{s}$ on the narrow ends, so $\vec{B} \cdot d\vec{s} = 0$. The field at *all* points on the left edge, at position x, is $\vec{B}(x)$, and this field is parallel to $d\vec{s}$ to make $\vec{B} \cdot d\vec{s} = B_z(x)ds$. Similarly, $\vec{B} \cdot d\vec{s} = -B_z(x + \Delta x)ds$ at all points on the right edge, where $\vec{B}$ is opposite to $d\vec{s}$. Thus, if we let $\Delta x \to 0$,

$$\oint \vec{B} \cdot d\vec{s} = B_z(x)l - B_z(x + \Delta x)l = -[B_z(x + \Delta x) - B_z(x)]l$$
$$= -\frac{\partial B_z}{\partial x}l\Delta x \qquad (34.42)$$

Equations 34.41 and 34.42 can now be used in the Ampère-Maxwell law:

$$\oint \vec{B} \cdot d\vec{s} = -\frac{\partial B_z}{\partial x}l\Delta x = \epsilon_0\mu_0\frac{d\Phi_e}{dt} = \epsilon_0\mu_0\frac{\partial E_y}{\partial t}l\Delta x$$

The area of the rectangle cancels, and we're left with

$$\frac{\partial B_z}{\partial x} = -\epsilon_0\mu_0\frac{\partial E_y}{\partial t} \qquad (34.43)$$

Equation 34.43 is a second required condition that the fields must satisfy. If we again evaluate the partial derivatives, using Equation 34.43 for E_y and B_z, we find that

$$\frac{\partial E_y}{\partial t} = -2\pi f E_0 \cos(2\pi(x/\lambda - ft))$$

$$\frac{\partial B_z}{\partial x} = \frac{2\pi B_0}{\lambda} \cos(2\pi(x/\lambda - ft))$$

With these, Equation 34.43 becomes

$$\frac{\partial B_z}{\partial x} = \frac{2\pi B_0}{\lambda} \cos(2\pi(x/\lambda - ft)) = -\epsilon_0\mu_0\frac{\partial E_y}{\partial t} = 2\pi\epsilon_0\mu_0 f E_0 \cos(2\pi(x/\lambda - ft))$$

A final round of cancellations, and another use of $\lambda f = v_{em}$, leaves us with

$$E_0 = \frac{B_0}{\epsilon_0\mu_0\lambda f} = \frac{B_0}{\epsilon_0\mu_0 v_{em}} \tag{34.44}$$

The last of Maxwell's equations gives us another constraint between E_0 and B_0.

The Speed of Light

But how can Equation 34.40, which required $E_0 = v_{em}B_0$, and Equation 34.44 both be true at the same time? The one and only way is if

$$\frac{1}{\epsilon_0\mu_0 v_{em}} = v_{em}$$

from which we find

$$v_{em} = \frac{1}{\sqrt{\epsilon_0\mu_0}} = 3.00 \times 10^8 \text{ m/s} = c \tag{34.45}$$

This is a remarkable conclusion. The constants ϵ_0 and μ_0 are from electrostatics and magnetostatics, where they determine the size of $\vec{E}$ and $\vec{B}$ due to point charges. Coulomb's law and the Biot-Savart law, where ϵ_0 and μ_0 first appeared, have nothing to do with waves. Yet Maxwell's theory of electromagnetism ends up predicting that electric and magnetic fields can form a self-sustaining electromagnetic wave *if* that wave travels at the specific speed $v_{em} = 1/\sqrt{\epsilon_0\mu_0}$. No other speed will satisfy Maxwell's equations.

We've made no assumption about the frequency of the wave, so apparently all electromagnetic waves, regardless of their frequency, travel (in a vacuum) at the same speed $v_{em} = 1/\sqrt{\epsilon_0\mu_0}$. We now call this speed c, the "speed of light," but it applies equally well from low-frequency radio waves to ultrahigh-frequency x rays.

STOP TO THINK 34.4 An electromagnetic wave is propagating in the positive x-direction. At this instant of time, what is the direction of $\vec{E}$ at the center of the rectangle?

a. In the positive x-direction
b. In the negative x-direction
c. In the positive y-direction
d. In the negative y-direction
e. In the positive z-direction
f. In the negative z-direction

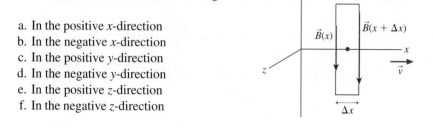

34.7 Properties of Electromagnetic Waves

We've demonstrated that one very specific sinusoidal wave is consistent with Maxwell's equations. It's possible to show that *any* electromagnetic wave, whether it's sinusoidal or not, must satisfy four basic conditions:

1. The fields $\vec{E}$ and $\vec{B}$ are perpendicular to the direction of propagation $\vec{v}_{em}$. Thus an electromagnetic wave is a transverse wave.
2. $\vec{E}$ and $\vec{B}$ are perpendicular to each other in a manner such that $\vec{E} \times \vec{B}$ is in the direction of $\vec{v}_{em}$.
3. The wave travels in a vacuum at speed $v_{em} = 1/\sqrt{\epsilon_0\mu_0} = c$.
4. $E = cB$ at any point on the wave.

In this section, we'll look at some other properties of electromagnetic waves.

Energy and Intensity

Waves transfer energy. Ocean waves erode beaches, sound waves set your eardrum to vibrating, and light from the sun warms the earth. The energy flow of an electromagnetic wave is described by the **Poynting vector** $\vec{S}$, defined as

$$\vec{S} \equiv \frac{1}{\mu_0}\vec{E} \times \vec{B} \qquad (34.46)$$

The Poynting vector, shown in Figure 34.31, has two important properties:

1. At any point, the Poynting vector points in the direction in which an electromagnetic wave is traveling. You can see this by looking back at Figure 34.26.
2. The magnitude S of the Poynting vector measures the rate of energy transfer per unit area of the wave. As a homework problem, you can show that the units of S are W/m², or power (joules per second) per unit area.

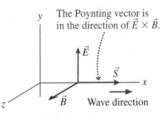

FIGURE 34.31 The Poynting vector.

Because $\vec{E}$ and $\vec{B}$ of an electromagnetic wave are perpendicular to each other, and $E = cB$, the magnitude of the Poynting vector is

$$S = \frac{EB}{\mu_0} = \frac{E^2}{c\mu_0}$$

The Poynting vector is a function of time, oscillating from zero to $S_{max} = E_0^2/c\mu_0$ and back to zero twice during each period of the wave's oscillation. That is, the energy flow in an electromagnetic wave is not smooth. It "pulses" as the electric and magnetic fields oscillate in intensity. We're unaware of this pulsing because the electromagnetic waves that we can sense—light waves—have such high frequencies.

Of more interest is the *average* energy transfer, averaged over one cycle of oscillation, which is the wave's **intensity** I. In our earlier study of waves, we defined the intensity of a wave to be $I = P/A$, where P is the power (energy transferred per second) of a wave that impinges on area A. Because $E = E_0\sin(2\pi(x/\lambda - ft))$, and the average over one period of $\sin^2(2\pi(x/\lambda - ft))$ is $\frac{1}{2}$, the intensity of an electromagnetic wave is

$$I = \frac{P}{A} = S_{avg} = \frac{1}{2c\mu_0}E_0^2 = \frac{c\epsilon_0}{2}E_0^2 \qquad (34.47)$$

Equation 34.47 relates the intensity of an electromagnetic wave, a quantity that is easily measured, to the amplitude of the wave's electric field.

EXAMPLE 34.6 The electric field of a laser beam

A helium-neon laser, the laser commonly used for classroom demonstrations, emits a 1.0-mm-diameter laser beam with a power of 1.0 mW. What is the amplitude of the oscillating electric field in the laser beam?

MODEL The laser beam is an electromagnetic plane wave. Assume that the energy is uniformly distributed over the diameter of the laser beam.

SOLVE 1.0 mW, or 1.0×10^{-3} J/s, is the energy transported per second by the light wave. This energy is carried within a 1.0-mm-diameter beam, so the light intensity is

$$I = \frac{P}{A} = \frac{P}{\pi r^2} = \frac{1.0 \times 10^{-3} \text{ W}}{\pi (0.00050 \text{ m})^2} = 1270 \text{ W/m}^2$$

We can use Equation 34.47 to relate this intensity to the electric field amplitude:

$$E_0 = \sqrt{\frac{2I}{c\epsilon_0}} = \sqrt{\frac{2(1270 \text{ W/m}^2)}{(3.00 \times 10^8 \text{ m/s})(8.85 \times 10^{-12} \text{ C}^2/\text{N m}^2)}}$$

$$= 978 \text{ V/m}$$

ASSESS This is a sizable electric field, comparable to the electric field near a charged glass or plastic rod.

The intensity of a plane wave, with constant electric field amplitude E_0, would not change with distance. But a plane wave is an idealization; there are no true plane waves in nature. You learned in Chapter 20 that, in order to conserve energy, the intensity of a wave far from its source decreases with the inverse square of the distance. If a source with power P_{source} emits electromagnetic waves *uniformly* in all directions, the electromagnetic wave intensity at distance r from the source is

$$I = \frac{P_{\text{source}}}{4\pi r^2} \tag{34.48}$$

Equation 34.48 simply expresses the recognition that the energy of the wave is spread over a sphere of surface area $4\pi r^2$.

Some sources, such as antennas, emit waves in some directions but not in others. Although Equation 34.48 does not apply to such nonuniform sources, it remains true that the intensity along any line from the source decreases with the inverse square of the distance.

STOP TO THINK 34.5 An electromagnetic wave is traveling in the positive y-direction. The electric field at one instant of time is shown at one position. The magnetic field at this position points

a. In the positive x-direction.
b. In the negative x-direction.
c. In the positive y-direction.
d. In the negative y-direction.
e. Toward the origin.
f. Away from the origin.

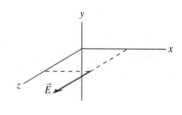

Radiation Pressure

Electromagnetic waves transfer not only energy but also momentum. An object gains momentum when it absorbs electromagnetic waves, much as a ball at rest gains momentum when struck by a ball in motion.

Suppose we shine a beam of light on an object that completely absorbs the light energy. If the object absorbs energy during a time interval Δt, its momentum changes by

$$\Delta p = \frac{\text{energy absorbed}}{c}$$

This is a consequence of Maxwell's theory that we'll state without proof.

The momentum change implies that the light is exerting a force on the object. Newton's second law, in terms of momentum, is $F = \Delta p/\Delta t$. The radiation force due to the beam of light is

$$F = \frac{\Delta p}{\Delta t} = \frac{(\text{energy absorbed})/\Delta t}{c} = \frac{P}{c}$$

where P is the power (joules per second) of the light.

It's more interesting to consider the force exerted on an object per unit area, which is called the **radiation pressure** p_{rad}. The radiation pressure on an object that absorbs all the light is

$$p_{rad} = \frac{F}{A} = \frac{(P/A)}{c} = \frac{I}{c} \qquad (34.49)$$

where I is the intensity of the light wave. The subscript on p_{rad} is important in this context to distinguish the radiation pressure from the momentum p.

EXAMPLE 34.7 Solar sailing

A low-cost way of sending spacecraft to other planets would be to use the radiation pressure on a solar sail. The intensity of the sun's electromagnetic radiation at distances near the earth's orbit is about 1300 W/m². What size sail would be needed to accelerate a 10,000 kg spacecraft toward Mars at 0.010 m/s²?

MODEL Assume that the solar sail is perfectly absorbing.

SOLVE The force that will create a 0.010 m/s² acceleration is $F = ma = 100$ N. We can use Equation 34.49 to find the sail area that, by absorbing light, will receive a 100 N force from the sun:

$$A = \frac{cF}{I} = \frac{(3.00 \times 10^8 \text{ m/s})(100 \text{ N})}{1300 \text{ W/m}^2} = 2.3 \times 10^7 \text{ m}^2$$

ASSESS If the sail is a square, it would need to be 4.8 km × 4.8 km, or roughly 3 mi × 3 mi. This is large, but not entirely out of the question with thin films that can be unrolled in space. But how will the crew return from Mars?

Antennas

We've seen that an electromagnetic wave is self-sustaining, independent of charges or currents. However, charges and currents are needed at the *source* of an electromagnetic wave. We'll take a brief look at how an electromagnetic wave is generated by an antenna.

Figure 34.32 is the electric field of an electric dipole. If the dipole is vertical, the electric field $\vec{E}$ at points along a horizontal line is also vertical. Reversing the dipole, by switching the charges, reverses $\vec{E}$. If the charges were to oscillate back and forth, switching position at frequency f, then $\vec{E}$ would oscillate in a vertical plane. The changing $\vec{E}$ would then create an induced magnetic field $\vec{B}$, which could then create an $\vec{E}$, which could then create a $\vec{B}$, ... and an electromagnetic wave at frequency f would radiate out into space.

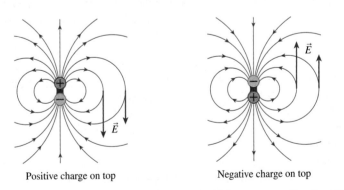

Positive charge on top   Negative charge on top

FIGURE 34.32 An electric dipole creates an electric field that reverses direction if the dipole charges are switched.

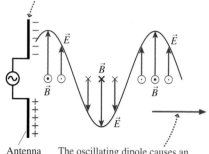

An oscillating voltage causes the dipole to oscillate.

Antenna wire

The oscillating dipole causes an electromagnetic wave to move away from the antenna at speed $v_{em} = c$.

FIGURE 34.33 An antenna generates a self-sustaining electromagnetic wave.

This is exactly what an **antenna** does. Figure 34.33 shows two metal wires attached to the terminals of an oscillating voltage source. The figure shows an instant when the top wire is negative and the bottom is positive, but these will reverse in half a cycle. The wire is basically an oscillating dipole, and it creates an oscillating electric field. The oscillating $\vec{E}$ induces an oscillating $\vec{B}$, and they take off as an electromagnetic wave at speed $v_{em} = c$. The wave does need oscillating charges as a *wave source,* but once created it is self-sustaining and independent of the source. The antenna might be destroyed, but the wave could travel billions of light years across the universe, bearing the legacy of James Clerk Maxwell.

> **STOP TO THINK 34.6**   The amplitude of the oscillating electric field at your cell phone is 4.0 μV/m when you are 10 km east of the broadcast antenna. What is the electric field amplitude when you are 20 km east of the antenna?
>
> a. 1.0 μV/m
> b. 2.0 μV/m
> c. 4.0 μV/m
> d. There's not enough information to tell.

34.8 Polarization

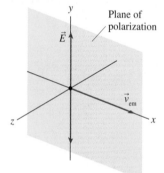

(a) Vertical polarization

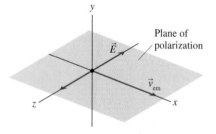

(b) Horizontal polarization

FIGURE 34.34 The plane of polarization is the plane in which the electric field vector oscillates.

16.9 Activ Physics

The plane of the electric field vector $\vec{E}$ and the Poynting vector $\vec{S}$ (the direction of propagation) is called the **plane of polarization** of an electromagnetic wave. Figure 34.34 shows just the electric field of two waves moving along the *x*-axis. The magnetic field, not shown, is perpendicular to $\vec{E}$. The electric field in Figure 34.34a oscillates vertically, so we would say that this wave is *vertically polarized.* Similarly the wave in Figure 34.34b is *horizontally polarized.* Other polarizations are possible, such as a wave polarized 30° away from horizontal.

NOTE ▶ This use of the term "polarization" is completely independent of the idea of *charge polarization* that you learned about in Chapter 25. ◀

Some wave sources, such as lasers and radio antennas, emit *polarized* electromagnetic waves with a well-defined plane of polarization. By contrast, most natural sources of electromagnetic radiation are unpolarized. Each atom in the sun's hot atmosphere emits light independently of all the other atoms, as does each tiny piece of metal in the incandescent filament of a light bulb. An electromagnetic wave that you see or measure is a superposition of waves from each of these tiny emitters. Although the wave from each individual emitter is polarized, it is polarized in a random direction with respect to the waves from all its neighbors. The net result is what we call an *unpolarized* wave, a wave whose electric field oscillates randomly with all possible orientations.

A few natural sources are *partially polarized,* meaning that one direction of polarization is more prominent than others. The light of the sky at right angles to the sun is partially polarized, because of how the sun's light scatters from air molecules to create skylight. Bees and other insects make use of this partial polarization to navigate. Light reflected from a flat, horizontal surface, such as a road or the surface of a lake, has a predominantly horizontal polarization. This is the rationale for using polarizing sunglasses.

The most common way of artificially generating polarized visible light is to send unpolarized light through a *polarizing filter.* The first widely used polarizing filter was invented by Edwin Land in 1928, while he was still an undergraduate student. He developed an improved version, called Polaroid, in 1938. Polaroid, as shown in Figure 34.35, is a plastic sheet containing very long organic molecules known as polymers. The sheets are formed in such a way that the polymers are all aligned to form a grid, rather like the metal bars in a barbecue grill. The sheet is then chemically treated to make the polymer molecules somewhat conducting.

As a light wave travels through Polaroid, the component of the electric field oscillating parallel to the polymer grid drives the conduction electrons up and down the molecules. The electrons absorb energy from the light wave, so the parallel component of $\vec{E}$ is absorbed in the filter. But the conduction electrons can't oscillate perpendicular to the molecules, so the component of $\vec{E}$ perpendicular to the polymer grid passes through without absorption. Thus the light wave emerging from a polarizing filter is polarized perpendicular to the polymer grid.

Malus's Law

Suppose a *polarized* light wave of intensity I_0 approaches a polarizing filter. What is the intensity of the light that passes through the filter? Figure 34.36 shows that an oscillating electric field can be decomposed into components parallel and perpendicular to the polarizer's axis (i.e., the polarization direction transmitted by the polarizer). If we call the polarizer axis the y-axis, then the incident electric field is

$$\vec{E}_{\text{incident}} = E_\perp \hat{i} + E_\parallel \hat{j} = E_0 \sin\theta \hat{i} + E_0 \cos\theta \hat{j} \qquad (34.50)$$

where θ is the angle between the incident plane of polarization and the polarizer axis.

If the polarizer is ideal, meaning that light polarized parallel to the axis is 100% transmitted and light perpendicular to the axis is 100% blocked, then the electric field of the light transmitted by the filter is

$$\vec{E}_{\text{transmitted}} = E_\parallel \hat{j} = E_0 \cos\theta \hat{j} \qquad (34.51)$$

Because the intensity depends on the square of the electric field amplitude, you can see that the transmitted intensity is related to the incident intensity by

$$I_{\text{transmitted}} = I_0 \cos^2\theta \qquad \text{(incident light polarized)} \qquad (34.52)$$

This result, which was discovered experimentally in 1809, is called **Malus's law.**

Figure 34.37a shows that Malus's law can be demonstrated with two polarizing filters. The first, called the *polarizer,* is used to produce polarized light of intensity I_0. The second, called the *analyzer,* is rotated by angle θ relative to the polarizer. As the photographs of Figure 34.37b show, the transmission of the analyzer is (ideally) 100% when $\theta = 0°$ and steadily decreases to zero when $\theta = 90°$. Two polarizing filters with perpendicular axes, called *crossed polarizers,* block all the light.

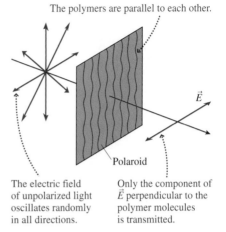

The polymers are parallel to each other.

Polaroid

The electric field of unpolarized light oscillates randomly in all directions.

Only the component of $\vec{E}$ perpendicular to the polymer molecules is transmitted.

FIGURE 34.35 A polarizing filter.

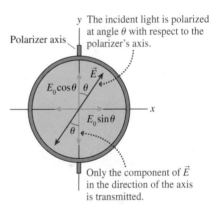

The incident light is polarized at angle θ with respect to the polarizer's axis.

Polarizer axis

$E_0 \cos\theta$

$E_0 \sin\theta$

Only the component of $\vec{E}$ in the direction of the axis is transmitted.

FIGURE 34.36 An incident electric field can be decomposed into components parallel and perpendicular to a polarizer's axis.

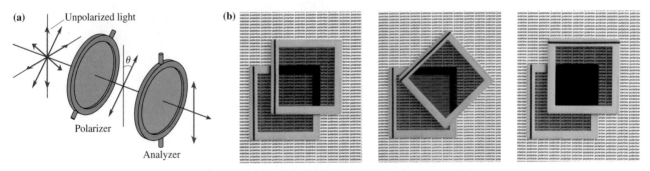

(a) Unpolarized light

θ

Polarizer

Analyzer

(b)

FIGURE 34.37 The intensity of the transmitted light depends on the angle between the polarizing filters.

Suppose the light incident on a polarizing filter is *unpolarized,* as is the light incident from the left on the polarizer in Figure 34.37a. The electric field of unpolarized light varies randomly through all possible values of θ. Because the *average* value of $\cos^2\theta$ is $\frac{1}{2}$, the intensity transmitted by a polarizing filter is

$$I_{\text{transmitted}} = \frac{1}{2}I_0 \text{ (incident light unpolarized)} \qquad (34.53)$$

In other words, a polarizing filter passes 50% of unpolarized light and blocks 50%.

In polarizing sunglasses, the polymer grid is aligned horizontally (when the glasses are in the normal orientation) so that the glasses transmit vertically polarized light. Most natural light is unpolarized, so the glasses reduce the light intensity by 50%. But glare—the reflection of the sun and the skylight from roads and other horizontal surfaces—has a strong horizontal polarization. This light is almost completely blocked by the Polaroid, so the sunglasses "cut glare" without affecting the main scene you wish to see.

You can test whether your sunglasses are polarized by holding them in front of you and rotating them as you look at the glare reflecting from a horizontal surface. Polarizing sunglasses will substantially reduce the glare when the glasses are "normal" but not when the glasses are 90° from normal. (You can also test them against a pair of sunglasses known to be polarizing by seeing if all light is blocked when the lenses of the two pairs are crossed.)

If you do have polarizing sunglasses, look at the sky about 90° from the sun in the early morning or late afternoon. You can detect the sky's polarization by rotating the glasses. Bees can automatically sense the polarization of skylight, but humans can't.

STOP TO THINK 34.7 Unpolarized light of equal intensity is incident on four pairs of polarizing filters. Rank in order, from largest to smallest, the intensities I_a to I_d transmitted through the second polarizer of each pair.

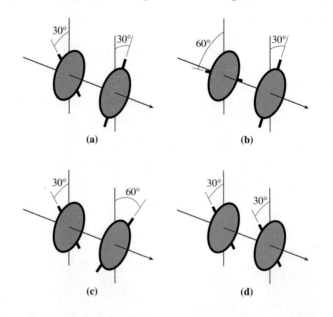

SUMMARY

The goal of Chapter 34 has been to study the properties of electromagnetic fields and waves.

GENERAL PRINCIPLES

Maxwell's Equations

These equations govern electromagnetic fields:

$$\oint \vec{E} \cdot d\vec{A} = \frac{Q_{in}}{\epsilon_0} \qquad \text{Gauss's law}$$

$$\oint \vec{B} \cdot d\vec{A} = 0 \qquad \text{Gauss's law for magnetism}$$

$$\oint \vec{E} \cdot d\vec{s} = -\frac{d\Phi_m}{dt} \qquad \text{Faraday's law}$$

$$\oint \vec{B} \cdot d\vec{s} = \mu_0 I_{through} + \epsilon_0 \mu_0 \frac{d\Phi_e}{dt} \qquad \text{Ampère-Maxwell law}$$

Maxwell's equations tell us that:

An electric field can be created by

- Charged particles
- A changing magnetic field

A magnetic field can be created by

- A current
- A changing electric field

Lorentz Force

This force law governs the interaction of charged particles with electromagnetic fields:

$$\vec{F} = q(\vec{E} + \vec{v} \times \vec{B})$$

- An electric field exerts a force on any charged particle.
- A magnetic field exerts a force on a moving charged particle.

Field Transformations

Fields measured in frame S to be $\vec{E}$ and $\vec{B}$ are found in frame S' to be

$$\vec{E}' = \vec{E} + \vec{V} \times \vec{B}$$

$$\vec{B}' = \vec{B} - \frac{1}{c^2}\vec{V} \times \vec{E}$$

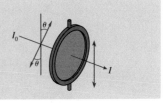

IMPORTANT CONCEPTS

Induced fields

An induced electric field is created by a changing magnetic field.

An induced magnetic field is created by a changing electric field.

These fields can exist independently of charges and currents.

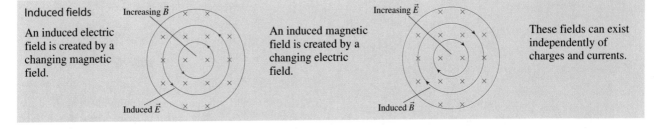

An electromagnetic wave is a self-sustaining electromagnetic field.

- An em wave is a transverse wave with $\vec{E}$, $\vec{B}$, and $\vec{v}$ mutually perpendicular.
- An em wave propagates with speed $v_{em} = c = 1/\sqrt{\epsilon_0 \mu_0}$.
- The electric and magnetic field strengths are related by $E = cB$.
- The **Poynting vector** $\vec{S} = (\vec{E} \times \vec{B})/\mu_0$ is the energy transfer in the direction of travel.
- The wave **intensity** is $I = P/A = (1/2c\mu_0)E_0^2 = (c\epsilon_0/2)E_0^2$.

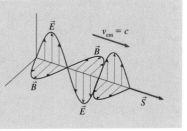

APPLICATIONS

Polarization

The electric field and the Poynting vector define the **plane of polarization.** The intensity of polarized light transmitted through a polarizing filter is given by Malus's law

$$I = I_0 \cos^2\theta$$

where θ is the angle between the electric field and the polarizer axis.

TERMS AND NOTATION

Lorentz force law	Maxwell's equations	antenna
Galilean field transformation equations	Poynting vector, $\vec{S}$	plane of polarization
displacement current	intensity, I	Malus's law
	radiation pressure, p_{rad}	

EXERCISES AND PROBLEMS

Exercises

Section 34.1 Electromagnetic Fields and Forces

1. The magnetic field is uniform over each face of the box shown in Figure Ex34.1. What are the magnetic field strength and direction on the front surface?

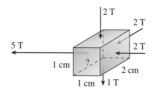

FIGURE EX34.1

2. What is the force (magnitude and direction) on the proton in Figure Ex34.2?

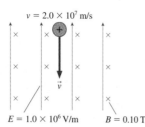

FIGURE EX34.2 $E = 1.0 \times 10^6$ V/m $B = 0.10$ T

3. An electron travels with $\vec{v} = 5.0 \times 10^6 \hat{\imath}$ m/s through a point in space where $\vec{E} = (2.0 \times 10^5 \hat{\imath} - 2.0 \times 10^5 \hat{\jmath})$ V/m and $\vec{B} = -0.10\hat{k}$ T. What is the force on the electron?

4. What electric field strength and direction will allow the electron in Figure Ex34.4 to pass through this region of space without being deflected?

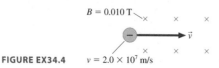

FIGURE EX34.4 $v = 2.0 \times 10^7$ m/s

5. What are the electric field strength and direction at the position of the proton in Figure Ex34.5?

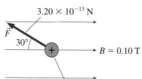

FIGURE EX34.5   Moving out of page at 1.0×10^7 m/s

6. An electron travels with $\vec{v} = 5.0 \times 10^6 \hat{\imath}$ m/s through a point in space where $\vec{B} = 0.10\hat{\jmath}$ T. The force on the electron at this point is $\vec{F} = (9.6 \times 10^{-14}\hat{\imath} - 9.6 \times 10^{-14}\hat{k})$ N. What is the electric field?

Section 34.2 *E* or *B*? It Depends on Your Perspective

7. A rocket cruises past a laboratory at 1.0×10^6 m/s in the positive x-direction just as a proton is launched with velocity (in the laboratory frame) $\vec{v} = (1.41 \times 10^6 \hat{\imath} + 1.41 \times 10^6 \hat{\jmath})$ m/s. What are the proton's speed and its angle from the y-axis (or y'-axis) in (a) the laboratory frame and (b) the rocket frame?

8. Figure Ex34.8 shows the electric and magnetic field in Frame S. A rocket travels parallel to one of the axes of the S coordinate system. Along which axis must the rocket travel, and in which direction (or directions), in order for the rocket scientists to measure (a) $B' > B$, (b) $B' = B$, and (c) $B' < B$?

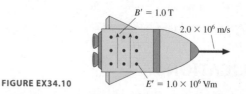

FIGURE EX34.8

9. Scientists in the laboratory create a uniform electric field $\vec{E} = -1.0 \times 10^6 \hat{\jmath}$ V/m in a region of space where $\vec{B} = \vec{0}$. What are the fields in the reference frame of a rocket traveling in the positive x-direction at 1.0×10^6 m/s?

10. A rocket zooms past the earth at $v = 2.0 \times 10^6$ m/s. Scientists on the rocket have created the electric and magnetic fields shown in Figure Ex34.10. What are the fields measured by an earthbound scientist?

FIGURE EX34.10

11. Laboratory scientists have created the electric and magnetic fields shown in Figure Ex34.11. These fields are also seen by scientists that zoom past in a rocket traveling in the x-direction at 1.0×10^6 m/s. According to the rocket scientists, what angle does the electric field make with the axis of the rocket?

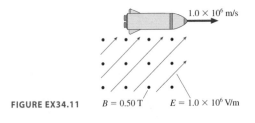

FIGURE EX34.11 $B = 0.50$ T $E = 1.0 \times 10^6$ V/m

Section 34.3 Faraday's Law Revisited

12. Figure Ex34.12 shows the current as a function of time through a 20-cm-long, 4.0-cm-diameter solenoid with 400 turns. Draw a graph of the induced electric field strength as a function of time at a point 1.0 cm from the axis of the solenoid.

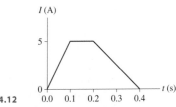

FIGURE EX34.12

13. The magnetic field inside a 5.0-diameter solenoid is 2.0 T and decreasing at 4.0 T/s. What is the electric field strength inside the solenoid at point (a) on the axis and (b) 2.0 cm from the axis?

14. The magnetic field in Figure Ex34.14 is decreasing at the rate 0.10 T/s. What is the acceleration (magnitude and direction) of a proton at rest at points a to d?

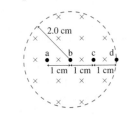

FIGURE EX34.14

Section 34.4 The Displacement Current

15. Show that the quantity $\epsilon_0 (d\Phi_e/dt)$ has units of current.
16. Show that the displacement current inside a parallel-plate capacitor can be written $C(dV_C/dt)$.
17. At what rate must the potential difference increase across a 1.0 μF capacitor to create a 1.0 A displacement current in the capacitor?
18. A 10-cm-diameter parallel-plate capacitor has a 1.0 mm spacing. The electric field between the plates is increasing at the rate 1.0×10^6 V/m s. What is the magnetic field strength (a) on the axis, (b) 3.0 cm from the axis, and (c) 7.0 cm from the axis?
19. A square parallel-plate capacitor 5.0 cm on a side has a 0.50 mm gap. What is the displacement current in the capacitor if the potential difference across the capacitor is increasing at 500,000 V/s?

Section 34.6 Electromagnetic Waves

20. What is the magnetic field amplitude of an electromagnetic wave whose electric field amplitude is 10 V/m?
21. What is the electric field amplitude of an electromagnetic wave whose magnetic field amplitude is 2.0 mT?

22. The magnetic field of an electromagnetic wave in a vacuum is $B_z = (3.0\ \mu\text{T})\sin((1.00 \times 10^7)x - \omega t)$, where x is in m and t is in s. What are the wave's (a) wavelength, (b) frequency, and (c) electric field amplitude?
23. The electric field of an electromagnetic wave in a vacuum is $E_y = (20\ \text{V/m})\cos((6.28 \times 10^8)x - \omega t)$, where x is in m and t is in s. What are the wave's (a) wavelength, (b) frequency, and (c) magnetic field amplitude?

Section 34.7 Properties of Electromagnetic Waves

24. A radio wave is traveling in the negative y-direction. What is the direction of $\vec{E}$ at a point where $\vec{B}$ is in the positive x-direction?
25. Show that:
    a. The quantity cB has the same units as E.
    b. The Poynting vector has units W/m^2.
26. a. What is the magnetic field amplitude of an electromagnetic wave whose electric field amplitude is 100 V/m?
    b. What is the intensity of the wave?
27. A radio receiver can detect signals with electric field amplitudes as small as 300 μV/m. What is the intensity of the smallest detectable signal?
28. A 200 MW laser pulse is focused with a lens to a diameter of 2.0 μm.
    a. What is the laser beam's electric field amplitude at the focal point?
    b. How does this electric field compare to the electric field that keeps the electron bound to the proton of a hydrogen atom? The radius of the electron's orbit is 0.053 nm.
29. A radio antenna broadcasts a 1.0 MHz radio wave with 25 kW of power. Assume that the radiation is emitted uniformly in all directions.
    a. What is the wave's intensity 30 km from the antenna?
    b. What is the electric field amplitude at this distance?
30. At what distance from a 10 W point source of electromagnetic waves is the electric field amplitude (a) 100 V/m and (b) 0.010 V/m?
31. A 1000 W carbon-dioxide laser emits light with a wavelength of 10 μm into a 3.0-mm-diameter laser beam. What force does the laser beam exert on a completely absorbing target?

Section 34.8 Polarization

32. Figure Ex34.32 shows a vertically polarized radio wave of frequency 1.0×10^6 Hz traveling into the page. The maximum electric field strength is 1000 V/m. What are
    a. The maximum magnetic field strength?
    b. The magnetic field strength and direction at a point where $\vec{E} = (500\ \text{V/m, down})$?
    c. The smallest distance between a point on the wave having the magnetic field of part b and a point where the magnetic field is at maximum strength?

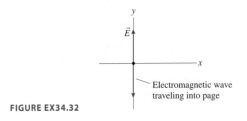

FIGURE EX34.32

33. Only 25% of the intensity of a polarized light wave passes through a polarizing filter. What is the angle between the electric field and the axis of the filter?

34. A 200 mW horizontally polarized laser beam passes through a polarizing filter whose axis is 25° from vertical. What is the power of the laser beam as it emerges from the filter?

35. Unpolarized light with intensity 350 W/m² passes first through a polarizing filter with its axis vertical, then through a polarizing filter with its axis 30° from vertical. What light intensity emerges from the second filter?

Problems

36. Figure P34.36 is an overhead view of children on a playground. Mary (M), who is at the origin of reference frame S, throws a ball horizontally to Tom (T) at 5.0 m/s. At the instant she releases the ball, Carlos (C) runs past her at 3.0 m/s.
    a. Write the ball's velocity vector $\vec{v}$ in component form, $\vec{v} = v_x\hat{\imath} + v_y\hat{\jmath}$.
    b. What are Carlos's xy-coordinates in frame S at the instant Mary tosses the ball and at the instant Tom catches it?
    c. Carlos is at the origin of reference frame S′. Draw a picture of Carlos's coordinate axes, then show the position of the ball, as seen by Carlos, at the instant Mary tosses it and at the instant Tom catches it.
    d. Use Carlos's measurements of position and time to find the ball's velocity $\vec{v}'$ in Carlos's reference frame.
    e. Use the Galilean velocity transformation to find $\vec{v}'$ and show that it agrees with your answer to part d.

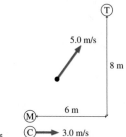

FIGURE P34.36

37. A proton is fired with a speed of 1.0×10^6 m/s through the parallel-plate capacitor shown in Figure P34.37. The capacitor's electric field is $\vec{E} = (1.0 \times 10^5$ V/m, down).
    a. What magnetic field $\vec{B}$, both strength and direction, must be applied to allow the proton to pass through the capacitor with no change in speed or direction?
    b. Find the electric and magnetic fields in the proton's reference frame.
    c. How does an experimenter in the proton's frame explain that the proton experiences no force as the charged plates fly by?

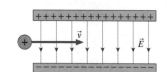

FIGURE P34.37

38. In Figure P34.38, a circular loop of radius r travels with speed v along a charged wire having linear charge density λ. The wire is at rest in the laboratory frame, and it passes through the center of the loop.
    a. What are $\vec{E}$ and $\vec{B}$ at a point on the loop as measured by a scientist in the laboratory? Include both strength and direction.
    b. What are the fields $\vec{E}'$ and $\vec{B}'$ at a point on the loop as measured by a scientist in the frame of the loop?
    c. Show that an experimenter in the loop's frame sees a current $I = \lambda v$ passing through the center of the loop.
    d. What electric and magnetic fields would an experimenter in the loop's frame calculate at distance r from the current of part c?
    e. Show that your field of parts b and d are the same.
    f. If the loop is made of a conducting material, will it have an induced current? Explain.

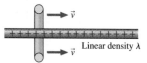

FIGURE P34.38

39. A very long, 1.0-mm-diameter wire carries a 2.5 A current from left to right. Thin plastic insulation on the wire is positively charged with linear charge density 2.5 nC/cm. A mosquito 1.0 cm from the center of the wire would like to move in such a way as to experience an electric field but no magnetic field. How fast and which direction should she fly?

40. The magnetic field inside a 4.0-cm-diameter superconducting solenoid varies sinusoidally between 8.0 T and 12.0 T at a frequency of 10 Hz.
    a. What is the maximum electric field strength at a point 1.5 cm from the solenoid axis?
    b. What is the value of B at the instant E reaches its maximum value?

41. Equation 34.27 is an expression for the induced electric field inside a solenoid ($r < R$). Find an expression for the induced electric field outside a solenoid ($r > R$) in which the magnetic field is changing at the rate dB/dt.

42. A simple series circuit consists of a 150 Ω resistor, a 25 V battery, a switch, and a 2.5 pF parallel-plate capacitor (initially uncharged) with plates 5.0 mm apart. The switch is closed at $t = 0$ s.
    a. After the switch is closed, find the maximum electric flux and the maximum displacement current through the capacitor.
    b. Find the electric flux and the displacement current at $t = 0.50$ ns.

43. A wire with conductivity σ carries current I. The current is increasing at the rate dI/dt.
    a. Show that there is a displacement current in the wire equal to $(\epsilon_0/\sigma)(dI/dt)$.
    b. Evaluate the displacement current for a copper wire in which the current is increasing at 1.0×10^6 A/s.

44. A 10 A current is charging a 1.0-cm-diameter parallel-plate capacitor.
    a. What is the magnetic field strength at a point 2.0 mm from the center of the wire leading to the capacitor?
    b. What is the magnetic field strength at a point 2.0 mm from the center of the capacitor?

45. Figure P34.45 shows the electric field inside a cylinder of radius $R = 3.0$ mm. The field strength is increasing with time as $E = 1.0 \times 10^8 t^2$ V/m, where t is in s. The electric field

outside the cylinder is always zero, and the field inside the cylinder was zero for $t < 0$.

a. Find an expression for the electric flux Φ_e through the entire cylinder as a function of time.

b. Draw a picture showing the magnetic field lines inside and outside the cylinder. Be sure to include arrowheads showing the field's direction.

c. Find an expression for the magnetic field strength as a function of time at a distance $r < R$ from the center. Evaluate the magnetic field strength at $r = 2.0$ mm, $t = 2.0$ s.

d. Find an expression for magnetic field strength as a function of time at a distance $r > R$ from the center. Evaluate the magnetic field strength at $r = 4.0$ mm, $t = 2.0$ s.

FIGURE P34.45

46. Assume that a 100 W light bulb radiates all its energy as a single wavelength of visible light. Estimate the electric and magnetic field strength at the surface of the bulb.

47. The intensity of sunlight reaching the earth is 1360 W/m².
    a. What is the power output of the sun?
    b. What is the intensity of sunlight on Mars?

48. When the Voyager 2 spacecraft passed Neptune in 1989, it was 4.5×10^9 km from the earth. Its radio transmitter, with which it sent back data and images, broadcast with a mere 21 W of power. Assuming that the transmitter broadcast equally in all directions,
    a. What signal intensity was received on the earth?
    b. What electric field amplitude was detected?
    The received signal was somewhat stronger than your result because the spacecraft used a directional antenna, but not by much.

49. In reading the instruction manual that came with your garage-door opener, you see that the transmitter unit in your car produces a 250 mW signal and that the receiver unit is supposed to respond to a radio wave of the correct frequency if the electric field amplitude exceeds 0.10 V/m. You wonder if this is really true. To find out, you put fresh batteries in the transmitter and start walking away from your garage while opening and closing the door. Your garage door finally fails to respond when you're 42 m away. Are the manufacturer's claims true?

50. The intensity of sunlight reaching the earth is 1360 W/m². Assuming all the sunlight is absorbed, what is the radiation-pressure force on the earth? Give your answer in newtons and as a percentage of the sun's gravitational force on the earth.

51. A laser beam shines straight up onto a flat, black foil with a mass of 25 μg. What laser power is needed to levitate the foil?

52. For a science project, you would like to horizontally suspend an 8.5 by 11 inch sheet of black paper in a vertical beam of light whose dimensions exactly match the paper. If the mass of the sheet is 1.0 g, what light intensity will you need?

53. You've recently read about a chemical laser that generates a 20-cm-diameter, 25 MW laser beam. One day, after physics class, you start to wonder if you could use the radiation pres-

sure from this laser beam to launch small payloads into orbit. To see if this might be feasible, you do a quick calculation of the acceleration of a 20-cm-diameter, 100 kg, perfectly absorbing block. What speed would such a block have if pushed *horizontally* 100 m along a frictionless track by such a laser?

54. An 80 kg astronaut has gone outside his space capsule to do some repair work. Unfortunately, he forgot to lock his safety tether in place, and he has drifted 5.0 m away from the capsule. Fortunately, he has a 1000 W portable laser with fresh batteries that will operate it for 1.0 hr. His only chance is to accelerate himself toward the space capsule by firing the laser in the opposite direction. He has a 10-hr supply of oxygen. Can he make it?

55. Unpolarized light of intensity I_0 is incident on three polarizing filters. The axis of the first is vertical, that of the second is 45° from vertical, and that of the third is horizontal. What light intensity emerges from the third filter?

Challenge Problems

56. A 4.0-cm-diameter parallel-plate capacitor with a 1.0 mm spacing is charged to 1000 V. A switch closes at $t = 0$ s, and the capacitor is discharged through a wire with 0.20 Ω resistance.
    a. Find an expression for the magnetic field strength inside the capacitor at $r = 1.0$ cm as a function of time.
    b. Draw of graph of B versus t.

57. a. Show that u_E and u_B, the energy densities of the electric and magnetic fields, are equal to each other in an electromagnetic wave. In other words, show that the wave's energy is divided equally between the electric field and the magnetic field.
    b. What is the total energy density in an electromagnetic wave of intensity 1000 W/m²?

58. Large quantities of dust should have been left behind after the creation of the solar system. Larger dust particles, comparable in size to soot and sand grains, are common. They create shooting stars when they collide with the earth's atmosphere. But very small dust particles are conspicuously absent. Astronomers believe that the very small dust particles have been blown out of the solar system by the sun. By comparing the forces on dust particles, determine the diameter of the smallest dust particles that can remain in the solar system over long periods of time. Assume that the dust particles are spherical, black, and have a density of 2000 kg/m³. The sun emits electromagnetic radiation with power 3.9×10^{26} W.

59. Consider current I passing through a resistor of radius r, length L, and resistance R.
    a. Determine the electric and magnetic fields at the surface of the resistor. Assume that the electric field is uniform throughout, including at the surface.
    b. Determine the strength and direction of the Poynting vector at the surface of the resistor.
    c. Show that the flux of the Poynting vector (i.e., the integral of $\vec{S} \cdot d\vec{A}$) over the surface of the resistor is I^2R. Then give an interpretation of this result.

60. Unpolarized light of intensity I_0 is incident on a stack of 7 polarizing filters, each with its axis rotated 15° cw with respect to the previous filter. What light intensity emerges from the last filter?

Stop to Think 34.1: a. The charge is negative, so the electric force is to the left. $\vec{v} \times \vec{B}$ is to the right, so the magnetic force on a negative charge is also to the left.

Stop to Think 34.2: b. $\vec{V}$ is parallel to $\vec{B}$, hence $\vec{V} \times \vec{B}$ is zero. Thus $\vec{E}' = \vec{E}$ and points in the positive z-direction. $\vec{V} \times \vec{E}$ points down, in the negative y direction, so $-\vec{V} \times \vec{E}/c^2$ points in the positive y-direction and causes $\vec{B}'$ to be angled upward.

Stop to Think 34.3: $B_c > B_a > B_d > B_b$. The induced magnetic field strength depends on the *rate dE/dt* at which the electric field is changing. Steeper slopes on the graph correspond to larger magnetic fields.

Stop to Think 34.4: e. $\vec{E}$ is perpendicular to $\vec{B}$ and to $\vec{v}$, so it can only be along the z-axis. According to the Ampère-Maxwell law, $d\Phi_e/dt$ has the same sign as the line integral of $\vec{B} \cdot d\vec{s}$ around the closed curve. The integral is positive for a cw integration. Thus, from the right-hand rule, $\vec{E}$ is either into the page (negative z-direction) and increasing, or out of the page (positive z-direction) and decreasing.

We can see from the figure that B is decreasing as the wave moves left to right, so E must also be decreasing. Thus $\vec{E}$ points along the positive z-axis.

Stop to Think 34.5: a. The Poynting vector $\vec{S} = (\vec{E} \times \vec{B})/\mu_0$ points in the direction of travel, which is the positive y-direction. $\vec{B}$ must point in the positive x-direction in order for $\vec{E} \times \vec{B}$ to point upward.

Stop to Think 34.6: b. The intensity along a line from the antenna decreases inversely with the square of the distance, so the intensity at 20 km is $\frac{1}{4}$ that at 10 km. But the intensity depends on the square of the electric field amplitude, or, conversely, E_0 is proportional to $I^{1/2}$. Thus E_0 at 20 km is $\frac{1}{2}$ that at 10 km.

Stop to Think 34.7: $I_d > I_a > I_b = I_c$. The intensity depends upon $\cos^2\theta$, where θ is the angle *between* the axes of the two filters. The filters in d have $\theta = 0°$. The two filters in both b and c are crossed ($\theta = 90°$) and transmit no light at all.

35 AC Circuits

Transmission lines carry
alternating current at voltages
as high as 500,000 V.

▶ **Looking Ahead**

The goal of Chapter 35 is to
understand and apply basic
techniques of AC circuit analysis.
In this chapter you will learn to:

- Use phasors to analyze an AC
  circuit with resistors, capacitors,
  and inductors.
- Understand *RC* filter circuits.
- Understand resonance in an
  RLC circuit.
- Calculate the power loss in
  an AC circuit.

◀ **Looking Back**

The material in this chapter
depends on the fundamentals of
circuits and on the properties of
resistors, capacitors, and inductors.
The mathematical representation
of AC circuits is based on simple
harmonic motion. Please review:

- Sections 14.1–14.2 and 14.8
  Simple harmonic motion and
  resonance.
- Section 30.6 Capacitors.
- Sections 31.1–31.4
  Fundamentals of circuit
  analysis.
- Sections 33.8–33.10 Inductors.

Thomas Edison built the first large-scale electric generating station in 1882, in
New York City. His entrepreneurial motive was to sell light bulbs, which he had
invented a few years earlier. Edison's company, which later became General
Electric, is still one of the largest manufacturers of electrical equipment.

Edison soon had competition from George Westinghouse, another name that
probably looks familiar. Whereas Edison's system used *direct current* (DC), Westinghouse favored *alternating current* (AC). The technological debate between
these two men and their companies lasted nearly 20 years, but eventually alternating current proved to be superior for the long-distance transmission of electric
energy.

Today, more than a century later, a "grid" of AC electrical distribution systems
spans the United States and other countries. Any device that plugs into an electric outlet uses an AC circuit. In this chapter, you will learn some of the basic
techniques for analyzing AC circuits. However, these ideas are not limited to

1121

(a)

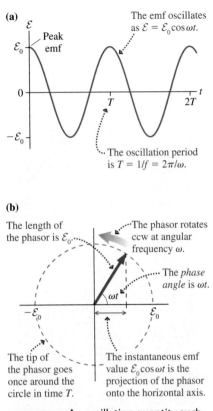

The emf oscillates as $\mathcal{E} = \mathcal{E}_0 \cos \omega t$.

Peak emf

The oscillation period is $T = 1/f = 2\pi/\omega$.

(b)

The length of the phasor is $\mathcal{E}_0$.

The phasor rotates ccw at angular frequency ω.

The *phase angle* is ωt.

The tip of the phasor goes once around the circle in time T.

The instantaneous emf value $\mathcal{E}_0 \cos \omega t$ is the projection of the phasor onto the horizontal axis.

FIGURE 35.1 An oscillating quantity such as emf can be represented either as a graph or as a phasor diagram.

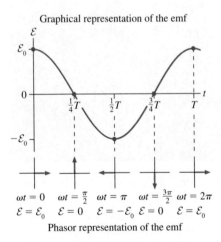

Graphical representation of the emf

$$\omega t = 0 \quad \omega t = \tfrac{\pi}{2} \quad \omega t = \pi \quad \omega t = \tfrac{3\pi}{2} \quad \omega t = 2\pi$$
$$\mathcal{E} = \mathcal{E}_0 \quad \mathcal{E} = 0 \quad \mathcal{E} = -\mathcal{E}_0 \quad \mathcal{E} = 0 \quad \mathcal{E} = \mathcal{E}_0$$

Phasor representation of the emf

FIGURE 35.2 The correspondence between a rotating phasor and points on a graph.

power-line circuits. Audio, radio, television, and telecommunication electronics are based on circuits that use oscillating voltages and currents. Any practical understanding of modern electronics is grounded, so to speak, in AC circuit analysis.

35.1 AC Sources and Phasors

One of the examples of Faraday's law cited in Chapter 33 was an electric generator. A turbine, which might be powered by expanding steam or falling water, causes a coil of wire to rotate in a magnetic field. As the coil spins, the emf and the induced current oscillate sinusoidally. The emf is alternately positive and then negative, causing the charges to flow in one direction and then, a half cycle later, in the other. The oscillation frequency in North and South America is $f = 60$ Hz, whereas most of the rest of the world uses a 50 Hz oscillation.

The generator's emf—the voltage—is determined by the magnetic field strength and the number of turns in the generator coil. The emf is a fixed, unvarying quantity, so it might seem logical to call a generator an *alternating voltage source*. Nonetheless, circuits powered by a sinusoidal emf are called **AC circuits,** where AC stands for *alternating current*. By contrast, the steady-current circuits you studied in Chapter 31 are called **DC circuits,** for *direct current*.

AC circuits are not limited to the use of 50 Hz or 60 Hz power-line voltages. Audio, radio, television, and telecommunication equipment all make extensive use of AC circuits, with frequencies ranging from approximately 10^2 Hz in audio circuits to approximately 10^9 Hz in cell phones. These devices use *electrical oscillators* rather than generators to produce a sinusoidal emf, but the basic principles of circuit analysis are the same.

You can think of an AC generator or oscillator as a battery whose output voltage undergoes sinusoidal oscillations. The instantaneous emf of an AC generator or oscillator, shown graphically in Figure 35.1a, can be written

$$\mathcal{E} = \mathcal{E}_0 \cos \omega t \tag{35.1}$$

where $\mathcal{E}_0$ is the peak or maximum emf and $\omega = 2\pi f$ is the angular frequency in radians per second. Recall that the units of emf are volts. As you can imagine, the mathematics of AC circuit analysis are going to be very similar to the mathematics of simple harmonic motion.

An alternative way to represent the emf and other oscillatory quantities is with the *phasor diagram* of Figure 35.1b. A **phasor** is a vector that rotates *counterclockwise* (ccw) around the origin at angular frequency ω. The length or magnitude of the phasor is the maximum value of the quantity. For example, the length of an emf phasor is $\mathcal{E}_0$. The angle ωt is the *phase angle,* an idea you learned about in Chapter 14, where we made a connection between circular motion and simple harmonic motion.

The quantity's instantaneous value, the value you would measure at time t, is the projection of the phasor onto the horizontal axis. This is also analogous to the connection between circular motion and simple harmonic motion. Figure 35.2 helps you visualize the phasor rotation by showing how the phasor corresponds to the more familiar graph at several specific points in the cycle.

STOP TO THINK 35.1 The magnitude of the instantaneous value of the emf represented by this phasor is

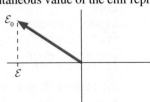

a. Increasing.
b. Decreasing.
c. Constant.
d. It's not possible to tell without knowing t.

Resistor Circuits

In Chapter 31 you learned to analyze a circuit in terms of the current I, voltage V, and potential difference ΔV. Now, because the current and voltage are oscillating, we will use lowercase i to represent the *instantaneous* current through a circuit element and v for the circuit element's *instantaneous* voltage.

Figure 35.3 shows the instantaneous current i_R through a resistor R. The potential difference across the resistor, which we call the *resistor voltage* v_R, is given by Ohm's law:

$$v_R = i_R R \qquad (35.2)$$

The potential *decreases* in the direction of the current.

Figure 35.4 shows a resistor R connected across an AC emf $\mathcal{E}$. Notice that the circuit symbol for an AC generator is —⊙—. We can analyze this circuit in exactly the same way we analyzed a DC resistor circuit. Kirchhoff's loop law says that the sum of all the potential differences around a closed path is zero:

$$\sum \Delta V = \Delta V_{\text{source}} + \Delta V_R = \mathcal{E} - v_R = 0 \qquad (35.3)$$

The minus sign appears, just as it did in the equation for a DC circuit, because the potential *decreases* when we travel through a resistor in the direction of the current. We find from the loop law that $v_R = \mathcal{E} = \mathcal{E}_0 \cos \omega t$. This isn't surprising because the resistor is connected directly across the terminals of the emf.

The resistor voltage is a sinusoidal voltage at angular frequency ω. It will be useful to write

$$v_R = V_R \cos \omega t \qquad (35.4)$$

where V_R is the peak or maximum voltage. You can see that $V_R = \mathcal{E}_0$ in the single-resistor circuit of Figure 35.4. Thus the current through the resistor is

$$i_R = \frac{v_R}{R} = \frac{V_R \cos \omega t}{R} = I_R \cos \omega t \qquad (35.5)$$

where $I_R = V_R/R$ is the peak current.

NOTE ▶ Ohm's law applies to both the instantaneous *and* peak currents and voltages. ◀

The resistor's instantaneous current and voltage are in phase, both oscillating as $\cos \omega t$. Figure 35.5 shows the voltage and the current simultaneously on a graph and as a phasor diagram. The fact that the current phasor is shorter than the voltage phasor has no significance. Current and voltage are measured in different units, so you can't compare the length of one to the length of the other. Showing the two different quantities on a single graph—a tactic that can be misleading if you're not careful—is simply to show that they oscillate in phase and that their phasors rotate together at the same angle and frequency.

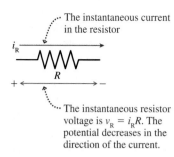

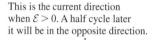

FIGURE 35.3 Instantaneous current i_R through a resistor.

This is the current direction when $\mathcal{E} > 0$. A half cycle later it will be in the opposite direction.

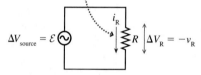

FIGURE 35.4 An AC resistor circuit.

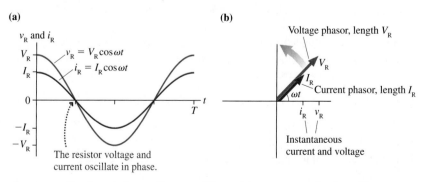

FIGURE 35.5 Graph and phasor diagram of the resistor current and voltage. The current and voltage are in phase.

EXAMPLE 35.1 Finding resistor voltages

In the circuit of Figure 35.6, what are (a) the peak voltage across each resistor and (b) the instantaneous resistor voltages at $t = 20$ ms?

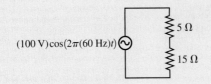

FIGURE 35.6 An AC resistor circuit.

VISUALIZE Figure 35.6 shows the circuit diagram. The two resistors are in series.

SOLVE

a. The equivalent resistance of the two series resistors is $R_{eq} = 5\ \Omega + 15\ \Omega = 20\ \Omega$. The instantaneous current through the equivalent resistance is

$$i_R = I_R \cos\omega t = \frac{v_R}{R_{eq}} = \frac{\mathcal{E}_0 \cos\omega t}{R_{eq}}$$

$$= \frac{(100\ \text{V})\cos(2\pi(60\ \text{Hz})t)}{20\ \Omega}$$

$$= (5.0\ \text{A})\cos(2\pi(60\ \text{Hz})t)$$

The peak current is $I_R = 5.0$ A, and this is also the peak current through the two resistors that form the 20 Ω equivalent resistance. Hence the peak voltage across each resistor is

$$V_R = I_R R = \begin{cases} 25\ \text{V} & 5\ \Omega\ \text{resistor} \\ 75\ \text{V} & 15\ \Omega\ \text{resistor} \end{cases}$$

b. The instantaneous current at $t = 0.020$ s is

$$i_R = (5.0\ \text{A})\cos(2\pi(60\ \text{Hz})(0.020\ \text{s})) = 1.545\ \text{A}$$

The resistor voltages at this time are

$$v_R = i_R R = \begin{cases} 7.7\ \text{V} & 5\ \Omega\ \text{resistor} \\ 23.2\ \text{V} & 15\ \Omega\ \text{resistor} \end{cases}$$

ASSESS The sum of the instantaneous voltages, 30.9 V, is what you would find by calculating $\mathcal{E}$ at $t = 20$ ms. This self-consistency gives us confidence in the answer.

STOP TO THINK 35.2 The resistor whose voltage and current phasors are shown here has resistance R

a. $>1\ \Omega$.
b. $<1\ \Omega$.
c. It's not possible to tell.

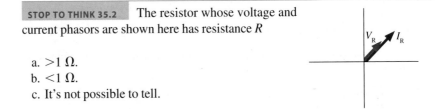

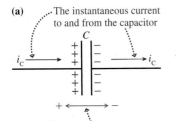

(a) The instantaneous current to and from the capacitor

The instantaneous capacitor voltage is $v_C = q/C$. The potential decreases from + to −.

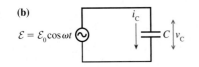

(b)

FIGURE 35.7 An AC capacitor circuit.

35.2 Capacitor Circuits

Figure 35.7a shows current i_C charging a capacitor with capacitance C. The instantaneous capacitor voltage is $v_C = q/C$, where $\pm q$ is the charge on the two capacitor plates at this instant of time. It is useful to compare Figure 35.7a to Figure 35.3 for a resistor.

Figure 35.7b, where capacitance C is connected across an AC source of emf $\mathcal{E}$, is the most basic capacitor circuit. The capacitor is in parallel with the source, so the capacitor voltage equals the emf: $v_C = \mathcal{E} = \mathcal{E}_0 \cos\omega t$. It will be useful to write

$$v_C = V_C \cos\omega t \tag{35.6}$$

where V_C is the peak or maximum voltage across the capacitor. You can see that $V_C = \mathcal{E}_0$ in this single-capacitor circuit.

To find the current to and from the capacitor, first write the charge

$$q = Cv_C = CV_C \cos\omega t \tag{35.7}$$

The current is the *rate* at which charge flows through the wires, $i_C = dq/dt$, thus

$$i_C = \frac{dq}{dt} = \frac{d}{dt}(CV_C\cos\omega t) = -\omega CV_C\sin\omega t \qquad (35.8)$$

We can most easily see the relationship between the capacitor voltage and current if we use the trigonometric identity $-\sin(x) = \cos(x + \pi/2)$ to write

$$i_C = \omega CV_C\cos\left(\omega t + \frac{\pi}{2}\right) \qquad (35.9)$$

In contrast to a resistor, a capacitor's current and voltage are *not* in phase. In Figure 35.8a, which shows a graph of the instantaneous voltage v_C and current i_C, you can see that the current peaks one-quarter of a period *before* the voltage peaks. The phase angle of the current phasor on the phasor diagram of Figure 35.8b is $\pi/2$ rad—a quarter of a circle—larger than the phase angle of the voltage phasor.

We can summarize this finding by saying

The AC current through a capacitor *leads* the capacitor voltage by $\pi/2$ rad, or 90°.

The current reaches its peak value I_C at the instant the capacitor is fully discharged and $v_C = 0$. The current is zero at the instant the capacitor is fully charged. You saw a similar behavior in the oscillation of an *LC* circuit in Chapter 33.

A simple harmonic oscillator provides a mechanical analogy of the 90° phase difference between current and voltage. You learned in Chapter 14 (refer to Section 14.1 and Figure 14.5) that the position and velocity of a simple harmonic oscillator are

$$x = A\cos\omega t$$

$$v = \frac{dx}{dt} = -\omega A\sin\omega t = -v_{max}\sin\omega t = v_{max}\cos\left(\omega t + \frac{\pi}{2}\right)$$

You can see in Figure 35.9 that the velocity leads the position by 90° in the same way that the capacitor current (which is proportional to the charge velocity) leads the voltage.

Capacitive Reactance

We can use Equation 35.9 to see that the peak current to and from a capacitor is $I_C = \omega CV_C$. This relationship between the peak voltage and peak current looks much like Ohm's law for a resistor if we define the **capacitive reactance** X_C to be

$$X_C \equiv \frac{1}{\omega C} \qquad (35.10)$$

With this definition,

$$I_C = \frac{V_C}{X_C} \quad \text{or} \quad V_C = I_CX_C \qquad (35.11)$$

The units of reactance, like those of resistance, are ohms.

NOTE ▶ Reactance relates the *peak* voltage V_C and current I_C. But reactance differs from resistance in that it does *not* relate the instantaneous capacitor voltage and current because they are out of phase. That is, $v_C \neq i_CX_C$. ◀

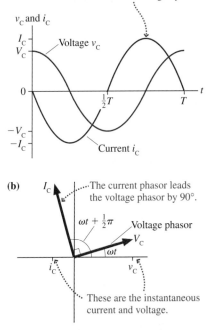

(a) i_C peaks $\frac{1}{4}T$ before v_C peaks. We say that the current *leads* the voltage by 90°.

(b) The current phasor leads the voltage phasor by 90°.

These are the instantaneous current and voltage.

FIGURE 35.8 Graph and phasor diagrams of the capacitor current and voltage.

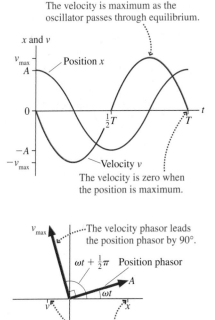

The velocity is maximum as the oscillator passes through equilibrium.

The velocity is zero when the position is maximum.

The velocity phasor leads the position phasor by 90°.

These are the instantaneous velocity and position.

FIGURE 35.9 In a mechanical analogy, the velocity of a simple harmonic oscillator leads the position by 90°.

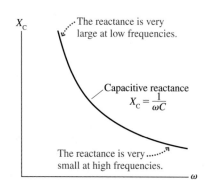

FIGURE 35.10 The capacitive reactance as a function of frequency.

A resistor's resistance R is independent of the emf frequency. In contrast, as Figure 35.10 shows, a capacitor's reactance X_C depends inversely on the frequency. The reactance becomes very large at low frequencies (i.e., the capacitor is a large impediment to current). This makes sense, because $\omega = 0$ would be a nonoscillating, DC circuit, and we know that a steady DC current cannot pass through a capacitor. The reactance decreases as the frequency increases until, at very high frequencies, $X_C \approx 0$ and the capacitor begins to act like an ideal wire. This result has important consequences for how capacitors are used in many circuits.

EXAMPLE 35.2 Capacitive reactance

What is the capacitive reactance of a 0.10 μF capacitor at a 100 Hz audio frequency and at a 100 MHz FM-radio frequency?

SOLVE At 100 Hz,

$$X_C(\text{at } 100 \text{ Hz}) = \frac{1}{\omega C} = \frac{1}{2\pi(100 \text{ Hz})(1.0 \times 10^{-7} \text{ F})}$$

$$= 15{,}900 \ \Omega$$

Increasing the frequency by a factor of 10^6 decreases X_C by a factor of 10^6, giving

$$X_C(\text{at } 100 \text{ MHz}) = 0.0159 \ \Omega$$

ASSESS A capacitor with a substantial reactance at audio frequencies has virtually no reactance at FM-radio frequencies.

EXAMPLE 35.3 Capacitor current

A 10 μF capacitor is connected to a 1000 Hz oscillator with a peak emf of 5.0 V. What is the peak current to the capacitor?

VISUALIZE Figure 35.7b showed the circuit diagram. It is a simple one-capacitor circuit.

SOLVE The capacitive reactance at $\omega = 2\pi f = 6280$ rad/s is

$$X_C = \frac{1}{\omega C} = \frac{1}{(6280 \text{ rad/s})(10 \times 10^{-6} \text{ F})} = 15.9 \ \Omega$$

The peak voltage across the capacitor is $V_C = \mathcal{E}_0 = 5.0$ V, hence the peak current is

$$I_C = \frac{V_C}{X_C} = \frac{5.0 \text{ V}}{15.9 \ \Omega} = 0.314 \text{ A}$$

ASSESS Using reactance is just like using Ohm's law, but don't forget it applies only to the *peak* current and voltage, not the instantaneous values.

STOP TO THINK 35.3 What is the capacitive reactance of "no capacitor," just a continuous wire?

a. 0
b. ∞
c. Undefined

35.3 *RC* Filter Circuits

You learned in Chapter 31 that a resistance R causes a capacitor to be charged or discharged with time constant $\tau = RC$. We called this an *RC* circuit. Now that we've looked at resistors and capacitors individually, let's explore what happens if an *RC* circuit is driven continuously by an alternating current source.

Figure 35.11 shows a circuit in which a resistor R and capacitor C are in series with an emf $\mathcal{E}$ that oscillates at angular frequency ω. Before launching into a formal analysis, let's try to understand qualitatively how this circuit will respond as the frequency is varied. If the frequency is very low, the capacitive reactance will be very large, and thus the peak current I_C will be very small. The peak current through the resistor is the same as the peak current to and from the capacitor (conservation of current requires $I_R = I_C$), hence we expect the resistor's peak voltage $V_R = I_R R$ to be very small at very low frequencies.

On the other hand, suppose the frequency is very high. Then the capacitive reactance approaches zero and the peak current, determined by the resistance alone, will be $I_R = \mathcal{E}_0/R$. The resistor's peak voltage $V_R = IR$ will approach the peak source voltage $\mathcal{E}_0$ at very high frequencies.

This reasoning leads us to expect that V_R will *increase* steadily from 0 to $\mathcal{E}_0$ as ω is increased from 0 to very high frequencies. Kirchhoff's loop law has to be obeyed, so the capacitor voltage V_C will *decrease* from $\mathcal{E}_0$ to 0 during the same change of frequency. A quantitative analysis will show us how this behavior can be used as a *filter*.

The goal of a quantitative analysis is to determine the peak current I and the two peak voltages V_R and V_C as functions of the emf amplitude $\mathcal{E}_0$ and frequency ω. Although this goal can be reached with a purely algebraic analysis, using a phasor diagram is easier and more informative. Our analytic procedure is based on the fact that the current i is the same for two circuit elements in series.

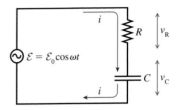

FIGURE 35.11 An *RC* circuit driven by an AC source.

Analyzing an *RC* circuit

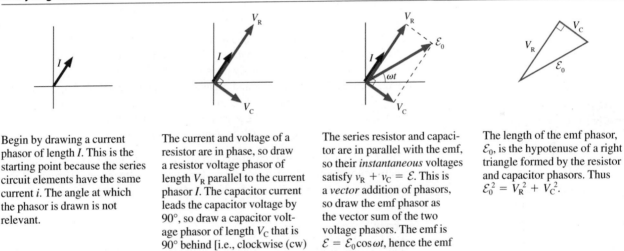

| Begin by drawing a current phasor of length I. This is the starting point because the series circuit elements have the same current i. The angle at which the phasor is drawn is not relevant. | The current and voltage of a resistor are in phase, so draw a resistor voltage phasor of length V_R parallel to the current phasor I. The capacitor current leads the capacitor voltage by 90°, so draw a capacitor voltage phasor of length V_C that is 90° behind [i.e., clockwise (cw) from] the current phasor. | The series resistor and capacitor are in parallel with the emf, so their *instantaneous* voltages satisfy $v_R + v_C = \mathcal{E}$. This is a *vector* addition of phasors, so draw the emf phasor as the vector sum of the two voltage phasors. The emf is $\mathcal{E} = \mathcal{E}_0 \cos \omega t$, hence the emf phasor is at angle ωt. | The length of the emf phasor, $\mathcal{E}_0$, is the hypotenuse of a right triangle formed by the resistor and capacitor phasors. Thus $\mathcal{E}_0^2 = V_R^2 + V_C^2$. |

The relationship $\mathcal{E}_0^2 = V_R^2 + V_C^2$ is based on the peak values, not the instantaneous values, because the peak values are the lengths of the sides of the right triangle. The peak voltages are related to the peak current I via $V_R = IR$ and $V_C = IX_C$, thus

$$\mathcal{E}_0^2 = V_R^2 + V_C^2 = (IR)^2 + (IX_C)^2 = (R^2 + X_C^2)I^2$$
$$= (R^2 + 1/\omega^2 C^2)I^2 \tag{35.12}$$

Consequently, the peak current in the *RC* circuit is

$$I = \frac{\mathcal{E}_0}{\sqrt{R^2 + X_C^2}} = \frac{\mathcal{E}_0}{\sqrt{R^2 + 1/\omega^2 C^2}} \tag{35.13}$$

Knowing I gives us the two peak voltages:

$$V_R = IR = \frac{\mathcal{E}_0 R}{\sqrt{R^2 + X_C^2}} = \frac{\mathcal{E}_0 R}{\sqrt{R^2 + 1/\omega^2 C^2}}$$

$$V_C = IX_C = \frac{\mathcal{E}_0 X_C}{\sqrt{R^2 + X_C^2}} = \frac{\mathcal{E}_0/\omega C}{\sqrt{R^2 + 1/\omega^2 C^2}}$$

(35.14)

Frequency Dependence

Our goal was to see how the peak current and voltages varied as functions of the frequency ω. Equations 35.13 and 35.14 are rather complex and best interpreted by looking at graphs. Figure 35.12 is a graph of V_R and V_C versus ω.

You can see that our qualitative predictions have been borne out. That is, V_R increases from 0 to $\mathcal{E}_0$ as ω is increased while V_C decreases from $\mathcal{E}_0$ to 0. The explanation for this behavior is that the capacitive reactance X_C decreases as ω increases. For low frequencies, where $X_C \gg R$, the circuit is primarily capacitive. For high frequencies, where $X_C \ll R$, the circuit is primarily resistive.

The frequency at which $V_R = V_C$ is called the **crossover frequency** ω_c. The *crossover* frequency is easily found by setting the two expressions in Equation 35.14 equal to each other. The denominators are the same and cancel, as does $\mathcal{E}_0$, leading to

$$\omega_c = \frac{1}{RC}$$

(35.15)

In practice, $f_c = \omega_c/2\pi$ is also called the crossover frequency.

We'll leave it as a homework problem to show that $V_R = V_C = \mathcal{E}_0/\sqrt{2}$ when $\omega = \omega_c$. This may seem surprising. After all, shouldn't V_R and V_C add up to $\mathcal{E}_0$?

No! V_R and V_C are the *peak values* of oscillating voltages, not the instantaneous values. The instantaneous values do, indeed, satisfy $v_R + v_C = \mathcal{E}$ at all instants of time. But the resistor and capacitor voltages are out of phase with each other, as the phasor diagram shows, so the two circuit elements don't reach their peak values at the same time. The peak values are related by $\mathcal{E}_0^2 = V_R^2 + V_C^2$, and you can see that $V_R = V_C = \mathcal{E}_0/\sqrt{2}$ satisfies this equation.

NOTE ▶ It's very important in AC circuit analysis to make a clear distinction between instantaneous values and peak values of voltages and currents. Relations that are true for one set of values may not be true for the other. ◀

Filters

Figure 35.13a is the circuit we've just analyzed, the only difference is that the capacitor voltage v_C is now identified as the *output voltage* v_{out}. This is a voltage you might measure or, perhaps, send to an amplifier for use elsewhere in an electronic instrument. You can see from the capacitor voltage graph in Figure 35.12 that the peak output voltage is $V_{out} \approx \mathcal{E}_0$ if $\omega \ll \omega_c$, but $V_{out} \approx 0$ if $\omega \gg \omega_c$. In other words,

■ If the frequency of an input signal is well below the crossover frequency, the input signal is transmitted with little loss to the output.
■ If the frequency of an input signal is well above the crossover frequency, the input signal is strongly attenuated and the output is very nearly zero.

This circuit is called a **low-pass filter.**

The circuit of Figure 35.13b, which instead uses the resistor voltage v_R for the output v_{out}, is a **high-pass filter.** The output is $V_{out} \approx 0$ if $\omega \ll \omega_c$, but $V_{out} \approx \mathcal{E}_0$ if $\omega \gg \omega_c$. That is, an input signal whose frequency is well above the crossover frequency is transmitted without loss to the output.

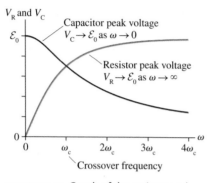

FIGURE 35.12 Graph of the resistor and capacitor peak voltages as functions of the emf frequency ω.

(a) Low-pass filter

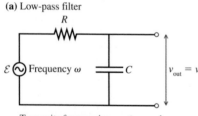

Transmits frequencies $\omega < \omega_c$ and blocks frequencies $\omega > \omega_c$.

(b) High-pass filter

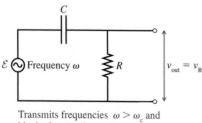

Transmits frequencies $\omega > \omega_c$ and blocks frequencies $\omega < \omega_c$.

FIGURE 35.13 Low-pass and high-pass filter circuits.

Filter circuits are widely used in electronics. For example, a high-pass filter designed to have $f_c = 100$ Hz would pass the audio frequencies associated with speech ($f > 200$ Hz) while blocking 60 Hz "noise" that can be picked up from power lines. Similarly, the high-frequency hiss from old vinyl records can be attenuated with a low-pass filter that allows the lower-frequency audio signal to pass through.

A simple RC filter suffers from the fact that the crossover region where $V_R \approx V_C$ is fairly broad. More sophisticated filters have a sharper transition from off ($V_{out} \approx 0$) to on ($V_{out} \approx \mathcal{E}_0$), but they're based on the same principles as the RC filter analyzed here.

EXAMPLE 35.4 Designing a filter

For a science project, you've built a radio to listen to AM radio broadcasts at frequencies near 1 MHz. The basic circuit is an antenna, which produces a very small oscillating voltage when it absorbs the energy of an electromagnetic wave, and an amplifier. Unfortunately, your neighbor's short-wave radio broadcast at 10 MHz interferes with your reception. Having just finished physics, you decide to solve this problem by placing a filter between the antenna and the amplifier. You happen to have a 500 pF capacitor. What frequency should you select as the filter's crossover frequency? What value of resistance will you need to build this filter?

MODEL You want to block signals at 10 MHz while passing the lower-frequency AM signal at 1 MHz. Thus you need a low-pass filter.

VISUALIZE The circuit will look like the low-pass filter in Figure 35.13a. The oscillating voltage generated by the antenna will be the emf, and v_{out} will be sent to the amplifier.

SOLVE You might think that a crossover frequency near 5 MHz, about halfway between 1 MHz and 10 MHz, would work best. But 5 MHz is a factor of 5 higher than 1 MHz while only a factor of 2 less than 10 MHz. A crossover frequency that is the same factor above 1 MHz as it is below 10 MHz will give the best results. In practice, choosing $f_c = 3$ MHz would be sufficient. You can then use Equation 35.15 to select the proper resistor value:

$$R = \frac{1}{\omega_c C} = \frac{1}{2\pi(3 \times 10^6 \text{ Hz})(500 \times 10^{-12} \text{ F})}$$

$$= 106 \ \Omega \approx 100 \ \Omega$$

ASSESS Rounding to 100 Ω is appropriate because the crossover frequency was determined only to one significant figure. Such "sloppy design" is quite adequate when the two frequencies you need to distinguish are well separated.

STOP TO THINK 35.4 Rank in order, from largest to smallest, the crossover frequencies $(\omega_c)_a$ to $(\omega_c)_d$ of these four circuits.

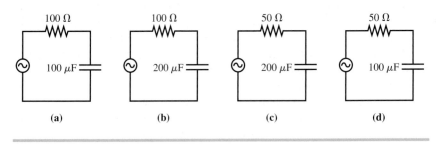

(a)   (b)   (c)   (d)

35.4 Inductor Circuits

Figure 35.14a on the next page shows the instantaneous current i_L through an inductor. If the current is changing, the instantaneous inductor voltage is

$$v_L = L\frac{di_L}{dt} \tag{35.16}$$

You learned in Chapter 33 that the potential decreases in the direction of the current if current is increasing ($di_L/dt > 0$) and increases if current is decreasing ($di_L/dt < 0$).

(a) The instantaneous current through the inductor

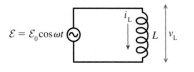

i_L

L

$+$ $-$

The instantaneous inductor voltage is $v_L = L(di_L/dt)$.

(b)

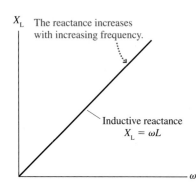

$\mathcal{E} = \mathcal{E}_0 \cos\omega t$ i_L L v_L

FIGURE 35.14 Using an inductor in an AC circuit.

Figure 35.14b, where inductance L is connected across an AC source of emf $\mathcal{E}$, is the simplest inductor circuit. The inductor is in parallel with the source, so the inductor voltage equals the emf: $v_L = \mathcal{E} = \mathcal{E}_0 \cos\omega t$. We can write

$$v_L = V_L \cos\omega t \qquad (35.17)$$

where V_L is the peak or maximum voltage across the inductor. You can see that $V_L = \mathcal{E}_0$ in this single-inductor circuit.

We can find the inductor current i_L by integrating Equation 35.17. First, use Equation 35.17 to write Equation 35.16 as

$$di_L = \frac{v_L}{L}dt = \frac{V_L}{L}\cos\omega t\, dt \qquad (35.18)$$

Integrating gives

$$i_L = \frac{V_L}{L}\int \cos\omega t\, dt = \frac{V_L}{\omega L}\sin\omega t = \frac{V_L}{\omega L}\cos\left(\omega t - \frac{\pi}{2}\right)$$
$$= I_L\cos\left(\omega t - \frac{\pi}{2}\right) \qquad (35.19)$$

where $I_L = V_L/\omega L$ is the peak or maximum inductor current.

> **NOTE** ▶ Mathematically, Equation 35.19 could have an integration constant i_0. An integration constant would represent a constant DC current through the inductor, but there is no DC source of potential in an AC circuit. Hence, on physical grounds, we set $i_0 = 0$ for an AC circuit. ◀

Define the **inductive reactance**, analogous to the capacitive reactance, to be

$$X_L \equiv \omega L \qquad (35.20)$$

Then the peak current $I_L = V_L/\omega L$ and the peak voltage are related by

$$I_L = \frac{V_L}{X_L} \quad \text{or} \quad V_L = I_L X_L \qquad (35.21)$$

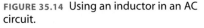

X_L The reactance increases with increasing frequency.

Inductive reactance $X_L = \omega L$

ω

FIGURE 35.15 The inductive reactance as a function of frequency.

Figure 35.15 shows that the inductive reactance increases as the frequency increases. This makes sense. Faraday's law tells us that the induced voltage across a coil increases as the time rate of change of $\vec{B}$ increases, and $\vec{B}$ is directly proportional to the inductor current. For a given peak current I_L, $\vec{B}$ changes more rapidly at higher frequencies than at lower frequencies, and thus V_L is larger at higher frequencies than at lower frequencies.

Figure 35.16a is a graph of the inductor voltage and current. You can see that the current peaks one-quarter of a period *after* the voltage peaks. The angle of the current phasor on the phasor diagram of Figure 35.16b is $\pi/2$ rad less than the angle of the voltage phasor. We can summarize this finding by saying

The AC current through an inductor *lags* the inductor voltage by $\pi/2$ rad, or 90°.

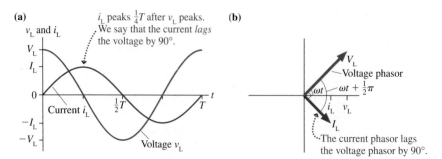

(a) v_L and i_L

i_L peaks $\frac{1}{4}T$ after v_L peaks. We say that the current *lags* the voltage by 90°.

V_L

I_L

0 $\frac{1}{2}T$ T t

Current i_L

$-I_L$

$-V_L$

Voltage v_L

(b)

V_L Voltage phasor

ωt $\omega t + \frac{1}{2}\pi$

i_L v_L

I_L

The current phasor lags the voltage phasor by 90°.

FIGURE 35.16 Graphs and phasor diagrams of the inductor current and voltage.

EXAMPLE 35.5 Current and voltage of an inductor

A 25 μH inductor is used in a circuit that oscillates at 100 kHz . The current through the inductor reaches a peak value of 20 mA at $t = 5.0\ \mu$s. What is the peak inductor voltage, and when, closest to $t = 5.0\ \mu$s, does it occur?

MODEL The inductor current lags the voltage by 90°, or, equivalently, the voltage reaches its peak value one-quarter period *before* the current.

VISUALIZE The circuit looks like Figure 35.14b.

SOLVE The inductive reactance at $f = 100$ kHz is

$$X_C = \omega L = 2\pi(1.0 \times 10^5\ \text{Hz})(25 \times 10^{-6}\ \text{H}) = 16\ \Omega$$

Thus the peak voltage is $V_L = I_L X_L = (20\ \text{mA})(16\ \Omega) = 320$ mV. The voltage peak occurs one-quarter period before the current peaks, and we know that the current peaks at $t = 5.0\ \mu$s. The period of a 100 kHz oscillation is 10.0 μs, so the voltage peaks at

$$t = 5.0\ \mu\text{s} - \frac{10.0\ \mu\text{s}}{4} = 2.5\ \mu\text{s}$$

35.5 The Series *RLC* Circuit

The circuit of Figure 35.17, where a resistor, inductor, and capacitor are in series, is called a **series *RLC* circuit.** The series *RLC* circuit has many important applications because, as you will see, it exhibits resonance behavior.

The analysis, which is very similar to our analysis of the *RC* circuit in Section 35.3, will be based on a phasor diagram. Notice that the three circuit elements are in series with each other and, together, are in parallel with the emf. We can draw two conclusions that form the basis of our analysis:

1. The instantaneous current of all three elements is the same: $i = i_R = i_L = i_C$.
2. The sum of the instantaneous voltages matches the emf: $\mathcal{E} = v_R + v_L + v_C$.

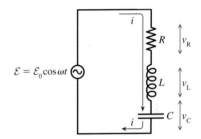

FIGURE 35.17 A series *RLC* circuit.

Analyzing an *RLC* circuit

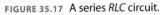

Begin by drawing a current phasor of length I. This is the starting point because the series circuit elements have the same current i.

The current and voltage of a resistor are in phase, so draw a resistor voltage phasor parallel to the current phasor I. The capacitor current leads the capacitor voltage by 90°, so draw a capacitor voltage phasor that is 90° behind the current phasor. Finally, the inductor current *lags* the voltage by 90°, so draw an inductor voltage phasor 90° ahead of the current phasor.

The instantaneous voltages satisfy $\mathcal{E} = v_R + v_L + v_C$. In terms of phasors, this is a *vector* addition. We can do the addition in two steps. Because the capacitor and inductor phasors are in opposite directions, their vector sum has length $V_L - V_C$. Adding the resistor phasor, at right angles, then gives the emf phasor $\mathcal{E}$ at angle ωt.

The length $\mathcal{E}_0$ of the emf phasor is the hypotenuse of a right triangle. Thus $\mathcal{E}_0^2 = V_R^2 + (V_L - V_C)^2$.

14.2, 14.3   Activ
Physics

If $V_L > V_C$, which we've assumed, then the instantaneous current i lags the emf by a phase angle ϕ. We can write the current, in terms of ϕ, as

$$i = I\cos(\omega t - \phi) \tag{35.22}$$

Of course, there's no guarantee that V_L will be larger than V_C. If the opposite is true, $V_L < V_C$, the emf phasor is on the other side of the current phasor. Our analysis is still valid if we consider ϕ to be negative when i is ccw from $\mathcal{E}$. Thus ϕ can be anywhere between $-90°$ and $+90°$.

Now we can continue much as we did with the *RC* circuit. Based on the right triangle, $\mathcal{E}_0^2$ is

$$\mathcal{E}_0^2 = V_R^2 + (V_L - V_C)^2 = [R^2 + (X_L - X_C)^2]I^2 \tag{35.23}$$

where we wrote each of the peak voltages in terms of the peak current I and a resistance or a reactance. Consequently, the peak current in the *RLC* circuit is

$$I = \frac{\mathcal{E}_0}{\sqrt{R^2 + (X_L - X_C)^2}} = \frac{\mathcal{E}_0}{\sqrt{R^2 + (\omega L - 1/\omega C)^2}} \tag{35.24}$$

The three peak voltages, if you need them, are then found from $V_R = IR$, $V_L = IX_L$ and $V_C = IX_C$.

Impedance

The denominator of Equation 35.24 is called the **impedance** Z of the circuit:

$$Z = \sqrt{R^2 + (X_L - X_C)^2} \tag{35.25}$$

Impedance, like resistance and reactance, is measured in ohms. The circuit's peak current can be written in terms of the source emf and the circuit impedance as

$$I = \frac{\mathcal{E}_0}{Z} \tag{35.26}$$

Equation 35.26 is a compact way to write I, but it doesn't add anything new to Equation 35.24.

Phase Angle

It is often useful to know the phase angle ϕ between the emf and the current. You can see from Figure 35.18 that

$$\tan\phi = \frac{V_L - V_C}{V_R} = \frac{(X_L - X_C)I}{RI}$$

The current I cancels, and we're left with

$$\phi = \tan^{-1}\left(\frac{X_L - X_C}{R}\right) \tag{35.27}$$

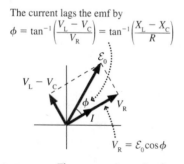

The current lags the emf by
$$\phi = \tan^{-1}\left(\frac{V_L - V_C}{V_R}\right) = \tan^{-1}\left(\frac{X_L - X_C}{R}\right)$$

$V_R = \mathcal{E}_0\cos\phi$

FIGURE 35.18 The current is not in phase with the emf.

We can check that Equation 35.27 agrees with our analyses of single-element circuits. A resistor-only circuit has $X_L = X_C = 0$ and thus $\phi = \tan^{-1}(0) = 0$ rad. In other words, as we discovered previously, the emf and current are in phase. An AC inductor circuit has $R = X_C = 0$ and thus $\phi = \tan^{-1}(\infty) = \pi/2$ rad, agreeing with our earlier finding that the inductor current lags the voltage by 90°. A homework problem will let you check that Equation 35.27 gives the correct result for an AC capacitor circuit.

Other relationships can be found from the phasor diagram and written in terms of the phase angle. For example, it is frequently useful to write the peak resistor voltage as

$$V_R = \mathcal{E}_0 \cos\phi \tag{35.28}$$

Notice that the resistor voltage oscillates in phase with the emf only if $\phi = 0$ rad.

Resonance

Suppose we vary the emf frequency ω while keeping everything else constant. There is very little current at very low frequencies because the capacitive reactance $X_C = 1/\omega C$ is very large. Similarly, there is very little current at very high frequencies because the inductive reactance $X_L = \omega L$ becomes very large.

If I approaches zero at very low and very high frequencies, there should be some intermediate frequency where I is a maximum. Indeed, you can see from Equation 35.24 that the denominator will be a minimum, making I a maximum, when $X_L = X_C$, or

$$\omega L = \frac{1}{\omega C} \tag{35.29}$$

The frequency ω_0 that satisfies Equation 35.39 is called the **resonance frequency:**

$$\omega_0 = \frac{1}{\sqrt{LC}} \tag{35.30}$$

This is the frequency for *maximum current* in the series *RLC* circuit. The maximum current

$$I_{max} = \frac{\mathcal{E}_0}{R} \tag{35.31}$$

is that of a purely resistive circuit because the impedance is $Z = R$ at resonance.

You'll recognize ω_0 as the oscillation frequency of the *LC* circuit that we analyzed in Chapter 33. The current in an ideal *LC* circuit oscillates forever as energy is transferred back and forth between the capacitor and the inductor. This is analogous to an ideal, frictionless simple harmonic oscillator in which the energy is transformed back and forth between kinetic and potential.

Adding a resistor to the circuit is like adding damping to a mechanical oscillator. The emf is then a sinusoidal driving force, and the series *RLC* circuit is directly analogous to the driven, damped oscillator that you studied in Chapter 14. A mechanical oscillator exhibits *resonance* by having a large-amplitude response when the driving frequency matches the system's natural frequency. Equation 35.30 is the natural frequency of the series *RLC* circuit, the frequency at which the current would like to oscillate. Consequently, the circuit has a large current response when the oscillating emf matches this frequency.

Figure 35.19 shows the peak current I of a series *RLC* circuit as the emf frequency ω is varied. Notice how the current increases until reaching a maximum at frequency ω_0, then decreases. This is the hallmark of a resonance.

As R decreases, causing the damping to decrease, the maximum current becomes larger and the curve in Figure 35.19 becomes narrower. You saw exactly the same behavior for a driven mechanical oscillator. The emf frequency must be very close to ω_0 in order for a lightly damped system to respond, but the response at resonance is very large.

For a different perspective, Figure 35.20 on the next page graphs the instantaneous emf $\mathcal{E} = \mathcal{E}_0 \cos\omega t$ and current $i = I\cos(\omega t - \phi)$ for frequencies below, at, and above ω_0. The current and the emf are in phase at resonance ($\phi = 0$ rad)

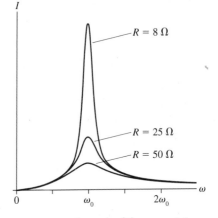

FIGURE 35.19 A graph of the current I versus emf frequency for a series *RLC* circuit.

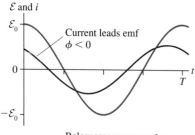

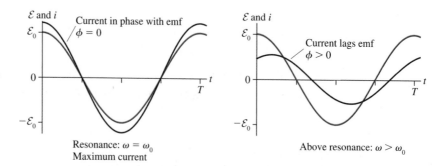

Below resonance: $\omega < \omega_0$

Resonance: $\omega = \omega_0$
Maximum current

Above resonance: $\omega > \omega_0$

FIGURE 35.20 Graphs of the emf $\mathcal{E}$ and the current I at frequencies below, at, and above the resonance frequency ω_0.

because the capacitor and inductor essentially cancel each other to give a purely resistive circuit. Away from resonance, the current decreases *and* begins to get out of phase with the emf. You can see, from Equation 35.27, that the phase angle ϕ is negative when $X_L < X_C$ (i.e., the frequency is below resonance) and positive when $X_L > X_C$ (the frequency is above resonance).

Resonance circuits are widely used in radio, television, and communication equipment because of their ability to respond to one particular frequency (or very narrow range of frequencies) while suppressing others. The selectivity of a resonance circuit improves as the resistance decreases, but the inherent resistance of the wires and the inductor coil keep R from being 0 Ω.

EXAMPLE 35.6 Designing a radio receiver

An AM radio antenna picks up a 1000 kHz signal with a peak voltage of 5.0 mV. The tuning circuit consists of a 60 μH inductor in series with a variable capacitor. The inductor coil has a resistance of 0.25 Ω, and the resistance of the rest of the circuit is negligible.

a. To what value should the capacitor be tuned to listen to this radio station?
b. What is the peak current through the circuit at resonance?
c. A stronger station at 1050 kHz produces a 10 mV antenna signal. What is the current at this frequency when the radio is tuned to 1000 MHz?

MODEL The inductor's 0.25 Ω resistance can be modeled as a resistance in series with the inductance, hence we have a series *RLC* circuit. The antenna signal at $\omega = 2\pi \times 1000$ kHz is the emf.

VISUALIZE The circuit looks like Figure 35.17.

SOLVE

a. The capacitor needs to be tuned to where it and the inductor are resonant at $\omega_0 = 2\pi \times 1000$ kHz. The appropriate value is

$$C = \frac{1}{L\omega_0^2} = \frac{1}{(60 \times 10^{-6}\,\text{H})(6.28 \times 10^6\,\text{rad/s})^2}$$
$$= 4.23 \times 10^{-10}\,\text{F} = 423\,\text{pF}$$

b. $X_L = X_C$ at resonance, so the peak current is

$$I = \frac{\mathcal{E}_0}{R} = \frac{5.0 \times 10^{-3}\,\text{V}}{0.25\,\Omega} = 0.020\,\text{A} = 20\,\text{mA}$$

c. The 1050 kHz signal is "off resonance," so we need to compute $X_L = \omega L = 396\,\Omega$ and $X_C = 1/\omega C = 358\,\Omega$ at $\omega = 2\pi \times 1050$ kHz. The peak voltage of this signal is $\mathcal{E}_0 = 10$ mV. With these values, Equation 35.24 for the peak current is

$$I = \frac{\mathcal{E}_0}{\sqrt{R^2 + (X_L - X_C)^2}} = 0.26\,\text{mA}$$

ASSESS These are realistic values for the input stage of an AM radio. You can see that the signal from the 1050 kHz station is strongly suppressed when the radio is tuned to 1000 kHz.

STOP TO THINK 35.5 A series *RLC* circuit has $V_C = 5.0$ V, $V_R = 7.0$ V, and $V_L = 9.0$ V. Is the frequency above, below, or equal to the resonance frequency?

35.6 Power in AC Circuits

A primary role of the emf is to supply energy. Some circuit devices, such as motors and lightbulbs, use the energy to perform useful tasks. Other circuit devices dissipate the energy as an increased thermal energy in the components and the surrounding air. Chapter 31 examined the topic of power in DC circuits. Now we can perform a similar analysis for AC circuits.

The emf supplies energy to a circuit at the rate

$$p_{\text{source}} = i\mathcal{E} \tag{35.32}$$

where i and $\mathcal{E}$ are the instantaneous current from and potential difference across the emf. We've used a lowercase p to indicate that this is the instantaneous power. We need to look at the power losses in individual circuit elements.

Resistors

You learned in Chapter 31 that the current through a resistor causes the resistor to dissipate energy at the rate

$$p_R = i_R v_R = i_R^2 R \tag{35.33}$$

We can use $i_R = I_R \cos \omega t$ to write the resistor's instantaneous power loss as

$$p_R = i_R^2 R = I_R^2 R \cos^2 \omega t \tag{35.34}$$

Figure 35.21 shows the instantaneous power graphically. You can see that, because the cosine is squared, the power oscillates twice during every cycle of the emf. The energy dissipation peaks both when $i_R = I_R$ and when $i_R = -I_R$. This shows that energy dissipation doesn't depend on the current's direction through the resistor, a result that is hardly surprising.

In practice, we're usually more interested in the *average power* than in the instantaneous power. The **average power** P is the total energy dissipated per second. We can find P_R for a resistor by using the trigonometric identity $\cos^2(x) = \frac{1}{2}(1 + \cos 2x)$ to write

$$P_R = I_R^2 R \cos^2 \omega t = I_R^2 R \left[\frac{1}{2}(1 + \cos 2\omega t) \right] = \frac{1}{2} I_R^2 R + \frac{1}{2} I_R^2 R \cos 2\omega t$$

The $\cos 2\omega t$ term oscillates positive and negative twice during each cycle of the emf. Its average, over one cycle of the emf, is therefore zero. Thus the average power loss in a resistor is

$$P_R = \frac{1}{2} I_R^2 R \qquad \text{(average power loss in a resistor)} \tag{35.35}$$

It is useful to write Equation 35.35 as

$$P_R = \left(\frac{I_R}{\sqrt{2}} \right)^2 R = (I_{\text{rms}})^2 R \tag{35.36}$$

where the quantity

$$I_{\text{rms}} = \frac{I_R}{\sqrt{2}} \tag{35.37}$$

is called the **root-mean-square current,** or rms current, I_{rms}. Technically, an rms quantity is the square root of the average, or mean, of the quantity squared. For quantities that oscillate sinusoidally, the rms value turns out to be the peak value divided by $\sqrt{2}$.

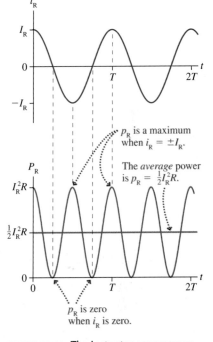

FIGURE 35.21 The instantaneous power loss in a resistor.

The rms current allows us to compare Equation 35.36 directly to the energy dissipated by a resistor in a DC circuit: $P = I^2R$. You can see that the average power loss of a resistor in an AC circuit with $I_{\text{rms}} = 1$ A is the same as in a DC circuit with $I = 1$ A. **As far as power is concerned, an rms current is equivalent to an equal DC current.**

Similarly, we can define the root-mean-square voltage and emf:

$$V_{\text{rms}} = \frac{V_R}{\sqrt{2}} \qquad \mathcal{E}_{\text{rms}} = \frac{\mathcal{E}_0}{\sqrt{2}} \tag{35.38}$$

The resistor's average power loss in terms of the rms quantities is

$$P_R = (I_{\text{rms}})^2 R = \frac{(V_{\text{rms}})^2}{R} = I_{\text{rms}} V_{\text{rms}} \tag{35.39}$$

and the average power supplied by the emf is

$$P_{\text{source}} = I_{\text{rms}} \mathcal{E}_{\text{rms}} \tag{35.40}$$

The single-resistor circuit that we analyzed in Section 35.1 had $V_R = \mathcal{E}$ or, equivalently, $V_{\text{rms}} = \mathcal{E}_{\text{rms}}$. You can see from Equations 35.39 and 35.40 that the power loss in the resistor exactly matches the power supplied by the emf. This must be the case in order to conserve energy.

NOTE ▶ Voltmeters, ammeters, and other AC electronic measuring instruments are almost always calibrated to give the rms value. An AC voltmeter would show that the "line voltage" of an electrical outlet in the United States is 120 V. This is $\mathcal{E}_{\text{rms}}$. The peak voltage $\mathcal{E}_0$ is larger by a factor of $\sqrt{2}$, or $\mathcal{E}_0 = 170$ V. The power-line voltage is sometimes specified as "120 V/60 Hz," showing the rms voltage and the frequency. ◀

EXAMPLE 35.7 Lighting a bulb

A 100 W incandescent lightbulb is plugged into a 120 V/60 Hz outlet. What is the resistance of the bulb's filament? What is the peak current through the bulb?

MODEL The filament in a lightbulb acts as a resistor.

VISUALIZE Figure 35.22 is a diagram of a simple one-resistor circuit.

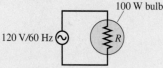

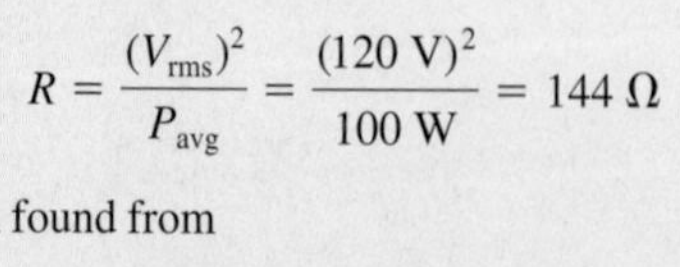

100 W bulb

120 V/60 Hz

R

FIGURE 35.22 An AC circuit with a lightbulb as a resistor.

SOLVE A bulb labeled 100 W is designed to dissipate an average 100 W at $V_{\text{rms}} = 120$ V. We can use Equation 35.39 to find

$$R = \frac{(V_{\text{rms}})^2}{P_{\text{avg}}} = \frac{(120 \text{ V})^2}{100 \text{ W}} = 144 \ \Omega$$

The rms current is then found from

$$I_{\text{rms}} = \frac{P_{\text{avg}}}{V_{\text{rms}}} = \frac{100 \text{ W}}{120 \text{ V}} = 0.833 \text{ A}$$

The peak current is $I_R = \sqrt{2} I_{\text{rms}} = 1.18$ A.

ASSESS Calculations with rms values are just like the calculations you did for DC circuits.

Capacitors and Inductors

In Section 35.2, we found that the instantaneous current to a capacitor is $i_C = -\omega C V_C \sin \omega t$. Thus the instantaneous energy dissipation in a capacitor is

$$p_C = v_C i_C = (V_C \cos \omega t)(-\omega C V_C \sin \omega t) = -\frac{1}{2}\omega C V_C^2 \sin 2\omega t \quad (35.41)$$

where we used $\sin(2x) = 2\sin(x)\cos(x)$.

Figure 35.23 shows Equation 35.41 graphically. Energy is transferred into the capacitor (positive power) as it is charged, but, instead of being dissipated, as it would be by a resistor, the energy is stored as potential energy in the capacitor's electric field. Then, as the capacitor discharges, this energy is given back to the circuit. Power is the rate at which energy is *removed* from the circuit, hence p is negative as the capacitor transfers energy back into the circuit.

Returning to our mechanical analogy, a capacitor is like an ideal, frictionless simple harmonic oscillator. Kinetic and potential energy are constantly being exchanged, but there is no dissipation because none of the energy is transformed into thermal energy. The important conclusion is that **a capacitor's average power loss is zero: $P_C = 0$.**

The same is true of an inductor. An inductor alternately stores energy in the magnetic field, as the current is increasing, then transfers energy back to the circuit as the current decreases. The instantaneous power oscillates between positive and negative, but **an inductor's average power loss is zero: $P_L = 0$.**

> **NOTE** ▶ We're assuming ideal capacitors and inductors. Real capacitors and inductors inevitably have a small amount of resistance and dissipate a small amount of energy. However, their energy dissipation is negligible compared to that of the resistors in most practical circuits. ◀

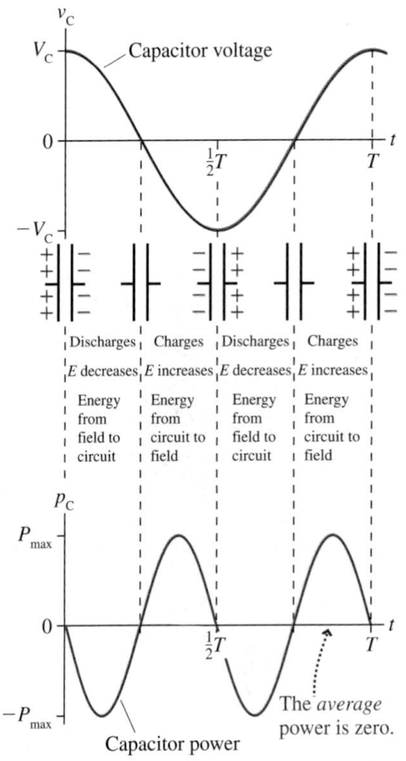

FIGURE 35.23 Energy flows into and out of a capacitor as it is charged and discharged.

The Power Factor

In an *RLC* circuit, energy is supplied by the emf and dissipated by the resistor. But an *RLC* circuit is unlike a purely resistive circuit in that the current is not in phase with the potential difference of the emf.

We found in Equation 35.22 that the instantaneous current in an *RLC* circuit is $i = I\cos(\omega t - \phi)$, where ϕ is the angle by which the current lags the emf. Thus the instantaneous power supplied by the emf is

$$p_{\text{source}} = i\mathcal{E} = (I\cos(\omega t - \phi))(\mathcal{E}_0 \cos \omega t) = I\mathcal{E}_0 \cos \omega t \cos(\omega t - \phi) \quad (35.42)$$

We can use the expression $\cos(x - y) = \cos(x)\cos(y) + \sin(x)\sin(y)$ to write the power as

$$p_{\text{source}} = (I\mathcal{E}_0 \cos \phi)\cos^2 \omega t + (I\mathcal{E}_0 \sin \phi)\sin \omega t \cos \omega t \quad (35.43)$$

In our analysis of the power loss in a resistor and a capacitor, we found that the average of $\cos^2 \omega t$ is $\frac{1}{2}$ and the average of $\sin \omega t \cos \omega t$ is zero. Thus we can immediately write that the *average* power supplied by the emf is

$$P_{\text{source}} = \frac{1}{2}I\mathcal{E}_0 \cos \phi = I_{\text{rms}}\mathcal{E}_{\text{rms}} \cos \phi \quad (35.44)$$

The rms values, you will recall, are $I/\sqrt{2}$ and $\mathcal{E}_0/\sqrt{2}$.

The term $\cos \phi$, which is called the **power factor,** arises because the current and the emf in a series *RLC* circuit are not in phase. Because the current and the emf aren't pushing and pulling together, the source delivers less energy to the circuit.

We'll leave it as a homework problem for you to show that the peak current in an RLC circuit can be written

$$I = I_{max} \cos\phi \tag{35.45}$$

where $I_{max} = \mathcal{E}_0/R$ was given in Equation 35.31. In other words, the current term in Equation 35.44 is a function of the power factor. Consequently, the average power is

$$P_{source} = P_{max} \cos^2\phi \tag{35.46}$$

where $P_{max} = \frac{1}{2}I_{max}\mathcal{E}_0$ is the *maximum* power the source can deliver to the circuit.

The source delivers maximum power only when $\cos\phi = 1$. This is the case when $X_L - X_C = 0$, requiring either a purely resistive circuit or an RLC circuit operating at the resonance frequency ω_0. The average power loss is zero for a purely capacitive or purely inductive load with, respectively, $\phi = -90°$ or $\phi = +90°$, which is in agreement with our analysis above.

Industrial motors use a significant fraction of the electric energy generated in the United States.

Motors of various types, especially large industrial motors, use a significant fraction of the electric energy generated in industrialized nations. Motors operate most efficiently, doing the maximum work per second, when the power factor is as close to 1 as possible. But motors are inductive devices, due to their electromagnet coils, and if too many motors are attached to the electric grid, the power factor is pulled away from 1. To compensate, the electric company places large capacitors throughout the transmission system. The capacitors dissipate no energy, but they allow the electric system to deliver energy more efficiently by keeping the power factor close to 1.

Finally, we found in Equation 35.28 that the resistor's peak voltage in an RLC circuit is related to the emf peak voltage by $V_R = \mathcal{E}_0\cos\phi$ or, dividing both sides by $\sqrt{2}$, $V_{rms} = \mathcal{E}_{rms}\cos\phi$. We can use this result to write the energy loss in the resistor as

$$P_R = I_{rms}V_{rms} = I_{rms}\mathcal{E}_{rms}\cos\phi \tag{35.47}$$

But this expression is P_{source}, as we found in Equation 35.44. Thus we see that the energy supplied to an RLC circuit by the emf is ultimately dissipated by the resistor.

EXAMPLE 35.8 The power used by a motor

A motor attached to a 120 V/60 Hz power line uses 600 W of power at a power factor of 0.80.

a. What is the rms current to the motor?
b. What is the motor's resistance?

MODEL The motor has inductance. Otherwise, the power factor would be 1. Treat it as a series RLC circuit without the capacitor ($X_C = 0$).

SOLVE

a. The average power is $P_{source} = I_{rms}\mathcal{E}_{rms}\cos\phi$. Thus

$$I_{rms} = \frac{P_{avg}}{\mathcal{E}_{rms}\cos\phi} = \frac{600 \text{ W}}{(120 \text{ V})(0.80)} = 6.25 \text{ A}$$

b. We can use $V_{rms} = \mathcal{E}_{rms}\cos\phi$ and Ohm's law to find that the resistance is

$$R = \frac{V_{rms}}{I_{rms}} = \frac{\mathcal{E}_{rms}\cos\phi}{I_{rms}} = \frac{(120 \text{ V})(0.80)}{6.25 \text{ A}} = 15.4 \text{ }\Omega$$

STOP TO THINK 35.6   The emf and the current in a series *RLC* circuit oscillate as shown. Which of the following (perhaps more than one) would increase the rate at which energy is supplied to the circuit?

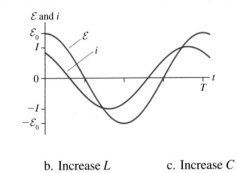

a. Increase $\mathcal{E}_0$

b. Increase L

c. Increase C

d. Decrease $\mathcal{E}_0$

e. Decrease L

f. Decrease C

SUMMARY

The goal of Chapter 35 has been to understand and apply basic techniques of AC circuit analysis.

IMPORTANT CONCEPTS

AC circuits are driven by an emf

$$\mathcal{E} = \mathcal{E}_0 \cos \omega t$$

that oscillates with angular frequency $\omega = 2\pi f$.

Phasors can be used to represent the oscillating emf, current, and voltage.

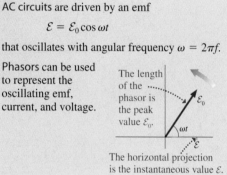

The length of the phasor is the peak value $\mathcal{E}_0$.

The horizontal projection is the instantaneous value $\mathcal{E}$.

Basic circuit elements

Element	i and v	Resistance/ reactance	I and V	Power
Resistor	In phase	R is fixed	$V = IR$	$V_{rms}I_{rms}$
Capacitor	i leads v by 90°	$X_C = 1/\omega C$	$V = IX_L$	0
Inductor	i lags v by 90°	$X_L = \omega L$	$V = IX_C$	0

For many purposes, especially calculating power, the **root-mean-square** (rms) quantities

$$V_{rms} = V/\sqrt{2} \qquad I_{rms} = I/\sqrt{2} \qquad \mathcal{E}_{rms} = \mathcal{E}/\sqrt{2}$$

are equivalent to the corresponding DC quantities.

KEY SKILLS

Phasor diagrams

- Start with a phasor (v or i) common to two or more circuit elements.

- The sum of instantaneous quantities is vector addition.

- Use the Pythagorean theorem to relate peak quantities.

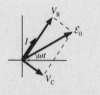

For an RC circuit, shown here,

$$v_R + v_C = \mathcal{E}$$

$$V_R^2 + V_C^2 = \mathcal{E}_0^2$$

Kirchhoff's laws

Loop law The sum of the potential differences around a loop is zero.

Junction law The sum of currents entering a junction equals the sum leaving the junction.

Instantaneous and peak quantities

Instantaneous quantities v and i generally obey different relationships than peak quantities V and I.

APPLICATIONS

RC filter circuits

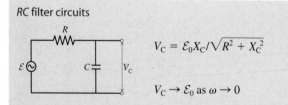

$$V_C = \mathcal{E}_0 X_C/\sqrt{R^2 + X_C^2}$$

$$V_C \rightarrow \mathcal{E}_0 \text{ as } \omega \rightarrow 0$$

A **low-pass filter** transmits low frequencies and blocks high frequencies.

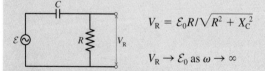

$$V_R = \mathcal{E}_0 R/\sqrt{R^2 + X_C^2}$$

$$V_R \rightarrow \mathcal{E}_0 \text{ as } \omega \rightarrow \infty$$

A **high-pass filter** transmits high frequencies and blocks low frequencies.

Series *RLC* circuits

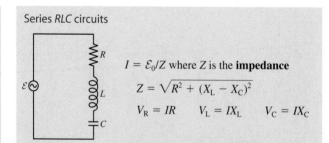

$I = \mathcal{E}_0/Z$ where Z is the **impedance**

$$Z = \sqrt{R^2 + (X_L - X_C)^2}$$

$$V_R = IR \qquad V_L = IX_L \qquad V_C = IX_C$$

When $\omega = \omega_0 = 1/\sqrt{LC}$ (the **resonance frequency**), the current in the circuit is a maximum $I_{max} = \mathcal{E}_0/R$.

In general, the current i lags behind $\mathcal{E}$ by the **phase angle** $\phi = \tan^{-1}((X_L - X_C)/R)$.

The power supplied by the emf is $P_{source} = I_{rms}\mathcal{E}_{rms} \cos\phi$, where $\cos\phi$ is called the **power factor.**

The power lost in a resistor is $P_R = I_{rms}V_{rms} = (I_{rms})^2 R$.

TERMS AND NOTATION

AC circuit	low-pass filter	resonance frequency, ω_0
DC circuit	high-pass filter	average power, P
phasor	inductive reactance, X_L	root-mean-square current, I_{rms}
capacitive reactance, X_C	series RLC circuit	power factor, $\cos\phi$
crossover frequency, ω_c	impedance, Z	

EXERCISES AND PROBLEMS

Exercises

Section 35.1 AC Sources and Phasors

1. The emf phasor in Figure Ex35.1 is shown at $t = 15$ ms.
   a. What is the angular frequency ω? Assume this is the first rotation.
   b. What is the instantaneous value of the emf?

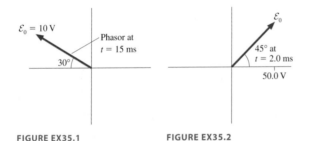

FIGURE EX35.1   FIGURE EX35.2

2. The emf phasor in Figure Ex35.2 is shown at $t = 2.0$ ms.
   a. What is the angular frequency ω? Assume this is the first rotation.
   b. What is the peak value of the emf?
3. A 111 Hz source of emf has a peak voltage of 50 V. Draw the emf phasor at $t = 3.0$ ms.
4. Draw the phasor for the emf $\mathcal{E} = (170\text{ V})\cos((2\pi \times 60\text{ Hz})t)$ at $t = 60$ ms.
5. A 200 Ω resistor is connected to an AC source with $\mathcal{E}_0 = 10$ V. What is the peak current through the resistor if the emf frequency is (a) 100 Hz? (b) 100 kHz?
6. Figure Ex35.6 shows voltage and current graphs for a resistor.
   a. What is the value of the resistance R?
   b. What is the emf frequency f?
   c. Draw the resistor's voltage and current phasors at $t = 15$ ms.

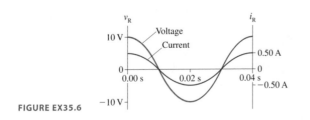

FIGURE EX35.6

Section 35.2 Capacitor Circuits

7. A 0.30 μF capacitor is connected across an AC generator that produces a peak voltage of 10.0 V. What is the peak current through the capacitor if the emf frequency is (a) 100 Hz? (b) 100 kHz?
8. The peak current through a capacitor is 10.0 mA. What is the current if
   a. The emf frequency is doubled?
   b. The emf peak voltage is doubled (at the original frequency)?
   c. The frequency is halved and, at the same time, the emf is doubled?
9. A 20 nF capacitor is connected across an AC generator that produces a peak voltage of 5.0 V.
   a. At what frequency f is the peak current 50 mA?
   b. What is the instantaneous value of the emf at the instant when $i_C = I_C$?
10. A capacitor is connected to a 15 kHz oscillator. The peak current is 65 mA when the rms voltage is 6.0 V. What is the value of the capacitance C?
11. The peak current through a capacitor is 8.0 mA when connected to an AC source with a peak voltage of 1.0 V. What is the capacitive reactance of the capacitor?

Section 35.3 RC Filter Circuits

12. A high-pass RC filter is connected to an AC source with a peak voltage of 10.0 V. The peak capacitor voltage is 6.0 V. What is the peak resistor voltage?
13. What are V_R and V_C if the emf frequency in Figure Ex35.13 is 10 kHz?

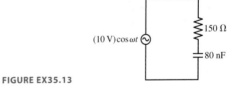

FIGURE EX35.13

14. A high-pass RC filter with a crossover frequency of 1000 Hz uses a 100 Ω resistor. What is the value of the capacitor?
15. A low-pass RC filter with a crossover frequency of 1000 Hz uses a 100 Ω resistor. What is the value of the capacitor?
16. A low-pass RC filter has a crossover frequency of 6280 rad/s. What is the crossover frequency if the value of the capacitor is doubled?

Section 35.4 Inductor Circuits

17. A 20 mH inductor is connected across an AC generator that produces a peak voltage of 10.0 V. What is the peak current through the inductor if the emf frequency is (a) 100 Hz? (b) 100 kHz?

18. The peak current through an inductor is 10.0 mA. What is the current if
    a. The emf frequency is doubled?
    b. The emf peak voltage is doubled (at the original frequency)?
    c. The frequency is halved and, at the same time, the emf is doubled?

19. A 500 μH inductor is connected across an AC generator that produces a peak voltage of 5.0 V.
    a. At what frequency f is the peak current 50 mA?
    b. What is the instantaneous value of the emf at the instant when $i_L = I_L$?

20. An inductor is connected to a 15 kHz oscillator. The peak current is 65 mA when the rms voltage is 6.0 V. What is the value of the inductance L?

21. The peak current through an inductor is 12.5 mA when connected to an AC source with a peak voltage of 1.0 V. What is the inductive reactance of the inductor?

Section 35.5 The Series *RLC* Circuit

22. At what frequency f do a 1.0 μF capacitor and a 1.0 μH inductor have the same reactance? What is the value of the reactance at this frequency?

23. What capacitor in series with a 100 Ω resistor and a 20 mH inductor will give a resonance frequency of 1000 Hz?

24. What inductor in series with a 100 Ω resistor and a 2.5 μF capacitor will give a resonance frequency of 1000 Hz?

25. A series *RLC* circuit has a 200 kHz resonance frequency. What is the resonance frequency if
    a. The resistor value is doubled?
    b. The capacitor value is doubled?

26. A series *RLC* circuit has a 200 kHz resonance frequency. What is the resonance frequency if
    a. The resistor value is doubled?
    b. The capacitor value is doubled and, at the same time, the inductor value is halved?

27. What is the phase angle of the current, in degrees, when the emf frequency in Figure Ex35.27 is (a) 14 kHz? (b) 18 kHz?

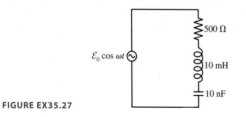

FIGURE EX35.27

Section 35.6 Power in AC Circuits

28. A resistor dissipates 2.0 W when the rms voltage of the emf is 10.0 V. At what rms voltage will the resistor dissipate 10.0 W?

29. The heating element of a hair drier dissipates 1500 W when connected to a 120 V/60 Hz power line. What is its resistance?

30. A 100 Ω resistor is connected to a 120 V/60 Hz power line. What is its average power loss?

31. A series *RLC* circuit attached to a 120 V/60 Hz power line draws 2.4 A of current with a power factor of 0.87. What is the value of the resistor?

32. A series *RLC* circuit with a 100 Ω resistor dissipates 80 W when attached to a 120 V/60 Hz power line. What is the power factor?

33. The motor of an electric drill draws a 3.5 A current at the power-line voltage of 120 V rms. What is the motor's power if the current lags the voltage by 20°?

Problems

34. a. For an *RC* circuit, find an expression for the angular frequency ω_{cap} at which $V_C = \frac{1}{2}\mathcal{E}_0$.
    b. What is V_R at this frequency?
    c. What is ω_{cap} if the crossover frequency is 6280 rad/s?

35. a. For an *RC* circuit, find an expression for the angular frequency ω_{res} at which $V_R = \frac{1}{2}\mathcal{E}_0$.
    b. What is V_C at this frequency?
    c. What is ω_{res} if the crossover frequency is 6280 rad/s?

36. a. Evaluate V_R in Figure P35.36 at emf frequencies 100, 300, 1000, 3000, and 10,000 Hz.
    b. Graph V_R versus frequency. Draw a smooth curve through your five points.

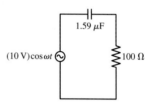

FIGURE P35.36

37. a. Evaluate V_C in Figure P35.37 at emf frequencies 1, 3, 10, 30, and 100 kHz.
    b. Graph V_C versus frequency. Draw a smooth curve through your five points.

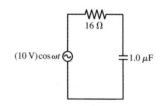

FIGURE P35.37

38. When two capacitors are connected in parallel across a 10.0 V rms, 1.0 kHz oscillator, the oscillator supplies a total rms current of 545 mA. When the same two capacitors are connected to the oscillator in series, the oscillator supplies an rms current of 126 mA. What are the values of the two capacitors?

39. For an *RC* filter circuit, show that $V_R = V_C = \mathcal{E}_0/\sqrt{2}$ at $\omega = \omega_c$.

40. Show that Equation 35.27 for the phase angle ϕ of a series *RLC* circuit gives the correct result for a capacitor-only circuit.

41. A low-pass filter consists of a 100 μF capacitor in series with a 159 Ω resistor. The circuit is driven by an AC source with a peak voltage of 5.0 V.
    a. What is the crossover frequency f_c?
    b. What is V_C when $f = \frac{1}{2}f_c$, f_c, and $2f_c$?

42. A high-pass filter consists of a 1.59 μF capacitor in series with a 100 Ω resistor. The circuit is driven by an AC source with a peak voltage of 5.0 V.
    a. What is the crossover frequency f_c?
    b. What is V_R when $f = \frac{1}{2}f_c$, f_c, and $2f_c$?
43. a. What is the peak current supplied by the emf in Figure P35.43?
    b. What is the peak voltage across the 3 μF capacitor?

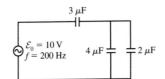

FIGURE P35.43

44. a. What is the average power supplied by the emf in Figure P35.44?
    b. What is the energy dissipated by each resistor?

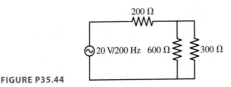

FIGURE P35.44

45. You have a resistor and a capacitor of unknown values. First, you charge the capacitor and discharge it through the resistor. By monitoring the capacitor voltage on an oscilloscope, you see that the voltage decays to half its initial value in 2.5 ms. You then use the resistor and capacitor to make a low-pass filter. What is the crossover frequency f_c?
46. Figure P35.46 shows a parallel RC circuit.
    a. Use a phasor-diagram analysis to find expressions for the peak currents I_R and I_C.
    Hint: What do the resistor and capacitor have in common? Use that as the initial phasor.
    b. Complete the phasor analysis by finding an expression for the peak emf current I.

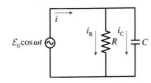

FIGURE P35.46

47. Use a phasor diagram to analyze the RL circuit of Figure P35.47. In particular,
    a. Find expressions for I, V_R, and V_L.
    b. What is V_R in the limits $\omega \to 0$ and $\omega \to \infty$?
    c. If the output is taken from the resistor, is this a low-pass or a high-pass filter? Explain.
    d. Find an expression for the crossover frequency ω_c.

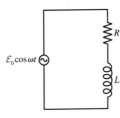

FIGURE P35.47

48. A series RLC circuit consists of a 100 Ω resistor, a 0.10 H inductor, and a 100 μF capacitor. It is attached to a 120 V/60 Hz power line. What are (a) the peak current I, (b) the phase angle ϕ, and (c) the average power loss?
49. A series RLC circuit consists of a 100 Ω resistor, a 0.15 H inductor, and a 30 μF capacitor. It is attached to a 120 V/60 Hz power line. What are (a) the peak current I, (b) the phase angle ϕ, and (c) the average power loss?
50. For the circuit of Figure P35.50,
    a. What is the resonance frequency, in both rad/s and Hz?
    b. Find V_R and V_L.
    c. How can V_L be larger than $\mathcal{E}_0$? Explain.

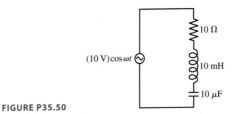

FIGURE P35.50

51. For the circuit of Figure P35.51,
    a. What is the resonance frequency, in both rad/s and Hz?
    b. Find V_R and V_C at resonance.
    c. How can V_C be larger than $\mathcal{E}_0$? Explain.

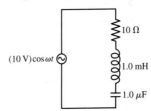

FIGURE P35.51

52. In Figure P35.52, what is the current supplied by the emf when (a) the frequency is very small and (b) the frequency is very large?

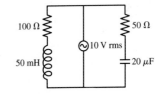

FIGURE P35.52

53. A series RLC circuit consists of a 50 Ω resistor, a 3.3 mH inductor, and a 480 nF capacitor. It is connected to an oscillator with a peak voltage of 5.0 V. Determine the impedance, the peak current, and the phase angle at frequencies (a) 3000 Hz, (b) 4000 Hz, and (c) 5000 Hz.
54. A series RLC circuit consists of a 50 Ω resistor, a 3.3 mH inductor, and a 480 nF capacitor. It is connected to a 5.0 kHz oscillator with a peak voltage of 5.0 V. What is the instantaneous current i when
    a. $\mathcal{E} = \mathcal{E}_0$?
    b. $\mathcal{E} = 0$ V and is decreasing?
55. A series RLC circuit consists of a 50 Ω resistor, a 3.3 mH inductor, and a 480 nF capacitor. It is connected to a 3.0 kHz oscillator with a peak voltage of 5.0 V. What is the instantaneous emf $\mathcal{E}$ when
    a. $i = I$?
    b. $i = 0$ A and is decreasing?
    c. $i = -I$?

56. A series *RLC* circuit consists of a 100 Ω resistor, a 10 mH inductor, and a 1.0 nF capacitor. It is connected to an oscillator with an rms voltage of 10 V. What is the power supplied to the circuit if (a) $\omega = \frac{1}{2}\omega_0$? (b) $\omega = \omega_0$? (c) $\omega = 2\omega_0$?

57. Show that the impedance of a series *RLC* circuit can be written

$$Z = \sqrt{R^2 + \omega^2 L^2 (1 - \omega_0^2/\omega^2)^2}$$

58. For a series *RLC* circuit, show that
    a. The peak current can be written $I = I_{max} \cos\phi$.
    b. The average power dissipation can be written $P_{avg} = P_{max} \cos^2\phi$.

59. The tuning circuit in an FM radio receiver is a series *RLC* circuit with a 0.200 μH inductor.
    a. The receiver is tuned to a station at 104.3 MHz. What is the value of the capacitor in the tuning circuit?
    b. FM radio stations are assigned frequencies every 0.2 MHz, but two nearby stations cannot use adjacent frequencies. What is the maximum resistance the tuning circuit can have if the peak current at a frequency of 103.9 MHz, the closest frequency that can be used by a nearby station, is to be no more than 0.10% of the peak current at 104.3 MHz? The radio is still tuned to 104.3 MHz, and you can assume the two stations have equal strength.

60. A television channel is assigned the frequency range from 54 MHz to 60 MHz. A series *RLC* tuning circuit in a TV receiver resonates in the middle of this frequency range. The circuit uses a 16 pF capacitor.
    a. What is the value of the inductor?
    b. In order to function properly, the current throughout the frequency range must be at least 50% of the current at the resonance frequency. What is the minimum possible value of the circuit's resistance?

61. Lightbulbs labeled 40 W, 60 W, and 100 W are connected to a 120 V/60 Hz power line as shown in Figure P35.61. What is the rate at which energy is dissipated in each bulb?

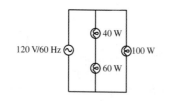

FIGURE P35.61

62. Commercial electricity is generated and transmitted as *three-phase electricity*. Instead of a single emf, three separate wires carry currents for the emfs $\mathcal{E}_1 = \mathcal{E}_0\cos\omega t$, $\mathcal{E}_2 = \mathcal{E}_0\cos(\omega t + 120°)$, and $\mathcal{E}_3 = \mathcal{E}_0\cos(\omega t - 120°)$ over three parallel wires, each of which supplies one-third of the power. This is why the long-distance transmission lines you see in the countryside have three wires. Suppose the transmission lines into a city supply a total of 450 MW of electric power, a realistic value.
    a. What would be the current in each wire if the transmission voltage were $\mathcal{E}_0 = 120$ V rms?
    b. In fact, transformers are used to step the transmission-line voltage up to 500 kV rms. What is the current in each wire?
    c. Big transformers are expensive. Why does the electric company use step-up transformers?

63. A motor attached to a 120 V/60 Hz power line draws an 8.0 A current. Its average energy dissipation is 800 W.
    a. What is the power factor?
    b. What is the rms resistor voltage?
    c. What is the motor's resistance?
    d. How much series capacitance needs to be added to increase the power factor to 1.0?

Challenge Problems

64. Commercial electricity is generated and transmitted as *three-phase electricity*. Instead of a single emf $\mathcal{E} = \mathcal{E}_0\cos\omega t$, three separate wires carry currents for the emfs $\mathcal{E}_1 = \mathcal{E}_0\cos\omega t$, $\mathcal{E}_2 = \mathcal{E}_0\cos(\omega t + 120°)$, and $\mathcal{E}_3 = \mathcal{E}_0\cos(\omega t - 120°)$. This is why the long-distance transmission lines you see in the countryside have three parallel wires, as do many distribution lines within a city.
    a. Draw a phasor diagram showing phasors for all three phases of a three-phase emf.
    b. Show that the sum of the three phases is zero, producing what is referred to as *neutral*. In *single-phase* electricity, provided by the familiar 120 V/60 Hz electric outlets in your home, one side of the outlet is neutral, as established at a nearby electrical substation. The other, called the *hot side*, is one of the three phases. (The round opening is connected to ground.)
    c. Show that the potential difference between any two of the phases has the rms value $\sqrt{3}\,\mathcal{E}_{rms}$, where $\mathcal{E}_{rms}$ is the familiar single-phase rms voltage. Evaluate this potential difference for $\mathcal{E}_{rms} = 120$ V. Some high-power home appliances, especially electric clothes dryers and hot-water heaters, are designed to operate between two of the phases rather than between one phase and neutral. Heavy-duty industrial motors are designed to operate from all three phases, but full three-phase power is rare in residential or office use.

65. The small transformers that power many consumer products produce a 12.0 V rms, 60 Hz emf. Design a circuit using resistors and capacitors that uses the transformer voltage as an input and produces a 6.0 V rms output that leads the input voltage by 45°.

66. You're the operator of a 15,000 V rms, 60 Hz electrical substation. When you get to work one day, you see that the station is delivering 6.0 MW of power with a power factor of 0.90.
    a. What is the rms current leaving the station?
    b. How much series capacitance should you add to bring the power factor up to 1.0?
    c. How much power will the station then be delivering?

67. a. Show that the peak inductor voltage in a series *RLC* circuit is maximum at frequency

$$\omega_L = \left(\frac{1}{\omega_0^2} - \frac{1}{2}R^2C^2\right)^{-1/2}$$

    b. A series *RLC* circuit with $\mathcal{E}_0 = 10.0$ V consists of a 1.0 Ω resistor, a 1.0 μH inductor, and a 1.0 μF capacitor. What is V_L at $\omega = \omega_0$ and at $\omega = \omega_L$?

68. a. Show that the average power loss in a series *RLC* circuit is

$$P_{\text{avg}} = \frac{\omega^2 \mathcal{E}_{\text{rms}}^2 R}{\omega^2 R^2 + L^2(\omega^2 - \omega_0^2)^2}$$

  b. Prove that the energy dissipation is a maximum at $\omega = \omega_0$.

69. Consider the parallel *RLC* circuit shown in Figure CP35.69.

  a. Show that the current drawn from the emf is

$$I = \mathcal{E}_0 \sqrt{\frac{1}{R^2} + \left(\frac{1}{\omega L} - \omega C\right)^2}$$

  Hint: Start with a phasor that is common to all three circuit elements.

  b. What is *I* in the limits $\omega \rightarrow 0$ and $\omega \rightarrow \infty$?

  c. Find the frequency for which *I* is a minimum.

  d. Sketch a graph of *I* versus ω.

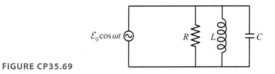

FIGURE CP35.69

70. The telecommunication circuit shown in Figure CP35.70 has a parallel inductor and capacitor in series with a resistor.

  a. Use a phasor diagram to show that the peak current through the resistor is

$$I = \frac{\mathcal{E}_0}{\sqrt{R^2 + \left(\dfrac{1}{X_L} - \dfrac{1}{X_C}\right)^{-2}}}$$

  Hint: Start with the inductor phasor v_L. Which voltages are shared? What does the junction law tell you about the currents? How do currents and voltages add on a phasor diagram?

  b. What is *I* in the limits $\omega \rightarrow 0$ and $\omega \rightarrow \infty$?

  c. What is the resonance frequency ω_0? What is *I* at this frequency?

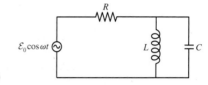

FIGURE CP35.70

STOP TO THINK ANSWERS

Stop to Think 35.1: a. The instantaneous emf value is the projection down onto the horizontal axis. The emf is negative but increasing in magnitude as the phasor, which rotates ccw, approaches the horizontal axis.

Stop to Think 35.2: c. Voltage and current are measured using different scales and units. You can't compare the length of a voltage phasor to the length of a current phasor.

Stop to Think 35.3: a. There is "no capacitor" when the separation between the two capacitor plates becomes zero and the plates touch. Capacitance *C* is inversely proportional to the plate spacing *d*, hence $C \rightarrow \infty$ as $d \rightarrow 0$. The capacitive reactance is inversely proportional to *C*, so $X_C \rightarrow 0$ as $C \rightarrow \infty$.

Stop to Think 35.4: $(\omega_c)_d > (\omega_c)_c = (\omega_c)_a > (\omega_c)_b$. The crossover frequency is $1/RC$.

Stop to Think 35.5: Above. $V_L > V_C$ tells us that $X_L > X_C$. This is the condition above resonance, where X_L is increasing with ω while X_C is decreasing.

Stop to Think 35.6: a, b, and **f.** You can always increase power by turning up the voltage. The current leads the emf, telling us that the circuit is primarily capacitive. The current can be brought into phase with the emf, thus maximizing the power, by decreasing *C* or increasing *L*.

Electricity and Magnetism

Mass and charge are the two most fundamental properties of matter. The first five parts of this text were investigations of the properties and interactions of masses. Part VI has been a study of the physics of charge—what charge is and how charges interact.

Electric and magnetic fields were introduced to enable us to understand the long-range forces of electricity and magnetism. The field concept is subtle, but it is an essential part of our modern understanding of the physical universe. One charge—the source charge—alters the space around it by creating an electric field and, if the charge is moving, a magnetic field. Other charges experience forces exerted *by*

the fields. Thus the electric and magnetic fields are the agents by which charges interact.

Faraday's discovery of electromagnetic induction led scientists to recognize that the fields are *real* and can exist independently of charges. The most vivid confirmation of this reality was Maxwell's discovery of electromagnetic waves—the quintessential electromagnetic phenomenon.

Part VI has introduced many new phenomena, concepts and laws. The knowledge structure table draws together the major ideas about charges and fields, and it briefly summarizes some of the most important applications of electricity and magnetism.

KNOWLEDGE STRUCTURE VI **Electricity and Magnetism**

ESSENTIAL CONCEPTS Charge, dipole, field, potential, emf
BASIC GOALS How do charged particles interact?
What are the properties and characteristics of electromagnetic fields?

GENERAL PRINCIPLES

Coulomb's law
$$\vec{E}_{\text{point charge}} = \frac{1}{4\pi\epsilon_0}\frac{q}{r^2}\hat{r} = \left(\frac{1}{4\pi\epsilon_0}\frac{q}{r^2}, \text{away from } q\right)$$

Biot-Savart law
$$\vec{B}_{\text{point charge}} = \frac{\mu_0}{4\pi}\frac{q\vec{v}\times\hat{r}}{r^2} = \left(\frac{\mu_0}{4\pi}\frac{qv\sin\theta}{r^2}, \text{direction of right-hand rule}\right)$$

Faraday's law
$\mathcal{E} = |d\Phi_m/dt|$ $I_{\text{induced}} = \mathcal{E}/R$ in the direction of Lenz's law

Lenz's law
An induced current flows around a conducting loop in the direction such that the induced magnetic field opposes the *change* in the magnetic flux.

Lorentz force law
$\vec{F}_{\text{on } q} = q(\vec{E} + \vec{v}\times\vec{B})$

Superposition
The electric or magnetic field due to multiple charges is the vector sum of the field of each charge. This principle was used to derive the fields of many special charge distributions, such as wires, planes, and loops.

FIELD AND POTENTIAL The electric field of charges can also be described in terms of an electric potential *V*.

$$V_{\text{point charge}} = \frac{q}{4\pi\epsilon_0 r}$$

- The electric field is perpendicular to equipotential surfaces and in the direction of decreasing potential.

- The potential energy of charge *q* is $U = qV$. The total energy $K + U$ of a group of charges is conserved.

ELECTROMAGNETIC WAVES All the properties of electromagnetic fields are summarized mathematically in four equations called *Maxwell's equations*. From Maxwell's equations we learn that electromagnetic fields can exist independently of charges as an *electromagnetic wave*.

- An em wave travels at speed $c = 1/\sqrt{\epsilon_0\mu_0}$.

- $\vec{E}$ and $\vec{B}$ are perpendicular to each other and to the direction of travel, with $E = cB$.

Electric and magnetic properties of materials

- Charges move through conductors but not through insulators.

- Conductors and insulators are *polarized* in an electric field.

- A magnetic moment in a magnetic field experiences a torque.

Model of current and conductivity

- The charge carriers in metals are electrons.

- Emf → electric field → current density $J = \sigma E \rightarrow I = JA$.

Applications to circuits

- Circuits obey Kirchhoff's loop law (conservation of energy) and junction law (conservation of current).

- Resistors control the current: $I = \Delta V/R$ (Ohm's law).

- Capacitors store charge $Q = C\Delta V$ and energy $V_C = \frac{1}{2}C(\Delta V_C)^2$.

The Telecommunications Revolution

In 1800, the year that Volta invented the battery and that Thomas Jefferson was elected president, the fastest a message could travel was the speed of a man or woman on horseback. News took three days to travel from New York to Boston, and well over a month to reach the frontier outpost of Cincinnati.

But Oersted's 1820 discovery that a current creates a magnetic field soon introduced revolutionary changes to communications. The American scientist Joseph Henry, who shares with Faraday the credit for the discovery of electromagnetic induction, saw a simple electromagnet in 1825. Inspired, he set about improving the device. By 1830, Henry was able to send current through more than a mile of wire to activate an electromagnet and strike a bell.

In 1835, Henry met an entrepreneur interested in the commercial development of electric technology—Samuel F. B. Morse. Morse was one of the most prominent American artists of the early 19th century, but he also had an abiding interest in technology. In the 1830s he invented the famous code that bears his name—Morse code—and began to experiment with electromagnets.

With advice and encouragement from Henry, Morse developed the first practical telegraph. The first telegraph line, between Washington, D.C., and Baltimore, began operating in 1844; the first message sent was, "What hath God wrought?" For the first time, long-distance communications could take place essentially instantaneously.

Telegraph communication advanced as quickly as wire could be strung, and a worldwide network had been established by 1875. But the telegraph didn't hold its monopoly for long, as other inventors began to think about using electromagnetic devices to transmit speech. The first to succeed was Alexander Graham Bell, who invented the telephone in 1876.

The telegraph and telephone provided electromagnetic communication over wires, but the discovery of electromagnetic waves opened up another possibility—wireless communication at the speed of light. Radio technology developed rapidly in the late 19th century, and in 1901 the Italian inventor Guglielmo Marconi sent and received the first transatlantic radio message. World War I prompted further development of radio, because of the need to communicate with military units as they moved about, and by 1925 more than 1000 radio stations were operating in the United States.

Radio and, later, television spanned the globe by 1960, but radio stations reached a few hundred miles at best, and television transmission was pretty much limited to each city. National broadcasts within the United States required the signal to be transmitted via microwave relays to local stations for rebroadcast. This system made possible network television shows, but not live-from-the-scene broadcasts. News journalists had to film events, then return the film to the studio for broadcast. Television images from overseas could only be seen the next day, after film was flown back to the United States.

The first communications satellite was launched by NASA in 1960, followed two years later by a more practical satellite, Telstar, that used solar power to amplify signals received from earth and beam them back down. The first live transatlantic television transmission was made on July 11, 1962, and was broadcast throughout the United States.

Plans were made for a system of roughly 100 satellites, so that one would always be overhead, but another idea soon proved more practical. In 1945, 12 years before space flight began, the science-fiction writer Arthur C. Clarke proposed placing satellites in orbits 22,300 miles above the earth. A satellite at this altitude orbits with a 24-hour period, so from the ground it appears to hang stationary in space. We now call this a *geosynchronous orbit*. One such satellite would allow microwave communication between two points one-third of a world apart, so just three geosynchronous satellites would span the entire earth.

Much more energy is required to reach geosynchronous orbit than to reach low-earth orbit, but rocket technology was advancing faster than NASA could build Telstar satellites. The first commercial communications satellite was placed in geosynchronous orbit in 1965, and, for the first time, television images could be broadcast live to anywhere in the world. Today all of the world's intercontinental television and much of the intercontinental telephone traffic travels via microwaves to and from a cluster of these artificial stars floating high above the earth.

As we enter the 21st century, information and images span the world as quickly as or more quickly than they once moved through a small village. You can pick up the phone and talk to friends or relatives anywhere around the globe, and each evening's news brings live images from remote places. Telecommunication unites our world, and the technologies of telecommunications are direct descendants of Coulomb, Ampère, Oersted, Henry, and—most of all—Michael Faraday.

Relativity

Contemporary Physics

Our journey into physics is nearing its end. We began roughly 350 years ago with Newton's discovery of the laws of motion. The conclusion of Part VI brought us to the end of the 19th century, just over 100 years ago. Along the way you've learned about the motion of particles, the conservation of energy, the physics of waves, and the electromagnetic interactions that hold atoms together and generate light waves. We can begin the last phase of our journey with confidence.

Newton's mechanics and Maxwell's electromagnetism were the twin pillars of science at the end of the 19th century and the basis for much of engineering and applied science in the 20th century. Despite the successes of these theories, a series of discoveries starting around 1900 and continuing into the first few decades of the 20th century profoundly altered our understanding of the universe at the most fundamental level. These discoveries forced scientists to reconsider the very nature of space and time and to develop new models of light and matter.

The discoveries and new ideas of the early 20th century led to two new theories: relativity and quantum physics. These two theories form the basis for physics as it is practiced today and are already having a significant impact on 21st-century engineering. We will end our journey into physics with a look at these contemporary topics and some of their applications.

Relativity

The idea of measuring distance with a meter stick and time with a clock or stopwatch has been with us since Chapter 1. The basic notions of space and time seem so self-evident that no one had seriously questioned them. No one, that is, until an unknown young scientist named Albert Einstein began to ponder these issues in the years right around 1900.

It wasn't space and time that first troubled Einstein. Instead, he was bothered by what he saw as paradoxes and difficulties in Maxwell's theory of electromagnetism. Einstein was able to show that electromagnetism is a self-consistent theory only if the speed of electromagnetic waves—the speed of light—is the same in all inertial reference frames, no matter how the reference frames might be moving with respect to each other or to the source of the wave. But this strange behavior of a traveling wave can be true only if *space and time are different* for two experimenters moving relative to each other.

We'll need to explore what it means for space and time to be different for different experimenters. In doing so we'll discover some of the well-known puzzles of relativity, such as length contraction, the twin paradox, and a cosmic speed limit for particles. Our exploration of these fascinating ideas will end with what is perhaps the most famous equation in physics: Einstein's $E = mc^2$.

1149

36 Relativity

These are the fundamental tools with which we learn about space and time.

▶ **Looking Ahead**
The goal of Chapter 36 is to understand how Einstein's theory of relativity changes our concepts of space and time. In this chapter you will learn to:

- Use the principle of relativity.
- Understand how time dilation and length contraction change our concepts of space and time.
- Use the Lorentz transformations of positions and velocities.
- Calculate relativistic momentum and energy.
- Understand how mass and energy are equivalent.

◀ **Looking Back**
The material in this chapter depends on an understanding of relative motion in Newtonian mechanics. Please review:

- Section 6.4 Inertial reference frames and the Galilean transformations.

Space and time seem like straightforward ideas. You can measure lengths with a ruler or meter stick. You can time events with a stop watch. Nothing could be simpler.

So it seemed to everyone until 1905, when an unknown young scientist had the nerve to suggest that this simple view of space and time was in conflict with other principles of physics. In the century since, Einstein's theory of relativity has radically altered our understanding of some of the most fundamental ideas in physics.

Relativity, despite its esoteric reputation, has very real implications for modern technology. Global positioning system (GPS) satellites depend on relativity, as do the navigation systems used by airliners. Nuclear reactors make tangible use of Einstein's famous equation $E = mc^2$ to generate 20% of the electricity used in the United States. The annihilation of matter in positron-emission tomography (PET scanners) has given neuroscientists an unprecedented ability to monitor activity within the brain.

The theory of relativity is fascinating, perplexing, and challenging. It is also vital to our contemporary understanding of the universe in which we live.

Albert Einstein (1879–1955) was one of the most influential thinkers in history.

36.1 Relativity: What's It All About?

What do you think of when you hear the phrase "theory of relativity"? A white-haired Einstein? $E = mc^2$? Black holes? Time travel? Perhaps you've heard that the theory of relativity is so complicated and abstract that only a handful of people in the whole world really understand it.

There is, without doubt, a certain mystique associated with relativity, an aura of the strange and exotic. The good news is that understanding the ideas of relativity is well within your grasp. Einstein's *special theory of relativity,* the portion of relativity we'll study, is not mathematically difficult at all. The challenge is conceptual because relativity questions deeply held assumptions about the nature of space and time. In fact, that's what relativity is all about—space and time.

In one sense, relativity is not a new idea at all. Certain ideas about relativity are part of Newtonian mechanics. You had an introduction to these ideas in Chapter 6, where you learned about reference frames and the Galilean transformations. Einstein, however, thought that relativity should apply to *all* the laws of physics, not just mechanics. The difficulty, as you'll see, is that some aspects of relativity appear to be incompatible with the laws of electromagnetism, particularly the laws governing the propagation of light waves.

Lesser scientists might have concluded that relativity simply doesn't apply to electromagnetism. Einstein's genius was to see that the incompatibility arises from *assumptions* about space and time, assumptions no one had ever questioned because they seem so obviously true. Rather than abandon the ideas of relativity, Einstein changed our understanding of space and time.

Fortunately, you need not be a genius to follow a path that someone else has blazed. However, we will have to exercise the utmost care with regard to logic and precision. We will need to state very precisely just how it is that we know things about the physical world, then ruthlessly follow the logical consequences. The challenge is to stay on this path, not to let our prior assumptions—assumptions that are deeply ingrained in all of us—lead us astray.

What's Special About Special Relativity?

Einstein's first paper on relativity, in 1905, dealt exclusively with inertial reference frames, reference frames that move relative to each other with constant velocity. Ten years later, Einstein published a more encompassing theory of relativity that considers accelerated motion and its connection to gravity. The second theory, because it's more general in scope, is called *general relativity.* General relativity is the theory that describes black holes, curved spacetime, and the evolution of the universe. It is a fascinating theory but, alas, very mathematical and outside the scope of this textbook. If you're interested, many popular science books provide a nontechnical introduction to general relativity.

Motion at constant velocity is a "special case" of motion; namely, motion for which the acceleration is zero. Hence Einstein's first theory of relativity has come to be known as **special relativity.** It is special in the sense of being a restricted, special case of his more general theory, not special in the everyday sense of meaning distinctive or exceptional. Special relativity, with its conclusions about time dilation and length contraction, is what we will study.

36.2 Galilean Relativity

A firm grasp of Galilean relativity is necessary if we are to appreciate and understand what is new in Einstein's theory. Thus we begin with the ideas of relativity that are embodied in Newtonian mechanics.

Reference Frames

Suppose you're passing me as we both drive in the same direction along a freeway. My car's speedometer reads 55 mph while your speedometer shows 60 mph. Is 60 mph your "true" speed? That is certainly your speed relative to someone standing beside the road, but your speed relative to me is only 5 mph. Your speed is 120 mph relative to a driver approaching from the other direction at 60 mph.

An object does not have a "true" speed or velocity. The very definition of velocity, $v = \Delta x/\Delta t$, assumes the existence of a coordinate system in which, during some time interval Δt, the displacement Δx is measured. The best we can manage is to specify an object's velocity relative to, or with respect to, the coordinate system in which it is measured.

Let's define a **reference frame** to be a coordinate system in which experimenters equipped with meter sticks, stopwatches, and any other needed equipment make position and time measurements on moving objects. Three ideas are implicit in our definition of a reference frame:

- A reference frame extends infinitely far in all directions.
- The experimenters are at rest in the reference frame.
- The number of experimenters and the quality of their equipment are sufficient to measure positions and velocities to any level of accuracy needed.

The first two bullets are especially important. It is often convenient to say "the laboratory reference frame" or "the reference frame of the rocket." These are shorthand expressions for "a reference frame, infinite in all directions, in which the laboratory (or the rocket) and a set of experimenters happen to be at rest."

NOTE ▶ A reference frame is not the same thing as a "point of view." That is, each person or each experimenter does not have his or her own private reference frame. **All experimenters at rest relative to each other share the same reference frame.** ◀

Figure 36.1 shows two reference frames called S and S′. The coordinate axes in S are x, y, z and those in S′ are x', y', z'. Reference frame S′ moves with velocity v relative to S or, equivalently, S moves with velocity $-v$ relative to S′. There's no implication that either reference frame is "at rest." Notice that the zero of time, when experimenters start their stopwatches, is the instant that the origins of S and S′ coincide.

We will restrict our attention to *inertial reference frames,* implying that the relative velocity v is constant. You should recall from Chapter 6 that an **inertial reference frame** is a reference frame in which Newton's first law, the law of inertia, is valid. In particular, an inertial reference frame is one in which an isolated particle, one on which there are no forces, either remains at rest or moves in a straight line at constant speed.

Any reference frame that moves at constant velocity with respect to an inertial reference frame is itself an inertial reference frame. Conversely, a reference frame that accelerates with respect to an inertial reference frame is *not* an inertial reference frame. Our restriction to reference frames moving with respect to each other at constant velocity—with no acceleration—is the "special" part of special relativity.

NOTE ▶ An inertial reference frame is an idealization. A true inertial reference would need to be floating in deep space, far from any gravitational influence. In practice, an earthbound laboratory is a good approximation of an inertial reference frame because the accelerations associated with the earth's rotation and motion around the sun are too small to influence most experiments. ◀

1. The axes of S and S′ have the same orientation.

2. Frame S′ moves with velocity v relative to frame S. The relative motion is entirely along the x- and x′-axes.

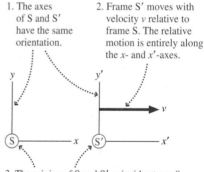

3. The origins of S and S′ coincide at $t = 0$. This is our definition of $t = 0$.

FIGURE 36.1 The standard reference frames S and S′.

Which of these is an inertial reference frame (or a very good approximation)?

a. Your bedroom
b. A car rolling down a steep hill
c. A train coasting along a level track
d. A rocket being launched
e. A roller coaster going over the top of a hill
f. A sky diver falling at terminal speed

The Galilean Transformations

Suppose a firecracker explodes at time t. The experimenters in reference frame S determine that the explosion happened at position x. Similarly, the experimenters in S′ find that the firecracker exploded at x' in their reference frame. What is the relationship between x and x'?

Figure 36.2 shows the explosion and the two reference frames. You can see from the figure that $x = x' + vt$, thus

$$
\begin{array}{lll}
x = x' + vt & & x' = x - vt \\
y = y' & \text{or} & y' = y \\
z = z' & & z' = z
\end{array} \tag{36.1}
$$

These equations, which you met in Chapter 6, are the *Galilean transformations of position*. If you know a position measured by the experimenters in one inertial reference frame, you can calculate the position that would be measured by experimenters in any other inertial reference frame.

Suppose the experimenters in both reference frames now track the motion of the object in Figure 36.3 by measuring its position at many instants of time. The experimenters in S find that the object's velocity is $\vec{u}$. During the *same time interval* Δt, the experimenters in S′ measure the velocity to be $\vec{u}'$.

NOTE ▶ In this chapter, we will use v to represent the velocity of one reference frame relative to another. We will use $\vec{u}$ and $\vec{u}'$ to represent the velocities of objects with respect to reference frames S and S′. This notation differs from the notation of Chapter 6, where we used V to represent the relative velocity. ◀

We can find the relationship between $\vec{u}$ and $\vec{u}'$ by taking the time derivatives of Equation 36.1 and using the definition $u_x = dx/dt$:

$$
u_x = \frac{dx}{dt} = \frac{dx'}{dt} + v = u'_x + v
$$

$$
u_y = \frac{dy}{dt} = \frac{dy'}{dt} = u'_y
$$

The equation for u_z is similar. The net result is

$$
\begin{array}{lll}
u_x = u'_x + v & & u'_x = u_x - v \\
u_y = u'_y & \text{or} & u'_y = u_y \\
u_z = u'_z & & u'_z = u_z
\end{array} \tag{36.2}
$$

Equations 36.2 are the *Galilean transformations of velocity*. If you know the velocity of a particle as measured by the experimenters in one inertial reference frame, you can use Equations 36.2 to find the velocity that would be measured by experimenters in any other inertial reference frame.

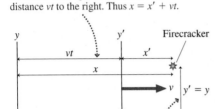

At time t, the origin of S′ has moved distance vt to the right. Thus $x = x' + vt$.

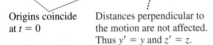

Origins coincide at $t = 0$

Distances perpendicular to the motion are not affected. Thus $y' = y$ and $z' = z$.

FIGURE 36.2 The position of an exploding firecracker is measured in reference frames S and S′.

The object's velocity in frame S is $\vec{u}$.

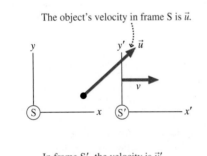

In frame S′, the velocity is $\vec{u}'$.

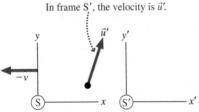

FIGURE 36.3 The velocity of a moving object is measured in reference frames S and S′.

EXAMPLE 36.1 **The speed of sound**

An airplane is flying at speed 200 m/s with respect to the ground. Sound wave 1 is approaching the plane from the front, sound wave 2 is catching up from behind. Both waves travel at 340 m/s relative to the ground. What is the speed of each wave relative to the plane?

MODEL Assume that the earth (frame S) and the airplane (frame S') are inertial reference frames. Frame S', in which the airplane is at rest, moves with velocity $v = 200$ m/s relative to frame S.

VISUALIZE Figure 36.4 shows the airplane and the sound waves.

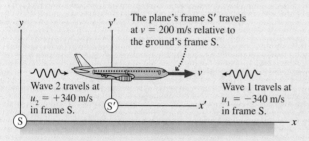

FIGURE 36.4 Experimenters in the plane measure different speeds for the sound waves than do experimenters on the ground.

SOLVE The speed of a mechanical wave, such as a sound wave or a wave on a string, is its speed *relative to its medium.* Thus the *speed of sound* is the speed of a sound wave through a reference frame in which the air is at rest. This is reference frame S, where wave 1 travels with velocity $u_1 = -340$ m/s and wave 2 travels with velocity $u_2 = +340$ m/s. Notice that the Galilean transformations use *velocities,* with appropriate signs, not just speeds.

The airplane travels to the right with reference frame S' at velocity v. We can use the Galilean transformations of velocity to find the velocities of the two sound waves in frame S':

$$u_1' = u_1 - v = -340 \text{ m/s} - 200 \text{ m/s} = -540 \text{ m/s}$$

$$u_2' = u_2 - v = 340 \text{ m/s} - 200 \text{ m/s} = 140 \text{ m/s}$$

ASSESS This isn't surprising. If you're driving 50 mph, a car coming the other way at 55 mph is approaching you at 105 mph. A car coming up behind you at 55 mph seems to be gaining on you at the rate of only 5 mph. Wave speeds behave the same. Notice that a mechanical wave would appear to be stationary to a person moving at the wave speed. To a surfer, the crest of the ocean wave remains at rest under his or her feet.

STOP TO THINK 36.2 Ocean waves are approaching the beach at 10 m/s. A boat heading out to sea travels at 6 m/s. How fast are the waves moving in the boat's reference frame?

a. 16 m/s   b. 10 m/s   c. 6 m/s   d. 4 m/s

The Galilean Principle of Relativity

Experimenters in reference frames S and S' measure different values for position and velocity. What about the force on and the acceleration of the particle in Figure 36.5? The strength of a force can be measured with a spring scale. The experimenters in reference frames S and S' both see the *same reading* on the scale (we'll assume the scale has a bright digital display easily seen by all experimenters), leading them to conclude that the force is the same in both frames. That is, $F' = F$.

We can compare the accelerations measured in the two reference frames by taking the time derivative of the velocity transformation equation $u' = u - v$. (We'll assume, for simplicity, that the velocities and accelerations are all in the x-direction.) The relative velocity v between the two reference frames is *constant,* thus

$$a' = \frac{du'}{dt} = \frac{du}{dt} = a \qquad (36.3)$$

Experimenters in reference frames S and S' measure different values for an object's position and velocity, but they *agree* on its acceleration.

If $F = ma$ in reference frame S, then $F' = ma'$ in reference frame S'. Stated another way, if Newton's second law is valid in one inertial reference frame, then it is valid in all inertial reference frames. Because other laws of mechanics, such

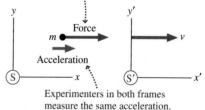

FIGURE 36.5 Experimenters in both reference frames test Newton's second law by measuring the force on a particle and its acceleration.

as the conservation laws, follow from Newton's laws of motion, we can state this conclusion as the *Galilean principle of relativity:*

> **Galilean principle of relativity** The laws of mechanics are the same in all inertial reference frames.

The Galilean principle of relativity is easy to state, but to understand it we must understand what is and is not "the same." To take a specific example, consider the law of conservation of momentum. Figure 36.6a shows two particles about to collide. Their total momentum in frame S, where particle 2 is at rest, is $P_i = 9$ kg m/s. This is an isolated system, hence the law of conservation of momentum tells us that the momentum after the collision will be $P_f = 9$ kg m/s.

Figure 36.6b has used the velocity transformation to look at the same particles in frame S′ in which particle 1 is at rest. The initial momentum in S′ is $P'_i = -18$ kg m/s. Thus it is not the *value* of the momentum that is the same in all inertial reference frames. Instead, the Galilean principle of relativity tells us that the *law* of momentum conservation is the same in all inertial reference frames. If $P_f = P_i$ in frame S, then it must be true that $P'_f = P'_i$ in frame S′. Consequently, we can conclude that P'_f will be -18 kg m/s after the collision in S′.

Using Galilean Relativity

The principle of relativity is concerned with the laws of mechanics, not with the values that are needed to satisfy the laws. If momentum is conserved in one inertial reference frame, it is conserved in all inertial reference frames. Even so, a problem may be easier to solve in one reference frame than in others.

Elastic collisions provide a good example of using reference frames. You learned in Chapter 10 how to calculate the outcome of a perfectly elastic collision between two particles in the reference frame in which particle 2 is initially at rest. We can use that information together with the Galilean transformations to solve elastic-collision problems in any inertial reference frame.

(a) Collision seen in frame S

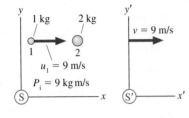

(b) Collision seen in frame S′

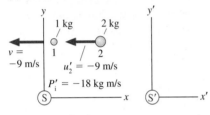

FIGURE 36.6 Total momentum measured in two reference frames.

> **TACTICS BOX 36.1 Analyzing elastic collisions**
>
> ❶ Use the Galilean transformations to transform the initial velocities of particles 1 and 2 from frame S to a reference frame S′ in which particle 2 is at rest.
>
> ❷ The outcome of the collision in S′ is given by
>
> $$u'_{1f} = \frac{m_1 - m_2}{m_1 + m_2} u'_{1i}$$
>
> $$u'_{2f} = \frac{2m_1}{m_1 + m_2} u'_{1i}$$
>
> ❸ Transform the two final velocities from frame S′ back to frame S.

EXAMPLE 36.2 An elastic collision
A 300 g ball moving to the right at 2 m/s has a perfectly elastic collision with a 100 g ball moving to the left at 4 m/s. What are the direction and speed of each ball after the collision?

MODEL The velocities are measured in the laboratory frame, which we call frame S.

VISUALIZE Figure 36.7a shows both the balls and reference frame S′ in which ball 2 is at rest.

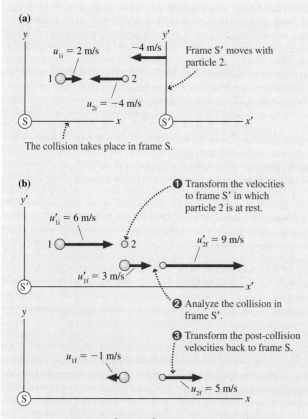

(a)

$u_{1i} = 2$ m/s

-4 m/s

Frame S′ moves with particle 2.

1 ⟶ ⟵ 2

$u_{2i} = -4$ m/s

The collision takes place in frame S.

(b)

❶ Transform the velocities to frame S′ in which particle 2 is at rest.

$u'_{1i} = 6$ m/s

1 ⟶ 2

$u'_{2f} = 9$ m/s

$u'_{1f} = 3$ m/s

❷ Analyze the collision in frame S′.

❸ Transform the post-collision velocities back to frame S.

$u_{1f} = -1$ m/s

$u_{2f} = 5$ m/s

FIGURE 36.7 Using reference frames to solve an elastic-collision problem.

SOLVE The three steps of Tactics Box 36.1 are illustrated in Figure 36.7b. We're given u_{1i} and u_{2i}. The Galilean transformations of these velocities to frame S′, using $v = -4$ m/s, are

$$u'_{1i} = u_{1i} - v = (2 \text{ m/s}) - (-4 \text{ m/s}) = 6 \text{ m/s}$$

$$u'_{2i} = u_{2i} - v = (-4 \text{ m/s}) - (-4 \text{ m/s}) = 0 \text{ m/s}$$

The 100 g ball is at rest in frame S′, which is what we wanted. The velocities after the collision are

$$u'_{1f} = \frac{m_1 - m_2}{m_1 + m_2} u'_{1i} = 3 \text{ m/s}$$

$$u'_{2f} = \frac{2m_1}{m_1 + m_2} u'_{1i} = 9 \text{ m/s}$$

We've finished the collision analysis, but we're not done because these are the post-collision velocities in frame S′. Another application of the Galilean transformations tells us that the post-collision velocities in frame S are

$$u_{1f} = u'_{1f} + v = (3 \text{ m/s}) + (-4 \text{ m/s}) = -1 \text{ m/s}$$

$$u_{2f} = u'_{2f} + v = (9 \text{ m/s}) + (-4 \text{ m/s}) = 5 \text{ m/s}$$

Thus the 300 g ball rebounds to the left at a speed of 1 m/s and the 100 g ball is knocked to the right at a speed of 5 m/s.

ASSESS You can easily verify that momentum is conserved: $P_f = P_i = 0.20$ kg m/s. The calculations in this example were easy. The important point of this example, and one worth careful thought, is the *logic* of what we did and why we did it.

36.3 Einstein's Principle of Relativity

The 19th century was an era of optics and electromagnetism. Thomas Young demonstrated in 1801 that light is a wave, and by midcentury scientists had devised techniques for measuring the speed of light. Faraday discovered electromagnetic induction in 1831, setting in motion a train of events leading to Maxwell's conclusion, in 1864, that light is an electromagnetic wave.

If light is a wave, what is the medium in which it travels? This was perhaps *the* most important scientific question of the second half of the 19th century. The medium in which light waves were assumed to travel was called the **ether.** Experiments to measure the speed of light were assumed to be measuring its speed through the ether. But just what *is* the ether? What are its properties? Can we collect a jar full of ether to study? Despite the significance of these questions, experimental efforts to detect the ether or measure its properties kept coming up empty handed.

Maxwell's theory of electromagnetism didn't help the situation. The crowning success of Maxwell's theory was his prediction that light waves travel with speed

$$c = \frac{1}{\sqrt{\epsilon_0 \mu_0}} = 3.00 \times 10^8 \text{ m/s}$$

This is a very specific prediction with no wiggle room. The difficulty with such a specific prediction was the implication that Maxwell's laws of electromagnetism

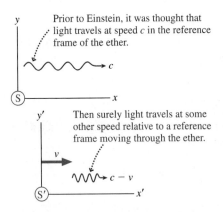

FIGURE 36.8 It seems as if the speed of light should differ from c in a reference frame moving through the ether.

are valid *only* in the reference frame of the ether. After all, as Figure 36.8 shows, the light speed should certainly be larger or smaller than c in a reference frame moving through the ether, just as the sound speed is different to someone moving through the air.

As the 19th century closed, it appeared that Maxwell's theory did not obey the classical principle of relativity. There was just one reference frame, the reference frame of the ether, in which the laws of electromagnetism seemed to be true. And to make matters worse, the fact that no one had been able to detect the ether meant that no one could identify the one reference frame in which Maxwell's equations "worked."

It was in this muddled state of affairs that a young Albert Einstein made his mark on the world. Even as a teenager, Einstein had wondered how a light wave would look to someone "surfing" the wave, traveling alongside the wave at the wave speed. You can do that with a water wave or a sound wave, but light waves seemed to present a logical difficulty. An electromagnetic wave sustains itself by virtue of the fact that a changing magnetic field induces an electric field and a changing electric field induces a magnetic field. But to someone moving with the wave, *the fields would not change.* How could there be an electromagnetic wave under these circumstances?

Several years of thinking about the connection between electromagnetism and reference frames led Einstein to the conclusion that *all* the laws of physics, not just the laws of mechanics, should obey the principle of relativity. In other words, the principle of relativity is a fundamental statement about the nature of the physical universe. Thus we can remove the restriction in the Galilean principle of relativity and state a much more general principle:

> **Principle of relativity** All the laws of physics are the same in all inertial reference frames.

All of the results of Einstein's theory of relativity flow from this one simple statement.

The Constancy of the Speed of Light

If Maxwell's equations of electromagnetism are laws of physics, and there's every reason to think they are, then, according to the principle of relativity, Maxwell's equations must be true in *every* inertial reference frame. On the surface this seems to be an innocuous statement, equivalent to saying that the law of conservation of momentum is true in every inertial reference frame. But follow the logic:

1. Maxwell's equations are true in all inertial reference frames.
2. Maxwell's equations predict that electromagnetic waves, including light, travel at speed $c = 3.00 \times 10^8$ m/s.
3. Therefore, **light travels at speed c in all inertial reference frames.**

Figure 36.9 shows the implications of this conclusion. *All* experimenters, regardless of how they move with respect to each other, find that *all* light waves, regardless of the source, travel in their reference frame with the *same* speed c. If Cathy's velocity toward Bill and away from Amy is $v = 0.9c$, Cathy finds, by making measurements in her reference frame, that the light from Bill approaches her at speed c, not at $c + v = 1.9c$. And the light from Amy, which left Amy at speed c, catches up from behind at speed c *relative to Cathy,* not the $c - v = 0.1c$ you would have expected.

Although this prediction goes against all shreds of common sense, the experimental evidence for it is strong. Laboratory experiments are difficult because

This light wave leaves Amy at speed c relative to Amy. It approaches Cathy at speed c relative to Cathy.

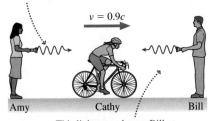

This light wave leaves Bill at speed c relative to Bill. It approaches Cathy at speed c relative to Cathy.

FIGURE 36.9 Light travels at speed c in all inertial reference frames, regardless of how the reference frames are moving with respect to the light source.

even the highest laboratory speed is insignificant in comparison to c. In the 1930s, however, the physicists R. J. Kennedy and E. M. Thorndike realized that they could use the earth itself as a laboratory. The earth's speed as it circles the sun is about 30,000 m/s. The *relative* velocity of the earth in January differs by 60,000 m/s from its velocity in July, when the earth is moving in the opposite direction. Kennedy and Thorndike were able to use a very sensitive and stable interferometer to show that the numerical values of the speed of light in January and July differ by less than 2 m/s.

More recent experiments have used unstable elementary particles, called π mesons, that decay into high-energy photons of light. The π mesons, created in a particle accelerator, move through the laboratory at 99.975% the speed of light, or $v = 0.99975c$ as they emit photons at the speed c in the π meson's reference frame. As Figure 36.10 shows, you would expect the photons to travel through the laboratory with speed $c + v = 1.99975c$. Instead, the measured speed of the photons in the laboratory was, within experimental error, 3.00×10^8 m/s.

In summary, *every* experiment designed to compare the speed of light in different reference frames has found that light travels at 3.00×10^8 m/s in every inertial reference frame, regardless of how the reference frames are moving with respect to each other.

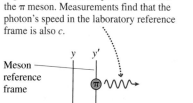

A photon is emitted at speed c relative to the π meson. Measurements find that the photon's speed in the laboratory reference frame is also c.

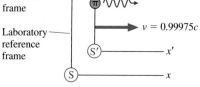

FIGURE 36.10 Experiments find that the photons travel through the laboratory with speed c, not the speed $1.99975c$ that you might expect.

How Can This Be?

You're in good company if you find this impossible to believe. Suppose I shot a ball forward at 50 m/s while driving past you at 30 m/s. You would certainly see the ball traveling at 80 m/s relative to you and the ground. What we're saying with regard to light is equivalent to saying that the ball travels at 50 m/s relative to my car and *at the same time* travels at 50 m/s relative to the ground, even though the car is moving across the ground at 30 m/s. It seems logically impossible.

You might think that this is merely a matter of semantics. If we can just get our definitions and use of words straight, then the mystery and confusion will disappear. Or perhaps the difficulty is a confusion between what we "see" versus what "really happens." In other words, a better analysis, one that focuses on what really happens, would find that light "really" travels at different speeds in different reference frames.

Alas, what "really happens" is that light travels at 3.00×10^8 m/s in every inertial reference frame, regardless of how the reference frames are moving with respect to each other. It's not a trick. There remains only one way to escape the logical contradictions.

The definition of velocity is $u = \Delta x / \Delta t$, the ratio of a distance traveled to the time interval in which the travel occurs. Suppose you and I both make measurements on an object as it moves, but you happen to be moving relative to me. Perhaps I'm standing on the corner, you're driving past in your car, and we're both trying to measure the velocity of a bicycle. Further, suppose we have agreed in advance to measure the bicycle as it moves from the tree to the lamppost in Figure 36.11 on the next page. Your $\Delta x'$ differs from my Δx because of your motion relative to me, causing you to calculate a bicycle velocity u' in your reference frame that differs from its velocity u in my reference frame. This is just the Galilean transformations showing up again.

Now let's repeat the measurements, but this time let's measure the velocity of a light wave as it travels from the tree to the lamppost. Once again, your $\Delta x'$ differs from my Δx, although the difference will be pretty small unless your car is moving at well above the legal speed limit. The obvious conclusion is that your light speed u' differs from my light speed u. But it doesn't. The experiments show that, for a light wave, we'll get the *same* values: $u' = u$.

The only way this can be true is if your Δt is not the same as my Δt. If the time it takes the light to move from the tree to the lamppost in your reference frame, a

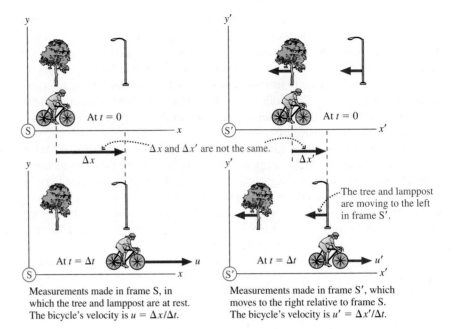

FIGURE 36.11 Measuring the velocity of an object by appealing to the basic definition $u = \Delta x/\Delta t$.

time we'll now call $\Delta t'$, differs from the time Δt it takes the light to move from the tree to the lamppost in my reference frame, then we might find that $\Delta x'/\Delta t' = \Delta x/\Delta t$. That is, $u' = u$ even though you are moving with respect to me.

We've assumed, since the beginning of this textbook, that time is simply time. It flows along like a river, and all experimenters in all reference frames simply use it. For example, suppose the tree and the lamppost both have big clocks that we both can see. Shouldn't we be able to agree on the time interval Δt the light needs to move from the tree to the lamppost?

Perhaps not. It's demonstrably true that $\Delta x' \neq \Delta x$. It's experimentally veri- fied that $u' = u$ for light waves. Something must be wrong with *assumptions* that we've made about the nature of time. The principle of relativity has painted us into a corner, and our only way out is to reexamine our understanding of time.

36.4 Events and Measurements

To question some of our most basic assumptions about space and time requires extreme care. We need to be certain that no assumptions slip into our analysis unnoticed. Our goal is to describe the motion of a particle in a clear and precise way, making the barest minimum of assumptions.

Events

The fundamental entity of relativity is called an **event.** An event is a physical activity that takes place at a definite point in space and at a definite instant of time. A firecracker exploding is an event. A collision between two particles is an event. A light wave hitting a detector is an event.

Events can be observed and measured by experimenters in different reference frames. An exploding firecracker is as clear to you as you drive by in your car as it is to me standing on the street corner. We can quantify where and when an event occurs with four numbers: the coordinates (x, y, z) and the instant of time t. These four numbers, illustrated in Figure 36.12, are called the **spacetime coordinates** of the event.

An event has spacetime coordinates (x, y, z, t) in frame S and different spacetime coordinates (x', y', z', t') in frame S'.

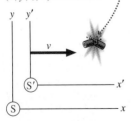

FIGURE 36.12 The location and time of an event are described by its spacetime coordinates.

The spatial coordinates of an event measured in reference frames S and S′ may differ. It now appears that the instant of time recorded in S and S′ may also differ. Thus the spacetime coordinates of an event measured by experimenters in frame S are (x, y, z, t) and the spacetime coordinates of the *same event* measured by experimenters in frame S′ are (x', y', z', t').

The motion of a particle can be described as a sequence of two or more events. We introduced this idea in the previous section when we agreed to measure the velocity of a bicycle and then of a light wave by comparing the object passing the tree (first event) to the object passing the lamppost (second event).

Measurements

Events are what "really happen," but how do we learn about an event? That is, how do the experimenters in a reference frame determine the spacetime coordinates of an event? This is a problem of *measurement*.

We defined a reference frame to be a coordinate system in which experimenters can make position and time measurements. That's a good start, but now we need to be more precise as to *how* the measurements are made. Imagine that a reference frame is filled with a cubic lattice of meter sticks, as shown in Figure 36.13. At every intersection is a clock, and all the clocks in a reference frame are *synchronized*. We'll return in a moment to consider how to synchronize the clocks, but assume for the moment it can be done.

Now, with our meter sticks and clocks in place, we can use a two-part measurement scheme:

- The (x, y, z) coordinates of an event are determined by the intersection of meter sticks closest to the event.
- The event's time t is the time displayed on the clock nearest the event.

You can imagine, if you wish, that each event is accompanied by a flash of light to illuminate the face of the nearest clock and make its reading known.

Several important issues need to be noted:

1. The clocks and meter sticks in each reference frame are imaginary, so they have no difficulty passing through each other.
2. Measurements of position and time made in one reference frame must use only the clocks and meter sticks in that reference frame.
3. There's nothing special about the sticks being 1 m long and the clocks 1 m apart. The lattice spacing can be altered to achieve whatever level of measurement accuracy is desired.
4. We'll assume that the experimenters in each reference frame have assistants sitting beside every clock to record the position and time of nearby events.
5. Perhaps most important, t is the time at which the event *actually happens*, not the time at which an experimenter sees the event or at which information about the event reaches an experimenter.
6. All experimenters in one reference frame agree on the spacetime coordinates of an event. In other words, **an event has a unique set of spacetime coordinates in each reference frame.**

The spacetime coordinates of this event are measured by the nearest meter stick intersection and the nearest clock.

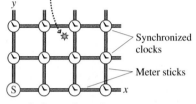

Reference frame S

Reference frame S′ has its own meter sticks and its own clocks.

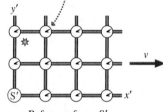

Reference frame S′

FIGURE 36.13 The spacetime coordinates of an event are measured by a lattice of meter sticks and clocks.

STOP TO THINK 36.3   A carpenter is working on a house two blocks away. You notice a slight delay between seeing the carpenter's hammer hit the nail and hearing the blow. At what time does the event "hammer hits nail" occur?

a. At the instant you hear the blow.
b. At the instant you see the hammer hit.
c. Very slightly before you see the hammer hit.
d. Very slightly after you see the hammer hit.

Clock Synchronization

It's important that all the clocks in a reference frame be **synchronized,** meaning that all clocks in the reference frame have the same reading at any one instant of time. We would not be able to use a sequence of events to track the motion of a particle if the clocks differed in their readings. Thus we need a method of synchronization. One idea that comes to mind is to designate the clock at the origin as the *master clock.* We could then carry this clock around to every clock in the lattice, adjust that clock to match the master clock, and finally return the master clock to the origin.

This would be a perfectly good method of clock synchronization in Newtonian mechanics, where time flows along smoothly, the same for everyone. But we've been driven to reexamine the nature of time by the possibility that time is different in reference frames moving relative to each other. Because the master clock would *move,* we cannot assume that the master clock keeps time in the same way as the stationary clocks.

We need a synchronization method that does not require moving the clocks. Fortunately, such a method is easy to devise. Each clock is resting at the intersection of meter sticks, so by looking at the meter sticks, the assistant knows, or can calculate, exactly how far each clock is from the origin. Once the distance is known, the assistant can calculate exactly how long a light wave will take to travel from the origin to each clock. For example, light will take 1.00 μs to travel to a clock 300 m from the origin.

NOTE ▶ It's handy for many relativity problems to know that the speed of light is $c = 300$ m/μs. ◀

To synchronize the clocks, the assistants begin by setting each clock to display the light travel time from the origin, but they don't start the clocks. Next, as Figure 36.14 shows, a light flashes at the origin and, simultaneously, the clock at the origin starts running from $t = 0$ s. The light wave spreads out in all directions at speed c. A photodetector on each clock recognizes the arrival of the light wave and, without delay, starts the clock. The clock had been preset with the light travel time, so each clock as it starts reads exactly the same as the clock at the origin. Thus all the clocks will be synchronized after the light wave has passed by.

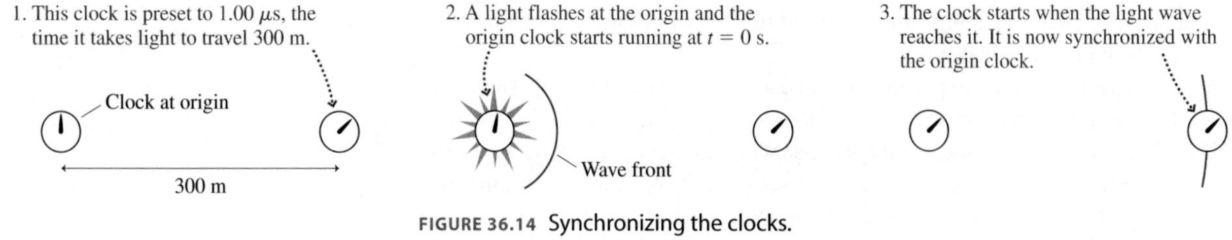

FIGURE 36.14 Synchronizing the clocks.

Events and Observations

We noted above that t is the time the event *actually happens.* This is an important point, one that bears further discussion. Light waves take time to travel. Messages, whether they're transmitted by light pulses, telephone, or courier on horseback, take time to be delivered. An experimenter *observes* an event, such as an exploding firecracker, only *at a later time* when light waves reach his or her eyes. But our interest is in the event itself, not the experimenter's observation of the event. The time at which the experimenter sees the event or receives information about the event is not when the event actually occurred.

Suppose at $t = 0$ s a firecracker explodes at $x = 300$ m. The flash of light from the firecracker will reach an experimenter at the origin at $t_1 = 1.0$ μs. The sound of the explosion will reach a sightless experimenter at $t_2 = 0.88$ s. Neither of these is the time t_{event} of the explosion, although the experimenter can work backward from these times, using known wave speeds, to determine t_{event}. In this example, the spacetime coordinates of the event—the explosion—are (300 m, 0 m, 0 m, 0 s).

EXAMPLE 36.3 Finding the time of an event

Experimenter A in reference frame S stands at the origin looking in the positive x-direction. Experimenter B stands at $x = 900$ m looking in the negative x-direction. A firecracker explodes somewhere between them. Experimenter B sees the light flash at $t = 3.00$ μs. Experimenter A sees the light flash at $t = 4.00$ μs. What are the spacetime coordinates of the explosion?

MODEL Experimenters A and B are in the same reference frame and have synchronized clocks.

VISUALIZE Figure 36.15 shows the two experimenters and the explosion at unknown position x.

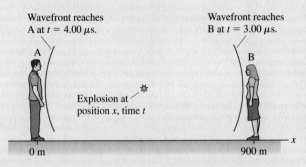

Wavefront reaches A at $t = 4.00$ μs.

Wavefront reaches B at $t = 3.00$ μs.

A

B

Explosion at position x, time t

0 m

900 m

x

FIGURE 36.15 The light wave reaches the experimenters at different times. Neither of these is the time at which the event actually happened.

SOLVE The two experimenters observe light flashes at two different instants, but there's only one event. Light travels 300 m/μs, so the additional 1.00 μs needed for the light to reach experimenter A implies that distance $(x - 0$ m) is 300 m longer than distance (900 m $- x$). That is,

$$(x - 0 \text{ m}) = (900 \text{ m} - x) + 300 \text{ m}$$

This is easily solved to give $x = 600$ m as the position coordinate of the explosion. The light takes 1.00 μs to travel 300 m to experimenter B, 2.00 μs to travel 600 m to experimenter A. The light is received at 3.00 μs and 4.00 μs, respectively, hence it was emitted by the explosion at $t = 2.00$ μs. The spacetime coordinates of the explosion are (600 m, 0 m, 0 m, 2.00 μs).

ASSESS Although the experimenters *see* the explosion at different times, they agree that the explosion actually *happened* at $t = 2.00$ μs.

Simultaneity

Two events 1 and 2 that take place at different positions x_1 and x_2 but at the *same time* $t_1 = t_2$, as measured in some reference frame, are said to be **simultaneous** in that reference frame. Simultaneity is determined by when the events actually happen, not when they are seen or observed. In general, simultaneous events are *not* seen at the same time because of the difference in light travel times from the events to an experimenter.

EXAMPLE 36.4 Are the explosions simultaneous?

An experimenter in reference frame S stands at the origin looking in the positive x-direction. At $t = 3.0$ μs she sees firecracker 1 explode at $x = 600$ m. A short time later, at $t = 5.0$ μs, she sees firecracker 2 explode at $x = 1200$ m. Are the two explosions simultaneous? If not, which firecracker exploded first?

MODEL Light from both explosions travels toward the experimenter at 300 m/μs.

SOLVE The experimenter *sees* two different explosions, but perceptions of the events are not the events themselves. When did the explosions *actually* occur? Using the fact that light travels 300 m/μs, it's easy to see that firecracker 1 exploded at $t_1 = 1.0$ μs and firecracker 2 also exploded at $t_2 = 1.0$ μs. The events *are* simultaneous.

A tree and a pole are 3000 m apart. Each is suddenly hit by a bolt of lightning. Mark, who is standing at rest midway between the two, sees the two lightning bolts at the same instant of time. Nancy is at rest under the tree. Define event 1 to be "lightning strikes tree" and event 2 to be "lightning strikes pole." For Nancy, does event 1 occur before, after, or at the same time as event 2?

36.5 The Relativity of Simultaneity

We've now established a means for measuring the time of an event in a reference frame, so let's begin to investigate the nature of time. The following "thought experiment" is very similar to one suggested by Einstein.

Figure 36.16 shows a long railroad car traveling to the right with a velocity v that may be an appreciable fraction of the speed of light. A firecracker is tied to each end of the car, right above the ground. Each firecracker is powerful enough that, when it explodes, it will make a burn mark on the ground at the position of the explosion.

Ryan is standing on the ground, watching the railroad car go by. Peggy is standing in the exact center of the car with a special box at her feet. This box has two light detectors, one facing each way, and a signal light on top. The box works as follows:

1. If a flash of light is received at the right detector before a flash is received at the left detector, then the light on top of the box will turn green.
2. If a flash of light is received at the left detector before a flash is received at the right detector, or if two flashes arrive simultaneously, the light on top will turn red.

The firecrackers explode as the railroad car passes Ryan, and he sees the two light flashes from the explosions simultaneously. He then measures the distances to the two burn marks and finds that he was standing exactly halfway between the marks. Because light travels equal distances in equal times, Ryan concludes that the two explosions were simultaneous in his reference frame, the reference frame of the ground. Further, because he was midway between the two ends of the car, he was directly opposite Peggy when the explosions occurred.

Figure 36.17a shows the sequence of events in Ryan's reference frame. Light travels at speed c in all inertial reference frames, so, although the firecrackers were moving, the light waves are spheres centered on the burn marks. Ryan determines that the light wave coming from the right reaches Peggy and the box before the light wave coming from the left. Thus, according to Ryan, the signal light on top of the box turns green.

How do things look in Peggy's reference frame, a reference frame moving to the right at velocity v relative to the ground? As Figure 36.17b shows, Peggy sees Ryan moving to the left with speed v. Light travels at speed c in all inertial reference frames, so the light waves are spheres centered on the ends of the car. If the explosions are simultaneous, as Ryan has determined, the two light waves reach her and the box simultaneously. Thus, according to Peggy, the signal light on top of the box turns red!

Now the light on top must be either green or red. *It can't be both!* Later, after the railroad car has stopped, Ryan and Peggy can place the box in front of them. Either it has a red light or a green light. Ryan can't see one color while Peggy sees the other. Hence we have a paradox. It's impossible for Peggy and Ryan both to be right. But who is wrong, and why?

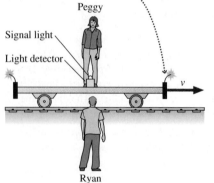

The firecrackers will make burn marks on the ground at the positions where they explode.

Peggy

Signal light

Light detector

v

Ryan

FIGURE 36.16 A railroad car traveling to the right with velocity v.

(a) The events in Ryan's frame

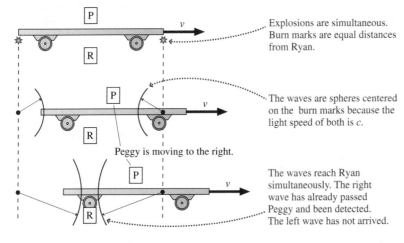

Explosions are simultaneous. Burn marks are equal distances from Ryan.

The waves are spheres centered on the burn marks because the light speed of both is c.

Peggy is moving to the right.

The waves reach Ryan simultaneously. The right wave has already passed Peggy and been detected. The left wave has not arrived.

(b) The events in Peggy's frame

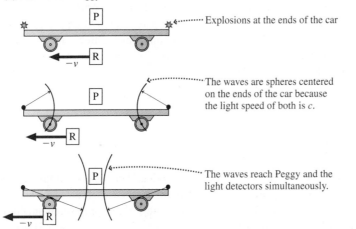

Explosions at the ends of the car

The waves are spheres centered on the ends of the car because the light speed of both is c.

The waves reach Peggy and the light detectors simultaneously.

FIGURE 36.17 Exploding firecrackers seen in two different reference frames.

What do we know with absolute certainty?

1. Ryan detected the flashes simultaneously.
2. Ryan was halfway between the firecrackers when they exploded.
3. The light from the two explosions traveled toward Ryan at equal speeds.

The conclusion that the explosions were simultaneous in Ryan's reference frame is unassailable. The light is green.

Peggy, however, made an assumption. It's a perfectly ordinary assumption, one that seems sufficiently obvious that you probably didn't notice, but an assumption nonetheless. Peggy assumed that the explosions were simultaneous.

Didn't Ryan find them to be simultaneous? Indeed, he did. Suppose we call Ryan's reference frame S, the explosion on the right event R, and the explosion on the left event L. Ryan found that $t_R = t_L$. But Peggy has to use a different set of clocks, the clocks in her reference frame S′, to measure the times t'_R and t'_L at which the explosions occurred. The fact that $t_R = t_L$ in frame S does *not* allow us to conclude that $t'_R = t'_L$ in frame S′.

In fact, the right firecracker must explode *before* the left firecracker in frame S′. Figure 36.17b, with its assumption about simultaneity, was incorrect. Figure 36.18 shows the situation in Peggy's reference frame with the right firecracker exploding first. Now the wave from the right reaches Peggy and the box first, as Ryan had concluded, and the light on top turns green.

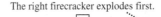
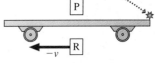

The right firecracker explodes first.

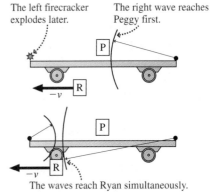

The left firecracker explodes later.

The right wave reaches Peggy first.

The waves reach Ryan simultaneously. The left wave has not reached Peggy.

FIGURE 36.18 The real sequence of events in Peggy's reference frame.

One of the most disconcerting conclusions of relativity is that **two events occurring simultaneously in reference frame S are *not* simultaneous in any reference frame S′ that is moving relative to S.** This is called the **relativity of simultaneity.**

The two firecrackers *really* explode at the same instant of time in Ryan's reference frame. And the right firecracker *really* explodes first in Peggy's reference frame. It's not a matter of when they see the flashes. Our conclusion refers to the times at which the explosions actually occur.

The paradox of Peggy and Ryan contains the essence of relativity, and it's worth careful thought. First, review the logic until you're certain that there *is* a paradox, a logical impossibility. Then convince yourself that the only way to resolve the paradox is to abandon the assumption that the explosions are simultaneous in Peggy's reference frame. If you understand the paradox and its resolution, you've made a big step toward understanding what relativity is all about.

STOP TO THINK 36.5   A tree and a pole are 3000 m apart. Each is suddenly hit by a bolt of lightning. Mark, who is standing at rest midway between the two, sees the two lightning bolts at the same instant of time. Nancy is flying her rocket at $v = 0.5c$ in the direction from the tree toward the pole. The lightning hits the tree just as she passes by it. Define event 1 to be "lightning strikes tree" and event 2 to be "lightning strikes pole." For Nancy, does event 1 occur before, after, or at the same time as event 2?

36.6 Time Dilation

17.1 Act|v Physics

The principle of relativity has driven us to the logical conclusion that time is not the same for two reference frames moving relative to each other. Our analysis thus far has been mostly qualitative. It's time to start developing some quantitative tools that will allow us to compare measurements in one reference frame to measurements in another reference frame.

Figure 36.19a shows a special clock called a **light clock.** The light clock is a box of height h with a light source at the bottom and a mirror at the top. The light source emits a very short pulse of light that travels to the mirror and reflects back to a light detector beside the source. The clock advances one "tick" each time the detector receives a light pulse, and it immediately, with no delay, causes the light source to emit the next light pulse.

Our goal is to compare two measurements of the interval between two ticks of the clock: one taken by an experimenter standing next to the clock and the other by an experimenter moving with respect to the clock. To be specific, Figure 36.19b shows the clock at rest in reference frame S′. We call this the **rest frame** of the clock. Reference frame S′ moves to the right with velocity v relative to reference frame S.

Relativity requires us to measure *events,* so let's define event 1 to be the emission of a light pulse and event 2 to be the detection of that light pulse. Experimenters in both reference frames are able to measure where and when these events occur *in their frame.* In frame S, the time interval $\Delta t = t_2 - t_1$ is one tick of the clock. Similarly, one tick in frame S′ is $\Delta t' = t_2' - t_1'$.

To be sure we have a clear understanding of the relativity result, let's first do a classical analysis. In frame S′, the clock's rest frame, the light travels straight up and down, a total distance $2h$, at speed c. The time interval is $\Delta t' = 2h/c$.

Figure 36.20a shows the operation of the light clock as seen in frame S. The clock is moving to the right at speed v in S, thus the mirror moves distance $\frac{1}{2}v(\Delta t)$

(a) A light clock

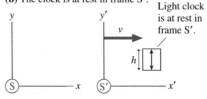

Mirror

h

Time display

Light source   Light detector

(b) The clock is at rest in frame S′.

Light clock is at rest in frame S′.

y y' v

h

(S) —— x (S′) —— x'

FIGURE 36.19 The ticking of a light clock can be measured by experimenters in two different reference frames.

during the time $\frac{1}{2}(\Delta t)$ in which the light pulse moves from the source to the mirror. The distance traveled by the light during this interval is $\frac{1}{2}u_{light}(\Delta t)$, where u_{light} is the speed of light in frame S. You can see from the vector addition in Figure 36.20b that the speed of light in frame S′ is $u_{light} = (c^2 + v^2)^{1/2}$. (Remember, this is a classical analysis in which the speed of light *does* depend on the motion of the reference frame relative to the light source.)

The Pythagorean theorem applied to the right triangle in Figure 36.20a is

$$h^2 + \left(\frac{1}{2}v\Delta t\right)^2 = \left(\frac{1}{2}u_{light}\Delta t\right)^2 = \left(\frac{1}{2}\sqrt{c^2 + v^2}\,\Delta t\right)^2$$

$$= \left(\frac{1}{2}c\,\Delta t\right)^2 + \left(\frac{1}{2}v\,\Delta t\right)^2 \qquad (36.4)$$

The term $(\frac{1}{2}v\Delta t)^2$ is common to both sides and cancels. Solving for Δt gives $\Delta t = 2h/c$, identical to $\Delta t'$. In other words, a classical analysis finds that the clock ticks at exactly the same rate in both frame S and frame S′. This shouldn't be surprising. There's only one kind of time in classical physics, measured the same by all experimenters independent of their motion.

The principle of relativity changes only one thing, but that change has profound consequences. According to the principle of relativity, light travels at the same speed in *all* inertial reference frames. In frame S′, the rest frame of the clock, the light simply goes straight up and back. The time of one tick,

$$\Delta t' = \frac{2h}{c} \qquad (36.5)$$

is unchanged from the classical analysis.

Figure 36.21 shows the light clock as seen in frame S. The difference from Figure 36.20a is that the light now travels along the hypotenuse at speed c. We can again use the Pythagorean theorem to write

$$h^2 + \left(\frac{1}{2}v\Delta t\right)^2 = \left(\frac{1}{2}c\,\Delta t\right)^2 \qquad (36.6)$$

Solving for Δt gives

$$\Delta t = \frac{2h/c}{\sqrt{1 - v^2/c^2}} = \frac{\Delta t'}{\sqrt{1 - v^2/c^2}} \qquad (36.7)$$

The time interval between two ticks in frame S is *not* the same as in frame S′.

It's useful to define $\beta = v/c$, the velocity as a fraction of the speed of light. For example, a reference frame moving with $v = 2.4 \times 10^8$ m/s has $\beta = 0.80$. In terms of β, Equation 36.7 is

$$\Delta t = \frac{\Delta t'}{\sqrt{1 - \beta^2}} \qquad (36.8)$$

NOTE ▶ The expression $(1 - v^2/c^2)^{1/2} = (1 - \beta^2)^{1/2}$ occurs frequently in relativity. The value of the expression is 1 when $v = 0$, and it steadily decreases to 0 as $v \rightarrow c$ (or $\beta \rightarrow 1$). The square root is an imaginary number if $v > c$, which would make Δt imaginary in Equation 36.8. Time intervals certainly have to be real numbers, suggesting that $v > c$ is not physically possible. One of the predictions of the theory of relativity, as you've undoubtedly heard, is that nothing can travel faster than the speed of light. Now you can begin to see why. We'll examine this topic more closely in Section 36.9. In the meantime, we'll require v to be less than c. ◀

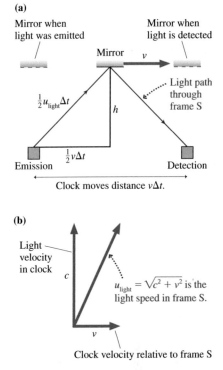

FIGURE 36.20 A classical analysis of the light clock.

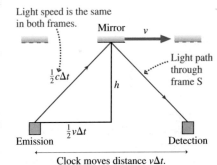

FIGURE 36.21 A light clock analysis in which the speed of light is the same in all reference frames.

Proper Time

Frame S′ has one important distinction. It is the *one and only* inertial reference frame in which the clock is at rest. Consequently, it is the one and only inertial reference frame in which the times of both events—the emission of the light and the detection of the light—are measured by the *same* clock. You can see that the light pulse in Figure 36.19, the rest frame of the clock, starts and ends at the same position and can be measured by one clock. In Figure 36.21, the emission and detection take place at different positions in frame S and must be measured by different clocks.

The time interval between two events that occur at the *same position* is called the **proper time** $\Delta\tau$. Only one inertial reference frame measures the proper time, and it does so with a single clock that is present at both events. An inertial reference frame moving with velocity $v = \beta c$ relative to the proper time frame must use two clocks to measure the time interval because the two events occur at different positions. The time interval in this frame is

$$\Delta t = \frac{\Delta\tau}{\sqrt{1 - \beta^2}} \geq \Delta\tau \qquad \text{(time dilation)} \qquad (36.9)$$

The "stretching out" of the time interval implied by Equation 36.9 is called **time dilation.** Time dilation is sometimes described by saying that "moving clocks run slow." This is not an accurate statement because it implies that some reference frames are "really" moving while others are "really" at rest. The whole point of relativity is that all inertial reference frames are equally valid, that all we know about reference frames is how they move relative to each other. A better description of time dilation is the statement that **the time interval between two ticks is the shortest in the reference frame in which the clock is at rest.** The time interval between two ticks is longer (i.e., the clock "runs slower") when it is measured in any reference frame in which the clock is moving.

NOTE ▶ Equation 36.9 was derived using a light clock because the operation of a light clock is clear and easy to analyze. But the conclusion is really about time itself. *Any* clock, regardless of how it operates, behaves the same. ◀

EXAMPLE 36.5 From the sun to Saturn
Saturn is 1.43×10^{12} m from the sun. A rocket travels along a line from the sun to Saturn at a constant speed of $0.9c$ relative to the solar system. How long does the journey take as measured by an experimenter on earth? As measured by an astronaut on the rocket?

MODEL Let the solar system be in reference frame S and the rocket be in reference frame S′ that travels with velocity $v = 0.9c$ relative to S. Relativity problems must be stated in terms of *events*. Let event 1 be "the rocket and the sun coincide" (the experimenter on earth says that the rocket passes the sun; the astronaut on the rocket says that the sun passes the rocket) and event 2 be "the rocket and Saturn coincide."

VISUALIZE Figure 36.22 shows the two events as seen from the two reference frames. Notice that the two events occur at the *same position* in S′, the position of the rocket, and consequently can be measured by *one* clock.

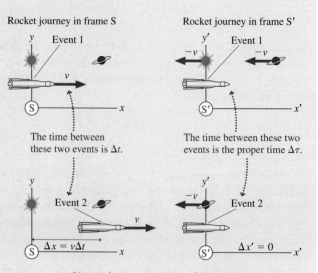
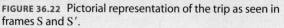

FIGURE 36.22 Pictorial representation of the trip as seen in frames S and S′.

SOLVE The time interval measured in the solar system reference frame, which includes the earth, is simply

$$\Delta t = \frac{\Delta x}{v} = \frac{1.43 \times 10^{12} \text{ m}}{0.9 \times (3.00 \times 10^8 \text{ m/s})} = 5300 \text{ s}$$

Relativity hasn't abandoned the basic definition $v = \Delta x/\Delta t$, although we do have to be sure that Δx and Δt are measured in just one reference frame and refer to the same two events.

How are things in the rocket's reference frame? The two events occur at the *same position* in S′ and can be measured by *one* clock, the clock at the origin. Thus the time measured by the astronauts is the *proper time* $\Delta \tau$ between the two events. We can use Equation 36.9 with $\beta = 0.9$ to find

$$\Delta \tau = \sqrt{1 - \beta^2}\,\Delta t = \sqrt{1 - 0.9^2}\,(5300 \text{ s}) = 2310 \text{ s}$$

ASSESS The time interval measured between these two events by the astronauts is less than half the time interval measured by experimenters on earth. The difference has nothing to do with when earthbound astronomers *see* the rocket pass the sun and Saturn. Δt is the time interval from when the rocket actually passes the sun, as measured by a clock at the sun, until it actually passes Saturn, as measured by a synchronized clock at Saturn. The interval between *seeing* the events from earth, which would have to allow for light travel times, would be something other than 5300 s. Δt and $\Delta \tau$ are different because *time is different* in two reference frames moving relative to each other.

STOP TO THINK 36.6 Molly flies her rocket past Nick at constant velocity v. Molly and Nick both measure the time it takes the rocket, from nose to tail, to pass Nick. Which of the following is true?

a. Both Molly and Nick measure the same amount of time.
b. Molly measures a shorter time interval than Nick.
c. Nick measures a shorter time interval than Molly.

Experimental Evidence

Is there any evidence for the crazy idea that clocks moving relative to each other tell time differently? Indeed, there's plenty. An experiment in 1971 sent an atomic clock around the world on a jet plane while an identical clock remained in the laboratory. This was a difficult experiment because the traveling clock's speed was so small compared to c, but measuring the small differences between the time intervals was just barely within the capabilities of atomic clocks. It was also a more complex experiment than we've analyzed because the clock accelerated as it moved around a circle. Nonetheless, the traveling clock, upon its return, was 200 ns behind the clock that stayed at home, which was exactly as predicted by relativity.

Very detailed studies have been done on unstable particles called *muons* that are created at the top of the atmosphere, at a height of about 60 km, when high-energy cosmic rays collide with air molecules. It is well known, from laboratory studies, that stationary muons decay with a *half-life* of 1.5 μs. That is, half the muons decay within 1.5 μs, half of those remaining decay in the next 1.5 μs, and so on. The decays can be used as a clock.

The muons travel down through the atmosphere at very nearly the speed of light. The time needed to reach the ground, assuming $v \approx c$, is $\Delta t \approx (60,000 \text{ m})/(3 \times 10^8 \text{ m/s}) = 200 \text{ }\mu$s. This is 133 half lives, so the fraction of muons reaching the ground should be $\approx (1/2)^{133} = 10^{-40}$. That is, only 1 out of every 10^{40} muons should reach the ground. In fact, experiments find that about 1 in 10 muons reach the ground, an experimental result that differs by a factor of 10^{39} from our prediction!

The discrepancy is due to time dilation. In Figure 36.23, the two events "muon is created" and "muon hits ground" take place at two different places in the earth's reference frame. However, these two events occur at the *same position* in the muon's reference frame. (The muon is like the rocket in Example 36.5.) Thus

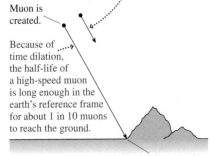

A muon travels ≈450 m in 1.5 μs. We would not detect muons at ground level if the half-life of a moving muon were 1.5 μs.

Muon is created.

Because of time dilation, the half-life of a high-speed muon is long enough in the earth's reference frame for about 1 in 10 muons to reach the ground.

Muon hits ground.

FIGURE 36.23 We wouldn't detect muons at the ground if not for time dilation.

the muon's internal clock measures the proper time. The time-dilated interval $\Delta t = 200\ \mu s$ in the earth's reference frame corresponds to a proper time $\Delta\tau \approx 5\ \mu s$ in the muon's reference frame. That is, in the muon's reference frame it takes only 5 μs from creation at the top of the atmosphere until the ground runs into it. This is 3.3 half-lives, so the fraction of muons reaching the ground is $(1/2)^{3.3} = 0.1$, or 1 out of 10. We wouldn't detect muons at the ground at all if not for time dilation.

The details are beyond the scope of this textbook, but dozens of high-energy particle accelerators around the world that study quarks and other elementary particles have been designed and built on the basis of Einstein's theory of relativity. The fact that they work exactly as planned is strong testimony to the reality of time dilation.

The Twin Paradox

The most well-known relativity paradox is the twin paradox. George and Helen are twins. On their 25th birthday, Helen departs on a starship voyage to a distant star. Let's imagine, to be specific, that her starship accelerates almost instantly to a speed of $0.95c$ and that she travels to a star that is 9.5 light years (9.5 ly) from earth. Upon arriving, she discovers that the planets circling the star are inhabited by fierce aliens, so she immediately turns around and heads home at $0.95c$.

A **light year,** abbreviated ly, is the distance that light travels in one year. A light year is vastly larger than the diameter of the solar system. The distance between two neighboring stars is typically a few light years. For our purpose, we can write the speed of light as $c = 1$ ly/year. That is, light travels 1 light year per year.

This value for c allows us to determine how long, according to George and his fellow earthlings, it takes Helen to travel out and back. Her total distance is 19 ly and, due to her rapid acceleration and rapid turn around, she travels essentially the entire distance at speed $v = 0.95c = 0.95$ ly/year. Thus the time she's away, as measured by George, is

$$\Delta t_G = \frac{19\ \text{ly}}{0.95\ \text{ly/year}} = 20\ \text{years} \tag{36.10}$$

George will be 45 years old when his sister Helen returns with tales of adventure.

While she's away, George takes a physics class and studies Einstein's theory of relativity. He realizes that time dilation will make Helen's clocks run more slowly than his clocks, which are at rest relative to him. Her heart—a clock—will beat fewer times and the minute hand on her watch will go around fewer times. In other words, she's aging more slowly than he is. Although she is his twin, she will be younger than he is when she returns.

Calculating Helen's age is not hard. We simply have to identify Helen's clock, because it's always with Helen as she travels, as the clock that measures proper time $\Delta\tau$. From Equation 36.9,

$$\Delta t_H = \Delta\tau = \sqrt{1 - \beta^2}\,\Delta t_G = \sqrt{1 - 0.95^2}\,(20\ \text{years}) = 6.25\ \text{years} \tag{36.11}$$

George will have just celebrated his 45th birthday as he welcomes home his 31-year-and-3-month-old twin sister.

This may be unsettling, because it violates our commonsense notion of time, but it's not a paradox. There's no logical inconsistency in this outcome. So why is it called "the twin paradox"? Read on.

Helen, knowing that she had quite of bit of time to kill on her journey, brought along several physics books to read. As she learns about relativity, she begins to think about George and her friends back on earth. Relative to her, they are all moving away at $0.95c$. Later they'll come rushing toward her at $0.95c$. Time dilation will cause their clocks to run more slowly than her clocks, which are at rest relative to her. In other words, as Figure 36.24 shows, Helen concludes that people on earth are aging more slowly than she is. Alas, she will be much older than they when she returns.

Finally, the big day arrives. Helen lands back on earth and steps out of the starship. George is expecting Helen to be younger than he is. Helen is expecting George to be younger than she is.

Here's the paradox! It's logically impossible for each to be younger than the other at the time when they are reunited. Where, then, is the flaw in our reasoning? It seems to be a symmetrical situation—Helen moves relative to George and George moves relative to Helen—but symmetrical reasoning has led to a conundrum.

But are the situations really symmetrical? George goes about his business day after day without noticing anything unusual. Helen, on the other hand, experiences three distinct periods during which the starship engines fire, she's crushed into her seat, and free dust particles that had been floating inside the starship are no longer, in the starship's reference frame, at rest or traveling in a straight line at constant speed. In other words, George spends the entire time in an inertial reference frame, *but Helen does not.* The situation is *not* symmetrical.

The principle of relativity applies *only* to inertial reference frames. Our discussion of time dilation was for inertial reference frames. Thus George's analysis and calculations are correct. Helen's analysis and calculations are *not* correct because she was trying to apply an inertial reference frame result to a noninertial reference frame.

Helen is younger than George when she returns. This is strange, but not a paradox. It is a consequence of the fact that time flows differently in two reference frames moving relative to each other.

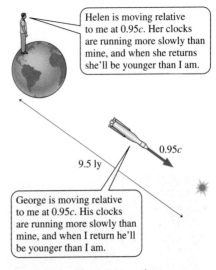

Helen is moving relative to me at 0.95c. Her clocks are running more slowly than mine, and when she returns she'll be younger than I am.

9.5 ly

0.95c

George is moving relative to me at 0.95c. His clocks are running more slowly than mine, and when I return he'll be younger than I am.

FIGURE 36.24 The twin paradox.

36.7 Length Contraction

We've seen that relativity requires us to rethink our idea of time. Now let's turn our attention to the concepts of space and distance. Consider the rocket that traveled from the sun to Saturn in Example 36.5. Figure 36.25a shows the rocket moving with velocity v through the solar system reference frame S. Define $L = \Delta x = x_{\text{Saturn}} - x_{\text{sun}}$ as the distance between the sun and Saturn in frame S or, more generally, the *length* of the spatial interval between two points. The rocket's speed is $v = L/\Delta t$, where Δt is the time measured in frame S for the journey from the sun to Saturn.

Activ Physics   17.2

(a) Reference frame S: The solar system is stationary.

(b) Reference frame S′: The rocket is stationary.

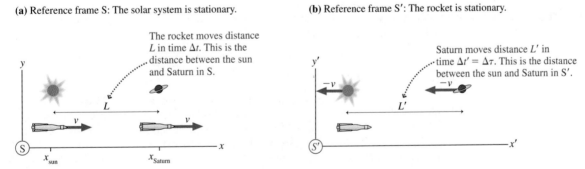

The rocket moves distance L in time Δt. This is the distance between the sun and Saturn in S.

L

v v

x_{sun} x_{Saturn}

Saturn moves distance L' in time $\Delta t' = \Delta \tau$. This is the distance between the sun and Saturn in S′.

$-v$ $-v$

L'

FIGURE 36.25 L and L' are the distances between the sun and Saturn in frames S and S′.

Figure 36.25b shows the situation in reference frame S′, where the rocket is at rest. The sun and Saturn move to the left at speed $v = L'/\Delta t'$, where $\Delta t'$ is the time measured in frame S′ for Saturn to travel distance L'.

Speed v is the relative speed between S and S′ and is the same for experimenters in both reference frames. That is,

$$v = \frac{L}{\Delta t} = \frac{L'}{\Delta t'} \qquad (36.12)$$

The time interval $\Delta t'$ measured in frame S′ is the proper time $\Delta \tau$ because both events occur at the same position in frame S′ and can be measured by one clock. We can use the time-dilation result, Equation 36.9, to relate $\Delta \tau$ measured by the astronauts to Δt measured by the earthbound scientists. Then Equation 36.12 becomes

$$\frac{L}{\Delta t} = \frac{L'}{\Delta \tau} = \frac{L'}{\sqrt{1 - \beta^2}\,\Delta t} \qquad (36.13)$$

The Δt cancels, and the distance L' in frame S′ is

$$L' = \sqrt{1 - \beta^2}\,L \qquad (36.14)$$

Surprisingly, we find that **the distance between two objects in reference frame S′ is *not the same* as the distance between the same two objects in reference frame S.**

Frame S, in which the distance is L, has one important distinction. It is the *one and only* inertial reference frame in which the objects are at rest. Experimenters in frame S can take all the time they need to measure L because the two objects aren't going anywhere. The distance L between two objects or two points in space measured in the reference frame in which the objects are at rest is called the **proper length** ℓ. Only one inertial reference frame can measure the proper length.

We can use the proper length ℓ to write Equation 36.14 as

$$L' = \sqrt{1 - \beta^2}\,\ell \leq \ell \qquad (36.15)$$

This "shrinking" of the distance between two objects, as measured by an experiment moving with respect to the objects, is called **length contraction.** Although we derived length contraction for the distance between two distinct objects, it applies equally well to the length of any physical object that stretches between two points along the x- and x'-axes. The length of an object is greatest in the reference frame in which the object is at rest. The object's length is less (i.e., the length is contracted) when it is measured in any reference frame moving relative to the object.

The Stanford Linear Accelerator (SLAC) is a 2-mi-long electron accelerator. The accelerator's length is less than 1 m in the reference frame of the electrons.

EXAMPLE 36.6 **The distance from the sun to Saturn**
In Example 36.5 a rocket traveled along a line from the sun to Saturn at a constant speed of 0.9c relative to the solar system. The Saturn-to-sun distance was given as 1.43×10^{12} m. What is the distance between the sun and Saturn in the rocket's reference frame?

MODEL Saturn and the sun are, at least approximately, at rest in the solar system reference frame S. Thus the given distance is the proper length ℓ.

SOLVE We can use Equation 36.15 to find the distance in the rocket's frame S′:

$$L' = \sqrt{1 - \beta^2}\,\ell = \sqrt{1 - 0.9^2}\,(1.43 \times 10^{12} \text{ m})$$
$$= 0.62 \times 10^{12} \text{ m}$$

ASSESS The sun-to-Saturn distance measured by the astronauts is less than half the distance measured by experimenters on earth. L' and ℓ are different because *space is different* in two reference frames moving relative to each other.

The conclusion that space is different in reference frames moving relative to each other is a direct consequence of the fact that time is different. Experimenters in both reference frames agree on the relative velocity v, leading to Equation 36.12: $v = L/\Delta t = L'/\Delta t'$. We had already learned that $\Delta t' < \Delta t$ because of time dilation. Thus L' *has* to be less than L. That is the only way experimenters in the two reference frames can reconcile their measurements.

To be specific, the earthly experimenters in Examples 36.5 and 36.6 find that the rocket takes 5300 s to travel the 1.43×10^{12} m between the sun and Saturn. The rocket's speed is $v = L/\Delta t = 2.7 \times 10^6$ m/s $= 0.9c$. The astronauts in the rocket find that it takes only 2310 s for Saturn to reach them after the sun has passed by. But there's no conflict, because they also find that the distance is only 0.62×10^{12} m. Thus Saturn's speed toward them is $v = L'/\Delta t' = (0.62 \times 10^{12}$ m$)/(2310$ s$) = 2.7 \times 10^6$ m/s $= 0.9c$.

Another Paradox?

Carmen and Dan are in their physics lab room. They each select a meter stick, lay the two side by side, and agree that the meter sticks are exactly the same length. Then, for an extra-credit project, they go outside and run past each other, in opposite directions, at a relative speed $v = 0.9c$. Figure 36.26 shows their experiment and a portion of their conversation.

Now, Dan's meter stick can't be both longer and shorter than Carmen's meter stick. Is this another paradox? No! Relativity allows us to compare the *same* events as they're measured in two different reference frames. This did lead to a real paradox when Peggy rolled past Ryan on the train. There the signal light on the box turns green (a single event) or it doesn't, and Peggy and Ryan have to agree about it. But the events by which Dan measures the length (in Dan's frame) of Carmen's meter are *not the same events* as those that Carmen uses to measure the length (in Carmen's frame) of Dan's meter stick.

There's no conflict between their measurements. In Dan's reference frame, Carmen's meter stick has been length contracted and is less than 1 m in length. In Carmen's reference frame, Dan's meter stick has been length contracted and is less than 1 m in length. If this weren't the case, if both agreed that one of the meter sticks was shorter than the other, then we could tell which reference frame was "really" moving and which was "really" at rest. But the principle of relativity doesn't allow us to make that distinction. Each is moving relative to the other, so each should make the same measurement for the length of the other's meter stick.

The Spacetime Interval

Forget relativity for a minute and think about ordinary geometry. Figure 36.27 shows two ordinary coordinate systems. They are identical except for the fact that one has been rotated relative to the other. A student using the xy-system would measure coordinates (x_1, y_1) for point 1 and (x_2, y_2) for point 2. A second student, using the $x'y'$-system, would measure (x_1', y_1') and (x_2', y_2').

The students soon find that none of their measurements agree. That is, $x_1 \neq x_1'$ and so on. Even the intervals are different: $\Delta x \neq \Delta x'$ and $\Delta y \neq \Delta y'$. Each is a perfectly valid coordinate system, giving no reason to prefer one over the other, but each yields different measurements.

Is there *anything* on which the two students can agree? Yes, there is. The distance d between points 1 and 2 is independent of the coordinates. We can state this mathematically as

$$d^2 = (\Delta x)^2 + (\Delta y)^2 = (\Delta x')^2 + (\Delta y')^2 \tag{36.16}$$

The quantity $(\Delta x)^2 + (\Delta y)^2$ is called an **invariant** in geometry because it has the same value in any Cartesian coordinate system.

Returning to relativity, is there an invariant in the spacetime coordinates, some quantity that has the *same value* in all inertial reference frames? There is, and to find it let's return to the light clock that we analyzed in Figure 36.21. Figure 36.28 on the next page shows the light clock as seen in reference frames S′ and S″. The speed of light is the same in both frames, even though both are moving with respect to each other and with respect to the clock.

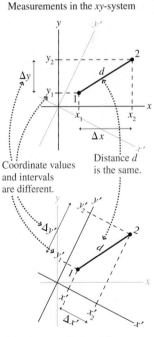

FIGURE 36.26 Carmen and Dan each measure the length of the other's meter stick as they move relative to each other.

Your meter stick is shorter than mine. Its length contracted because you're moving relative to me.

Carmen   Meter sticks

That can't be. Your meter stick is the one whose length contracted. *Your* meter stick is the shorter one.   Dan

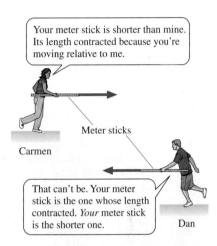

Measurements in the xy-system

Coordinate values and intervals are different.

Distance d is the same.

Measurements in the $x'y'$-system

FIGURE 36.27 Distance d is the same in both coordinate systems.

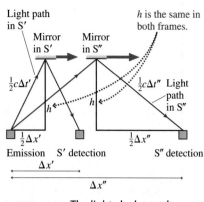

FIGURE 36.28 The light clock seen by experimenters in reference frames S′ and S″.

Notice that the clock's height h is common to both reference frames. Thus

$$h^2 = \left(\frac{1}{2}c\,\Delta t'\right)^2 - \left(\frac{1}{2}\Delta x'\right)^2 = \left(\frac{1}{2}c\,\Delta t''\right)^2 - \left(\frac{1}{2}\Delta x''\right)^2 \tag{36.17}$$

The factor $\frac{1}{2}$ cancels, allowing us to write

$$c^2(\Delta t')^2 - (\Delta x')^2 = c^2(\Delta t'')^2 - (\Delta x'')^2 \tag{36.18}$$

Let us define the **spacetime interval** s between two events to be

$$s^2 = c^2(\Delta t)^2 - (\Delta x)^2 \tag{36.19}$$

What we've shown in Equation 36.18 is that **the spacetime interval s has the same value in all inertial reference frames.** That is, the spacetime interval between two events is an invariant. It is a value that all experimenters, in all reference frames, can agree upon.

EXAMPLE 36.7 Using the spacetime interval
A firecracker explodes at the origin of an inertial reference frame. Then, 2.0 μs later, a second firecracker explodes 300 m away. Astronauts in a passing rocket measure the distance between the explosions to be 200 m. According to the astronauts, how much time elapses between the two explosions?

MODEL The spacetime coordinates of two events are measured in two different inertial reference frames. Call the reference frame of the ground S and the reference frame of the rocket S′. The spacetime interval between these two events is the same in both reference frames.

SOLVE The spacetime interval (or, rather, its square) in frame S is

$$s^2 = c^2(\Delta t)^2 - (\Delta x)^2 = (600 \text{ m})^2 - (300 \text{ m})^2$$
$$= 270{,}000 \text{ m}^2$$

where we used $c = 300$ m/μs to determine that $c\,\Delta t = 600$ m. The spacetime interval has the same value in frame S′. Thus

$$s^2 = 270{,}000 \text{ m}^2 = c^2(\Delta t')^2 - (\Delta x')^2$$
$$= c^2(\Delta t')^2 - (200 \text{ m})^2$$

This is easily solved to give $\Delta t' = 1.85$ μs.

ASSESS The two events are closer together in both space and time in the rocket's reference frame than in the reference frame of the ground.

Einstein's legacy, according to popular culture, was the discovery that "everything is relative." But it's not so. Time intervals and space intervals may be relative, as were the intervals Δx and Δy in the purely geometric analogy with which we opened this section, but some things are *not* relative. In particular, the spacetime interval s between two events is not relative. It is a well-defined number, agreed to by experimenters in each and every inertial reference frame.

STOP TO THINK 36.7 Beth and Charles are at rest relative to each other. Anjay runs past at velocity v while holding a long pole parallel to his motion. Anjay, Beth, and Charles each measure the length of the pole at the instant Anjay passes Beth. Rank in order, from largest to smallest, the three lengths L_A, L_B, and L_C.

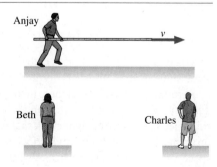

36.8 The Lorentz Transformations

The Galilean transformation $x' = x - vt$ of classical relativity lets us calculate the position x' of an event in frame S' if we know its position x in frame S. Classical relativity, of course, assumes that $t' = t$. Is there a similar transformation in relativity that would allow us to calculate an event's spacetime coordinates (x', t') in frame S' if we know their values (x, t) in frame S? Such a transformation would need to satisfy three conditions. It must

1. Agree with the Galilean transformations in the low-speed limit $v \ll c$.
2. Transform not only spatial coordinates but also time coordinates.
3. Ensure that the speed of light is the same in all reference frames.

We'll continue to use reference frames in the standard orientation of Figure 36.29. The motion is parallel to the x- and x'-axes, and we *define* $t = 0$ and $t' = 0$ as the instant when the origins of S and S' coincide.

The requirement that a new transformation agree with the Galilean transformation when $v \ll c$ suggests that we look for a transformation of the form

$$x' = \gamma(x - vt) \quad \text{and} \quad x = \gamma(x' + vt') \qquad (36.20)$$

where γ is a dimensionless function of velocity that satisfies $\gamma \to 1$ as $v \to 0$.

To determine γ, consider the following two events:

Event 1: A flash of light is emitted from the origin of both reference frames $(x = x' = 0)$ at the instant they coincide $(t = t' = 0)$.

Event 2: The light strikes a light detector. The spacetime coordinates of this event are (x, t) in frame S and (x', t') in frame S'.

Light travels at speed c in both reference frames, so the positions of event 2 are $x = ct$ in S and $x' = ct'$ in S'. Substituting these expressions for x and x' into Equation 36.20 gives

$$ct' = \gamma(ct - vt) = \gamma(c - v)t$$
$$ct = \gamma(ct' + vt') = \gamma(c + v)t' \qquad (36.21)$$

Solve the first for t', by dividing by c, then substitute this result for t' into the second:

$$ct = \gamma(c + v)\frac{\gamma(c - v)t}{c} = \gamma^2(c^2 - v^2)\frac{t}{c}$$

The t cancels, leading to

$$\gamma^2 = \frac{c^2}{c^2 - v^2} = \frac{1}{1 - v^2/c^2}$$

Thus the γ that "works" in the proposed transformation of Equation 36.20 is

$$\gamma = \frac{1}{\sqrt{1 - v^2/c^2}} = \frac{1}{\sqrt{1 - \beta^2}} \qquad (36.22)$$

You can see that $\gamma \to 1$ as $v \to 0$, as expected.

The transformation between t and t' is found by requiring that $x = x$ if you use Equation 36.20 to transform a position from S to S' and then back to S. The details will be left for a homework problem. Another homework problem will let you demonstrate that the y and z measurements made perpendicular to the relative motion are not affected by the motion. We tacitly assumed this condition in our analysis of the light clock.

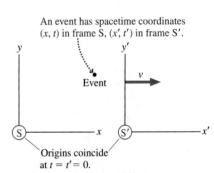

An event has spacetime coordinates (x, t) in frame S, (x', t') in frame S'.

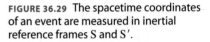

FIGURE 36.29 The spacetime coordinates of an event are measured in inertial reference frames S and S'.

The full set of equations are called the **Lorentz transformations.** They are

$$
\begin{aligned}
x' &= \gamma(x - vt) & x &= \gamma(x' + vt') \\
y' &= y & y &= y' \\
z' &= z & z &= z' \\
t' &= \gamma(t - vx/c^2) & t &= \gamma(t' + vx'/c^2)
\end{aligned}
\tag{36.23}
$$

The Lorentz transformations transform the spacetime coordinates of *one* event. You should compare these to the Galilean transformation equations in Equation 36.1.

NOTE ▶ These transformations are named after the Dutch physicist H. A. Lorentz, who derived them prior to Einstein. Lorentz was close to discovering special relativity, but he didn't recognize that our concepts of space and time have to be changed before these equations can be properly interpreted. ◀

Using Relativity

Relativity is phrased in terms of *events,* hence relativity problems are solved by interpreting the problem statement in terms of specific events.

(MP) PROBLEM-SOLVING STRATEGY 36.1 Relativity

MODEL Frame the problem in terms of events, things that happen at a specific place and time.

VISUALIZE A pictorial representation defines the reference frames.

- Sketch the reference frames, showing their motion relative to each other.
- Show events. Identify objects that are moving with respect to the reference frames.
- Identify any proper time intervals and proper lengths. These are measured in an object's rest frame.

SOLVE The mathematical representation is based on the Lorentz transformations, but not every problem requires the full transformation equations.

- Problems about time intervals can often be solved using time dilation: $\Delta t = \gamma \Delta \tau$.
- Problems about distances can often be solved using length contraction: $L = \ell/\gamma$.

ASSESS Are the results consistent with Galilean relativity when $v \ll c$?

EXAMPLE 36.8 Ryan and Peggy revisited

Peggy is standing in the center of a long, flat railroad car that has firecrackers tied to both ends. The car moves past Ryan, who is standing on the ground, with velocity $v = 0.8c$. Flashes from the exploding firecrackers reach him simultaneously 1.0 μs after the instant that Peggy passes him, and he later finds burn marks on the track 300 m to either side of where he had been standing.

a. According to Ryan, what is the distance between the two explosions and at what times do the explosions occur relative to the time that Peggy passes him?

b. According to Peggy, what is the distance between the two explosions and at what times do the explosions occur relative to the time that Ryan passes her?

MODEL Let the explosion on Ryan's right, the direction in which Peggy is moving, be event R. The explosion on his left is event L.

VISUALIZE Peggy and Ryan are in inertial reference frames. Figure 36.30 shows Peggy's frame S′ moving with $v = 0.8c$ relative to Ryan's frame S. We've defined the reference frames such that Peggy and Ryan are at the origins. The instant they pass, by definition, is $t = t' = 0$ s. The two events are shown in Ryan's reference frame.

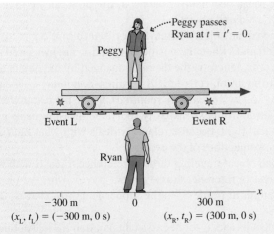

FIGURE 36.30 A pictorial representation of the reference frames and events.

SOLVE

a. The two burn marks tell Ryan that the distance between the explosions was $L = 600$ m. Light travels at $c = 300$ m/μs, and the burn marks are 300 m on either side of him, so Ryan can determine that each explosion took place 1.0 μs before he saw the flash. But this was the instant of time that Peggy passed him, so Ryan concludes that the explosions were simultaneous with each other and with Peggy's passing him. The spacetime coordinates of the two events in frame S are $(x_R, t_R) = (300$ m, 0 μs$)$ and $(x_L, t_L) = (-300$ m, 0 μs$)$.

b. We already know, from our qualitative analysis in Section 36.5, that the explosions are *not* simultaneous in Peggy's reference frame. Event R happens before event L in S′, but we don't know how they compare to the time at which Ryan passes Peggy. We can now use the Lorentz transformations to relate the spacetime coordinates of these events as measured by Ryan to the spacetime coordinates as measured by Peggy. Using $v = 0.8c$, we find that γ is

$$\gamma = \frac{1}{\sqrt{1 - v^2/c^2}} = \frac{1}{\sqrt{1 - 0.8^2}} = 1.667$$

For event L, the Lorentz transformations are

$$x_L' = 1.667((-300 \text{ m}) - (0.8c)(0 \text{ } \mu s)) = -500 \text{ m}$$

$$t_L' = 1.667((0 \text{ } \mu s) - (0.8c)(-300 \text{ m})/c^2) = 1.33 \text{ } \mu s$$

And for event R,

$$x_R' = 1.667((300 \text{ m}) - (0.8c)(0 \text{ } \mu s)) = 500 \text{ m}$$

$$t_R' = 1.667((0 \text{ } \mu s) - (0.8c)(300 \text{ m})/c^2) = -1.33 \text{ } \mu s$$

According to Peggy, the two explosions occur 1000 m apart. Furthermore, the first explosion, on the right, occurs 1.33 μs before Ryan passes her at $t' = 0$ s. The second, on the left, occurs 1.33 μs after Ryan goes by.

ASSESS Events that are simultaneous in frame S are *not* simultaneous in frame S′. The results of the Lorentz transformations agree with our earlier qualitative analysis.

A follow-up discussion of Example 36.8 is worthwhile. Because Ryan moves at speed $v = 0.8c = 240$ m/μs relative to Peggy, he moves 320 m during the 1.33 μs between the first explosion and the instant he passes Peggy, then another 320 m before the second explosion. Gathering this information together, Figure 36.31 shows the sequence of events in Peggy's reference frame.

The firecrackers define the ends of the railroad car, so the 1000 m distance between the explosions in Peggy's frame is the car's length L' in frame S′. The car is at rest in frame S′, hence length L' is the proper length: $\ell = 1000$ m. Ryan is measuring the length of a moving object, so he should see the car length contracted to

$$L = \sqrt{1 - \beta^2}\ell = \frac{\ell}{\gamma} = \frac{1000 \text{ m}}{1.667} = 600 \text{ m}$$

And, indeed, that is exactly the distance Ryan measured between the burn marks.

Finally, we can calculate the spacetime interval s between the two events. According to Ryan,

$$s^2 = c^2(\Delta t^2) - (\Delta x)^2 = c^2(0 \text{ } \mu s)^2 - (600 \text{ m})^2 = -(600 \text{ m})^2$$

Peggy computes the spacetime interval to be

$$s^2 = c^2(\Delta t')^2 - (\Delta x')^2 = c^2(2.67 \text{ } \mu s)^2 - (1000 \text{ m})^2 = -(600 \text{ m})^2$$

Their calculations of the spacetime interval agree, showing that s really is an invariant, but notice that s itself is an imaginary number.

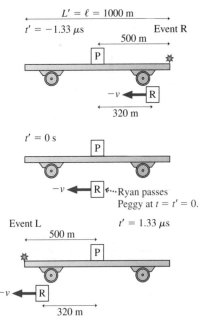

FIGURE 36.31 The sequence of events as seen in Peggy's reference frame.

Length

We've already introduced the idea of length contraction, but we didn't precisely define just what we mean by the *length* of a moving object. The length of an object at rest is clear because we can take all the time we need to measure it with

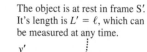

The object is at rest in frame S′. It's length is $L' = \ell$, which can be measured at any time.

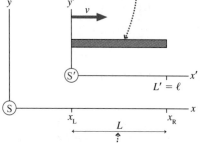

Since the object is moving in frame S, simultaneous measurements of its ends must be made in order to find its length L in frame S.

FIGURE 36.32 The length of an object is the distance between *simultaneous* measurements of the positions of the end points.

The binomial approximation

If $x \ll 1$, then $(1 + x)^n \approx 1 + nx$

meter sticks, surveying tools, or whatever we need. But how can we give clear meaning to the length of a moving object?

A reasonable definition of an object's length is the distance $L = \Delta x = x_R - x_L$ between the right and left ends when the positions x_R and x_L are measured *at the same time t*. In other words, length is the distance spanned by the object at *one instant* of time. Measuring an object's length requires *simultaneous* measurements of two positions (i.e., two events are required), hence the result won't be known until the information from two spatially separated measurements can be brought together. That's all right, because relativity deals with what *really happens* rather than with our perceptions of the events.

Figure 36.32 shows an object traveling through reference frame S with velocity v. The object is at rest in reference frame S′ that travels with the object at velocity v, hence the length in frame S′ is the proper length ℓ. That is, $\Delta x' = x_R' - x_L' = \ell$ in frame S′.

At time t, an experimenter (and his or her assistants) in frame S makes simultaneous measurements of the positions x_R and x_L of the ends of the object. The difference $\Delta x = x_R - x_L = L$ is the length in frame S. The Lorentz transformations of x_R and x_L are

$$x_R' = \gamma(x_R - vt)$$
$$x_L' = \gamma(x_L - vt)$$
(36.24)

where, it is important to note, t is the *same* for both because the measurements are simultaneous.

Subtracting the second equation from the first, we find

$$x_R' - x_L' = \ell = \gamma(x_R - x_L) = \gamma L = \frac{L}{\sqrt{1 - \beta^2}}$$

Solving for L, we find, in agreement with Equation 36.15, that

$$L = \sqrt{1 - \beta^2}\,\ell$$
(36.25)

This analysis has accomplished two things. First, by giving a precise definition of length, we've put our length-contraction result on a firmer footing. Second, we've had good practice at relativistic reasoning using the Lorentz transformation.

NOTE ▶ Length contraction does not tell us how an object would *look*. The visual appearance of an object is determined by light waves that arrive simultaneously at the eye. These waves left points on the object at different times (i.e., *not* simultaneously) because they had to travel different distances to the eye. The analysis needed to determine an object's visual appearance is considerably more complex. Length and length contraction are concerned only with the *actual* length of the object at one instant of time. ◀

The Binomial Approximation

You've met the binomial approximation earlier in this text and in your calculus class. The binomial approximation is useful when we need to calculate a relativistic expression for a nonrelativistic velocity $v \ll c$. Because $v^2/c^2 \ll 1$ in these cases, we can write

$$\sqrt{1 - \beta^2} = (1 - v^2/c^2)^{1/2} \approx 1 - \frac{1}{2}\frac{v^2}{c^2}$$
$$\gamma = \frac{1}{\sqrt{1 - \beta^2}} = (1 - v^2/c^2)^{-1/2} \approx 1 + \frac{1}{2}\frac{v^2}{c^2}$$
(36.26)

The following example illustrates the use of the binomial approximation.

EXAMPLE 36.9 The shrinking school bus

An 8.0-m-long school bus drives past at 30 m/s. By how much is its length contracted?

MODEL The school bus is at rest in an inertial reference frame S′ moving at velocity $v = 30$ m/s relative to the ground frame S. The given length, 8.0 m, is the proper length ℓ in frame S′.

SOLVE In frame S, the school bus is length contracted to

$$L = \sqrt{1 - \beta^2}\,\ell$$

The bus's velocity v is much less than c, so we can use the binomial approximation to write

$$L \approx \left(1 - \frac{1}{2}\frac{v^2}{c^2}\right)\ell = \ell - \frac{1}{2}\frac{v^2}{c^2}\ell$$

The *amount* of the length contraction is

$$\ell - L = \frac{1}{2}\frac{v^2}{c^2}\ell = \left(\frac{30 \text{ m/s}}{3.0 \times 10^8 \text{ m/s}}\right)^2 (4.0 \text{ m})$$

$$= 4.0 \times 10^{-14} \text{ m} = 40 \text{ fm}$$

where 1 fm = 1 femtometer = 10^{-15} m.

ASSESS The amount the bus "shrinks" is only slightly larger than the diameter of the nucleus of an atom. It's no wonder that we're not aware of length contraction in our everyday lives. If you had tried to calculate this number exactly, your calculator would have shown $\ell - L = 0$. The difficulty is that the difference between ℓ and L shows up only in the 14th decimal place. A scientific calculator determines numbers to 10 or 12 decimal places, but that isn't sufficient to show the difference. The binomial approximation provides an invaluable tool for finding the very tiny difference between two numbers that are nearly identical.

The Lorentz Velocity Transformations

Figure 36.33 shows an object that is moving in both reference frame S and reference frame S′. Experimenters in frame S determine that the object's velocity is u while experimenters in S′ find it to be $u′$. For simplicity, we'll assume that the object moves parallel to the x- and x′-axes.

The Galilean velocity transformation $u′ = u - v$ was found by taking the time derivative of the position transformation. We can do the same with the Lorentz transformation if we take the derivative with respect to the time in each frame. Velocity $u′$ in frame S′ is

$$u' = \frac{dx'}{dt'} = \frac{d(\gamma(x - vt))}{d(\gamma(t - vx/c^2))} \tag{36.27}$$

where we've used the Lorentz transformations for position $x′$ and time $t′$.

Carrying out the differentiation gives

$$u' = \frac{\gamma(dx - v\,dt)}{\gamma(dt - v\,dx/c^2)} = \frac{dx/dt - v}{1 - v(dx/dt)/c^2} \tag{36.28}$$

But dx/dt is u, the object's velocity in frame S, leading to

$$u' = \frac{u - v}{1 - uv/c^2} \tag{36.29}$$

You can see that Equation 36.29 reduces to the Galilean transformation $u′ = u - v$ when $v \ll c$, as expected.

The reverse transformation, from S′ to S, is found by reversing the sign of v. Altogether,

$$u' = \frac{u - v}{1 - uv/c^2} \quad \text{and} \quad u = \frac{u' + v}{1 + u'v/c^2} \tag{36.30}$$

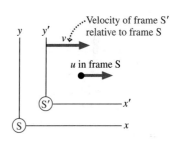
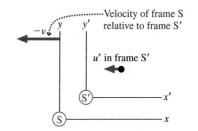

FIGURE 36.33 The velocity of a moving object is measured to be u in frame S and $u′$ in frame S′.

Equations 36.30 are the Lorentz velocity transformation equations.

NOTE ▶ It is important to distinguish carefully between v, which is the relative velocity of the reference frames in which measurements are carried out, and u and $u′$, which are the velocities of an *object* as measured in two different reference frames. ◀

EXAMPLE 36.10 A really fast bullet

A rocket flies past the earth at $0.9c$. As it goes by, the rocket fires a bullet in the forward direction at $0.95c$ with respect to the rocket. What is the bullet's speed with respect to the earth?

MODEL The rocket and the earth are inertial reference frames. Let the earth be frame S and the rocket be frame S'. The velocity of frame S' relative to frame S is $v = 0.9c$. The bullet's velocity in frame S' is $u' = 0.95c$.

SOLVE We can use the Lorentz velocity transformation to find

$$u = \frac{u' + v}{1 + u'v/c^2} = \frac{0.95c + 0.90c}{1 + (0.95c)(0.90c)/c^2} = 0.997c$$

The bullet's speed with respect to the earth is 99.7% of the speed of light.

NOTE ▶ Many relativistic calculations are much easier when velocities are specified as a fraction of c. ◀

ASSESS In Newtonian mechanics, the Galilean transformation of velocity would give $u = 1.85c$. Now, despite the very high speed of the rocket and of the bullet with respect to the rocket, the bullet's speed with respect to the earth remains less than c. This is yet more evidence that objects cannot exceed the speed of light.

Suppose the rocket in Example 36.10 fired a laser beam in the forward direction as it traveled past the earth at velocity v. The laser beam would travel away from the rocket at speed $u' = c$ in the rocket's reference frame S'. What is the laser beam's speed in the earth's frame S? According to the Lorentz velocity transformation, it must be

$$u = \frac{u' + v}{1 + u'v/c^2} = \frac{c + v}{1 + cv/c^2} = \frac{c + v}{1 + v/c} = \frac{c + v}{(c + v)/c} = c \quad (36.31)$$

Light travels at speed c in both frame S and frame S'. This important consequence of the principle of relativity is "built into" the Lorentz transformations.

36.9 Relativistic Momentum

In Newtonian mechanics, the total momentum of a system is a conserved quantity. Further, as we've seen, the law of conservation of momentum, $P_f = P_i$, is true in all inertial reference frames *if* the particle velocities in different reference frames are related by the Galilean velocity transformations.

The difficulty, of course, is that the Galilean transformations are not consistent with the principle of relativity. It is a reasonable approximation when all velocities are much less than c, but the Galilean transformations fail dramatically as velocities approach c. It's not hard to show that $P_f' \neq P_i'$ if the particle velocities in frame S' are related to the particle velocities in frame S by the Lorentz transformations.

There are two possibilities:

1. The so-called law of conservation of momentum is not really a law of physics. It is approximately true at low velocities but fails as velocities approach the speed of light.
2. The law of conservation of momentum really is a law of physics, but the expression $p = mu$ is not the correct way to calculate momentum when the particle velocity u becomes a significant fraction of c.

Momentum conservation is such a central and important feature of mechanics that it seems unlikely to fail in relativity. How else might the momentum of a particle be defined?

The classical momentum, for one-dimensional motion, is $p = mu = m(\Delta x/\Delta t)$. Δt is the time needed to move distance Δx. That seemed clear enough within a Newtonian framework, but now we've learned that experimenters in different reference frames disagree about the amount of time needed. So whose Δt should we use?

One possibility is to use the time measured *by the particle*. This is the proper time $\Delta \tau$ because the particle is at rest in its own reference frame and needs only one clock. With this in mind, let's redefine the momentum of a particle of mass m moving with velocity $u = \Delta x/\Delta t$ to be

$$p = m\frac{\Delta x}{\Delta \tau} \tag{36.32}$$

We can relate this new expression for p to the familiar Newtonian expression by using the time-dilation result $\Delta \tau = (1 - u^2/c^2)^{1/2}\Delta t$ to relate the proper time interval measured by the particle to the more practical time interval Δt measured by experimenters in frame S. With this substitution, Equation 36.32 becomes

$$p = m\frac{\Delta x}{\Delta \tau} = m\frac{\Delta x}{\sqrt{1 - u^2/c^2}\,\Delta t} = \frac{mu}{\sqrt{1 - u^2/c^2}} \tag{36.33}$$

You can see that Equation 36.33 reduces to the classical expression $p = mu$ when the particle's speed $u \ll c$. That is an important requirement, but whether this is the "correct" expression for p depends on whether the total momentum P is conserved when the velocities of a system of particles are transformed with the Lorentz velocity transformation equations. The proof is rather long and tedious, so we will assert, without actual proof, that the momentum defined in Equation 36.33 does, indeed, transform correctly. **The law of conservation of momentum is still valid in all inertial reference frames *if* the momentum of each particle is calculated with Equation 36.33.**

The factor that multiplies mu in Equation 36.33 looks much like the factor γ in the Lorentz transformation equations for x and t, but there's one very important difference. The v in the Lorentz transformation equations is the velocity of a *reference frame*. The u in Equation 36.33 is the velocity of a particle moving *in a* reference frame.

With this distinction in mind, let's define the quantity

$$\gamma_p = \frac{1}{\sqrt{1 - u^2/c^2}} \tag{36.34}$$

where the subscript p indicates that this is γ for a particle, not for a reference frame. In frame S′, where the particle moves with velocity u', the corresponding expression would be called γ_p'. With this definition of γ_p, the momentum of a particle is

$$p = \gamma_p mu \tag{36.35}$$

EXAMPLE 36.11 Momentum of a subatomic particle
Electrons in a particle accelerator reach a speed of $0.999c$ relative to the laboratory. One collision of an electron with a target produces a muon that moves forward with a speed of $0.95c$ relative to the laboratory. The electron mass is $m_e = 9.11 \times 10^{-31}$ m/s and the muon mass is 1.90×10^{-28} m/s. What is the muon's momentum in the laboratory frame and in the frame of the electron beam?

MODEL Let the laboratory be reference frame S. The reference frame S′ of the electron beam (i.e., a reference frame in which the electrons are at rest) moves in the direction of the electrons at $v = 0.999c$. The muon velocity in frame S is $u = 0.95c$.

SOLVE γ_p for the muon in the laboratory reference frame is

$$\gamma_p = \frac{1}{\sqrt{1 - u^2/c^2}} = \frac{1}{\sqrt{1 - 0.95^2}} = 3.20$$

Thus the muon's momentum in the laboratory is

$$p = \gamma_p mu = (3.20)(1.90 \times 10^{-28}\text{ kg})(0.95 \times 3.00 \times 10^8\text{ m/s})$$
$$= 1.73 \times 10^{-19}\text{ kg m/s}$$

The momentum is a factor of 3.2 larger than the Newtonian momentum mu. To find the momentum in the electron-beam reference frame, we must first use the velocity transformation equation to find the muon's velocity in frame S′:

$$u' = \frac{u - v}{1 - uv/c^2} = \frac{0.95c - 0.999c}{1 - (0.95c)(0.999c)/c^2} = -0.962c$$

In the laboratory frame, the faster electrons are overtaking the slower muon. Hence the muon's velocity in the electron-beam frame is negative. γ_p' for the muon in frame S′ is

$$\gamma_p' = \frac{1}{\sqrt{1 - u'^2/c^2}} = \frac{1}{\sqrt{1 - 0.962^2}} = 3.66$$

The muon's momentum in the electron-beam reference frame is

$$p' = \gamma_p' m u'$$

$$= (3.66)(1.90 \times 10^{-28}\,\text{kg})(-0.962 \times 3.00 \times 10^8\,\text{m/s})$$

$$= -2.01 \times 10^{-19}\,\text{kg m/s}$$

ASSESS From the laboratory perspective, the muon moves only slightly slower than the electron beam. But it turns out that the muon moves faster with respect to the electrons, although in the opposite direction, than it does with respect to the laboratory.

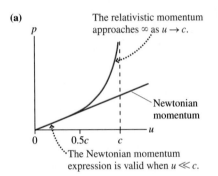

(a)

The relativistic momentum approaches ∞ as $u \to c$.

Newtonian momentum

The Newtonian momentum expression is valid when $u \ll c$.

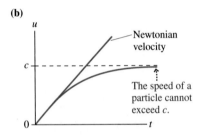

(b)

Newtonian velocity

The speed of a particle cannot exceed c.

FIGURE 36.34 The speed of a particle cannot reach the speed of light.

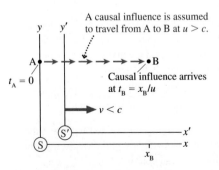

A causal influence is assumed to travel from A to B at $u > c$.

Causal influence arrives at $t_B = x_B/u$

$v < c$

FIGURE 36.35 Assume that a causal influence can travel from A to B at a speed $u > c$.

The Cosmic Speed Limit

Figure 36.34a is a graph of momentum versus velocity. For a Newtonian particle, with $p = mu$, the momentum is directly proportional to the velocity. The relativistic expression for momentum agrees with the Newtonian value if $u \ll c$, but p approaches ∞ as $u \to c$.

The implications of this graph become clear when we relate momentum to force. Consider a particle subjected to a constant force, such as a rocket that never runs out of fuel. If F is constant, we can see from $F = dp/dt$ that the momentum is $p = Ft$. If Newtonian physics were correct, a particle would go faster and faster as its velocity $u = p/m = (F/m)t$ increased without limit. But the relativistic result, shown in Figure 36.34b, is that the particle's velocity asymptotically approaches the speed of light ($u \to c$) as p approaches ∞. Relativity gives a very different outcome than Newtonian mechanics.

The speed c is a "cosmic speed limit" for material particles. A force cannot accelerate a particle to a speed higher than c because the particle's momentum becomes infinitely large as the speed approaches c. The amount of effort required for each additional increment of velocity becomes larger and larger until no amount of effort can raise the velocity any higher.

Actually, at a more fundamental level, c is a speed limit for *any* kind of **causal influence.** If I throw a rock and break a window, my throw is the *cause* of the breaking window and the rock is the *causal influence.* If I shoot a laser beam at a light detector that is wired to a firecracker, the light wave is the *causal influence* that leads to the explosion. A causal influence can be any kind of particle, wave, or information that travels from A to B and allows A to be the cause of B.

For two unrelated events—a firecracker explodes in Tokyo and a balloon bursts in Paris—the relativity of simultaneity tells us that they may be simultaneous in one reference frame but not in others. Or in one reference frame the firecracker may explode before the balloon bursts but in some other reference frame the balloon may burst first. These possibilities violate our commonsense view of time, but they're not in conflict with the principle of relativity.

For two causally related events—A *causes* B—it would be nonsense for an experimenter in any reference frame to find that B occurs before A. No experimenter in any reference frame, no matter how it is moving, will find that you are born before your mother is born. If A causes B, then it must be the case that $t_A < t_B$ in *all* reference frames.

Suppose there exists some kind of causal influence that *can* travel at speed $u > c$. Figure 36.35 shows a reference frame S in which event A is at the origin ($x_A = 0$). The faster-than-light causal influence—perhaps some yet-to-be-discovered "z ray"—leaves A at $t_A = 0$ and travels to the point at which it will cause event B. It arrives at x_B at time $t_B = x_B/u$.

How do events A and B appear in a reference frame S′ that travels at an ordinary speed $v < c$ relative to frame S? We can use the Lorentz transformations to find out. Because $x_A = 0$ and $t_A = 0$, it's easy to see that $x_A' = 0$ and $t_A' = 0$. That is, the origins of S and S′ overlap at the instant the causal influence leaves event A. More interesting is the time at which this influence reaches B in frame S′. The Lorentz time transformation for event B is

$$t'_B = \gamma\left(t_B - \frac{vx_B}{c^2}\right) = \gamma t_B\left(1 - \frac{v(x_B/t_B)}{c^2}\right) = \gamma t_B\left(1 - \frac{vu}{c^2}\right) \qquad (36.36)$$

where we first factored out t_B, then made use of the fact that $u = x_B/t_B$ in frame S.

We're assuming $u > c$, so let $u = \alpha c$ where $\alpha > 1$ is a constant. Then $vu/c^2 = \alpha v/c$. Now follow the logic:

1. If $v > c/\alpha$, which is possible because $\alpha > 1$, then $vu/c^2 > 1$.
2. If $vu/c^2 > 1$, then the term $(1 - vu/c^2)$ is negative and $t'_B < 0$.
3. If $t'_B < 0$, then event B happens *before* event A in reference frame S′.

In other words, if a causal influence can travel faster than c, then there exist reference frames in which the effect happens before the cause. We know this can't happen, so our assumption $u > c$ must be wrong. **No causal influence of any kind—particle, wave, or yet-to-be-discovered z rays—can travel faster than c.**

The existence of a cosmic speed limit is one of the most interesting consequences of the theory of relativity. "Warp drive," in which a spaceship suddenly leaps to faster-than-light velocities, is simply incompatible with the theory of relativity. Rapid travel to the stars will remain in the realm of science fiction unless future scientific discoveries find flaws in Einstein's theory and open the doors to yet-undreamed-of theories. While we can't say with certainty that a scientific theory will never be overturned, there is currently not even a hint of evidence that disagrees with the special theory of relativity.

36.10 Relativistic Energy

Energy is our final topic in this chapter on relativity. Space, time, velocity, and momentum are changed by relativity, so it seems inevitable that we'll need a new view of energy.

In Newtonian mechanics, a particle's kinetic energy $K = \frac{1}{2}mu^2$ can be written in terms of its momentum $p = mu$ as $K = p^2/2m$. This suggests that a relativistic expression for energy will likely involve both the square of p and the particle's mass. We also hope that energy will be conserved in relativity, so a reasonable starting point is with the one quantity we've found that is the same in all inertial reference frames: the spacetime interval s.

Let a particle of mass m move through distance Δx during a time interval Δt, as measured in reference frame S. The spacetime interval is

$$s^2 = c^2(\Delta t)^2 - (\Delta x)^2 = \text{invariant}$$

We can turn this into an expression involving momentum if we multiply by $(m/\Delta\tau)^2$, where $\Delta\tau$ is the proper time (i.e., the time measured by the particle). Doing so gives

$$(mc)^2\left(\frac{\Delta t}{\Delta\tau}\right)^2 - \left(\frac{m\Delta x}{\Delta\tau}\right)^2 = (mc)^2\left(\frac{\Delta t}{\Delta\tau}\right)^2 - p^2 = \text{invariant} \qquad (36.37)$$

where we used $p = m(\Delta x/\Delta\tau)$ from Equation 36.32.

Now Δt, the time interval in frame S, is related to the proper time by the time-dilation result $\Delta t = \gamma_p\Delta\tau$. With this change, Equation 36.37 becomes

$$(\gamma_p mc)^2 - p^2 = \text{invariant}$$

Finally, for reasons that will be clear in a minute, multiply by c^2, to get

$$(\gamma_p mc^2)^2 - (pc)^2 = \text{invariant} \qquad (36.38)$$

To say that the right side is an *invariant* means it has the same value in all inertial reference frames. We can easily determine the constant by evaluating it in the reference frame in which the particle is at rest. In that frame, where $p = 0$ and $\gamma_p = 1$, we find that

$$(\gamma_p mc^2)^2 - (pc)^2 = (mc^2)^2 \tag{36.39}$$

Let's reflect on what this means before taking the next step. The spacetime interval s has the same value in all inertial reference frames. In other words, $c^2(\Delta t)^2 - (\Delta x)^2 = c^2(\Delta t')^2 - (\Delta x')^2$. Equation 36.39 was derived from the definition of the spacetime interval, hence the quantity mc^2 is also an invariant having the same value in all inertial reference frames. In other words, if experimenters in frames S and S' both make measurements on this particle of mass m, they will find that

$$(\gamma_p mc^2)^2 - (pc)^2 = (\gamma_p' mc^2)^2 - (p'c)^2 \tag{36.40}$$

Experimenters in different reference frames measure different values for the momentum, but experimenters in all reference frames agree that momentum is a conserved quantity. Equations 36.39 and 36.40 suggest that the quantity $\gamma_p mc^2$ is also an important property of the particle, a property that changes along with p in just the right way to satisfy Equation 36.39. But what is this property?

The first clue comes from checking the units. γ_p is dimensionless and c is a velocity, so $\gamma_p mc^2$ has the same units as the classical expression $\frac{1}{2}mv^2$; namely, units of energy. For a second clue, let's examine how $\gamma_p mc^2$ behaves in the low-velocity limit $u \ll c$. We can use the binomial approximation expression for γ_p to find

$$\gamma_p mc^2 = \frac{mc^2}{\sqrt{1 - u^2/c^2}} \approx \left(1 + \frac{1}{2}\frac{u^2}{c^2}\right)mc^2 = mc^2 + \frac{1}{2}mu^2 \tag{36.41}$$

The second term, $\frac{1}{2}mu^2$, is the low-velocity expression for the kinetic energy K. This is an energy associated with motion. But the first term suggests that the concept of energy is more complex than we originally thought. It appears that **there is an inherent energy associated with mass itself.**

With that as a possibility, subject to experimental verification, let's define the **total energy** E of a particle to be

$$E = \gamma_p mc^2 = E_0 + K = \text{rest energy} + \text{kinetic energy} \tag{36.42}$$

This total energy consists of a **rest energy**

$$E_0 = mc^2 \tag{36.43}$$

and a relativistic expression for the *kinetic energy*

$$K = (\gamma_p - 1)mc^2 = (\gamma_p - 1)E_0 \tag{36.44}$$

This expression for the kinetic energy is very nearly $\frac{1}{2}mu^2$ when $u \ll c$ but, as Figure 36.36 shows, differs significantly from the classical value for very high velocities.

Equation 36.43 is, of course, Einstein's famous $E = mc^2$, perhaps the most famous equation in all of physics. Before discussing its significance, we need to tie up some loose ends. First, notice that the right-hand side of Equation 36.39 is the square of the rest energy E_0. Thus we can write a final version of that equation:

$$E^2 - (pc)^2 = E_0^2 \tag{36.45}$$

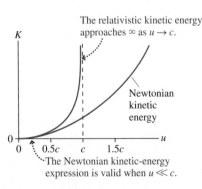

The relativistic kinetic energy approaches ∞ as $u \to c$.

Newtonian kinetic energy

The Newtonian kinetic-energy expression is valid when $u \ll c$.

FIGURE 36.36 The relativistic kinetic energy.

The quantity E_0 is an *invariant* with the same value mc^2 in *all* inertial reference frames.

Second, notice that we can write

$$pc = (\gamma_{\mathrm{p}}mu)c = \frac{u}{c}(\gamma_{\mathrm{p}}mc^2)$$

But $\gamma_{\mathrm{p}}mc^2$ is the total energy E and $u/c = \beta_{\mathrm{p}}$, where the subscript p, as on γ_{p}, implies that we're referring to the motion of a particle within a reference frame, not the motion of two reference frames relative to each other. Thus

$$pc = \beta_{\mathrm{p}}E \qquad (36.46)$$

Figure 36.37 shows the "velocity-energy-momentum triangle," a convenient way to remember the relationships between the three quantities.

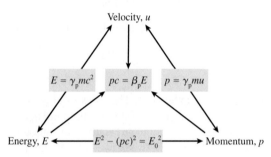

FIGURE 36.37 The velocity-energy-momentum triangle.

EXAMPLE 36.12 Kinetic energy and total energy
Calculate the rest energy and the kinetic energy of (a) a 100 g ball moving with a speed of 100 m/s and (b) an electron with a speed of 0.999c.

MODEL The ball, with $u \ll c$, is a classical particle. We don't need to use the relativistic expression for its kinetic energy. The electron is highly relativistic.

SOLVE

a. For the ball, with $m = 0.10$ kg,

$$E_0 = mc^2 = 9.0 \times 10^{15} \text{ J}$$

$$K = \frac{1}{2}mu^2 = 500 \text{ J}$$

b. For the electron, we start by calculating

$$\gamma_{\mathrm{p}} = 1/(1 - u^2/c^2)^{1/2} = 22.4$$

Then, using $m_{\mathrm{e}} = 9.11 \times 10^{-31}$ kg,

$$E_0 = mc^2 = 8.2 \times 10^{-14} \text{ J}$$

$$K = (\gamma_{\mathrm{p}} - 1)E_0 = 170 \times 10^{-14} \text{ J}$$

ASSESS The ball's kinetic energy is a typical kinetic energy. Its rest energy, by contrast, is a staggeringly large number. For a relativistic electron, on the other hand, the kinetic energy is more important than the rest energy.

STOP TO THINK 36.8 An electron moves through the lab at 99% the speed of light. The lab reference frame is S and the electron's reference frame is S′. In which reference frame is the electron's rest mass larger?

a. Frame S, the lab frame
b. Frame S′, the electron's frame
c. It is the same in both frames.

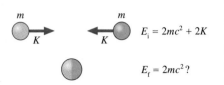

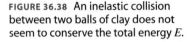

FIGURE 36.38 An inelastic collision between two balls of clay does not seem to conserve the total energy E.

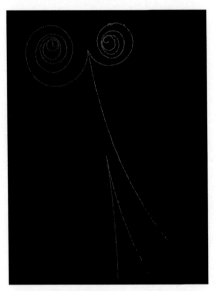

The tracks of elementary particles in a bubble chamber show the creation of an electron-positron pair. The negative electron and positive positron spiral in opposite directions in the magnetic field.

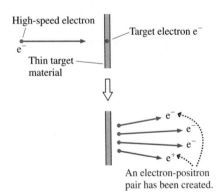

FIGURE 36.39 An inelastic collision between electrons can create an electron-positron pair.

Mass-Energy Equivalence

Now we're ready to explore the significance of Einstein's famous equation $E = mc^2$. Figure 36.38 shows two balls of clay approaching each other. They have equal masses, equal kinetic energies, and slam together in a perfectly inelastic collision to form one large ball of clay at rest. In Newtonian mechanics, we would say that the initial energy $2K$ is dissipated by being transformed into an equal amount of thermal energy, raising the temperature of the coalesced ball of clay. But Equation 36.42, $E = E_0 + K$, doesn't say anything about thermal energy. The total energy before the collision is $E_i = 2mc^2 + 2K$, with the factor of 2 appearing because there are two masses. It seems like the total energy after the collision, when the clay is at rest, should be $2mc^2$, but this value doesn't conserve total energy.

There's ample experimental evidence that energy is conserved, so there must be a flaw in our reasoning. The statement of energy conservation is

$$E_f = Mc^2 = E_i = 2mc^2 + 2K \qquad (36.47)$$

where M is the mass of clay after the collision. But, remarkably, this requires

$$M = 2m + \frac{2K}{c^2} \qquad (36.48)$$

In other words, **mass is not conserved.** The mass of clay after the collision is larger than the mass of clay before the collision. Total energy can be conserved only if kinetic energy is transformed into an "equivalent" amount of mass.

The mass increase in a collision between two balls of clay is incredibly small, far beyond any scientist's ability to detect. So how do we know if such a crazy idea is true?

Figure 36.39 shows an experiment that has been done countless times in the last 50 years at particle accelerators around the world. An electron that has been accelerated to $u \approx c$ is aimed at a target material. When a high-energy electron collides with an atom in the target, it can easily knock one of the electrons out of the atom. Thus we would expect to see two electrons leaving the target: the incident electron and the ejected electron. Instead, *four* particles emerge from the target: three electrons and a positron. A *positron,* or positive electron, is the antimatter version of an electron, identical to an electron in all respects other than having charge $q = +e$.

In chemical-reaction notation, the collision is

$$e^-(\text{fast}) + e^-(\text{at rest}) \rightarrow e^- + e^- + e^- + e^+$$

An electron and a positron have been *created,* apparently out of nothing. Mass $2m_e$ before the collision has become mass $4m_e$ after the collision. (Notice that charge has been conserved in this collision.)

Although the mass has increased, it wasn't created "out of nothing." This is an inelastic collision, just like the collision of the balls of clay, because the kinetic energy after the collision is less than before. In fact, if you measured the energies before and after the collision, you would find that the decrease in kinetic energy is exactly equal to the energy equivalent of the two particles that have been created: $\Delta K = 2m_e c^2$. The new particles have been created *out of energy!*

Particles can be created from energy and particles can return to energy. Figure 36.40 shows an electron colliding with a positron, its antimatter partner. When a particle and its antiparticle meet, they *annihilate* each other. The mass disappears, and the energy equivalent of the mass is transformed into two high-energy photons of light. Momentum conservation requires two photons, rather than one, and specifies that the two photons have equal energies and be emitted back to back.

If the electron and positron are fairly slow, so that $K \ll mc^2$, then $E_i \approx E_0 = mc^2$. In that case, energy conservation requires

$$E_f = 2E_{photon} = E_i \approx 2m_e c^2 \tag{36.49}$$

You learned in Chapter 24 that the energy of a photon of light is $E_{photon} = hc/\lambda$, where h is Planck's constant. (Photons and their properties will be discussed again in Chapter 38.) Hence the wavelength of the emitted photons is

$$\lambda = \frac{hc}{m_e c^2} \approx 0.0024 \text{ nm} \tag{36.50}$$

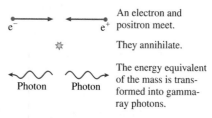

An electron and positron meet.

They annihilate.

The energy equivalent of the mass is transformed into gamma-ray photons.

FIGURE 36.40 The annihilation of an electron-positron pair.

This is an extremely short wavelength, even shorter than the wavelengths of x rays. Photons in this wavelength range are called *gamma rays*. And, indeed, the emission of 0.0024 nm gamma rays is observed in many laboratory experiments in which positrons are able to collide with electrons and thus annihilate. In recent years, with the advent of gamma-ray telescopes on satellites, astronomers have found 0.0024 nm photons coming from many places in the universe, especially galactic centers—evidence that positrons are abundant throughout the universe.

Positron-electron annihilation is also the basis of the medical procedure known as a positron-emission tomography, or PET scans. A patient ingests a very small amount of a radioactive substance that decays by the emission of positrons. This substance is taken up by certain tissues in the body, especially those tissues with a high metabolic rate. As the substance decays, the positrons immediately collide with electrons, annihilate, and create two gamma-ray photons that are emitted back to back. The gamma rays, which easily leave the body, are detected, and their trajectories are traced backward into the body. The overlap of many such trajectories shows quite clearly the tissue where the positron emission is occurring. The results are usually shown as false-color photographs, with redder areas indicating regions of higher positron emission.

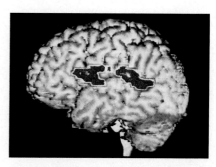

Positron-electron annihilation (a PET scan) provides a noninvasive look into the brain.

Conservation of Energy

The creation and annihilation of particles with mass, processes strictly forbidden in Newtonian mechanics, are vivid proof that neither mass nor the Newtonian definition of energy are conserved. Even so, the *total* energy—the kinetic energy *and* the energy equivalent of mass—remains a conserved quantity.

> **Law of conservation of total energy** The energy $E = \sum E_i$ of an isolated system is conserved, where $E_i = (\gamma_p)_i m_i c^2$ is the total energy of particle i.

Mass and energy are not the same thing, but, as the last few examples have shown, they are *equivalent* in the sense that mass can be transformed into energy and energy can be transformed into mass as long as the total energy is conserved.

Probably the most well-known application of the conservation of total energy is nuclear fission. The uranium isotope ^{236}U, containing 236 protons and neutrons, does not exist in nature. It can be created when a ^{235}U nucleus absorbs a neutron, increasing its atomic mass from 235 to 236. The ^{236}U nucleus quickly fragments into two smaller nuclei and several extra neutrons, a process known as **nuclear fission.** The nucleus can fragment in several ways, but one is

$$n + {}^{235}U \rightarrow {}^{236}U \rightarrow {}^{144}Ba + {}^{89}Kr + 3n$$

Ba and Kr are the atomic symbols for barium and krypton.

This reaction seems like an ordinary chemical reaction—until you check the masses. The masses of atomic isotopes are known with great precision from many decades of measurement in instruments called mass spectrometers. If you add up

The mass of the reactants is 0.185 u more than the mass of the products.

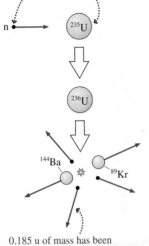

0.185 u of mass has been converted into kinetic energy.

FIGURE 36.41 In nuclear fission, the energy equivalent of lost mass is converted into kinetic energy.

the masses on both sides, you find that the mass of the products is 0.185 u smaller than the mass of the initial neutron and ^{235}U, where, you will recall, 1 u = 1.66 × 10^{-27} kg is the unified atomic mass unit. Converting to kilograms gives us the mass loss of 3.07 × 10^{-28} kg.

Mass has been lost, but the energy equivalent of the mass has not. As Figure 36.41 shows, the mass has been converted to kinetic energy, causing the two product nuclei and three neutrons to be ejected at very high speeds. The kinetic energy is easily calculated: $\Delta K = m_{\text{lost}}c^2 = 2.8 \times 10^{-11}$ J.

This is a very tiny amount of energy, but it is the energy released from *one* fission. The number of nuclei in a macroscopic sample of uranium is on the order of N_A, Avogadro's number. Hence the energy available if *all* the nuclei fission is enormous. This energy, of course, is the basis for both nuclear power reactors and nuclear weapons.

We started this chapter with an expectation that relativity would challenge our basic notions of space and time. We end by finding that relativity changes our understanding of mass and energy. Most remarkable of all is that each and every one of these new ideas flows from one simple statement: The laws of physics are the same in all inertial reference frames.

SUMMARY

The goal of Chapter 36 has been to understand how Einstein's theory of relativity changes our concepts of space and time.

GENERAL PRINCIPLES

Principle of Relativity All the laws of physics are the same in all inertial reference frames.

• The speed of light c is the same in all inertial reference frames.

• No particle or causal influence can travel at a speed greater than c.

IMPORTANT CONCEPTS

Space

Spatial measurements depend on the motion of the experimenter relative to the events. An object's length is the difference between *simultaneous* measurement of the positions of both ends.

Proper length ℓ is the length of an object measured in a reference frame in which the object is at rest. The L in a frame in which the object moves with velocity v is

$$L = \sqrt{1 - \beta^2}\, \ell \leq \ell$$

This is called **length contraction**.

Time

Time measurements depend on the motion of the experimenter relative to the events. Events that are simultaneous in reference frame S are not simultaneous in frame S' moving relative to S.

Proper time $\Delta\tau$ is the time interval between two events measured in a reference frame in which the events occur at the same position. The time interval Δt in a frame moving with relative velocity v is

$$\Delta t = \Delta\tau/\sqrt{1 - \beta^2} \geq \Delta\tau$$

This is called **time dilation**.

Momentum

The law of conservation of momentum is valid in all inertial reference frames if the momentum of a particle with velocity u is $p = \gamma_\mathrm{p} m u$, where

$$\gamma_\mathrm{p} = 1/\sqrt{1 - u^2/c^2}$$

The momentum approaches ∞ as $u \to c$.

Energy

The law of conservation of energy is valid in all inertial reference frames if the energy of a particle with velocity u is $E = \gamma_\mathrm{p} m c^2 = E_0 + K$

Rest energy $E_0 = m c^2$

Kinetic energy $K = (\gamma_\mathrm{p} - 1) m c^2$

Invariants are quantities that have the same value in all inertial reference frames.

Spacetime interval: $s^2 = (c\Delta t)^2 - (\Delta x)^2$

Particle rest energy: $E_0^2 = (mc^2)^2 = E^2 - (pc)^2$

Mass-energy equivalence

Mass m can be transformed into energy $E = mc^2$.

Energy can be transformed into mass $m = \Delta E/c^2$.

APPLICATIONS

An event happens at a specific place in space and time. Spacetime coordinates are (x, t) in frame S and (x', t') in frame S'.

A reference frame is a coordinate system with meter sticks and clocks for measuring events. Experimenters at rest relative to each other share the same reference frame.

The Lorentz transformations transform spacetime coordinates and velocities between reference frames S and S'.

$$x' = \gamma(x - vt) \qquad x = \gamma(x' + vt')$$
$$y' = y \qquad y = y'$$
$$z' = z \qquad z = z'$$
$$t' = \gamma(t - vx/c^2) \qquad t = \gamma(t' + vx'/c^2)$$
$$u' = \frac{u - v}{1 - uv/c^2} \qquad u = \frac{u' + v}{1 + u'v/c^2}$$

where u and u' are the x- and x'-components of velocity.

$$\beta = \frac{v}{c} \quad \text{and} \quad \gamma = 1/\sqrt{1 - v^2/c^2} = 1/\sqrt{1 - \beta^2}$$

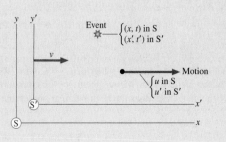

TERMS AND NOTATION

special relativity	simultaneous	invariant
reference frame	relativity of simultaneity	spacetime interval, s
inertial reference frame	light clock	Lorentz transformations
Galilean principle of relativity	rest frame	causal influence
ether	proper time, $\Delta\tau$	total energy, E
principle of relativity	time dilation	rest energy, E_0
event	light year, ly	law of conservation of total energy
spacetime coordinates, (x, y, z, t)	proper length, ℓ	nuclear fission
synchronized	length contraction	

EXERCISES AND PROBLEMS

Exercises

Section 36.2 Galilean Relativity

1. At $t = 1$ s, a firecracker explodes at $x = 10$ m in reference frame S. Four seconds later, a second firecracker explodes at $x = 20$ m. Reference frame S′ moves in the x-direction at a speed of 5 m/s. What are the positions and times of these two events in frame S′?

2. A firecracker explodes in reference frame S at $t = 1$ s. A second firecracker explodes at the same position at $t = 3$ s. In reference frame S′, which moves in the x-direction at speed v, the first explosion is detected at $x' = 4$ m and the second at $x' = -4$ m.
   a. What is the speed of frame S′ relative to frame S?
   b. What is the position of the two explosions in frame S?

3. A sprinter crosses the finish line of a race. The roar of the crowd in front approaches her at a speed of 360 m/s. The roar from the crowd behind her approaches at 330 m/s. What are the speed of sound and the speed of the sprinter?

4. A baseball pitcher can throw a ball with a speed of 40 m/s. He is in the back of a pickup truck that is driving away from you. He throws the ball in your direction, and it floats toward you at a lazy 10 m/s. What is the speed of the truck?

5. A boy on a skateboard coasts along at 5 m/s. He has a ball that he can throw at a speed of 10 m/s. What is the ball's speed relative to the ground if he throws the ball
   a. forward?
   b. backward?
   c. to the side?

Section 36.3 Einstein's Principle of Relativity

6. An out-of-control alien spacecraft is diving into a star at a speed of 1×10^8 m/s. At what speed, relative to the spacecraft, is the starlight approaching?

7. A starship blasts past the earth at 2×10^8 m/s. Just after passing the earth, it fires a laser beam out the back of the starship. With what speed does the laser beam approach the earth?

8. A positron moving in the positive x-direction at 2×10^8 m/s collides with an electron at rest. The positron and electron annihilate, producing two gamma-ray photons. Photon 1 travels in the positive x-direction and photon 2 travels in the negative x-direction. What is the speed of each photon?

Section 36.4 Events and Measurements

Section 36.5 The Relativity of Simultaneity

9. Your job is to synchronize the clocks in a reference frame. You are going to do so by flashing a light at the origin at $t = 0$ s. To what time should the clock at $(x, y, z) = (30$ m, 40 m, 0 m$)$ be preset?

10. Bjorn is standing at $x = 600$ m. Firecracker 1 explodes at the origin and firecracker 2 explodes at $x = 900$ m. The flashes from both explosions reach Bjorn's eye at $t = 3$ μs. At what time did each firecracker explode?

11. Bianca is standing at $x = 600$ m. Firecracker 1, at the origin, and firecracker 2, at $x = 900$ m, explode simultaneously. The flash from firecracker 1 reaches Bianca's eye at $t = 3$ μs. At what time does she see the flash from firecracker 2?

12. You are standing at $x = 9$ km. Lightning bolt 1 strikes at $x = 0$ km and lightning bolt 2 strikes at $x = 12$ km. Both flashes reach your eye at the same time. Your assistant is standing at $x = 3$ km. Does your assistant see the flashes at the same time? If not, which does she see first and what is the time difference between the two?

13. You are standing at $x = 9$ km and your assistant is standing at $x = 3$ km. Lightning bolt 1 strikes at $x = 0$ km and lightning bolt 2 strikes at $x = 12$ km. You see the flash from bolt 2 at $t = 10$ μs and the flash from bolt 1 at $t = 50$ μs. According to your assistant, were the lightning strikes simultaneous? If not, which occurred first and what was the time difference between the two?

14. Jose is looking to the east. Lightning bolt 1 strikes a tree 300 m from him. Lightning bolt 2 strikes a barn 900 m from him in the same direction. Jose sees the tree strike 1 μs before he sees the barn strike. According to Jose, were the lightning strikes simultaneous? If not, which occurred first and what was the time difference between the two?

15. You are flying your personal rocketcraft at $0.9c$ from Star A toward Star B. The distance between the stars, in the stars' reference frame, is 1 ly. Both stars happen to explode simultaneously in your reference frame at the instant you are exactly halfway between them. Do you see the flashes simultaneously? If not, which do you see first and what is the time difference between the two?

Section 36.6 Time Dilation

16. A cosmic ray travels 60 km through the earth's atmosphere in $400 \mu s$, as measured by experimenters on the ground. How long does the journey take according to the cosmic ray?

17. At what speed, as a fraction of c, does a moving clock tick at half the rate of an identical clock at rest?

18. An astronaut travels to a star system 4.5 ly away at a speed of $0.9c$. Assume that the time needed to accelerate and decelerate is negligible.
    a. How long does the journey take according to Mission Control on earth?
    b. How long does the journey take according to the astronaut?
    c. How much time elapses between the launch and the arrival of the first radio message from the astronaut saying that she has arrived?

19. A starship voyages to a distant planet 10 ly away. The explorers stay 1 yr, return at the same speed, and arrive back on earth 26 yr after they left. Assume that the time needed to accelerate and decelerate is negligible.
    a. What is the speed of the starship?
    b. How much time has elapsed on the astronauts' chronometers?

20. You fly 5000 km across the United States on an airliner at 250 m/s. You return two days later at the same speed.
    a. Have you aged more or less than your friends at home?
    b. By how much?
    Hint: Use the binomial approximation.

21. Two clocks are synchronized. One is placed in a race car that drives around a 2.0-km-diameter track at 100 m/s for 24 hours. Afterward, by how much do the two clocks differ?
    Hint: Use the binomial approximation.

Section 36.7 Length Contraction

22. At what speed, as a fraction of c, will a moving rod have a length 60% that of an identical rod at rest?

23. Jill claims that her new rocket is 100 m long. As she flies past your house, you measure the rocket's length and find that it is only 80 m. Should Jill be cited for exceeding the $0.5c$ speed limit?

24. A muon travels 60 km through the atmosphere at a speed of $0.9997c$. According to the muon, how thick is the atmosphere?

25. The Stanford Linear Accelerator (SLAC) accelerates electrons to $c = 0.99999997c$ in a 3.2-km-long tube. If they travel the length of the tube at full speed (they don't, because they are accelerating), how long is the tube in the electrons' reference frame?

26. Our Milky Way galaxy is 100,000 ly in diameter. A spaceship crossing the galaxy measures the galaxy's diameter to be a mere 1 ly.
    a. What is the speed of the spaceship relative to the galaxy?
    b. How long is the crossing time as measured in the galaxy's reference frame?

27. An optical interferometer can detect a displacement of 0.1λ, or ≈ 50 nm. At what speed would a meter stick "shrink" by 50 nm?
    Hint: Use the binomial approximation.

Section 36.8 The Lorentz Transformations

28. An event has spacetime coordinates $(x, t) = (1200 \text{ m}, 2.0 \mu s)$ in reference frame S. What are the event's spacetime coordinates (a) in reference frame S' that moves in the positive x-direction at $0.8c$ and (b) in reference frame S'' that moves in the negative x-direction at $0.8c$?

29. A rocket travels in the x-direction at speed $0.6c$ with respect to the earth. An experimenter on the rocket observes a collision between two comets and determines that the spacetime coordinates of the collision are $(x', t') = (3 \times 10^{10} \text{ m}, 200 \text{ s})$. What are the spacetime coordinates of the collision in earth's reference frame?

30. In the earth's reference frame, a tree is at the origin and a pole is at $x = 30$ km. Lightning strikes both the tree and the pole at $t = 10 \mu s$. The lightning strikes are observed by a rocket traveling in the x-direction at $0.5c$.
    a. What are the spacetime coordinates for these two events in the rocket's reference frame?
    b. Are the events simultaneous in the rocket's frame? If not, which occurs first?

31. A rocket cruising past earth at $0.8c$ shoots a bullet out the back door, opposite the rocket's motion, at $0.9c$ relative to the rocket. What is the bullet's speed relative to the earth?

32. A laboratory experiment shoots an electron to the left at $0.9c$. What is the electron's speed relative to a proton moving to the right at $0.9c$?

33. A distant quasar is found to be moving away from the earth at $0.8c$. A galaxy closer to the earth and along the same line of sight is moving away from us at $0.2c$. What is the recessional speed of the quasar as measured by astronomers in the other galaxy?

Section 36.9 Relativistic Momentum

34. A proton is accelerated to $0.999c$.
    a. What is the proton's momentum?
    b. By what factor does the proton's momentum exceed its Newtonian momentum?

35. A 1 g particle has momentum 400,000 kg m/s. What is the particle's speed?

36. At what speed is a particle's momentum twice its Newtonian value?

37. What is the speed of a particle whose momentum is mc?

Section 36.10 Relativistic Energy

38. What are the kinetic energy, the rest energy, and the total energy of a 1 g particle with a speed of $0.8c$?

39. A quarter-pound hamburger with all the fixings has a mass of 200 g. The food energy of the hamburger (480 food calories) is 2 MJ.
    a. What is the energy equivalent of the mass of the hamburger?
    b. By what factor does the energy equivalent exceed the food energy?

40. How fast must an electron move so that its total energy is 10% more than its rest mass energy?
41. At what speed is a particle's kinetic energy twice its rest energy?
42. At what speed is a particle's total energy twice its rest energy?

Problems

43. A 50 g ball moving to the right at 4.0 m/s overtakes and collides with a 100 g ball moving to the right at 2.0 m/s. The collision is perfectly elastic. Use reference frames and the Chapter 10 result for perfectly elastic collisions to find the speed and direction of each ball after the collision.
44. A 300 g ball moving to the right at 2 m/s has a perfectly elastic collision with a 100 g ball moving to the left at 8 m/s. Use reference frames and the Chapter 10 result for perfectly elastic collisions to find the speed and direction of each ball after the collision.
45. A billiard ball has a perfectly elastic collision with a second billiard ball of equal mass. Afterward, the first ball moves to the left at 2.0 m/s and the second to the right at 4.0 m/s. Use reference frames and the Chapter 10 result for perfectly elastic collisions to find the speed and direction of each ball before the collision.
46. A 9.0 kg artillery shell is moving to the right at 100 m/s when suddenly it explodes into two fragments, one twice as heavy as the other. Measurements reveal that 900 J of energy are released in the explosion and that the heavier fragment was in front of the lighter fragment. Find the velocity of each fragment relative to the ground by analyzing the explosion in the reference frame of (a) the ground and (b) the shell. (c) Is the problem easier to solve in one reference frame?
47. The diameter of the solar system is 10 light hours. A spaceship crosses the solar system in 15 hours, as measured on earth. How long, in hours, does the passage take according to passengers on the spaceship?
    Hint: $c = 1$ light hour per hour.
48. A 30-m-long rocket train car is traveling from Los Angeles to New York at $0.5c$ when a light at the center of the car flashes. When the light reaches the front of the car, it immediately rings a bell. Light reaching the back of the car immediately sounds a siren.
    a. Are the bell and siren simultaneous events for a passenger seated in the car? If not, which occurs first and by how much time?
    b. Are the bell and siren simultaneous events for a bicyclist waiting to cross the tracks? If not, which occurs first and by how much time?
49. The star Alpha goes supernova. Ten years later and 100 ly away, as measured by astronomers in the galaxy, star Beta explodes.
    a. Is it possible that the explosion of Alpha is in any way responsible for the explosion of Beta? Explain.
    b. An alien spacecraft passing through the galaxy finds that the distance between the two explosions is 120 ly. According to the aliens, what is the time between the explosions?
50. Two events in reference frame S occur 10 μs apart at the same point in space. The distance between the two events is 2400 m in reference frame S'.
    a. What is the time interval between the events in reference frame S'?
    b. What is the velocity of S' relative to S?

51. a. How fast must a rocket travel on a journey to and from a distant star so that the astronauts age 10 years while the Mission Control workers on earth age 120 years?
    b. As measured by Mission Control, how far away is the distant star?
52. In Section 36.6 we explained that muons can reach the ground because of time dilation. But how do things appear in the muon's reference frame, where the muon's half-life is only 1.5 μs? How can a muon travel the 60 km to reach the earth's surface before decaying? Resolve this apparent paradox. Be as quantitative as you can in your answer.
53. A cube has a density of 2000 kg/m³ while at rest in the laboratory. What is the cube's density as measured by an experimenter in the laboratory as the cube moves through the laboratory at 90% of the speed of light in a direction perpendicular to one of its faces?
54. In an attempt to reduce the extraordinarily long travel times for voyaging to distant stars, some people have suggested traveling at close to the speed of light. Suppose you wish to visit the red giant star Betelgeuse, which is 430 ly away, and that you want your 20,000 kg rocket to move so fast that you age only 20 years during the round trip.
    a. How fast must the rocket travel relative to earth?
    b. How much energy is needed to accelerate the rocket to this speed?
    c. Compare this amount of energy to the total energy used by the United States in the year 2000, which was roughly 1.0×10^{20} J.
55. A rocket traveling at $0.5c$ sets out for the nearest star, Alpha Centauri, which is 4.25 ly away from earth. It will return to earth immediately after reaching Alpha Centauri. What distance will the rocket travel and how long will the journey last according to (a) stay-at-home earthlings and (b) the rocket crew? (c) Which answers are the correct ones, those in part a or those in part b?
56. The star Delta goes supernova. One year later and 2 ly away, as measured by astronomers in the galaxy, star Epsilon explodes. Let the explosion of Delta be at $x_D = 0$ and $t_D = 0$. The explosions are observed by three spaceships cruising through the galaxy in the direction from Delta to Epsilon at velocities $v_1 = 0.3c$, $v_2 = 0.5c$, and $v_3 = 0.7c$.
    a. What are the times of the two explosions as measured by scientists on each of the three spaceships?
    b. Does one spaceship find that the explosions are simultaneous? If so, which one?
    c. Does one spaceship find that Epsilon explodes before Delta? If so, which one?
    d. Do your answers to parts b and c violate the idea of causality? Explain.
57. Two rockets approach each other. Each is traveling at $0.75c$ in the earth's reference frame. What is the speed of one rocket relative to the other?
58. A military jet traveling at 1500 m/s has engine trouble and the pilot must bail out. Her ejection seat shoots her forward at 300 m/s relative to the jet. How fast is she traveling relative to the ground according to the relativistic velocity transformation? Would the pilot or anyone else be able to tell the difference between the relativistic and the nonrelativistic result?
59. What is the speed of an electron after being accelerated from rest through a 20×10^6 V potential difference?

60. What is the speed of a proton after being accelerated from rest through a 50×10^6 V potential difference?

61. The half-life of a muon at rest is 1.5 μs. Muons that have been accelerated to a very high speed and are then held in a circular storage ring have a half-life of 7.5 μs.
    a. What is the speed of the muons in the storage ring?
    b. What is the total energy of a muon in the storage ring? The mass of a muon is 207 times the mass of an electron.

62. A solar flare blowing out from the sun at $0.9c$ is overtaking a rocket as it flies away from the sun at $0.8c$. According to the crew on board, with what speed is the flare gaining on the rocket?

63. This chapter has assumed that lengths perpendicular to the direction of motion are not affected by the motion. That is, motion in the x-direction does not cause length contraction along the y- or z-axes. To find out if this is really true, consider two spray-paint nozzles attached to rods perpendicular to the x-axis. It has been confirmed that, when both rods are at rest, both nozzles are exactly 1 m above the base of the rod. One rod is placed in the S reference frame with its base on the x-axis; the other is placed in the S' reference frame with its base on the x'-axis. The rods then swoop past each other and, as Figure P36.63 shows, each paints a stripe across the other rod.

    We will use proof by contradiction. Assume that objects perpendicular to the motion *are* contracted. An experimenter in frame S finds that the S' nozzle, as it goes past, is less than 1 m above the x-axis. The principle of relativity says that an experiment carried out in two different inertial reference frames will have the same outcome in both.
    a. Pursue this line of reasoning and show that you end up with a logical contradiction, two mutually incompatible situations.
    b. What can you conclude from this contradiction?

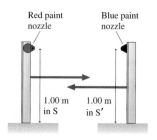

Red paint nozzle   Blue paint nozzle

1.00 m in S   1.00 m in S'

FIGURE P36.63

64. Derive the Lorentz transformations for t and t'.
    Hint: See the comment following Equation 36.22.

65. a. Derive a velocity transformation equation for u_y and u'_y. Assume that the reference frames are in the standard orientation with motion parallel to the x- and x'-axes.
    b. A rocket passes the earth at $0.8c$. As it goes by, it launches a projectile at $0.6c$ perpendicular to the direction of motion. What is the projectile's speed in the earth's reference frame?

66. What is the momentum of a particle with speed $0.95c$ and total energy 2.0×10^{-10} J?

67. What is the momentum of a particle whose total energy is four times its rest energy? Give your answer as a multiple of mc.

68. a. What are the momentum and total energy of a proton with speed $0.99c$?
    b. What is the proton's momentum in a different reference frame in which $E' = 5.0 \times 10^{-10}$ J?

69. At what speed is the kinetic energy of a particle twice its Newtonian value?

70. What is the speed of an electron whose total energy equals the rest mass of a proton?

71. A typical nuclear power plant generates electricity at the rate of 1000 MW. The efficiency of transforming thermal energy into electrical energy is $\frac{1}{3}$ and the plant runs at full capacity for 80% of the year. (Nuclear power plants are down about 20% of the time for maintenance and refueling.)
    a. How much thermal energy does the plant generate in one year?
    b. What mass of uranium is transformed into energy in one year?

72. The sun radiates energy at the rate 3.8×10^{26} W. The source of this energy is fusion, a nuclear reaction in which mass is transformed into energy. The mass of the sun is 2.0×10^{30} kg.
    a. How much mass does the sun lose each year?
    b. What percentage is this of the sun's total mass?
    c. Estimate the lifetime of the sun.

73. The radioactive element radium (Ra) decays by a process known as *alpha decay*, in which the nucleus emits a helium nucleus. (These high-speed helium nuclei were named alpha particles when radioactivity was first discovered, long before the identity of the particles was established.) The reaction is ^{226}Ra $\rightarrow$ ^{222}Rn $+$ ^{4}He, where Rn is the element radon. The accurately measured atomic masses of the three atoms are 226.025, 222.017, and 4.003. How much energy is released in each decay? (The energy released in radioactive decay is what makes nuclear waste "hot.")

74. The nuclear reaction that powers the sun is the fusion of four protons into a helium nucleus. The process involves several steps, but the net reaction is simply $4p \rightarrow$ ^{4}He $+$ energy. The mass of a helium nucleus is known to be 6.64×10^{-27} kg.
    a. How much energy is released in each fusion?
    b. What fraction of the initial rest mass energy is this energy?

75. An electron moving to the right at $0.9c$ collides with a positron moving to the left at $0.9c$. The two particles annihilate and produce two gamma-ray photons. What is the wavelength of the photons?

76. Section 36.10 looked at the inelastic collision e^- (fast) $+$ e^- (at rest) $\rightarrow$ $e^- + e^- + e^- + e^+$.
    a. What is the threshold kinetic energy of the fast electron? That is, what minimum kinetic energy must the electron have to allow this process to occur?
    b. What is the speed of an electron with the threshold kinetic energy?

Challenge Problems

77. Two rockets, A and B, approach the earth from opposite directions at speed $0.8c$. The length of each rocket measured in its rest frame is 100 m. What is the length of rocket A as measured by the crew of rocket B?

78. Two rockets are each 1000 m long in their rest frame. Rocket Orion, traveling at $0.8c$ relative to the earth, is overtaking rocket Sirius, which is poking along at a mere $0.6c$. According to the crew on Sirius, how long does Orion take to completely pass? That is, how long is it from the instant the nose of Orion is at the tail of Sirius until the tail of Orion is at the nose of Sirius?

79. Some particle accelerators allow protons (p^+) and antiprotons (p^-) to circulate at equal speeds in opposite directions in a device called a *storage ring*. The particle beams cross each other at various points to cause $p^+ + p^-$ collisions. In one collision, the outcome is $p^+ + p^- \rightarrow e^+ + e^- + \gamma + \gamma$, where γ represents a high-energy gamma-ray photon. The electron and positron are ejected from the collision at $0.9999995c$ and the gamma-ray photon wavelengths are found to be 1.0×10^{-6} nm. What were the proton and antiproton speeds prior to the collision?

80. The rockets of the Goths and the Huns are each 1000 m long in their rest frame. The rockets pass each other, virtually touching, at a relative speed of $0.8c$. The Huns have a laser cannon at the rear of their rocket that shoots a deadly laser beam at right angles to the motion. The captain of the Hun rocket wants to send a threatening message to the Goths by "firing a shot across their bow." He tells his first mate, "The Goths' rocket is length contracted to 600 m. Fire the laser cannon at the instant the nose of our rocket passes the tail of their rocket. The laser beam will cross 400 m in front of them." But

things are different in the Goths' reference frame. The Goth captain muses, "The Huns' rocket is length contracted to 600 m, 400 m shorter than our rocket. If they fire the laser cannon as their nose passes the tail of our rocket, the lethal laser blast will go right through our side."

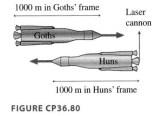

1000 m in Goths' frame

Laser cannon

Goths

Huns

1000 m in Huns' frame

FIGURE CP36.80

The first mate on the Hun rocket fires as ordered. Does the laser beam blast the Goths or not? Resolve this paradox. Show

that, when properly analyzed, the Goths and the Huns agree on the outcome. Your analysis should contain both quantitative calculations and written explanation.

81. A very fast pole vaulter lives in the country. One day, while practicing, he notices a 10-m-long barn with the doors open at both ends. He decides to run through the barn at $0.866c$ while carrying his 16-m-long pole. The farmer, who sees him coming, says. "Aha! This guy's pole is length contracted to 8 m. There will be a short interval of time when the pole is entirely inside the barn. If I'm quick, I can simultaneously close both barn doors while the pole vaulter and his pole are inside." The pole vaulter, who sees the farmer beside the barn, thinks to himself, "That farmer is crazy. The barn is length contracted and is only 5 m long. My 16-m-long pole cannot fit into a 5-m-long barn. If the farmer closes the doors just as the tip of my pole reaches the back door, the front door will break off the last 11 m of my pole."

Can the farmer close the doors without breaking the pole? Show that, when properly analyzed, the farmer and the pole vaulter agree on the outcome. Your analysis should contain both quantitative calculations and written explanation. It's obvious that the pole vaulter cannot stop quickly, so you can assume that the doors are paper thin and that the pole breaks through without slowing down.

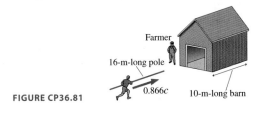

Farmer

16-m-long pole

0.866c   10-m-long barn

FIGURE CP36.81

Stop to Think 36.1: a, c, and **f.** These move at constant velocity, or very nearly so. The others are accelerating.

Stop to Think 36.2: a. $u' = u - v = -10$ m/s $- 6$ m/s $= -16$ m/s. The *speed* is 16 m/s.

Stop to Think 36.3: c. Even the light has a slight travel time. The event is the hammer hitting the nail, not your seeing the hammer hit the nail.

Stop to Think 36.4: At the same time. Mark is halfway between the tree and the pole, so the fact that he *sees* the lightning bolts at the same time means they *happened* at the same time. It's true that Nancy *sees* event 1 before event 2, but the events actually occurred before she sees them. Mark and Nancy share a reference frame, because they are at rest relative to each other, and all experimenters in a reference frame, after correcting for any signal delays, *agree* on the spacetime coordinates of an event.

Stop to Think 36.5: After. This is the same as the case of Peggy and Ryan. In Mark's reference frame, as in Ryan's, the events are simultaneous. Nancy *sees* event 1 first, but the time when an event is seen is not when the event actually happens. Because all experi-

menters in a reference frame agree on the spacetime coordinates of an event, Nancy's position in her reference frame cannot affect the order of the events. If Nancy had been passing Mark at the instant the lightning strikes occur in Mark's frame, then Nancy would be equivalent to Peggy. Event 2, like the firecracker at the front of Peggy's railroad car, occurs first in Nancy's reference frame.

Stop to Think 36.6: c. Nick measures proper time because Nick's clock is present at both the "nose passes Nick" event and the "tail passes Nick" event. Proper time is the smallest measured time interval between two events.

Stop to Think 36.7: $L_A > L_B = L_C$. Anjay measures the pole's proper length because it is at rest in his reference frame. Proper length is the longest measured length. Beth and Charles may *see* the pole differently, but they share the same reference frame and their *measurements* of the length agree.

Stop to Think 36.8: c. The rest energy E_0 is an invariant, the same in all inertial reference frames. Thus $m = E_0/c^2$ is independent of speed.

Mathematics Review

Algebra

Using exponents:

$$a^{-x} = \frac{1}{a^x} \qquad a^x a^y = a^{(x+y)} \qquad \frac{a^x}{a^y} = a^{(x-y)} \qquad (a^x)^y = a^{xy}$$

$$a^0 = 1 \qquad a^1 = a \qquad a^{1/n} = \sqrt[n]{a}$$

Fractions:

$$\left(\frac{a}{b}\right)\left(\frac{c}{d}\right) = \frac{ac}{bd} \qquad \frac{a/b}{c/d} = \frac{ad}{bc} \qquad \frac{1}{1/a} = a$$

Logarithms:

If $a = e^x$, then $\ln(a) = x$ $\qquad$ $\ln(e^x) = x$ $\qquad$ $e^{\ln(x)} = x$

$$\ln(ab) = \ln(a) + \ln(b) \qquad \ln\left(\frac{a}{b}\right) = \ln(a) - \ln(b) \qquad \ln(a^n) = n\ln(a)$$

The expression $\ln(a + b)$ cannot be simplified.

Linear equations: The graph of the equation $y = ax + b$ is a straight line. a is the slope of the graph. b is the y-intercept.

Proportionality: To say that y is proportional to x, written $y \propto x$, means that $y = ax$, where a is a constant. Proportionality is a special case of linearity. A graph of a proportional relationship is a straight line that passes through the origin. If $y \propto x$, then

$$\frac{y_1}{y_2} = \frac{x_1}{x_2}$$

(graph) Slope $a = \dfrac{\text{rise}}{\text{run}} = \dfrac{\Delta y}{\Delta x}$; Δy; Δx; y-intercept $= b$

Quadratic equation: The quadratic equation $ax^2 + bx + c = 0$ has the two solutions $x = \dfrac{-b \pm \sqrt{b^2 - 4ac}}{2a}$.

Geometry and Trigonometry

Area and volume:

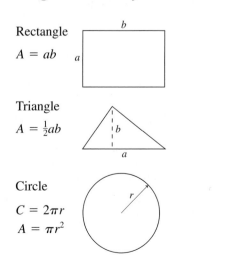

Rectangle
$A = ab$

Triangle
$A = \frac{1}{2}ab$

Circle
$C = 2\pi r$
$A = \pi r^2$

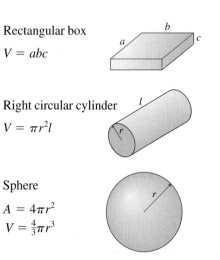

Rectangular box
$V = abc$

Right circular cylinder
$V = \pi r^2 l$

Sphere
$A = 4\pi r^2$
$V = \frac{4}{3}\pi r^3$

A

APPENDIX

Arc length and angle: The angle θ in radians is defined as $\theta = s/r$.

The arc length that spans angle θ is $s = r\theta$.

2π rad $= 360°$

Right triangle: Pythagorean theorem $\quad c = \sqrt{a^2 + b^2} \text{ or } a^2 + b^2 = c^2$

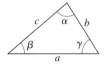

$$\sin\theta = \frac{b}{c} = \frac{\text{far side}}{\text{hypotenuse}} \qquad \theta = \sin^{-1}\left(\frac{b}{c}\right)$$

$$\cos\theta = \frac{a}{c} = \frac{\text{adjacent side}}{\text{hypotenuse}} \qquad \theta = \cos^{-1}\left(\frac{a}{c}\right)$$

$$\tan\theta = \frac{b}{a} = \frac{\text{far side}}{\text{adjacent side}} \qquad \theta = \tan^{-1}\left(\frac{b}{a}\right)$$

General triangle: $\alpha + \beta + \gamma = 180° = \pi$ rad

Law of cosines $c^2 = a^2 + b^2 - 2ab\cos\gamma$

Identities:

$$\tan\alpha = \frac{\sin\alpha}{\cos\alpha} \qquad\qquad \sin^2\alpha + \cos^2\alpha = 1$$

$$\sin(-\alpha) = -\sin\alpha \qquad\qquad \cos(-\alpha) = \cos\alpha$$

$$\sin(\alpha \pm \beta) = \sin\alpha\cos\beta \pm \cos\alpha\sin\beta \qquad\qquad \cos(\alpha \pm \beta) = \cos\alpha\cos\beta \mp \sin\alpha\sin\beta$$

$$\sin(2\alpha) = 2\sin\alpha\cos\alpha \qquad\qquad \cos(2\alpha) = \cos^2\alpha - \sin^2\alpha$$

$$\sin(\alpha \pm \pi/2) = \pm\cos\alpha \qquad\qquad \cos(\alpha \pm \pi/2) = \mp\sin\alpha$$

$$\sin(\alpha \pm \pi) = -\sin\alpha \qquad\qquad \cos(\alpha \pm \pi) = -\cos\alpha$$

Expansions and Approximations

Binomial expansion: $\quad (1 + x)^n = 1 + nx + \dfrac{n(n-1)}{2}x^2 + \ldots$

Binomial approximation: $\quad (1 + x)^n \approx 1 + nx \quad \text{if} \quad x \ll 1$

Trigonometric expansions: $\quad \sin\alpha = \alpha - \dfrac{\alpha^3}{3!} + \dfrac{\alpha^5}{5!} - \dfrac{\alpha^7}{7!} + \ldots \text{ for } \alpha \text{ in rad}$

$$\cos\alpha = 1 - \dfrac{\alpha^2}{2!} + \dfrac{\alpha^4}{4!} - \dfrac{\alpha^6}{6!} + \ldots \text{ for } \alpha \text{ in rad}$$

Small-angle approximation: If $\alpha \ll 1$ rad, then $\sin\alpha \approx \tan\alpha \approx \alpha$ and $\cos\alpha \approx 1$.

The small-angle approximation is excellent for $\alpha < 5°$ (≈ 0.1 rad) and generally acceptable up to $\alpha \approx 10°$.

Periodic Table of Elements

Key:

Atomic number →	27
	Co ← Symbol
Atomic mass →	58.9

Transition elements

Inner transition elements

Period	Group																	
1	1 H 1.0																	2 He 4.0
2	3 Li 6.9	4 Be 9.0											5 B 10.8	6 C 12.0	7 N 14.0	8 O 16.0	9 F 19.0	10 Ne 20.2
3	11 Na 23.0	12 Mg 24.3											13 Al 27.0	14 Si 28.1	15 P 31.0	16 S 32.1	17 Cl 35.5	18 Ar 39.9
4	19 K 39.1	20 Ca 40.1	21 Sc 45.0	22 Ti 47.9	23 V 50.9	24 Cr 52.0	25 Mn 54.9	26 Fe 55.8	27 Co 58.9	28 Ni 58.7	29 Cu 63.5	30 Zn 65.4	31 Ga 69.7	32 Ge 72.6	33 As 74.9	34 Se 79.0	35 Br 79.9	36 Kr 83.8
5	37 Rb 85.5	38 Sr 87.6	39 Y 88.9	40 Zr 91.2	41 Nb 92.9	42 Mo 95.9	43 Tc 96.9	44 Ru 101.1	45 Rh 102.9	46 Pd 106.4	47 Ag 107.9	48 Cd 112.4	49 In 114.8	50 Sn 118.7	51 Sb 121.8	52 Te 127.6	53 I 126.9	54 Xe 131.3
6	55 Cs 132.9	56 Ba 137.3	57 La 138.9	72 Hf 178.5	73 Ta 180.9	74 W 183.9	75 Re 186.2	76 Os 190.2	77 Ir 192.2	78 Pt 195.1	79 Au 197.0	80 Hg 200.6	81 Tl 204.4	82 Pb 207.2	83 Bi 209.0	84 Po 209.0	85 At 210.0	86 Rn 222.0
7	87 Fr 223.0	88 Ra 226.0	89 Ac 227.0	104 Rf 261	105 Db 262	106 Sg 263	107 Bh 264	108 Hs 269	109 Mt 268	110 Ds 271	111 272	112 285						

Lanthanides 6

58 Ce 140.1	59 Pr 140.9	60 Nd 144.2	61 Pm 144.9	62 Sm 150.4	63 Eu 152.0	64 Gd 157.3	65 Tb 158.9	66 Dy 162.5	67 Ho 164.9	68 Er 167.3	69 Tm 168.9	70 Yb 173.0	71 Lu 175.0

Actinides 7

90 Th 232.0	91 Pa 231.0	92 U 238.0	93 Np 237.0	94 Pu 239.1	95 Am 241.1	96 Cm 244.1	97 Bk 249.1	98 Cf 252.1	99 Es 257.1	100 Fm 257.1	101 Md 258.1	102 No 259.1	103 Lr 262.1

Answers

Answers to Odd-Numbered Exercises and Problems

Solutions to questions posed in the Part Overview captions can be found at the end of this answer list.

Chapter 25

1. a. Electrons removed from glass b. 3.13×10^{10}
3. 3.04×10^{-11}
7. Right negatively charged, left positively charged
11. a. 9.0×10^9 N b. 9.0×10^9 m/s^2
13. -10 nC
15. $\vec{F}_{\text{B on A}} = 4.50 \times 10^{-3} \hat{j}$ N, $\vec{F}_{\text{A on B}} = -4.50 \times 10^{-3} \hat{j}$ N
17. 30 N/kg
19. a. (9.83 N/kg, toward earth) b. (2.70×10^{-3} N/kg, toward earth)
21. 0.111 nC
23. -8.0 nC
25. (3.27×10^6 N/C, downward)
27. a. $3.6 \times 10^4 \hat{i}$ N/C, $(-1.27 \times 10^4 \hat{i} + 1.27 \times 10^4 \hat{j})$ N/C, $(-1.27 \times 10^4 \hat{i} - 1.27 \times 10^4 \hat{j})$ N/C
    b.

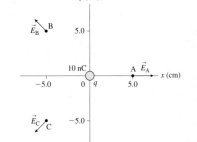

29. 1.36×10^5 C, -1.36×10^5 C
31. a. Electrons removed from sphere and added to rod b. 2.5×10^{10}
33. -160 nC and 0 nC
35. a. 498 N b. 2.98×10^{29} m/s^2
37. a. 0.45 N b. 1.0×10^{-6} C, 5.0×10^{-7} C c. 4.5 m/s^2
39. 1.80×10^{-4} N to the right
41. 4.74×10^{-3} N, 71.6° above $-x$-axis
43. 1.74×10^{-4} N, 51.75° below $+x$-axis
45. $-1.02 \times 10^{-3} \hat{i}$ N
47. $(1.02 \times 10^{-5} \hat{i} + 2.16 \times 10^{-5} \hat{j})$ N
49. 0.68 nC
51. $-2KQqa/(y^2 + a^2)^{3/2}$
53. a.

b. $(2 - \sqrt{2})KQq/L^2$

55. a. 243 N b. Yes. Any difference must therefore be smaller than 1 part in 10^9.
57. 3.2×10^{15}
59. 4.42°
61. $1.0 \times 10^5 \hat{j}$ N/C, $(2.88 \times 10^4 \hat{i} + 2.16 \times 10^4 \hat{j})$ N/C, $5.63 \times 10^4 \hat{i}$ N/C
63. $(4.02 \times 10^4 \hat{i} + 8.05 \times 10^4 \hat{j})$ N/C, $4.5 \times 10^5 \hat{i}$ N/C, $(4.02 \times 10^4 \hat{i} - 8.05 \times 10^4 \hat{j})$ N/C
65. a. $(-1$ cm, 2 cm) b. (3 cm, 3 cm) c. (4 cm, -2 cm)
67. a. $(3.20 \hat{i} + 6.40 \hat{j}) \times 10^{-17}$ N b. $(-3.20 \hat{i} - 6.40 \hat{j}) \times 10^{-17}$ N c. 4.28×10^{10} m/s^2 d. 7.85×10^{13} m/s^2
69. 14.3°
71. b. 22.4 nC
73. b. 5.13 nC
75. 4.06 g

Chapter 26

1. (2550 N/C, 0° above horizontal)
3. (3975 N/C, 9.3° above horizontal)
5. a. 18.0 N/C b. 36.0 N/C
7. 2.28×10^5 N/C, 1.67×10^5 N/C, 2.28×10^5 N/C
9. (8.78×10^{-4} N, toward rod)
11. -0.056 nC
13. a. 0 N/C b. 4110 N/C
15. a. 0 N/C b. 1.49×10^5 N/C
17. 1.39×10^{-3} nC
19. 1.41×10^5 N/C
21. 1.86 cm
23. 6.13×10^5 N/C, down
25. 5.93×10^5 N/C
27. 0.185 m
29. (9.0×10^{-13} N, direction opposite $\vec{p}$)
31. $(132,600 \hat{i} - 12,130 \hat{j})$ N/C; (133,200 N/C, 5.23° below the $+x$-axis)
33. $(675 \hat{i} - 78,400 \hat{j})$ N/C; (78,400 N/C, 89.5° below the $+x$-axis)
35. a. $\vec{E}_1 = [q/(4\pi\epsilon_0)5\sqrt{5}a^2](-3\hat{i} + 2\hat{j})$; $\vec{E}_2 = [7q/(4\pi\epsilon_0)9a^2]\hat{i}$; $\vec{E}_3 = [17q/(4\pi\epsilon_0)9a^2]\hat{i}$; $\vec{E}_4 = [q/(4\pi\epsilon_0)5\sqrt{5}a^2](-3\hat{i} - 2\hat{j})$
    b.

39.
1.08×10^5 N/C

41. a. $\dfrac{8\lambda d}{4\pi\epsilon_0(4y^2 + d^2)}$

    b.

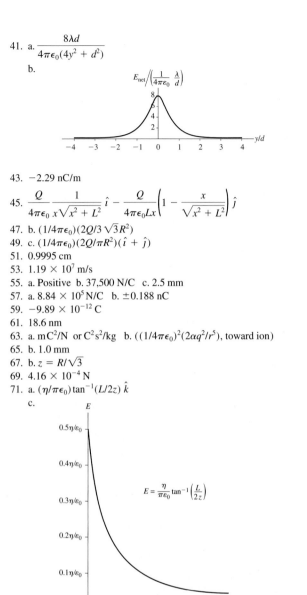

43. -2.29 nC/m

45. $\dfrac{Q}{4\pi\epsilon_0}\dfrac{1}{x\sqrt{x^2 + L^2}}\,\hat{i} - \dfrac{Q}{4\pi\epsilon_0 Lx}\left(1 - \dfrac{x}{\sqrt{x^2 + L^2}}\right)\hat{j}$

47. b. $(1/4\pi\epsilon_0)(2Q/3\sqrt{3}R^2)$

49. c. $(1/4\pi\epsilon_0)(2Q/\pi R^2)(\hat{i} + \hat{j})$

51. 0.9995 cm

53. 1.19×10^7 m/s

55. a. Positive b. 37,500 N/C c. 2.5 mm

57. a. 8.84×10^5 N/C b. ± 0.188 nC

59. -9.89×10^{-12} C

61. 18.6 nm

63. a. mC^2/N or C^2s^2/kg b. $((1/4\pi\epsilon_0)^2(2\alpha q^2/r^5)$, toward ion)

65. b. 1.0 mm

67. b. $z = R/\sqrt{3}$

69. 4.16×10^{-4} N

71. a. $(\eta/\pi\epsilon_0)\tan^{-1}(L/2z)\,\hat{k}$

    c.

$$E = \frac{\eta}{\pi\epsilon_0}\tan^{-1}\left(\frac{L}{2z}\right)$$

73. c. 2.0×10^{12} Hz

Chapter 27

1.

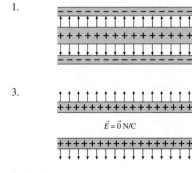

3.

$$\vec{E} = \vec{0}\text{ N/C}$$

5. No charge

7. 5 N/C, pointing in

9. -1.0 N m^2/C

11. a. 6.0×10^{-2} N m^2/C b. 0 N m^2/C

13. 1.26 N m^2/C

15. a. 0 b. $2\pi R^2 E$

17. -1.0 N m^2/C

19. 5.31 nC

21.

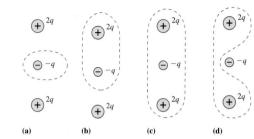

| (a) | (b) | (c) | (d) |

23. 113 N m^2/C

25. 2.66×10^{-5} C/m^2

27. a. -0.390 N m^2/C, 0.225 N m^2/C, 0.390 N m^2/C, -0.225 N m^2/C

    b. 0 N m^2/C

29. a. -3.46 N m^2/C b. 1.15 N m^2/C

31. $-2Q/\epsilon_0$

33. a. 2000 N/C b. 251 N m^2/C c. 2.22 nC

35. a. -100 nC b. $+50$ nC

37. a. 2.39×10^{-6} C/m^3 b. 1.25 nC, 10.0 nC, 80.0 nC

    c. 4500 N/C, 9000 N/C, 18,000 N/C

39. a. -1.068×10^{-8} C b. $+1.068 \times 10^{-8}$ C c. 4.82×10^{-8} C

41.

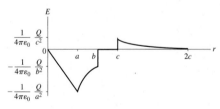

43. a. $(q/4\pi\epsilon_0 r^2)\,\hat{r}$ b. $-(q/4\pi\epsilon_0 r^2)\,\hat{r}$

45. a. $(-Qr/4\pi\epsilon_0 a^3)\,\hat{r}$, $-(Q/4\pi\epsilon_0 r^2)\,\hat{r}$, $\vec{0}$, $(Q/4\pi\epsilon_0 r^2)\,\hat{r}$

    b.

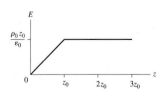

47. a. $\rho_0 z/\epsilon_0$ b. $\rho_0 z_0/\epsilon_0$

    c.

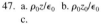

49. a. $3Q/2A\epsilon_0$, 0, $Q/2A\epsilon_0$, 0, $3Q/2A\epsilon_0$ b. $\frac{3}{2}Q/A$, $-\frac{1}{2}Q/A$, $\frac{1}{2}Q/A$, $\frac{3}{2}Q/A$

51. a. $(\lambda/2\pi\epsilon_0 r)\,\hat{r}$ b. $(\lambda r/2\pi\epsilon_0 R^2)\,\hat{r}$

53. b. 0 N/C c. 4.64×10^{13} N/C

55. a.

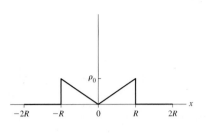

c. $(\lambda r^2/2\pi\epsilon_0 R^3)\,\hat{r}$

Chapter 28

1. 7.5×10^{-5} m/s
3. 2.62×10^5 s
5. Aluminum
7. a. 4.63×10^{21} b. 4.32×10^{-12} m
9. 0.31 N/C
11. 9.4×10^{18}
13. 3.2 mA
15. a. 6.25×10^5 A/m^2 b. 6.51×10^{-5} m/s
17. a. 1.73×10^7 A/m^2 b. 5.31×10^{18} s^{-1}
19. 0.141 mA
21. 2.08×10^{-14} s, 4.19×10^{-15} s
23. 1.68 A
25. 5.01×10^{-8} Ωm
27. a. 1.64×10^{-3} N/C b. 1.10×10^{-5} m/s
31. a. Doubled b. Unchanged c. Unchanged d. Doubled
33. a. 3.12×10^{14} b. 398 A/m^2 c. 9.11×10^5 N/C d. 0.227 W
35. 22.6 mA
37. a. 120 C b. 0.449 mm
39. 1/4
41. 10.4 A
43. a. $I/4\pi\sigma r^2$ b. 3.32×10^{-4} N/C, 5.31×10^{-5} N/C
45. a.

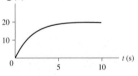

b. $(10\text{ A})e^{-t/(2.0\text{s})}$ c. 10.0 A
d.

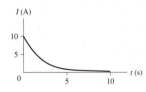

47. 2.43 A
49. 5.6×10^{-6} m/s
51. 1/2
53. 1.01×10^{23}
55. a. 1.15×10^5 m/s b. 1.5 nm
57. a. $(\epsilon_0 I/A)(1/\sigma_2 - 1/\sigma_1)$ b. 3.68×10^{-18} C

Chapter 29

1. 1.38×10^5 m/s
3. 7.07×10^4 m/s
5. 2.82×10^{-6} J
7. -2.24×10^{-19} J
9. 1.61×10^8 N/C
11. 1.87×10^7 m/s
13. -2.09×10^4 V
15. a. Lower b. -0.712 V
19. a. 200 V b. 3.54×10^{-10} C
21. a. Right plate b. 1.0×10^5 V/m c. 2.40×10^{-17} J
23. a. 1800 V, 1800 V, 900 V b. -2.88×10^{-16} J, -2.88×10^{-16} J, -1.44×10^{-16} J c. 0 V, -900 V
25. 4.17×10^{-10} C
27. $+1410$ V
29. a. 3140 V b. 5.02×10^{-16} J
31. $x = 3$ cm and 6 cm
33. a. q_a is positive, q_b is negative with the same magnitude
b.

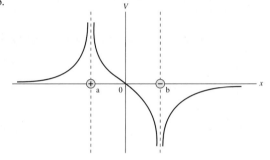

35. 0 V
37. 1.44×10^{-3} N
39. a. $x = \pm\infty$ b. $x = \pm\infty$ and 0
c.

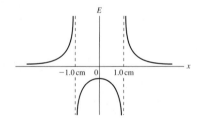

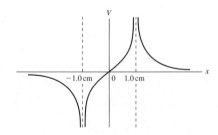

41. a. 0.720 J b. 14.4 N c. 21.9 m/s and 10.95 m/s
43. 25.3×10^{-6} J
45. a. 1000 V b. 1.39×10^{-9} C c. 7.0×10^6 m/s

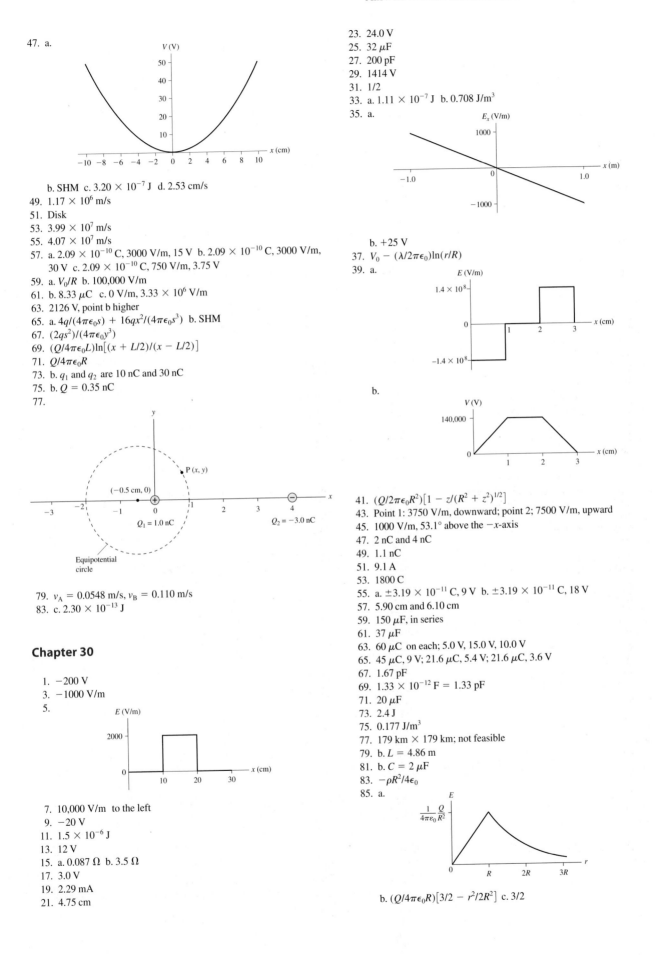

47. a.

b. SHM c. 3.20×10^{-7} J d. 2.53 cm/s
49. 1.17×10^6 m/s
51. Disk
53. 3.99×10^7 m/s
55. 4.07×10^7 m/s
57. a. 2.09×10^{-10} C, 3000 V/m, 15 V b. 2.09×10^{-10} C, 3000 V/m, 30 V c. 2.09×10^{-10} C, 750 V/m, 3.75 V
59. a. V_0/R b. 100,000 V/m
61. b. 8.33 μC c. 0 V/m, 3.33×10^6 V/m
63. 2126 V, point b higher
65. a. $4q/(4\pi\epsilon_0 s) + 16qx^2/(4\pi\epsilon_0 s^3)$ b. SHM
67. $(2qs^2)/(4\pi\epsilon_0 y^3)$
69. $(Q/4\pi\epsilon_0 L)\ln[(x + L/2)/(x - L/2)]$
71. $Q/4\pi\epsilon_0 R$
73. b. q_1 and q_2 are 10 nC and 30 nC
75. b. $Q = 0.35$ nC
77.

79. $v_A = 0.0548$ m/s, $v_B = 0.110$ m/s
83. c. 2.30×10^{-13} J

Chapter 30

1. -200 V
3. -1000 V/m
5.

7. 10,000 V/m to the left
9. -20 V
11. 1.5×10^{-6} J
13. 12 V
15. a. 0.087 Ω b. 3.5 Ω
17. 3.0 V
19. 2.29 mA
21. 4.75 cm

23. 24.0 V
25. 32 μF
27. 200 pF
29. 1414 V
31. 1/2
33. a. 1.11×10^{-7} J b. 0.708 J/m^3
35. a.

b. +25 V
37. $V_0 - (\lambda/2\pi\epsilon_0)\ln(r/R)$
39. a.

b.

41. $(Q/2\pi\epsilon_0 R^2)[1 - z/(R^2 + z^2)^{1/2}]$
43. Point 1: 3750 V/m, downward; point 2; 7500 V/m, upward
45. 1000 V/m, 53.1° above the $-x$-axis
47. 2 nC and 4 nC
49. 1.1 nC
51. 9.1 A
53. 1800 C
55. a. $\pm 3.19 \times 10^{-11}$ C, 9 V b. $\pm 3.19 \times 10^{-11}$ C, 18 V
57. 5.90 cm and 6.10 cm
59. 150 μF, in series
61. 37 μF
63. 60 μC on each; 5.0 V, 15.0 V, 10.0 V
65. 45 μC, 9 V; 21.6 μC, 5.4 V; 21.6 μC, 3.6 V
67. 1.67 pF
69. 1.33×10^{-12} F = 1.33 pF
71. 20 μF
73. 2.4 J
75. 0.177 J/m^3
77. 179 km $\times$ 179 km; not feasible
79. b. $L = 4.86$ m
81. b. $C = 2$ μF
83. $-\rho R^2/4\epsilon_0$
85. a.

b. $(Q/4\pi\epsilon_0 R)[3/2 - r^2/2R^2]$ c. 3/2

d.

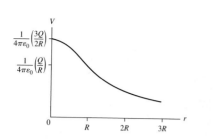

Chapter 31

1. $5.5 \, \Omega$
3. $2.0 \, A$
5. 0.64 mm
7.

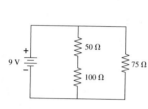

9. $5 \, A$, toward the junction
11. a. $0.50 \, A$, left to right
b.

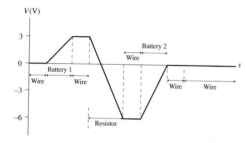

13. $9.6 \, \Omega$, $12.5 \, A$
15. $23.6 \, \mu m$
17. a. $11.6 \, A$ b. $10.4 \, \Omega$
19. $78.4 \, m\Omega$
21. $25 \, \Omega$
23. $93.4 \, W$
25. 3.23%
27. $R/4$
29. $12.0 \, \Omega$
31. $24 \, \Omega$
33. $183.3 \, \Omega$
35. $13 \, V$, $9 \, V$, $0 \, V$, $-2 \, V$
37. 2 ms
39. 6.93 ms
41. $869 \, \Omega$
43. $A > D = E > B = C$
45. Increase
47. $8.4 \times 10^{-8} \, \Omega$
49. $7 \, \Omega$
51. $60 \, V$, $10 \, \Omega$
53. $9.00 \, V$, $0.50 \, \Omega$
55. $2.0 \, V$ for each
57. $1.0 \, A$, $2.0 \, A$, $15.0 \, V$
59. $3.0 \, A$
61. a. \$14.40 b. 34.7 months
63. a. $8 \, A$, $8 \, V$ b. $9.14 \, A$, $0 \, V$
65. a. $0.505 \, \Omega$ b. $0.50 \, \Omega$

67.

$R \, (\Omega)$	$I \, (A)$	$\Delta V \, (V)$
6	2.0	12.0
15	0.8	12.0
6	1.2	7.2
4	1.2	4.8

69.

$R \, (\Omega)$	$I \, (A)$	$\Delta V \, (V)$
4	2	8
6	$\frac{4}{3}$	8
8	1	8
24	$\frac{1}{3}$	8
24	$\frac{2}{3}$	16

71. a. $10 \, A$ b. $80 \, W$ c. $60 \, V$
73. $36.4 \, \Omega$
75. 0.69 ms
77. a. $80 \, \mu C$ b. 0.23 ms
81. b. $5140 \, \Omega$

Chapter 32

1. Out of the page
3. $(2.0 \, mT$, into the page$)$, $(4.0 \, mT$, into the page$)$
5. a. $1.60 \times 10^{-15} \, \hat{k} \, T$ b. $0 \, T$ c. $0 \, T$
7. $1.13 \times 10^{-15} \, \hat{k} \, T$
9. 6.25×10^{6} m/s in the $+z$-direction
11. 4.0 cm, 0.4 mm, $20 \, \mu m$ to $2 \, \mu m$, $0.20 \, \mu m$
13. a. $20 \, A$ b. 1.60×10^{-3} m
15. $2.0 \times 10^{-4} \, \hat{i} \, T$, $4.0 \times 10^{-4} \, \hat{i} \, T$, and $2.0 \times 10^{-4} \, \hat{i} \, T$
17. a. $0.025 \, A \, m^2$ b. $5.0 \, \mu T$
19. $8.75 \times 10^{-5} \, m^2$
21. $0.0707 \, T \, m$
23. $1.0 \, A$
25. $2390 \, A$
27. a. Into the page b. No deflection
29. a. In the plane of the paper, $45°$ cw from straight up
b. In the plane of the paper, $45°$ ccw from straight down
31. a. $-8.0 \times 10^{-13} \, \hat{k} \, N$ b. $5.66 \times 10^{-13} \, (-\hat{j} - \hat{k}) \, N$
33. $1.61 \times 10^{-3} \, T$
35. $2.9 \times 10^{28} \, m^{-3}$
37. $0.025 \, N$, to the right
39. $3.0 \, \Omega$
41. $7.5 \times 10^{-4} \, N \, m$
43. a. $1.26 \times 10^{-11} \, N \, m$ b. Rotated $90°$
45. a. $1.0 \, \mu T$ b. 2.0% c. $2.0 \, \mu T$ d. $2.0 \, \mu T$; twice field in (a)
47. $(5.2 \times 10^{-5} \, T$, out of the page$)$; $0 \, T$
49.

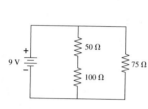

51. $750 \, A$
53. a. $1.13 \times 10^{10} \, A$ b. $0.014 \, A/m^2$
c. The current density in the earth is much less than the current density in the wire.

55. #18 gauge; 4.06 A
57. a. $(1.25)^{-3/2}\mu_0 NI/R$ b. 1.80×10^{-4} T
59. $\mu_0 I/4R$
61. 0; $(\mu_0 I/2\pi r)[(r^2 - R_1^2)/(R_2^2 - R_1^2)]$; $\mu_0 I/2\pi r$
63. 2.9×10^{-3} T
65. a. $(2.4 \times 10^{10}$ m/s^2, down) b. $(2.2 \times 10^{11}$ m/s^2, up)
67. a. 4.6×10^{-13} J b. 2850
69. 2.10 T
71. 2.0 A
73. (0.00864 T, down)
75. 12.5 T
77. 0.036 mm
81. a. $\mu_0 IL/4\pi d \sqrt{(L/2)^2 + d^2}$ b. $\sqrt{2}\mu_0 I/\pi R$ c. 90.0%
83. a. $\mu_0 Ir/2\pi R_1^2$, $\mu_0 I/2\pi r$, 0
    b.

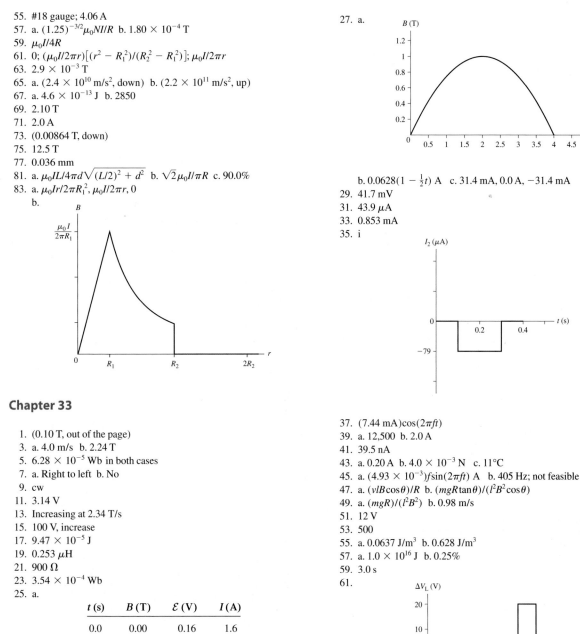

Chapter 33

1. (0.10 T, out of the page)
3. a. 4.0 m/s b. 2.24 T
5. 6.28×10^{-5} Wb in both cases
7. a. Right to left b. No
9. cw
11. 3.14 V
13. Increasing at 2.34 T/s
15. 100 V, increase
17. 9.47×10^{-5} J
19. 0.253 μH
21. 900 Ω
23. 3.54×10^{-4} Wb
25. a.

t (s)	B (T)	$\mathcal{E}$ (V)	I (A)
0.0	0.00	0.16	1.6
0.5	1.50	0.08	0.8
1.0	2.00	0.00	0.0
1.5	1.50	−0.08	−0.8
2.0	2.00	−0.16	−1.6

   b.

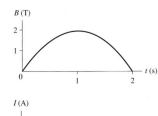

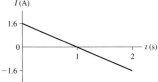

27. a.

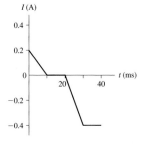

    b. $0.0628(1 - \frac{1}{2}t)$ A c. 31.4 mA, 0.0 A, −31.4 mA
29. 41.7 mV
31. 43.9 μA
33. 0.853 mA
35. i
37. $(7.44$ mA$)\cos(2\pi ft)$
39. a. 12,500 b. 2.0 A
41. 39.5 nA
43. a. 0.20 A b. 4.0×10^{-3} N c. 11°C
45. a. $(4.93 \times 10^{-3})f\sin(2\pi ft)$ A b. 405 Hz; not feasible
47. a. $(vlB\cos\theta)/R$ b. $(mgR\tan\theta)/(l^2 B^2 \cos\theta)$
49. a. $(mgR)/(l^2 B^2)$ b. 0.98 m/s
51. 12 V
53. 500
55. a. 0.0637 J/m^3 b. 0.628 J/m^3
57. a. 1.0×10^{16} J b. 0.25%
59. 3.0 s
61.

63.

65. a. $(LI_0/\tau)e^{-t/\tau}$ b. 1.0 V, 0.37 V, 0.13 V, 0.05 V
c.

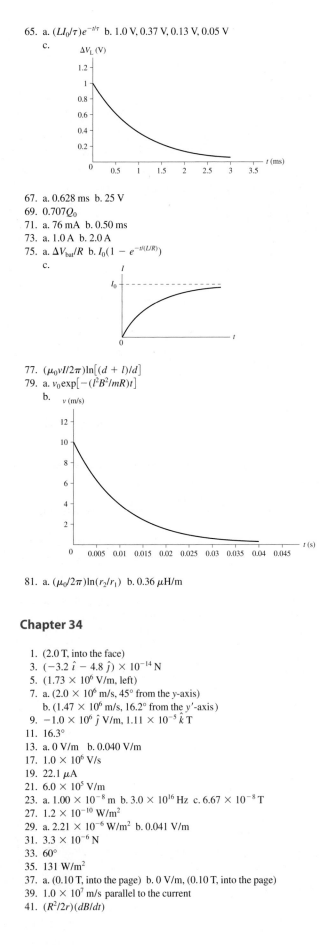

67. a. 0.628 ms b. 25 V
69. $0.707Q_0$
71. a. 76 mA b. 0.50 ms
73. a. 1.0 A b. 2.0 A
75. a. $\Delta V_{bat}/R$ b. $I_0(1 - e^{-t/(L/R)})$
c.

77. $(\mu_0 vI/2\pi)\ln[(d + l)/d]$
79. a. $v_0\exp[-(l^2B^2/mR)t]$
b.

81. a. $(\mu_0/2\pi)\ln(r_2/r_1)$ b. 0.36 μH/m

Chapter 34

1. (2.0 T, into the face)
3. $(-3.2\,\hat{\imath} - 4.8\,\hat{\jmath}) \times 10^{-14}$ N
5. $(1.73 \times 10^6$ V/m, left)
7. a. $(2.0 \times 10^6$ m/s, 45° from the y-axis)
   b. $(1.47 \times 10^6$ m/s, 16.2° from the y'-axis)
9. $-1.0 \times 10^6\,\hat{\jmath}$ V/m, $1.11 \times 10^{-5}\,\hat{k}$ T
11. 16.3°
13. a. 0 V/m b. 0.040 V/m
17. 1.0×10^6 V/s
19. 22.1 μA
21. 6.0×10^5 V/m
23. a. 1.00×10^{-8} m b. 3.0×10^{16} Hz c. 6.67×10^{-8} T
27. 1.2×10^{-10} W/m^2
29. a. 2.21×10^{-6} W/m^2 b. 0.041 V/m
31. 3.3×10^{-6} N
33. 60°
35. 131 W/m^2
37. a. (0.10 T, into the page) b. 0 V/m, (0.10 T, into the page)
39. 1.0×10^7 m/s parallel to the current
41. $(R^2/2r)(dB/dt)$

43. b. 1.48×10^{-13} A
45. a. $(2.83 \times 10^3 t^2)$ V m
b.

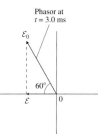

c. $1.11 \times 10^{-9}rt$ T; 4.44×10^{-12} T
d. $1.00 \times 10^{-14}t/r$ T; 5.0×10^{-12} T
47. a. 3.85×10^{26} W b. 589 W/m^2
49. Yes
51. 73.5 W
53. 0.408 m/s
55. $I_0/8$
57. b. 6.67×10^{-6} J/m^3
59. a. IR/L; $\mu_0 I/2\pi r$ b. ($I^2 R/2\pi rL$, radially inward)

Chapter 35

1. a. 175 rad/s b. -8.66 V
3.

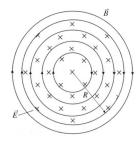

5. a. 50 mA b. 50 mA
7. a. 1.88 mA b. 1.88 A
9. a. 79.6 kHz b. 0 V
11. 125 Ω
13. 6.02 V, 7.99 V
15. 1.59 μF
17. a. 0.796 A b. 0.796 mA
19. a. 3.18×10^4 Hz b. 0 V
21. 80 Ω
23. 1.27 μF
25. a. 200 kHz b. 141 kHz
27. a. $-27.2°$ b. $+26.3°$
29. 9.6 Ω
31. 43.5 Ω
33. 395 W
35. a. $1/\sqrt{3}RC$ b. $(\sqrt{3}/2)\mathcal{E}_0$ c. 3630 rad/s
37. a. 9.95 V, 9.57 V, 7.05 V, 3.15 V, 0.990 V
b.

41. a. 10.0 Hz b. 4.47 V, 3.45 V, 2.24 V
43. a. 25.1 mA b. 6.67 V
45. 44.1 Hz
47. a. $\mathcal{E}_0/\sqrt{R^2 + \omega^2 L^2}$, $\mathcal{E}_0 R/\sqrt{R^2 + \omega^2 L^2}$, $\mathcal{E}_0 \omega L/\sqrt{R^2 + \omega^2 L^2}$
    c. Low d. R/L
49. a. 1.62 A b. $-17.7°$ c. 131 W
51. a. 3.16×10^4 rad/s $= 5.03 \times 10^3$ Hz b. 10.0 V, 31.6 V
53. a. 69.53 Ω, 0.072 A, $-44.0°$ b. 50.0 Ω, 0.100 A, 0°
    c. 62.42 Ω, 0.080 A, 36.8°
55. a. 3.60 V b. 3.47 V c. -3.60 V
59. a. 11.6 pF b. 1.5×10^{-3} Ω
61. 40 W: 14.4 W; 60 W: 9.6 W; 100 W: 100 W
63. a. 0.833 b. 100 V c. 12.5 Ω d. 320 μF
65.

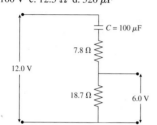

67. b. 10.0 V, 11.55 V
69. c. $\sqrt{1/LC}$
    d.

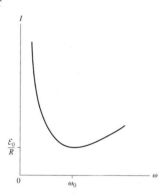

Part Overview Solutions

PART VI **Overview**

Anakin's and Obi-Wan's speeder discharges the power coupler on the planet Coruscant by flying through it. If the discharge current through the speeder is 5000 A, what voltage is developed across the speeder? To find out, what properties of the speeder do you need to estimate?

MODEL The speeder's shell is built from carbon-nanotube fibers, an advanced material that is both lightweight and extremely rigid. Carbon-nanotube fibers are poor conductors, with resistivity 5.0×10^{-6} Ω m. To estimate the resistance of the speeder's shell, model it as a 5.0-m-long tube with a 2.0 m × 5.0 m rectangular cross section. The shell's thickness is 2 mm.

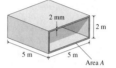

VISUALIZE The figure shows the simplified geometry of the speeder's shell.

SOLVE The cross-section area of the material is the perimeter of the shell multiplied by the thickness, or

$$A = (5.0 \text{ m} + 2.0 \text{ m} + 5.0 \text{ m} + 2.0 \text{ m}) \times (0.002 \text{ m})$$
$$= 0.028 \text{ m}^2$$

With this information, we can calculate that the speeder's resistance is

$$R = \frac{\rho L}{A} = \frac{(5.0 \times 10^{-6} \text{ Ω m})(5.0 \text{ m})}{0.028 \text{ m}^2} = 9.0 \times 10^{-4} \text{ Ω}$$

According to Ohm's law, the potential difference produced across the speeder by a 5000 A current though the speeder is

$$\Delta V = IR = (5000 \text{ A})(9.0 \times 10^{-4} \text{ Ω}) = 4.5 \text{ V}$$

ASSESS This potential difference is easily measured, but it is not sufficient to disrupt the speeder's electronic systems.

Index

For users of the five-volume edition: pages 1–481 are in Volume 1; pages 482–607 are in Volume 2; pages 608–779 are in Volume 3; pages 780–1194 are in Volume 4; and pages 1148–1383 are in Volume 5.
Pages 1195–1383 are not in the Standard Edition of this textbook.

A

Absolute temperature, 492
Absolute zero, 492, 555
Absorption
    excitation by, 1335–36
    in hydrogen, 1336
    of light, 1212–13
    in sodium, 1336
Absorption spectra, 1237–38, 1245, 1294–95, 1303
Acceleration, 2, 15–17
    angular, 371–73
    average, 16, 54, 155
    centripetal, 19, 183–85, 346–47
    constant, 53–59, 107, 156
    force and, 107–8
    in free fall, 60, 131–32
    Galilean transformation of, 169, 1089
    due to gravity, 20, 60–61, 352
    instantaneous, 67–69, 155
    nonuniform, 68–69
    in planar motion, 155–57
    of projectile, 159
    sign of, 36–37, 54
    in simple harmonic motion, 424–25
    tangential, 197, 371–72
    in two dimensions, 152, 155–57
    in uniform circular motion, 183–85
    in uniform electric field, 836
Acceleration constraints, 216
Acceleration vector, 16–17, 20, 87
AC circuits, 1121–45
    AC sources, 1122–24
    capacitor circuits, 1124–26, 1137
    inductor circuits, 1129–31, 1137
    phasors, 1122–24
    power in, 1135–39
    RC filter circuits, 1126–29
    resistor circuits, 981–82, 1123–24
    series RLC circuit, 1131–34
Action at a distance, 802
Action/reaction pair, 208. See also Newton's third law of motion
    identifying, 209–13
    propulsion, 211–13
Activity, 1367
AC voltage, 1063
Adiabatic process, 536–39
Adiabats, 537
Agent of force, 98, 113
Air conditioner, 580–81
Air pressure, 114
Air resistance, 59–60, 103, 139, 432

Allowed transitions, 1336
Alpha decay, 929, 1193, 1369–70
Alpha particle, 929, 1209
Alpha rays, 1206, 1353, 1363
Alternating current (AC), 1121. See also AC circuits
Ammeters, 975
Amorphous solids, 486
Ampère-Maxwell law, 1101, 1102, 1107–8
Ampère's law, 1011–14, 1099, 1100–1101
Amplitude, 415, 621, 652
Amplitude function ($A(x)$), 650, 1254
Angle of incidence, 718, 722
Angle of reflection, 718
Angle of refraction, 722
Angle of repose, 137
Angular acceleration, 371–73, 382
Angular displacement, 179–80
Angular frequency, 416, 418, 623
Angular momentum, 258–60, 360, 401, 402–4, 1242–43
    conservation of, 401–2
    of hydrogen, 1319–20
    inherent, 1353
    of a particle, 399–400
    quantization of, 1242–43, 1319–20
    of rigid body, 400–404
Angular position, 4, 178–79
Angular resolution, 748–49
Angular velocity, 179–81, 196, 371–73
    angular momentum and, 402–4
    average, 180
Angular velocity vector, 397
Anode, 1197
Antennas, 1111–12
Antibonding molecular orbital, 1307
Antielectron (positron), 847, 1022, 1186
Antimatter, 1186
Antineutrino, 1371
Antinodal lines, 668–69
Antinodes, 649–50
Antireflection coating, 665, 666
Apertures, 716–17
Apparent weight, 130–31, 132, 351
Archimedes' principle, 459, 480
Arc length, 179
Area vector, 856, 1051
Aristotle, 3, 60, 110–11
Atmosphere, scale height of, 478
Atmospheric pressure, 449–50, 455
Atom(s), 101, 487–89, 495–96, 1150, 1196.
        See also Electron(s); Nuclear physics;
        Nucleus; Proton(s)
    Bohr model of, 1235

electricity and, 788–89
    hard-sphere model of, 496
    nuclear model of, 1207–9
    nucleus of. See Nucleus
    plum-pudding or raisin-cake model of, 1205
    properties of, 1382
    shell model of, 1298, 1324, 1360–62
    structure of, 758–60, 782, 789–90
Atomic beam, 1252
Atomic clock, 26
Atomic magnets, 1028–29
Atomic mass, 488–89, 1354–55
Atomic mass number (A), 488, 489
Atomic mass unit (u), 253, 488, 1354–55
Atomic model, 101
Atomic number, 759, 1210, 1353
Atomic physics, 1317–51
    electron's spin, 1029, 1325–28
    emission spectra, 1212, 1237–38, 1244–45, 1305, 1337–38
    excited states, 1235, 1334–37, 1339–42
    hydrogen atom, 1318–24
    lasers, 1343–46
    multielectron atoms, 1328–30
    periodic table of the elements, 1330–34
    stimulated emission, 1343
Atom interferometer, 1231–32
Avogadro's number, 489

B

Ballistic pendulum, 277–78
Balmer, Johann, 759–60
Balmer formula, 1213
Balmer series, 759–60, 1244–45
Bandwidth, 1266–67
Bar charts, 245–46, 273–74
Barometers, 454–55
Battery(ies), 889–90, 932, 942–43, 1197
    charge escalator model of, 942
    emf of, 943
    ideal, 943
    real, 976–78
    short-circuited, 977–78
    as source of potential, 916
Beam splitter, 704
Beat frequency, 674
Beats, 672–74
Bernoulli, Daniel, 495, 1196
Bernoulli's equation, 465–69, 480
Beta decay, 235, 265, 929, 1203, 1362, 1370–71
Beta particle, 929
Beta rays, 1206, 1353, 1363

Big Bang, 235, 638
Binding energy, 1242, 1355, 1357–59
Binding energy per nucleon, 1358
Binomial approximation, 356, 831
Biologically equivalent dose, 1373
Biot-Savart law, 1001, 1091, 1146
  in terms of current, 1004
Blood pressure, 456
Blue shift, 637
Bohr, Niels, 1150, 1234–35, 1291
Bohr model of atomic quantization, 1234–36
  basic assumption, 1235
  quantum jumps, 1235
  stationary states, 1235
Bohr model of hydrogen atom, 1238–43
  binding energy, 1242
  energy levels, 1241–42
  ionization energy, 1242
  quantization of angular momentum, 1242–43
  stationary states, 1238–40, 1242
Bohr radius, 1240
Boiling point, 493
Boltzmann's constant, 500
Bonding molecular orbital, 1306–7
Boundary conditions, 652, 1281–82, 1318
Bound states, 1294
Bound system, 907, 1357
Boyle's law, 478
Bragg condition, 761–63, 766
Brahe, Tycho, 344
Brayton cycle, 586–88
Breakdown field strength, 816
Brownian motion, 1225
Bubble chamber, 1022
Bulk modulus, 472
Bulk properties, 485
Buoyancy, 458–62
  of boats, 461–62
  neutral, 460
Buoyant force, 458–59

C

Caloric, 2, 520
Calorimetry, 530–33
  with a gas and a solid, 534
  with a phase change, 532–33
Camera obscura, 716–17
Capacitance, 946–48, 949
Capacitive reactance, 1125–26
Capacitor(s), 946–53. *See also* Parallel-plate
    capacitor
  charging, 948, 987
  combinations of, 949–51
  conservation of energy inside, 903–4
  discharged, 880
  energy stored in, 951–53
  fields inside, 1101–2
  parallel plate, 949–50
  particle in, 1303–5
  power in, 1137
  series, 949, 950
  spherical, 948–49

Capacitor circuits, 1124–26, 1137
Carbon dating, 1367–68
Carnot, Sadi, 590
Carnot cycle, 594–97
Carnot engine, 591–93, 595
Carnot thermal efficiency, 595
Cartesian coordinate system, 84
Cathode, 1197
Cathode glow, 1198, 1212
Cathode rays, 760, 1198–1202
Cathode-ray tube (CRT) devices, 837
Causal influence, 1182–83
Cavendish, Henry, 352–53
Celsius (centigrade) scale, 491
Center of mass, 374–76
Central force, 906
Central maximum, 688, 696, 699, 700
Centrifugal force, 193–94
Centripetal acceleration, 19, 183–85, 346–47
Charge(s), 781, 788–91, 891. *See also*
    Point charge
  of capacitor, 987
  within conductor, 871–72
  conservation of, 790–91
  fundamental unit of, 789
  like, 786
  micro/macro connection, 789–90
  mobility of, 787
  moving, 1001–4, 1017–23
  opposite, 786
  separation of, 941
  source, 806, 835
  transferring, 788
Charge carriers, 791, 880–81
Charge density, 824–25
Charge diagrams, 790–91
Charge distribution. *See also* Electric field(s)
  continuous, 824–28, 921–24
  symmetric, 850
Charged particles, 782. *See also* Electron(s);
    Proton(s)
Charged sphere, 919–20
Charge escalator, 890, 942
Charge polarization, 794–95
Charge quantization, 790
Charge-to-mass ratio, 836, 1020, 1038
Charging, 785, 786
  by friction, 790
  by induction, 796
  insulators and conductors, 792–93
Chemical energy, 339, 514
Chromatic aberration, 746
Circuit(s), 961–95. *See also* AC circuits
  basic, 967–69
  direct current (DC), 961
  elements of, 964–65
  energy and power in, 970–73
  grounded, 983–85
  induced current in, 1046
  Kirchhoff's laws and, 965–66
  LC, 1069–72
  LR, 1072–74
  Ohm's law and, 962–64

  oscillator, 995
  parallel resistors, 978–81
  RC, 985–87
  RC filter, 1126–29
  resistor circuits, 981–83, 1123–24
  series resistors, 973–75
  single-resistor, 968
Circuit diagrams, 949, 964–65
Circular-aperture diffraction, 700–702
Circular motion, 4, 177–206, 370–71
  angular momentum and, 259
  fictitious forces and apparent weight in,
    193–96
  nonuniform, 196–200
  orbits, 190–93
  period of, 358
  problem-solving strategies for, 200
  simple harmonic motion and, 418–21
  slowing, 199
  uniform. *See* Uniform circular motion
Circular wave, 626, 667
Classically forbidden regions, 1293, 1295–96
Classical physics
  breakdown of, 559–60, 685, 1214
  equations of, 1103
Clock synchronization, 1161, 1162
Closed-closed tube, 656, 657
Closed-cycle device, 577
Closed shell, 1330
Closed system, 238
Coaxial cable, 960, 1083
Coefficient of kinetic friction, 133
Coefficient of performance, 581
  Carnot, 596
Coefficient of rolling friction, 134
Coefficient of static friction, 133
Coils
  metal detector, 1064–65
  transformer, 1064
Cold reservoir, 575
Collisional excitation, 1235, 1336–37
Collisions, 239, 240
  elastic, 255, 287–91, 1156–57
  inelastic, 255–56
  mean time between, 889, 894, 895
  molecular, 548–50
Color, 728–29, 730
  in solids, 1338–39
Compasses, 998, 999
Complete circuit, 967
Components of vectors, 85–88
Component vectors, 85, 88
Compression ratio, 605
Compression stroke, 605
Condensation point, 493
Conduction
  electrical, 1197–98
  model of, 887–89
Conductivity, 893–95
Conductor(s), 787, 791–96
  charge polarization, 794–95
  charging, 792–93, 796
  discharging, 793–94

Conductor(s) *cont.*
  electric dipole, 795–96
  in electrostatic equilibrium, 870–72, 939–41
  isolated, 792
Conservation laws, 237–38, 338–39
Conservation of angular momentum, 259–60, 401–2
Conservation of charge, 790–91
Conservation of current, 883–84, 892–93
Conservation of energy, 238, 270, 323–28, 339. *See also* Thermodynamics: first law of
  inside capacitor, 903–4
  in charge interactions, 912
  in fluid flow, 465–67
  Kirchhoff's loop law, 938–39
  in motional emf, 1047
  relativity and, 1187–88
  in simple harmonic motion, 423
Conservation of mass, 237–38
Conservation of mechanical energy, 278–79, 318
Conservation of momentum, 246–52, 1180
Conservative forces, 316–18, 901, 902, 906
Constant-pressure (isobaric) processes, 503, 518
Constant-temperature (isothermal) processes, 503–5, 518–19, 525–26
Constant-volume gas thermometer, 492
Constant-volume (isochoric) process, 502–3, 518, 525, 533
Constructive interference, 649, 661, 667, 668–69, 696
Contact forces, 98, 104, 130
Continuity, equation of, 463–65
Continuous spectrum, 758
Contour map, 670–71, 915–17, 918
Converging lens, 732–33, 736
Coordinate axes, 85
Coordinate systems, 7–8, 36, 84–88
  acceleration and, 164
  axes of, 85
  Cartesian, 84
  inertial reference frame, 112–13
  origin of, 7, 10
  with tilted axes, 91–92
Copernicus, Nicholas, 344, 481
Corpuscles, 685, 1197
Correspondence principle, 1291–93
Coulomb electric field, 1061, 1085, 1091
Coulomb's law, 796–802, 1146
  electrostatic forces and, 799
  Gauss's law vs., 861–62
  units of charge, 797–98
  using, 798–802
Couples, 380, 840
Covalent bond, 1305–7
Crests, 626
Critical angle, 725
Critical point, 494
Critical speed, 195
Crookes tubes, 1199–1200
Crossed-field experiment, 1201–2
Crossed polarizers, 1113
Crossover frequency, 1128

Cross product, 397–99, 1003–4
Crystal lattice, 761, 1316
Crystals, 486, 712
Current, 781, 782, 791, 890. *See also* Induced current
  batteries, 889–90
  conservation of, 883–84
  conventional, 891
  creating, 884–89
  direction of, 891
  displacement, 1099–1102
  eddy, 1048–49
  effect on compass, 999
  electric field in a wire, 885–87
  electric potential and, 943–46
  electron, 880–84
  of inductor, 1131
  magnetic field of, 1004–8
  model of conduction, 887–89
  peak, 1125
  root-mean-square (rms), 1135–36
  in series *RLC* circuit, 1133
  tunneling, 1310
Current balance, 1026
Current density, 890–93
Current loop, 1006, 1008–9
  forces and torques on, 1026–28
  as magnetic dipole, 1009
  magnetic field of, 1006–9
  parallel, 1026–27
Curve of binding energy, 1358–59
Cyclotron, 1021–22
Cyclotron frequency, 1019–21
Cyclotron motion, 1019–21
Cylindrical symmetry, 850, 852

D

Damped oscillation, 432–34
Damped systems, energy in, 433–34
Damping constant, 432
Daughter nucleus, 1369
Davisson, Clinton, 766, 768
Davisson-Germer experiment, 766
DC circuits, 961, 1122
De Broglie, Louis-Victor, 767
De Broglie wavelength, 767–68, 773, 1231
  Schrödinger equation and, 1278–79
Decay. *See also* Radioactivity
  exponential, 433–34
  nuclear, 1364–66
  rate of, 1341, 1364
Decay equation, 1340–42
Decay mechanisms, 1369–73. *See also* Radioactivity
  alpha decay, 929, 1193, 1369–70
  beta decay, 235, 265, 929, 1203, 1362, 1370–71
  decay series, 1372–73
  gamma decay, 1372
  weak interaction, 1371–72
Defibrillator, 953
Degrees, 179

Degrees of freedom, 557, 558–59
Density(ies), 487, 824
  average, 460
  of fluids, 445–47
Destructive interference, 650, 661, 662, 667, 668, 690, 697
Deuterium, 488, 1355
Diatomic gases, 489, 584
Diatomic molecules, thermal energy of, 558–61
Dichroic glass, 646, 665
Diesel cycle, 586, 605
Diffraction, 685, 695
  circular-aperture, 700–702
  of electrons, 768–69
  of matter, 768–70
  of neutron, 769
  order of, 693
  single-slit, 695–99
  x-ray, 760–63
Diffraction grating, 693–95
  reflection gratings, 695, 713
  resolution of, 712
  transmission grating, 695
Diffraction-limited lens, 747
Diffuse reflection, 719
Diode, 963
  resonant tunneling, 1310–11
Dipole(s)
  acceleration of, 840
  electric, 795–96
  electric field of, 821–22, 823
  electric potential of, 920
  induced, 847
  induced magnetic, 1030–31
  magnetic, 998, 1008–11
  in nonuniform electric field, 840–41
  potential energy of, 909–10
  torque on, 839–40
  in uniform electric field, 839–40
Dipole moment, 822
Direct current (DC), 1121
Direct current (DC) circuits, 961, 1122
Discharging, 787, 793–94
Discrete spectrum, 758, 1212
Disk of charge
  electric fields of, 830–32
  electric potential of, 923, 934–35
Disorder, 565–66
Disordered systems, 566–67
Displaced fluid, 459
Displacement, 9–11, 13, 80–83
  algebraic addition to find, 90
  angular, 179–80
  from equilibrium, 281
  graphical addition to find, 81
  net, 80–81, 82, 83
  of sinusoidal wave, 622–23
  velocity and, 49–53, 83
  of wave, 618–20
Displacement current, 1099–1102
  Ampère's law and, 1100–1101
  induced magnetic field and, 1101–2

Displacement vectors, 12, 80, 153
Dissipative forces, 322–24
Divergence angle of laser beam, 711
Diverging lens, 732, 737–38
Doppler effect, 634–38
Dose, radiation, 1373–74
Dot product, 312–14
Double-slit experiment. *See* Young's double-slit experiment
Double-slit interference, 764, 769–70
  interference fringes in, 1254–56
  of neutrons, 770
Drag, 103, 104
Drag force, 137–41
  model of, 138
  for slowly moving object, 432
  terminal speed, 140–41
Drift speed, 881, 882, 883, 888, 892
Driven oscillations, 435–36
Driving frequency, 435
Dry ice, 494
Dynamic equilibrium, 111, 116, 123, 124–25
Dynamics, 2, 97, 234
  of nonuniform circular motion, 198–200
  of simple harmonic motion, 424–27
  in two dimensions, 157–59
  of uniform circular motion, 185–90
Dynodes, 1251

E

Earthquakes, 282, 643
Eddy currents, 1048–49
Edison, Thomas, 1121
Efficiency, limits of, 590–93
Einstein, Albert, 113, 685, 1149, 1152, 1158, 1224–28
Elastic collisions, 255, 287–91, 1156–57
  perfectly elastic collision, 287
  reference frames in, 290–91
Elasticity, 280, 470–72, 480
  tensile stress, 470–71
  volume stress, 472
Elastic limit, 470
Elastic potential energy, 283–87, 291
Electrical conduction, 1197–98
Electrically neutral object, 789
Electric dipole, 795–96, 821
Electric field(s), 94, 803, 805, 817–48, 1088–94, 1146. *See also* Current; Gauss's law
  calculating, 809
  of charged wire, 868
  within conductor, 871–72
  of continuous charge distribution, 824–28
  Coulomb, 1061, 1085, 1091
  of dipole, 821–22, 823
  of disk of charge, 830–32
  electric flux of, 855, 857–59
  electric potential and, 933–39
  energy in, 953
  from equipotential surfaces, 938
  field model, 805–9
  geometry of, 937–38

induced, 1061–63, 1097–99
  of laser beam, 1110
  limiting cases, 818–19
  of line of charge, 826–28
  Maxwell's equations and, 1102–4
  motion of charged particle in, 835–39
  motion of dipole in, 839–41
  of multiple point charges, 819–24
  non-Coulomb, 1061
  nonuniform, 838–39, 840–41, 855, 857–59, 1052
  of parallel-plate capacitor, 834–35
  picturing, 823–24
  of plane of charge, 818, 832–33, 834, 869
  of point charge, 807–8
  of ring of charge, 829–30, 936
  of sphere of charge, 818, 833, 866–67
  symmetry of, 850–53
  torque by, 839
  transformation of, 1089–93
  typical field strengths, 818–19
  uniform, 835, 836–38, 839–40, 902, 1015
  unit vector notation, 808–9
  in current-carrying wire, 885–87
Electric field lines, 823
Electric field strength, 806
Electric field vectors, 823, 916
Electric flux, 853–61
  Ampère's law and, 1100
  calculating, 855–61
  through closed surface, 860–61
  definition of, 856–57
  independence from surface shape and radius, 862–63
  of nonuniform electric field, 855, 857–59
  inside parallel-plate capacitor, 857
Electricity, 104, 235, 783–816, 906, 1146, 1196–97
  atoms and, 788–89
  charge, 788–91
  charge model, 784–88
  Coulomb's law, 796–802
  insulators and conductors, 787, 791–96
  magnetism and, 998–99
  phenomenon of, 781–82
Electric potential energy, 901–4, 914
  mechanical energy and, 901–2
  uniform field and, 902–4
Electric potential, 910–31
  of continuous distribution of charge, 921–24
  current and, 943–46
  of dipole, 920
  of disk of charge, 934–35
  electric field and, 933–39
  geometry of, 937–38
  of many charges, 920–24
  inside parallel-plate capacitor, 914–17
  of point charge, 918–20
  sources of, 941–43
  superposition principle and, 920
Electrodes, 832, 1197
  geometry of, 948
  parallel, 837

Electrolysis, 1197–98
Electromagnetic fields, 612, 1198
  forces and, 1085–88
  transformations of, 1089–93
Electromagnetic field strength, 618
Electromagnetic force, 235
Electromagnetic induction, 1041–83
  in circular loop, 1058
  Faraday's law, 1056–60
  induced currents, 1041–43, 1063–65
  induced electric field, 1061–63, 1097–99
  inductors, 1065–69
  LC circuits, 1069–72
  Lenz's law, 1053–56, 1146
  LR circuits, 1072–74
  magnetic flux, 1050–52
  motional emf, 1043–49, 1056–57
  problem-solving strategy for, 1058
  in solenoid, 1059
Electromagnetic spectrum, 631, 685
Electromagnetic waves, 235, 612, 618, 1062, 1104–14, 1146
  Ampère-Maxwell law and, 1107–8
  antenna-generated, 1111–12
  energy of, 1109–10
  intensity of, 1109–10
  polarization of, 1112–14
  radiation pressure, 1110–11
  speed of light, 1108
  standing, 654–55
  traveling, 630–31
Electromagnetism, 782. *See also* Maxwell's equations
  theory of, 1102–3
Electromagnets, 1009
Electron(s), 1203
  acceleration of, 184–85
  charge of, 788
  de Broglie wavelength of, 768, 1231
  diffraction of, 768–69
  discovery of, 1200–1203
  drift speed of, 892
  energy levels of, 1234
  magnetic force on, 1018
  minimum energy of, 772
  sea of, 791, 792, 881–82, 884
  secondary, 1230
  spin of, 1325–28
  structure of, 789
  valence, 791
  wave-like properties of, 766–67
  wave properties of, 757
Electron beam, 837–38
Electron capture (EC), 1371
Electron cloud, 789, 1321
Electron configuration, 1329
Electron current, 880–84. *See also* Current
  charge carriers, 791, 880–81
  conservation of, 883–84, 892–93
  field strength and, 889
Electron-electron interaction, 1328
Electron gun, 837
Electron spin, 1029, 1325–28

Electron volt, 1208–9
Electro-optic crystals, 712
Electroscope, 792–93, 794, 796
Electrostatic constant, 797
Electrostatic equilibrium, 792
    conductors in, 870–72, 939–41
Electrostatic forces, Coulomb's law and, 799
Electroweak force, 235
Elements, spectrum of, 1236. *See also*
        Periodic table
Emf
    of battery, 943
    chemical, 1044
    induced, 1057
    motional. *See* Motional emf
Emission, spontaneous and stimulated, 1343
Emission spectra, 1212, 1237–38, 1337–38
    of electron, 1305
    of hydrogen, 1244–45
Energy(ies), 268–303, 401, 513–16. *See also*
        Kinetic energy; Mechanical energy;
        Potential energy; Thermal energy;
        Work
    basic model of, 279–80, 304–6, 338
    binding, 1242, 1355, 1357–59
    chemical, 339, 514
    in circuits, 970–73
    conservation of. *See* Conservation of
        energy
    in damped systems, 433–34
    in electric field, 953
    of electromagnetic waves, 1109–10
    forms of, 270
    heat and, 238
    internal, 514
    ionization, 1242, 1321, 1334, 1357
    nuclear, 514
    of photon, 765
    problem-solving strategy for, 326–27
    quantization of, 770–73, 1232–34
    relativistic, 1183–88
    rest, 1184
    rotational, 392–93
    in simple harmonic motion, 421–23
    stored in capacitors, 951–53
    of system, 238, 305
    total, 1184
    transfer of, 515
    work and, 514–15
Energy bar charts, 273–74
    thermal energy in, 324–26
    work done by external forces in, 324–26
Energy diagram(s), 291–96, 1281
    equilibrium positions, 293–95
    molecular bonds and, 295–96
Energy equation, 324
Energy-level diagrams, 1237–38, 1243
Energy levels, 772, 1233, 1234, 1241–42
    of hydrogen atom, 1320–21
    vibrational, 1303
Energy reservoir, 575
Energy storage, modes of, 557–58
Energy transfer, 305

Energy-transfer diagrams, 575–76
Energy transformations, 305
Engine. *See also* Heat engines
    Carnot, 591–93, 595
    gas turbine, 586
    steam, 573
English units
    of force, 110, 130
    of power, 328
Entropy, 565–66, 607
Environment, 208–9, 305, 342
Equation of continuity, 463–65
Equation of motion, 425–27
Equilibrium, 123–25
    displacement from, 281
    dynamic, 111, 116, 123, 124–25
    electrostatic, 792
    hydrostatic, 452
    mechanical, 111, 514
    phase, 493
    problem-solving strategies, 123
    rigid-body, 390–91
    rotational, 390
    stable, 294
    static, 111, 123–24
    thermal, 487, 501, 521, 562
    total, 390
    translational, 390
    unstable, 294
Equipartition theorem, 557–58
Equipotential surfaces, 915, 918, 938
Equivalence, principle of, 349–50
Equivalent capacitance, 949
Equivalent focal length, 745
Equivalent resistance, 974, 978
Escape speed, 355, 907, 908
Estimate, order-of-magnitude, 30
Ether, 1157
Events, 1160–61
    observations and, 1162–63
Excitation, 1334–37
    by absorption, 1335–36
    collisional, 1235, 1336–37
Excitation transfer, 1346
Excited states, 1235, 1334–37
    lifetimes of, 1339–42
Explosions, 239, 252–54
Exponential decay, 433–34
Exponential function, 433
Extended object, 370
External forces, 209

F

Fahrenheit scale, 491
Faraday, Michael, 802–3, 1042–43, 1197–98
Faraday's law, 1056–60, 1094–99, 1102,
        1103, 1146
    electromagnetic waves and, 1106–7
    using, 1058–59
Far focal point, 733
Fermat's principle, 756
Ferromagnetism, 1029–30

Fiber optics, 726
Fictitious forces, 193–96
    centrifugal force, 193–94
Field(s), 781, 782, 798, 802–9, 1042. *See also*
        Electric field(s); Electromagnetic fields
    inside charging capacitor, 1101–2
    fringe, 834
    gravitational, 803–5, 902–4
    particle vs., 803
    vector, 806
Field diagram, 807–8
Field lines, 1085–86
Filter(s), 1128–29
    high-pass, 1128
    low-pass, 1128
    polarizing, 1112–13
    RC circuit, 1126–29
Finite potential wells, 1293–98
    classically forbidden region, 1293, 1295–96
    nuclear physics, 1297–98
    quantum-well devices, 777, 1277, 1288–89,
        1296–97
Flame test, 1338
Flashlamp, 953, 1345
Flow tube, 463–64
    energy conservation in, 465–66
Fluid(s), 341, 444–69, 480
    buoyancy, 458–62
    displaced, 459
    gases and liquids, 445
    pressure, 447–58
    volume and density of, 445–47
Fluid dynamics, 462–69
    applications of, 468–69
    Bernoulli's equation, 465–69, 480
    equation of continuity, 463–65
Fluid mechanics, 444
Fluid statics, 444
Fluorescence, 1199
Focal length of lens, 743
    equivalent, 745
Focal point, 732–33
Force(s), 98–121
    acceleration and, 107–8, 185–86
    agent of, 98, 113
    buoyant, 458–59
    centrifugal, 193–94
    combining, 100
    conservative, 316–18, 901, 902, 906
    constant, 309–11, 901–2
    contact, 98, 104, 130
    on current loops, 1026–28
    dissipative, 322–24
    drag, 103, 104, 432
    electromagnetic fields and, 1085–88
    external, 209, 248, 323
    fictitious, 193–96
    friction, 102–3, 104
    in gas pressure, 550–52
    gravitational, 235, 348–49, 351–52
    identifying, 104–6, 113
    impulsive, 240–41
    inertial reference frame and, 112–13

interaction, 248
long-range, 98, 802, 803
Lorentz, 1087–88
misconceptions about, 113–14
of nature, 235
net, 96, 100, 109, 111, 112, 185, 189, 551
nonconservative, 316–18
normal, 102, 104
polarization, 296, 795, 839–40
from potential energy, 319–20
restoring, 280–83, 352, 425, 431
spring, 101, 104
strong, 302, 1297–98, 1359–60
superposition of, 100
tension, 101, 104
thrust, 103–4, 212
units of, 110
weight, 100–101, 104
work and, 316–19
Free-body diagrams, 114–16
Free fall, 59–62, 129. *See also* Projectile motion
acceleration in, 131–32
energy diagram of, 291–92
orbiting projectile in, 191
projectile motion and, 166
Freezing point, 493
Frequency
angular, 416, 418, 623
beat, 674
crossover, 1128
cyclotron, 1019–21
driving, 435
fundamental, 653, 658
modulation, 673
natural, 435
of oscillation, 414, 416
resonance, 436, 1133
of sinusoidal wave, 621
threshold, 1221, 1224, 1226, 1228
Frequency-time relationship, 1266–67
Friction, 102–3, 104, 132–37
atomic-level model of, 322
causes of, 137
charging by, 790
coefficients of, 133
kinetic, 103, 133, 134
model of, 134–37
rolling, 133–34
static, 103, 132–33, 134, 143–44, 187–88, 212
Fringe field, 834
Fringe spacing, 690
Fuel cells, 1197
Fundamental frequency, 653, 658
Fundamental quantum of energy, 1233
Fundamental unit of charge, 789, 1203–5

G

Galilean field transformation equations, 1092
Galilean relativity, 168–70, 1089, 1152–57
reference frames and, 1153–54
using, 1156–57

Galilean transformation, 1154–55
of acceleration, 169, 1089
of position, 165, 1154
of velocity, 167, 290, 1089, 1154
Galileo, 3–4, 35, 60, 111, 345, 481
Gamma decay, 1372
Gamma rays, 776, 1187, 1298, 1353, 1363
Gas(es), 445, 480, 486–87, 496
diatomic, 489, 584
ideal, 491, 495–505
inert, 1334
molecular collisions in, 548–50
monatomic, 489, 556–57, 584
pressure in, 448–50, 454, 550–54
sound waves in, 629
specific heats of, 533–39
Gas discharge tube, 1198, 1212
Gas turbine engine, 586
Gauge pressure, 453–54
Gaussian surface, 854, 865, 1086–87
multiple charges on, 864
point charge outside, 863–64
at surface of conductor, 870–71
Gauss's law, 861–69, 1086–87, 1102, 1103
charged wire field and, 868
charge outside surface, 863–64, 866–67
Coulomb's law vs., 861–62
electric flux independence from surface shape and radius, 862–63
for magnetic fields, 1086–87
for magnetism, 1102, 1103
plane of charge field and, 869
problem-solving strategy for, 866
sphere of charge field and, 866–67
using, 865–69
Geiger counter, 1364
General relativity, 1152
Generators, 1047, 1063–64
Geomagnetism, 998
Geosynchronous orbits, 203, 359, 1147
Grand unified theory, 235
Graphical representation, 23, 24, 64–67
Gravitational collapse, 366–67, 412
Gravitational constant, 337, 348, 352–53
Gravitational field, 803–5
uniform, 902–4
Gravitational force, 235, 348–49
decrease with distance, 351–52
weight and, 348–49
Gravitational mass, 349
Gravitational potential energy, 270–80, 291, 353–57, 902
conservation of mechanical energy and, 278–79
flat-earth approximation of, 356–57
zero of, 274–75
Gravitational torque, 379–80
Gravity, 98, 100, 341, 343–68
acceleration due to, 20, 60–61, 352
circular orbits and, 192–93
cosmology and, 344–45
little *g* (gravitational force) and big *G* (gravitational constant), 350–53

Newton's law of, 193, 347–50
Newton's theory of, 350, 480
pressure and, 449
satellite orbits and energies, 357–62
universality of, 346
Gravity waves, 704
Grounded circuits, 983–85
Grounding, 793–94
Ground state, 1235, 1329–35
of hydrogen, 1319, 1320, 1321, 1327
Ground symbol, 983
Ground wire, 983

H

Half-lives, 1169, 1364–66
Hall effect, 1022–23
Hall probe, 1023
Hall voltage, 1023
Harmonics, 653, 681
Heat, 305, 515, 521
energy and, 238
path followed in *pV* diagram, 536
properties of, 522
specific, 527–29
temperature vs., 523
thermal energy vs., 523
thermal interactions and, 561–63
waste, 578
into work, 574–77
Heat engines, 484, 573–88
Brayton cycle, 586–88
first law of thermodynamics for, 577–78
ideal-gas, 583–88
perfect, 577, 578, 581–82
perfectly reversible, 591–93
problem-solving strategy for, 584–86
thermal efficiency of, 578, 590–93, 596–97
Heat exchanger, 586
Heat of fusion, 529
Heat of transformation, 529–30
Heat of vaporization, 529
Heat pump, 601
Heat, 327, 520–23
thermal interactions, 521–22
trouble with, 522–23
units of, 522
Heavy water (deuterium), 488, 1355
Heisenberg uncertainty principle, 1268–70, 1382
Helium, 1329–30
Helium-neon laser, 1345–46
Henry, Joseph, 1042, 1147
Hertz, Heinrich, 1220–21
High-pass filter, 1128
High-temperature superconductors, 895
High-*Z* nuclei, 1362
History graph, 615–17, 1268
Holography, 706–7
Hooke, Robert, 346
Hooke's law, 281–83, 470
Hot reservoir, 575
Hot side, 1144

Huygens' principle, 696, 722
Hydraulic lift, 457–58
Hydrogen, 759–60, 1318–24. *See also* Bohr
    model of hydrogen atom
    absorption in, 1336
    angular momentum of, 1319–20
    energy levels of, 1320–21
    excitation of, 1337
    ground state of, 1319, 1320, 1321, 1327
    radial wave functions of, 1322–24
    stationary states of, 1318–19
Hydrogen-like ions, 1245–46
Hydrogen spectrum, 1243–46
    visible, 759
Hydrostatic pressure, 448–49, 451–53
Hydrostatics, 454

I

Ideal battery, 943
Ideal-fluid model, 462
Ideal gases, 491, 495–505
Ideal-gas heat engines, 583–88
    Brayton cycle, 586–88
Ideal-gas law, 498–501
Ideal-gas model, 496–97
Ideal-gas processes, 501–5, 583–84
    adiabatic, 536–39
    constant-pressure (isobaric), 503, 518
    constant-temperature (isothermal), 503–5,
        518–19
    constant-volume (isochoric), 502–3, 518,
        525, 533
    problem-solving strategy for, 518
    pV diagram of, 501
    quasi-static, 501–2
    work in, 516–20
Ideal-gas refrigerators, 588–90
Ideal inductor, 1065
Ideal insulator, 964
Ideal solenoid, 1015
Ideal wire, 964
Image distance, 719, 727
Impulse, 242–46, 338
Impulse approximation, 244
Impulse-momentum theorem, 242–43, 245–46
    work-kinetic energy theorem and, 308
Impulsive force, 240–41
Inclined plane, 62–67
Independent particle approximation (IPA), 1328
Index of refraction, 631–33, 722–23
    measuring, 705–6, 724
Induced current, 1041–43
    applications of, 1063–65
    in circuit, 1046
    eddy currents, 1048–49
    energy considerations, 1047–48
Induced dipole, 847
Induced electric field, 1061–63, 1097–99
    Maxwell's theory, 1062–63
Induced emf, 1057
Induced magnetic dipoles, 1030–31
Induced magnetic field, 1053, 1062, 1101–2

Inductance, 1065–66
Induction, 796. *See also* Electromagnetic
    induction
Inductive reactance, 1130
Inductor circuits, 1129–31, 1137
Inductors, 1065–69
    energy in, 1068–69
    ideal, 1065
    potential difference across, 1066–68
    power in, 1137
Inelastic collisions, 255–56
Inert gases, 1334
Inertia, 108
    moment of, 381–85
    thermal, 528
Inertial mass, 108, 349, 382
Inertial reference frames, 112–13, 165, 169,
    1089, 1153
Initial conditions, 418, 419
Ink-jet printer, electrostatic, 846
Insulator(s), 787, 791–96, 964
    charge polarization, 794–95
    charging, 792–93
    discharging, 793–94
    electric dipole, 795–96
    ideal, 964
Integration, limits of, 52
Intensity, 1255
    of wave, 633–34, 650, 1109–10
Interacting systems, 208–9, 223–27. *See also*
    Newton's third law of motion
    revised problem-solving strategy for, 216–18
Interference, 609, 659, 684, 685, 686–92, 778.
    See also Double-slit interference;
    Wave optics
    constructive, 649, 661, 667, 668–69, 696
    destructive, 650, 661, 662, 667, 668, 690, 697
    of light, 687–92
    mathematics of, 663–66
    of matter, 768–70, 1259
    in one dimension, 659–66
    phase difference, 661–63
    photon analysis of, 1257
    picturing, 670–72
    problem-solving strategy for, 669–70
    between sound waves, 662–63, 664
    thin-film optical coatings, 665–66
    in two and three dimensions, 667–72
    of two waves, 669–70
    wave analysis of, 1254–55
Interference fringes, 688
Interferometer(s), 702–7
    acoustical, 702
    indices of refraction measured with, 705–6
    Michelson, 703–6
Internal energy, 514
Internal resistance, 976
Inverse-square law, 797
Inverted image, 734
Ion cores, 791
Ionization, 790, 1209–10
Ionization energy, 1242, 1321, 1334, 1357
Ionization limit, 1243

Ionizing radiation, 1363–64
Ions, 1197, 1200
    hydrogen-like, 1245–46
    molecular, 790
Iridescence, 695
Irreversible processes, 563–66
    equilibrium direction and, 564–65
    order, disorder, and entropy, 565–66
Irrotational flow, 462
Isobaric (constant-pressure) processes, 503,
    518, 533
Isobars, 1354
Isochoric (constant-volume) process, 502–3,
    518, 525, 533
Isolated system, 248–49, 324, 338
    second law of thermodynamics and, 607
Isothermal (constant-temperature) processes,
    503–5, 518–19, 525–26
Isotherms, 504
Isotopes, 1211, 1354

J

Joule, James, 520–21

K

Kelvin scale, 491, 492, 494–95
Kepler, Johannes, 344–45, 360–61
Kepler's laws of planetary orbits, 344–45, 358–60
Kinematics, 1, 35–77, 234. *See also* Planar
    motion
    with constant acceleration, 53–59
    free fall, 59–62
    instantaneous acceleration, 67–69, 155
    instantaneous velocity, 43–49
    position from velocity, 49–53
    rotational, 370–73
    of simple harmonic motion, 415–17, 480
    in two dimensions, 152–57
    uniform motion, 38–42
Kinetic energy, 270–75, 304–5
    at microscopic level, 321–22
    relativistic expression for, 1184
    of rolling object, 394–95
    rotational, 392–93, 480
    of satellite, 361–62
    in terms of momentum, 308
    translational, 554, 557
    work and, 306–8
Kinetic friction, 103, 133, 134
Kinetic theory, 547–72
    of gas pressure, 550–54
    of irreversible processes, 563–66
    molecular collisions and, 548–50
    second law of thermodynamics, 328, 566–67
    of temperature, 554–56
    of thermal energy and specific heat, 556–61
    of thermal interactions and heat, 561–63
Kirchhoff's laws, 965–66
    junction law, 893, 966
    loop law, 938–39, 966, 1123
Knowledge structure, 234

L

Laminar flow, 462, 463
Laser(s), 1343–46
    helium-neon, 1345–46
    quantum-well, 1220, 1297
    ruby, 1345
    semiconductor diode, 1297
    single-slit diffraction of, 698
Laser beam, 654, 711, 1110
Laser cavity, 654–55
Laser cooling, 556
Laser range-finding, 711
Lasing medium, 1344
Launch angle, 159
LC circuits, 1069–72
Length(s), 30
    units of, 26, 27
Length contraction, 1171–74
Lens(es), 731, 742–44
    aberrations, 745–46
    converging, 732–33, 736
    diffraction-limited, 747
    diverging, 732
    focal length of, 743
    meniscus, 743
    thick, 742
    thin. *See* Thin lenses
Lens maker's equation, 743
Lens plane, 732, 739
Lenz's law, 1053–56, 1146
Lever arm (moment arm), 378, 379, 381
Lifetimes of excited states, 1339–42
    decay equation, 1340–42
Lift, 469
Light, 685–92, 778, 1197. *See also*
        Electromagnetic waves
    absorption of, 1212–13
    coherent, 1344
    corpuscular theory of, 685, 1197
    as electromagnetic wave, 685, 1062–63
    emission of, 1212–13
    filtering, 730
    interference of, 684, 685, 686–92
    models of, 609, 684, 686, 701–2
    particle-like behavior of, 763–64
    photon model of, 609, 764–66, 1257–59
    properties of, 1382
    quanta of, 686, 1225–27
    as ray. *See* Ray optics
    scattering of, 731
    spectrum of, 630, 729, 758
    speed of, 26, 170, 630, 722, 1104, 1108,
        1158–59
    as wave. *See* Wave optics
    wavelength of, 705
Light clock, 1166
Light ray, 715. *See also* Ray optics
Light waves, 630–31
    Doppler effect for, 637–38
    reflection of, 651–52
Light year (ly), 72, 368, 1170
Like charges, 786

Linear charge density, 824–25
Linear density, 614–15
Linear motion, 122–50, 480
    drag, 137–41
    equilibrium and, 123–25
    friction, 132–37
    mass and weight, 129–32
    Newton's second law, 125–28, 141–44
Linear restoring force, 431
Line integrals, 1011–12, 1095
Line of action, 378
Line of charge, 826–27, 828
Line of stability, 1357
Line spectrum, 758
Lines per millimeter, 693
Liquid crystals, 486
Liquid-drop model, 1356
Liquids, 445, 480, 486–87, 496
    pressure in, 448–49, 451–53
    sound waves in, 629
Load, 967
Longitudinal wave, 613, 617–18
Long-range forces, 98, 785, 802, 803
Lorentz force, 1087–88
Lorentz force law, 1103, 1146
Lorentz transformations, 1175–80
    binomial approximation, 1178–79
    equations of, 1176
    length and, 1177–78
    velocity transformations, 1179–80
Loschmidt number, 569
Low-pass filter, 1128
Low-Z nuclei, 1361–62
LR circuits, 1072–74
Lyman series, 760, 1244–45

M

Macrophysics, 321, 484
Macroscopic systems, 483, 484, 485
Magnet(s)
    atomic, 1028–29
    permanent, 1030
Magnetic dipole moment, 1009–11
Magnetic dipoles, 998, 1008–11
    induced, 1030–31
Magnetic domains, 1029–30
Magnetic energy density, 1069
Magnetic field, 996–1040, 1042, 1088–94, 1146
    Ampère's law and, 1011–14
    of current, 1004–8
    of current loop, 1006–9
    discovery of, 997, 998–1001
    energy in, 1068–69
    Galilean relativity and, 1089, 1093–94
    Gauss's law for, 1086–87
    induced, 1053, 1062, 1101–2
    Maxwell's equations and, 1102–4
    moving charges as source of, 1001–4
    of solenoid, 1015–17
    transformation of, 1089–93
    uniform, 1015, 1016, 1027
Magnetic field lines, 1000

Magnetic flux, 1050–52, 1053
    over closed Gaussian surface, 1086–87
    Lenz's law and, 1053
    in nonuniform field, 1052
    sign of, 1096–97
Magnetic force, 235
    on current-carrying wires, 1024–26
    on moving charge, 1017–23
Magnetic materials, 998
Magnetic moment, 1325–26
Magnetic monopole, 998
Magnetic quantum number, 1319
Magnetic resonance imaging (MRI), 1016,
        1080, 1375–76
Magnetism, 98, 104, 997–98, 1146
    atomic structure and, 782
    electricity and, 998–99
    kinds of, 1000–1001
    phenomenon of, 781–82
    theory of, 781–82
Magnetron, 1019, 1038
Magnification, 717, 735–36
Malus's law, 1113–14
Manometers, 454–55
Maser, 1343
Mass, 30, 129–32
    atomic, 488, 1354–55
    center of, 374–76
    conservation of, 237–38
    force and, 108
    gravitational, 349
    inertial, 108, 349, 382
    measuring, 129–30
    molar, 489–90
    molecular, 488
    weight vs., 101, 129
Mass density, 446, 487
Mass-energy equivalence, 1186–87
Massless string approximation, 221–23
Mass number, 1211, 1353
Mass spectrometer, 1038, 1211
Matter, 485–511, 1196, 1353. *See also* Atom(s)
    atomic mass and atomic mass number,
        488–89
    ideal gases, 491, 495–505
    interference and diffraction of, 768–70
    magnetic properties of, 1028–31
    moles and molar mass, 489–90
    phase changes in, 486, 493–95, 529–30, 652
    phases of, 485, 486
    solids, liquids, and gases, 486–87
    temperature and, 490–92
    thermal properties of, 527–30
Matter waves, 612, 766–70, 778
    de Broglie wavelength, 767–68, 773
    interference and diffraction of matter,
        768–70, 1259
    quantization of, 1231–32
Maxwell, James Clerk, 235, 803, 1062, 1102
Maxwell's equations, 803, 1102–4, 1146
    source-free, 1104
Maxwell's theory of electromagnetism,
        1062–63, 1157–58

Mean free path, 548–49
Mean time between collisions, 889, 894, 895
Mechanical energy, 305, 318–19, 513
    conservation of, 278–79, 318
    electric potential energy and, 901–2
Mechanical equilibrium, 111, 514
Mechanical interaction, 514
Mechanical waves, 612
Medium, 721
    of light wave, 630
    of mechanical wave, 612
    motion in, 167
Melting point, 493
Meniscus lens, 743
Metal detectors, 1064–65
Metals, 881
Michelson interferometer, 703–6
Micro/macro connection, 484, 485–86. *See also*
        Kinetic theory
Microphysics, 321, 484
Microwave radiation, 655
Millikan, Robert, 1203–5
Millikan oil-drop experiment, 1204–5
Minimum spot size, 747
Mirror, 718
    plane, 719–20
Mks units, 26
Model(s), 2
    atomic, 101, 1207
    Bohr model of atomic quantization,
        1234–36
    Bohr model of the hydrogen atom, 1238–43
    charge, 786–87
    light, 686
    particle, 6–7
    wave, 609–10
Mode number, 760
Modulation frequency, 673
Modulation of wave, 673
Molar mass, 489–90
Molar specific heat, 528–29
    at constant pressure, 533
    at constant volume, 533
    of solid, 558
Molecular beam, 497
Molecular bond(s), 101, 104
    bond length, 296
    covalent, 1305–7
    energy diagrams and, 295–96
    friction and, 137
    potential energy of, 1302–3
Molecular collisions, 548–50
    mean free path between, 548–49
Molecular mass, 488
Molecular orbitals, 1306–7
Molecular speeds, 497
Molecular springs, 102
Molecular vibration, 296, 1302–3
Molecules, 1196
    polarizability of, 847
Moles, 489–90
Moment arm (lever arm), 378, 379, 381
Moment of inertia, 381–85

Momentum, 238, 239–60, 401. *See also*
        Angular momentum
    basic model of, 338
    conservation of, 246–52, 1180
    impulse and, 240–41
    of isolated system, 249
    kinetic energy in terms of, 308
    relativistic, 1180–83
    total, 248
    in two dimensions, 257–58
Momentum bar charts, 245–46
Monatomic gases, 489, 556–57, 584
Monopoles, 998
Moon, orbit of, 192–93
Motion, 3–25. *See also* Acceleration; Circular
        motion; Force(s); Kinematics; Linear
        motion; Newton's laws of motion
        (general); Oscillation(s); Planar motion;
        Rotational motion; Simple harmonic
        motion (SHM); Uniform circular motion;
        Velocity
    along a line, notation for, 23
    with constant acceleration, 53–59
    cyclotron, 1019–21
    graphical representations of, 64–67
    on inclined plane, 62–67
    in one dimension, 36–38
    particle model of, 6–7
    relative, 164–70
    signs of quantities of, 36–37
    types of, 4, 6, 7
    uniform (constant velocity), 38–42, 56,
        111, 112
Motional emf, 1043–49, 1056–57, 1094
    calculating, 1095–96
    induced current in circuit, 1046
Motion diagram(s), 3–6
    acceleration vectors on, 16–17
    analyzing, 11–13
    complete, 17
    with displacement vectors, 12
    for uniform circular motion, 183
    velocity vectors on, 14
Motor(s)
    electric, 1027–28
    power used by, 1138–39
MRI, 1016, 1080, 1375–76
Muons, 1169–70, 1252

N

Nanostructures, 342, 1277
Natural abundance, 1354
Natural frequency, 435
Near focal point, 733
Near point, 755, 756
Negative ion, 790
Neutral, 1144
Neutral buoyancy, 460
Neutral objects, 785, 786, 789
Neutrino, 265, 929, 1371
Neutron number, 1353
Neutrons, 265, 769–70, 789, 1211, 1353

Neutron star, 367, 412, 1379
Newton, Isaac, 4, 46, 97, 345–47, 360–61, 685
Newtonian physics. *See* Classical physics
Newtonian synthesis, 481
Newton's first law of motion (law of inertia),
        110–14
Newton's law of gravity, 193, 347–50
Newton's laws of motion (general), 97, 234–35
    inertial reference frames and, 169
Newton's second law of motion, 108–10,
        141–44, 480
    with drag included, 432
    for oscillations, 424–26
    for rotation of rigid body, 381–83
    stopping distances, 141–42
    in terms of momentum, 241
    using, 125–28
Newton's theory of gravity, 350, 480
Newton's theory of motion, 1260–61
Newton's third law of motion, 207–8, 213–18.
        See also Interacting systems
    acceleration constraints, 216
    conservation of momentum and, 246–47, 249
    reasoning with, 214–15
Nodal lines, 669
Nodes of standing waves, 649–50
Non-Coulomb electric field, 1061
Normal force, 102, 104
Normalization, 1261–64
Normal modes, 653–54
North pole, 997
Nuclear decay, 1364–66. *See also* Decay
        mechanisms
Nuclear energy, 514
Nuclear fission, 1187–88
Nuclear force. *See* Strong force
Nuclear magnetic resonance (nmr), 1375–76
Nuclear model of atom, 1207–9
Nuclear physics, 1297–98, 1352–84
    biological applications of, 1373–76
    decay mechanisms, 1369–73
    first experiment in, 1206–8
    nuclear stability, 1356–59
    nuclear structure, 1352–56
    radiation and radioactivity, 1353, 1362–68
    shell model, 1298, 1324, 1360–62
    strong force, 235, 302, 1297–98, 1359–60
Nuclear potential well, 1298
Nucleons, 302, 1353
    potential energy of interacting, 1359–60
Nucleus, 235, 789, 1150, 1207
    daughter, 1369
    discovery of, 1205–10
    parent, 1369
    properties of, 1382
    size and density of, 1355–56
Number density, 488

O

Object distance, 727
Objective lens, 738
Object plane, 734

Objects (sources of light rays), 715, 716
Observations, events and, 1162–63
Ohmic materials, 963–64
Ohm's law, 962–64
One-dimensional quantum mechanics. *See* Quantum mechanics
One-dimensional waves, 614–20
 displacement of, 618–20
 longitudinal, 617–18
 snapshot graphs and history graphs of, 615–17
 on a string, 614–15
Open-closed tube, 657
Open-open tube, 657
Operational definitions, 5–6
Opposite charges, 786
Optical amplifier, 1344
Optical axis, 727
Optical cavity, 1344
Optical instruments, resolution of, 746–49
Optical lever, 753
Optics, 684, 685–86, 757–77, 778. *See also* Wave optics
 photons, 763–66, 770
 spectroscopy, 695, 758–60
 x-ray diffraction, 760–63
Orbital quantum number, 1318
Orbitals, molecular, 1306–7
Orbits, 191
 circular, 190–93
 elliptical, 344–45, 357
 geosynchronous, 203
 satellite, 357–62
Ordered systems, 566–67
Order of diffraction, 693
Orders of magnitude, 30
Oscillation(s), 283, 293, 341, 413–43, 480. *See also* Simple harmonic motion (SHM)
 amplitude of, 415
 damped, 432–34
 driven, 435–36
 envelope of, 433, 653
 frequency of, 414, 416, 435
 initial conditions of, 418, 419
 Newton's second law for, 424–26
 pendulum, 429–31
 period of, 414
 phase of, 419
 resonance and, 435–36
 turning points in, 415, 422
 vertical, 427–29
Oscillator circuit, 995
Oscillators, 414, 1300–1303
Otto cycle, 586, 605

P

Parallel-axis theorem, 386–87
Parallel capacitors, 949–50
Parallel-plate capacitor, 857, 946–48
 electric field of, 834–35
 electric flux inside, 857
 electric potential inside, 914–17
 potential difference of, 947

Parallel resistors, 978–81
Paraxial rays, 727–28, 739–40
Parent nucleus, 1369
Particle(s), 803
 alpha, 929
 angular momentum of, 399–400
 beta, 929
 charged, 782
 field vs., 803
 system of, 321
 torque on, 399
Particle accelerator, 1021
"Particle in a box," 770–71, 1232–33, 1283–91
 allowed energies for, 1286–87
 boundary conditions for, 1284–85
 energy-level diagram for, 1286
 interpreting solution to, 1287–91
 potential-energy function for, 1284
 wave functions of, 1285–86
 zero-point motion, 1289–91
Particle model, 6–7, 612
Particle-wave duality, 1220
Pascal's principle, 453
Paschen series, 760
Path-length difference, 661–62, 668
Pauli exclusion principle, 1329–30, 1382
Pendulum
 ballistic, 277–78
 conical, 203
 damped, 434
 motion of, 429–31
Penetration distance, 1296
Perfect destructive interference, 661, 662, 668
Perfectly elastic collision, 287
Perfectly inelastic collision, 255
Perfectly reversible engine (Carnot engine), 591–93, 595
Period
 of circular motion, 358
 of oscillation, 414
 of sinusoidal wave, 621–22
 of uniform circular motion, 178, 181
Periodic table, 1210–11, 1330–34
 elements with $Z > 10$, 1332–33
 first two rows of, 1331–32
 ionization energies, 1334
Permanent magnet, 1030
Permeability constant, 1002
Permittivity constant, 798
Phase angle, 1122, 1132–33
Phase changes, 486, 493–95, 529–30
 upon reflection, 652
Phase constant, 419–21, 622
Phased array, 683
Phase diagram, 494
Phase difference, 628, 661–63
 inherent, 661
Phase equilibrium, 493
Phase of oscillation, 419
Phase of wave, 627–28
Phases of matter, 485, 486
Phasor, 1122–24
Phasor diagram, 1122

Photodetectors, 1229
Photodissociation, 296
Photoelectric effect, 685, 764, 767, 1150, 1220–28
 characteristics of, 1221–22
 classical interpretation of, 1222, 1224
 Einstein's explanation of, 1224–28
 stopping potential, 1223–24
Photoelectrons, 1221
Photolithography, 747
Photomultiplier tube (PMT), 1229–30
Photon(s), 685, 686, 763–66, 770, 1150, 1228–30, 1257
 double-slit interference experiment and, 764
 energy of, 765
 wavelength of, 1236
Photon model of light, 609, 764–66, 1257–59
Physical representation, 23, 24, 104, 114
Pinhole camera, 717
Planar motion, 151–76. *See also* Projectile motion
 acceleration in, 155–57
 momentum in, 257–58
 position and velocity in, 153–55
 relative motion, 164–70
Planck's constant, 764, 767, 1225
Plane mirror, 719–20
Plane of charge, 818
 electric fields of, 832–33, 869
Plane of polarization, 1112
Planetary orbits, Kepler's laws of, 344–45
Planets, extrasolar, 359–60
Plane wave, 627
Plum-pudding model of atom, 1205–6
Point charge, 797, 818, 834
 electric field of, 807–8
 electric field of multiple, 819–24
 electric potential of, 918–20
 magnetic field of, 1003
 multiple, 908–9
 potential energy of, 904–9
Point source, 696, 716
Polarization of electromagnetic waves, 1112–14
Polarization force, 296, 795, 839–40
Polarizer, 1113
Polarizing filter, 1112–13
Polaroid, 1113
Population inversion, 1345
Position vector, 8
Positron (antielectron), 847, 1022, 1186
Positron-emission tomography (PET) scans, 1187
Potassium-argon dating, 1380
Potential, electric, 910–31
Potential difference, 912–13, 914
 across battery, 942–43
 across capacitor, 914–16
 across inductor, 1066–68
 across resistor, 962, 966
Potential energy, 304–5, 317
 of dipole, 909–10

Potential energy *cont.*
    elastic, 283–87, 291
    electric, 914
    force from, 319–20
    gravitational, 270–80, 291, 353–57, 902
    at microscopic level, 321–22
    of molecular bond, 1302–3
    negative, 354
    of point charges, 904–9
    of satellite, 361–62
    of two nucleons interacting via the strong
        force, 1359–60
    zero of, 274–75, 907–8
Potential-energy curve, 291–92
Potential graph, 915, 918
Potential well, 1293. *See also* Finite
        potential wells
    nuclear, 1298
Power, 328–30, 1047
    average, 1135
    in circuits, 970–73, 1135–39
    used by motor, 1138–39
    of wave, 633–34
Power factor, 1137
Power loss in resistors and capacitors, 1136–37
Power stroke, 605
Poynting vector, 1109
Prefixes used in science, 27
Pressure, 447–58
    atmospheric, 449–50, 455
    blood, 456
    causes of, 448–49
    in gases, 448–50, 454, 550–54
    in liquids (hydrostatic pressure), 448–49,
        451–53
    measuring, 453–56
    radiation, 1110–11
    units of, 456
    using, 456–58
    vapor, 448, 455
Pressure gauge, 453
Prime symbol, 164
Principal quantum number, 1318
Probability, 1255–56
Probability density, 1258–59
Probability wave, 779
Projectile motion, 19–20, 159–64, 190–91
    acceleration vector for, 20
    free-fall motion and, 166
    independent motions making up, 160
    problem-solving strategies for, 161–64
    reasoning about, 160–61
Proper length, 1172
Proper time, 1168–69
Proportionality constant, 107–8
Propulsion, 211–13
Proton(s), 104, 789, 790, 1210, 1353
Psi, law of, 1261. *See also* Schrödinger's
        equation
Ptolemy, 344
Pulleys, 222–23
    massless, frictionless approximation,
        222–23
    rotation constraints, 389

Pulsars, 367, 412, 1379
Pulse train, 1273
pV diagram, 501

Q

Quadrants of coordinate system, 84–85
Quanta of light, 686, 1225–27
Quantization, 1150, 1220–52, 1282
    of angular momentum, 1242–43, 1319–20
    Bohr hydrogen atom and, 1238–43
    Bohr's model of, 1234–38
    charge, 790
    of energy, 770–73, 1232–34
    hydrogen spectrum, 1243–46
    of matter waves, 1231–32
    photoelectric effect, 1220–28
    photons, 1228–30
    Schrödinger's equation and, 1282
Quantum computers, 1383–84
Quantum corral, 1220
Quantum jump (transition), 1235
Quantum mechanics, 559–60, 685, 757–58,
        1150, 1253, 1254, 1277–1316, 1382–84.
        See also Atomic physics
    correspondence principle, 1291–93
    covalent bonds, 1305–7
    finite potential wells, 1293–98
    particle in a rigid box, 1283–91
    problem solving in, 1283
    quantum harmonic oscillator, 1300–1303
    quantum-mechanical tunneling, 1307–11
    Schrödinger's equation, 1277–83
    wave-function shapes, 1298–1300
Quantum number, 772, 1233
    magnetic, 1319
    orbital, 1318
    principal, 1318
    spin, 1327
Quantum-well devices, 777, 1277, 1288–89,
        1296–97
    laser, 1220, 1297
Quarks, 235
Quasar, 638
Quasi-static processes, 501–2

R

Rad, 1373
Radial axis, 182
Radial probability density, 1323
Radial wave functions, 1322–24
Radians, 179, 419
Radiation, 1362–68
    alpha, 929, 1193, 1363, 1369–70
    beta, 235, 265, 929, 1203, 1362, 1363,
        1370–71
    gamma, 1363, 1372
    ionizing, 1363–64
    medical uses of, 1374–75
    microwave, 655
Radiation dose, 1373–74
Radiation exposure, 1374
Radiation pressure, 1110–11

Radiation therapy, 1374
Radioactive dating, 1367–68
Radioactivity, 253–54, 1206, 1353, 1354,
        1362–68. *See also* Decay; Decay
        mechanisms
Raisin-cake model of atom, 1205–6
Rarefactions, 629
Rate equation, 1341
Ray diagrams, 716
Rayleigh scattering, 731
Rayleigh's criterion, 713, 748
Ray optics, 714–56
    dispersion, 729
    ray model of light, 609, 701–2, 715–17
    reflection. *See* Reflection
    refraction. *See* Refraction
    resolution of optical instruments, 746–49
    thin lenses. *See* Thin lenses
Ray tracing, 732, 734, 737
RC circuits, 985–87
RC filter circuits, 1126–29
Real images, 733–35
Recoil, 253
Red shift, 637, 638
Reference frame(s), 164–65, 290–91, 1153–54
    accelerating, 113
    inertial, 112–13, 165, 169, 1089, 1153
Reflection, 651–52, 718–21, 850
    diffuse, 719
    law of, 176
    left and right "reversal," 720–21
    phase change upon, 652
    from plane mirror, 719–20
    specular, 718
    total internal, 725–26
Reflection gratings, 695, 713
Refraction, 721–28
    analyzing, 723
    image formation by, 727–28
    index of, 631–33, 705–6, 722–23, 724
    Snell's law of, 722
    theory of, 739–46
Refrigerators, 573, 580–83
    Brayton-cycle, 588–89, 591
    Carnot, 596
    ideal-gas, 588–90
Relative biological effectiveness (RBE), 1373
Relative motion, 164–70
    Galilean principle of relativity, 168–70,
        1155–56
    position, 164–66
    velocity, 164, 166–68
Relativity, 170, 1149, 1151–94, 1382–83.
        See also Galilean relativity
    clock synchronization, 1161, 1162
    Einstein's principle of, 1157–60
    energy and, 1183–88
    events, 1160–61, 1162–63
    general, 1152
    length contraction, 1171–74
    Lorentz transformations, 1175–80
    measurements in, 1161
    momentum and, 1180–83
    problem-solving strategy in, 1176

simultaneity, 1163–66
special, 1152
time dilation, 1166–71
Rem, 1373
Resistance, 944–45, 962
equivalent, 974, 978
internal, 976
resistivity vs., 894, 945
Resistive force, 103
Resistivity, 893–95, 944, 945
Resistor(s), 962–64, 1135–36
parallel, 978–81
power dissipated by, 972
series, 973–75
Resistor circuits, 981–83, 1123–24
Resistor voltage, 1123–24
Resolution, 746–49
angular, 748–49
of diffraction grating, 712
Resonance, 435–36, 1072
in AC circuits, 1133–34
standing-wave, 777
Resonance frequency, 436, 1133
Resonant tunneling diode, 1310–11
Rest energy, 1184
Rest frame, 1166
Restoring force, 280–83, 352, 425
Hooke's law and, 281–83
linear, 431
Resultant vector, 81
Revolutions per minute (rpm), 180
Right-hand rule, 397–98, 1018
Rigid bodies, 370
rotation of. *See* Rotational motion
Rigid-body equilibrium, 390–91
Ring of charge
electric field of, 829–30, 936
electric potential of, 922–23
Rocket propulsion, 212, 254
Rolling constraint, 394
Rolling friction, 133–34
air resistance compared to, 139
coefficient of, 134
Rolling motion, 393–96
Röntgen, Wilhelm, 760–61
Root-mean-square (rms) current, 1135–36
Root-mean-square (rms) speed, 552–54, 555
Ropes, 219–22
massless string approximation, 221–23
tension, 219–20
Rotation, 341, 850
Rotational equilibrium, 390
Rotational kinetic energy, 480
Rotational motion, 4, 7, 369–412, 480
angular momentum of rigid body, 400–404
about center of mass, 374–76
constraints due to ropes and
pulleys, 389–90
dynamics of, 380–87
energetics of, 392–93
about a fixed axis, 387–90
kinematics of, 370–73
moment of inertia, 381–85

Newton's second law for, 381–83, 390
parallel-axis theorem, 386–87
problem-solving strategies for, 387, 390
rolling motion, 393–96
torque. *See* Torque
vector description of, 396–400
rtz-coordinate system, 182
Rutherford, Ernest, 1205–10

S

Satellite orbits and energies, 357–62
angular momentum, 360
geosynchronous orbits, 203, 359, 1147
Kepler's second law and, 360
Kepler's third law and, 358–60
orbital energetics, 361–62
Satellites, 192
s-axis, 40–41
Scalars, 8, 78–79
vector multiplication by, 82–83, 90–91
Scanning tunneling microscope (STM), 757,
1309–10
Scattering, light, 731
Schrödinger, Erwin, 1150, 1277–78
Schrödinger's equation, 1277–83, 1382.
See also Atomic physics
de Broglie wavelength and, 1278–79
justifying, 1278–80
for particle in rigid box, 1285–86
quantization and, 1282
for quantum harmonic oscillator,
1300–1301
quantum-mechanical models and, 1280–81
restrictions and boundary conditions in,
1281–82
solving, 1281–83
three-dimensional, 1321–22
Scientific notation, 30
Scientific revolution, 342
Screening of electric fields, 871
Sea of electrons, 791, 792, 881–82, 884
Second, 26
Second law of thermodynamics, 328, 566–67,
574, 581, 582, 592–93, 596, 607
Seismic waves, 643
Selection rules, 1336
Self-energy, 931
Self-inductance, 1065
Semimajor-axis length, 345
Series capacitors, 949, 950
Series limit wavelength, 759
Series resistors, 973–75
Series *RLC* circuit, 1131–34
analyzing, 1131
impedance of, 1132
phase angle of, 1132–33
resonance in, 1133–34
Shear waves, 629
Shell model of atom, 1298, 1324, 1360–62
Short circuit, 977–78
Sign convention
for electric and magnetic flux, 1096

for motion in one dimension, 36
for rotational motion, 372
for thin lenses, 743
Significant figures, 28–30
Simple harmonic motion (SHM), 414–27
circular motion and, 418–21
conditions for, 431
conservation of energy in, 423
dynamics of, 424–27
energy in, 421–23
kinematics of, 415–17, 480
Simultaneity, 1163–66
Single-slit diffraction, 695–99
analyzing, 696–98
Huygens' principle, 696
width of, 699
Sinusoidal oscillation. *See* Simple harmonic
motion (SHM)
Sinusoidal waves, 620–26, 634
fundamental relationship for, 621–22
mathematics of, 622–24
on string, 624–26
SI unit(s), 26–27
of activity, 1367
of angle, 179
of angular acceleration, 371
of angular momentum, 259, 400
of angular velocity, 180
of capacitance, 947
of charge, 797
of current, 890
of force, 110
of gravitational constant, 348
of heat, 522
of impulse, 242
of inductance, 1065
of inertia, 381
of intensity, 633
of kinetic energy, 272
of magnetic dipole moment, 1010
of magnetic field, 1001
of magnetic flux, 1050
of mass density, 446
of momentum, 241
of number density, 488
of oscillation, 414
of power, 328
of pressure, 447
primary, 110
of resistance, 945
secondary, 110
of temperature, 492
of torque, 377
of universal gas constant, 498
of velocity, 39
of volume, 445.
of volume flow rate, 464
of work, 307
Slipping, 134–37
rolling without, 394
Small-angle approximation, 429–31
Snapshot graph, 615–17, 1268
Snell's law, 722

Sodium
    absorption in, 1336
    emission spectrum of, 1338
Solenoid
    electromagnetic induction in, 1059
    ideal, 1015
    inductance of, 1065–66
    magnetic field of, 1015–17
Solid(s), 486–87, 496
    amorphous, 486
    color in, 1338–39
    molar specific heat of, 558
    sound waves in, 629
    thermal energy of, 558
Sound waves, 629–30, 656–59, 662–63, 664
    interference between, 662–63, 664
    power of, 972
    speed of, 321, 1155
    standing, 656–59
Source charges, 806, 835
Spacetime coordinates, 1160–61
Spacetime interval, 1173–74
Special relativity, 1152
Specific heat, 527–29
    of gases, 533–39
    micro/macro connection and, 556–61
    molar, 528–29, 534–35
Specific heat ratio, 537
Spectral analysis, 695
Spectral line, 758, 1212
Spectrometer, 758, 1212
Spectroscopy, 695, 758–60
    hydrogen atom and, 759–60
Spectrum/spectra, 758. *See also* Absorption
    spectra; Emission spectra
Specular reflection, 718
Speed, 40, 79, 154
    of alpha particle, 1209
    average, 12–13
    escape, 355
    of flowing gas, 468
    of light, 26, 170, 630, 722, 1104, 1108,
        1158–59
    molecular, 497
    of projectile, 159
    root-mean-square, 552–54, 555
    of sound, 321, 629, 1155
    terminal, 122, 140–41
    wave, 612, 614, 615
Sphere of charge, 818, 833, 866–67
Spherical aberration, 746
Spherical symmetry, 852
Spherical waves, 627, 634, 667
Spin quantum number, 1327
Spontaneous emission, 1343
Spring constant, 281
Spring force, 101, 104
Springs, molecular, 102, 240
Spring scales, 129–30
Stable equilibrium, 294
Stable isotopes, 1354
Standard atmosphere (atm), 449–50, 455
Standard temperature and pressure (STP), 500

Standing-wave resonance, 777
Standing waves, 609, 648–59, 760, 771,
    778, 1234
    mathematics of, 650–51
    nodes and antinodes, 649–50
    in phase, 649
    out of phase, 650
    sound, 656–59
    transverse, 651–55
    wavelength of, 771
State variables, 487, 498
Static equilibrium, 111, 123–24
Static friction, 103, 132–33, 134, 143–44,
    187–88, 212
    coefficient of, 133
Statics, 390
Stationary state(s), 1235, 1237, 1238–40,
    1242, 1246
    energies of, 1241
    of hydrogen, 1318–19
Steam engine, 573
Stern-Gerlach experiment, 1325–27
Stick-slip motion, 282
Stimulated emission, 1343
Stopping distances, 141–42
Stopping potential, 1223–24
STP, 500
Strain, 471, 472
Streamline, 463
Strong force, 235, 302, 1297–98, 1359–60
Subatomic particle, 1203
Sublimation, 494
Subshell, 1331–33
Superconductivity, 895
Supernova, 366–67
Superposition, 646–83, 818, 1146
    beats, 672–74
    electric potential and, 920
    of forces, 100
    interference in one dimension, 659–66
    interference in two and three dimensions,
        667–72
    magnetic fields and, 1002–3
    quantum systems in, 1383
    standing waves, 648–59
Surface charge density, 825
Surface integral, 858–59
Symbols, 21–23
Symmetry, 850–53
    cylindrical, 850, 852
    field shape and, 852
    planar, 852
    spherical, 852
System(s), 208–9
    closed, 238
    disordered, 566–67
    elastic, 280
    energy of, 238, 305
    isolated, 248–49, 324, 338
    momentum conservation and, 251–52, 256
    ordered, 566–67
    of particles, 321
    self-organizing, 607

    total momentum of, 248
    uniform, 488
    work done by, 574–75
Système Internationale d'Unités, le.
    See SI unit(s)

T

Tangential acceleration, 197, 371–72
Tangential axis, 182
Telescope, Galilean, 738–39
Temperature, 305, 490–92, 530
    absolute, 492
    average translational kinetic energy and, 555
    change in, 527–29
    heat vs., 523
    micro/macro connection and, 554–56
    pressure and, 449
    scales, 491
    thermal energy and, 523, 557
Tensile strength, 496
Tensile stress, 470–71
Tension force, 101, 104
    in massless string, 221
Terminal speed, 122, 140–41
Terminal velocity, 1218
Terminal voltage, 943, 976
Thermal efficiency, 578
    Carnot, 595
    maximum, 596–97
Thermal emission, 880, 1222
Thermal energy, 305, 321–23, 339, 490–91,
    513–14, 560
    of diatomic molecules, 558–61
    dissipative forces, 322–24
    in energy bar charts, 324–26
    heat vs., 523
    kinetic and potential energy at the
        microscopic level and, 321–22
    micro/macro connection and, 556–61
    of monatomic gas, 556–57
    of solids, 558
    temperature vs., 523, 557
Thermal equilibrium, 487, 501, 521, 562
Thermal inertia, 528
Thermal interactions, 515, 521–22, 561–63
Thermal properties of matter, 527–30
    heat of transformation, 529–30
    phase change, 486, 493–95, 529–30
    temperature change and specific heat,
        527–29
Thermocouples, 491
Thermodynamic energy model, 524
Thermodynamics, 483–84, 574, 606–7
    first law of, 327, 524–26, 574, 575, 577
    nonequilibrium, 607
    second law of, 328, 566–67, 574, 581, 582,
        592–93, 596, 607
Thermometers, 491, 492
Thick lens, 742
Thin-film optical coatings, 665–66
Thin lenses, 731–46
    aberrations, 745–46

combinations of, 738–39
converging, 732–33, 734
diverging, 737–38
image formation from, 744–45
magnification, 735–36
real images, 733–35
refraction theory, 739–46
thin-lens equation, 743
virtual images, 736–38
Thompson, Benjamin (Count Rumford), 520
Thomson, J.J., 1200–1203
Three-phase electricity, 1144
Threshold frequency in photo electric effect,
    1221, 1224, 1226, 1228
Thrust, 103–4, 212
Tides, 208
Time
    change in, 11
    measurement of, 8–9
Time constant, 433–34, 986, 1073, 1365
Time dilation, 1166–71
    experimental evidence of, 1169–70
    proper time, 1168–69
    twin paradox, 1170–71
Time-frequency relationship, 1266–67
Tolman-Stewart experiment, 880–81
Toroid, 1037
Torque, 377–80
    angular acceleration and, 382
    couples, 380
    on current loops, 1026–28
    on dipole, 839–40
    gravitational, 379–80
    interpreting, 378–79
    net, 379, 380, 401
    on a particle, 399
    signs and strengths of, 377
Torque vector, 399
Torsion balance, 352–53
Total energy, 1184, 1187–88
Total equilibrium, 390
Total internal reflection (TIR), 725–26
Townes, Charles, 1343
Tracers, radioactive, 1374–75
Trajectory, 4, 153
    parabolic, 158, 159
Transformers, 1064
Transition elements, 1332–33
Transition (quantum jump), 1235
Translational equilibrium, 390
Translational kinetic energy, 554, 557
Translational motion, 4, 6, 370
Transmission grating, 695
Transverse standing waves, 651–55
    electromagnetic waves, 654–55
    on a string, 652–54
Transverse wave, 613
Traveling waves, 611–45, 778. *See also*
    Superposition
    amplitude of, 621
    displacement of, 618–20
    Doppler effect, 634–38
    electromagnetic, 630–31

frequency of, 621
index of refraction of, 631–33
intensity of, 633–34
one-dimensional, 614–20
period of, 621–22
in phase, 661
out of phase, 661, 662
power of, 633–34
sinusoidal, 620–26, 634
sound, 629–30
spherical, 634
in two and three dimensions, 626–28
wavelength of, 621–22
wave model, 612–14
Triple point, 494
Tritium, 929, 1379
Tunneling current, 1310
Turbulent flow, 462
Turning point, 48, 51, 62, 292–93, 415, 422
Twin paradox, 1170–71

U

Ultrasonic frequencies, 629
Uncertainty, 1267–68
    fractional, 1275
Uncertainty principle, 1268–70, 1382
Unified atomic mass unit, 253
Uniform circular motion, 178–90
    acceleration in, 183–85
    angular position, 178–79
    angular velocity, 179–81
    dynamics of, 185–90
    period of, 178, 181
    velocity in, 182–83
Uniform electric fields, 835, 1015
    dipoles in, 839–40
    motion in, 836–38
Uniformly charged object, 825
Uniform magnetic field, 1015, 1016, 1027
Uniform motion, 38–42, 56. *See also* Uniform
    circular motion
    mathematics of, 40–42
    as "natural state" of object, 111, 112
Unit(s), 26–28. *See also* English units;
    SI unit(s)
Unit vector notation, 808
Unit vectors, 88–89, 399
Unit volume, 446
Universal constant, 353, 764
Universal gas constant, 498
Unstable equilibrium, 294
Upright image, 736

V

Vacuum, 114
    perfect, 449
Valence electrons, 791
Van Allen radiation belt, 1021
Van de Graaff generator, 930, 941–42
Vapor pressure, 448, 455
Variable capacitor, 1072

Vector(s), 8–9, 78–92
    addition of, 9–10, 80–82, 89–90
    algebra, 88–92
    area, 856, 1051
    components of, 85–88, 91–92
    component vector, 85, 88
    coordinate description of, 84–88
    cross product, 397–99, 1003–4
    decomposition of, 85, 86, 89
    direction of, 79
    displacement, 80, 153
    dot product of two, 312–14
    geometric representation of, 79
    magnitude of, 8, 79
    multiplication by scalar, 82–83, 90–91
    negative of, 10
    notation for, 8–9
    position, 8
    properties of, 80–84
    resultant, 81
    subtraction, 10–11, 84, 90–91
    unit, 88–89, 399
    working with, 89–91
    zero, 10, 83
Vector algebra, 88–92
    addition, 89–90
    multiplication, 90–91
    subtraction, 90–91
    tilted axes of coordinate system and, 91–92
    unit vectors and, 88–89
Vector equations, 91, 123
Vector field, 806
Vector product, 397
Velocity, 12–15, 16. *See also* Acceleration;
    Angular velocity
    acceleration related to, 54
    average, 13, 39, 43, 153
    displacement and, 83
    escape, 907, 908
    finding from acceleration, 68
    Galilean transformation of, 167, 290,
        1089, 1154
    instantaneous, 43–49, 67, 153, 154
    Lorentz transformations of, 1179–80
    negative, 66
    in planar motion, 153–55
    position from, 49–53
    relating position to, 15
    relative, 164, 166–68
    signs of, 36–37
    terminal, 1218
    in two dimensions, 152
    in uniform circular motion, 182–83
Velocity-energy-momentum triangle, 1185
Velocity selector, 497
Velocity vectors, 13, 14
Velocity-versus-time graph for constant
    acceleration, 54
Venturi tube, 468–69
Vibrational energy levels, 1303
Virial theorem, 1218
Virtual images, 719, 727, 734, 736–38
Viscosity, 462

Visible spectrum, 631, 759
Voltage, 912
    AC, 1063
    Hall, 1023
    of inductor, 1131
    peak, 1125
    resistor, 1123–24
    terminal, 943, 976
Voltage difference, 793–94
Voltage divider, 993
Voltage drop, 964
Voltmeters, 980–81
Volume flow rate, 464–65
Volume strain, 472
Volume stress, 472. *See also* Pressure

W

Waste heat, 578
Watt, 328
Wave(s), 1254–55. *See also* Electromagnetic
        waves; Matter waves; One-dimensional
        waves; Sinusoidal waves; Standing waves;
        Traveling waves
    circular, 626, 667
    deep-water, 682
    electromagnetic, 612, 618
    gravity, 704
    intensity of, 650
    longitudinal, 613, 617–18
    mechanical, 612
    modulation of, 673
    phase of, 627–28
    plane, 627
    probability, 779
    seismic, 643
    shear, 629
    sound, 629–30, 656–59, 662–63, 664
    spherical, 627, 667
    transverse, 613
Wave fronts, 626
Wave function, 1150, 1253, 1259–61. *See also*
        Schrödinger's equation
    drawing, 1299
    of hydrogen atom, 1322–24

interpreting, 1262–63
    normalized, 1262–63, 1287
    of particle in a box, 1285–86
    radial, 1322–24
    shapes of, 1298–1300
Wavelength, 621–22. *See also* De Broglie
        wavelength
    of light, 631, 705
    measurement of, 758
    series limit, 759
    of standing wave, 771
Wave number, 623
Wave optics, 609, 684–713, 1197, 1257–59.
        See also Light
    circular-aperture diffraction, 700–702
    diffraction grating, 693–95
    holography, 706–7
    interferometers, 702–7
    single-slit diffraction, 695–99
Wave packets, 1229, 1264–68
    bandwidth, 1266–67
    uncertainty, 1267–68
Wave-particle duality, 610, 779, 1220
Wave pulse, 613–14
Wave speed, 612, 614, 615
Weak force, 235
Weak interaction, 1371–72
Weight, 100–101, 104, 129–32
    apparent, 130–31, 132, 351
    gravitational force and, 348–49
    mass vs., 101, 129
    measuring, 129–30
Weighted average, 374
Weightlessness, 131–32, 192
Wire(s), 964
    charged, 818, 868
    current density in, 892
    electric field of, 868, 885–87
    magnetic field of, 1005–6, 1014
    magnetic flux from current in, 1052
    magnetic forces on, 1024–26
    parallel, 1025–26
Work function, 1222, 1310
Work-kinetic energy theorem, 308, 324
Work, 304–39, 514

calculating and using, 309–14
    by constant force, 309–11, 313, 901–2
    done by system, 574–75
    dot product to calculate, 313–14
    energy and, 514–15
    force and, 316–19
    by force perpendicular to direction of motion,
        311–12
    heat into, 574–77
    in ideal-gas processes, 516–20
    kinetic energy and, 306–8
    potential energy and, 316–19
    power and, 2
    properties of, 522
    sign of, 514, 515
    by variable force, 314–16

X

x-axis, 7
x-component of vector, 85
X-ray diffraction, 760–63
X-ray diffraction spectrum, 762
X-ray monochromator, 763
X rays, 761
xy-plane, 152

Y

y-axis, 7
y-component of vector, 85
Young, Thomas, 685
Young's double-slit experiment, 687–92, 1197
    analyzing, 688–90
    intensity of interference pattern, 691–92
Young's modulus, 470–71

Z

Zero of energy, 1335
Zero of potential energy, 274–75, 907–8
Zero-point motion, 1289–91
Zero vector, 10, 83